Recommended Dietary Allowances (RDA) and Adequate Intakes (AI) for Vitamins

Age (yr)	Thiamin RDA (mg/day)	Riboflavin RDA (mg/day)	Niacin RDA (mg/day)[a]	Biotin AI (µg/day)	Pantothenic acid AI (mg/day)	Vitamin B6 RDA (mg/day)	Folate RDA (µg/day)[b]	Vitamin B12 RDA (µg/day)	Choline AI (mg/day)	Vitamin C RDA (mg/day)	Vitamin A RDA (µg/day)[c]	Vitamin D RDA (IU/day)[d]	Vitamin E RDA (mg/day)[e]	Vitamin K AI (µg/day)
Infants														
0–0.5	0.2	0.3	2	5	1.7	0.1	65	0.4	125	40	400	400 (10 µg)	4	2.0
0.5–1	0.3	0.4	4	6	1.8	0.3	80	0.5	150	50	500	400 (10 µg)	5	2.5
Children														
1–3	0.5	0.5	6	8	2	0.5	150	0.9	200	15	300	600 (15 µg)	6	30
4–8	0.6	0.6	8	12	3	0.6	200	1.2	250	25	400	600 (15 µg)	7	55
Males														
9–13	0.9	0.9	12	20	4	1.0	300	1.8	375	45	600	600 (15 µg)	11	60
14–18	1.2	1.3	16	25	5	1.3	400	2.4	550	75	900	600 (15 µg)	15	75
19–30	1.2	1.3	16	30	5	1.3	400	2.4	550	90	900	600 (15 µg)	15	120
31–50	1.2	1.3	16	30	5	1.3	400	2.4	550	90	900	600 (15 µg)	15	120
51–70	1.2	1.3	16	30	5	1.7	400	2.4	550	90	900	600 (15 µg)	15	120
>70	1.2	1.3	16	30	5	1.7	400	2.4	550	90	900	800 (20 µg)	15	120
Females														
9–13	0.9	0.9	12	20	4	1.0	300	1.8	375	45	600	600 (15 µg)	11	60
14–18	1.0	1.0	14	25	5	1.2	400	2.4	400	65	700	600 (15 µg)	15	75
19–30	1.1	1.1	14	30	5	1.3	400	2.4	425	75	700	600 (15 µg)	15	90
31–50	1.1	1.1	14	30	5	1.3	400	2.4	425	75	700	600 (15 µg)	15	90
51–70	1.1	1.1	14	30	5	1.5	400	2.4	425	75	700	600 (15 µg)	15	90
>70	1.1	1.1	14	30	5	1.5	400	2.4	425	75	700	800 (20 µg)	15	90
Pregnancy														
≤18	1.4	1.4	18	30	6	1.9	600	2.6	450	80	750	600 (15 µg)	15	75
19–30	1.4	1.4	18	30	6	1.9	600	2.6	450	85	770	600 (15 µg)	15	90
31–50	1.4	1.4	18	30	6	1.9	600	2.6	450	85	770	600 (15 µg)	15	90
Lactation														
≤18	1.4	1.6	17	35	7	2.0	500	2.8	550	115	1200	600 (15 µg)	19	75
19–30	1.4	1.6	17	35	7	2.0	500	2.8	550	120	1300	600 (15 µg)	19	90
31–50	1.4	1.6	17	35	7	2.0	500	2.8	550	120	1300	600 (15 µg)	19	90

Note: For all nutrients, values for infants are AI. The glossary on the following pages and Appendix B define units of nutrient measure.
[a]Niacin recommendations are expressed as niacin equivalents (NE), except for recommendations for infants younger than 6 months, which are expressed as preformed niacin.
[b]Folate recommendations are expressed as dietary folate equivalents (DFE).
[c]Vitamin A recommendations are expressed as retinol activity equivalents (RAE).
[d]Vitamin D recommendations are expressed as cholecalciferol and assume an absence of adequate exposure to sunlight.
[e]Vitamin E recommendations are expressed as α-tocopherol.

SOURCE: Adapted with permission from the *Dietary Reference Intakes* series, National Academies Press. Copyright 1997, 1998, 2000, 2001, 2002, 2005, 2011 by the National Academies of Sciences.

Recommended Dietary Allowances (RDA) and Adequate Intakes (AI) for Minerals

Age (yr)	Sodium AI (mg/day)	Chloride AI (mg/day)	Potassium AI (mg/day)	Calcium RDA (mg/day)	Phosphorus RDA (mg/day)	Magnesium RDA (mg/day)	Iron RDA (mg/day)	Zinc RDA (mg/day)	Iodine RDA (µg/day)	Selenium RDA (µg/day)	Copper RDA (µg/day)	Manganese AI (mg/day)	Fluoride AI (mg/day)	Chromium AI (µg/day)	Molybdenum RDA (µg/day)
Infants															
0–0.5	120	180	400	200	100	30	0.27	2	110	15	200	0.003	0.01	0.2	2
0.5–1	370	570	700	260	275	75	11	3	130	20	220	0.6	0.5	5.5	3
Children															
1–3	1000	1500	3000	700	460	80	7	3	90	20	340	1.2	0.7	11	17
4–8	1200	1900	3800	1000	500	130	10	5	90	30	440	1.5	1.0	15	22
Males															
9–13	1500	2300	4500	1300	1250	240	8	8	120	40	700	1.9	2	25	34
14–18	1500	2300	4700	1300	1250	410	11	11	150	55	890	2.2	3	35	43
19–30	1500	2300	4700	1000	700	400	8	11	150	55	900	2.3	4	35	45
31–50	1500	2300	4700	1000	700	420	8	11	150	55	900	2.3	4	35	45
51–70	1300	2000	4700	1000	700	420	8	11	150	55	900	2.3	4	30	45
>70	1200	1800	4700	1200	700	420	8	11	150	55	900	2.3	4	30	45
Females															
9–13	1500	2300	4500	1300	1250	240	8	8	120	40	700	1.6	2	21	34
14–18	1500	2300	4700	1300	1250	360	15	9	150	55	890	1.6	3	24	43
19–30	1500	2300	4700	1000	700	310	18	8	150	55	900	1.8	3	25	45
31–50	1500	2300	4700	1000	700	320	18	8	150	55	900	1.8	3	25	45
51–70	1300	2000	4700	1200	700	320	8	8	150	55	900	1.8	3	20	45
>70	1200	1800	4700	1200	700	320	8	8	150	55	900	1.8	3	20	45
Pregnancy															
≤18	1500	2300	4700	1300	1250	400	27	12	220	60	1000	2.0	3	29	50
19–30	1500	2300	4700	1000	700	350	27	11	220	60	1000	2.0	3	30	50
31–50	1500	2300	4700	1000	700	360	27	11	220	60	1000	2.0	3	30	50
Lactation															
≤18	1500	2300	5100	1300	1250	360	10	13	290	70	1300	2.6	3	44	50
19–30	1500	2300	5100	1000	700	310	9	12	290	70	1300	2.6	3	45	50
31–50	1500	2300	5100	1000	700	320	9	12	290	70	1300	2.6	3	45	50

Note: For all nutrients, values for infants are AI. The glossary on the following pages and Appendix B define units of nutrient measure.
SOURCE: Adapted with permission from the *Dietary Reference Intakes* series, National Academies Press. Copyright 1997, 1998, 2000, 2001, 2002, 2005, 2011 by the National Academies of Sciences.

Tolerable Upper Intake Levels (UL) for Vitamins

Age (yr)	Niacin (mg/day)[a]	Vitamin B_6 (mg/day)	Folate (µg/day)[a]	Choline (mg/day)[a]	Vitamin C (mg/day)	Vitamin A (µg/day)[b]	Vitamin D (IU/day)	Vitamin E (mg/day)[c]
Infants								
0–0.5	–	–	–	–	–	600	1000 (25 µg)	–
0.5–1	–	–	–	–	–	600	1500 (38 µg)	–
Children								
1–3	10	30	300	1000	400	600	2500 (63 µg)	200
4–8	15	40	400	1000	650	900	3000 (75 µg)	300
9–13	20	60	600	2000	1200	1700	4000 (100 µg)	600
Adolescents								
14–18	30	80	800	3000	1800	2800	4000 (100 µg)	800
Adults								
19–70	35	100	1000	3500	2000	3000	4000 (100 µg)	1000
>70	35	100	1000	3500	2000	3000	4000 (100 µg)	1000
Pregnancy								
≤18	30	80	800	3000	1800	2800	4000 (100 µg)	800
19–50	35	100	1000	3500	2000	3000	4000 (100 µg)	1000
Lactation								
≤18	30	80	800	3000	1800	2800	4000 (100 µg)	800
19–50	35	100	1000	3500	2000	3000	4000 (100 µg)	1000

[a]The UL for niacin and folate apply to synthetic forms obtained from supplements, fortified foods, or a combination of the two.
[b]The UL for vitamin A applies to the preformed vitamin only.
[c]The UL for vitamin E applies to any form of supplemental α-tocopherol, fortified foods, or a combination of the two.
Note: An Upper Level was not established for vitamins not listed and for those age groups listed with a dash (—) because of a lack of data, not because these nutrients are safe to consume at any level of intake. All nutrients can have adverse effects when intakes are excessive. See Module 5 for toxicity signs and symptoms.

SOURCE: Adapted with permission from the *Dietary Reference Intakes* series, National Academies Press. Copyright 1997, 1998, 2000, 2001, 2002, 2005, 2011 by the National Academies of Sciences.
© Cengage Learning 2013

Tolerable Upper Intake Levels (UL) for Minerals

Age (yr)	Sodium (mg/day)	Chloride (mg/day)	Calcium (mg/day)	Phosphorus (mg/day)	Magnesium (mg/day)[d]	Iron (mg/day)	Zinc (mg/day)	Iodine (µg/day)	Selenium (µg/day)	Copper (µg/day)	Manganese (mg/day)	Fluoride (mg/day)	Molybdenum (µg/day)	Boron (mg/day)	Nickel (mg/day)	Vanadium (mg/day)
Infants																
0–0.5	–	–	1000	–	–	40	4	–	45	–	–	0.7	–	–	–	–
0.5–1	–	–	1500	–	–	40	5	–	60	–	–	0.9	–	–	–	–
Children																
1–3	1500	2300	2500	3000	65	40	7	200	90	1000	2	1.3	300	3	0.2	–
4–8	1900	2900	2500	3000	110	40	12	300	150	3000	3	2.2	600	6	0.3	–
9–13	2200	3400	3000	4000	350	40	23	600	280	5000	6	10	1100	11	0.6	–
Adolescents																
14–18	2300	3600	3000	4000	350	45	34	900	400	8000	9	10	1700	17	1.0	–
Adults																
19–50	2300	3600	2500	4000	350	45	40	1100	400	10,000	11	10	2000	20	1.0	1.8
51–70	2300	3600	2000	4000	350	45	40	1100	400	10,000	11	10	2000	20	1.0	1.8
>70	2300	3600	2000	3000	350	45	40	1100	400	10,000	11	10	2000	20	1.0	1.8
Pregnancy																
≤18	2300	3600	3000	3500	350	45	34	900	400	8000	9	10	1700	17	1.0	–
19–50	2300	3600	2500	3500	350	45	40	1100	400	10,000	11	10	2000	20	1.0	–
Lactation																
≤18	2300	3600	3000	4000	350	45	34	900	400	8000	9	10	1700	17	1.0	–
19–50	2300	3600	2500	4000	350	45	40	1100	400	10,000	11	10	2000	20	1.0	–

[d]The UL for magnesium applies to synthetic forms obtained from supplements or drugs only.
Note: An Upper Level was not established for minerals not listed and for those age groups listed with a dash (—) because of a lack of data, not because these nutrients are safe to consume at any level of intake. All nutrients can have adverse effects when intakes are excessive. See Module 5 for toxicity signs and symptoms.

SOURCE: Adapted with permission from the *Dietary Reference Intakes* series, National Academies Press. Copyright 1997, 1998, 2000, 2001, 2002, 2005, 2011 by the National Academies of Sciences.
© Cengage Learning 2013

Acceptable Macronutrient Distribution Ranges (AMDRs) for Healthy Diets as a Percent of Energy						
Age (years)	Carbohydrate	Sugar	Total Fat	Linoleic Acid	α-Linolenic Acid	Protein
1–3	45–65	≤ 25	30–40	5–10	0.6–1.2	5–20
4–18	45–65	≤ 25	25–35	5–10	0.6–1.2	10–30
≥ 19	45–65	≤ 25	20–35	5–10	0.6–1.2	10–35

Assumes adequate Calorie intake determined by the EER.

SOURCE: Adapted with permission from the *Dietary Reference Intakes* series, National Academies Press. Copyright 1997, 1998, 2000, 2001, 2002, 2005, 2011 by the National Academies of Sciences.
© Cengage Learning 2013

Daily Values for Food Labels

The Daily Values are standard values developed by the Food and Drug Administration (FDA) for use on food labels. Daily Values for protein, vitamins, and minerals reflect the highest allowance based on the 1968 RDA. Daily Values for nutrients and food components, such as fat and fiber, that do not have an established RDA but do have important relationships with health are based on recommended calculation factors as noted.

Nutrient	Amount
Protein[a]	50 g
Thiamin	1.5 mg
Riboflavin	1.7 mg
Niacin	20 mg NE
Biotin	300 μg
Pantothenic acid	10 mg
Vitamin B_6	2 mg
Folate	400 μg
Vitamin B_{12}	6 μg
Vitamin C	60 mg
Vitamin A	5000 IU[b]
Vitamin D	400 IU[b]
Vitamin E	30 IU[b]
Vitamin K	80 μg
Calcium	1000 mg
Iron	18 mg
Zinc	15 mg
Iodine	150 μg
Copper	2 mg
Chromium	120 μg
Selenium	70 μg
Molybdenum	75 μg
Manganese	2 mg
Chloride	3400 mg
Magnesium	400 mg
Phosphorus	1000 mg

[a]The Daily Values for protein vary for different groups of people: pregnant women, 60 g; nursing mothers, 65 g; infants under 1 year, 14 g; children 1 to 4 years, 16 g.

[b]Equivalent values for nutrients expressed as IU are: vitamin A, 1500 RAE (assumes a mixture of 40% retinol and 60% beta-carotene); vitamin D, 10 μg; vitamin E, 20 mg.
© Cengage Learning 2013

Food Component	Amount	Calculation Factors
Fat	65 g	30% of Calories
Saturated fat	20 g	10% of Calories
Cholesterol	300 mg	Same regardless of Calories
Carbohydrate (total)	300 g	60% of Calories
Fiber	25 g	11.5 g per 1000 Calories
Protein	50 g	10% of Calories
Sodium	2400 mg	Same regardless of Calories
Potassium	3500 mg	Same regardless of Calories

Note: Daily Values were established for adults and children over 4 years old. The values for energy-yielding nutrients are based on 2000 Calories a day.

SOURCE: Adapted with permission from the *Dietary Reference Intakes* series, National Academies Press. Copyright 1997, 1998, 2000, 2001, 2002, 2005, 2011 by the National Academies of Sciences.
© Cengage Learning 2013

Glossary of Nutrient Measures

Calories: a unit by which energy is measured (Module 1 provides more details).

g: grams; a unit of weight equivalent to about 0.03 ounces.

mg: milligrams; one-thousandth of a gram.

μg: micrograms; one-millionth of a gram.

IU: international units; an old measure of vitamin activity determined by biological methods (as opposed to new measures that are determined by direct chemical analyses). Many fortified foods and supplements use IU on their labels.
- For vitamin A, 1 IU = 0.3 μg retinol, 3.6 μg β-carotene, or 7.2 μg other vitamin A carotenoids.
- For vitamin D, 1 IU = 0.025 μg cholecalciferol.
- For vitamin E, 1 IU = 0.67 natural α-tocopherol (other conversion factors are used for different forms of vitamin E).

mg NE: milligrams niacin equivalents; a measure of niacin activity (Module 5 provides more details).
- 1 NE = 1 mg niacin.
  = 60 mg tryptophan (an amino acid).

μg DFE: micrograms dietary folate equivalents; a measure of folate activity (Module 5 provides more details).
- 1 μg DFE = 1 μg food folate.
  = 0.6 μg fortified food or supplement folate.
  = 0.5 μg supplement folate taken on an empty stomach.

μg RAE: micrograms retinol activity equivalents; a measure of vitamin A activity (Module 5 provides more details).
- 1 μg RE = 1 μg retinol.
  = 12 μg β-carotene.
  = 24 μg other vitamin A carotenoids.

© Cengage Learning 2013

Body Mass Index (BMI)

Height	18	19	20	21	22	23	24	25	26	27	28	29	30	31	32	33	34	35	36	37	38	39	40
									Body Weight (pounds)														
4'10"	86	91	96	100	105	110	115	119	124	129	134	138	143	148	153	158	162	167	172	177	181	186	191
4'11"	89	94	99	104	109	114	119	124	128	133	138	143	148	153	158	163	168	173	178	183	188	193	198
5'0"	92	97	102	107	112	118	123	128	133	138	143	148	153	158	163	168	174	179	184	189	194	199	204
5'1"	95	100	106	111	116	122	127	132	137	143	148	153	158	164	169	174	180	185	190	195	201	206	211
5'2"	98	104	109	115	120	126	131	136	142	147	153	158	164	169	175	180	186	191	196	202	207	213	218
5'3"	102	107	113	118	124	130	135	141	146	152	158	163	169	175	180	186	191	197	203	208	214	220	225
5'4"	105	110	116	122	128	134	140	145	151	157	163	169	174	180	186	192	197	204	209	215	221	227	232
5'5"	108	114	120	126	132	138	144	150	156	162	168	174	180	186	192	198	204	210	216	222	228	234	240
5'6"	112	118	124	130	136	142	148	155	161	167	173	179	186	192	198	204	210	216	223	229	235	241	247
5'7"	115	121	127	134	140	146	153	159	166	172	178	185	191	198	204	211	217	223	230	236	242	249	255
5'8"	118	125	131	138	144	151	158	164	171	177	184	190	197	203	210	216	223	230	236	243	249	256	262
5'9"	122	128	135	142	149	155	162	169	176	182	189	196	203	209	216	223	230	236	243	250	257	263	270
5'10"	126	132	139	146	153	160	167	174	181	188	195	202	209	216	222	229	236	243	250	257	264	271	278
5'11"	129	136	143	150	157	165	172	179	186	193	200	208	215	222	229	236	243	250	257	265	272	279	286
6'0"	132	140	147	154	162	169	177	184	191	199	206	213	221	228	235	242	250	258	265	272	279	287	294
6'1"	136	144	151	159	166	174	182	189	197	204	212	219	227	235	242	250	257	265	272	280	288	295	302
6'2"	141	148	155	163	171	179	186	194	202	210	218	225	233	241	249	256	264	272	280	287	295	303	311
6'3"	144	152	160	168	176	184	192	200	208	216	224	232	240	248	256	264	272	279	287	295	303	311	319
6'4"	148	156	164	172	180	189	197	205	213	221	230	238	246	254	263	271	279	287	295	304	312	320	328
6'5"	151	160	168	176	185	193	202	210	218	227	235	244	252	261	269	277	286	294	303	311	319	328	336
6'6"	155	164	172	181	190	198	207	216	224	233	241	250	259	267	276	284	293	302	310	319	328	336	345

Under-weight (<18.5)	Healthy Weight (18.5–24.9)	Overweight (25–29.9)	Obese (≥ 30)

Find your height along the left-hand column and look across the row until you find the number that is closest to your weight. The number at the top of that column identifies your BMI. Module 4 describes how BMI correlates with disease risks and defines obesity. The area shaded in green represents healthy weight ranges.

NUTRITION
Your Life Science

Jennifer Turley

WEBER STATE UNIVERSITY

Joan Thompson

WEBER STATE UNIVERSITY

 WADSWORTH
CENGAGE Learning

Australia • Brazil • Japan • Korea • Mexico • Singapore • Spain • United Kingdom • United States

Nutrition: Your LIfe Science
Jennifer Turley, Joan Thompson

Publisher: Yolanda Cossio

Acquisitions Editor: Peggy Williams

Developmental Editor: Nedah Rose

Assistant Editor: Elesha Feldman

Editorial Assistant: Sean Cronin

Media Editor: Miriam Myers

Marketing Manager: Laura McGinn

Marketing Assistant: Jing Hu

Marketing Communications Manager: Mary Anne Payumo

Content Project Manager: Carol Samet

Creative Director: Rob Hugel

Art Director: John Walker

Print Buyer: Judy Inouye

Rights Acquisitions Specialist: Roberta Broyer

Production Service: Lachina Publishing Services

Text Designer: Diane Beasley

Photo Researcher: Bill Smith Group

Copy Editor: Ginjer Clarke

Cover Designer: Diane Beasley

Cover Image: Alamy Images

Compositor: Lachina Publishing Services

Library of Congress Control Number: 2011933490
ISBN-13: 978-0-538-49484-7
ISBN-10: 0-538-49484-0

Wadsworth
20 Davis Drive
Belmont, CA 94002-3098
USA

Cengage Learning is a leading provider of customized learning solutions with office locations around the globe, including Singapore, the United Kingdom, Australia, Mexico, Brazil, and Japan. Locate your local office at **www.cengage.com/global**.

Cengage Learning products are represented in Canada by Nelson Education, Ltd.

To learn more about **Wadsworth**, visit **www.cengage.com/Wadsworth.**

Purchase any of our products at your local college store or at our preferred online store **www.CengageBrain.com.**

Unless otherwise noted, all art is © Cengage Learning.

Printed in the United States of America
1 2 3 4 5 6 7 15 14 13 12 11

About the Authors

Jennifer Turley, Ph.D

Dr. Jennifer Turley is a professor of nutrition and online nutrition program director at Weber State University in Ogden, Utah, where she has been actively involved in teaching, scholarship, and service for almost 13 years. She received her Ph.D. in nutritional science from the University of Texas at Austin and held a four-year postdoctoral research fellowship at the National Cancer Institute in Frederick, Maryland.

Dr. Turley's nutrition specialty areas are cancer and immunity. Her basic laboratory experiences centered on investigating vitamin E as an anticancer agent in human cancer cells and determining novel mechanisms of action for this essential nutrient. Nutrition and immunity, especially as they relate to food allergies and intolerances, are areas of both personal and professional interest. She is also interested the areas of food system sustainability and the personal and environmental benefits of organic foods.

She centers her personal life on faith, family, and friendship and her professional life on education and empowerment for the enhancement of self and life. Dr. Turley pursues a healthy, active, and balanced lifestyle and hopes that students who study nutrition are able to apply the concepts learned to experience the most valuable thing in life, personal health and long-lived well being.

Joan Thompson, Ph.D., M.S., R.D., C.D.

Dr. Joan Thompson is an associate professor of nutrition and director of the nutrition program at Weber State University, where she has been for 24 years. This career opportunity has given her a chance to reach out to young adults and share sound strategies for promoting health and human performance.

Many experiences that Dr. Thompson had during her undergraduate education at the University of California, Berkeley, motivated her to pursue career opportunities in the clinical dietetics area. After completing a master's degree and doctorate degree from the University of Arizona, Tucson, and doing clinical nutrition research, she was convinced that preventive medicine and lifestyle management are among the keys to lifelong health and happiness.

As an athlete, dietitian, professor, and parent, she has had abundant opportunities to speak, write, research, teach, and share her knowledge in the areas of nutrition, health, and human performance to a wide variety of audiences.

Brief Contents

Contents

Valentyn Volkov/Shutterstock.com

MODULE **2** Tools to Plan, Manage, and Evaluate Diets   45

Valentyn Volkov/Shutterstock.com

Valentyn Volkov/Shutterstock.com

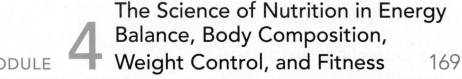

MODULE **4** The Science of Nutrition in Energy
Balance, Body Composition,
Weight Control, and Fitness   169

Valentyn Volkov/Shutterstock.com

MODULE 5 The Vitamins and Minerals   223

LockStockBob/Shutterstock.com

Fedorov Oleksiy/Shutterstock.com

MODULE **6** Nutrition Information and the Food Industry 287

Preface

Hello and welcome to *Nutrition: Your Life Science!* We hope you will enjoy learning the principles of nutrition science using the combination of printed and online learning tools that we have prepared for you. Your instructor will explain how you can easily gain access to the online resources.

Courtesy Jennifer Turley

Courtesy Joan Thompson

How to Use This Book

When you open the textbook, you will see that there are seven modules, each of which has a corresponding eBook unit, as well as a variety of other activities. Each module begins and ends with learning objectives that define the key concepts you will be focusing on; in addition, we have included personal improvement goals that demonstrate how you can apply the information to your own life.

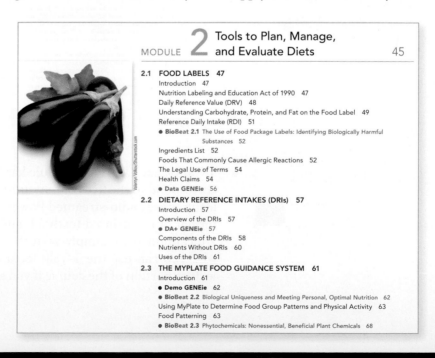

Valentyn Volkov/Shutterstock.com

Cengage Learning

Each module supplies five lectures, an assessment, an exam, and several learning tools that prepare you to answer questions (demonstrate knowledge and cognitive learning), apply the information (demonstrate skills and competencies), and master the terminology.

When you are progressing through each of the seven modules, we recommend you begin by: reading the text material for each of the five lectures, then listening to or going to class lectures and highlighting the material in the text or taking notes as you follow along. The information in the lectures is the "need-to-know" content. After covering all of the lectures in the module, it is best to complete the Total Recall (located at the end of each module in the text) and submit the answers online. These exercises help you answer questions, apply knowledge you have gained in reading and listening, master the terminology, and improve your exam grade. The next step is to complete the Homework Assessment and submit the answers online, reread the module, use the downloadable online study activities in each module, take the practice exam available online that mimics the actual exam testing format, study the material for the questions that were missed, then take the real exam! You will be ready for success.

Features and Learning Tools

The learning journey is substantial, yet fully supported by teaching and learning aids and high-interest features, including:

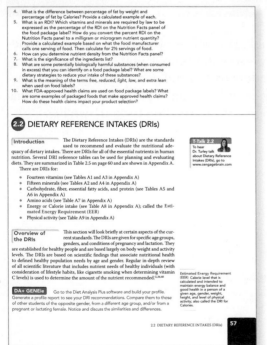

- **Textbook and eBook:** You can choose to use the printed textbook, the electronic version (eBook), or both. While the textbook is portable and easy to use, the eBook has hotlinks and embedded animations and videos.

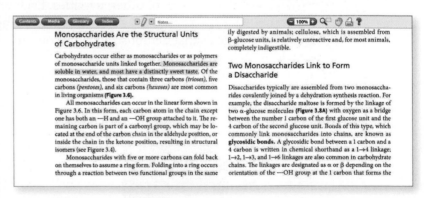

- **T-Talks:** Each learning module has five "T-Talk" lectures, which correspond to the five content units in each module. The T-Talk lectures are audio-streamed PowerPoint lectures that follow the format of the printed textbook and eBook. All T-Talk lectures are captioned to comply with the Americans with Disabilities Act. You can use the T-Talk lectures as a study tool or as the lecture portion of the course if you are an online student. T-Talk audio can also be downloaded to a mobile device. Following each T-Talk lecture content area are "Summary Points" and "Take Ten on Your Knowledge Know-How," which are reflective questions to apply and assess learning.

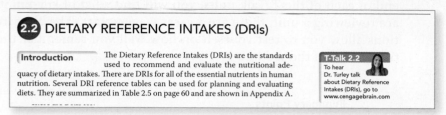

2.2 DIETARY REFERENCE INTAKES (DRIs)

Introduction The Dietary Reference Intakes (DRIs) are the standards used to recommend and evaluate the nutritional adequacy of dietary intakes. There are DRIs for all of the essential nutrients in human nutrition. Several DRI reference tables can be used for planning and evaluating diets. They are summarized in Table 2.5 on page 60 and are shown in Appendix A.

T-Talk 2.2
To hear
Dr. Turley talk
about Dietary Reference
Intakes (DRIs), go to
www.cengagebrain.com

- **BioBeats:** BioBeats address major themes and organizing principles important to the study of nutrition. They often deal with the biological or chemical underpinnings of how our bodies work, how food becomes fuel, how one's genes affect one's health risks, and so on.

BioBeat 1.2

Food, Energy, and Me

All living things abide by the laws of **energy** and **thermodynamics** as manifested in biochemical reactions. Energy can be **kinetic** (the ability to do work) or **potential** (stored energy) in nature. The laws of thermodynamics say that energy is neither created nor destroyed; it can only be transferred from one form to another. The way that energy is transferred to sustain life is through nutrition. Food nourishes living systems by providing energy-producing nutrients and non-energy-producing nutrients, both of which can provide **essential nutrients**. The energy-producing nutrients—carbohydrates, proteins, and fats—are used to produce **adenosine triphosphate (ATP)** inside **mitochondria**, also known as the powerhouses of cells (see Figure 1.5).

ATP functions as the body's short-term energy storage center. ATP is generated by converting adenosine diphosphate (ADP) to ATP. This chemical reaction requires energy. The addition of the third phosphate to adenosine results in an energy transfer from the carbon bonds in energy-producing nutrients. Hence, the potential energy in carbohydrates, proteins, and fats is used to produce ATP, and ATP can be used kinetically to supply the energy required to do cellular work. The non-energy-producing essential nutrients, though not metabolized directly to yield ATP, participate in metabolism.[25,29]

What is the ultimate form of energy used by the body to do any type of work?

- **GENEies:** GENEies are the places in the module where you can solve problems. There are three different kinds: Diet Analysis Plus (DA+) GENEie allows you to use the online DA+ software program to analyze your own diet, Data GENEie asks you to calculate a dietary fact and work with numbers relevant to nutrition science, and Demo GENEie outlines a demonstration that lets you see a point made in the module using everyday items or foods.

DA+ GENEie Go to Diet Analysis Plus and enter some foods like coconut oil, olive oil, flax seeds, fish, and processed foods that have crunchy or flaky aspects to them, such as pot pie, fast-food breakfast potatoes, crackers, and so forth. Create a source analysis spreadsheet report. Study the fatty acid content by food entered for sat-fat, mono-fat, poly-fat, *trans*-fat, omega-3, and omega-6. Note your observations for discussion.

MyPlate An icon identifying the portions of the major food groups to consume each day; accompanied by a Web-based, Calorie-controlled, personalized food guidance system based on the age, gender, height, weight, and physical activity level of a person; sponsored by the U.S. Department of Agriculture.

- **Glossary Terms:** There are running glossary terms posted throughout each module to help you learn the unique nutrition terminology used in the module.
- **Summary Points and Take Ten on Your Knowledge Know-How:** At the end of each of the T-Talk units, you will find a list of key points. If you are reviewing a unit, these are a great way to start. The Take Ten questions allow you to test yourself on how well you remember the important points of the section.

Summary Points

- DRIs are the nutrient intake standards used for many reasons. They are based on age, gender, and special conditions.
- The exact DRI value may be as an RDA, AI, or EAR, depending on scientific information available for the nutrient.
- DRIs include ULs and AMDRs.
- For Calories, the DRI is called the EER.
- There is a DRI recommendation for PA.
- Tables A1 through A8 in Appendix A are referred to collectively as the DRIs and are all needed to perform a thorough dietary analysis.

Take Ten on Your Knowledge Know-How

1. What are DRIs and how are they determined and expressed?
2. What are the various components of the DRIs?
3. What specifically are the adult DRIs for protein, physical activity, and fiber?
4. Based on your current body weight and typical Calorie intake, calculate your DRI for protein and fiber.
5. Which nutrients have and don't have DRIs?
6. What are all of your DRI values?
7. After converting the percent RDI for vitamin C and calcium to the milligram value per serving in your favorite food, calculate the percent DRI provided by one serving of the food for you.
8. What is your calculated EER?
9. How many grams of carbohydrate, sugar, protein, and fat do you consume each day? Using those values, calculate your percentage of Calories from these energy-producing nutrients. Does your intake pattern meet the AMDRs?
10. How are DRIs used? How can you use them to improve your diet and health?

FUN-DUH-MENTAL PUZZLE

ACROSS
4. A risk factor for heart disease.
5. Organ that makes bile and stores glycogen.
6. Disease caused by excess sugar in the blood stream.
9. Process causing protein shape to change in response to heat.
10. Food group that decreases cancer risk.
12. Clinical name for protein deficiency.
15. Type of food that may contain trans fatty acids.
17. Transportation vessels for water-soluble substances.
19. Anabolic process by which plant cells make carbohydrates.
20. Action that moves food along the gastrointestinal tract.
22. Fiber does what to transit time?
24. Energy source of the brain.
25. People with type 2 diabetes usually have insulin _____.

DOWN
1. Amino acid intermediate that causes arterial damage.
2. Enzyme that breaks down starch.
3. Digestion involving muscles and nerves.
7. Protein type that has a uniform alpha-helix or beta-sheet formation.
8. Process generating sticky lipid fragments that increase atherosclerosis.
11. An excellent food source of fiber.
13. Leading cause of arterial wall injury and atherosclerosis.
14. A soluble fiber found in apples.
16. Gallbladder secretion that emulsifies fat.
18. Monounsaturated fatty acids decrease this type of serum cholesterol.
21. Hormone made in the pancreas to reduce blood sugar levels when needed.
23. Common allergic food.

- **Study Activities:** A variety of online or small-screen device activities have been created from the Glossary. Terms and their definitions have been used to create flashcards, fill-in-the-blank questions, matching games, and crossword puzzles. We have created these activities so that you have a fun way to master the terminology and learn the material.
- **Homework Assessment:** Each learning module has a homework assessment that assesses mastery of the module content. There are two different versions of the assessment by module—one printed at the end of each module and one available in the online Instructor's Resources. Submitted homework is instantaneously graded and reviewable, a huge benefit for students and instructors.
- **Total Recall:** At the conclusion of each module, there are learning exercises called "Total Recall." The Total Recall section includes 10 questions, a case study with 10 questions, and a crossword puzzle with 25 questions about the content of the module. These have proved to be excellent tools for reinforcing the skills, knowledge,

and competencies of the module. Students can submit the answers to the questions, case study, and crossword puzzle for extra credit or for required coursework, whichever the instructor desires, from the online platform. Like the Homework Assessments, submitted Total Recall work is instantaneously graded and reviewable. Author research has demonstrated that those students who use Total Recall score significantly higher (typically a grade and a half better) on exams.

- **Exams:** Each learning module has an exam. Exam questions are equally distributed among the content of the five different T-Talk lectures in each module. None of the questions in Homework Assessments, Total Recalls, or exam versions are repeated. Each exam incorporates application of knowledge through use of a case study.
- **Appendices:** Appendices are very valuable resource tools. *Nutrition: Your Life Science* thus incorporates a number of valuable and frequently used appendices.
  - **Appendix A:** Reference Tables (*DRI, UL, AMDR, EER, DRV, and RDI*) and Chemical Structures of Nutrients.
  - **Appendix B:** Facts, Formulas, Conversions, and Sample Calculations
  - **Appendix C:** The MyPlate Food Guidance System
  - **Appendix D:** The Exchange Lists
  - **Appendix E:** Nutrition Resources
  - **Appendix F:** CDC Growth Charts
- A **Glossary** and **Index** are also included at the end of the book.
- **The Online Platform:** This is the virtual textbook home and doubles as the nutrition classroom, whether face-to-face, enhanced, hybrid, or fully online. The online platform hosts:
  - **The eBook**
  - **T-Talk Lectures**
  - **Homework Assessment Submission**
  - **Total Recall Submission**
  - **The Exam Bank**
  - **Weblinks:** The content of the textbook is heavily researched and referenced. Most of the references are from the primary nutrition science literature. There are a number of supportive Web references in each module. The Web references have weblinks that can be easily accessed from the eBook. The online platform of the textbook also has weblink pages by learning module.
  - **Animations and Videos:** Some topics warrant an animation or short instructional video. When available and applicable, these resources are available by module in the online platform.
  - **Communication:** Vital to success in any learning environment is communication. The online platform for the textbook offers e-mail, discussions, and chats.
  - **Diet Analysis Plus Software:** Diet Analysis Plus software is used in some Homework Assessments, the DA+ GENEie in each learning module, and the Module 4 exam. The online platform for the textbook has the gateway to the most current version of the Diet Analysis Plus software and database.
  - **Study Activities:** A variety of study activities, including flashcards and puzzles, have been made from the complete glossary by learning module. Many study activities can be downloaded to a small-screen device for on-the-go learning entertainment.

- **Syllabus:** An extensive and comprehensive syllabus is available from the online platform. The syllabus speaks to each attribute of the course and integrates the printed and online textbook components, which were designed to teach a three-credit-hour, college-level, general education class on foundations in nutrition life science. Since a syllabus is, in essence, a course contract, the syllabus reflects the estimated 140 hours of student time needed to successfully complete such a class.

Ancillaries

A number of ancillaries are available for instructors online in the Instructors Resources folder and on the accompanying PowerLecture DVD. Most of these ancillaries are described above under "The Online Platform" and include:

- Five PowerPoint lectures per module, which coincide with the five T-Talk content areas
- Tables, figures, and other images used throughout the hybrid textbook
- Animations of key concepts
- Short video clips
- Clicker response card activities
- Learning exam review games
- Weblinks
- Exam banks
- Study activities
- Syllabus
- Weekly e-mail communication to help students stay on task
- Version B of the Homework Assessments

Acknowledgments

The authors would like to thank their loving and wonderful families. The Turley and Thompson family support systems have been entertained for many years with stories of the progress of this project, and the continual updating and integration of their food and fitness lives are seen in many of the photographs used. We have been devoted to creating and producing materials that not only develop academic skills, but dramatically change lives for the better. We celebrate the benefactors of our efforts, including faculty, students, and their families.

We also wish to thank the 30,000 students who have been touched by this product during its development at Weber State University.

This work would not be possible without the vision, trust, and commitment of the Cengage team.

Influential Reviewers and Focus Group Participants A number of colleagues around the country have reviewed these materials at various stages. Their comments—critical, occasionally enthusiastic, and always professional—have helped shape the version you see here. We appreciate and value their time and contributions.

Bernice Adeleye
University of Louisiana—Lafayette

Andrea E. Altice
Florida State College at Deerwood

Charalee Allen
Cincinnati State University

Emily Anschlowar
Molloy College

Pat Baird
Norwalk Community College

Patsy Beffa-Negrini
University of Massachusetts—Amherst

Virginia Bennett
Central Washington University

Rita Berthelsen
Iowa Western Community College

Joye M. Bond
Minnesota State University—Mankato

Anne Bridges
University of Alaska—Anchorage

Ardith Brunt
North Dakota State University

Cinda Catchings
Alcorn State University

Barbra Cerio-Loco
Rochester Institute of Technology

Melissa Chabot
University at Buffalo

Katie Clark
University of San Diego
University of California—San Francisco

Nicholle Clark
College of the Desert

Paula K. Cochrane
Central New Mexico Community College

Tammie Collum
University of Northern Iowa

Cathy Cunningham
Tennessee Technological Institute

Joanne DeMarchi
Saddleback College

Mary Ann Donnell
State University of New York—Oneonta

Rachelle D. Duncan
Oklahoma State University—Institute of Technology

Betty J. Forbes
West Virginia University

William Forsythe
East Carolina University

Crista Galvin-Cox
John Tyler Community College

Ying Gao-Balch
University of Arkansas

Melodye Gold
Bellevue College

Shelby Goldberg
Pima Community College

Jill Goode Englett
University of North Alabama

April Graveman
Ellsworth Community College

Michele Grodner
William Paterson University

Sandra S. Haggard
University of Maine

Nancy Harris
East Carolina University

Lee Harrison
Marywood University

Dawn Hedges
Bowling Green State University

Patricia Henry
Lone Star College—North Harris

Lisa Herzig
California State University—Fresno

Suzanne Hewitt
Saddleback College

Dennis Hunt
Florida Gulf Coast University

Ann Hunter
Wichita State University

Teresa Johnson
Troy University School of Nursing

Walt Justice
Southwestern College

Stephen Kabrhel
Community College of Baltimore County

Kathleen Laquale
Bridgewater State University

Kris Levy
Columbus State University

Scott T. Macdowall
Quinnipiac University

Sarah Murray
Missouri State University

Katherine Musgrave
University of Maine

Carmen Nochera
Grand Valley State University

Esther Okeiyi
North Carolina Central University

Yi-Ling Pan
North Carolina A&T State University

Elizabeth Quintana
West Virginia University

Peggy Ramsay
Fuller Community College

Yeong Rhee
North Dakota State University

Carol Ruwe
Barton College

Claire Schmelzer
Eastern Kentucky University

Matt Schmidt
Southern Utah University

Karen Schuster
Florida State College at Jacksonville

Claudia Sealey-Potts
University of North Florida

Debra Sheats
St. Catherine University

Tammy J. Stephenson
University of Kentucky

Judy Swanson
Minot State University

Priya Venkatesan
Pasadena City College

Janelle Walter
Baylor University

Nancy Zwick
Northern Kentucky University

Nutrition Basics

Valentyn Volkov/Shutterstock.com

LEARNING OBJECTIVES

When you complete this learning module, you will:

- Know basic nutrition terminology.
- Understand the order of life.
- Be able to differentiate good nutrition from malnutrition and identify the characteristics of a sound diet.
- Know the fundamentals of each of the six categories of nutrients in human nutrition: carbohydrates, proteins, lipids/fats, vitamins, minerals, and water.
- Be introduced to basic information about each nutrient category, including simple chemical characteristics, functions, general food sources, human needs for the essential nutrients, and the energy yield of each nutritional substance.

PERSONAL IMPROVEMENT GOALS

This learning module is designed to empower you to:

- Adopt and practice the essentials of eating a sound diet.
- Choose to eat a balanced diet.
- Practice dietary moderation.
- Drink plenty of fluid each day.
- Eat more whole, fresh foods each day.
- Take responsibility for your daily dietary intake.

Scientific research has firmly established the link between a healthy diet and good overall health; likewise, the link between a poor diet and a variety of chronic diseases is clear. Growing older is inevitable, and as a person's bodily functions decline, chronic disease may develop. But a healthy lifestyle can slow the rate of aging, and eating healthy foods is a critical component of that lifestyle. The diet a person chooses to eat has a profound daily effect on his or her function, performance, and health.[1]

In this learning module, you will be introduced to the basics of nutrition and will begin to learn about the components of a healthy diet. You will be exposed to many new terms used regularly in science and in the field of nutrition. Furthermore, you will learn about the six categories of essential nutrients and will see how many nutrients are required by the human body each day. You will also learn how many substances that are consumed in a typical American diet are harmful to human health. By learning about the nutritional value of food, choosing foods that promote health, and avoiding those foods that undermine health, you can maintain your quality of life at a higher level.[31]

1.1 NUTRITION BASICS AND TERMINOLOGY

Introduction

Every discipline has its own vocabulary or terminology, and nutrition is no exception. The following section provides a quick introduction to basic nutrition terms; next we will look at the characteristics and results of consuming a sound diet. The purpose is to become knowledgeable about the nutrients in food, and then you can eat a good diet and make wise food choices so that you are better able to enjoy a long and healthy life. We will describe how sound dietary practices can have profound effects on an individual's health and longevity. It is one thing to understand factors that affect a person's longevity but another to understand why a person chooses to eat or behave in particular ways. Thus, the factors that affect food choice will also be addressed.

T-Talk 1.1
To hear Dr. Turley talk about nutrition basics and terminology, go to www.cengagebrain.com

Common Nutrition Terms

Becoming familiar with the following terminology and concepts will help you understand the basics of the study of nutrition:[3,4,29]

Diet: The kind and amount of food consumed each day.

Food: Anything edible that nourishes the body.

Nourish: To provide food or other substances necessary to sustain life and support growth. All parts of the body need to be nourished; from the individual cells to tissues, organs, organ systems, and the organism (see BioBeat 1.1).

Nutrition: The act or process of nourishing or being nourished. The processes of ingestion, digestion, absorption, assimilation, and excretion of food are part of nutrition. It is also the science or study of the nourishment of humans or other creatures.[60]

Nutritional sciences: The study of nutrition, including dietary components and **metabolism**.

Nutrients: Molecular substances that provide nourishment to cells and thus every multicelluar component of the human organism (see Table 1.1 and BioBeat 1.1). Nutrients can be essential (nutritional deficiency signs and symptoms occur without intake) or nonessential (nutritional deficiency signs and symptoms do not occur without intake). They can also be energy-producing or non-energy-producing.[35,45]

Energy-producing: A substance that provides Calories when metabolized (converted from food to energy) by the body. The energy-producing and non-energy-producing nutrients are shown in Table 1.1.

Calorie: The unit used to measure energy. One Calorie is the amount of heat energy required to raise 1 kilogram of water 1 degree Celsius from 36 to 37 degrees Celsius (actually a kilocalorie, Kcal, or Calorie denoted with a capital C). Calorie values in food are estimated by the bomb calorimeter (see Figure 1.1).

Nutritious: A food that provides a high degree of nourishment.

Nutrient density: The amount of nutrients relative to the number of Calories in a given quantity of food (see Figure 1.2).

Alison Wright/CORBIS

Deficiency of an essential nutrient causes specific signs and symptoms. Here a woman deficient in iodine has developed a goiter (enlarged thyroid gland).

DA+ GENEie Go to Diet Analysis Plus software and enter 12 ounces of regular soda, 1% milk, and real orange juice. Use the intake spreadsheet to compare the nutrient analysis results. Draw your conclusions about the nutrient density of these beverages.

metabolism The conversion (anabolic or catabolic) of a substance from one form to another by a living organism.

TABLE 1.1

The Six Categories of Nutrients in Human Nutrition

Energy-Producing Caloric Nutrients: Function and Values	Non-Energy-Producing Noncaloric Nutrients: Functions and Values
Carbohydrates:	**Vitamins:**
Preferred fuel	Regulate metabolism
4 Calories per gram	0 Calories per gram
Proteins:	**Minerals:**
Tissue repair, maintenance, and growth	Structural and regulate metabolism
4 Calories per gram	0 Calories per gram
Fats:	**Water:**
Sustaining fuel	Medium for metabolism and nutrient support
9 Calories per gram	0 Calories per gram

© Cengage Learning 2013

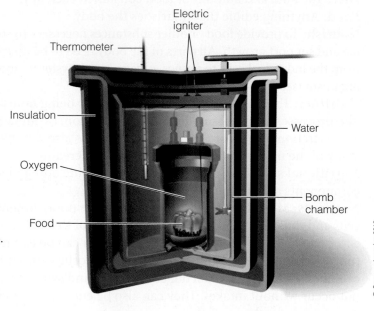

FIGURE 1.1 The bomb calorimeter is used to estimate the Calorie value of food. The transfer of heat energy from the combusted (burnt) bell pepper to the water raises the water temperature; a Calorie value is estimated based on how many degrees the water temperature goes up.

© Cengage Learning 2013

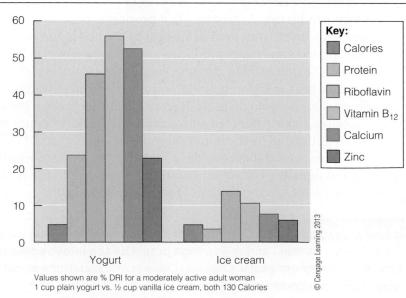

FIGURE 1.2 Nutrient density of yogurt and ice cream compared. With less fat and sugar to add Calories, yogurt has a higher nutrient density than ice cream.

Key:
- Calories
- Protein
- Riboflavin
- Vitamin B$_{12}$
- Calcium
- Zinc

Values shown are % DRI for a moderately active adult woman
1 cup plain yogurt vs. ½ cup vanilla ice cream, both 130 Calories

© Cengage Learning 2013

biological classification The scientific organization of living organisms on earth into the categories kingdom, phylum, class, order, family, genus, and species.

cell The smallest structural and functional unit of life in all known living organisms.

BioBeat 1.1

Molecules to Cells to Organisms in the Order of Life

In biology, the scientific classification of life forms on earth is known as **biological classification**. Biological classification is a form of scientific taxonomy that follows the classification through domain, kingdom, phylum, class, order, family, genus, and species. The **cell** is the smallest structural and functional unit of life among all known living organisms. It is often called the building block of life and comes into life by division of a pre-existing cell. Cells have the capacity for metabolism, homeostasis, growth, and reproduction. **Molecules** from the six categories of biological molecular nutrients in human nutrition provide the raw materials for cell structure. More complex living systems have several types of cells and are referred to as **multicellular organisms**. Cells are individually alive while contributing to a multicellular organism (see Figures 1.3 and 1.4). Humans have an estimated 100 trillion or 10^{14} cells. There are 210 different cell types in the human body. Figure 1.4 illustrates the parts of a cell and describes the many functions that occur within each one.[59]

How many different cell types can you think of?

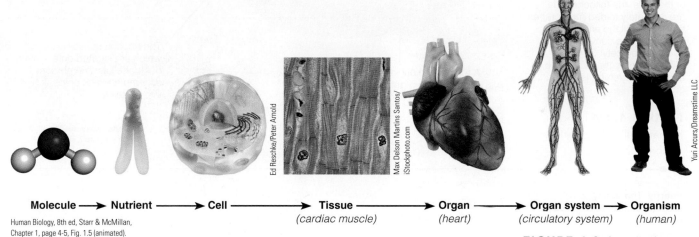

Molecule → Nutrient → Cell → Tissue *(cardiac muscle)* → Organ *(heart)* → Organ system *(circulatory system)* → Organism *(human)*

Human Biology, 8th ed, Starr & McMillan, Chapter 1, page 4-5, Fig. 1.5 (animated).

FIGURE 1.3 An overview of the levels of organization in humans.

Substances That Nourish the Body: An Overview of the Nutrients

Individuals consume foods based on a variety of factors and personal preferences. The combination of foods eaten in a particular day makes up an individual's diet and has the potential to greatly impact his or her **health**. This next section will introduce several features of the diet and foods. The components in food that participate in sustaining life can provide energy (Calories) directly or participate in necessary chemical reactions or be used for tissue structures (see BioBeat 1.2). These food components are carbohydrates, proteins, lipids/fats, vitamins, minerals, and water (Table 1.1). Carbohydrates, proteins, and fats are the energy-producing nutrients and are termed **macronutrients**, because they are required in large amounts. The vitamins and minerals are the non-energy-producing nutrients and are considered **micronutrients**, because they are required in very small amounts. Water is required based on energy expenditure in an amount necessary to stay in **water balance** (intake is balanced with loss via excretion); achieving water balance is not the amount of water needed to optimize health and body function, because function may depend on a greater amount to support metabolism, exercise levels, and environmental conditions.[25,28,53]

molecule Two or more atoms (the smallest component of an element) held together or stabilized by a chemical bond.

multicellular organisms Complex living systems that have several types of cells in the life form.

health Complete physical, mental, and social well-being; not just the absence of infirmity.

macronutrient Any of the categories of energy-producing nutrients: carbohydrates, proteins, and fats.

micronutrient Any of the categories of non-energy-producing nutrients: vitamins and minerals.

water balance When the amount of water consumed equals the amount lost by the body, and equilibrium is achieved.

Inside the cell membrane lies the cytoplasm, a lattice-type structure that supports and controls the movement of the cell's structures. A protein-rich jelly-like fluid called cytosol fills the spaces within the lattice. The cytosol contains the enzymes involved in glycolysis.

A separate inner membrane encloses the cell's nucleus.

Inside the nucleus are the chromosomes, which contain the genetic material DNA.

This network of membranes is known as smooth endoplasmic reticulum—the site of lipid synthesis.

Known as the "powerhouses" of the cells, the mitochondria are intricately folded membranes that house all the enzymes involved in the conversion of pyruvate to acetyl CoA, fatty acid oxidation, the TCA cycle, and the electron transport chain.

A membrane encloses each cell's contents and regulates the passage of molecules in and out of the cell.

The ribosomes, some of which are located on a system of intracellular membranes, assemble amino acids into proteins.

FIGURE 1.4 A living human cell with its variety of components.

Understanding Nutrition, 11th ed, Whitney & Rolfes, Chapter 7, Fig. 7-1, p. 214.

---

energy The ability to do work (chemical, mechanical, or osmotic).

thermodynamics The processes involved in the conversion of energy to work (mechanical, chemical, or osmotic) and the energy lost as heat.

kinetic The motion of molecules, related to the ability to do work.

potential Stored energy, in the context of the ability to do work.

essential nutrients Substances that are found in food and are needed by the body, but are not made by the body in amounts sufficient to meet physiological needs.

adenosine triphosphate (ATP) The ultimate form of energy and a short-term energy store generated by converting adenosine diphosphate (ADP) to ATP.

mitochondria The cellular organelles that generate most of the cell's supply of adenosine triphosphate (ATP), which is the ultimate source of chemical energy used by cells to do work.

BioBeat 1.2

Food, Energy, and Me

All living things abide by the laws of **energy** and **thermodynamics** as manifested in biochemical reactions. Energy can be **kinetic** (the ability to do work) or **potential** (stored energy) in nature. The laws of thermodynamics say that energy is neither created nor destroyed; it can only be transferred from one form to another. The way that energy is transferred to sustain life is through nutrition. Food nourishes living systems by providing energy-producing nutrients and non-energy-producing nutrients, both of which can provide **essential nutrients**. The energy-producing nutrients—carbohydrates, proteins, and fats—are used to produce **adenosine triphosphate (ATP)** inside **mitochondria**, also known as the powerhouses of cells (see Figure 1.5).

ATP functions as the body's short-term energy storage center. ATP is generated by converting adenosine diphosphate (ADP) to ATP. This chemical reaction requires energy. The addition of the third phosphate to adenosine results in an energy transfer from the carbon bonds in energy-producing nutrients. Hence, the potential energy in carbohydrates, proteins, and fats is used to produce ATP, and ATP can be used kinetically to supply the energy required to do cellular work. The non-energy-producing essential nutrients, though not metabolized directly to yield ATP, participate in metabolism.[25,29]

What is the ultimate form of energy used by the body to do any type of work?

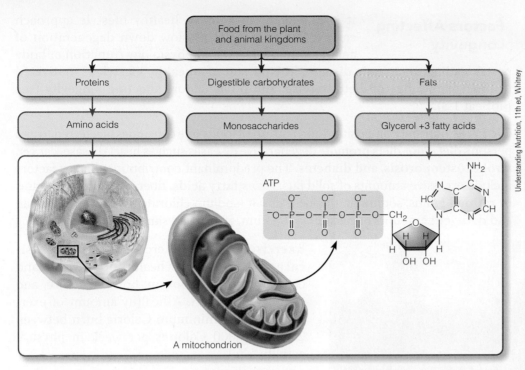

Understanding Nutrition, 11th ed, Whitney & Rolfes, Chapter 7, Fig. 7-1, p. 214

FIGURE 1.5 Food to ATP energy from carbohydrates, proteins, and fats. Inside the mitochondria of the cell, carbon bonds are broken to form carbon dioxide and water while ATP is produced.

Characteristics of a Sound Diet The following terms are often used to describe a sound diet. A sound diet is identified by:

Calorie control: Selecting foods that are nutrient dense so that food and energy intake maintain a healthy body weight.

Adequacy: A diet that provide essential nutrients, **fiber**, and energy (Calories) in amounts adequate or sufficient to maintain health.

Balance: A diet that provides an appropriate number of servings from a variety of food types that complement one another.

Moderation: A diet that avoids excesses of unhealthy substances.

Variety: A diet in which different foods that provide the same nutrients are chosen.

Results of a Sound Diet To stay healthy, the human body requires many different nutrients. To sustain life minimally, the body needs energy to do work, building materials to maintain form and structure, and agents to control these processes. A diet that is based on the principles of Calorie control, adequacy, balance, moderation, and variety promotes physiological health, but there is more to health than that.

The World Health Organization defines health as the state of complete physical, mental, and social well-being; health is not just the absence of infirmity. Our philosophical statement about health is that, "Healthy lifestyle behaviors promote health, and unhealthy lifestyle behaviors promote disease." This means that a person whose lifestyle is unhealthy, even if he or she is disease-free at the moment, should not be labeled as a healthy person.

Results of a Poor Diet When the diet and lifestyle are not sound, the result is **malnutrition**. This simply means bad nutrition. Malnutrition is the impairment of health resulting from deficiency, toxicity, or imbalance of nutrient intake or body utilization. It includes both **overnutrition** and **undernutrition**. Overnutrition can result from a single nutrient toxicity or from excess Calorie intake leading to obesity. Undernutrition can result in a single nutrient deficiency or lack of sufficient Calories or starvation.[8,48,61,65]

fiber Plant polysaccharides, such as cellulose, that are composed predominately of repeating units of glucose hooked together by beta bonds, and are indigestible by humans.

malnutrition Bad nutrition; can include over- and undernutrition related to the intake of too much or too little of a nutrient, energy, and/or Calories.

overnutrition Excess intake of energy and/or nutrients.

undernutrition A lack of (or deficiency in) energy and/or essential nutrients.

Factors Affecting Longevity

It is important to adopt a healthy lifestyle approach early in life to prevent or slow down degeneration of anatomy (body parts) and physiology (function of body parts) leading to **chronic disease**. The pursuit of a healthy lifestyle is motivated through knowledge of diet, exercise, and other factors affecting longevity (see BioBeat 1.3).[6,39,42]

Diet The nutrient composition and pattern of foods one consumes is referred to as a person's diet. Poor diets promote degenerative diseases such as heart disease, cancer, **stroke, osteoporosis**, and **diabetes**. The predominant contributing dietary factors include excessive amounts of solid fats, *trans* **fatty acids**, fiber-poor carbohydrates or **refined** grains, sodium in the form of salt (sodium chloride), **alcohol**, and sugar, and **deficient** amounts of calcium, potassium, and magnesium.[A,B,C,2,21]

Exercise affects longevity.

Exercise Adequate exercise (also called **physical activity**) promotes healthy body weight and composition, metabolism, bone structure, and cognitive function. A healthy amount of exercise creates a minimum Calorie burn between 2,000 to 3,000 Calories per week in physical activity.[40,58,71]

Other Factors Smoking, other habits (e.g., inadequate sleep, alcohol and drug use, and unsafe sex), accidents, and disease-promoting microorganisms (such as hepatitis, which is caused by viruses) affect the development of degenerative diseases. Genetics is also a factor in the development of degenerative diseases. Certain genetic predispositions can increase the importance of maintaining a healthy diet and lifestyle.[41]

Factors Affecting Food Choices

Food choices impact health, both now and later. Food choices affect disease and stress resistance as well. Some influences are **physiological**, some are **psychological**, some are **sociological**, and others are cultural. Food selection is actually very complex (see Figure 1.6 and BioBeat 1.4).[9,11,12,27,31,34,43,46,62] Nutritional knowledge can maximize the physiological benefits of nourishment, and the purpose of eating (your food choices) will be to fuel and nourish your body optimally.[14] The following factors affect your food choices:

Hunger: The physiological need for food. The physical body sends signals indicating a need for food caused by nutrient store depletion and the need to replenish.

Satiety: The physiological feedback mechanism that terminates food intake. The stomach is full and the person should stop eating. Many signals are sent to the brain that make it easy to stop food intake.

Appetite: The psychological desire for food. The brain sends signals indicating a desire for food because of sensory inputs such as seeing, smelling, or thinking about food. There is no need to replenish nutrients.

Personal preferences: The food likes and dislikes of individual people that often have sociological and cultural meaning.

Availability: Food supply, geographic area, climate, soil, and labor affect food availability. Americans have the most abundant food supply in the

world. This abundance can lead to overnutrition, which leads to some degenerative diseases, especially heart disease, stroke, diabetes, and cancer.
Economics: Social status, income, and misperceptions pertaining to food, such as the notion that healthy food costs more, all determine what kinds of foods are consumed.
Social factors: Family, friends, holidays, celebrations, weight-loss groups for support, or simply what a person is raised on or surrounded by determines what that person will eat in a given setting.

FIGURE 1.6 How a variety of factors can affect hunger, appetite, and satiety.

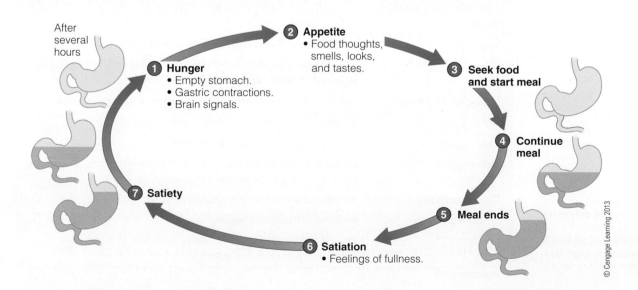

After several hours

1 **Hunger**
 • Empty stomach.
 • Gastric contractions.
 • Brain signals.

2 **Appetite**
 • Food thoughts, smells, looks, and tastes.

3 **Seek food and start meal**

4 **Continue meal**

5 **Meal ends**

6 **Satiation**
 • Feelings of fullness.

7 **Satiety**

© Cengage Learning 2013

Family and ethnic traditions: Beliefs, values, customs, morals, and cultural variation all affect food intake.

Advertising: Some of the American population is becoming more health conscious. Advertising via television, radio, magazines, newspapers, and food manufacturers (product promotion) can influence food purchase and thus food intake.

Other: Habits, feelings associated with certain foods, time, health beliefs, and nutrition knowledge can influence food intake.

Knowing some of the factors that affect food choice, one should keep in mind the nutrition-educated person's goal for eating, which is to consume foods that fuel and nourish the body optimally.

BioBeat 1.4

Eating Styles: All Creatures Eat Something

All forms of life must be nourished. The biological molecules used to nourish life forms come from the six categories of nutrients. These are the molecules of life and are commonly utilized by a wide variety of life forms through metabolism. Metabolism is the concert of biochemical processes that support life. Animals consume the molecules of life in the form of food. Based on religious preferences, cultural norms, ethical issues, or personal choice, many people exclude certain domains of food sources. A few eating styles are worth noting: **vegan**, lacto-ovo vegetarian, **omnivore**, and a wide variety of semivegetarians (see Table 1.2). Vegans will not consume any animal products in their diets. Neither honey nor gelatin are consumed. Gelatin is omitted because it is made from the collagen that was boiled out of animal bones. Lacto-ovo vegetarians eat no muscle meats but will include milk products and eggs in their diets. Omnivores eat both plant and animal foods. Most Americans are omnivores. Semivegetarians include pollo-vegetarians, who eat only chicken as their source of muscle meat, and pesco-vegetarians, who eat only seafood as their source of muscle meat.[2,33,66,68]

What eating style do you have?

TABLE 1.2	
Types of Vegetarianism	
Term	**Dietary Style**
Vegetarian	Includes plant foods and limits or restricts animal foods
Vegan	Includes plant foods only (grains, vegetables, fruits, nuts, seeds, legumes); a strict vegetarian
Lacto-ovo vegetarian	Includes dairy products and eggs within a vegetarian diet
Lacto-vegetarian	Includes dairy products within a vegetarian diet
Ovo-vegetarian	Includes eggs within a vegetarian diet
Semivegetarian	Includes a vegetarian diet with restrictions on certain animal foods or amounts of animal foods
Pollo-vegetarian	Includes chicken within a vegetarian diet
Pesco-vegetarian	Includes fish within a vegetarian diet
Fruitarian	Includes raw or dried fruits, seeds, and nuts in the diet; a nutritionally inadequate eating style

© Cengage Learning 2013

vegan An eating style that includes consuming only plant foods.

omnivore An eating style that includes consuming plant and animal foods.

Summary Points

- Diet is the collection of food and fluids consumed by an individual within a 24-hour period.
- Food nourishes the body by providing nutrients that enable cellular functioning; nutrients can be essential, nonessential, caloric, or noncaloric.
- Nutrition is the study of how food nourishes and affects body function throughout the day and a person's health over several years.
- The six categories of nutrients include the energy-producing macronutrients (carbohydrates, proteins, and fats), the non-energy-producing micronutrients (vitamins and minerals), and water.
- The goal of eating should be to fuel and nourish the body optimally, noting that selecting nutrient-dense foods is key for Calorie control.
- It is important to consume a healthy diet in order to promote health and prevent chronic disease.
- Many factors affect food choice.

Take Ten on Your Knowledge Know-How

1. What do the terms *food, diet, nutrition, nutrient, Calorie, nutrient density, caloric density, health, malnutrition, hunger, appetite,* and *satiety* mean? How does knowing the aforementioned terms affect your view of food in today's society?
2. What is the basic unit of life?
3. What is the difference between an essential and a nonessential nutrient? What are some examples of each in your body? How do they support cell structure and function?
4. What is energy, and how is food energy converted to usable energy inside your body?
5. What are the energy-producing (caloric) and non-energy-producing (noncaloric) nutrients? What are their primary functions in your body?
6. What are the characteristics of a sound diet?
7. What are the results of a sound versus a poor diet? What changes can you make so that your diet is sound?
8. How do diet, exercise, and other factors affect longevity and the development of degenerative diseases? What can you do to age more gracefully?
9. Why do you eat what you eat? List several factors that affect your food choices.
10. What type of eating style do you have? What are some other eating styles practiced by those around you?

1.2 CARBOHYDRATES

T-Talk 1.2

To hear Dr. Turley talk about carbohydrates, go to www.cengagebrain.com

Introduction   The class of nutrients known as **carbohydrates** provides energy for the brain so that you can read this text, as well as energy for the muscles so that you can run a mile or walk up stairs. We will explore this important family of nutrients in terms of its composition, energy yield, functions, categories, recommended intake levels, and food sources.[37,72]

Composition of Carbohydrates   Carbohydrates are made of the elements carbon, hydrogen, and oxygen in a typical molecular relationship. For every carbon in the molecule, there are twice the number of hydrogen molecules and the same number of oxygen molecules. Six-carbon sugars such as glucose dominate in food. One glucose molecule ($C_6H_{12}O_6$) plus six oxygen

carbohydrates Organic compounds composed of carbon, oxygen, and hydrogen that provide the preferred fuel of the body; categorized as simple and complex; many provide 4 Calories per gram (except fibers, which are indigestible and thus noncaloric).

molecules ($6 O_2$) break down to six carbon dioxide molecules ($6 CO_2$) plus six water molecules ($6 H_2O$) plus 36 adenosine triphosphate (ATP):

$$C_6H_{12}O_6 + 6 O_2 \leftrightarrow 6 CO_2 + 6 H_2O + 36 \text{ ATP}$$

This is a perfectly balanced chemical reaction with no toxic waste products or **metabolites** produced. Thus, carbohydrate as an energy source for the body is excellent. It is a clean-burning, high-performance fuel (the fastest fuel to use for ATP production).

By scientific definition, **organic** means that the chemical compound contains the element carbon. A chemical compound is a structure with several elements linked together. Therefore, carbohydrate is an organic compound. Other lay meanings of the term *organic* can be a substance that is derived from a living organism or a food that is grown or raised without synthetic fertilizers, pesticides, hormones, or medications.

Energy Yield and Functions of Carbohydrates

Carbohydrates such as **simple sugars** and **starch** provide 4 Calories per gram. Carbohydrates have the fastest rate of energy production, and they are the preferred fuel for the body. It is the primary fuel used in high-intensity exercise and the only fuel used to make ATP by the brain, central nervous system, and red blood cells. Even though it has a low caloric density of 4 Calories per gram, it is quick to be converted to ATP.

Categories of Carbohydrates

Carbohydrates are divided into simple sugars and **complex carbohydrates** from a nutrition perspective.[63]

Simple Sugars Common dietary sugars are single (mono) or double (di) units of six-carbon sugars (saccharides). These simple sugars are thus **monosaccharides** and **disaccharides** (see Figure 1.7). Let's study and learn the six most common simple sugars consumed in the diet and used by humans.

Monosaccharides Single basic molecular units of sugar are called monosaccharides. The three most common single units of dietary carbohydrate include **glucose**, **galactose**, and **fructose** (see Figure 1.7). It will be apparent that the majority of carbohydrates consumed in the diet consist mostly of glucose, galactose, and fructose.

Disaccharides Double units of sugar are called disaccharides. The three most common double units of dietary carbohydrate include **sucrose**, **lactose**, and **maltose** (see Figure 1.7). Sucrose is made up of glucose and fructose. Lactose is made up of glucose and galactose. Maltose is made up of two units of glucose.

Glucose is a very important biological molecule. It is the form of carbohydrate in the blood, and it is referred to as blood sugar. Fructose occurs naturally in fruit, so it is known as fruit sugar, but high-fructose corn sweetener is artificially produced and is often added to processed foods. Sucrose is table sugar and is the molecule that dominates in any type of sugar product, such as brown sugar, granulated sugar, powdered sugar, honey, or raw sugar. Maltose, found in fermented grain products and processed foods, is known as malt sugar, and lactose is the type of carbohydrate in milk and dairy products that is known as milk sugar.

Complex Carbohydrates There are two types of complex carbohydrates common in whole foods: starch and fiber. Both starch and fiber are **polysaccharides**. This

metabolites The products (molecules) of biochemical reactions or metabolism.

organic A carbon-containing substance or molecule; in lay terms or on food labels, means organically produced.

simple sugars Carbohydrates that are monosaccharides or disaccharides.

starch Plant polysaccharides, such as amylose and amylopectin, that are made up of repeating units of glucose hooked together by alpha bonds, that are digestible by humans.

complex carbohydrates Polysaccharides composed of straight or branched chains of monosaccharides.

monosaccharide Single units of sugar that typically form a single ring and include glucose, fructose, and galactose.

disaccharide Double units of sugar including sucrose, maltose, and lactose; two monosaccharides chemically bound together.

glucose A monosaccharide, sometimes called blood sugar.

galactose A monosaccharide that is part of the disaccharide lactose.

fructose A monosaccharide, sometimes called fruit sugar.

sucrose A disaccharide that is made up of glucose and fructose; commonly called table sugar.

lactose A disaccharide that is made up of glucose and galactose, commonly called milk sugar.

maltose A disaccharide that is composed of two units of glucose; can be referred to as malt sugar.

polysaccharides Organic compounds composed of many monosaccharides chemically linked together; also referred to as complex carbohydrates (such as starch and fiber).

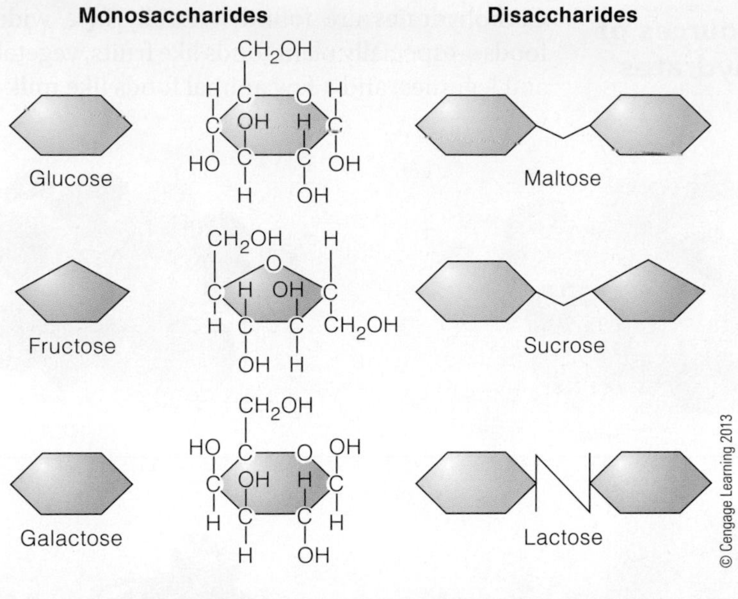

FIGURE 1.7 The chemistry of the most common simple sugars, the monosaccharides and disaccharides.

means they are made of many units of sugar (see Figure 1.8). Eating combinations of these carbohydrates is important for maintaining good health.[23]

Starch Amylose and amylopectin are types of starches present in foods. Starches are large molecules made of repeating glucose units. The type of bond that links the glucose units together distinguishes the complex carbohydrate. Starch contains alpha-linked glucose and bonded molecules. The alpha-amylase **enzyme** is the enzyme that can break the alpha bond and thus make the glucose molecule available for absorption and utilization. Then when glucose is utilized, energy is produced.

Fiber There are many types of **dietary fiber**. The most common fiber type consumed in the diet, cellulose, consists of large molecules made of repeating glucose units. Unlike the starch amylose, all fibers, including the fiber cellulose, are made of beta-bond-linked glucose units. The beta-link between the glucose molecules is impervious to acid and enzymes. This means that the bonds between the glucose molecules in fiber are not broken, so the glucose cannot be absorbed or utilized. Thus, fiber cannot be used for energy production. Because the glucose units cannot be broken apart or absorbed, fiber is indigestible and noncaloric. Although it cannot be used for energy production, fiber is important for digestive health. Fiber is abundant in whole plant foods like **legumes**, fruits, vegetables, and whole grains.[63]

enzyme A protein synthesized by cells to catalyze, or facilitate, a specific chemical reaction involving other substances without itself being altered.

dietary fiber The nonstarch polysaccharides in plant foods that are not digested by human digestive enzymes, although some are digested by bacteria in the gastrointestinal tract.

legumes Plant foods from the bean and pea family, with seeds that are rich in protein compared to other plant-derived foods.

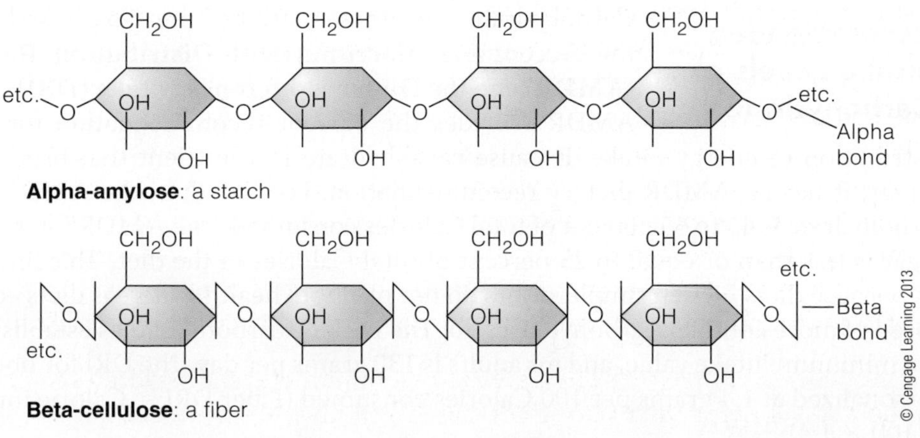

FIGURE 1.8 The chemistry of the complex carbohydrates (polysaccharides), starch and fiber.

Food Sources of Carbohydrates

Carbohydrates are found naturally in a wide variety of foods—especially plant foods like fruits, vegetables, grains, and legumes, and a few animal foods like milk and honey.

Plant sources of simple sugars.

Plant sources of complex carbohydrates.

Plant Sources The Calories from plant foods are mostly provided by carbohydrates, though plant foods can also provide protein and fat. The food sources rich in carbohydrates include grains, cereals, legumes, vegetables, and fruits. Plant food sources of simple carbohydrate (sugars) include fruits, fruit juices, sweets, sugars, and sugar-sweetened cereals. Plant food sources of complex carbohydrates (starch) include pastas, breads, cereals, potatoes, corn, grains, vegetables, and legumes. Fiber is found in whole plant foods rather than refined plant products.[10,13]

Animal Sources Two carbohydrate-rich foods come from the animal kingdom: milk and honey. Both milk and honey provide simple carbohydrate. There are no significant sources of complex carbohydrates from animal foods. Most animal food sources provide Calories from protein and fat rather than from carbohydrate.

Animal sources of simple sugars.

Acceptable Macronutrient Distribution Range (AMDR) A recommended range of Calories expressed as percents for carbohydrate, sugar, protein, fat, and essential fatty acid dietary intake.

Dietary Reference Intake (DRI) A recommended intake value for an essential nutrient, fiber, and Calories; also exists for physical activity.

Recommended Intake Levels of Carbohydrates

Carbohydrate recommendations are expressed as the **Acceptable Macronutrient Distribution Range (AMDR)** and the **Dietary Reference Intake (DRI)**. The AMDR provides the current recommendation for the distribution of energy intake. Because carbohydrate is a nutrient that produces energy, it has an AMDR dietary recommendation. For an adult, the AMDR for carbohydrate is 45 to 65 percent of total Calories consumed. The AMDR for added sugar is less than or equal to 25 percent of total Calories in the diet. This limit is set because diets high in simple sugars do not promote health, whereas diets composed of more complex carbohydrates do. The DRI for carbohydrate is established as a minimum intake value, and for adults is 130 grams per day. The DRI for fiber is personalized at 1.4 grams per 100 Calories consumed (Fiber DRI = Calorie intake ÷ 100 × 1.4).[22,55,56,63]

Alcohol

Alcohol is an organic energy-producing compound containing two carbons and a hydroxyl group. Its chemical formula more closely matches that of a carbohydrate compared to other energy-producing macronutrients. Alcohol provides 7 Calories per gram. Because of its lack of nutritional value and its central nervous system depressant effects, alcohol is considered a drug; thus, one can find the most information on alcohol in pharmacology or advanced nutrition books. Both positive and negative health effects are associated with alcohol use (see BioBeat 1.5). These will be reflected in the dietary recommendations mentioned in Module 2.[7,19,41]

BioBeat 1.5

Alcoholic Beverages from Health to Drunkenness and Disease

Alcoholic beverages contain a chemical compound called ethanol (CH_3CH_2OH). Ethanol, also called ethyl alcohol, is a caloric (7 Calories per gram), carbohydrate-related molecule found in commonly marketed alcoholic beverages like beer, wine, and distilled spirits (liquor). Small amounts of ethanol are present in a variety of products such as mouth rinses, flavorings like vanilla extract, and some fermented products. Most alcoholic beverages provide their Calories from ethanol and carbohydrates with very little protein or fat. Alcoholic beverages provide empty Calories, meaning physiologically insignificant amounts of essential nutrients are provided for the Calories consumed. Because alcoholic beverages do not belong in any food group and provide empty Calories, alcoholic beverages are not nutritious. The caloric values among the alcoholic beverages vary greatly depending on the type and additional ingredients used to make mixed drinks.[D]

A standard drink contains about 14 grams of ethanol, and note the varying percentage concentration of ethanol in alcoholic beverages (see Table 1.3). At the present time, the grams of ethanol are not given on the food package label of alcoholic products. Rather, the percentage of "alcohol" by volume is listed on the container, in small print either on the side or bottom of the container. Many consumers feel that there is a need to improve the nutrition labeling on alcoholic beverages, such as to include the Calorie and alcohol con-

tent per serving, the number of drinks per container, guidance information about how many servings can be consumed per hour safely, as well as any ingredients that may stimulate allergic responses. The standard drink or volume of fluid to deliver the roughly 14 grams of alcohol per serving includes 12 ounces of beer, 5 ounces of wine, or 1.5 ounces of distilled spirits. The recommendation for alcohol consumption by U.S. governmental agencies is that women should consume no more than one standard drink per day, and men should consume no more than two standard drinks per day.[G] This is because alcohol has many detrimental effects in the body. Its most notable effects include a central nervous system depressant effect, a diuretic effect on the renal system (kidney), an inhibitor of muscle and liver carbohydrate storage, and altered nutrient metabolism.

An average adult (150 pounds) can metabolize less than one drink (14 grams of ethanol) per hour, so typically if the alcohol ingestion exceeds two drinks per hour, the blood alcohol concentration (BAC) increases.[E] A BAC of 0.08 indicates that the person is legally impaired, and it is illegal to drive at this level in any state in the United States. It takes only two to four drinks (based on gender and body weight) to achieve a BAC of 0.08 and be legally intoxicated and significantly impaired. With lower BAC (0.02 to 0.06) there are feelings of slight euphoria, reduced inhibitions, the sensation of warmth, minor

impairment of reasoning and memory, and intensified emotions; good emotions are better and bad emotions are worse. With a BAC of 0.07 to 0.09, there is happiness with impairment of balance, coordination, speech, vision, reaction time, hearing, memory, and reasoning. Judgment and self-control are reduced, and caution, reasoning, and memory are notably impaired. Increasing the BAC above 0.10 leads to greater dysfunction, including slurred speech, blurred vision, staggered walking, dysphoria (an emotional state of anxiety, depression, or unease), and nausea. Above a BAC of 0.20 there is total disorientation, few pain sensations, and vomiting, while above a BAC of 0.30 a state of stupor exists, with increasing risk of coma and death from respiratory arrest.

In small amounts, alcohol can actually benefit heart health. Too much alcohol promotes a variety of liver diseases, because ethanol requires a special system available only in the liver to break it down. In large amounts, alcohol promotes cancer, especially stomach and esophageal cancer, because several cancer-causing chemicals are generated in the metabolism of ingested ethanol. Additionally, the empty Calories consumed and metabolic byproducts produced from alcohol intake promote body weight gain and obesity and can displace the ingestion of healthier foods and beverages in the diet.[E,F,19,44,57,70]

What is the alcohol intake recommendation for yourself?

TABLE 1.3

The Chemical Analysis of Selected Alcoholic Beverages[E]

Beverage	Serving (ounces)	Cals	Carb (g)	Pro (g)	Fat (g)	% Vol. Alcohol	Ethanol (g)
Regular beer	12	146	13.2	1.1	0	4.6	12.8
Light beer	12	99	4.6	0.7	0	4.0	11.3
Distilled Spirits							
80 proof	1.5	97	0	0	0	31.1	14
86 proof	1.5	105	0	0	0	33.6	15.1
90 proof	1.5	110	0	0	0	35.3	15.9
Liqueur (53 proof)	1.5	175	24.3	0.1	0.2	25.1	11.3
Mixed Drinks							
Daiquiri	2	114	4	0.1	0.1	23.2	13.9
Pina colada	4.5	235	39.9	0.6	2.4	10.4	14
Dessert dry wine	3	179	2.2	0.2	0	15	14
Dessert sweet wine	3	246	10.9	0.2	0	15	14
Table red wine	5	93	2.3	0.25	0	9.1	14
Table white wine	5	88	1.0	0.13	0	9.1	14

© Cengage Learning 2013

Summary Points

- Carbohydrates are organic compounds.
- Carbohydrates are categorized as simple and complex.
- Six different mono- and disaccharides largely comprise the simple sugars consumed in the diet.
- Digestible polysaccharides (starch) provide the majority of complex carbohydrates consumed in the diet.
- Digestible carbohydrates (starch and simple sugars) provide 4 Calories per gram.
- Indigestible complex carbohydrate, cellulose, and other fiber types provide 0 Calories per gram (i.e., they are noncaloric).
- Carbohydrate is the preferred fuel of the body.
- Consume 45 to 65 percent of Calories from carbohydrate (AMDR for total carbohydrate).
- Consume 25 percent of Calories or less from simple sugars (AMDR for sugars).
- Consume 1.4 grams of fiber per 100 Calories consumed (DRI for fiber).
- Carbohydrate is found in plant foods, except animal sources of milk and honey.
- The ethanol in alcoholic beverages provides 7 Calories per gram.

Take Ten on Your Knowledge Know-How

1. What elements are found in carbohydrates?
2. How are carbohydrates categorized?
3. What are the names of the carbohydrates in each category?
4. What are the differences between starch and fiber?
5. What are the functions of carbohydrate in your body?
6. How much energy do digestible carbohydrates provide per gram? How much energy does alcohol provide per gram?

7. What are the recommended intake levels for total carbohydrate, sugar, and dietary fiber? Do you think your diet meets these recommendations?
8. What are food sources of simple and complex carbohydrates? How can you increase your intake of starch and fiber and decrease your intake of added sugar?
9. Chemically, what is alcohol?
10. What are your views on alcohol and its role in society as a consumable substance?

1.3 PROTEINS

Introduction

Proteins are important structural and working substances in the cells of the body. We will explore proteins in terms of their composition, energy yield, functions, categories, recommended intake levels, and food sources.[36]

T-Talk 1.3
To hear Dr. Turley talk about proteins, go to **www.cengagebrain.com**

Composition of Proteins

Proteins are made from **amino acids** (see Figure 1.9). There are 20 known amino acids that are used to make proteins. As shown in Table 1.4, amino acids are categorized as essential and nonessential. The human body cannot make essential amino acids. The human body can make nonessential amino acids given nitrogen and carbohydrate intermediates. Amino acids, and thus proteins, contain the elements nitrogen, carbon, oxygen, and hydrogen. However, two of the 20 amino acids contain the essential mineral sulfur. Because proteins are made of amino acids, which contain carbon and have several elements linked together, they are organic compounds.

proteins Organic, energy-producing compounds made of amino acids for tissue repair and maintenance, as well as for growth; classified as complete or incomplete, high-quality or low-quality, or high or low biological value.

amino acids Organic and nitrogen containing compounds that can be essential and nonessential; there are 20 used to make proteins.

complete protein A high-quality or high biological value protein that is from the animal kingdom and provides all the essential amino acids.

Energy Yield and Functions of Proteins

Protein and its component amino acids yield 4 Calories per gram. In adults, the body's first function and use for dietary protein is to provide the amino acids for enzymes, hormones, nutrient transport, cellular tissue and vital organ maintenance, growth, and repair of lean body mass. However, if the body is stressed or energy intake is inadequate, then it can use the amino acids in protein for energy.

Categories of Proteins

Even though dietary protein is made up of essential and non-essential amino acids (which is how amino acids are categorized), the categories and quality of protein are determined by the protein's essential amino acid composition (amounts and proportions of the essential amino acids in the context of all the amino acids that make up the protein). Dietary proteins are categorized as high or low quality based on these amounts.[69]

High-Quality Proteins Proteins that contain all of the essential amino acids are called **complete proteins**. They are also referred to as having high biological value. Animal proteins provide dietary sources of high-quality proteins.

TABLE 1.4

Categories of Amino Acids

Essential Amino Acids	Nonessential Amino Acids
Histidine	Alanine
Isoleucine	Arginine
Leucine	Asparagine
Lysine	Aspartic acid
Methionine	Cysteine
Phenylalanine	Glutamic acid
Tryptophan	Glutamine
Threonine	Glycine
Valine	Proline
	Serine
	Tyrosine

© Cengage Learning 2013

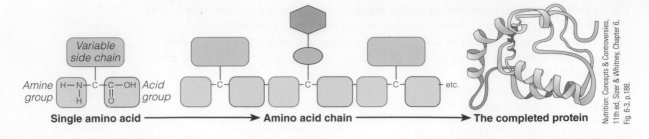

Nutrition: Concepts & Controversies, 11th ed. Sizer & Whitney, Chapter 6, Fig. 6-3, p.188.

Single amino acid ⟶ Amino acid chain ⟶ The completed protein

Variable side chain

Amine group H—N—C—C—OH *Acid group*

FIGURE 1.9 Amino acids are the building blocks of proteins. Each amino acid contains a central carbon, an amine group (NH₂), an acid group (COOH), and a variable side chain. Amino acids are combined to build a completed protein.

Low-Quality Proteins Proteins that lack one or more essential amino acids are called **incomplete proteins**. They are also referred to as having low biological value. Plant proteins provide dietary sources of low-quality proteins.

Proteins can be combined in such a way that the essential amino acids that are missing in one incomplete protein are provided by the other incomplete protein. This plant protein combination system is called **complementation** (see Figure 1.10). Table 1.5 shows how complementation can occur by combining two foods from two different columns: for example, brown rice and vegetables or peanut butter and whole-wheat bread. Some of the most complete amino acid complementation occurs by crossing grains with legumes or grains with vegetables.

TABLE 1.5

Some Plant Protein Complements

Grains	Legumes	Nuts and Seeds	Vegetables
Wheat	Lentils	Walnuts	Broccoli
Barley	Peanuts	Pecans	Carrots
Rye	Soybeans	Cashews	Leafy greens
Oats	Pinto beans	Other tree nuts	Green beans
Rice	Kidney beans	Sesame seeds	Squash
Quinoa	Lima beans	Sunflower seeds	Tomatoes
Other grains	Other dried beans	Other seeds	Other vegetables

© Cengage Learning 2013

Food Sources of Proteins

Both plant and animal foods can provide protein.

Plant Sources Legumes (dried beans), including soybean products (like tofu); nuts, seeds, grains, and grain products like bread, cereal, and pasta; and vegetables all provide low-quality, low biological value, incomplete proteins.[16]

incomplete protein A low-quality or low biological value protein from the plant kingdom that is missing or limited in one or more of the essential amino acids.

complementation Combining protein-containing plant foods so that all the essential amino acids are present with the food combination; examples include rice and black beans, whole-wheat bread and peanut butter, and soy milk and cereal.

studiogi/Shutterstock.com

Plant sources of low-quality proteins.

Biology: The Dynamic Science, 1st ed, Fussell, Chapter 45, Fig. 45-6, p.1021.

	Isoleucine	Lysine	Methionine	Tryptophan
Peanut Butter *(a legume)*				
Bread *(a grain product)*				
Complementation *(legume + grain)*				

A

Essential amino acids

Rice, corn, or other grains

Methionine
Tryptophan
Leucine
Phenylalanine
Threonine
Valine
Isoleucine
Lysine

Lentils, soybeans (for example, tofu), or other legumes

B

C

FIGURE 1.10 Protein complementation with plant foods. A: The essential amino acids missing in peanut butter are provided by the wheat bread and vice versa. B: The combination of grains with legumes results in protein complementation. C: Protein complementation with rice and vegetables.

Andi Berger/Shutterstock.com

Animal Sources Milk, yogurt, cheese, eggs, and meats such as beef, poultry, pork, seafood, and wild game all provide high-quality, high biological value, complete proteins.[16]

Recommended Intake Levels of Proteins

The dietary goal for protein intake is to provide the amino acids for lean body mass, growth, and repair. For an adult, the AMDR for protein is 10 to 35 percent of total Calories consumed. The DRI for protein is personalized. It is 0.8 grams of high-quality protein per kilogram (kg) of body weight for adults (ages 19 and older).[63] There are 2.2 pounds per kilogram, so the protein requirement is calculated as:

$$\text{Body weight in kg} \times 0.8 \text{ g/kg}$$

See Appendix B for conversions and sample calculations.

Data GENEie Solah weighs 140 pounds. What is her DRI for protein? How much do you weigh, and what is your DRI for protein?

Eric Isselée/Shutterstock.com

Animal sources of high-quality proteins.

Summary Points

- Dietary protein is made of 20 different amino acids, which are categorized as essential and nonessential.
- Amino acids are composed of the elements carbon, hydrogen, oxygen, and nitrogen.
- Dietary protein is categorized as complete (high-quality, high biological value) or incomplete (low-quality, low biological value).
- Complete proteins or those of high biological value come from animal proteins.
- Incomplete proteins or those of low biological value come from plant proteins.

- Combining plant proteins so that all of the essential amino acids are present is protein complementation.
- The primary function of dietary protein is tissue repair and functional protein maintenance.
- Protein provides 4 Calories per gram when used to generate ATP energy in the body.
- Adults need 0.8 grams of protein per kilogram of body weight per day to meet the DRI for protein.
- Adults may consume 10 to 35 percent of their total Calorie intake from protein and meet the AMDR for protein.

Take Ten on Your Knowledge Know-How

1. What elements are amino acids composed of?
2. How are amino acids and proteins categorized from a nutrition perspective?
3. What is the primary function of proteins in your body?
4. How much energy do amino acids and protein provide per gram?
5. What are some sources of complete (high-quality, high biological value) proteins that you consume?
6. What are some sources of incomplete (low-quality, low biological value) proteins that you consume?
7. What is protein complementation?
8. Why is protein complementation important in the diet? How can you incorporate protein complementation into your diet?
9. What are the recommended intake levels for protein?
10. Do you think your diet meets these recommendations? Why or why not?

1.4 LIPIDS/FATS

T-Talk 1.4
To hear Dr. Turley talk about lipids/fats, go to www.cengagebrain.com

fats Triglycerides or dietary fat.

triglyceride An organic, energy-producing compound (commonly called fat) that is made up of three fatty acids attached to a glycerol backbone and provides 9 Calories per gram.

phospholipids A compound (such as lecithin) composed of a glycerol backbone and a phosphate group with choline (a B vitamin), which is water-soluble, and two fatty acids, which are fat-soluble; these compounds are used as emulsifying agents, to build cell membranes, and as a precursor for acetylcholine; provides 0 Calories per gram.

sterols Fat-soluble compounds containing a four-ring carbon structure with any of a variety of side chains attached that provide 0 Calories per gram.

Introduction

Fat fits into the classification of lipids. As you will see, the class of lipids contains three categories of fat-soluble substances: **triglycerides** (dietary fats), **phospholipids**, and **sterols**. Although we sometimes think of fat or lipids as all bad, in fact a variety of lipids perform some important roles in the body, from being an energy source to insulating the body to helping maintain cell membranes. In foods, **lipids** usually are present in combination; therefore, when we consume fats in foods, we are consuming a variety of lipids. Each category of lipids will be explored in terms of its composition, energy yield, functions, categories, recommended intake levels, and food sources.[3,5,25,47]

Composition of Lipids

Lipids are a chemical class of organic compounds that are fat-soluble, meaning they associate with other fat-soluble substances, not water. They contain mostly carbon and hydrogen molecules and very small amounts of oxygen. Dietary fats are greasy substances, whereas sterols are waxy. Phospholipids are natural soaps; these molecules contain phosphorus in addition to carbon, hydrogen, and oxygen.

Categories of Lipids

The three categories of lipids include triglycerides, phospholipids, and sterols.

Triglycerides (Dietary Fats) Most dietary fats naturally exist as chemical structures of triglycerides. Triglycerides are made up of three fatty acids that are attached to a **glycerol backbone** (see Figure 1.11). The triglyceride can

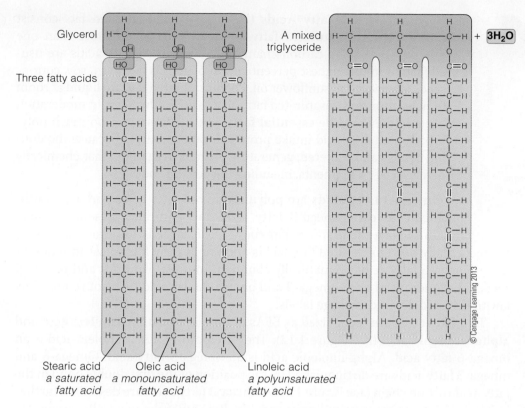

Glycerol

A mixed triglyceride + $3H_2O$

Three fatty acids

Stearic acid
a saturated fatty acid

Oleic acid
a monounsaturated fatty acid

Linoleic acid
a polyunsaturated fatty acid

FIGURE 1.11 Glycerol plus three fatty acids makes a triglyceride (the common form of Calorie-dense fat in food). The three fatty acids shown in this triglyceride are stearic acid, oleic acid, and linoleic acid.

© Cengage Learning 2013

be broken down into its smaller components, fatty acids, and then the fatty acids are broken down to produce energy. When the fatty acids are broken down into carbon dioxide plus water, ATP is released. Triglycerides are the only energy-producing lipid. The type of fatty acid present in the triglyceride determines the characteristics of the fat. Fatty acids are categorized and described in the following pages.[37,51]

Energy Yield   Dietary triglycerides are the most **calorically dense** of all the energy producing nutrients. Triglycerides and their composite fatty acids provide 9 Calories per gram, while protein and carbohydrate provide 4 Calories per gram, and alcohol provides 7 Calories per gram.

Categories of Fatty Acids in Triglycerides   The fatty acids that make up triglycerides and also phospholipids are categorized as saturated, monounsaturated, and polyunsaturated (see Figure 1.11).[3]

- **Saturated Fatty Acids (SFAs):** Saturated fats consist of mostly saturated fatty acids, which have no double bonds between the carbon molecules. They are usually found in the highest percentages in animal products, hydrogenated vegetable fats, and tropical oils such as palm and coconut oil. Saturated fats are solid at room temperature. Saturated fatty acids are unhealthy to eat because they promote heart disease. This is because a diet high in saturated fat raises the bad kind of cholesterol in the blood that tends to narrow the arteries that supply oxygenated blood to the heart muscle.

- **Monounsaturated Fatty Acids (MUFAs):** Monounsaturated fats consist mostly of monounsaturated fatty acids, which have only one double bond in the carbon chain. There are a few food sources of fats that are rich in monounsaturated fatty acids, such as olive oil, canola oil, almonds, and avocados. Some monounsaturated fats can become semisolid when refrigerated; at room temperature, they are liquid. Monounsaturated fatty acids are healthy to eat in moderation. They are not linked to promoting disease.

lipids A family of fat-soluble organic compounds that includes triglycerides, phospholipids, and sterols.

glycerol backbone The three-carbon sugar alcohol that forms the three sites for the three fatty acids in a triglyceride to attach to (in an esterification process).

calorically dense An energy-producing substance that contains a lot of energy per gram weight, such as triglycerides (9 Calories per gram) and alcohol (7 Calories per gram), compared to starches, sugars, and proteins (4 Calories per gram).

saturated fatty acids (SFAs) A fatty acid with no double bonds between the carbon molecules; found in abundance in animal meats, high-fat dairy products, tropical oils (coconut and palm), and hydrogenated oils.

monounsaturated fatty acids (MUFAs) A fatty acid with one double bond in the carbon chain, found in abundance in olive and canola oils, almonds, and avocados.

A pan full of fat. The liquid sunflower seed oil is polyunsaturated, whereas the solid coconut oil and butter are saturated.

polyunsaturated fatty acids (PUFAs) A fatty acid with more than one double bond in the carbon chain; found in abundance in nuts, seeds, and most liquid plant oils.

essential fatty acids (EFAs) Linoleic acid and alpha-linolenic acid, which are needed by the body but are not made by the body in amounts sufficient to meet physiological needs.

omega-3 fatty acid Long-chained, polyunsaturated fatty acids (PUFAs) in which the first double bond is three carbons away from the methyl (CH_3) end of the carbon chain.

FIGURE 1.12 The two essential polyunsaturated fatty acids in human nutrition. The methyl end (CH_3) of the fatty acid (highlighted in light purple) represents the first carbon in the omega system. The first double bond of two in linoleic acid is at carbon number six; thus, it is an omega-6 polyunsaturated fatty acid. The first double bond of three in alpha-linolenic acid is at carbon number three; thus, it is an omega-3 polyunsaturated fatty acid.

- **Polyunsaturated Fatty Acids (PUFAs):** Polyunsaturated fats consist mostly of polyunsaturated fatty acids, which contain more than one double bond in the carbon chain. Polyunsaturated fatty acids are usually found in the highest percentages in plant oils like corn, soy, cottonseed, safflower, and sunflower oil. Polyunsaturated fats are liquid at room temperature. Polyunsaturated fatty acids are healthy to eat in moderation because they provide **essential fatty acids (EFAs)**, but too much polyunsaturated fatty acid intake promotes cancer. This is because the double bonds can be altered, generating unstable fragments that chemically damage cell components, including the genetic code.

Omega-3 fatty acids are polyunsaturated fatty acids and are health-enhancing to eat. Omega-3 fatty acids are naturally present in small amounts in fish and in some plant foods like flax seeds. Omega-6 fatty acids are technically polyunsaturated fatty acids and are healthy fats to eat in moderation. Omega-6 fatty acids are naturally abundant in most plant oils and nuts and seeds. The amounts of both omega-3 and omega-6 fatty acids are not required by law to be listed on food package labels.

Two fatty acids are recognized as EFAs in human nutrition: **linoleic acid** and **alpha-linolenic acid** (see Figure 1.12). They are both PUFAs. Linoleic acid is an omega-6 fatty acid. Alpha-linolenic acid is an omega-3 fatty acid. Omega-6 and omega-3 fatty acids are distinguished by the position of the first double bond in the fatty acid carbon chain (see Figure 1.12). Omega-3 fatty acids are described further in Module 3. Good sources of linoleic acid include all of the plant oils, except for olive and canola oils. Good sources of alpha-linolenic acid include canola oil, walnuts, flax seeds, primrose oil, and borage oil.[3,15,47,64,67]

Food Sources of Triglycerides   Triglycerides are found in plant and animal foods.

Plant Sources: Oils, margarine, mayonnaise, salad dressings, olives, avocados, nuts, and seeds are all foods that provide plant fat. Coconut and palm oils are also plant fats, but they have a high percentage of saturated fatty acids and are not conducive to good health. Plant fat sources used to make liquid-at-room-temperature oils provide good sources of healthy essential fatty acids and other PUFAs as well as MUFAs (see Figure 1.13 on page 24 and Table 1.6). Plant food sources that are modified by food manufacturing techniques to make them more solid at room temperature, such as stick margarine and shortening, are disease promoting and unhealthy.

Animal Sources: In general, animal sources of fat include butter, cream, milk fat, cream cheese, ice cream, lard, chicken skin, pork rind, fish oil, and fat marbled in muscle meat. Animal fat sources, with the exception of fish, provide mostly solid saturated fats, which are potentially unhealthy because they promote obesity, heart disease, and some cancers and thus should be limited in the diet (that will be elaborated on in Module 3). Fish provides a good source of the healthy omega-3 fatty acids called eicosapentaenoic acid and docosahexaenoic acid.

Functions of Dietary Fats   Fat has many important functions: it plays structural and metabolic roles in the body. It increases the satiety value of a meal; improves the texture, flavor, and aroma of food; is required for the absorption of fat-soluble vitamins; cushions vital organs; is an essential structural component of cell membranes; provides

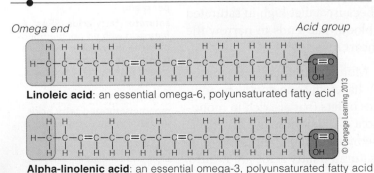

Omega end   *Acid group*

Linoleic acid: an essential omega-6, polyunsaturated fatty acid

Alpha-linolenic acid: an essential omega-3, polyunsaturated fatty acid

TABLE 1.6

Fatty Acids in Foods

Fatty Acid Name	Carbon Length	Double Bonds	Type	Omega	Food Sources
Butryic	4	0	SFA	N/A	Butter fat
Caproic	6	0	SFA	N/A	Butter fat
Capryic	8	0	SFA	N/A	Coconut oil
Capric	10	0	SFA	N/A	Palm oil
Lauric	12	0	SFA	N/A	Coconut and palm oil
Myristic	14	0	SFA	N/A	Coconut and palm oil
Palmitic	16	0	SFA	N/A	Palm oil
Stearic	18	0	SFA	N/A	Animal fats
Arachidic	20	0	SFA	N/A	Peanut oil
Behenic	22	0	SFA	N/A	Seeds
Lignoceric	24	0	SFA	N/A	Peanut oil
Palmitoleic	16	1	MUFA	7	Seafood, beef
Oleic	18	1	MUFA	9	Olive and canola oil
Elaidic	18	1	MUFA/TFA	9	Partially hydrogenated oil
Linoleic	18	2	PUFA/EFA	6	Plant oils, nuts, and seeds
Linolenic	18	3	PUFA/EFA	3	Flax seeds and primrose oil
Arachidonic	20	4	PUFA	6	Eggs and animal fat
Eicosapentanoic	20	5	PUFA	3	Fatty fish
Docosahexaenoic	22	6	PUFA	3	Fatty fish

Key:

SFA: Saturated Fatty Acid

MUFA: Monounsaturated Fatty Acid

PUFA: Polyunsaturated Fatty Acid

EFA: Essential Fatty Acid

TFA: *Trans* Fatty Acid

© Cengage Learning 2013

insulation; and provides the major energy stores in the body. When energy is stored, fat cells are filled with triglycerides. All excess Calories consumed in the diet, whether from carbohydrates, alcohol, proteins, or fats, are converted into triglycerides and stored as body fat. As a person gains body weight, it is usually fat weight resulting from excess Calorie intake that has increased the volume of triglycerides inside fat cells (fat cell fill).

Recommended Intake Levels of Dietary Fats For an adult, the AMDR for total fat intake is 20 to 35 percent of total Calories consumed. Of this total fat intake, the adult AMDR for linoleic acid is 5 to 10 percent of total Calories consumed, and for alpha-linolenic acid, the adult AMDR is 0.6 to 1.2 percent of Calories. There is no DRI for total fat; however, there is a DRI for the essential fatty acids. For adult men ages 19–30, the DRI for omega-6 linoleic acid is 17 grams per day, and for omega-3 alpha-linolenic acid, it is 1.6 grams per day. For adult women ages 19–30, the DRI for linoleic acid is 12 grams per day, and for alpha-linolenic acid, it is 1.1 grams per day. There is no AMDR or DRI for saturated fatty acids, but for good health, less than 7 percent of the total Calories should come from saturated fatty acids.[28,63]

linoleic acid An essential fatty acid with 18 carbons and two double bonds, with the first double bond occurring at the sixth carbon from the methyl end, making it an omega-6 fatty acid.

alpha-linolenic acid An essential fatty acid with 18 carbons and three double bonds, with the first double bond occurring at the third carbon from the methyl end, making it an omega-3 fatty acid.

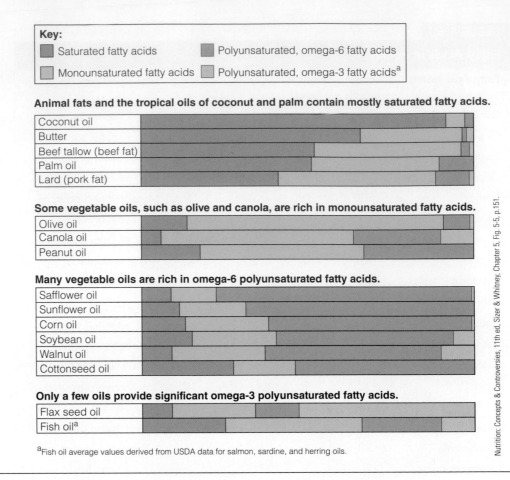

Key:

■ Saturated fatty acids	■ Polyunsaturated, omega-6 fatty acids
■ Monounsaturated fatty acids	■ Polyunsaturated, omega-3 fatty acids[a]

Animal fats and the tropical oils of coconut and palm contain mostly saturated fatty acids.

Coconut oil
Butter
Beef tallow (beef fat)
Palm oil
Lard (pork fat)

Some vegetable oils, such as olive and canola, are rich in monounsaturated fatty acids.

Olive oil
Canola oil
Peanut oil

Many vegetable oils are rich in omega-6 polyunsaturated fatty acids.

Safflower oil
Sunflower oil
Corn oil
Soybean oil
Walnut oil
Cottonseed oil

Only a few oils provide significant omega-3 polyunsaturated fatty acids.

Flax seed oil
Fish oil[a]

[a]Fish oil average values derived from USDA data for salmon, sardine, and herring oils.

FIGURE 1.13 Fatty acid composition of common fats.

Nutrition: Concepts & Controversies, 11th ed, Sizer & Whitney, Chapter 5, Fig. 5-5, p.151.

Trans fatty acids (TFAs) have been coined "*trans* fats" by the food industry. The majority of *trans* fats are technically monounsaturated fatty acids, but they can be found in any unsaturated fatty acid source that has undergone the food-processing technique called partial hydrogenation (see Figure 1.14). These types

FIGURE 1.14 The hydrogenation process. Hydrogenation adds hydrogen atoms to unsaturated fatty acids and thus removes double bonds generating saturated fatty acids. The partial hydrogenation of the same unsaturated fatty acid can generate *trans* fatty acids. The *trans* configuration in eladic acid is highlighted to show how the hydrogen molecules are on opposite sides of the double bond in the carbon backbone of the still unsaturated fatty acid.

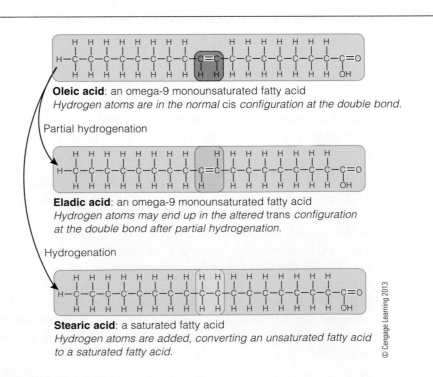

Oleic acid: an omega-9 monounsaturated fatty acid
Hydrogen atoms are in the normal cis *configuration at the double bond.*

Partial hydrogenation

Eladic acid: an omega-9 monounsaturated fatty acid
Hydrogen atoms may end up in the altered trans *configuration at the double bond after partial hydrogenation.*

Hydrogenation

Stearic acid: a saturated fatty acid
Hydrogen atoms are added, converting an unsaturated fatty acid to a saturated fatty acid.

of fats are very unhealthy to eat. *Trans* fatty acids are worse than saturated fatty acids, because they promote heart disease even more than saturated fatty acids. The food industry has taken steps to reduce *trans* fatty acids in margarine by altering the processing technique of hardened fats, but *trans* fatty acids can be present in partially hydrogenated fats used in **processed foods**. The grams of *trans* fatty acids per serving are included on food package labels if at least 0.5 gram per serving is present. The goal for *trans* fat intake is as little as possible, less than 1 percent of total Calorie intake.[17,32]

Phospholipids Small amounts of phospholipids are found in food. The most common dietary phospholipid is **lecithin** (see Figure 1.15). Phospholipids are the most water compatible of all of the lipids. One small end of the molecule is water-soluble, whereas the dominant end of the molecule is fat-soluble (see Figure 1.15). This chemical characteristic allows phospholipids to function as an **emulsifying agent**. In other words, phospholipids allow water-soluble substances to stay mixed up with fat-soluble substances.[25]

Non-Energy Yield and Functions of Phospholipids Phospholipids provide 0 Calories per gram; thus, phospholipids are noncaloric molecules that are used in the body for their chemical structure. Some of the functions of lecithin include using it as an emulsifier; the synthesis of the neurochemical **acetylcholine**; and using it to make cell membranes (see Figure 1.15).

processed food A food that has been manipulated to change its physical, chemical, microbiological, or sensory properties.

lecithin A phospholipid used by the food industry as an emulsifier in processed foods and by the body in cell membranes.

emulsifying agent (or emulsifier) A substance that associates water-soluble and fat-soluble substances such as water and oil together.

acetylcholine A neurotransmitter made from the water-soluble vitamin choline.

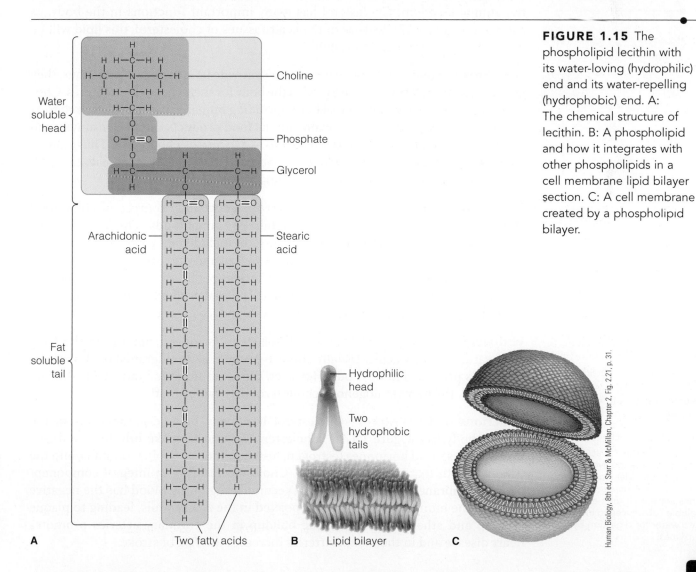

FIGURE 1.15 The phospholipid lecithin with its water-loving (hydrophilic) end and its water-repelling (hydrophobic) end. A: The chemical structure of lecithin. B: A phospholipid and how it integrates with other phospholipids in a cell membrane lipid bilayer section. C: A cell membrane created by a phospholipid bilayer.

Human Biology, 8th ed, Starr & McMillan, Chapter 2, Fig. 2.21, p. 31.

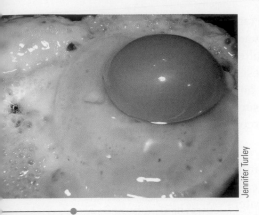

The egg yolk is an excellent food source of both lecithin and cholesterol.

Food Sources of Phospholipids   Lecithin is found in plant and animal foods.

Plant Sources: A natural plant source of lecithin is the soybean. Soy lecithin is added to many processed foods, such as salad dressing and sandwich spreads, as an emulsifying agent.

Animal Sources: A natural animal source of lecithin is egg yolk. Lecithin from egg yolk is commonly used as an emulsifying agent in processed foods like gravy and cream sauce.

Recommended Intake Level of Phospholipids   Because phospholipids like lecithin can be synthesized in the body and are not essential nutrients, there is no DRI for phospholipids. Choline is a water-soluble essential B vitamin that is a component of lecithin. Furthermore, because phospholipids are noncaloric, they have no AMDR.

Demo GENEie   The emulsifying power of an egg yolk. Try mixing oil and water together, and notice what happens. Next, beat an egg yolk into the mixture. Observe and explain what happens.

FIGURE 1.16 The chemical structure of cholesterol, a common sterol in the human body and animal foods.

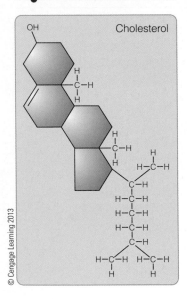

Sterols   Small amounts of sterols are present in both plant and animal foods. Sterols are used to build a variety of chemical structures in the body. Cholesterol (see Figure 1.16), a well-known sterol, is found in notable amounts in foods from the animal kingdom. Cholesterol has many important functions in the body, as shown in Figure 1.17. Because of the health issues of cholesterol, this lipid will be covered in greater detail in Module 3.[25]

Non-Energy Yield of Cholesterol   Sterols provide 0 Calories per gram; thus, they are noncaloric molecules that are used in the body for their chemical structures. Cholesterol can be made in any animal body, including humans, and this is referred to as **endogenous** cholesterol. Additionally, animal food products can be consumed, and this food source of cholesterol is referred to as exogenous cholesterol. Thus, animal sources of cholesterol can be endogenous or **exogenous**, because animals can both produce and consume cholesterol-containing animal products.

Food (Exogenous) Sources of Cholesterol   Exogenous sources of cholesterol come from outside the body, from the intake of animal foods. The most concentrated dietary sources of exogenous cholesterol include egg yolks (approximately 275 milligrams each), organ meats, and some crustaceans, such as shrimp, crab, crayfish, and lobster (approximately 190 milligrams per 3 ounces). Smaller amounts of cholesterol are present in the fat portions of animal products, such as beef, poultry, pork, fish, wild game, milk, yogurt, cheese, and ice cream.

Endogenous Cholesterol Sources   Cholesterol is made primarily in the **liver** from saturated fatty acids. Usually about 1 gram (1,000 milligrams) of cholesterol per day is produced in the body. However, the more saturated fatty acids that are consumed, the more endogenous cholesterol that can be made.

Functions of Cholesterol   Cholesterol has many beneficial functions in the human body (see Figure 1.17).[18] Cholesterol is used to make **bile** for the digestion of fats, steroid hormones (estrogen, testosterone, cortisol), vitamin D, and the myelin sheath that covers nerve cells. Cholesterol is also an integral component of cell membranes. Excess cholesterol accumulating in the blood has the negative effect in the human body of being deposited in the artery walls, leading to plaque buildup and atherosclerosis. Plaque buildup in the coronary arteries promotes heart disease and in the cranial arteries increases the risk of stroke.

endogenous From inside the body.

exogenous From outside the body.

liver An organ with multiple functions, including the synthesis of cholesterol and bile and detoxification of chemicals.

bile An emulsifier that is synthesized by the liver, stored in the gallbladder, and released into the small intestine with fat consumption. It then enables fat-soluble substances to integrate into water for digestion by lipase enzymes.

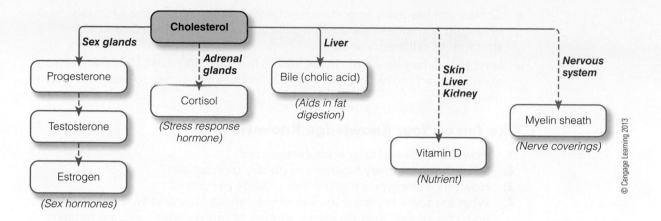

Recommended Intake Levels of Cholesterol   Because cholesterol is synthesized in animals and is not an essential nutrient, there is no DRI for cholesterol. Furthermore, because cholesterol is noncaloric, it has no AMDR. Because cholesterol intake is associated with an increased risk for heart disease, there is a recommendation to consume less than 300 milligrams of cholesterol per day to maintain health, and less than 200 milligrams per day to improve heart health.[63]

Plant Sterols and Stanols   Many **plant sterols** and **stanols** promote heart health. They are present naturally in small quantities in many plant foods, including vegetables, nuts, seeds, whole grains, legumes, and vegetable oils. They can be added to commercially prepared margarine-type spreads, as indicated on food package labels, and they are also available as dietary supplements.[18]

FIGURE 1.17 The metabolic functions and fates of cholesterol: steroid hormones (such as sex and stress hormones), vitamin D (a nutrient), bile (a fat digestion aid), and myelin sheaths (nerve coverings).

Summary Points

- All lipids are organic compounds.
- There are three categories of lipids: triglycerides, phospholipids, and sterols.
- Triglycerides are the only energy-producing lipids consumed in the diet and are commonly called fats.
- Triglycerides provide 9 Calories per gram.
- Triglycerides contain fatty acids that are categorized as saturated fatty acids, monounsaturated fatty acids, and polyunsaturated fatty acids.
- There are two essential fatty acids in human nutrition, and both are polyunsaturated fatty acids.
- Linoleic acid is an essential omega-6 polyunsaturated fatty acid.
- Alpha-linolenic acid is an essential omega-3 polyunsaturated fatty acid.
- Triglycerides have many important functions.
- Consume 20 to 35 percent of total Calories from fat to meet the adult AMDR for total fat.
- Limit saturated fatty acid intake to less than 7 percent of total Calories to meet the recommendation for saturated fat intake.
- *Trans* fatty acids are produced by partial hydrogenation of unsaturated fatty acids.
- Limit *trans* fatty acid intake to less than 1 percent of total Calories.
- Phospholipids are noncaloric.
- The most common phospholipid consumed in the diet is lecithin, which is used to emulsify fat and used in the chemical structure of acetylcholine, a neurotransmitter, and to make cell membranes.
- Cholesterol is noncaloric and is found only in animal foods.
- Cholesterol is nonessential because it can be made in the liver.

plant sterols and stanols Heart health-promoting chemicals from plant foods, including grains, vegetables, nuts, seeds, cereals, and legumes.

- Cholesterol has many important functions: It is needed for the synthesis of steroid hormones, bile acids, myelin sheaths, and vitamin D, and in the structure of cell membranes.
- Limit the dietary intake of cholesterol to less than 300 milligrams per day to maintain heart health and less than 200 milligrams per day to improve heart health.

Take Ten on Your Knowledge Know-How

1. How are lipids and fatty acids categorized?
2. What are the primary functions of dietary triglycerides?
3. How much energy do triglycerides provide per gram?
4. What are some key food sources of the various fatty acid types that make up triglycerides? How do animal sources of triglycerides compare to plant food sources?
5. What are some excellent food sources of the two essential fatty acids?
6. What are the recommended intake levels for dietary fat? Do you think your diet meets this recommendation? If so, why? If not, what dietary changes can you make to meet the recommended dietary fat intake level while increasing your intake of beneficial fatty acids and decreasing your intake of fatty acids that are potentially harmful when consumed in excess?
7. What are phospholipids? How do they function in the body and in food?
8. What are some plant and animal food sources of phospholipids?
9. How does the sterol cholesterol function in your body?
10. What are endogenous and exogenous sources of cholesterol? How much cholesterol should you limit your dietary intake to daily? What dietary changes could you make to meet this recommendation if you don't already?

1.5 VITAMINS, MINERALS, AND WATER

T-Talk 1.5
To hear Dr. Turley talk about vitamins, minerals, and water, go to www.cengagebrain.com

Introduction to Vitamins

Vitamins are chemical compounds that play several important roles in living systems. The essential vitamins are vital to sustain life. All vitamins are instrumental in catalyzing biochemical reactions. They are necessary for body function and good health.[52] Vitamins will be explored in terms of their composition, non-energy yield, functions, categories, recommended intake levels, and food sources.

Composition of Vitamins

Vitamins are organic compounds composed of carbon, hydrogen, and oxygen. Some vitamins (the B vitamins) also contain nitrogen in their chemical structure. All vitamins are organic compounds.

Recommended Intake Levels of Vitamins

The recommended amounts for daily consumption of the essential vitamins are established and collectively known as the Dietary Reference Intakes (DRIs). The DRI values are given as **Recommended Dietary Allowances (RDAs)** and **Adequate Intakes (AIs)**. They are used to recommend nutrient intake throughout the life span and to analyze dietary intakes. The essential vitamins are needed for cellular metabolism in tiny amounts—microgram or milligram—as compared to gram amounts for carbohydrate, protein, and fat. One microgram is ten to the minus sixth power (10^{-6}) or one-millionth of a gram, whereas a milligram is one-thousandth of a gram. Because vitamins are required in small amounts, they are referred to as micronutrients. (See Appendix B for

vitamins Organic, essential nutrients (categorized as fat- and water-soluble) that are needed in small amounts (milligrams or micrograms) by the body for health.

Recommended Dietary Allowances (RDAs) The dietary amount of a nutrient considered adequate to meet the known nutrient needs of practically all healthy people; also a nutrient intake goal for individuals.

Adequate Intake (AI) The average dietary amount of a nutrient that appears sufficient for health; used when an RDA cannot be determined.

calculations and conversion factors. The specific DRI values for the essential vitamins are discussed in Module 2 and are found in Appendix A.[28,53])

Non-Energy Yield and Functions of Vitamins

Vitamins provide 0 Calories per gram; thus, they are noncaloric molecules that are used in the body for their chemical structure. The essential vitamins have distinct functions at DRI levels, deficiency characteristics at low levels, and many have toxicity symptoms at high levels. Thus, too much or too little of essential nutrients is dangerous, can cause health problems, and is undesirable. Vitamins are commonly integrated into the chemical structures of compounds called enzymes that are needed to drive biochemical reactions. Thus, vitamins are known as **coenzymes** or cofactors (see Figure 1.18 and BioBeat 1.6). Most of the water-soluble vitamins play a role in energy metabolism as enzyme cofactors. Other vitamins are required for building connective tissue (vitamin C), blood clotting (vitamin K), antioxidant activity (vitamin C, E, and provitamin A), or sending hormone-like messages to regulate cell behavior (vitamins A and D).

Categories of Vitamins

Vitamins are categorized as fat-soluble or water-soluble. Fat-soluble vitamins dissolve in fat, and water-soluble vitamins dissolve in water.[69] Each vitamin fits into a category based on its chemical solubility.[49,73]

Fat-Soluble Vitamins The four fat-soluble vitamins that are essential in human nutrition include vitamins A (retinol, retinal, retinoic acid), D (1,25 dihydroxy-vitamin D_3), E (alpha, beta, gamma, and delta tocopherols and tocotrienols), and K (phylloquinone and naphthoquinone). **Beta-carotene** is a plant pigment and a precursor or a provitamin for vitamin A, but it is not an essential nutrient itself, though it has shown a wide variety of potentially beneficial health effects when obtained from food sources.

Water-Soluble Vitamins The 10 water-soluble vitamins that are essential in human nutrition include thiamin (B_1), riboflavin (B_2), niacin (B_3), vitamin B_6 (pyridoxine, pyridoxal, and pyridoxamine), vitamin B_{12} (cyanocobalamin), vitamin C (ascorbic acid), pantothenic acid, biotin, folic acid (also called folate and folacin), and choline.

Other Vitamins Hundreds of other compounds show vitamin-like activity. They are not considered essential vitamins and do not have a DRI established for them, because deficiencies have never been shown in humans without a dietary intake.

Food Sources of Vitamins

Vitamins are found in a wide variety of whole plant and animal foods and are added to many processed foods. The key to consuming an adequate level of the essential vitamins—not too much or not too little—is practicing the characteristics of a sound diet: moderation, variety, balance, and adequacy. These characteristics should be applied to each food group: grains, vegetables, fruits, milk and milk alternatives, meat and meat alternatives, and oils.

Introduction to Minerals

Minerals will be explored in terms of their composition, non-energy yield, functions, categories, recommended intake levels, and food sources.[25,37,69,73]

Composition of Minerals

Every mineral that is essential in human nutrition can be found on the periodic table (see Figure 1.19). All of the essential minerals have a low molecular weight, and

coenzyme A substance, such as a vitamin or mineral, that participates in metabolism as a structural part of an enzyme; also called a cofactor.

beta-carotene A phytochemical (plant-derived, health-promoting food component) that is a precursor or a "provitamin" for vitamin A.

minerals Inorganic elements naturally found in the earth that are categorized as major and trace in human nutrition; some minerals are essential nutrients needed in small amounts (milligrams or micrograms) by the body to function in a structural capacity, as coenzymes, and in fluid and pH balance.

inorganic Does not contain carbon and is not a living thing.

electrolytes Ions such as sodium, potassium, and chloride that govern fluid balance across semipermeable membranes (between the inside and outside of cells).

semipermeable membrane A membrane made of a lipid bilayer that will allow certain molecules or ions to pass through it by diffusion.

buffer A water-based solution that resists pH change.

pH Acid-base balance measured on a scale of 1 to14, with low numbers being acidic, middle numbers being neutral, and high numbers being basic (alkali).

major minerals Essential minerals found in the adult reference body (150 pounds) in quantities greater than 5 grams, including calcium, phosphorus, magnesium, sodium, potassium, chloride, and sulfur.

trace minerals Essential minerals found in the adult reference body (150 pounds) in quantities less than or equal to 5 grams, including iron, zinc, iodine, selenium, chromium, molybdenum, copper, manganese, fluoride, and cobalt.

every mineral is an element. That means minerals are naturally found in the earth, each having its unique set of physical and chemical properties, such as melting point, crystalline structure, boiling point, and molecular weight. Even though carbon is one of the elements, all minerals are **inorganic** substances. Minerals are not compounds containing carbon. Of all the minerals present in the body, the majority are used for structural integrity. However, a small amount of the total body mineral content is required for chemical processes.[24]

Non-Energy Yield and Functions of Minerals

Minerals provide 0 Calories per gram; thus, they are noncaloric molecules that are used for their chemical structure. Some examples of minerals as structural components include calcium, phosphorus, magnesium, and fluoride in bone and tooth structure and phosphorus in deoxyribonucleic acid (DNA) and ATP. Several minerals play chemical roles in cellular metabolism. Zinc, copper, manganese, iron, and selenium are a few minerals that are vital components of enzymes through their coenzymes (or cofactors) function. Other minerals play a role in body-fluid regulation. Sodium, potassium, and chloride are known as the **electrolytes** that govern fluid balance across **semipermeable membranes** (such as between the inside and outside of cells). Still other minerals are integral components of functional proteins that are vital to chemical structures, such as iron in hemoglobin and iodine in thyroxin, whereas other minerals like sodium and chloride function in chemical **buffer** reactions and thus help regulate **pH** (acid-base balance).

Categories of Minerals

The 17 minerals listed as follows are essential for proper growth and function of the human body. Minerals are categorized from a nutritional perspective as either **major minerals** or **trace minerals**, depending on the amount present in the 150-pound, reference adult body. The full name, chemical abbreviation, and category (major or trace) of the minerals is given below.[53,69]

BioBeat 1.6

Enzymes and Their Cofactors

Enzymes are proteins that **catalyze** chemical reactions. An enzyme can usually be recognized by its name. Most names of enzymes end in *-ase*. Biochemical reactions may build chemicals (**anabolic**), break apart compounds (**catabolic**), or rearrange molecules. When an enzyme catalyzes a biochemical reaction, the enzyme itself is not altered. Enzymes are necessary for most biochemical reactions to occur.

Many vitamins play their role in biochemistry as structural components of enzymes. This role earns them the title of coenzymes or cofactors. As such, they are essential for the chemical structure and function of enzymes. Vitamins are known as regulators of metabolism. For example, folic acid is a cofactor for **homocysteine transmethylase**. This means that folic acid plays an essential role in reducing homocysteine

levels. Without folic acid, homocysteine builds up and may contribute to heart disease. Some minerals also serve as cofactors to enzymes. An example of a mineral cofactor is selenium, which is required for the enzymatic antioxidant property of **glutathione peroxidase**.

What would happen to your biochemistry if you did not consume the needed vitamins?

FIGURE 1.18 The concept of an enzyme catalyzing a chemical reaction with its vitamin cofactor. Substance A is converted to product B by the enzyme/cofactor action.

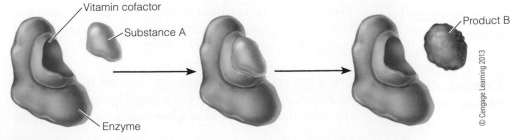

Vitamin cofactor

Substance A

Product B

Enzyme

© Cengage Learning 2013

1	2	3	4	5	6	7	8	9	10	11	12	13	14	15	16	17	18
H																	He
Li	Be											B	C	N	O	**F**	Ne
Na	**Mg**											Al	Si	**P**	**S**	**Cl**	Ar
K	**Ca**	Sc	Ti	V	**Cr**	**Mn**	**Fe**	**Co**	Ni	**Cu**	**Zn**	Ga	Ge	As	**Se**	Br	Kr
Rb	Sr	Y	Zr	Nb	**Mo**	Tc	Ru	Rh	Pd	Ag	Cd	In	Sn	Sb	Te	**I**	Xe
Cs	Ba	La	Hf	Ta	W	Re	Os	Ir	Pt	Au	Hg	Ti	Pb	Bi	Po	At	Rn
Fr	Ra	Ac	Rf	Ha	Sg	Ns	Hs	Mt									

Ce	Pr	Nd	Pm	Sm	Eu	Gd	Tb	Dy	Ho	Er	Tm	Yb	Lu
Th	Pa	U	Np	Pu	Am	Cm	Bk	Cf	Es	Fm	Md	No	Lr

© Cengage Learning 2013

FIGURE 1.19 The periodic table of elements with nutritional application for the essential minerals in human nutrition. The essential minerals are boldfaced and highlighted, including the major minerals [Calcium (**Ca**), Magnesium (**Mg**), Phosphorus (**P**), Sodium (**Na**), Potassium (**K**), Chloride (**Cl**), Sulfur (**S**)] and the trace minerals [Iron (**Fe**), Zinc (**Zn**), Iodine (**I**), Selenium (**Se**), Chromium (**Cr**), Molybdenum (**Mo**), Copper (**Cu**), Manganese (**Mn**), Fluoride (**F**), Cobalt (**Co**)].

Major Minerals The seven essential major minerals are found in the reference adult body in quantities greater than 5 grams (see Figure 1.20). These include calcium (Ca), phosphorus (P), magnesium (Mg), sodium (Na), chloride (Cl), potassium (K), and sulfur (S).

Trace Minerals The 10 essential trace minerals are found in the reference adult body in quantities less than or equal to 5 grams (see Figure 1.20). These include iron (Fe), zinc (Zn), iodine (I), selenium (Se), chromium (Cr), molybdenum (Mo), copper (Cu), manganese (Mn), fluoride (F), and cobalt (Co).

Other Minerals Other minerals are not currently recognized as essential for the human body, but they may gain essential status as research supports nutritional needs through the life span. Some examples include nickel, as a structural requirement for some metalloenzymes, and boron, which has been shown to function in calcium metabolism. Other minerals, such as mercury, lead, cadmium, and aluminum, can be found in human ash, but because no structural or metabolic roles have been identified for these minerals, they are believed to be contaminants in the body.[28]

Recommended Intake Levels of Minerals

The essential minerals have distinct functions at DRI levels, deficiency characteristics at low levels, and toxicity symptoms at high levels. Thus, too much or too little of an essential nutrient is dangerous, can cause health problems, and is undesirable. Minerals are needed in tiny amounts for cellular metabolism and structure. Specifically, they are needed in microgram or milligram amounts compared to gram amounts for carbohydrate, protein, and fat. Because minerals are needed in small amounts, they are known as micronutrients.[53]

Depending on age, gender, and condition, humans have differing needs for essential minerals; these special dietary needs are established in the DRIs (see Module 2 and Appendix A).[28] In general, major minerals have a DRI value greater than 100 milligrams per day, and trace minerals have a DRI value of less than or equal to 100 milligrams per day (see Figure 1.20). Note that there is no DRI for the essential minerals cobalt (a trace mineral) and sulfur (a major mineral). Vitamin B$_{12}$ has the mineral cobalt at the center of its chemical structure; thus, the nutritional need for cobalt is met through meeting the DRI for vitamin B$_{12}$. Sulfur is found as a component in two sulfur-containing amino acids, cysteine and methionine; thus, the nutritional need for sulfur is met through meeting the DRI for dietary protein.

catalyze To facilitate a biochemical reaction to occur.

anabolic When small molecules are put together to build larger ones through metabolic reactions; requires a condensation chemical reaction.

catabolic When large molecules are broken apart to yield smaller ones through metabolic reactions; requires a hydrolysis chemical reaction.

homocysteine transmethylase A folic acid-dependent enzyme that is necessary for the detoxification of homocysteine.

glutathione peroxidase A selenium-dependent liver enzyme.

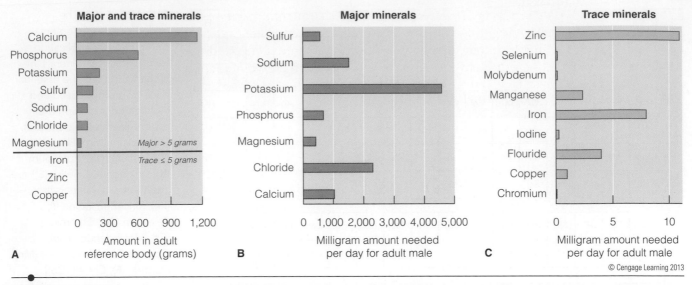

Major and trace minerals		
Calcium		
Phosphorus		
Potassium		
Sulfur		
Sodium		
Chloride		
Magnesium	*Major > 5 grams*	
Iron	*Trace ≤ 5 grams*	
Zinc		
Copper		

0   300   600   900   1,200

A Amount in adult reference body (grams)

Major minerals
Sulfur
Sodium
Potassium
Phosphorus
Magnesium
Chloride
Calcium

0   1,000 2,000 3,000 4,000 5,000

B Milligram amount needed per day for adult male

Trace minerals
Zinc
Selenium
Molybdenum
Manganese
Iron
Iodine
Flouride
Copper
Chromium

0   5   10

C Milligram amount needed per day for adult male

© Cengage Learning 2013

FIGURE 1.20 Major and trace minerals in human nutrition. A: Major and trace minerals in the adult reference person. B: Major minerals by the adult male DRI. C: Trace minerals by the adult male DRI.

water An inorganic compound that is made of hydrogen and oxygen (H_2O) and is essential for life; the medium for metabolism and nutrient transport; dietary sources include all fluids and fluid-rich foods, such as fruits and vegetables.

Water is the medium for metabolism, essential for thermal regulation, and the basis of excretion of most metabolic waste.

Food Sources of Minerals

Minerals, like the vitamins, are found in a wide variety of whole plant and animal foods and are added to many processed foods. The key to consuming an adequate level of the essential minerals—not too much, not too little—is practicing the characteristics of a sound diet: moderation, variety, balance, and adequacy. These characteristics should be applied to each food group: grains, vegetables, fruits, milk and milk alternatives, meat and meat alternatives, and oils.

Introduction to Water

Water will be explored in terms of its composition, non-energy yield, functions, recommended intake levels, and sources.[20,30,38,54]

Composition of Water

Water is the fluid of life; it is a compound made of the elements hydrogen and oxygen (H_2O). Because it does not contain carbon in its chemical structure, it is an inorganic compound.

Non-Energy Yield and Functions of Water

Water supplies 0 Calories per gram; thus, it is noncaloric. Water is the medium for metabolism and nutrient transport. Without an intake of water, dysfunction leading to death will result more quickly than with the limitation of any other of the essential nutrients in human nutrition.

Recommended Intake Level of Water

Although there is no one-size-fits-all recommendation that can be made accurately for water intake, a DRI value for individuals over 19 years of age has been set. The DRI is 2.7 liters (11 cups per day) for women and 3.7 liters (15 cups per day) for men. Individual needs for water vary greatly depending on activity and environmental conditions. To stay in water balance, 1 milliliter of water per 1 Calorie expended is needed. However, this is not an optimal level of water intake. Most metabolic waste is excreted from the body via the urine, and the kidneys can excrete more efficiently into dilute urine. Therefore, drinking more fluid promotes the excretion

Goodluz/Shutterstock.com

of toxic metabolic waste products. However, too much of a good thing can be fatal. Consuming excess water, though rare and deliberate, is termed water intoxication (see Module 5). Thirst mechanisms do not provide the motivation to drink until an individual is 2 percent dehydrated. At this point, optimal body function is decreased (see BioBeat 1.7). Thus, individuals need to learn to drink when they are not thirsty and should consume enough fluids to produce clear urine every 2 hours during the day.[20,38,50,70]

Sources of Water

Water is available in beverages of all sorts and from foods. Fluid is provided best by consuming water, non-caffeinated and nonalcoholic beverages, and water-rich foods such as fruits and vegetables. Another source of water in animals is metabolic water. For example, when glucose is broken down to carbon dioxide and water, the water produced is metabolic water.

BioBeat 1.7

Salt, Sweat, Exercise, and Human Performance

Americans are sedentary people. Fifty-one percent of Americans do not exercise. Only 17 percent exercise enough to yield a health benefit. An acceptable range of time to exercise is 30 to 90 minutes per day. When one exercises, one loses water and salts in sweat. The adult DRI for salt (sodium chloride) is 3,800 milligrams per day. This recommendation is made for the typical American and is targeted to control blood pressure. However, if a person exercises and is sweating while he or she is doing it, the sweat loss on average will be about a liter per hour. Although sweat is mostly water, on average 2,600 milligrams of salt are lost per liter of sweat. Because salt is 60 percent sodium and 40 percent chloride, this means that about 1,560 milligrams of sodium would be lost in

sweat per hour of exercise. Furthermore, based on these percentages, the 3,800-milligram-per-day DRI for sodium chloride equals a 1,500-milligram-per-day DRI for sodium. Avid exercisers need to make sure that they consume enough salt. About 1,500 milligrams of sodium should be ingested per hour during moderate exercise if the individual will be continuously sweating from 1 to 4 hours per day. If salt losses are too great during exercise, salt-depleted heat exhaustion will occur. Needless to say, the ability of the body to function properly—let alone to exercise—will deteriorate, and thermal injury may result.[20,38,50]

Is your salt and fluid intake optimal to sustain your health and performance?

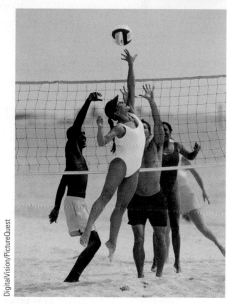

DigitalVision/PictureQuest

Fluid and electrolyte losses have to be replaced in physical activity and ordinary life.

Summary Points

- The noncaloric nutrients (0 Calories per gram) are vitamins, minerals, and water.
- Vitamins and minerals are referred to as the micronutrients.
- Vitamins are categorized as fat- or water-soluble.
- There are four fat-soluble vitamins essential in human nutrition.
- There are 10 water-soluble vitamins essential in human nutrition.
- Minerals are categorized as major and trace.

- There are seven major minerals essential in human nutrition.
- There are 10 trace minerals essential in human nutrition.
- Water is the fluid of life.
- It is healthier to consume more water than is required to achieve water balance, which is 1 milliliter of water per 1 Calorie expended. Care should be taken to prevent water intoxication.

Take Ten on Your Knowledge Know-How

1. How are vitamins categorized? What is the name for each essential vitamin in each category?
2. What is the primary function of vitamins in your body?
3. What are minerals, and how do they differ from vitamins?
4. How are minerals categorized? Name every essential mineral in each category.
5. What are the primary functions of minerals in your body?
6. What is the key to consuming enough of the essential vitamins and minerals?
7. How do the amounts of essential vitamins and minerals needed each day compare to those of carbohydrate, protein, and fat?
8. What elements are in water?
9. What are the functions of water in your body?
10. How much water or fluid should you consume each day? Do you think you reach this recommendation? If so, why? If not, what changes can you make in your diet to meet this recommendation?

 **SUMMARY**

CONTENT KNOWLEDGE

IN THIS MODULE, YOU HAVE LEARNED ABOUT:

- The relationships between your diet and your health.
- The six categories of biological nutrients, including their chemical composition, categories, functions, and food sources (see Table 1.7).
- The importance of consuming a sound diet for supporting cellular metabolism.
- How food keeps the organism, you, alive and well by supplying energy, building blocks of vital tissues, and providing essential nutrients that regulate metabolism.

PERSONAL IMPROVEMENT GOALS

IN THIS MODULE, YOU HAVE DISCOVERED THAT:

- You can achieve the goal of eating to fuel and nourish your body optimally.
- You can take responsibility for your daily dietary intake and physical activity patterns.
- Making small changes in the foods you select can have a huge impact on your health and body weight control over time.
- Your hunger and satiety mechanisms are highly tuned, and you can pay greater attention to the signals for initiating and terminating eating and drinking.

Here is a tip for you to adopt in your life that will reduce the risk of increasing fat cell fill: Exercise 1 hour per day (sweating while you do it) and eat a variety of low-fat, whole, fresh foods. This way hunger and satiety mechanisms are sharper.

You can assess if you met the learning objectives for this module by successfully completing the Homework Assessment and the Total Recall activities (sample questions, case study with questions, and crossword puzzle).

TABLE 1.7

Summary of Key Components Learned in Module 1

	Composition	Dietary Source by *Category*			Energy	Function	Intake Rec.
Carbo-hydrates	Carbon Hydrogen Oxygen	*Simple* Fruit, milk, honey, sweets	*Complex* Grains, vegetables, legumes, fruits (fiber)		4 Cal/g except fiber (0 Cal/g) and alcohol (7 Cal/g)	Preferred fuel	45–65% of Calories with ≤ 25% of Calories from simple
Proteins	Carbon Hydrogen Oxygen Nitrogen	*Complete* All animal foods	*Incomplete* All plant foods		4 Cal/g for protein and amino acids	Tissue repair, maintenance, growth	10–35% of Calories
Lipids	Carbon Hydrogen Oxygen	*Triglycerides* SFA: Animal fat, coconut oil, palm oil MUFA: Olive and canola oil PUFA: Plant oils, nuts, seeds EFA: Are PUFAs TFA: From partially hydrogenated oils in processed foods	*Phospho-lipids* Egg yolk, soy	*Sterols* Animal tissue for choles-terol; plants for plant stanols/ sterols	9 Cal/g for trigly-cerides and fatty acids	*Triglycerides* Energy *Phospho-lipids* Cell membranes, emulsify *Cholesterol* Synthesis of bile, hormones, and vitamin D; integration into myelin sheath and cell membranes	20–35% of Calories for triglycerides with < 7% SFA, < 1% TFA, 5–10% LA, and 0.6–1.2% ALA N/A < 300 mg/day
Essential vitamins	Carbon Hydrogen Oxygen Nitrogen (some vitamins)	*Fat Soluble and Water Soluble* *Many foods* Practice moderation, variety, balance, and adequacy			0 Cal/g	*Metabolism* Enzyme cofactors	Adequacy per the DRI
Essential minerals	Unique elements	*Major and Trace* *Many foods* Practice moderation, variety, balance, and adequacy			0 Cal/g	Enzyme cofactors, structure, fluid and pH balance	Adequacy per the DRI
Water	Hydrogen Oxygen	*No categories* Sources: Water, juice, milk, and fluid-rich foods like fruits and vegetables			0 Cal/g	Medium for metabolism and nutrient transport	Adequacy per the DRI or Calorie expenditure and fluid loss

Recall the order of life, terms, characteristics of a sound diet, and factors affecting longevity and food choices.

Key

SFA: Saturated Fatty Acid

MUFA: Monounsaturated Fatty Acid

PUFA: Polyunsaturated Fatty Acid

TFA: *Trans* Fatty Acid

EFA: Essential Fatty Acid

LA: Linoleic Acid

ALA: Alpha-Linolenic Acid

Homework Assessment

50 questions

1. The Calorie is the unit of measure used to describe the energy content of food.
   A. True   **B.** False

2. Polyunsaturated fatty acids contain more than one double bond in their carbon chain.
   A. True   **B.** False

3. The primary function of carbohydrates is to serve as cofactors for enzymes.
   A. True   **B.** False

4. Saturated fatty acids are the most abundant type of fatty acids in canola oil.
   A. True   **B.** False

5. The element nitrogen is a component of fatty acids.
   A. True   **B.** False

6. Cholesterol can be used to synthesize vitamin A inside the human body.
   A. True   **B.** False

7. Wheat bread provides a source of low-quality protein.
   A. True   **B.** False

8. Milk is a source of simple sugar.
   A. True   **B.** False

9. Alcohol provides 7 Calories per gram.
   A. True   **B.** False

10. Amino acids provide 4 Calories per gram.
    A. True   **B.** False

11. Triglycerides are a common form of dietary fats.
    A. True   **B.** False

12. The Acceptable Macronutrient Distribution Range (AMDR) for fat is less than 20 percent of total Calories consumed.
    A. True   **B.** False

13. The Acceptable Macronutrient Distribution Range (AMDR) for protein is 10 to 20 percent of total Calories consumed.
    A. True   **B.** False

14. Chicken is a source of dietary cholesterol.
    A. True   **B.** False

15. Oleic acid is an essential fatty acid for humans.
    A. True   **B.** False

16. Cholesterol is used to make bile.
    A. True   **B.** False

17. Tuna fish provides high-quality protein.
    A. True   **B.** False

18. High-fat animal products contain saturated fatty acids.
    A. True   **B.** False

19. Unsaturated fats are liquid at room temperature.
    A. True   **B.** False

20. Less than 300 milligrams of cholesterol should be consumed per day to maintain heart health.
    A. True   **B.** False

21. Fruit provides a source of carbohydrates.
    A. True   **B.** False

Match the term on the left with the definition on the right.

22. Hunger   **A.** Physiological need for food

23. Diet   **B.** Psychological desire for food

24. Appetite   **C.** The kind and amount of food consumed in a day

25. Food   **D.** Anything that nourishes the body

26. Cells are individually alive and contribute to multicellular organisms such as humans.
    A. True   **B.** False

27. Vitamins C and D are soluble in water.
    A. True   **B.** False

28. Zinc, iron, and manganese are essential trace minerals in human nutrition.
    A. True   **B.** False

29. Minerals are needed in very large amounts in the diet.
    A. True   **B.** False

30. Water is an organic compound that is essential for life.
    A. True   **B.** False

31. Iron is a major mineral.
    A. True   **B.** False

32. Biological classification is a method of categorizing living organisms.
    A. True   **B.** False

33. A lacto-ovo vegetarian consumes chicken.
    A. True   **B.** False

34. Which of the following best describes nutrition?
    A. The kind and amount of food consumed each day
    B. The provision of essential nutrients consumed from the diet
    C. The amount of a nutrient present in one serving of food in the context of the number of Calories provided
    D. The provision of anything that nourishes the body
    E. The processes involved with being nourished

35. Which of the following aspects of lifestyle management can slow the rate of aging or physiological decline?
    A. Diet
    B. Exercise
    C. Controlling certain behavioral risks
    D. All of the above
    E. None of the above

36. The energy stored between the carbon-carbon bonds in protein, carbohydrate, and fat is captured and used to produce which of the following molecules?
    A. Vitamins
    B. Minerals
    C. Amino acids
    D. Adenosine triphosphate (ATP)
    E. Cholesterol

37. Which of the following energy-producing nutrients generates adenosine triphosphate (ATP) at the fastest rate?
    A. Amino acids
    B. Fatty acids
    C. Starches
    D. Cellulose
    E. Alcohol

38. Which of the following foods provide(s) a plant source of complex carbohydrate?
    A. Legumes or dried beans
    B. Dark green and orange vegetables
    C. Whole-grain breads and cereals
    D. All of the above
    E. None of the above

39. Which of the following are polyunsaturated fatty acids?
    A. Omega-6 fatty acids
    B. Omega-3 fatty acids
    C. Alpha-linolenic acid
    D. Linoleic acid
    E. All of the above

40. Which of the following activities would increase sodium losses from the body and increase the daily need for sodium?
    A. Participation in a competitive sporting event
    B. Hiking in the heat
    C. Jogging
    D. Playing basketball
    E. All of the above

41. Sidney consumed 5 grams of sugar. How many Calories were consumed?
    A. 4
    B. 20
    C. 90
    D. 49
    E. None of the above

42. Terry consumed 10 grams of sunflower seed oil. How many Calories were consumed?
    A. 4
    B. 20
    C. 90
    D. 49
    E. None of the above

43. Stevie consumed 7 grams of alcohol. How many Calories were consumed?
    A. 4
    B. 20
    C. 90
    D. 49
    E. None of the above

44. A function of phospholipids is to:
    A. Make bile
    B. Repair tissue
    C. Make cell membranes
    D. Improve bone density
    E. All of the above

45. What is the Acceptable Macronutrient Distribution Range (AMDR) for linoleic acid?
    A. 20–35 percent of total Calories consumed per day
    B. 5–10 percent of total Calories consumed per day
    C. 0.6–1.2 percent of total Calories consumed per day
    D. Less than 7 percent of total Calories consumed per day
    E. None of the above

46. What is the recommended saturated fatty acid intake level to maintain health?
    A. 20–35 percent of total Calories consumed per day
    B. 5–10 percent of total Calories consumed per day
    C. 0.6–1.2 percent of total Calories consumed per day
    D. Less than 7 percent of total Calories consumed per day
    E. None of the above

47. Protein complementation would occur best by combining whole-grain cereal with:
    A. Dried fruit
    B. Whole fresh fruit
    C. Soy milk
    D. All of the above
    E. None of the above

48. Which of the following trace minerals function as electrolytes?
    A. Sodium, potassium, and water
    B. Calcium, phosphorus, and magnesium
    C. Zinc, iron, and molybdenum
    D. A and B
    E. None of the above

49. Which of the following major minerals are primarily concentrated in the bones?
    A. Sodium, potassium, and water
    B. Calcium, phosphorus, and magnesium
    C. Zinc, iron, and molybdenum
    D. A and C
    E. None of the above

50. Which of the following trace minerals is a structural component of vitamin B_{12}?
    A. Zinc
    B. Cobalt
    C. Molybdenum
    D. Boron
    E. None of the above

Total Recall

SAMPLE QUESTIONS

True/False Questions

1. Organic chemicals refer to chemical compounds that contain the element carbon.

2. Carbohydrates are organic compounds that produce 5 Calories per gram.

3. Chemically and structurally, alcohol is closely related to fatty acids.

4. Cholesterol is an essential nutrient for humans.

5. Water provides the medium for metabolism.

Multiple Choice Questions: Choose the best answer.

6. What are the three most common single-unit, simple sugars consumed in the diet?
    A. Sucrose, sugar, galactose
    B. Glucose, fructose, galactose
    C. Maltose, ribose, glucose
    D. Sucrose, maltose, lactose
    E. Starch, glucose, maltose

7. Which of the following answers contain all of the fat-soluble vitamins?
    A. Vitamin C, pantothenic acid, biotin, and folate
    B. Vitamins A, C, E, and biotin
    C. Zinc, iron, biotin, and vitamin K
    D. Vitamins A, D, E, and K
    E. Vitamins A and D, zinc, and calcium

8. Food provides nutrients that nourish the:
    A. Cells
    B. Organs
    C. Tissues
    D. Organism
    E. All of the above

9. Choose one of the following answers that includes only the trace minerals:
    A. Calcium, iron, and copper
    B. Iron, zinc, and iodine
    C. Magnesium, cobalt, and selenium
    D. Phosphorus, fluoride, and chromium
    E. Fluoride, cobalt, and sulfur

10. Which of the following is **not** a function of minerals in the human body?
    A. Bone structure
    B. Coenzyme
    C. pH balance
    D. Emulsifier
    E. Fluid regulation

CASE STUDY

Denise is a 24-year-old female. She has been following a high-protein weight-loss regimen for two months after watching an infomercial on this type of weight-loss diet. Denise loves to eat fast food and sweets, a diet that has caused her to gain excess body weight and fat. When asked about it, she says her usual way of eating needed to change, so she is "on a diet." The change in her food selection has caused her to eliminate the intake of all fruits, grain products, starchy vegetables, legumes, and refined baked goods, as well as nearly eliminate the intake of dairy products such as milk, yogurt, and ice cream. As a result, she is consuming a lot of meat, cheese, nuts, olives, and diet Jell-O. Because she doesn't like the taste of water, she is drinking mostly caffeinated diet soda. Most of her Calories are from fat (55 percent of total Calories are from fat, with a high intake of saturated fatty acids and cholesterol), a high level from protein (35 percent of total Calories), and a low intake level of carbohydrates (10 percent of Calories). Her diet was analyzed and found to be extremely deficient in Calories, dietary fiber, and many water-soluble vitamins and some major and trace minerals.

1. Denise has an accurate understanding of the meaning of the term *diet*.
   A. True
   B. False

2. Denise's diet should improve her health both now and in the future.
   A. True
   B. False

3. Which characteristic of a sound diet is Denise following?
   A. Adequacy
   B. Balance
   C. Moderation
   D. Variety
   E. None of the above

4. Based on the case study information provided, which factor do you think is affecting Denise's food intake the most?
   A. Appetite
   B. Availability
   C. Advertising
   D. Economics
   E. Professional advice

5. Denise's diet is providing a majority of Calories from the preferred energy source.
   A. True
   B. False

6. Which food that Denise is eliminating is also eliminating her fiber source?
   A. Milk
   B. Refined baked goods
   C. Yogurt
   D. Fruit
   E. All of the above

7. Denise's diet is based on the principle of protein complementation.
   A. True
   B. False

8. Which food that Denise is eating provides the best source of essential fatty acids?
   A. Meat
   B. Milk
   C. Nuts
   D. Diet Jell-O
   E. Olives

9. Denise has a fluid intake that is conducive to health promotion.
   A. True
   B. False

10. By consuming an inadequate level of vitamins and minerals, which body function could Denise be impairing the least?
    A. Enzyme function
    B. Cell membrane structure
    C. Bone structure
    D. Connective tissue building
    E. Fluid regulation

FUN-DUH-MENTAL PUZZLE

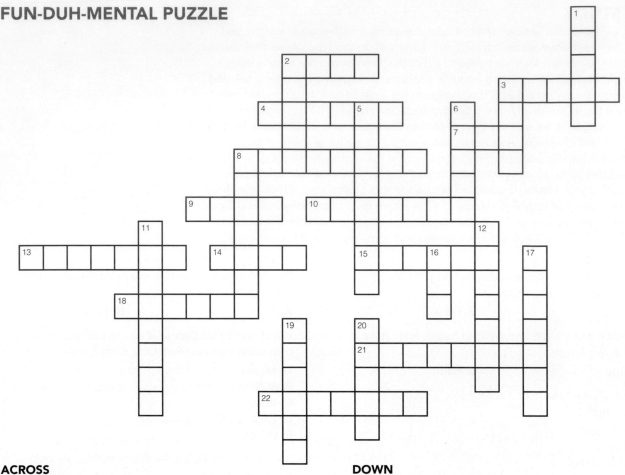

ACROSS

2. Anything that nourishes the body.
3. An animal source of carbohydrate.
4. Physiological need for food.
7. An exogenous source of cholesterol.
8. A substance that provides nourishment; can be essential or nonessential.
9. Alcohol is a carbohydrate but also a _____.
10. Milk sugar.
13. A unit used to measure energy in food.
14. The kind and amount of food typically consumed each day.
15. Mineral whose need is met by protein intake.
18. The result of a sound diet.
21. Psychological desire for food.
22. Feedback mechanism terminating food consumption.

DOWN

1. Noncaloric complex carbohydrate.
2. Food source of simple sugar.
3. Complete proteins have _____ biological value.
5. A factor affecting longevity that helps maintain a healthy body weight.
6. A food that supplies a large amount of nutrients relative to its number of Calories is a nutrient _____ food.
8. To keep alive by providing food.
11. Noncaloric essential nutrients providing structure to the body.
12. Used for tissue repair.
16. Energy-producing nutrient providing 9 Calories per gram.
17. When different foods are used for the same purpose.
19. Mineral whose need is met by vitamin B_{12} intake.
20. Medium for metabolism.

References

Web Resources

A. American Cancer Society: www.cancer.org
B. American Diabetes Association: www.diabetes.org
C. American Heart Association: www.heart.org
D. Nutrient Data Laboratory: www.nal.usda.gov/fnic/foodcomp/search
E. The Police Notebook, Blood Alcohol Calculator: www.ou.edu/oupd/bac.htm
F. The Athlete.org: www.theathlete.org
G. The United States Department of Agriculture: usda.gov

Works Cited

1. Aldana, S. G., Greenlaw, R. L., Diehl, H. A., Salberg, A., Merrill, R. M., Ohmine, S., & Thomas, C. (2005). Effects of an intensive diet and physical activity modification program on the health risks of adults. *Journal of the American Dietetic Association, 105*(3), 371–381.
2. American Dietetic Association. (2003). Position of the American Dietetic Association: Vegetarian diets. *Journal of the American Dietetic Association, 103,* 748–765.
3. American Dietetic Association. (2006). Nutrients. *Evidence Analysis Library*. Accessed at: http://adaevidencelibrary.com.
4. American Dietetic Association. (2007). Position of the American Dietetic Association: Total diet approach to communicating food and nutrition information. *Journal of the American Dietetic Association, 107*(7), 1224–1232.
5. American Dietetic Association. (2007). Position of the American Dietetic Association and Dietitians of Canada: Dietary fatty acids. *Journal of the American Dietetic Association, 107*(9), 1599–1611.
6. Aronson, D. (2009). Nutrition for health and longevity. *Today's Dietitian, 11*(2), 40–46.
7. Bellisle, F. (2005). Nutrition and health in France: Dissecting a paradox. *Journal of the American Dietetic Association, 105*(12), 1870–1880.
8. Blössner, M., & de Onis, M. (2005). Malnutrition: Quantifying the health impact at national and local levels. *World Health Organization (WHO) Environmental Burden of Disease Series, No. 12*, Geneva, Switzerland.
9. Borzekowski, D. L. G., & Robinson, T. N. (2001). The 30-second effect: An experiment revealing the impact of television commercials on food preferences of preschoolers. *Journal of the American Dietetic Association, 101,* 42–46.
10. Brannon, C. A. (2007, August). Ancient and alternative grains. *Today's Dietitian, 9*(5), 10–15.
11. Cason, K. L., & Wenrich, T. R. (2002). Health and nutrition beliefs, attitudes, and practices of undergraduate college students: A needs assessment. *Topics in Clinical Nutrition, 17,* 52–66.
12. Coll, A. P., Farooqi, I. S., & O'Rahilly, S. (2007). The hormonal control of food intake. *Cell, 129*(2), 251–262.
13. Coulston, A. M., & Johnson, R. K. (2002). Sugar and sugars: Myths and realities. *Journal of the American Dietetic Association, 102,* 351–353.
14. Cummings, D. E., & Overduin, J. (2007). Gastrointestinal regulation of food intake. *Journal of Clinical Investigation, 117*(1), 13–23.
15. Davis, B. (2010). Vegetarian's challenge: Optimizing essential fatty acid status. *Today's Dietitian, 12*(2), 22–28.
16. Dietrich, M., Brown, C. J. P., Cho, S., & Block, G. (2005, July/August). What is the average effect on energy, macronutrient, and micronutrient intake whenever Americans substitute plant protein for beef or pork? *Nutrition Today, 40*(4), 156–164.
17. Eckel, R. H., Borra, S., Lichtenstein, A. H., & Yin-Piazza, S. Y. (2007). Understanding the complexity of *trans* fatty acid reduction in the American diet: American Heart Association *Trans* Fat Conference, 2006 report of the *Trans* Fat Conference Planning Group. *Circulation, 115,* 2231–2246.
18. Ellegard, L. H., Anderson, S. W., Normen, A. L., & Andersson, H. A. (2007). Dietary plant sterols and cholesterol metabolism. *Nutrition Reviews, 65*(1), 39–45.
19. Ferreira, M. P., & Willoughby, D. (2008). Alcohol consumption: The good, the bad and the indifferent. *Applied Physiology and Nutrition Metabolism, 33,* 12–20.
20. Florentino, R. F. (2009). Hydration and human health: Critical issues update. *Nutrition Today, 44*(1), 6–13.
21. Fried, S. K., & Rao, S. P. (2003). Sugars, hypertriglycerideemia, and cardiovascular disease. *American Journal of Clinical Nutrition, 78,* 873S–880S.
22. Gaesser, G. A. (2007). Carbohydrate quantity and quality in relation to body mass index. *Journal of the American Dietetic Association, 107*(10), 1768–1780.
23. Griel, A. E., Ruder, E. H., & Kris-Etherton. P. M. (2006). The changing roles of dietary carbohydrates: From simple to complex. *Arteriosclerosis, Thrombosis, and Vascular Biology, 26,* 1958–1965.
24. Gropper, S. S. (2000). *The biochemistry of human nutrition: A desk reference* (2nd ed.). Belmont, CA: Wadsworth.
25. Gropper, S. S., Smith, J. L., & Groff, J. L. (2009). *Advanced nutrition and human metabolism* (5th ed.). Belmont, CA: Wadsworth.

26. Haber, D. (2007). *Health promotion and aging: Practical applications for health professionals* (4th ed.). New York: Springer.

27. Hart, A., Tinker, L. F., Bowen, D. J., & Satia-Abouta, J. (2004). Is religious orientation associated with fat and fruit/vegetable intake? *Journal of the American Dietetic Association, 104*(8), 1292–1296.

28. Institute of Medicine. (2006). *Dietary Reference Intakes: The essential guide to nutrient requirements.* Washington, DC: National Academies Press.

29. Jacobs, D. R., & Tapsell, L. C. (2007). Food, not nutrients, is the fundamental unit in nutrition. *Nutrition Reviews, 65*(10), 439–450.

30. Kleiner, S. M. (1999). Water: An essential but overlooked nutrient. *Journal of the American Dietetic Association, 99,* 200–206.

31. Kolodinsky, J., Harvey-Berino, J. R., Berlin, L., Johnson, R. K., & Reynolds, T. W. (2007). Knowledge of current dietary guidelines and food choice by college students: Better eaters have higher knowledge of dietary guidance. *Journal of the American Dietetic Association, 107*(8), 1409–1413.

32. Korver, O., & Katan, M. B. (2006). The elimination of trans fats from spreads: How science helped to turn an industry around. *Nutrition Reviews, 64*(6), 275–279.

33. Ledikwe, J. H., Blanck, H. M., Khan, L. K., Serdula, M. K., Seymour, J. D., Tohill, B. C., & Rolls, B. J. (2006). Low-energy-density diets are associated with high diet quality in adults in the United States. *Journal of the American Dietetic Association, 106*(8), 1172–1180.

34. Levine, A. S., Kotz, C. M., & Gosnell, B. A. (2003). Sugars and fats: The neurobiology of preference. *Journal of Nutrition, 133,* 831S–834S.

35. Lichtenstein, A. H., & Russell, R. M. (2005). Essential nutrients: Food or supplements? Where should the emphasis be? *Journal of the American Medical Association, 294,* 351–358.

36. Loma Linda University. (2003, April). Do you need more protein? *Nutrition and Health Letter, 6,* 1–7.

37. Mahan, K., & Escott-Stump, S. (2008). *Krause's food, nutrition, and diet therapy* (12th ed.). Philadelphia: Saunders.

38. McBurney, M. I. (2009). Drink fluids to maintain hydration and eat to obtain calories. *Nutrition Today, 44*(1), 14–16.

39. McKeown, R. E. (2009). The epidemiologic transition: Changing patterns of mortality and population dynamics. *American Journal of Lifestyle Medicine, 3*(S1), 19S–26S.

40. Mestek, M. L. (2009). Physical activity, blood lipids, and lipoproteins. *American Journal of Lifestyle Medicine, 3*(4), 279–283.

41. Miner, M. (2007). Fitness, antioxidants, and moderate drinking: All to lower cardiovascular risk. *American Journal of Lifestyle Medicine, 1*(2), 110–112.

42. Mokdad, A. H., Marks, J. S., Stroup, D. F., & Gerberdine, L. J. (2004). Actual causes of death in the United States, 2000. *Journal of the American Medical Association, 291,* 1238–1245.

43. Monsivais, P., & Drewnowski, A. (2007). The rising cost of low-energy-density foods. *Journal of the American Dietetic Association, 107*(12), 2071–2076.

44. Moore, B. J. (2010). The standard drink: Does anybody know what it is? Does anybody care? *Nutrition Today, 45*(2), 66–72.

45. Murphy, S. P., Barr, S. I., & Yates, A. A. (2006). The Recommended Dietary Allowance (RDA) should not be abandoned: An individual is both an individual and member of a group. *Nutrition Reviews, 64*(7), 313–348.

46. Newby, P. K. (2006). Examining energy density: Comments on diet quality, dietary advice, and the cost of healthful eating. *Journal of the American Dietetic Association, 106*(8), 1166–1169

47. Palmer, S. (2008). The fairest fats of them all (and those to avoid). *Today's Dietitian, 10*(10), 36–40.

48. Quatromoni, P. A., Pencina, M., Cobain, M. R., Jacques, P. F., & D'Agostino, R. B. (2006). Dietary quality predicts adult weight gain: Findings from the Framingham Offspring Study. *Obesity, 14*(8), 1383–1391.

49. Reichrath, J. (2007). Vitamin D and the skin: An ancient friend, revisited. *Experimental Dermatology, 16*(7), 618–625.

50. Ritz, P. (2005). The importance of good hydration for day-to-day health. *Nutrition Reviews, 63*(6), S6–S13.
51. Ronnett, G. V., Kleman, A. M., Kim, E. K., Landree, L. E., & Tu, Y. (2006). Fatty acid metabolism, the central nervous system, and feeding. *Obesity, 14*(S5), 201S–207S.
52. Rosenberg, I. H. (2007). Challenges and opportunities in the translation of the science of vitamins. *American Journal of Clinical Nutrition, 85*(1), 325S–3257S.
53. Russel, R. M. (2001). New micronutrient Dietary Reference Intakes from the National Academy of Sciences. *Nutrition Today, 36*, 163–171.
54. Sawka, M. N., Cheuvront, S. N., & Carter, R., III. (2005). Human water needs. *Nutrition Reviews, 63*(6), S30–S39.
55. Schorin, M. D. (2005, November/December). High fructose corn syrups, part 1. *Nutrition Today, 40*(6), 248–252.
56. Schorin, M. D. (2006, March/April). High fructose corn syrups, part 2. *Nutrition Today, 41*(2), 70–77.
57. Seitz, H. K., & Becker, P. (2007). Alcohol metabolism and cancer risk. *Alcohol Research and Health, 30*(1), 38–59.
58. Smith, D. (2006). Review: Increased physical activity and combined dietary changes reduce mortality in coronary artery disease: Commentary. *ACP Journal Club, 144*, 16.
59. Starr, C., Evers, C. A., & Starr, L. (2010). *Biology: Today and tomorrow with physiology* (3rd ed.). Belmont, CA: Brooks/Cole.
60. Stipanuk, M. H. (2000). *Biochemical and physiological aspects of human nutrition.* Philadelphia: Saunders.
61. Tanumihardjo, S. A., Anderson, C., Kaufer-Horwitz, M., Bode, L., Emenaker, N. J., Haqq, A. M., . . . & Stadler, D. D. (2007). Poverty, obesity, and malnutrition: An international perspective recognizing the paradox. *Journal of the American Dietetic Association, 107*(11), 1966–1972.
62. Tillotson, J. E. (2005, July/August). The heavy burden of eating out today. *Nutrition Today, 40*(4), 173–175.
63. Trumbo, P., Schlicker, S., Yates, A. A., & Poos, M. (2002). Dietary Reference Intakes for energy, carbohydrates, fiber, fat, fatty acids, cholesterol, protein and amino acids. *Journal of the American Dietetic Association, 102*, 1621–1631.
64. Uauy, R., & Dangour, A. D. (2006). Nutrition in brain development and aging: Role of essential fatty acids. *Nutrition Reviews, 64*(5), S24–S33.
65. Van Horn, L. (2007). Eating more . . . nourishing less. *Journal of the American Dietetic Association, 107*(7), 1087.
66. Venderley, A. M., & Campbell, W. W. (2006). Vegetarian diets: Nutritional considerations for athletes. *Sports Medicine, 36*, 293–305.
67. Watkins, B. A., & Hutchins, H. (2010). Omega-3 fatty acids: Past, present, and future. *Dietitians in Integrative and Functional Medicine, 13*(2), 21, 24–32.
68. Webb, D. (2010). Defending vegan diets: RDs clear up misconceptions about their completeness. *Today's Dietitian, 12*(9), 20–24.
69. Whitney, E., & Rolfes, S. R. (2011). *Understanding nutrition* (12th ed.). Belmont, CA: Wadsworth.
70. Wolf, A., Bray, G. A., & Popkin, B. M. (2008). A short history of beverages and how our body treats them. *Obesity Reviews, 9*, 151–164.
71. Wolfe, R. R. (2006). The underappreciated role of muscle in health and disease. *American Journal of Clinical Nutrition, 84*(3), 475–482.
72. Wylie, J. (2006, Fall). Does carbohydrate type impact weight management? *SCAN'S (A Publication for Sports, Cardiovascular, and Wellness Nutritionists) PULSE, 25*(4), 6–8.
73. Yetley, E. A. (2007). Multivitamin and multimineral dietary supplements: Definitions, characterization, bioavailability, and drug interactions. *American Journal of Clinical Nutrition, 85*(1), 269S–276S.

Tools to Plan, Manage, and Evaluate Diets

Valentyn Volkov/Shutterstock.com

MODULE GOAL
To understand how to use various dietary tools to plan, manage, and evaluate diets for nutritional adequacy.

LEARNING OBJECTIVES
When you complete this learning module, you will know:

- How to read and interpret the information displayed in the Nutrition Facts panel on food package labels.
- Your personal nutrient needs and how to consume a healthy diet.
- The MyPlate food guidance system.
- The dietary recommendations made to reduce the risk of chronic disease and promote health.
- How to use the Exchange Lists and tools for understanding food composition.

PERSONAL IMPROVEMENT GOALS
This learning module is designed to teach you to:

- Select or reject foods based on the nutritional information given in the Nutrition Facts panel.
- Identify and follow your personal MyPlate food plan.
- Put into action your knowledge of dietary guidelines, food composition information, and the Exchange List system.
- Use the tools to continue to plan, manage, and evaluate all aspects of your diet.

As you become more knowledgeable about your nutrient needs, what constitutes a healthy diet, and the nutritional value of foods, we hope you will be motivated to select and consume nutritious foods daily. Remember that diet impacts your health and disease risk. The top two leading causes of death in the United States are diseases of the heart and cancer. About one-third of the individuals who die from heart diseases die of a condition that might have been prevented if they had made healthy diet, exercise, and lifestyle choices. About 45 percent of the people who die from cancer die from a diet-related cancer. The third largest category of cause of death is stroke, which could be related to high blood pressure and is associated with a diet high in sodium and inadequate in potassium, calcium, and magnesium. After you have been exposed to the information in this learning module, we hope that you will find yourself eating foods that fulfill the food plan individually recommended to you, selecting foods based on their nutritional values, and planning your diet to optimize your health.[20,23,31,40,46,48,81]

2.1 FOOD LABELS

Introduction

By law, the **food label** provides consumers with a great deal of information about the nutrient content of the food within. Understanding the information in food package labels can help you make better food choices. In this section we will look at the legislation that governs food package labels and decode the nutritional information that is provided.

T-Talk 2.1
To hear Dr. Turley talk about food labels, go to www.cengagebrain.com

Nutrition Labeling and Education Act of 1990

As the food packaging industry has grown, laws regulating processing methods, vitamin supplementation, and labeling have been passed. A few of these laws are highlighted in Table 2.1.

In 1990, Congress passed the Nutrition Labeling and Education Act, which governs the information that food manufacturers must provide to consumers. The food package labeling law established a standard for the appearance and content of information on food package labels and in the **Nutrition Facts panel**. Some of the key items of information that appear on food labels include the **Daily Values (DVs)**, expressed as the **Daily Reference Value (DRV)** and the **Reference Daily Intake (RDI)**, **ingredients list**, terms, and **health claims** (see Figure 2.1).

As a result of the 1990 food labeling law, serving size was defined by the Food and Drug Administration (FDA) for more than 100 food categories. The FDA established set serving sizes so consumers could compare Nutrition Facts among products much more easily. It should be noted that if the food package size is smaller than 12 square inches in surface area, which is about the size of a pack age of chewing gum, then the food company is only required to include a phone number and does not have to include a complete Nutrition Facts panel.

food label The Nutrition Facts panel, which provides information according to law about the manufacturer, nutrients, ingredients, terms, health claims, and allergic foods in that item.

Nutrition Facts panel An area on the food package that shows the serving size, servings per container, Calories per serving, Calories from fat per serving, percent of the Daily Value (DV) including the DRVs and RDIs, and the ingredients.

Daily Values (DVs) Reference values including Daily Reference Values (DRVs) and Reference Daily Intakes (RDIs) used on food labels.

Daily Reference Values (DRVs) Daily Values for fat, saturated fat, cholesterol, carbohydrate, fiber, protein, sodium, and potassium that are based on a 2,000-Calorie diet and expressed as percents (except protein, sugar, and potassium) in the Nutrition Facts panel on food package labels.

Reference Daily Intake (RDI) The highest level of the essential vitamins and minerals for men or women based on the 1968 Recommended Dietary Allowances (RDAs); values are expressed as percentages on the Nutrition Facts panel of food labels, and only percents for vitamin A, vitamin C, calcium, and iron are required by law to be shown.

ingredients list A list of the components of a processed food product given in descending order by gram weight or volume.

health claims Statements approved by the Food and Drug Administration (FDA) linking the nutrition profile of a food to a reduced risk of a particular disease or health-related condition.

TABLE 2.1

Food Label Legislation Highlights

Year	Law
1942	The Enrichment Act of 1942 mandates that thiamin, niacin, riboflavin, and iron have to be added back to refined grain products at specified levels to prevent nutritional deficiencies in America.
1998	Amendment of the Enrichment Act of 1942 mandates folic acid to be added back into refined grain products to reduce the incidence of neural tube birth defects of the brain and spine (spina bifida) as well as to prevent elevated levels of homocysteine (hyperhomocystemia), a risk factor for heart disease.
1990	The Nutrition Labeling and Education Act of 1990 passed. By 1994, more than 300,000 packaged foods were relabeled. The Food and Drug Administration establishes set serving sizes for more than 100 food categories, making product comparison easier. Mandatory compliance by food manufacturers is required. Unprocessed foods and fast food are not required by law to label foods, although information may be optionally made available via brochures, posters, and/or Web sites.
2003	Legislation passed for *trans* fatty acids (TFAs) to appear on a separate line under saturated fatty acids (SFAs) in the Nutrition Facts panel starting January 1, 2006.
2004	The Food Allergen Labeling and Consumer Protection Act of 2004 passed. Manufacturers must plainly list milk, eggs, fish, crustacean shellfish, tree nuts, peanuts, wheat, and soybeans on the food package label.
2009	Country of Origin Labeling requires notice of the source of certain foods, including muscle cut and ground meats, wild and farm-raised fish and shellfish, fresh and frozen fruits and vegetables, peanuts, pecans, macadamia nuts, and ginseng.

© Cengage Learning 2013

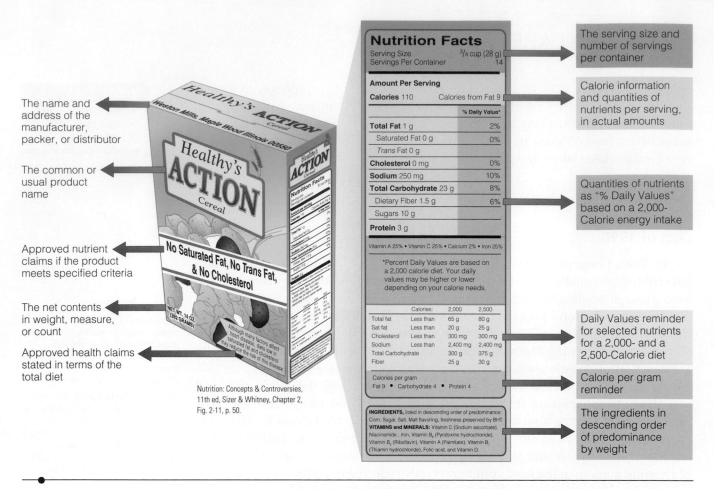

Nutrition: Concepts & Controversies, 11th ed, Sizer & Whitney, Chapter 2, Fig. 2-11, p. 50.

FIGURE 2.1 The anatomy of a food package label.

It is easy to misunderstand the food package label information unless you know something about the legal definitions of the terms used. So we will look at all the important information given on food labels to empower you to make better food choices.[E,I,8,40,47,49,65,78]

Daily Reference Value (DRV)

DRVs are based on a 2,000-Calorie diet for adults and children over 4 years old (see Table A11 in Appendix A). They are set for the nutrients listed in Table 2.2 and are expressed in the Nutrition Facts panel on food package labels as percent Daily Values, except for protein, sugars, *trans* fatty acids, and potassium.

TABLE 2.2

The Daily Reference Values Based on a 2,000-Calorie Diet	
Nutrient	**Daily Reference Value (DRV)**
Fat*	< 30% of Calories or 65 g (*equals 29% of Calories*)
Saturated fat	< 10% of Calories or 20 g (*equals 9% of Calories*)
Cholesterol	300 mg
Carbohydrate**	60% of Calories or 300 g
Fiber	12.5 g/1,000 Calories or 25 g
Sodium	2,400 mg
Potassium	3,500 mg
Protein	50 g high-quality or 65 g low-quality

*Total fat on a food label refers to triglycerides, which include SFAs, PUFAs, MUFAs, and/or TFAs.

**Total carbohydrate on a food label refers to digestible carbohydrates, which include starch, sugars, and alcohol.

© Cengage Learning 2013

These values can be somewhat misleading for consumers who are not aware of their meaning and not as applicable for consumers who do not consume a 2,000-Calorie diet.

Understanding Carbohydrate, Protein, and Fat on the Food Label

During the 1980s, public health organizations shifted their focus from concerns about vitamin and mineral deficiencies to the better management of macronutrients. Nutritionally related chronic diseases of the 21st century result largely from the mismanagement of total fat, saturated fatty acids, *trans* fatty acids, sodium, and sugars consumed in the American diet. These dietary components are included on the food package label, and you should take note of them in the effort to make better food choices.

Carbohydrates and Sugars on the Food Label The Acceptable Macronutrient Distribution Range (AMDR) for carbohydrate is 45 to 65 percent of Calories consumed in the diet, and the AMDR for sugars is 25 percent or less. The DRV for carbohydrate is based on 60 percent of the Calories from carbohydrate of a 2,000-Calorie diet. There is no DRV for sugars. In the DRV, total grams of carbohydrates are listed, as are grams of simple sugars. The grams of sugar include all simple sugars added or naturally occurring per serving. Dietary sugars largely present include lactose, sucrose, and fructose. **Added sugars** can include a wide variety of sources, as shown in Table 2.3. You can calculate the percentage of Calories from sugar by following the example given in Table 2.4 on page 53. The dietary fiber amount is also included in the carbohydrate section of the DRV. Like sugars, the grams of fiber per serving are listed separately.

TABLE 2.3

The Most Common Terms That Describe Added Sugar in the Ingredient List of Food Labels

Term	Definition
Concentrated fruit juice	Dehydrated fruit juice, such as grape juice, that is then used as an additive to sweeten food products that often claim to be all fruit.
Corn syrup	A syrupy sweetener that is made from corn by the enzyme action on cornstarch, generating mostly glucose and some maltose.
Dextrose	Another name for glucose that may have been produced from the hydrolysis of cornstarch.
High-fructose corn syrup	A commercially produced sweetener added to many processed food items, composed of high amounts of fructose generated from the chemical release and modification of glucose in cornstarch.
Honey	A sweetener generated by the enzymatic digestion of the sucrose in nectar to fructose and glucose by bees.
Molasses	A thick brown syrup left over from the refinement of sugar cane.
Sugar	Crystals of sucrose produced by dissolving, concentrating, and recrystallizing raw sugar harvested from sugar cane. Sugar can be found in many forms, including white, brown, powdered or confectionary, granulated, and invert.

© Cengage Learning 2013

Proteins on the Food Label The AMDR for protein is 10 to 35 percent of the Calories consumed in the diet, but the adult Dietary Reference Intake (DRI) for protein is 0.8 grams per kilogram of body weight. The DRV for protein is 50 grams in a 2,000-Calorie diet, which provides 12 percent of Calories from protein. The percent of the DRV for protein is never listed in the Nutrition Facts panel, only

added sugars Simple sugars and syrups used as an ingredient in the preparation of processed foods or added to foods by an individual.

the grams of protein per serving. The reason is that there is such a broad range of body weights, and thus such a large range of grams of protein for individual requirements, that any percentage given would be erroneous for most people. You can calculate the percentage of Calories from protein by following the example given in Table 2.4.

Fats on the Food Label Because of the Calorie density of dietary fats, the disease promotion of high-fat diets, and the health risks of consuming large amounts of saturated fatty acids and *trans* fatty acids, it is very important to understand the nutrition information about fats provided on the food package label (see BioBeat 2.1 on page 52). The AMDR for total fat is 20 to 35 percent of Calories; the goal for saturated fatty acid intake is less than 7 percent of Calories; and the goal for *trans* fatty acid intake is as low as possible. The DRV for total fat is 29 percent of Calories, while the DRV for saturated fat is 9 percent of Calories, both in a 2,000-Calorie diet. There is no DRV for *trans* fatty acids.

The Nutrition Facts panel on the food package label shows the grams of total fat contained in a single serving. Serving size is the first nutrition fact to appear in the Nutrition Facts panel. By law, the fatty acids connected with increased risk of disease are shown in gram amounts; thus, grams of saturated fatty acids and *trans* fatty acids are provided (see BioBeat 2.1). The total fat grams include the sum of all fatty acids. So this total fat gram number includes monounsaturated fatty acids and polyunsaturated fatty acids, which are not required by law to be shown separately on the food label, as well as saturated fatty acids and *trans* fatty acids, which are required to be shown. Food manufacturers often choose to include the amount of monounsaturated fatty acids and polyunsaturated fatty acids to underscore the healthy fat content of their food product.

The percent value shown on the Nutrition Facts panel for fat is the percent of the DRV. Thus, if a food label Nutrition Facts panel reads 10 percent for total fat under the DRV header, this means that 10 percent of the 65-gram DRV for fat is provided in a single serving of the food. The percent is not the percent of Calories from fat.[47,49,65]

Because most people don't eat exactly 2,000 Calories every day—and may not want to consume a diet of 29 percent of Calories from fat as their reference—looking at the percent of Calories from fat is a better way to assess individual foods for their fat content. You can make this calculation using the information on the Nutrition Facts panel, starting with the number of Calories from fat in the context of total Calories by following the example in Table 2.4.

You can also determine the percentage of Calories from fat by doing the following calculations. Take the grams of fat and multiply by 9 Calories per gram. The result equals the number of Calories from fat. Then take the Calories from fat and divide by the total Calories per serving, then multiply by 100. The result equals the percentage of Calories from fat (see Figure 2.2 and Table 2.4).

Classifying Food by Fat Content Once the percentage of Calories from fat has been determined, the food can be classified as high, moderate, or low fat. A **high-fat food** provides more than 35 percent of Calories from fat. A **moderate-fat food** provides between 25 and 35 percent of Calories from fat. A **low-fat food** provides less than 25 percent of Calories from fat. This same principle can be applied to assessing a day's worth of eating, or a diet, and determining whether the diet is high, moderate, or low fat. Individuals should avoid consuming an excess of high-fat foods so that the diet doesn't become a high-fat diet.

Fat Content by Weight The fat content of meats and dairy products are labeled as a percentage of fat by weight. The meat and dairy industries refer to the fat content by gram weight rather than by Calories. So 1 percent milk is 1 percent fat by

high-fat food Provides more than 35 percent of the Calories from fat.

moderate-fat food Provides 25 to 35 percent of the Calories from fat.

low-fat food Provides less than 25 percent of the Calories from fat.

HIGH FAT	MODERATE FAT	LOW FAT

% Calories from fat:

$$\frac{70 \text{ fat Calories}}{150 \text{ total Calories}} \times 100 = \mathbf{47\%}$$

% fat by weight:

$$\frac{8 \text{ fat grams}}{240 \text{ total grams}} \times 100 = \mathbf{3.3\%}$$

100% − 3% = **97% fat free**

*Note 1 mL weighs 1 gram

% Calories from fat:

$$\frac{42 \text{ fat Calories}}{120 \text{ total Calories}} \times 100 = \mathbf{35\%}$$

% fat by weight:

$$\frac{5 \text{ fat grams}}{240 \text{ total grams}} \times 100 = \mathbf{2\%}$$

100% − 2% = **98% fat free**

% Calories from fat:

$$\frac{0 \text{ fat Calories}}{80 \text{ total Calories}} \times 100 = \mathbf{0\%}$$

% fat by weight:

$$\frac{0 \text{ fat grams}}{240 \text{ total grams}} \times 100 = \mathbf{0\%}$$

100% − 0% = **100% fat free**

© Cengage Learning 2013

weight. Most of the milk's weight is from water, and it also provides protein and carbohydrate, along with fat. It is more valuable to know the percentage of fat by Calories, because evaluating fat content by the DRV, by weight, and by legal terms can be confusing and even misleading. (See Figure 2.2, Table 2.4, and Appendix B for calculated examples.[72,73])

FIGURE 2.2 Milk food labels provide examples of foods that are high fat (whole milk), moderate fat (2 percent milk), and low fat (nonfat milk). Calculations for percentage of Calories from fat and percent fat by weight are shown.

Reference Daily Intake (RDI)

The RDI values are set for the vitamins and minerals that are essential in human nutrition. RDIs are based on the highest level of nutrients needed by men or women recommended in the 1968 Recommended Dietary Allowances. The RDIs are expressed as percentages on food labels and by law indicate nutrient density for vitamins C and A as well as iron and calcium. Other nutrients may appear on the label if the food manufacturer chooses to include them. Table A12 in Appendix A shows the actual RDI values for the essential vitamin and mineral percentages on the food label.

The nutrient density of a food is determined by the amount of nutrient in relation to the number of Calories. On a food package label, a food is considered nutrient dense for a particular nutrient (another way of saying it is nutritious) if the food provides at least 20 percent of the RDI for that nutrient per serving.

BioBeat 2.1

The Use of Food Package Labels: Identifying Biologically Harmful Substances

Some substances that are included in the Nutrition Facts panel on food package labels are clearly detrimental to human health. Saturated fatty acids and *trans* fatty acids are two types of dietary fatty acids that have been noted to promote heart disease in the medical literature.[7,16,47,49] The grams of these fatty acid types are provided per serving on full-sized food package labels. Diets high in saturated fatty acids promote heart disease, cancer, and obesity. *Trans* fatty acids, although still unsaturated, promote heart disease, and it has been recently noted that *trans* fatty acids (with their altered hydrogen positioning on the carbons of a double bond) also promote inflammation and type 2 diabetes.[2,16]

Trans fatty acids are commonly consumed in processed foods that use partially hydrogenated oils. The natural food sources that contain *trans* fatty acids are ruminant animal (animals with more than one stomach) food products. For Americans, this means that foods coming from cows provide most natural sources of *trans* fatty acids.

The saturated fatty acids allowance for a 2,000-Calorie diet to maintain heart health is less than 20 grams per day. The saturated fatty acids allowance to comply with the less than 7 percent of Calories from saturated fatty acids for heart health would be less than 13 grams per day. Although there is no DRV for *trans* fatty acids, a heart-healthy recommendation would amount to less than 2 grams per day, which is less than 1 percent of Calories for a 2,000-Calorie diet. Use the information on the Nutrition Facts panel to control your intake of these unhealthy fats.[16,71]

Can you identify some high-fat foods to limit in your diet and provide some healthier alternatives?

iStockPhoto.com/diego cervo

Reading food labels helps you make healthier food selections.

Ingredients List

The order in which the ingredients are listed on a food package label is significant: They are listed in descending order by weight or volume. Food manufacturers are not required to provide how much of each ingredient is present in the food, just the proportion of that **ingredient** in relation to all the others.

Foods That Commonly Cause Allergic Reactions

Many food ingredients are known to cause allergic reactions, and food allergies can be life threatening. **Allergy** labeling has been mandated by law for the top eight **allergic foods**. The Food Allergen Labeling and Consumer Protection Act of 2004 was implemented on January 1,

ingredient A component of a processed food product.

allergy An immune-mediated reaction, usually to a protein component in food.

allergic foods Milk, eggs, fish, crustacean shellfish, tree nuts, peanuts, wheat, and soybeans are the most common foods that people have allergies to.

2006. By law, manufacturers must plainly state that the product contains milk, eggs, fish, crustacean shellfish, tree nuts, peanuts, wheat, and soybean ingredients if used as an ingredient in the processed food. These are the top eight food allergens (see Figure 2.3). Furthermore, if a food is processed in a manufacturing facility or with equipment shared with any of these food allergens, the manufacturer must note the risk of cross-contamination on the food package label (see Table 2.4 for an example).[39,68,75]

TABLE 2.4

A Food Package Label Dissected Mathematically and Legislatively

Smoked Chicken Sausage **Nutrition Facts** Serving size: 1 link (85 g) Servings per container: About 16			Dissecting the Label Mathematically and Legislatively
Amount per serving			**% Calories from Fat:**
Calories 150　　Calories from Fat 90			90 fat Calories ÷ 150 total Calories × 100 = 60%
		% Daily Value	This is a high-fat food.
Total Fat 10 g		**15%**	**% Fat by Weight:**
Saturated Fat 3.5 g		**18%**	10 g fat weight per serving ÷ 85 g total weight per serving × 100 = 12%
Trans Fatty Acids 0 g			This food is 88% fat free.
Cholesterol 80 mg		**27%**	This is not a low-cholesterol food.
Sodium 700 mg		**29%**	This is not a low-sodium food.
Total Carbohydrate 3 g			**% Calories from Total Carbohydrate:**
Dietary Fiber less than 1 g		**1%**	3 g carbohydrate × 4 Calories/g = 12 carbohydrate Calories ÷ 150 total Calories × 100 = 8%
Sugars less than 1 g		**0%**	**% Calories from Sugars:**
			1 g carbohydrate × 4 Calories/g = 4 sugar Calories ÷ 150 total Calories × 100 = 3%
			This food is not a good source of fiber.
Protein 12 g			**% Calories from Protein:**
			12 g protein × 4 Calories/g = 48 protein Calories ÷ 150 total Calories × 100 = 32% This food provides 32% of the total Calories as protein.
Vitamin A 10%　　**Vitamin C** 2%			**mg Iron per Serving:**
Calcium 10%　　**Iron** 5%			5% ÷ 100 × 18 mg = 0.9 mg
			This food is not nutrient dense for vitamin A, vitamin C, calcium, or iron.
Ingredients: Chicken, roasted red pepper, cheddar cheese culture, pasteurized milk, salt, enzymes, powdered cellulose, annatto, roasted poblano pepper, salt, tomato paste, vinegar, spices, cilantro, paprika, jalapeno, granulated garlic and celery powder, in a pork casing.			Ingredients are listed in descending order by weight. This food provides a source of complete protein (from chicken), complex carbohydrate (from vegetables), and simple carbohydrate (lactose from milk).
CONTAINS: MILK			Contains one of the eight leading allergens.

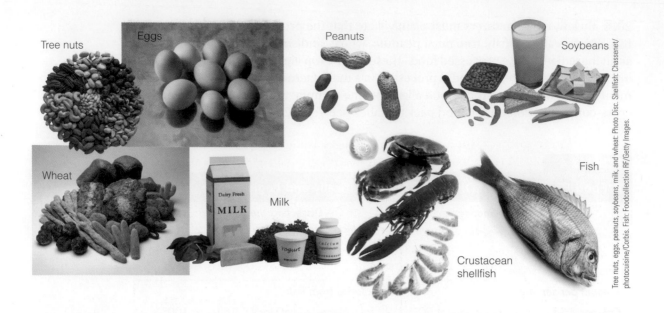

Tree nuts

Eggs

Peanuts

Soybeans

Wheat

MILK

Milk

Fish

Crustacean
shellfish

Tree nuts, eggs, peanuts, soybeans, milk, and wheat: Photo Disc. Shellfish: Chassenet/
photocuisine/Corbis. Fish: Foodcollection RF/Getty Images.

FIGURE 2.3 The top eight food allergens required by law to be plainly listed on packaged foods.

free Negligible amounts of fat, cholesterol, sodium, sugar, or Calories per serving in a food product.

reduced Twenty-five percent less of a nutrient is present as compared to the original food product; a food label must specifically state which nutrient is reduced (e.g., Calories, fat, or sodium).

light or lite One-third fewer Calories per serving as compared to the original product; one-half the fat or sodium as compared to the original product; or light in color or texture compared to the original product.

low 140 milligrams or fewer of sodium per serving; 20 milligrams or fewer of cholesterol per serving; 40 Calories or fewer per serving.

lean Ten grams of fat or less, 4.5 grams of saturated and *trans* fat or less, and 95 milligrams or less of cholesterol per 3.5-ounce serving of meat.

extra lean Five grams of fat or less, 2 grams of saturated and *trans* fat or less, and 95 milligrams or less of cholesterol per 3.5-ounce serving of meat.

The Legal Use of Terms

Many terms used on food package labels have strict legal definitions.[47] Following are some examples of terms that can only be used to mean certain, defined things. **Free** can only be used to refer to fat, cholesterol, sodium, sugar, or Calories. *Free* means that there are negligible amounts of fat, cholesterol, sodium, sugar, or Calories present per serving. The food label must specifically state what the food is free of (e.g., fat free or cholesterol free). When a food label says *sugar free*, it means *sucrose free*, even though the product may contain other types of simple sugars. The terms **reduced** or *less* must be used to indicate that 25 percent less of a nutrient is present as compared to the original product. The food label must specifically state what nutrient is reduced (e.g., reduced Calories, reduced fat, or reduced sodium).

Light or **lite** has several meanings depending on the context. When referring to Calories, *light* means that there are one-third fewer Calories per serving as compared to the original product. When referring to the fat or sodium content, *light* means that there is one-half the fat or sodium as compared to the original product. When referring to color or texture, *light* means simply that, compared to the original product, it is lighter in color or has a lighter texture.

The term **low** has multiple meanings as well, depending on the nutrient. *Low*, with reference to sodium, means that there are 140 milligrams or fewer of sodium per serving. *Low*, with reference to cholesterol, means that there are 20 milligrams or fewer of cholesterol per serving. *Low*, with reference to Calories, means that there are 40 Calories or fewer per serving.

Lean, with reference to the fat content of meats, means that there are 10 grams of fat or less, 4.5 grams of saturated fatty acids and *trans* fatty acids or less, and 95 milligrams or less of cholesterol per 3.5-ounce (100-gram) serving.

Extra lean, with reference to the fat content of meats, means that there are 5 grams of fat or less, 2 grams of saturated fatty acids and *trans* fatty acids or less, and 95 milligrams or less of cholesterol per 3.5-ounce serving.

Health Claims

By law, manufacturers can make certain statements or health claims on a food label linking the nutrition profile of the food to a reduced risk of a particular disease or health condition.[H,5,46,59,64,66,70,86,88] The statements have been approved by the FDA and are based on sound scientific evidence. For a manufacturer to make a claim that a food supplies a high amount of a nutrient, the food must provide at least 20 percent of the RDI per serving. Careful phrasing is required. For example, if a product provides at least 20 percent of the RDI for calcium (200 milligrams) per serving, then the health claim "this product *may* reduce the risk of osteoporosis" can legally be used. The company must also mention that other factors like exercise may reduce the risk of osteoporosis.

Claims can be made for the following:

1. *Calcium and osteoporosis:* If the product is high in calcium (nutrient dense), the claim can be made that consuming this product may reduce the risk of osteoporosis.

2. *Fat and cancer:* If the product is low in fat and, further, if the product is an extra-lean meat, the claim can be made that consuming this product may reduce the risk of certain types of cancers.

3. *Saturated fatty acids and* trans *fatty acids, cholesterol, and heart disease:* If the product is low in fat, saturated fatty acids and *trans* fatty acids, and cholesterol, and, further, if the product is an extra-lean meat, then the claim can be made that consuming this product may reduce the risk of heart disease or coronary heart disease.

4. *Fiber-containing fruits, vegetables, and whole grains and cancer:* If the product is a good fiber source (provides 20 percent of the DRV or more for fiber), and the fiber is provided from fruits, vegetables, and grains, the claim can be made that consuming this product may reduce the risk of certain types of cancer. The product must also be low in fat.[3,4,10]

5. *Fiber-containing fruits, vegetables, and whole grains and heart disease:* If the product is a good fiber source (provides 20 percent of the DRV or more for fiber, especially soluble fiber), and the fiber is provided from fruits, vegetables, or grains, the claim can be made that consuming this product may reduce the risk of heart disease. The product must also be low in fat, saturated fatty acids and *trans* fatty acids, and cholesterol. [3,4,10]

6. *Sodium and high blood pressure:* If the product is low in sodium (140 milligrams or less per serving), the claim can be made that consuming this product may reduce the risk of hypertension or high blood pressure.

7. *Fruits and vegetables (vitamin C and beta-carotene) and cancer:* If the product is nutrient dense for vitamin C and vitamin A as beta-carotene, and the vitamin C and beta-carotene are provided from fruits and vegetables, the claim can be made that consuming this product may reduce the risk of certain types of cancer. The product must also be low in fat.[3,4,10]

8. *Sugar alcohols and **dental caries**:* If **sugar alcohols** are used to sweeten the product and the product is sugar free, the statement can be made that sugar alcohols do not promote tooth decay.

9. *Folic acid and neural tube defects:* If the product provides 40 micrograms or more per serving of folic acid (also called folate and folacin), the claim can be made that consuming this product may reduce the risk of neural tube defects.[60,80]

10. *Soluble fiber from oats and barley and heart disease:* If the product is nutrient dense for fiber (provides 20 percent of the DRV or more for fiber), and the fiber is provided from whole oats or whole-grain barley, the claim can be made that consuming this product may reduce the risk of heart disease. The product must also be low in fat, saturated fatty acids and *trans* fatty acids, and cholesterol.

11. *Soy and heart disease:* If the product provides 6.25 grams of soy protein per serving, the claim can be made that consuming 25 grams of soy protein per day may reduce the risk of heart disease. The product must also be low in fat, saturated fatty acids and *trans* fatty acids, and cholesterol.

dental caries Tooth decay, cavities.

sugar alcohols Simple sugars that don't promote tooth decay, which may include any of the following chemicals: xylitol, mannitol, maltitol, lactitol, and erythritol.

12. *Potassium and blood pressure and stroke:* If the product is a good source of potassium, the claim can be made that consuming this product may reduce the risk of high blood pressure (hypertension) and stroke. The product must also be low in fat, saturated fatty acids and *trans* fatty acids, cholesterol, and sodium.

13. *Plant sterol/stanol esters and heart disease:* If the product contains significant amounts of **sterol esters** or **stanol esters** from plants, the claim can be made that consuming this product may reduce the risk of heart disease. The product must also be low in fat, saturated fatty acids and *trans* fatty acids, and cholesterol.[56,59]

14. *Fluoridated water and dental caries:* If the fluoridated water source provides more than 0.7 to 1.2 milligrams per liter of fluoride, the claim can be made that consuming this product may reduce the risk of dental caries. The new proposed recommendation for fluoridated water is up to 0.7 milligrams per liter.[12]

Data GENEie How many Calories are provided by sugars, and what percentage of Calories come from sugars in the photograph of the lite pears Nutrition Facts panel?

Example of a food label using a descriptive term and a health claim. The nutrition facts and ingredients are also provided.

Summary Points

- Food package labels contain the Nutrition Facts panel that uses DRVs and RDIs to provide nutrition information.
- Determining whether a food is high fat is best done by calculating the percentage of Calories from fat.
- A food is nutrient dense if 20 percent or more of the RDI for a nutrient is provided per serving.
- Ingredients are listed in descending order, by weight or volume.
- There are specific definitions for terms used on food labels.
- There are specific FDA-approved health claims for food package labels.

Take Ten on Your Knowledge Know-How

1. What key legislation governs food package labeling and education today?
2. What are DRVs? Which nutrients have DRVs?
3. How can you determine the number of Calories provided by total fat, saturated fatty acids, *trans* fatty acids, carbohydrate, sugar, and protein from the Nutrition Facts panel? Provide a calculated example of each.

sterol esters A group of chemically esterified phytosterol compounds found in plants that reduce the level of low-density lipoproteins in blood and thus are heart-healthy.

stanol esters A group of sterol compounds (chemically modified by the addition of hydrogen atoms) that are found in plants and reduce the level of low-density lipoproteins in blood and thus are heart-healthy.

4. What is the difference between percentage of fat by weight and percentage of fat by Calories? Provide a calculated example of each.
5. What is an RDI? Which vitamins and minerals are required by law to be expressed as the percentage of the RDI on the Nutrition Facts panel of the food package label? How do you convert the percent RDI on the Nutrition Facts panel to a milligram or microgram nutrient quantity? Provide a calculated example based on what the food manufacturer calls one serving of food. Then calculate for 2½ servings of food.
6. How can you determine nutrient density from the Nutrition Facts panel?
7. What is the significance of the ingredients list?
8. What are some potentially biologically harmful substances (when consumed in excess) that you can identify on a food package label? What are some dietary strategies to reduce your intake of these substances?
9. What is the meaning of the terms *free*, *reduced*, *light*, *low*, and *extra lean* when used on food labels?
10. What FDA-approved health claims are used on food package labels? What are some examples of packaged foods that make approved health claims? How do these health claims impact your product selection?

2.2 DIETARY REFERENCE INTAKES (DRIs)

Introduction
The Dietary Reference Intakes (DRIs) are the standards used to recommend and evaluate the nutritional adequacy of dietary intakes. There are DRIs for all of the essential nutrients in human nutrition. Several DRI reference tables can be used for planning and evaluating diets. They are summarized in Table 2.5 on page 60 and are shown in Appendix A. There are DRIs for:

- Fourteen vitamins (see Tables A1 and A3 in Appendix A)
- Fifteen minerals (see Tables A2 and A4 in Appendix A)
- Carbohydrate, fiber, essential fatty acids, and protein (see Tables A5 and A6 in Appendix A)
- Amino acids (see Table A7 in Appendix A)
- Energy or Calorie intake (see Table A8 in Appendix A); called the **Estimated Energy Requirement (EER)**
- Physical activity (see Table A9 in Appendix A)

Overview of the DRIs
This section will look briefly at certain aspects of the current standards. The DRIs are given for specific age groups, genders, and conditions of pregnancy and lactation. They are established for healthy people and are based largely on body weight and activity levels. The DRIs are based on scientific findings that associate nutritional health to defined healthy population needs by age and gender. Regular in-depth review of all scientific literature that includes nutrient needs of healthy individuals (with consideration of lifestyle habits, like cigarette smoking when determining vitamin C levels) is used to determine the amount of the nutrient recommended.[G,34,48]

T-Talk 2.2
To hear Dr. Turley talk about Dietary Reference Intakes (DRIs), go to www.cengagebrain.com

Estimated Energy Requirement (EER) Calorie level that is calculated and intended to maintain energy balance and good health in a person of a given age, gender, weight, height, and level of physical activity; also called the DRI for Calories.

DA+ GENEie Go to the Diet Analysis Plus software and build your profile. Generate a profile report to see your DRI recommendations. Compare them to those of other students of the opposite gender, from a different age group, and/or from a pregnant or lactating female. Notice and discuss the similarities and differences.

History of the DRIs The National Academy of Sciences is currently responsible for the revisions and updates of the dietary standards used in the United States. The first standards were released in 1943. There was a consistent updating process every four years until the 1980s, when a paradigm shift began to occur in the basis of the recommendations. Until this point, the goal of the recommendations was to prevent deficiency, but during the 1980s recommendations began to be based on the amounts of nutrients that could reduce the risk of chronic disease and optimize body function.

The current DRIs collectively reflect updates from the years 2001, 2002, and 2004. With each revision, a national committee of elite nutritional scientists evaluates scientific data, including clinical trials, case studies, laboratory experiments, and epidemiological (population-based) data, to premise their recommendations for nutrient intake. The federally funded committee draws its conclusions when several studies have shown consistent findings. The committee members determine whether the DRIs for essential vitamins, minerals, fatty acids, amino acids, protein, carbohydrate, fiber, and water are established as a Recommended Dietary Allowance (RDA), Adequate Intake (AI), or **Estimated Average Requirement (EAR)**. Furthermore, the most recent DRI committee also created the **Tolerable Upper Intake Level (UL)** for certain vitamins and minerals, Acceptable Macronutrient Distribution Ranges (AMDRs) for energy-producing nutrients, sugars, and essential fatty acids, the Estimated Energy Requirement (EER) for Calories, and a DRI for physical activity (PA) (see Tables A1 through A9 in Appendix A). An explanation of these terms appears below.[25-34]

2001 Photodisc, Inc.

Dietary Reference Intake soup.

Components of the DRIs

The RDA, AI, UL, AMDR, and EER are all components of the DRI that will be discussed in this section.

Recommended Dietary Allowances (RDAs) and Adequate Intakes (AIs) Taken together, all of the DRI components are intended to provide guidance for dietary assessment and prescription and to provide recommended safe and adequate levels—not minimum levels—for nutrient intake. The nutrient levels recommended are approximate, generous, and adequate (see Figure 2.4). For most of the nutrients, the recommended intake value is an RDA. The RDA value represents an average daily dietary nutrient intake level sufficient to meet the nutrient requirement of nearly all (97.5 percent) healthy individuals in a particular life stage and gender group. For some nutrients, the recommended intake value is set at an AI, which is the average (or mean) intake level for an apparently healthy life-stage group. Almost every essential vitamin or mineral has its particular signs and symptoms of toxicity (too much) and deficiency (not enough). Thus, a moderate intake of essential nutrients is recommended (see Tables A1 and A2 in Appendix A).[25-34,45,61,63,77,83,90]

Tolerable Upper Intake Levels (ULs) In the DRIs, the UL for eight vitamins and 16 minerals is included. The UL is the highest level of daily nutrient intake that is likely to pose no risk of adverse health effects for most individuals in the general population. As the intake increases above the UL, the potential risk of adverse side effects increases. It should be noted that for many of the nutrients that do have ULs, the majority of the population could consume higher levels than the UL without side effects (see Tables A3 and A4 in Appendix A).[26] Because common minerals present in the environment (e.g., boron, nickel, and vanadium) have potential toxicity, they have ULs established, but they are not essential.

Estimated Average Requirement (EAR) The average dietary amount of a nutrient that will maintain adequate function in half of the healthy people of a given age and gender group.

Tolerable Upper Intake Level (UL) The maximum dietary amount of a nutrient that can be consumed daily with little risk of illness; however, an intake at a higher level increases the risk of adverse health effects.

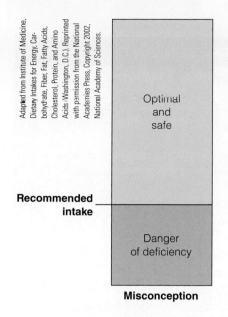

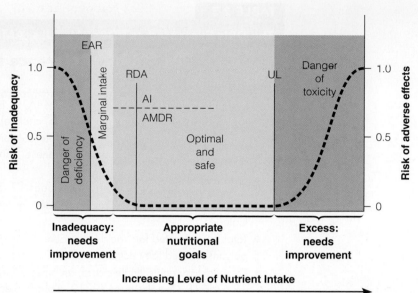

FIGURE 2.4 DRIs are appropriate, optimal, and safe nutrient levels to consume each day.

There are ULs for:

- Eight vitamins (see Table A3 in Appendix A)
- Sixteen minerals (three are not essential; see Table A4 in Appendix A)

Acceptable Macronutrient Distribution Ranges (AMDRs) The AMDR standards are used to prescribe and evaluate the distribution of energy intake specifically for carbohydrate including sugar, protein, and fat (including the essential fatty acids linoleic and alpha-linolenic acid). The AMDR values are expressed as a percent range of the total Calories consumed (see Figure 2.5, Table 2.5, and Table A6 in Appendix A).[76]

Estimated Energy Requirements (EERs) A person's Calorie needs used to be calculated using Calories per kilogram of body weight. Now it is specifically determined using formulas that are based on body mass (height and weight), age, physical activity level, and gender. The formulas are called the EERs and establish the DRI for energy or Calories as an EAR. This is the average daily nutrient intake level estimated to meet the requirement of half the healthy individuals in a particular life stage and gender group (see Table A8 in Appendix A).[76]

FIGURE 2.5 Acceptable Macronutrient Distribution Range intake amounts for carbohydrate, added sugars, protein, fat, and essential fatty acids.

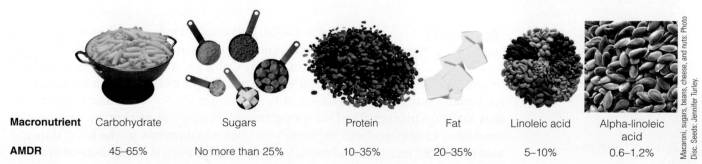

Macronutrient	Carbohydrate	Sugars	Protein	Fat	Linoleic acid	Alpha-linoleic acid
AMDR	45–65%	No more than 25%	10–35%	20–35%	5–10%	0.6–1.2%

TABLE 2.5

The DRIs Expressed as RDAs, AIs, ULs, AMDRs, EERs, and PA

The Dietary Reference Intake System

1. RDAs and AIs:

- Vitamins with specific DRIs: 14 vitamins (A, D, E, K, C, thiamin, riboflavin, niacin, folate, B_6, B_{12}, biotin, pantothenic acid, choline)
- Minerals with specific DRIs: 15 minerals (calcium, phosphorus, magnesium, iron, iodine, zinc, selenium, copper, manganese, chromium, molybdenum, fluoride, potassium, sodium, chloride)
- Carbohydrate (130 g/day for adults ages 19–30)
- Fiber (38 g/day for males and 25 g/day for females ages 19–30), 1.4 g/100 Calories
- Total fat, saturated fatty acids, monounsaturated fatty acids, and nonessential polyunsaturated fatty acids (not determined for adults)
- Essential fatty acids (linoleic acid, an omega-6 fatty acid: 17 g/day for males and 12 g/day for females ages 19–30; alpha-linolenic acid, an omega-3 fatty acid: 1.6 g/day for males and 1.1 g/day for females ages 19–30)
- Protein (0.8 grams of high-quality dietary protein per kilogram body weight for adults ≥ 19 years old)
- Water (3.7 L/day and 2.7 L/day for men and women ≥ 19 years old, respectively)

2. ULs:

- Vitamins with specific ULs: 8 vitamins (A, D, E, C, niacin, folate, B_6, choline)
- Minerals with specific ULs: 16 minerals (calcium, phosphorus, magnesium, iron, iodine, zinc, selenium, copper, manganese, molybdenum, fluoride, sodium, chloride, vanadium, nickel, boron)

3. AMDRs:

For an adult ≥ 19 years old:

- 45–65% of the Calories provided from carbohydrate
- 10–35% of the Calories provided from protein
- 20–35% of the Calories provided from fat
- ≤ 25% of the Calories provided from added sugars
- 0.6–1.2% of the Calories provided from alpha-linolenic acid
- 5–10% of the Calories provided from linoleic acid

4. EERs:

- Equations based on life-stage group in which age, gender, height, weight, and physical activity are used to determine an EAR for Calories per day

5. PA:

- ≥ 60 minutes of moderate physical activity per day

© Cengage Learning 2013

Nutrients Without DRIs

Cobalt, sulfur, and cholesterol, as well as nonessential dietary supplements, do not have DRI levels established for them. Cobalt is a structural component of vitamin B_{12}. The DRI for vitamin B_{12} will also cover the dietary amount needed for cobalt. Sulfur is a component of two amino acids (cysteine and methionine) that make up protein. The DRI for protein will cover the dietary amount needed for sulfur. Cholesterol can be made in sufficient quantities in the liver, thus it is a nonessential nutrient and doesn't require a DRI. Even though there are recommendations about dietary cholesterol intake, there is not a requirement for it from the diet.

DRIs are used for achieving good nutrition, evaluating the food supply, feeding the military, planning meals, conducting nutrition research and dietary analysis, and developing food programs and public policy. They are also used in the food and health care industries.[27,31,48]

Summary Points

- DRIs are the nutrient intake standards used for many reasons. They are based on age, gender, and special conditions.
- The exact DRI value may be as an RDA, AI, or EAR, depending on scientific information available for the nutrient.
- DRIs include ULs and AMDRs.
- For Calories, the DRI is called the EER.
- There is a DRI recommendation for PA.
- Tables A1 through A8 in Appendix A are referred to collectively as the DRIs and are all needed to perform a thorough dietary analysis.

Take Ten on Your Knowledge Know-How

1. What are DRIs and how are they determined and expressed?
2. What are the various components of the DRIs?
3. What specifically are the adult DRIs for protein, physical activity, and fiber?
4. Based on your current body weight and typical Calorie intake, calculate your DRI for protein and fiber.
5. Which nutrients have and don't have DRIs?
6. What are all of your DRI values?
7. After converting the percent RDI for vitamin C and calcium to the milligram value per serving in your favorite food, calculate the percent DRI provided by one serving of the food for you.
8. What is your calculated EER?
9. How many grams of carbohydrate, sugar, protein, and fat do you consume each day? Using those values, calculate your percentage of Calories from these energy-producing nutrients. Does your intake pattern meet the AMDRs?
10. How are DRIs used? How can you use them to improve your diet and health?

2.3 THE MYPLATE FOOD GUIDANCE SYSTEM

T-Talk 2.3
To hear Dr. Turley talk about the MyPlate food guidance system, go to www.cengagebrain.com

Introduction In 2011, the U.S. Department of Agriculture (USDA) for the first time released the **MyPlate** food guidance system (see Figure 2.6).[K] The icon of MyPlate helps consumers to understand the proportions of the major food groups to consume each day. The three overreaching messages associated with MyPlate include balancing Calories (enjoy your food, but eat less and avoid oversized portions), suggesting foods to increase consumption of (make half your plate fruits and vegetables, make at least half your grains whole grains, and switch to fat-free or low-fat (1%) milk), and suggesting foods to reduce consumption of (compare sodium in foods like soup, bread, and frozen meals—then choose the foods with lower numbers and drink water instead of sugary drinks). The Web site that supports MyPlate provides an individualized diet approach that considers age, gender, and physical activity level. It suggests a caloric intake level and then recommends a food pattern that will promote dietary Calorie control, adequacy, moderation, variety, and balance. It is built on the understanding that the diet is for an individual and that users can track their progress and make gradual improvements to be healthier (see BioBeat 2.2). On the Web site, www.choosemyplate.gov, consumers

MyPlate An icon identifying the portions of the major food groups to consume each day; accompanied by a Web-based, Calorie-controlled, personalized food guidance system based on the age, gender, height, weight, and physical activity level of a person; sponsored by the U.S. Department of Agriculture.

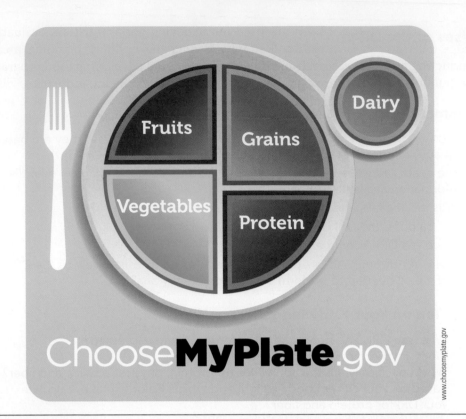

www.choosemyplate.gov

FIGURE 2.6 The MyPlate food guidance system icon.[K]

Demo GENEie The MyPlate Food Guidance System. Go to www .choosemyplate.gov. Create your personalized MyPlate plan by entering your age, gender, activity level, height, and weight. Explore each food group and visit the food gallery to observe portions of food by food group. Recall what you ate yesterday and see if you met your MyPlate recommendation. If not, which areas do you need to work on for improvement?

BioBeat 2.2

Biological Uniqueness and Meeting Personal, Optimal Nutrition

There is no such thing as a "one size fits all" type of diet. Each individual is a unique biological system, and the diet that supports that biological uniqueness is not specifically known at this time. However, now that the human genome has been sequenced, it is possible that in the near future dietitians will be prescribing diets based on one's genetic composition. Food allergies, food intolerances, and food sensitivities are common. It is left up to the individual's food intake experiences to understand what foods are compatible with his or her system.[35] You will note that the MyPlate personal dietary plan provides a Calorie-sensitive pattern of food intake; however, it is up to each person to make the specific food choices within each food group that best fits his or her needs.

Certain foods may be rich in nutrients but unsuitable for a particular person because of his or her dislikes, allergies, or intolerances. Adverse responses typically involve an immune system response and result in mucus membrane secretions, gastrointestinal disturbances, skin eruptions, or respiratory failure. Other subtle symptoms of intolerance may be nausea, headache, and itching.

Can you describe your MyPlate personal biological uniqueness for food choices in your diet plan?

can access in-depth information about each of the major food groups, as well as the related topic categories for oils, **empty Calories**, and physical activity. Serving sizes are simplified by using a "one MyPlate serving equivalent." Basic information on using MyPlate dietary planning and evaluation tools can be found in Appendix C.

Using MyPlate to Determine Food Group Patterns and Physical Activity

The amount of food needed from each MyPlate food group depends on age, gender, and physical activity level. The first step in getting a personalized MyPlate plan is to determine Calorie need. Estimated daily Calorie needs are shown in Appendix C, Table C1. These values are based on the average body mass for age and gender. These Calorie levels are not as accurate as those calculated using the EER included in the DRIs.

The next step is to determine the food intake pattern by Calorie level, as shown in Appendix C, Table C2. Here consumers are shown how much food to eat from each food group: grains, vegetables, fruits, protein foods, dairy and milk-alternatives, **oils**, and allowable empty Calories (see Appendix C, Table C3.G for foods providing empty Calories). In addition to providing Calories and food quantities, MyPlate educates users, including giving advice on food choices, common foods and serving sizes, some health benefits, and nutrient highlights (see the summary in Table 2.6 and Figure 2.7). Appendix C, Tables C3.A–G, show detailed food lists for each MyPlate food group and are used as a reference for food patterning. Please note at the end of Table C3.B the weekly vegetable subgroup recommendations based on Calorie level. Also please notice the superscripted antioxidant-rich food choices of fruits and vegetables that you can make in Tables C3.B and C3.C. Understanding what constitutes one MyPlate equivalent and learning to pattern foods sheds light on the extreme portion distortion to which Americans have become accustomed over time. What we perceive as a serving on a typical dinner plate may be two, three, or even more MyPlate equivalents when it comes to grains and protein foods, but half the serving size for fruits and vegetables (see Figure 2.8 on page 67).[17,22,67,74,79,84,89]

Food Patterning

It is important to learn how to pattern foods and diets to comply with the MyPlate model. The process of **patterning** a food begins first by identifying the correct food group and then determining the number of MyPlate serving equivalents that the food provides. A simple formula to determine the number of MyPlate serving equivalents is to take the amount

empty Calories Calories provided by solid fat and added sugars, neither of which are health promoting; an allowance is given based on Calorie need.

oils Dietary fats that are liquid at room temperature.

patterning Determining how an amount of food eaten quantifies as an equivalent amount from the MyPlate food guidance system or the Exchange List system.

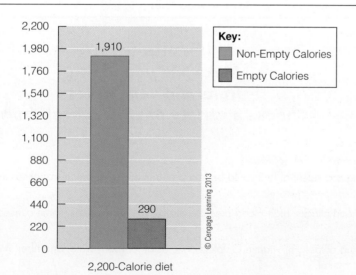

2,200-Calorie diet

FIGURE 2.7 The limited proportion of empty Calories from solid fat and added sugars compared to the total Calories consumed from health-promoting foods.

TABLE 2.6

Inside the MyPlate Food Guidance System

Grains (G)
Make at least half of your grains whole grain (WG)

Includes: Whole grains such as amaranth, barley, brown rice, buckwheat, bulgur (cracked wheat), cornmeal, millet, oatmeal, popcorn, quinoa, rye, sorghum, triticale, whole wheat, wild rice; and whole-grain bread, cereal, tortilla, and pasta products. **Refined grains** may include products such as breads, crackers, cereals, flour tortillas, noodles, processed grains, and bakery goods. There may be whole grain versions available. Consumers need to read the food label ingredients and look for the word "whole" by the grain type used to make the product.

Servings in General: A 1-ounce MyPlate serving equivalent of grain could be 1 slice of bread, 1 cup of ready-to-eat cereal, or ½ cup of cooked rice, pasta, or cereal (*approximately 80 Calories*).

Health Benefits: Grains reduce heart disease, high blood pressure, some cancers, type 2 diabetes, neural tube defects during fetal development, and both constipation and obesity (useful in weight management) when eaten as whole grains.

Nutrients: Grains provide many nutrients, including several B vitamins (thiamin, riboflavin, niacin, and folate), minerals (iron, magnesium, and selenium), carbohydrate, fiber (as whole grains), and protein.

Vegetables (V)
Make half your plate fruits and vegetables
Choose a variety of vegetables from the five subgroups

Includes: All fresh, frozen, canned, and dried vegetables and vegetable juices.

Servings in General: A 1 cup MyPlate serving equivalent of vegetables could be 1 cup raw or cooked vegetables or vegetable juice, or 2 cups of raw leafy greens (*approximately 50 Calories or 120 Calories for starchy vegetables*).

Health Benefits: Vegetables reduce heart disease, heart attacks, high blood pressure, stroke, type 2 diabetes, some cancers, kidney stones, obesity, and bone loss. Eating vegetables that are low in Calories instead of higher-Calorie foods may be useful in helping to lower Calorie intake.

Nutrients: Vegetables provide many nutrients, including potassium, vitamin A, vitamin C, folate (folic acid), carbohydrate, fiber, and protein. Most are low in fat and Calories. None have cholesterol.

Subgroups: Dark green, red-orange, beans and peas, starchy, and other vegetables.

Fruits (F)
Make half your plate fruits and vegetables
Choose a variety of whole, fresh fruit

Includes: All fresh, frozen, canned, and dried fruits and fruit juices.

Servings in General: A 1 cup MyPlate serving equivalent of fruit could be 1 cup of fruit or 100% fruit juice, or ½ cup of dried fruit (*approximately 100 Calories*).

Health Benefits: Fruits reduce heart disease, heart attacks, high blood pressure, stroke, type 2 diabetes, some cancers, kidney stones, obesity, and bone loss.

Nutrients: Fruits provide many nutrients, including potassium, vitamin C, folate (folic acid), carbohydrate, and fiber. Most are low in fat, sodium, and Calories. None have cholesterol.

Dairy (D)
Choose fat free or low fat (1%)

Includes: All fluid milk products and many foods made from milk that retain their calcium. Foods made from milk that have little to no calcium, such as cream cheese, cream, and butter, are not part of this group.

Servings in General: A 1 cup MyPlate serving equivalent of dairy could be 1 cup of milk or yogurt, 1½ ounces of natural cheese, or 2 ounces of processed cheese (*approximately 90 Calories when fat free or low fat*).

Health Benefits: Dairy products reduce the risk of low bone mass throughout the life cycle and may prevent osteoporosis.

Nutrients: Dairy products provide calcium, potassium, vitamin D, and protein. Low-fat or fat-free forms provide little or no solid fat.

Protein Foods (PF)
Choose lean or low-fat meats; choose fish (8 ounces per week), nuts, and seeds frequently instead of meat or poultry

Includes: All foods made from meat, poultry, fish, beans or peas, eggs, nuts, and seeds are considered part of this group. Beans and peas can be counted either as vegetables (beans and peas subgroup), or in the protein foods group. Generally, individuals who regularly eat meat, poultry, and fish would count beans and peas in the vegetable group. Individuals who seldom eat meat, poultry, or fish (vegetarians) would count some of the beans and peas they eat in the protein foods group.

Servings in General: A 1-ounce MyPlate serving equivalent of protein foods could be 1 ounce of lean meat, poultry, or fish, 1 egg, 1 tablespoon peanut butter, ¼ cup cooked beans, or ½ ounce of nuts or seeds (*approximately 55 Calories when lean*).

Health Implications: Foods in the protein foods group provide nutrients that are vital for health and body maintenance. However, choosing foods from this group that are high in saturated fat and cholesterol may increase the risk for heart disease.

Nutrients: Protein foods provide many nutrients including protein, B vitamins (niacin, thiamin, riboflavin, and B_6), vitamin E, iron, zinc, and magnesium.

Oils (O) Category
Consume the recommended amount of healthy liquid fats

Includes: Oils that are liquid at room temperature, like vegetable oils. Liquid oils come from plants (except coconut and palm) and from some fish. Foods that are mainly oil include mayonnaise, certain salad dressings, and soft margarine with no *trans* fats.

Servings in General: A 1 teaspoon MyPlate serving equivalent of oil could be 1 teaspoon of liquid plant or fish oil at room temperature (*approximately 40 Calories*). Most Americans consume enough oil in the foods they eat, such as nuts, fish, cooking oil, and salad dressing. Some oil is needed for health. Because it is a fat source, the amount should be limited to the recommendation to balance total Calorie intake.

Nutrients: Oils provide vitamin E, monounsaturated fatty acids, and polyunsaturated fatty acids, which contain essential fatty acids.

Health Benefits and Implications: Plant and fish oils promote heart health; however, overconsuming linoleic acid, which is dominant in most plant oils, can increase cancer risk.

continued

Empty Calorie Foods Category (EC)
Limit foods and beverages with solid fat and added sugars
Empty Calorie foods are discretionary food choices that promote malnutrition

Includes: Solid fats and added sugars that, when consumed in excess, promote obesity, which is associated with heart disease, type 2 diabetes, and cancer.

Solid Fats: Solid fats are solid at room temperature, like butter and shortening. Solid fats come from many animal foods, can be made from vegetable oils through hydrogenation, and are found naturally in coconut and palm plant foods.

- Common solid fats are butter, shortening, stick margarine, animal fat, and coconut oil.
- Foods high in solid fats include many cheeses, creams, ice creams, well-marbled cuts of meats, regular ground beef, bacon, sausages, poultry skin, and many baked goods (such as cookies, crackers, doughnuts, pastries, and croissants).

Added Sugars: Added sugars are sugars and syrups that are added to foods or beverages during processing or preparation. This does not include naturally occurring sugars, such as those that occur in milk and fruits.

- Foods that contain added sugars are regular soft drinks, candy, cakes, cookies, pies, fruit drinks (such as fruitades and fruit punch), milk-based desserts and products (such as ice cream, sweetened yogurt, and sweetened milk), and grain products (such as sweet rolls and cinnamon toast).
- Ingredients shown on the food labels of processed foods that indicate added sugar are brown sugar, corn sweetener, corn syrup, dextrose, fructose, fruit juice concentrates, glucose, high-fructose corn syrup, honey, invert sugar, lactose, maltose, malt syrup, molasses, raw sugar, sucrose, sugar, and syrup.

Allowance: The **empty Calorie allowance** is the remaining amount of Calories needed to meet the food intake pattern (after accounting for the Calories needed for all food groups and oils, using forms of foods that are fat free or low fat and with no added sugars). The empty Calorie allowance can also be used to eat more whole, fresh foods from the major food groups.

Physical Activity
Expend energy through body movement and exercise

Physical Activity: Physical activity and nutrition work together for better health. Being active increases the amount of Calories burned. As people age their metabolism slows, so maintaining energy balance requires moving more and eating less. Physical activity simply means movement of the body that uses energy. Walking, gardening, briskly pushing a baby stroller, climbing the stairs, playing soccer, or dancing the night away are all good examples of being active. For health benefits, physical activity should be moderate or vigorous and amount to ***30 minutes*** minimally each day. Increasing the intensity or the amount of time of activity can have additional health benefits and may be needed to control body weight. Some physical activities, like walking at a casual pace while grocery shopping and doing light household chores, are not intense enough to help meet the recommendations. Although the body is moving, these activities do not increase the heart rate, so they are not counted toward the 30 or more minutes a day that should minimally be achieved. About ***60 minutes*** a day of moderate physical activity may be needed to prevent weight gain. For those who have lost weight, at least ***60 to 90 minutes*** a day may be needed to maintain the weight loss. At the same time, Calorie needs should not be exceeded. Children and teenagers should be physically active for ***at least 60 minutes*** every day, or most days.

- Moderate physical activities include walking briskly (about 3½ miles per hour), hiking, gardening/yard work, dancing, golf (walking and carrying clubs), bicycling (less than 10 miles per hour), and weight training (general light workout).
- Vigorous physical activities include running/jogging (5 miles per hour), bicycling (more than 10 miles per hour), swimming (freestyle laps), aerobics, walking very fast (4½ miles per hour), heavy yard work such as chopping wood, weight lifting (vigorous effort), and basketball (competitive).

Health Benefits: Regular physical activity can produce long-term health benefits. People of all ages, shapes, sizes, and abilities can benefit from being physically active. The more physical activity you do (within the 2008 Physical Activity Guidelines for Americans), the greater the health benefits. Being physically active can help you increase your chances of living longer, feel better about yourself, decrease your chances of becoming depressed, sleep well at night, move around more easily, have stronger muscles and bones, stay at or get to a healthy weight, be with friends or meet new people, and enjoy yourself and have fun.

Health Implications: When you are not physically active, you are more likely to get heart disease, get type 2 diabetes, have high blood pressure, have high blood cholesterol, or have a stroke.

© Cengage Learning 2013

Jennifer Turley

Sandwich ingredients: 2 slices whole wheat bread, 3 ounces of 90% fat-free turkey meat, 1 leaf (1 cup) dark green lettuce, ¼ tomato (large), ¼ avocado, and 1 tablespoon mayonnaise

Patterning:

2 slices whole-wheat bread	÷ 1 slice bread per equivalent	= 2 ounce equivalents whole grains
3 ounces meat	÷ 1 ounce meat per equivalent	= 3 ounce equivalents protein foods
1 cup dark green lettuce	÷ 2 cups lettuce per equivalent	= ½ cup equivalent dark green vegetable
¼ large tomato	÷ 1 large tomato per equivalent	= ¼ cup other vegetable
¼ avocado provides 1.5 teaspoons oil		
10 ompty Calories per ounce of 90% fat-free meat	× 3 ounces	= 30 empty Calories

FIGURE 2.8 The patterning of a turkey sandwich according to the MyPlate food guidance system.

eaten and divide that by the serving size as given in the MyPlate food group information (see Appendix C). The result is the number of MyPlate serving equivalents. For example, if you consumed 2 cups of cooked brown rice, this would equal four MyPlate whole-grain ounce equivalents (2 cups eaten ÷ 0.5 cups per equivalent = 4 1-ounce serving equivalents). For additional patterning, see Figure 2.8 and Appendix C. To see what equivalents look like size-wise, go to the MyPlate food gallery online at www.choosemyplate.gov, and visit Appendix C.[K,6]

Once you determine the number of MyPlate serving equivalents eaten from each of the food groups throughout the day, you can assess whether your diet met your MyPlate recommendations and which food groups you consumed were adequate, deficient, or excessive. Your diet is considered deficient in a food group if less than the required amount was consumed. Your diet is considered adequate in a food group if the required amounts were consumed, or more than the required amount for whole grains, vegetables, and fruits (see BioBeat 2.3). Your diet is considered excessive in a food group if more than the required amounts were consumed for dairy, protein, oils, and especially empty Calories.

empty Calorie allowance
The Calories allotted for solid fat and added sugars in a person's energy allowance after consuming enough nutrient-dense foods from the MyPlate food groups to meet all nutrient needs for a day.

BioBeat 2.3

Phytochemicals: Nonessential, Beneficial Plant Chemicals

Phytochemicals are nonessential chemical compounds produced in plants that provide a health benefit. Even though the study of phytochemicals is in its scientific infancy, thousands of phytochemicals have been identified and studied. Their roles in the healthy functioning of the human body are varied: Some have shown antioxidant effects, some have been shown to positively affect the detoxification processes in the liver, and some are instrumental in gene regulation. Many of the phytonutrients complement each other chemically in their antioxidant effects.

Most phytochemicals are plant pigments, and thus give color to whole, fresh food. Fruits and vegetables that are dark green, orange, blue/purple, or red are great sources of these beneficial molecules. It has been documented in the literature that super-supplementing one powerful antioxidant nutrient, like beta-carotene, does not provide health-enhancing qualities. It appears that the natural assortment of antioxidants present in fruits and vegetables is the best way to derive the health-enhancing benefits from phytonutrients. Sadly, the diets of many people include few fresh, whole foods and a greater proportion of refined grains and added sugar, salt, and solid fats.[23,42,62]

Beata Becla/Shutterstock.com

Fruits and vegetables are sources of essential antioxidants and are rich in thousands of phytochemicals.

What are some colorful, phytochemical-rich plant foods that you enjoy eating?

Summary Points

- MyPlate is a personalized diet plan based on age, gender, and activity.
- It encourages the consumption of whole grains, vegetables, fruits, fat-free or low-fat milk or milk-alternatives, low-fat protein foods, and healthy plant oils.
- It discourages the intake of added sugar and solid fats as well as a sedentary lifestyle.
- Evaluating food intake by patterning the food consumed is a tool to determine nutritional adequacy.

Take Ten on Your Knowledge Know-How

1. What diet and health principles underlie the MyPlate food guidance system?
2. Within the MyPlate model, what are the units of the serving equivalents for each food group?
3. Provide some examples of foods and their equivalent amount in the MyPlate food groups.
4. What are the health benefits or implications for each of the MyPlate food groups?
5. In general, which nutrients are provided by each MyPlate food group?
6. What is the MyPlate recommendation for physical activity?
7. What are empty Calories?
8. Name 10 food items that contain empty Calories.
9. What is your personal MyPlate plan?
10. Pattern the foods that you consume for one day according to MyPlate. Does your diet follow the MyPlate plan? If so, why? If not, what dietary and lifestyle changes can you make to comply?

phytochemical A plant-derived, nonnutrient chemical that has biological activity and health-promoting properties in the body.

2.4 DIETARY GUIDELINES AND RECOMMENDATIONS

Introduction

In early times, most humans were hunters and gatherers, eating fruits, vegetables, nuts, berries, and occasionally meat. Today, many people are sedentary and eat a high proportion of processed foods. As a result, many health issues are related to diet. Scientific research has yielded information that is used to improve health and function and prevent disease (see BioBeat 2.4). Today, the emergence of public health and nutrition messages (communicated as **dietary guidelines** from several health agencies in the United States) offer guidance for optimizing gene–nutrient interactions.

Current dietary guidelines and recommendations, such as those reflected in the MyPlate food guidance system, come from several credible organizations. Most of these guidelines and recommendations are supported by following MyPlate and provide guidance for healthy eating and disease prevention, as well as meeting the DRIs. Four sets of dietary recommendations (Dietary Guidelines 2010, American Heart Association, American Cancer Society, and Healthy People 2020) will be highlighted in this section.[15,18–19,69,81,82] You will notice that the common themes of the dietary recommendations discussed here are to promote nutritional adequacy and reduce the risk for major diet-related chronic diseases: atherosclerotic coronary artery (heart) disease, diet-related cancers, obesity, type 2 diabetes, and osteoporosis.[18,24,37,38,41,43,44]

T-Talk 2.4
To hear Dr. Turley talk about dietary guidelines and recommendations, go to www.cengagebrain.com

dietary guidelines Modern society's instructional messages for reducing the risks for diet-related diseases, including U.S. Dietary Guidelines 2010; American Heart Association guidelines; American Cancer Society guidelines; and Healthy People 2020.

BioBeat 2.4

Adopting the Dietary Guidelines for Disease Prevention

With the emergence of modern society comes the need to communicate healthy eating behavior messages. Advice on eating behavior, food choices, and food preparation has been expressed to people through many avenues.[58] However, the recommendations have gone from being philosophical and religious in nature to more science- and medicine-based. There has been a shift from a focus on sanitation and prevention of nutrient deficiencies to prevention of chronic disease due to the dietary excesses that have increased disease incidences, and the incorporation of a healthy lifestyle. The manner in which dietary guideline messages are fashioned has changed as well. Instead of putting the main focus on food

components like saturated fat (present in high-fat animal foods), the emphasis now is on whole, fresh food choices.[15,35] For example, instead of telling consumers to reduce saturated fat intake, the message now is to increase their intake of fresh fruits and vegetables.

The development of dietary recommendations made by any national health organization is based on years of consistent, research-based scientific evidence that supports the relationship of diet to health or to disease prevention. The goal of the recommendations made by these health organizations is to reduce the risk of chronic disease.[9,15,18,37,38,41,43,51,82]

What changes are needed for you to adopt the dietary guidelines?

Dietary Guidelines 2010

Every five years, the **Dietary Guidelines for Americans** provide science-based advice to promote health and reduce risk for major chronic diseases through diet and physical activity. The first version of the Dietary Guidelines for Americans was released in 1990. An important component of each five-year revision is the analysis of new scientific information by the Dietary Guidelines Advisory Committee appointed by the Secretaries of the U.S. Department of Health and Human Services and the U.S. Department of Agriculture (USDA). A revised version of the Dietary Guidelines for Americans was released in the year 2010.[F,9,15,36,50,51,55,57,82,85]

The 2010 Dietary Guidelines are summarized in Table 2.7 on page 73 and are significantly different from those of 2005. The guidelines address the fact that the majority of Americans are overweight or obese and undernourished in several key essential nutrients. It is recognized that ensuring that all Americans consume a health-promoting diet and achieve and maintain energy balance requires far more than individual behavior change. The Dietary Guidelines Advisory Committee considers the total diet and personal food choices in the context of improving related environmental and societal issues.

The new Dietary Guidelines speak to individuals, society, and the environment (see Figure 2.9). A coordinated plan is directed toward individual Americans at all ages, of both genders, and of any racial and ethnic group, along with all sectors of society, including individuals, families, educators, communities, physicians and allied health professionals, public health advocates, policy makers, scientists, and small and large businesses (farmers, agricultural producers, food scientists, food manufacturers, and food retailers of all kinds). There are four simple yet effective directives to individuals and society in the 2010 Dietary Guidelines:

1. Reduce the incidence and prevalence of overweight and obesity of the U.S. population by reducing overall Calorie intake and increasing physical activity.

2. Shift food intake patterns to a more plant-based diet that emphasizes vegetables, cooked dried beans and peas, fruits, whole grains, nuts, and seeds. In addition, increase the intake of seafood and fat-free and low-fat milk and milk products and consume only moderate amounts of lean meats, poultry, and eggs.

3. Significantly reduce intake of foods containing added sugars and solid fats, because these dietary components contribute excess Calories and few, if any, nutrients. In addition, reduce sodium intake and lower intake of refined grains, especially refined grains that are coupled with added sugar, solid fat, and sodium.

4. Meet the 2008 Physical Activity Guidelines for Americans. For adults, substantial health benefits are gained by doing at least 150 minutes per week of moderate-intensity exercise, and more extensive health benefits come from doing 300 minutes per week. Adults should also do muscle-strengthening activities that are moderate or high intensity and involve all major muscle groups on two or more days per week, as these activities provide additional health benefits.[F]

Dietary Guidelines for Americans Science-based advice for Americans to promote health and to reduce risk for major chronic diseases through diet and physical activity; updated every five years by the Department of Health and Human Services in conjunction with the U.S. Department of Agriculture.

FIGURE 2.9 A summary of the Dietary Guidelines 2010: a call to action to value and enjoy good nutrition, physical activity, and a healthy lifestyle.

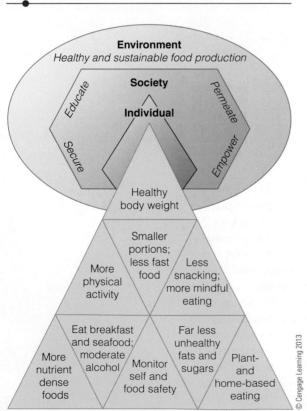

© Cengage Learning 2013

The 2010 Dietary Guidelines Advisory Committee states that change is needed in the overall food environment to support the efforts of all Americans to meet the four directives. Adopting the following sustainable recommendations will help Americans to eat well, be physically active, and maintain good health and function:

- Improve nutrition literacy and cooking skills, including safe food-handling skills, and empower and motivate the population, especially families with children, to prepare and consume healthy foods at home.
- Increase comprehensive health, nutrition, and physical education programs and curricula in U.S. schools and preschools, including food preparation, food safety, cooking, and physical education classes and improved quality of recess.
- Create greater financial incentives for consumers to purchase, prepare, and consume whole, fresh food: vegetables, fruit, whole grains, seafood, fat-free or low-fat milk and milk products, lean meats, and other healthy foods.
- Improve the availability of affordable fresh produce through greater access to grocery stores, produce trucks, and farmers' markets.
- Increase environmentally sustainable production of vegetables, fruits, and fiber-rich whole grains.
- Ensure household food security through measures that provide access to adequate amounts of foods that are nutritious and safe to eat.
- Develop safe, effective, and sustainable practices to expand aquaculture and increase the availability of seafood to all segments of the population. Enhance access to publicly available, user-friendly risk-benefit information that helps consumers make informed seafood choices.
- Encourage restaurants and the food industry to offer health-promoting foods that are low in salt (sodium); limited in added sugars, refined grains, and solid fats; and served in smaller portions.
- Implement the U.S. National Physical Activity Plan, which will increase physical activity and reduce sedentary activity. Develop efforts across all sectors of society, including health care and public health; education; business and industry; mass media; parks, recreation, fitness, and sports; transportation, land use, and community design; and volunteer and nonprofit. Reducing screen time, especially television, for all Americans also will be important.[F]

Several nutritional issues were also specifically identified in the 2010 Dietary Guidelines:

Energy balance and weight management: The American eating environment presents temptations in the form of tasty, energy-dense, micronutrient-poor foods and beverages and contributes to the present epidemic of obesity occurring in America. The macronutrient distribution of a person's diet is not the driving force behind the current obesity epidemic. Americans (children, adolescents, and adults) overconsume Calories (especially from solid fat and added sugars), are physically inactive, and spend too much time in sedentary behavior. Mindful or conscious eating in correct portion sizes of whole foods—along with increasing physical activity, breakfast consumption, and reductions in TV watching, eating out at fast-food restaurants, and snacking on sugar- and fat-filled foods and beverages—will prevent inappropriate weight gain. For special populations, the committee noted that maternal obesity before pregnancy and excessive weight gain during pregnancy pose health problems such as obesity and type 2 diabetes for the mother and fetus. As with the other age groups, older adults can derive health benefits, including reduced disabilities and chronic disease risk, from achieving and maintaining a healthy body weight.

Nutrient adequacy: Americans should lower their overall energy intake by replacing foods high in saturated fat and added sugar with vegetables, fruits, whole grains, and fluid milk and milk products. This will increase nutrient intake shortfalls for vitamin D, calcium, potassium, and dietary fiber. Women of reproductive age should consume foods rich in folate and iron.

Older adults should consume whole foods rich in vitamin B_{12} and consume vitamin B_{12}-fortified foods and vitamin B_{12} supplements if needed. Taking multivitamins or mineral supplements does not offer health benefits to healthy Americans, but multivitamin and mineral supplements can benefit some populations with known deficiencies.

Fatty acids and cholesterol: For several decades, American society has been burdened by cardiovascular disease and type 2 diabetes, two major causes of death and disease resulting from eating excess fat and cholesterol. Americans should limit saturated fatty acid intake to less than 7 percent of total Calories and instead substitute food sources containing monounsaturated fatty acids or polyunsaturated fatty acids. They should also limit cholesterol intake to less than 300 milligrams per day. Avoiding the intake of *trans* fat from industrial sources and consuming only small amounts (0.5 percent of total Calories) from ruminant animal sources will be beneficial, along with consuming two 4-ounce servings of cooked seafood per week rich in eicosapentaenoic acid and docosahexaenoic acid (omega-3 fatty acids). Individuals with (or at high risk for) cardiovascular disease and type 2 diabetes should limit cholesterol-raising fats and cholesterol even more.

Protein: Adult protein needs are based on 0.8 grams of protein per kilogram of body weight per day. Animal food sources provide the highest-quality proteins, yet planned plant-based diets are able to meet protein requirements for essential amino acids and offer other potential benefits, such as fiber and nutrients important for health.

Carbohydrates: Carbohydrates are the primary energy source for active people. However, most Americans are sedentary and should decrease their consumption of energy- and sugar-dense carbohydrate sources (like sweetened beverages and desserts) and choose to eat fiber-rich carbohydrate foods (such as whole grains, vegetables, fruits, and cooked dry beans and peas) as staple foods.

Sodium, potassium, and water: Americans consume an excess of sodium (usually in the form of salt) and not enough potassium (from fruits, vegetables, and whole grains), causing substantial health problems, including increased blood pressure and its consequences of heart disease and stroke. Americans should reduce sodium intake to less than 2,300 milligrams per day and then gradually to 1,500 milligrams per day. Individuals should increase their consumption of potassium to dietary adequacy (4,700 milligrams per day for adults). Water intake among Americans is neither excessive nor insufficient under normal circumstances. Americans should continue to consume water daily to prevent dehydration.

Alcohol: An average intake of one or two alcoholic beverages per day is associated with the lowest all-cause death and disease risk in middle-aged and older adults. If alcohol is consumed, then it should be consumed in moderation and only by adults. For women this equals up to one drink per day, and for men this is up to two drinks per day. A drink is 12 fluid ounces of beer, 5 fluid ounces of wine, or 1.5 fluid ounces of distilled spirits. Excess alcohol intake (more than three drinks per day for women and four drinks per day for men) has harmful effects. There are many situations that call for the complete avoidance of alcoholic beverages.

Food safety and technology: Food safety concerns have escalated as a result of disease-causing bacteria contamination and food adulterated with nonfood substances. This affects

5 oz wine (12% alcohol)

12 oz beer

10 oz wine cooler

1½ oz hard liquor (80 proof whiskey, gin, brandy, rum, vodka)

© Polara Studios, Inc.

One alcoholic drink equivalents defined.

commercial food products and those prepared in the home. Government policies and responsible food industry practices can help prevent foodborne illness. Americans should use the principles of "clean, separate, cook, and chill" to prevent foodborne illnesses. Federal and local advisories communicate risks associated with exposure to methylmercury and persistent organic pollutants. Americans—including women who may become pregnant or who are pregnant, nursing mothers, and children 12 and younger—can safely eat at least 12 ounces of a variety of cooked seafood per week. They should pay attention to local seafood advisories and limit their intake of large, predatory fish.[F]

TABLE 2.7

A Summary of the 2010 Dietary Guidelines

Four directives in a plan for all Americans to eat well, be physically active, and maintain good health and function.	1. Reduce the incidence and prevalence of overweight and obesity by reducing Calorie intake and increasing physical activity 2. Shift food intake patterns to a more plant-based diet. 3. Reduce the intake of foods devoid of nutrients and excessive in sodium and Calories from solid fat, added sugars, and refined grains. 4. Meet the 2008 Physical Activity Guidelines for Americans.
Nine sustainable food environment changes needed to support all Americans to meet the four directives.	1. Improve nutrition, food, and cooking literacy and skills. 2. Expand nutrition and physical education knowledge and programs. 3. Create financial incentives for consumers to purchase, prepare, and consume whole fresh foods. 4. Make fresh produce available and affordable. 5. Increase environmentally sustainable production of vegetables, fruits, and whole grains. 6. Ensure household food security. 7. Expand aquaculture and provide risk-benefit information pertaining to seafood. 8. Encourage restaurants to serve smaller food portions and foods low in sodium, solid fats, added sugars, and refined grains. 9. Implement the U.S. National Physical Activity Plan.
Eight topic-specific finding areas	1. Promote energy balance and weight control. 2. Ensure nutrient adequacy. 3. Limit saturated fatty acid ($<$ 7% of Calories) and cholesterol ($<$ 300 mg per day) intake. 4. Consume adequate protein within a plant-based diet. 5. Increase whole grain and fiber intake and reduce refined grain and added sugar intake. 6. Reduce sodium to $<$ 2,300 mg per day, then to 1,500 mg per day; increase potassium to adequacy (4,700 mg per day for adults), and consume adequate water. 7. Limit alcohol intake. 8. Practice food safety.

This call to action to value and enjoy good nutrition, physical activity, and a healthy lifestyle requires individual and societal shifts in the following areas:

Less . . .	More . . .
• Sedentary behaviors	• Physical activity
• Large portions	• Appropriate portions
• Saturated fat and *trans* fat	• Foods high in essential nutrients
• Added sugars and sodium	• Home cooking
• Snacking	• Plant-based and whole foods
• Fast food	• Seafood
• Mindless eating	• Nutrition and physical education
• Heavy alcohol consumption	• Financial incentives to eat right
• Foodborne illness	• Food safety and security
• ***Obesity and chronic disease***	• ***Good health and function***

American Heart Association (AHA) Dietary Guidelines

AHA recommendations regarding diet and related life-style practices for the general population are based on evidence indicating that modification of specific risk factors will decrease the incidence of heart disease. The risk factors include cigarette smoking, elevated levels of blood cholesterol (particularly **low-density lipoprotein (LDL)** cholesterol), low levels of **high-density lipoprotein (HDL)** cholesterol, increased blood pressure, type 2 diabetes, obesity (especially central [visceral] adiposity), and physical inactivity. The AHA identifies seven health and behavior factors that impact health and quality of life, including don't smoke; maintain a healthy weight; engage in regular physical activity; eat a healthy diet; manage blood pressure; take charge of cholesterol; and keep blood sugar, or glucose, at healthy levels. These seven factors can be managed by addressing four diet and lifestyle AHA recommendation areas, including use up at least as many Calories as you take in; eat a variety of nutritious foods from all the food groups; eat less of the nutrient-poor foods; and don't smoke tobacco—stay away from tobacco smoke. Use the AHA dietary recommendations shown in Table 2.8 as a guide to achieve heart health through dietary and lifestyle factors.[D,7,24,43,44]

TABLE 2.8

AHA Diet and Lifestyle Goals

AHA Dietary Recommendations

Consume an Overall Healthy Diet:

- Consume a diet rich in vegetables and fruits.
- Choose whole-grain, high-fiber foods.
- Consume fish, especially oily fish, at least twice a week.
- Minimize intake of beverages and foods with added sugars.

Be Physically Active and Aim for a Healthy Body Weight:

- All adults should minimally accumulate at least 30 minutes of physical activity most days of the week. Children and adults who are attempting to lose weight or maintain weight loss should accumulate at least 60 minutes of physical activity most days of the week.
- Match intake of energy (Calories) to overall energy needs with appropriate changes to achieve weight loss when indicated.

Aim for a Desirable Lipid Profile:

- Limit foods high in saturated fat, *trans* fat, and cholesterol.
  - Cholesterol intake should be less than 300 mg per day.
  - Saturated fatty acid intake should be less than 7% of Calories.
  - *Trans* fatty acid intake should be less than 1% of Calories.
  - Polyunsaturated fat intake should be up to 10% of Calories.
  - Monounsaturated fat can make up to 15% of total Calories.
- Select fat-free (skim), 1%, and low-fat dairy products; limit fatty meats and tropical oils.
- Choose lean meats and vegetable alternatives.

Aim for a Normal Blood Pressure:

- Choose and prepare foods with little or no salt.
  - Sodium intake should be between 1,500 and 2,300 mg per day, which is about 1 teaspoon of sodium chloride (salt).
- Limit alcohol.
  - If you drink, then do so in moderation. Have no more than one drink per day (for women) or two drinks per day (for men) of wine, beer, or liquor, and only when caloric limits allow.

© Cengage Learning 2013

---

low-density lipoprotein (LDL) A class of lipoproteins made of lipids including cholesterol, phospholipids, triglycerides, and protein that is known as the "bad" kind of cholesterol because it delivers cholesterol to tissues and can promote atherosclerosis and contribute to heart disease when elevated in the blood.

high-density lipoprotein (HDL) A class of lipoproteins made of lipids including cholesterol, phospholipids, triglycerides, and protein that is known as the "good" kind of cholesterol because it scavenges cholesterol from tissues and returns it to the liver for processing.

- Increase potassium intake to an adequate level.
- Follow the Dietary Approach to Stop Hypertension (DASH) Diet.

Aim for a Normal Blood Glucose Level of less than 100 mg/dL.

Avoid use of and exposure to tobacco products.

© Cengage Learning 2013

American Cancer Society (ACS) Dietary Guidelines

The ACS publishes nutrition guidelines to advise the public about dietary practices that reduce cancer risk. Reviews of the guidelines occur approximately every few years (1984, 1991, 1996, 2002, 2006, and 2008). This time frame allows the ACS to review and discuss the scientific findings that emerge during the interim years. See Table 2.9 for the current guidelines.[A,41]

Cigarette smoking causes one-third of the cancer deaths occurring annually in the United States; 25 percent of Americans smoke. Don't smoke! If you don't smoke, the most important risk factors that you control are a combination of diet and physical activity. Introducing healthful diet and exercise practices at any time from childhood to old age can promote health and reduce cancer risk.

One-third or more of cancer deaths are caused by a variety of dietary factors; 100 percent of Americans eat. Thus, it follows that diet is an important and modifiable risk factor to most cancers. More important than this simple deduction is the vast amount of scientific evidence linking unsound dietary patterns and practices to risk for certain types of cancer. To reduce your cancer risk, follow an overall dietary pattern that includes eating a high proportion of plant foods (fruits, vegetables, grains, and legumes/dried beans), limiting amounts of high-fat meat and dairy foods, limiting alcohol intake, and balancing caloric intake and physical activity.[13,23,41]

TABLE 2.9

ACS Nutrition and Physical Activity Guidelines (2008)

Eat a variety of healthful foods, with an emphasis on plant sources.

Maintain a healthy weight throughout life.

- Balance caloric intake with physical activity.
- Avoid excessive weight gain throughout the life cycle.
- Achieve and maintain a healthy weight if currently overweight or obese.

Adopt a physically active lifestyle.

- Adults: Engage in at least 30 minutes of moderate to vigorous physical activity, above usual activities, on five or more days of the week. Forty-five to 60 minutes of intentional physical activity are preferable.
- Children and adolescents: Engage in at least 60 minutes per day of moderate to vigorous physical activity at least five days per week.

Consume a healthy diet, with an emphasis on plant sources.

- Choose foods and beverages in amounts that help achieve and maintain a healthy weight.
- Eat five or more servings of a variety of vegetables and fruits each day.
- Choose whole grains in preference to processed (refined) grains.
- Limit consumption of processed and red meats.

If you drink alcoholic beverages, limit consumption.

- Drink no more than one drink per day for women or two per day for men.

Specific recommendations are made for breast, prostate, colorectal, endometrial, lung, stomach, and oral-esophageal cancers.

© Cengage Learning 2013

Eating a plant-based diet helps in the prevention of cancer.

Healthy People 2020

There are many public health issues that continue to evolve over time. Projecting the public objectives for the next 10 years gives public health providers guidance for reshaping their programs and setting new goals. In the Healthy People 2020 objectives, all public health issues are included. Only the objectives with nutritional issues are summarized in Table 2.10.[J]

TABLE 2.10

A Summary of Healthy People 2020 Nutrition and Weight Status Objectives

- Increase the proportion of adults who are at a healthy weight.
- Reduce the proportion of adults who are obese.
- Prevent inappropriate weight gain in youth and adults.
- Reduce the proportion of children and adolescents who are overweight or obese.
- Increase the proportion of worksites that offer nutrition or weight management classes or counseling.
- Increase the proportion of physician office visits that include counseling or education related to nutrition or weight.
- Increase the proportion of primary care physicians who regularly measure the body mass index of their patients.
- Reduce iron deficiency among young children and females of childbearing age.
- Reduce iron deficiency among pregnant females.
- Increase the contribution of fruits to the diets of the population aged 2 years and older.
- Increase the variety and contribution of vegetables to the diets of the population aged 2 years and older.
- Increase the contribution of whole grains to the diets of the population aged 2 years and older.
- Reduce consumption of saturated fat in the population aged 2 years and older.
- Reduce consumption of Calories from solid fats and added sugars in the population aged 2 years and older.
- Reduce consumption of sodium in the population aged 2 years and older.
- Increase consumption of calcium in the population aged 2 years and older.
- Eliminate very low food security among children in U.S. households.
- Increase the number of states that have state-level policies that incentivize food retail outlets to provide foods that are encouraged by the Dietary Guidelines.
- Increase the number of states with nutrition standards for foods and beverages provided to preschool-aged children in child care.
- Increase the percentage of schools that offer nutritious foods and beverages outside of school meals.

Healthy People 2010, www.healthypeople.gov

Summary Points

- The Dietary Guidelines 2010 are a call to action to value and enjoy good nutrition, physical activity, and a healthy lifestyle. They include four directives, nine sustainable food environment changes, and eight specific areas to reduce and prevent obesity and chronic disease.
- The American Heart Association includes a daily healthy eating plan of five or more fruits and vegetables, six or more grains with an emphasis on whole grains, and the inclusion of low-fat or nonfat dairy products, fish, legumes, poultry, and lean meats.

- The American Heart Association recommends maintaining an appropriate body weight, eating a desirable distribution of fat (saturated fatty acids 7 percent or less of daily consumption, polyunsaturated fatty acids up to 10 percent, monounsaturated fatty acids up to 15 percent, avoid *trans* fatty acids) and amount of cholesterol (less than 300 milligrams per day), maintaining a healthy blood pressure level by limiting sodium intake to less than 2,300 milligrams per day, and moderating alcohol intake.
- The American Cancer Society includes a recommendation to eat a plant-based diet with five or more fruits and vegetables daily and an emphasis on whole grains, limiting high-fat red meats, maintaining a healthy body weight, being physically active, and limiting alcohol intake.
- There are many nutritional objectives in Healthy People 2020.

Take Ten on Your Knowledge Know-How

1. What is the purpose of dietary guidelines and recommendations?
2. What are the Dietary Guidelines 2010?
3. Do you have a diet and lifestyle that strives to meet the Dietary Guidelines 2010? What is one thing you struggle with that you could change to become healthier and start to meet one of the guidelines?
4. What dietary guidelines does the American Heart Association recommend?
5. Which nutritional objectives from Healthy People 2020 could positively impact you?
6. If you knew your total Calorie intake and the grams of saturated fatty acids consumed in one day, how would you calculate the percentage of Calories from saturated fatty acids to see if it complied with the American Heart Association dietary guidelines?
7. What dietary guidelines does the American Cancer Society make?
8. In general, what is the key to consuming a diet that reduces your risk for developing cancer?
9. What is the importance of being physically active?
10. What is the importance of not using tobacco in cancer prevention?

2.5 FOOD COMPOSITION INFORMATION AND THE EXCHANGE LISTS

Introduction

Two necessary tools for performing diet analysis and creating a healthy diet are the nutrient composition of foods database and the Exchange List system. The average nutrient values of foods consumed in the United States form the foundation of diet analysis software and food composition tables and databases. The Exchange List system is often used to prescribe an individual eating plan with a wide variety of food options in meal planning.

T-Talk 2.5
To hear Dr. Turley talk about food composition and the Exchange Lists, go to www.cengagebrain.com

Food Composition Database

The USDA has compiled a massive database of information about the nutrient composition of foods (see BioBeat 2.5). The chemicals shown in the **food composition** tables and databases include energy-producing nutrients (carbohydrates, proteins, and fats) and non-energy-producing nutrients (vitamins, minerals, water). The tables quantify other bioactive substances in food, such as phytochemicals, alcohol, and caffeine, and give specific details on the amount of fatty acids, amino acids, and simple sugars in each of the foods. The

food composition The chemical composition of nutrients in foods, including carbohydrate, starch, fiber, sugars, fat, fatty acids, protein, amino acids, vitamins, minerals, water, and other bioactive substances such as phytochemicals, alcohol, and caffeine.

BioBeat 2.5

Food Composition: The Chemicals in Foods

The USDA is responsible for maintaining the national food composition database. The database is updated at least once a year. The USDA analyzes many food samples from all over the United States and maintains the nutritional composition database of food used in food composition tables. Diet analysis software packages also draw on the USDA database for nutrient values for the foods included in their software database. Learning the nutrient composition of foods can be overwhelming and a lifelong learning adventure. However, learning the composition of the foods you consume is easy. You will find that you do not have 365 different days of eating; rather, you consistently consume your favorite foods. By taking the time to become familiar with the nutrient composition of the foods you eat, it is possible to select foods you consume based on their ability to meet your nutritional needs and/or the food's health-promoting qualities.[L,14,21,52–54]

Is the food composition of your most common daily intake pattern meeting your nutritional needs?

Search the USDA National Nutrient Database for Standard Reference

Enter up to 5 keywords which best describe the food item. To further limit the search, select a specific Food Group.

Certain codes can also be searched: NDB number (the USDA 5-digit Nutrient Databank identifier); the USDA commodity code; and the URMIS number for specific cuts of meat (enter the # symbol followed without a space by the URMIS code).

Keyword(s): _____ Help
Select Food Group: All Food Groups ▾
Submit

To view reports on foods by single nutrients, such as calcium or niacin, go to Nutrient Lists.

Use these links to access SR23 datasets or SR23 documentation.

Home | How to get information

United States Department of Agriculture

The USDA online Nutrient Data Laboratory.

nutrient values provided in food composition tables and databases are based on portions of food commonly consumed. The USDA database allows the user to select a food, specify a quantity, and then generate the food composition value based on that quantity of food.[L,21,52–54]

The values in the food composition tables and databases are obtained through government research and sources. The values are important for performing dietary analysis. Dietary analysis is a critical tool for determining nutrient adequacy. A person's dietary intake is compared to his or her DRI to create the basis of nutritional adequacy. Updates for food composition values are released through the USDA, with the update indicated by the release number. New foods are constantly being added to the database.

Food composition tables and databases are used to perform dietary analysis. Diet analysis software can also be used to determine food composition. Most diet analysis software programs, including DA+, contain the chemical composition for thousands of foods. To access the USDA National Nutrient Database for Standard Reference, go to www.nal.usda.gov/fnic/foodcomp/search. You can use this Web site to plan, manage, and/or evaluate diets according to the DRIs and dietary guidelines and recommendations.

The Exchange Lists

Exchange Lists A diet-planning tool created by the American Diabetes Association and Academy of Nutrition and Dietetics that organizes foods by their proportions of carbohydrate, fat, and protein (and thus Calories); there are nine lists: Starch, Nonstarchy Vegetables, Fruit, Milk, Meat and Meat Alternatives, Fat, Free Foods, Combination Foods, and Other Carbohydrates, that are used to plan diets.

The **Exchange Lists** were created by the American Diabetes Association in conjunction with the Academy of Nutrition and Dietetics. There are nine different Exchange Lists by food group. The Exchange List system is an excellent tool for meal planning, Calorie control, and meeting the AMDRs. Historically, diabetic diet planning hinged on using the Exchange Lists so that carbohydrates in a meal were at a level compatible with a person's insulin prescription.[B,C,1,87]

Today, the Exchange Lists are widely used for weight control and human performance.[11] The portion size of each food in the Exchange Lists is based on grams of carbohydrates, proteins, and fats, and the total number of Calories, not micronutrient content of the food. It is important to distinguish the Exchange Lists as separate and more precise tools than the MyPlate food guidance system for measuring and prescribing diets.

Table 2.11 demonstrates how the Exchange Lists can be used for prescribing diabetic diets at specific Calorie levels. Then the individual makes specific food choices using the Exchange Lists (see Table 2.12). Specific Exchange Lists, as well as the guidelines and parameters of these Exchange Lists for caloric contribution by carbohydrate, protein, and fat, are found in Appendix D. For example, an individual can use the Starch Exchange List to make food choices (exchanges) to fulfill their starch diet prescription. The Starch Exchange List in Appendix D shows how the nutritional contributions are controlled by serving size depending on the food choice. One starch exchange (15 grams of carbohydrate, 3 grams of protein, little or no fat, and about 80 Calories) can be provided by 2 slices of reduced-Calorie bread, 3 cups of puffed rice cereal, ⅓ cup of cooked rice, or 1 cup of peas, for example.

TABLE 2.11

Dietary Calorie Levels and Exchange System Diet Prescription

Exchange List	1,000 Cals.	1,200 Cals.	1,500 Cals.	1,800 Cals.	2,000 Cals.	2,200 Cals.	3,000 Cals.	3,300 Cals.	3,800 Cals.	4,400 Cals.
Nonfat milk	2	2	2	2	2	2	3	4	5	6
Vegetables	2	2	4	4	4	6	9	10	11	13
Fruits	3	3	4	5	5	5	8	8	10	12
Starch	3	4	6	8	10	11	16	17	20	22
Lean meats	4	5	5	5	6	6	7	8	8	11
Fats	3	4	5	7	7	8	8	10	10	10

© Cengage Learning 2013

The Exchange Lists offer an opportunity to learn about the nutritional values of foods. There are Exchange Lists for Starch, Milk Products (broken into Fat-Free/Low-Fat, Reduced-Fat, and Whole), Fruits, Other Carbohydrates (including sweets and desserts), Nonstarchy Vegetables, Meat and Meat Substitutes (broken into Very Lean, Lean, Medium-Fat, and High-Fat), Fat (broken into Monounsaturated, Polyunsaturated, and Saturated), Free Foods, and Combination Foods. See Table 2.12 for a sample diet plan and an example of how to follow a breakfast Exchange List pattern.

TABLE 2.12

A Sample Exchange List Diet Plan

Exchange	Calories	Breakfast	Lunch	Snack	Dinner	Snack
9 starch	720	2	2	1	3	1
4 vegetables	100				4	
3 fruit	180	1	1	1		
6 lean meat	330		2		4	
2 fat-free milk	180	1				1
5 fat	225		1		4	

© Cengage Learning 2013

continued

A Sample Exchange List Diet Plan (*continued*)

Accompanying Meal Plan: Breakfast of cereal with fruit and milk, lunch of a ham sandwich and fruit, snack of popcorn and juice, dinner of pasta with meat sauce, salad sprinkled with seeds and dressing, green beans, and corn on the cob, and snack of milk and crackers. *This diet plan provides 1,735 Calories with 55–60% of its Calories from carbohydrate, 15–20% from protein, and < 30% from fat.*

Breakfast Variety: Given a breakfast plan of 2 starch, 1 fruit, and 1 fat-free milk exchanges, the Exchange Lists from Appendix D (Tables D3, D4, and D6) could be used to create the following three breakfasts:

1. 1 cup cooked oatmeal with 2 tablespoons raisins and 1 cup fat-free milk
2. ½ cup muesli cereal with ¾ cup blueberries and ⅔ cup low-fat yogurt
3. ¾ cup corn flakes with 1 cup low-fat soy milk and ½ cup orange juice

© Cengage Learning 2013

Summary Points

- Food composition tables and databases show the chemical composition of food based on a specified food portion and are used in diet planning and diet analysis.
- The Exchange Lists are used in diet planning and evaluation. They are based on grams of carbohydrates, proteins, and fats, as well as total Calories provided per food exchange.

Take Ten on Your Knowledge Know-How

1. What kind of information do you find in food composition tables and databases?
2. How can this information be used?
3. What is the primary Web site in the United States where you can access food composition information?
4. What is the composition of your favorite food?
5. What are the Exchange Lists?
6. Which Exchange List has sublists related to fat content?
7. Why were the Exchange Lists created?
8. How are food exchanges determined?
9. How many exchanges of meat do you typically consume each day?
10. How many grams of fat are in one fat exchange?

 2.6 SUMMARY

CONTENT KNOWLEDGE

THE CONTENT IN THIS MODULE IS BROADLY SUMMARIZED IN TABLE 2.13. IN THIS MODULE, YOU HAVE LEARNED HOW TO:

- Read and interpret the information displayed in the Nutrition Facts panel on the food package labels.
- Determine nutrient needs.
- Consume a healthy diet.
- Reduce the risks of heart disease, diet-related cancers, and other chronic diseases by using dietary recommendations.
- Use Exchange Lists and other tools for understanding food composition.

PERSONAL IMPROVEMENT GOALS

IF YOU:

- Adopt your personalized MyPlate eating pattern into your life, you will be consuming a healthy diet that meets the DRIs and is calorically appropriate for you.
- Fill up on whole, fresh food and have very little solid fat and added sugar in your diet, your body weight will be much easier to manage.
- Follow the dietary guidelines you've learned and apply food composition knowledge, you will be optimizing your nutritional status and reducing your risks for developing heart disease, cancer, hypertension, stroke, and other diet-related diseases and conditions.

Here is a tip for you to adopt when you do your grocery shopping: Shop around the periphery of the grocery store. That is where you will find the whole, fresh food. When you begin to go up and down the aisles, you get into the processed food, which is generally laden with solid fat, sugars, and salt. If you purchase packaged and processed foods, then use your food label skills to choose wisely.

You can assess if you met the learning objectives for this module by successfully completing the Homework Assessment and the Total Recall activities (sample questions, case study with questions, and crossword puzzle).

TABLE 2.13

Summary of Key Components Learned in Module 2

Nutrient	Dietary Recommendation
Carbohydrate  Jennifer Turley	• According to the DRV, 60% of total Calories should come from carbohydrates. Sugars in the Nutrition Facts panel include all simple sugars. • The AMDR for carbohydrate is 45–65% of Calories. • The AMDR for sugars is ≤ 25% of Calories • The DRI for carbohydrate is at least 130 g per day. • The DRI for fiber is 1.4 g per 100 Calories consumed. • These levels of complex carbohydrate intake, if energy intake level is adequate, will preserve lean body mass, provide fuel for the red blood cells, brain, and central nervous system, maintain optimal energy, promote nutrient density in the diet, and reduce the risk of many chronic diseases. • MyPlate and Dietary Guidelines 2010 recommend consuming carbohydrates from whole grains, vegetables, and fruits, while limiting added sugars that provide empty Calories and refined grain products. • Carbohydrates are present in the Starch, Fruit, Milk, Other Carbohydrate, and Vegetable Exchange Lists.
Protein  Jennifer Turley	• According to the DRV, 12% of total Calories should come from protein. • The AMDR for protein is 10–35% of total Calories for adults. • The adult DRI for protein is 0.8 g of protein per kg body weight. • These levels of protein, if energy intake level is adequate, may reduce the risk of many chronic diseases. • MyPlate and Dietary Guidelines 2010 recommend consuming protein from lean meats, low-fat milk and milk-alternatives, and in greater amounts from legumes (beans and peas), whole grains, vegetables, and nuts. • Proteins are present in the Meat, Milk, Starch, and Vegetable Exchange Lists.

continued

Fat

Jennifer Turley

- According to the DRV, < 30% of total Calories should come from fat and < 10% should be from saturated fatty acids.
- The AMDR for fat is 20–35% of total Calories.
- The DRI for the essential fatty acids is 17 g per day linoleic acid and 1.6 g per day alpha-linolenic acid (adult male). The AMDR is 5–10% for linoleic acid and 0.6–1.2% for alpha-linolenic acid.
- There is no DRI for total fat.
- The AHA recommends < 7% saturated fatty acids and < 1% *trans* fatty acids of total Calories consumed.
- These levels of fat may reduce the risk of obesity, heart disease, and cancer.
- MyPlate and Dietary Guidelines 2010 recommend consuming fats from liquid plant oils, nuts, and seeds, as well as from fish, while reducing the intake of solid fats that provide empty Calories.
- Fats are present in the Fat, Meat, Other Carbohydrate, and Milk Exchange Lists.

Cholesterol

Jennifer Turley

- According to the DRV, ≤ 300 mg of cholesterol should be consumed daily.
- This level may reduce the risk of heart disease.
- There is no DRI for cholesterol.
- Dietary Guidelines 2010 recommend limiting cholesterol intake to < 300 mg per day to maintain heart health.

Sodium

Jennifer Turley

- According to the DRV, ≤ 2,400 mg of sodium should be consumed daily.
- This level will support 1 hour of moderate exercise and may prevent elevations in blood pressure and risk of stroke.
- The adult DRI for sodium is 1,500 mg per day.
- The adult UL for sodium is 2,300 mg per day.
- Dietary Guidelines 2010 recommend limiting sodium intake to ≤ 2,300 mg per day and then gradually reducing it to 1,500 mg per day.

General Diet

Jennifer Turley

- Diets that are low in fat, saturated fatty acids, *trans* fatty acids, cholesterol, sodium, sugars, and alcohol, diets high in whole grains, legumes, fruits, and vegetables (especially those rich in vitamin C and beta-carotene), and diets adequate in potassium, calcium, essential fatty acids, and protein are associated with the lowest risk of nutrition related chronic disease.

© Cengage Learning 2013

Homework Assessment

50 questions

1. The Daily Reference Values (DRVs) are based on a 2,000-Calorie diet, and they are used to describe nutritional information in a Nutrition Facts panel on a food package label.
   A. True   **B.** False

2. A natural source of *trans* fatty acids in the food supply is vegetable oil.
   A. True   **B.** False

3. There are two FDA-approved health claims that can be made based on a good source of fiber from grains, fruits, and vegetables: one for reducing the risk for developing heart disease and the other for reducing the risk of cancer.
   A. True   **B.** False

4. A refined grain product contains homocysteine.
   A. True   **B.** False

5. On the food package label, total fat is the sum of saturated, monounsaturated, and polyunsaturated fatty acids.
   A. True   **B.** False

6. The percentage of Calories from fat for the olives is (see Figure 2.10):
   A. 40%
   B. 58%
   C. 80%
   D. 114%

7. In regards to fat content, the olives are (see Figure 2.10):
   A. High fat
   B. Low fat
   C. Moderately fat
   D. Cannot determine from the label

8. The olives can be labeled as a low-sodium food (see Figure 2.10).
   A. True   **B.** False

9. Which of the following statements is correct for using the term *light* on a food package label?
   A. The product has one-third fewer Calories per serving compared to the original product.
   B. The product has one-half the amount of sodium per serving compared to the original product.
   C. The product has one-half the amount of fat per serving compared to the original product.
   D. The product is a lighter color or texture per serving compared to the original product.
   E. All of the above.

10. The amount of sugars in one serving of the cereal is (see Figure 2.11):
    A. 2 g
    B. 9 g
    C. 32 g
    D. 3 g

Canned Olives Nutrition Facts

Serving Size	5 Olives (15 g)

Amount Per Serving	
Calories 25	Calories from Fat 20

	% Daily Value*
Total Fat 2.5 g	4%
Saturated Fat 0 g	0%
Trans Fat 0 g	
Cholesterol 0 mg	0%
Sodium 190 mg	8%
Total Carbohydrate 1 g	0%
Dietary Fiber <1 g	3%
Sugars 1 g	
Protein 0 g	

Vitamin A 2%	Vitamin C 0%
Calcium 2%	Iron 4%

*Percent Daily Values are based on a 2,000 calorie diet.

Ingredients: Olives, water, salt, ferrous gluconate

© Cengage Learning 2013

FIGURE 2.10
Canned olives food label for Assessment 2A, questions 6–8.

Oats Cereal Nutrition Facts

Serving Size	1 Cup (41 g)

Amount Per Serving	
Calories 170	Calories from Fat 30

	% Daily Value*
Total Fat 3.5 g	5%
Saturated Fat 0 g	0%
Trans Fat 1 g	
Cholesterol 0 mg	0%
Sodium 250 mg	10%
Total Carbohydrate 32 g	11%
Dietary Fiber 2 g	7%
Sugars 9 g	
Protein 3 g	

Vitamin A 20%	Vitamin C 0%
Calcium 2%	Iron 60%

*Percent Daily Values are based on a 2,000 calorie diet.

Ingredients: Whole grain oats, whole grain rolled wheat, sugar syrup, brown sugar, honey roasted almonds, sugar, corn syrup, honey, whole wheat flour, oat flour, salt, rice flour, partially hydrogenated cottonseed oil, calcium carbonate, corn syrup solids, natural and artificial flavors, mono and diglycerides, reduced iron, vitamin A palmitate, and BHT.

© Cengage Learning 2013

FIGURE 2.11
Oats cereal food label for Assessment 2A, questions 10–16.

11. The cereal is nutrient dense for (see Figure 2.11):
   A. Calcium
   B. Vitamin A and iron
   C. Vitamin A
   D. Vitamins A and C

12. What percent of the Reference Daily Intake (RDI) for vitamin A is provided if 1½ cups (61.5 grams) of the cereal are eaten (see Figure 2.11)?
   A. 25%
   B. 30%
   C. 50%
   D. 60%

13. One cup (41 grams) of the cereal provides 2 percent of the RDI for calcium. This is equal to (see Figure 2.11):
   A. 2 mg
   B. 20 mg
   C. 200 mg
   D. 2,000 mg

14. In the cereal, less than 20 percent of the Calories from carbohydrate are provided by simple sugars (see Figure 2.11).
   A. True   B. False

15. The manufacturer of the cereal could legally make the health claim (see Figure 2.11):
   A. "Our product may protect against osteoporosis."
   B. "Our product may protect against heart disease."
   C. "Our product may protect against night blindness."
   D. A and B
   E. B and C

16. What is the most abundant ingredient listed in the cereal (see Figure 2.11)?
   A. Honey roasted almonds
   B. Oat flour
   C. Whole-grain oats
   D. Brown sugar

17. Based on the percentage of fat by Calories, the Fakin Bakin could be called (see Figure 2.12):
   A. High fat
   B. Low fat
   C. Moderately fat
   D. Cannot determine from the label

18. The percentage of fat by weight for two pieces of Fakin Bakin is ___, thus the product could be labeled ___% fat free (see Figure 2.12).
   A. 4%, 96%
   B. 30%, 70%
   C. 1%, 99%
   D. 10%, 90%

19. A directive included in the Dietary Guidelines 2010 states, "Shift food intake patterns to a more plant-based diet."
   A. True   B. False

20. The Dietary Reference Intakes (DRIs) are established for the state of pregnancy and lactation.
   A. True   B. False

21. The Tolerable Upper Intake Level (UL) is a daily amount of a nutrient that is not likely to pose any adverse health effect.
   A. True   B. False

22. The adult Dietary Reference Intakes (DRIs) for energy or Calories called the Estimated Energy Requirement (EER) can be calculated using factors for various physical activity levels, age, height, and weight.
   A. True   B. False

23. Foods from the dairy group of the MyPlate food guidance system may reduce the risk of:
   A. Cancer
   B. Stroke
   C. Heart disease
   D. Diabetes
   E. Bone loss

Fakin Bakin Nutrition Facts

Serving Size   2 pieces (38 g)

Amount Per Serving

Calories 50   Calories from Fat 15

	% Daily Value*
Total Fat 1.5 g	3%
Saturated Fat 0 g	0%
Trans Fat 0 g	
Cholesterol 0 mg	0%
Sodium 150 mg	6%
Total Carbohydrate 4 g	1%
Dietary Fiber <1 g	3%
Sugars 0 g	
Protein 5 g	

Vitamin A 2%	Vitamin C 0%
Calcium 6%	Iron 4%

*Percent Daily Values are based on a 2,000 calorie diet.

Ingredients: Texturized vegetable protein (soy flour, caramel color), soybean oil, salt, and natural flavors.

© Cengage Learning 2013

FIGURE 2.12
FakinBakin food label for Assessment 2A, questions 17–18.

24. Which of the following foods is a starchy vegetable, according to the MyPlate food guidance system?
    A. Broccoli
    B. Corn
    C. Spinach
    D. Carrots
    E. Soybeans

25. Which of the following foods is not included in the dairy group of the MyPlate food guidance system?
    A. Whole milk
    B. Yogurt
    C. Ice cream
    D. Cheese
    E. None of the above

26. Which MyPlate food group provides a good source of phytochemicals, vitamin C, and beta-carotene?
    A. Dairy
    B. Grains
    C. Protein foods
    D. Vegetables
    E. Oils

27. Which of the following MyPlate food groups provide a good source of iron?
    A. Dairy
    B. Vegetables
    C. Protein foods
    D. Fruits
    E. Oils

28. Which of the following foods would reduce the risk of developing some cancers, according to the MyPlate food guidance system?
    A. Dairy
    B. Empty Calorie foods
    C. Protein foods
    D. Fruits
    E. Oils

29. The U.S. Department of Agriculture (USDA) maintains the database on food composition in the Unites States of America.
    A. True   B. False

Tiffany is female, age 20, weighs 137 pounds, is 58 inches tall, and is physically "active". Use the dietary analysis case study information on Tiffany in Tables 2.14 through 2.17 to solve and answer questions 30 to 50.

TABLE 2.14

Tiffany's Calorie and Macronutrient Analysis

Food	Amount	Calories	Protein (g)	Fat (g)	Starch (g)	Sugar (g)
Egg	2 large	199	13.53	14.9	0.57	2.11
Bagel	1 each 3"	144	6.1	0.68	27.35	0.93
Orange juice	12 fluid ounces	169	2.6	0.74	6.74	31.25
Chicken	1 cup	265	41.72	10.89	0	0
Lettuce	1 cup	10	0.65	0.1	0	1.42
Dressing	2 tablespoons	88	0.11	8.34	0.62	2.45
Chips	1 ounce	137	2.21	6.62	16.8	0.28
Cola	12 fluid ounces	156	0	0	0	38.93
Ham	3 ounces	129	13.94	7.22	2.12	0
Potato	1 large	254	7.48	0.39	52.11	3.53
Butter	1 tablespoon	104	0.12	11.52	0	0.01
Broccoli	1 cup	45	3.71	0.64	3.93	2.17
Tap water	12 fluid ounces	0	0	0	0	0
TOTAL		**1,700**	**92.17**	**62.04**	**110.24**	**83.08**

© Cengage Learning 2013

TABLE 2.15

Tiffany's Partial Dietary Analysis

Food	Chol (mg)	Fiber (g)	Calcium (mg)	Iron (mg)	Sodium (mg)	Vit C (mg)	Vit A (µg)
Egg	429	0	87	1.46	342	0.2	174
Bagel	0	2.1	7	1.76	289	0.1	1
Orange juice	0	0.7	41	0.74	4	186	37
Chicken	118	0	20	1.5	99	0	39
Lettuce	0	0.9	13	0.3	7	2	18
Dressing	0	0	2	0.19	486	0	1
Chips	0	1.5	49	0.66	119	0	0
Cola	0	0	7	0.07	15	0	0
Ham	48	1.1	20	0.86	1,095	3.4	0
Potato	0	6.6	45	3.23	30	28.7	3
Butter	31	0	3	0	82	0	97
Broccoli	0	5.1	62	1.05	64	101	120
Tap water	0	0	11	0	14	0	0
TOTAL	**626**	**18**	**367**	**11.82**	**2,646**	**322**	**490**

© Cengage Learning 2013

TABLE 2.16

Tiffany's Percent Calories Calculations

	Tiffany's Intake (in grams)		Tiffany's Intake (in Calories)		Tiffany's Total Calories		Percent Calories
Protein	92	× 4 Cal/g =		÷	1,700	× 100 =	
Fat	62	× 9 Cal/g =		÷	1,700	× 100 =	
Carbo-hydrate	193	× 4 Cal/g =		÷	1,700	× 100 =	
Sugar	83	× 4 Cal/g =		÷	1,700	× 100 =	

© Cengage Learning 2013

TABLE 2.17

Tiffany's Percent Dietary Reference Intake (DRI) Calculations

	Tiffany's Intake		Tiffany's DRI		Percent DRI	
Protein (g)	92	÷	Calculated (0.8 g/kg/day)	× 100 =		
Fiber (g)	18	÷	Calculated (1.4 g/100 Calories consumed)	× 100 =		
Calcium (mg)	367	÷		× 100 =		
Iron (mg)	11.82	÷		× 100 =		
Sodium (mg)	2,646	÷		× 100 =		
Vitamin C (mg)	322	÷		× 100 =		
Vitamin A (µg)	490	÷		× 100 =		

g (grams), mg (milligrams), µg or mcg (micrograms).

Tiffany is female, age 20, weighs 137 pounds, is 58 inches tall, and is physically "active".

© Cengage Learning 2013

30. Based on percentage of Calories, Tiffany's diet complied with the Acceptable Macronutrient Distribution Range (AMDR) for protein.
   A. True   B. False

31. A majority (more than 50 percent) of Tiffany's protein came from high biological value sources.
   A. True   B. False

32. Tiffany's protein intake exceeded 150 percent of her Dietary Reference Intake (DRI) for protein.
   A. True   B. False

33. Based on percentage of Calories, Tiffany's diet complied with the Acceptable Macronutrient Distribution Range (AMDR) for carbohydrate.
   A. True   B. False

34. A majority (more than 50 percent) of Tiffany's carbohydrate sources were complex.
   A. True   B. False

35. Based on percentage of Calories, Tiffany's diet complied with the Acceptable Macronutrient Distribution Range (AMDR) for sugar.
   A. True   B. False

36. Based on percentage of Calories, Tiffany's diet complied with the Acceptable Macronutrient Distribution Range (AMDR) for fat.
   A. True   B. False

37. Tiffany's percent Dietary Reference Intake (DRI) for calcium was greater than 100 percent.
   A. True   B. False

38. Tiffany's percent Dietary Reference Intake (DRI) for iron was greater than 100 percent.
   A. True   B. False

39. Tiffany's percent Dietary Reference Intake (DRI) for vitamin C was greater than 100 percent.
   A. True   B. False

40. Tiffany's dietary intake of vitamin C exceeded the Tolerable Upper Intake Level (UL).
   A. True   B. False

41. Tiffany's percent Dietary Reference Intake (DRI) for vitamin A was greater than 100 percent.
   A. True   B. False

42. Tiffany's diet did not meet the Dietary Guidelines 2010 for sodium.
   A. True   B. False

43. Overall, Tiffany's diet met the American Cancer Society (ACS) dietary guidelines.
   A. True   B. False

44. Tiffany's cholesterol intake complied with the American Heart Association dietary guidelines.
   A. True   B. False

45. Tiffany's percent Dietary Reference Intake (DRI) for fiber was greater than 100 percent based on the 1.4 grams per 100 Calorie intake personalized DRI.
   A. True   B. False

46. Tiffany's 2 tablespoons of Italian salad dressing are patterned as six MyPlate teaspoon equivalents of oil.
   A. True   B. False

47. Tiffany's butter provided more than 100 empty Calories, according to MyPlate.
   A. True   B. False

48. Tiffany's broccoli is patterned as one MyPlate cup equivalent from the dark green vegetable subgroup, and broccoli is a vitamin C–rich food.
   A. True   B. False

49. Tiffany's 1 cup of cooked broccoli provided two vegetable exchanges.
   A. True   B. False

50. Tiffany's food record shows that her diet met her MyPlate food pattern recommendation.
   A. True   B. False

Total Recall

SAMPLE QUESTIONS

True/False Questions

1. Reference Daily Intakes (RDIs) are expressed as a percent of the most recent Dietary Reference Intake (DRI) values.

2. A Tolerable Upper Intake Level (UL) for fiber has been established.

3. Joseph ate the following in one day: breakfast egg sandwich, coffee with cream, doughnut, regular soda, hamburger, French fries, apple, steak sub sandwich. Joseph's diet met the American Cancer Society guideline for consumption of fruits and vegetables.

4. The Acceptable Macronutrient Distribution Range (AMDR) for protein is 5 to 15 percent of total Calories.

5. American Heart Association and American Cancer Society guidelines both emphasize the importance of maintaining a healthy body weight.

Multiple Choice Questions: Choose the best answer.

6. Which information item is not included on the food package label?
   A. Ingredients
   B. Percent of Calories from fat
   C. Grams of *trans* fatty acids
   D. Percent of the Reference Daily Intake (RDI) for vitamin A
   E. Milligrams of sodium

7. Which disease or condition below does not have an approved food package label health claim established for it?
   A. Heart disease
   B. Dental caries
   C. Cancer
   D. Diabetes
   E. Osteoporosis

8. Which food below provides empty Calories?
   A. Pasta
   B. Bread
   C. Raw vegetables
   D. Nuts
   E. Whole milk

9. Which word below best describes the information provided in food composition tables and databases?
   A. Adequate
   B. Basal
   C. Chemical
   D. Dietary
   E. Essential

10. The MyPlate vegetable food group is rich in which nutrient(s)?
    A. Vitamin B_{12}
    B. Calcium
    C. Vitamin A
    D. Provitamin A and vitamin C
    E. Iron

CASE STUDY

Victor is a 23-year-old male who is physically "active" (engages in more than 60 minutes of moderate activity each day), is 6 feet 1 inch tall, and weighs 190 pounds. His Estimated Energy Requirement is 3,953 Calories. He ate the following foods in one day, which provided 2,540 Calories and the following nutrients:

TABLE 2.18

Victor's One-Day Diet and Diet Analysis Results

Food	Amount	Food	Amount
Whole milk	8 fluid ounces	Brown rice (cooked)	1.5 cup
Cooked oatmeal	1 cup	Broccoli	1¼ cups
Brown sugar	1 teaspoon	Steamed shrimp	4 ounces
Pear	1 each	Stick margarine	2 teaspoons
White bread, toasted	2 slices	Chocolate ice cream	2 cups
Peanut butter	2 tablespoons	Sugar cookies	3 each
Jelly	2 tablespoons		

© Cengage Learning 2013

Victor's One-Day Diet and Diet Analysis Results (*continued*)

Energy-Producing Nutrient Profile	Fatty Acid Profile	Other
Carbohydrate: 60%	SFA: 13%	Cholesterol: 365 mg
Protein: 13%	MUFA: 10%	Sugar: 230 g
Fat: 27%	PUFA: 5%	Fiber: 28 g

Vitamins and Minerals	Intake
Thiamin	1.28 mg
Riboflavin	1.71 mg
Niacin	17.81 mg
Vitamin B_6	1.49 mg
Vitamin B_{12}	3.49 μg
Folate	442.56 μg
Vitamin C	138.15 mg
Vitamin D	8.11 μg
Vitamin A	1,230.00 μg
Vitamin E	7.44 mg
Calcium	839.33 mg
Iron	13.46 mg
Magnesium	473.35 mg
Potassium	2,570.06 mg
Zinc	9.72 mg
Sodium	1,291.50 mg

© Cengage Learning 2013

1. In Victor's diet, half of his grains consumed were from whole-grain sources.
   A. True   B. False

2. Victor's diet is adequate in oils according to MyPlate.
   A. True   B. False

3. Victor's diet is promoting heart health, according to the American Heart Association (AHA) dietary guidelines.
   A. True   B. False

4. Victor's diet meets the Dietary Guidelines 2010 recommendation for protein.
   A. True   B. False

5. Victor's diet meets the American Cancer Society (ACS) dietary guidelines.
   A. True   B. False

6. Victor's diet does not meet the Acceptable Macronutrient Distribution Range (AMDR) for:
   A. Carbohydrate
   B. Protein
   C. Fat
   D. Sugar
   E. Cholesterol

7. Which fat-soluble vitamin is Victor not consuming enough of?
   A. Vitamin A
   B. Vitamin B_6
   C. Vitamin C
   D. Vitamin D
   E. Vitamin E

8. Which mineral is Victor consuming enough of?
   A. Calcium
   B. Iron
   C. Potassium
   D. Zinc
   E. Sodium

9. Which tool to plan, manage, and evaluate diets would provide the type of vitamin and mineral information in the table above?
   A. MyPlate
   B. Exchange Lists
   C. Food composition database
   D. Dietary Reference Intakes (DRIs)
   E. Dietary Guidelines 2010

10. Which Exchange List does Victor's ice cream fit into?
    A. Milk
    B. Starch
    C. Fat
    D. Other Carbohydrate
    E. Combination Foods

FUN-DUH-MENTAL PUZZLE

ACROSS

1. Standard used to prescribe and evaluate the distribution of energy intake for carbohydrate, protein, and fat.
8. To prevent cancer a person should eat a high proportion of _____ foods.
9. One of the Exchange Lists.
11. Portion sizes in the Exchange Lists are based on _____ of carbohydrate, protein, and fat.
13. A MyPlate food group that provides a source of both protein and fiber.
14. A mineral that is added back to refined grain products.
15. The name for the MyPlate healthy fat category.
17. The Dietary Guidelines 2010, AHA guidelines, and ACS guidelines recommend to moderate _____ intake.
19. Substance for which the DRI for adults is calculated at 0.8 grams per kilogram per day.
20. A food providing less than 140 milligrams of sodium per serving is a _____ sodium food.
21. Vitamins are expressed as a percentage of the _____ on the Nutrition Facts panel of a food package label.
22. A food providing 40 percent of its Calories from fat is a _____ fat food.
24. The DRI for physical activity is _____ minutes per day.
25. Meat, eggs, and _____ are included in the MyPlate protein foods group.

DOWN

2. A food that provides empty Calories is a _____ food choice in the MyPlate food guidance system.
3. An electrolyte that has a health claim for possibly reducing hypertension.
4. The MyPlate plan is an _____ diet prescription based on age, gender, and physical activity.
5. A fat-soluble substance that does not have a DRI because it is nonessential.
6. A sweet, carbohydrate- and fiber-rich food that Americans are encouraged to consume according to the Dietary Guidelines 2010.
7. The _____ composition of foods is provided in food composition tables and databases.
10. Consumption should be as low as possible of this type of fatty acid, according to the Dietary Guidelines 2010.
12. A use of the DRIs is for dietary _____.
16. The Dietary Guidelines 2010 recommend reducing the intake of this mineral-based substance.
18. A type of oil that has a health claim for possibly reducing the risk of heart disease.
23. A MyPlate food within the dairy group group that is associated with reduced risk of osteoporosis.

References

Web Resources

A. American Cancer Society: www.cancer.org
B. American Diabetes Association: www.diabetes.org
C. Academy of Nutrition and Dietetics (formerly the American Dietetic Association): www.eatright.org
D. American Heart Association: www.heart.org
E. U.S. Food and Drug Administration, Food Label Education Tools and General Information: www.fda.gov/Food/LabelingNutrition/ConsumerInformation/ucm121642.htm
F. Dietary Guidelines for Americans: www.health.gov/dietaryguidelines
G. Dietary Reference Intakes Tables and Application: www.iom.edu/Activities/Nutrition/SummaryDRIs/DRI-Tables.aspx
H. Food and Drug Administration: www.fda.gov
I. Food and Nutrition Information Center: fnic.nal.usda.gov
J. Healthy People 2020: healthypeople.gov
K. MyPlate: www.choosemyplate.gov
L. USDA Nutrient Data Laboratory: www.nal.usda.gov/fnic/foodcomp/search

Works Cited

1. American Diabetes Association & American Dietetic Association. (2008). Choose your foods: Exchange Lists for diabetes.
2. American Dietetic Association. (2007). Position of the American Dietetic Association and Dietitians of Canada: Dietary fatty acids. *Journal of the American Dietetic Association, 107*(9), 1599–1611.
3. Aronson, D. (2007, July). Fruits & veggies—More matters. *Today's Dietitian, 9*(7), 30–34.
4. Bazzano, L. A. (2006). The high cost of not consuming fruits and vegetables. *Journal of the American Dietetic Association, 106*(9), 1364–1368.
5. Brecher, S. J., Bender, M. M., Wilkening, V. L., McCabe, N. M., & Anderson, E. M. (2000). Status of nutrition labeling, health claims, and nutrient content claims for processed foods: 1997 food label and package survey. *Journal of the American Dietetic Association, 100*, 1057–1062.
6. Burger, K. S., Kern, M., & Coleman, K. J. (2007). Characteristics of self-selected portion size in young adults. *Journal of the American Dietetic Association, 107*(4), 611–618.
7. Burrowes, J. D. (2007, November/December). Preventing heart disease in women: What is new in diet and lifestyle recommendations? *Nutrition Today, 42*(6), 242–247.
8. Cappellano, K. L. (2009). Influencing food choices, nutrition labeling, health claims, and front-of-the-package labeling. *Nutrition Today, 44*(6), 269–273.
9. Carmona, R. H. (2006). Foundations for a healthier United States. *Journal of the American Dietetic Association, 106*(3), 341.
10. Celentano, J. C. (2009). Increased fruit and vegetable intake may reduce the nutrition-related health disparities in African Americans. *American Journal of Lifestyle Medicine, 3*(3), 185–187.
11. Daly, A., Franz, M., Holzmeister, L., Kulkarni, K., O'Connell, B., Wheeler, M., Dunbar, S., & Myers, E. (2003). Exchange Lists for weight management. *American Diabetes Association and the American Dietetic Association*, 1–47.
12. Department of Health and Human Services. (2011, January). *Federal Register, 76*(9), 2383–2384.
13. Divisi, D., Di Tommaso, S., Salvemini, S., Garramone, M., & Crisci, R. (2006). Diet and cancer. *Acta Biomedica, 77*(2), 118–23.
14. Dole Food Company & Mayo Clinic. (2002). *Encyclopedia of foods: A guide to healthy nutrition*. San Diego: Academic Press.

15. Dwyer, J. (2010). Dietary Guidelines 2010—Some things old, some things new, some things borrowed, and much that is true. *Nutrition Today, 45*(4), 144–146.

16. Eckel, R. H., Borra, S., Lichtenstein, A. H., & Yin-Piazza, S. Y. (2007). Understanding the complexity of *trans* fatty acid reduction in the American diet: American Heart Association *Trans* Fat Conference, 2006 report of the *Trans* Fat Conference Planning Group. *Circulation, 115*, 2231–2246.

17. Gall, S. (2007, Winter). Considerations in evaluation of nondairy milks. *Vegetarian Nutrition Update, XV*(II), 1, 8–10.

18. Garza, C., & Pelletier, D. L. (2007). Dietary Guidelines: Past, present, and future. In E. Kennedy & R. Deckelbaum (Eds.), *The nation's nutrition* (pp. 205–220). Washington, DC: International Life Sciences Institute.

19. Goldberg, J., & Folta, S. (2007). Dietary Guidelines 2005 and MyPyramid. In E. Kennedy & R. Deckelbaum (Eds.), *The nation's nutrition* (pp. 221–230). Washington, DC: International Life Sciences Institute.

20. Guenther, K. W., Dodd, J. R., & Krebs-Smith, S. M. (2006). Most Americans eat much less than recommended amounts of fruits and vegetables. *Journal of the American Dietetic Association, 106*(9), 1371–1379.

21. Hands, E. S. (2000). *Nutrients in foods* (D. Balado, Ed.). Baltimore, MD: Lippincott Williams & Wilkins.

22. Haven, J., & Britten, P. (2006). MyPyramid: The complete guide. *Nutrition Today, 41*(6), 253–259.

23. Heber, D., & Bowerman, S. (2001). Applying science to changing dietary patterns. *Journal of Nutrition, 131*, 3078S–3081S.

24. Hu, F. B. (2003). Plant-based foods and prevention of cardiovascular disease: An overview. *American Journal of Clinical Nutrition, 78*, 544S–551S.

25. Institute of Medicine. (1997). *Dietary Reference Intakes for calcium, phosphorus, magnesium, vitamin D, and fluoride.* Washington, DC: The National Academies Press.

26. Institute of Medicine. (1998). *Dietary Reference Intakes: A risk assessment model for establishing upper intake levels for nutrients.* Washington, DC: The National Academies Press.

27. Institute of Medicine. (2000). *Dietary Reference Intakes: Application in dietary assessment.* Washington, DC: The National Academies Press.

28. Institute of Medicine. (2000). *Dietary Reference Intakes for thiamin, riboflavin, niacin, vitamin B_6, folate, vitamin B_{12}, pantothenic acid, biotin, and choline.* Washington, DC: The National Academies Press.

29. Institute of Medicine. (2000). *Dietary Reference Intakes for vitamin C, vitamin E, selenium, and carotenoids.* Washington, DC: The National Academies Press.

30. Institute of Medicine. (2001). *Dietary Reference Intakes for vitamin A, vitamin K, arsenic, boron, chromium, copper, iodine, iron, manganese, molybdenum, nickel, silicon, vanadium, and zinc.* Washington, DC: The National Academies Press.

31. Institute of Medicine. (2003). *Dietary Reference Intakes: Applications in dietary planning.* Washington, DC: The National Academies Press.

32. Institute of Medicine. (2003). *Dietary Reference Intakes: Guiding principles for nutrition labeling and fortification.* Washington, DC: The National Academies Press.

33. Institute of Medicine. (2004). *Dietary Reference Intakes for water, potassium, sodium, chloride, and sulfate.* Washington, DC: The National Academies Press.

34. Institute of Medicine. (2006). *Dietary Reference Intakes: The essential guide to nutrient requirements.* Washington, DC: The National Academies Press.

35. Jacobs, D. R., & Tapsell, L. C. (2007). Food, not nutrients, is the fundamental unit in nutrition. *Nutrition Reviews, 65*(10), 439–450.

36. Johnson, M. A., & Kimlin, M. G. (2006). Vitamin D, aging, and the 2005 Dietary Guidelines for Americans. *Nutrition Reviews, 64*(9), 410–421.

37. King, J. C. (2007). An evidence-based approach for establishing dietary guidelines. *Journal of Nutrition, 137*, 480–483.

38. Kolodinsky, J., Harvey-Berino, J. R., Berlin, L., Johnson, R. K., & Reynolds, T. W. (2007). Knowledge of current dietary guidelines and food choice by college students: Better eaters have higher knowledge of dietary guidance. *Journal of the American Dietetic Association, 107*(8), 1409–1413.

39. Kupper, C. (2007). Gluten-free labeling: Decisions and dilemmas. *Today's Dietitian, 9*(4), 10–16.

40. Kupper, C. (2008). Too much information? New labeling regulations for 2008. *Today's Dietitian, 10*(5), 8–13.

41. Kushi, L. H., Byers, T., Doyle, C., Bandera, E. V., McCullough, M., McTiernan, A., . . . & Thun, M. J. (2006). American Cancer Society guidelines on nutrition and physical activity for cancer prevention: Reducing the risk of cancer with healthy food choices and physical activity. *CA: A Cancer Journal for Clinicians, 56,* 254–281.

42. Laquatra, I., Yeung, D. L., Storey, M., & Forshee, R. (2005, January/February). Health benefits of lycopene in tomatoes—Conference summary. *Nutrition Today, 40*(1), 29–36.

43. Lichtenstein, A. H., et al. (2006). Diet and lifestyle recommendations revision 2006: A scientific statement from the American Heart Association nutrition committee. *Circulation, 114,* 82–96.

44. Miner, M. (2007). Fitness, antioxidants, and moderate drinking: All to lower cardiovascular risk. *American Journal of Lifestyle Medicine, 1*(2), 110–112.

45. Monsen, E. R. (2000). Dietary Reference Intakes for the antioxidant nutrients: Vitamin C, vitamin E, selenium, and carotenoids. *Journal of the American Dietetic Association, 100,* 637–640.

46. Moon, M. (2008). Supplements: The role of a high potassium diet in managing blood pressure. *Nutrition in Complementary Care, 10*(4), 72–76.

47. Moriarty, P. (1993, Spring–Summer). Food labeling and the law. *Food News for Consumers,* 10.

48. Murphy, S. P., & Barr, S. I. (2005). Challenges in using the Dietary Reference Intakes to plan diets for groups. *Nutrition Reviews, 63*(8), 267–271.

49. Ollberding, N. J., Wolf, R. L., & Contento, I. (2010). Food label use and its relation to dietary intake among U.S. adults. *Journal of the American Dietetic Association, 110*(8), 1233–1237.

50. Palmer, S. (2009). Get Ready 2010 for the Dietary Guidelines. *Today's Dietitian, 11*(12), 20–27.

51. Pedersen, B. K. (2007). Health benefits related to exercise in patients with chronic low-grade systemic inflammation. *American Journal of Lifestyle Medicine, 1*(4), 289–298.

52. Pennington, J. A. T., & Douglass, J. S. (2005). *Bowes & Church's food values of portions commonly used* (18th ed.). Baltimore, MD: Lippincott Williams & Wilkins.

53. Pennington, J. A. T., Stumbo, P. J., Murphy, S. P., McNutt, S. W., Eldridge, A. L., McCabe-Sellers, B. J., & Chenard, C. A. (2007). Food composition data: The foundation of dietetic practice and research. *Journal of the American Dietetic Association, 107*(12), 2105–2113.

54. Pennington, J. A. T., & Young, B. (1990). Iron, zinc, manganese, selenium, and iodine in foods in the United States: Total diet study. *Journal of Food Composition, 3,* 166–184.

55. Peterson, S., Sigman-Grant, M., Eissenstat, B., & Kris-Etherton, P. (1999). Impact of adopting lower-fat food choices on energy and nutrient intakes of American adults. *Journal of the American Dietetic Association, 99,* 177–183.

56. Polagruto, J. A., Wang-Polagruto, J. F., Braun, M. M., Lee, L., Kwik-Uribe, K., & Keen, C. L. (2006). Cocoa flavanol-enriched snack bars containing phytosterols effectively lower total and low-density lipoprotein cholesterol levels. *Journal of the American Dietetic Association, 106*(11), 1804–1813.

57. Quatromoni, P. A., Pencina, M., Cobain, M. R., Jacques, P. F., & D'Agostino, R. B. (2006). Dietary quality predicts adult weight gain: Findings from the Framingham Offspring Study. *Obesity, 14*(8), 1383–1391.

58. Reedy, J., & Krebs-Smith, S. M. (2008). A comparison of food-based recommendations and nutrient values of three food guides: USDA's MyPyramid, NHLBI's Dietary Approaches to Stop Hypertension Eating Plan, and Harvard's Healthy Eating Pyramid. *Journal of the American Dietetic Association, 108*(3), 522–528.

59. Rideout, T. C., & Jones, P. J. H. (2010). Plant sterols: An essential component of preventive cardiovascular medicine. *SCAN'S (A Publication for Sports, Cardiovascular, and Wellness Nutritionists) PULSE, 29*(1), 1–6.

60. Rosenberg, I. H. (2007). Folic acid fortification. *Nutrition Reviews, 65*(11), 503.

61. Russel, R. M. (2001). New micronutrient Dietary Reference Intakes from the National Academy of Sciences. *Nutrition Today, 36,* 163–171.

62. Sacks, F. M., Lichtenstein, A., Van Horn, L., Harris, W., Kris-Etherton, P., & Winston, M. (2006). Soy protein, isoflavones, and cardiovascular health: An American Heart Association science advisory for professionals from the nutrition committee. *Circulation, 113,* 1034–1044.

63. Sanders, L. M., & Zeisel, S. H. (2007, July/August). Choline: Dietary requirements and role in brain development. *Nutrition Today, 42*(4), 181–186.

64. Schneeman, B. (2007). FDA's review of scientific evidence for health claims. *Journal of Nutrition, 137*(2), 493–494.

65. Schor, D., Maniscalco, S., Tuttle, M. M., Alligood, S., & Kapsak, W. R. (2010). Nutrition facts you can't miss: The evolution of front-of-pack labeling—Providing consumers with tools to help select foods and beverages to encourage more healthful diets. *Nutrition Today, 45*(1), 22–32.

66. Shuaibi, A. M., House, J. D., & Sevenhuysen, G. P. (2008). Folate status of young Canadian women after folic acid fortification of grain products. *Journal of the American Dietetic Association, 108*(12), 2090–2094.

67. Smith-Edge, M., Miller-Jones, J., & Marquart, L. (2005). A new life for whole grains. *Journal of the American Dietetic Association, 105*(12), 1856–1860.

68. Stein, K. (2009). Are food allergies on the rise, or is it misdiagnosis? *Journal of the American Dietetic Association, 109*(11), 1832–1837.

69. Story, M., & Forshee, R. (2008). The alcohol dietary guideline. *Nutrition Today, 43*(3), 91–97.

70. Talati, R., Sobieraj, D. M., Makanji, S. S., Phung, O. J., & Coleman, C. I. (2010). The comparative efficacy of plant sterols and stanols on serum lipids: A systematic review and meta-analysis. *Journal of the American Dietetic Association, 110*(5), 719–726.

71. Tarrago-Trani, M. T., Phillips, K. M., Lemar, L. E., & Holden, J. M. (2006). New and existing oils and fats used in products with reduced *trans*-fatty acid content. *Journal of the American Dietetic Association, 106*(6), 867–880.

72. Taylor, C. L., & Wilkening, V. L. (2008). How the nutrition food label was developed, part 1: The Nutrition Facts panel. *Journal of the American Dietetic Association, 108*(3), 437–442.

73. Tedford, K. (2005). Food labels, who is being educated? *Journal of the American Dietetic Association, 105*(3), 402–403.

74. Thompson, F. E., McNeel, T. S., Dowling, E. C., Midthune, D., Morrissette, M., & Zeruto, C. A. (2009). Interrelationships of added sugars intake, socioeconomic status, and race/ethnicity in adults in the United States: National Health Interview Survey, 2005. *Journal of the American Dietetic Association, 109*(8), 1376–1383.

75. Thompson, T., Kane, R. R., & Hager, M. H. (2006). Food allergen labeling and Consumer Protection Act of 2004 in effect. *Journal of the American Dietetic Association, 106*(11), 1742–1744.

76. Trumbo, P., Schlicker, S., Yates, A. A., & Poos, M. (2002). Dietary Reference Intakes for energy, carbohydrates, fiber, fat, fatty acids, cholesterol, protein and amino acids. *Journal of the American Dietetic Association, 102*, 1621–1631.

77. Trumbo, P., Yates, A., Schlicker, S., & Poos, M. (2001). Dietary Reference Intakes: Vitamin A, vitamin K, arsenic, chromium, copper, iodine, iron, manganese, molybdenum, nickel, vanadium, zinc. *Journal of the American Dietetic Association, 101*, 294–301.

78. Tufts University. (2004, September). How relevant are nutrition label numbers? *Tufts University Health & Nutrition Letter, 22*(7), 1.

79. Tufts University. (2009). Meat lovers' mortality risk is higher. *Tufts University Health & Nutrition Letter, 27*(4), 1–2.

80. Upton, J. (2007, August). Folate: An oft-neglected B vitamin. *Today's Dietitian, 9*(8), 58–59.

81. Van Horn, L. (2009). Dietary guidance: Targeting specific populations. *Journal of the American Dietetic Association, 109*(9), 1503.

82. Van Horn, L. (2010). Development of the 2010 U.S. Dietary Guidelines Advisory Committee Report: Perspectives from a registered dietitian. *Journal of the American Dietetic Association, 110*(11), 1638–1645.

83. Volpe, S. L. (2004). Serving on the Institute of Medicine's Dietary Reference Intake panel for electrolytes and water. *Journal of the American Dietetic Association, 104*, 1885–1887.

84. Wansink, B., & van Ittersum, K. (2007). Portion size me: Downsizing our consumption norms. *Journal of the American Dietetic Association, 107*(7), 1103–1106.

85. Watson, R. R., & Preddy, V. R. (2003). Nutrition and alcohol: Linking dietary interactions and dietary intake. New York: CRC Press.

86. Webb, D. (2010). Shedding light on soy. *Today's Dietitian, 12*(11), 28–34.

87. Wheeler, M. L., Daly, A., Evert, A., Franz, M. J., Geil, P., Holzmeister, L. A., . . . & Woolf, P. (2008). Choose your foods: Exchange Lists for diabetes, 2008: Description and guidelines for use (6th ed.). *Journal of the American Dietetic Association, 108*(5), 883–888.
88. Williams, P. (2005). Consumer understanding and use of health claims for foods. *Nutrition Reviews, 63*(7), 256–264.
89. Winham, D., Webb, D., & Barr, A. (2008). Beans and good health. *Nutrition Today, 43*(5), 201–209.
90. Yates, A. A., Schlicker, S. A., & Suitor, C. W. (1998). Dietary Reference Intakes: The new basis for recommendations for calcium and related nutrients, B-vitamins, and choline. *Journal of the American Dietetic Association, 98,* 699–706.

Nutrition in Health and Chronic Disease

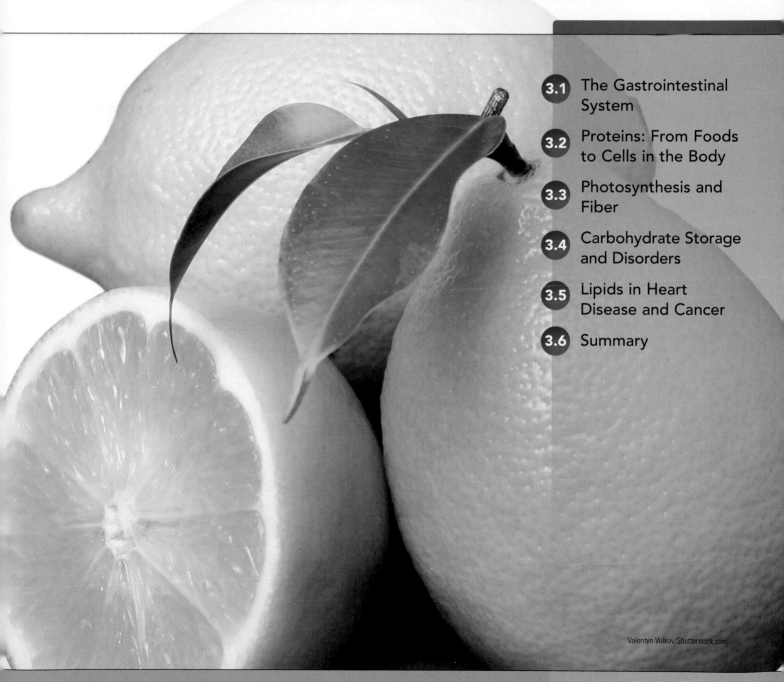

Valentyn Volkov/Shutterstock.com

97

MODULE GOAL

To understand how to manage the intake of energy to prevent the development of nutrition related chronic diseases.

LEARNING OBJECTIVES

After studying this learning module, you will:

- Encounter some major themes and organizing principles of the life sciences.
- Understand the biological processes involved in mechanical and chemical digestion.
- Be able to describe common abnormalities in carbohydrate metabolism.
- Describe how an unhealthy diet and lifestyle can promote heart disease, cancer, and type 2 diabetes.
- Understand how eating adequate amounts of protein, fiber-rich carbohydrates, and essential fatty acids supports optimal body function and health.

PERSONAL IMPROVEMENT GOALS

In this learning module, you will learn:

- How to promote better digestion and absorption of nutrients.
- The benefits of eating your meals and allowing digestion to occur when you are calm.
- To identify healthy carbohydrate, fat, and protein food sources.
- To choose low-fat, high-quality proteins and/or adequate amounts of plant proteins to meet your dietary protein needs.
- The benefits of consuming mostly low-glycemic-index carbohydrate food sources, such as whole grains, fruits, vegetables, and legumes.
- The reasons to avoid eating solid and *trans* fats and instead consume adequate amounts of the essential fatty acids (omega-3 and omega-6 fatty acids).

Did you ever take the time to think about what happens to the food you eat as it makes its way through your body? Which components of the food were absorbed by your body, and which were not? In this learning module, you will follow the fate of food as it is transformed in its journey through the systems of the body: digestive, integumentary (skin), endocrine, cardiovascular, lymphatic, reproductive, skeletal, muscular, respiratory, nervous, and urinary (see Figure 3.1). We will discuss the internal processing of nutrients from the breakdown of food components into their simplest forms of molecules, their entry into the body, the ways nutrients are taken up by cells and metabolized, and the excretion of waste products. You will see how types of carbohydrates, proteins, and fats can either promote health or promote disease depending on how much of each a person consumes and on that person's genetic makeup. Protein malnutrition, lactose intolerance, diabetes, heart disease, cancer, and allergies will also be addressed.[E,8]

3.1 THE GASTROINTESTINAL SYSTEM

Introduction

Because eating is such a natural process, digesting food may also appear simple. However, many chemical, mechanical, physiological, and psychological factors are involved in the process. Nourishing the body begins with the ingestion of food and ends with the excretion of waste. The ultimate purpose of food intake is to fuel and nourish the body optimally. In order to extract any nutrients or Calories from food, the food has to undergo some significant alterations. These alterations take place in the digestive or gastrointestinal (GI) system and affect all of the other organ systems in the body in some way (see Figure 3.1).[G,18,39]

T-Talk 3.1
To hear Dr. Turley talk about the gastrointestinal system, go to www.cengagebrain.com

FIGURE 3.1 The organ systems of the body.

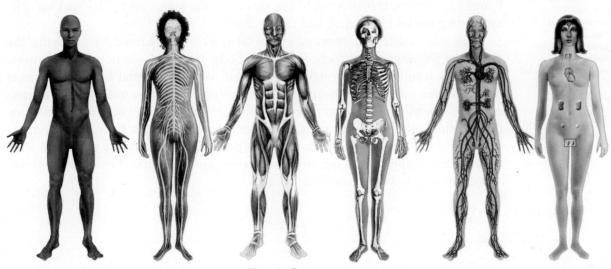

Integumentary System
Protects body from injury, dehydration, and some pathogens; controls its temperature; excretes certain wastes; receives some external stimuli.

Nervous System
Detects external and internal stimuli; controls and coordinates responses to stimuli; integrates all organ system activities.

Muscular System
Moves body and its internal parts; maintains posture; generates heat by increases in metabolic activity.

Skeletal System
Supports and protects body parts; provides muscle attachment sites; produces red blood cells; stores calcium, phosphorus.

Circulatory System
Rapidly transports many materials to and from cells; helps stabilize internal pH and temperature.

Endocrine System
Hormonally controls body functioning with nervous system; integrates short- and long-term activites. (Male testes added.)

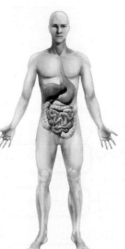

Lymphatic System
Collects and returns some tissue fluid to the blood stream; defends against infection and tissue damage.

Respiratory System
Rapidly delivers oxygen to the tissue fluid that bathes all living cells; removes carbon dioxide wastes of cells; helps regulate pH.

Digestive System
Ingests food and water; mechanically, chemically breaks down food and absorbs small molecules into internal environment; eliminates food residues.

Urinary System
Maintains the volume and solute composition of internal fluids; excretes excess fluid, solutes, and dissolved wastes.

Reproductive System
Male: Produces and transfers sperm to the female. Hormones of both systems also influence other organ systems. Female: Produces eggs; after fertilization, affords a protected, nutritive environment for the development of a new individual.

Biology Today and Tomorrow, 3rd ed, Starr, Chapter 19, Fig. 19.10 (animated), p. 396.

An Overview of the Digestive Process

Let's take a closer look at the anatomy of the digestive tract. We will focus on the steps in the metabolic processing of foods and their nutrients. Specifically, we will explore ingestion, **digestion**, **absorption**, **transportation**, utilization, cellular storage, and **excretion**.

Once food has been chewed up, it is referred to as **bolus**. The bolus is swallowed, passes through the **esophagus**, and enters the **stomach**, where it is mixed with stomach acid or **gastric acid**, which is largely hydrochloric acid, and other gastric secretions. As the smooth muscles around the stomach contract, the gastric contents are churned. This slushy mixture is called **chyme**. Food has to be systematically digested into its molecular components before absorption can occur.

After digestion has taken place—that is, the large molecules that make up food have been broken down into molecules or chemical compounds small enough to enter the intestinal cells—the nutrients are absorbed into the blood stream or the lymphatic system, transported to the various cells, taken up by the various cells making up the body systems, and metabolized.

The solid waste that remains in the GI tract is compacted by removing the water. As the water is absorbed into the body, the solid waste product known as **feces** is formed and excreted. The cellular waste generated by metabolism is secreted back into the blood stream and excreted via the respiratory system (**gaseous** emissions containing carbon dioxide), urinary system (**aqueous** secretions called urine), and the integumentary system (aqueous and **sebaceous** secretions from the skin). So you can see that each step of the process plays a part in systematically dismantling the food into its nutrient components, metabolizing them, and excreting the waste (see Figure 3.2).

FIGURE 3.2 The steps of digestion from intake to excretion.

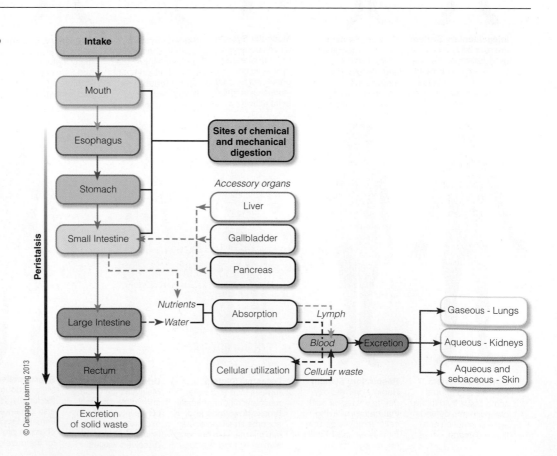

Mechanical and Chemical Digestion

The mechanical and chemical processes that participate in digestion efficiently break down the large food molecules into their molecular units. For example, 98 percent of carbohydrates, including starch (polysaccharides), lactose, maltose, and sucrose, is digested; 95 percent of fats (triglycerides) is digested; and 92 percent of proteins (plant and animal) is digested. This efficient **digestive system** accomplishes digestion by coordinating many organs that contribute the needed chemical and mechanical processes.

Mechanical Digestion The part of the digestive process that is mechanical uses muscles and nerves to create movement. Examples include chewing and muscular stomach contractions that mix and churn the chyme. Another mechanical process is **peristalsis**, a muscular wave action that is caused by the smooth muscle contractions in the digestive system. Peristalsis begins with swallowing and is controlled by the **autonomic nervous system**. Peristalsis propels bolus or chyme along the GI tract and is an automatic process—it happens without your consciously thinking about it. The ingestion, **mastication** (chewing), and swallowing of food are conscious processes.

Chemical Digestion Chemical digestion refers to the use of chemical substances that are needed to break down food such as enzymes, **hormones, hydrochloric acid**, and sodium bicarbonate. Chemical digestion is also controlled unconsciously by the autonomic nervous system. Approximately 7.5 liters of digestive juices are secreted into the GI tract over the course of the day. Think of almost four 2-liter bottles of fluids, such as **saliva**, mucus, gastric juice, intestinal cell secretions, bile, and pancreatic secretions pouring into the GI tract.

Let's take a closer look at the GI system's anatomy. We will focus on the steps in the metabolic processing of foods and their nutrients. Specifically, we will explore ingestion, digestion, absorption, transportation, utilization, cellular storage, and excretion of food and food components.

Anatomy of the Gastrointestinal System

The GI tract is a tube-like passage (often called the alimentary canal) that runs through the body, including the **mouth**, esophagus, stomach, **small intestine**, **large intestine**, and colon (see Figure 3.3). The **accessory organs** of the GI system, though not part of the tubular passage, are directly involved in digestion. The accessory organs of digestion produce and/or secrete chemicals into the GI tract that facilitate the digestion of chyme into its molecular components. The accessory organs include the liver, **gallbladder**, and **pancreas** (see Figure 3.3). The accessory organs are signaled to participate in digestion by many different hormones. A select few are presented.

Journey Through the GI Tract

The passage that food and drink make through the GI tract is long and winding, with many changes in the physiological environments. Let's make the journey one segment at a time.[18]

The Mouth After food is ingested, it immediately begins to be altered. In the mouth, mechanical digestion and some chemical digestion occur. Mechanical digestion occurs with mastication—the grinding and chewing of food. Chemical digestion occurs from the enzymes in saliva. Approximately 1.5 liters of saliva are produced each day by the salivary glands. This saliva is composed largely of water to moisten food. The enzyme salivary **amylase** digests starch. After partial digestion of food occurs in the mouth, the food substance is called a *bolus* and is swallowed.

---

stomach A large, muscular, sac-like organ of the digestive system where chemical and mechanical digestion occurs; the mixing of the bolus with gastric acid and enzymes and the churning due to muscle contractions produces the liquid mixture called *chyme*.

gastric acid An acidic secretion containing hydrochloric acid produced by the cells of the stomach.

chyme A liquefied, partly digested substance that is made and released from the stomach into the small intestine.

feces Solid waste that is compacted in the large intestine (colon) and then released from the anus; composed of bacteria, fiber, sloughed-off intestinal epithelial cells, undigested food, and small amounts of GI tract juices.

gaseous Of or pertaining to gas; a substance in vapor form.

aqueous Of or pertaining to water and water-soluble body secretions.

sebaceous Of or pertaining to fat and fat-soluble body secretions.

digestive system Collectively, the organs and processes associated with the ingestion, digestion, absorption, and excretion of food.

peristalsis Waves of circular muscular contractions in the GI tract, which are controlled by the autonomic nervous system and propel food along the entire GI tract.

autonomic nervous system The part of the central nervous system that regulates involuntary vital functions.

mastication The act of mechanical digestion, accomplished by chewing food to soften it for swallowing.

hormone An active chemical substance formed in one part of the body and carried to another part of the body to alter cellular behavior or activity.

hydrochloric acid An acid secreted by the stomach that causes protein denaturation and aids digestion.

saliva The secretions from the salivary glands in the mouth that contain the enzyme salivary amylase, which begins starch digestion.

FIGURE 3.3 The anatomy of the gastrointestinal tract.

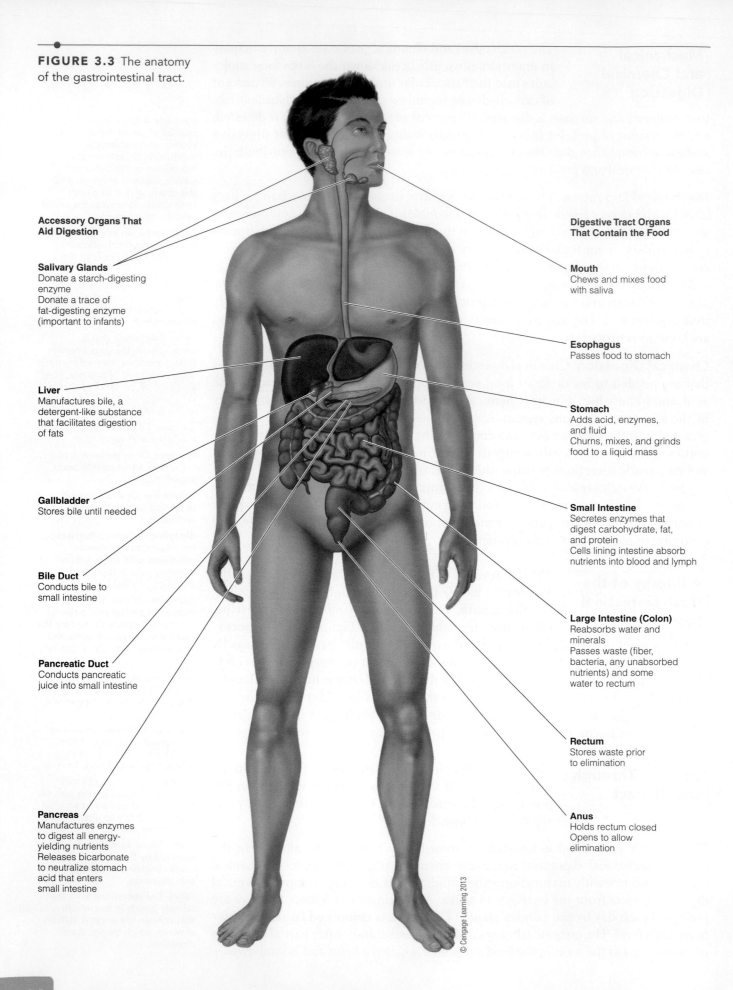

Accessory Organs That Aid Digestion

Salivary Glands
Donate a starch-digesting enzyme
Donate a trace of fat-digesting enzyme (important to infants)

Liver
Manufactures bile, a detergent-like substance that facilitates digestion of fats

Gallbladder
Stores bile until needed

Bile Duct
Conducts bile to small intestine

Pancreatic Duct
Conducts pancreatic juice into small intestine

Pancreas
Manufactures enzymes to digest all energy-yielding nutrients
Releases bicarbonate to neutralize stomach acid that enters small intestine

Digestive Tract Organs That Contain the Food

Mouth
Chews and mixes food with saliva

Esophagus
Passes food to stomach

Stomach
Adds acid, enzymes, and fluid
Churns, mixes, and grinds food to a liquid mass

Small Intestine
Secretes enzymes that digest carbohydrate, fat, and protein
Cells lining intestine absorb nutrients into blood and lymph

Large Intestine (Colon)
Reabsorbs water and minerals
Passes waste (fiber, bacteria, any unabsorbed nutrients) and some water to rectum

Rectum
Stores waste prior to elimination

Anus
Holds rectum closed
Opens to allow elimination

© Cengage Learning 2013

The Esophagus Once the bolus is swallowed, it enters the portion of the tube-like passage called the esophagus and is propelled to the stomach by the action of peristalsis (see BioBeat 3.1).

BioBeat 3.1

Digestion and the Nervous System

Much of the digestion process is controlled through the autonomic nervous system. Once you have chewed your food, you do not have to think about the chemical or mechanical actions of digestion; they are controlled automatically for the most part. Smooth muscles that line the GI tract wall contract and relax to propel the intestinal contents through the body. These muscles contract and relax without conscious guidance. The autonomic action that moves food along the entire GI tract is called peristalsis. The major concert of automatically controlled chemical secretions and actions begin in the stomach and continue through the first segment of the small intestine. Additionally, the release of hormones solicits the accessory organs' fluids for further chemical digestion, including the bile that is produced by the liver and stored in the gallbladder, as well as pancreatic juices that include a wide variety of digestive enzymes and sodium bicarbonate to neutralize gastric acid. The significant amount of chemical digestion that occurs in the small intestine produces a size of molecular components that are absorbed primarily in the first segment of the small intestine.

Which parts of digestion are under your conscious control?

mouth The opening of the oral cavity where food ingestion occurs.

small intestine The part of the digestive tract located between the stomach and the large intestine (colon) that has three sections—the duodenum, jejunum, and ileum—and is the part of the GI tract where most digestion and absorption of nutrients occurs.

large intestine The large bowel portion of the GI tract that completes the digestive process and includes the ascending, transverse, descending, and sigmoid colon.

accessory organs Relating to digestion, the liver, gallbladder, and pancreas.

gallbladder An accessory organ of digestion that stores bile produced by the liver.

pancreas An organ that has exocrine (secreting digestive enzymes and juices into the duodenum) and endocrine (secreting hormones into the blood that help to maintain glucose homeostasis) functions.

amylase An enzyme that hydrolyzes the starch amylose (cleaves the alpha bonds between the glucose molecules).

The Stomach The stomach is a large sac-like organ that plays an important role in digestion. A bizarre balance is maintained in this unique organ: The stomach has cells of the glandular epithelium (see Figure 3.4) that release hydrochloric acid and cells that release thick mucus that protects other stomach cells from the damaging acid. Further, there are three layers of muscle cells: horizontal, vertical, and oblique (see Figure 3.4). The bolus that has passed from the esophagus turns into a liquefied slurry as it is mixed and churned with all the gastric juices. Approximately 2 liters of gastric juice, composed of hydrochloric acid, enzymes, hormones, and mucus, is used each day. The hydrochloric acid plays a big role in the chemical digestion of food. The acid denatures proteins and prepares the chyme for enzymatic action. Some enzymes that digest protein, such as the

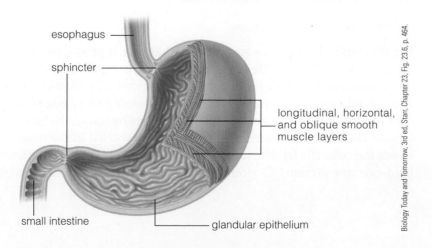

esophagus

sphincter

longitudinal, horizontal, and oblique smooth muscle layers

small intestine

glandular epithelium

Biology Today and Tomorrow, 3rd ed, Starr, Chapter 23, Fig. 23.6, p. 464.

FIGURE 3.4 The musculature of the stomach.

protease enzyme **pepsin**, do their work in the stomach. The stomach also serves as a food reservoir and controls the slow release of chyme from the stomach into the small intestine. The release of small amounts of chyme is controlled by the autonomic nervous system and hormones (see BioBeats 3.1 and 3.2). Having small amounts of chyme entering the small intestine increases the efficiency of chemical digestion in the small intestine.

BioBeat 3.2

Coordinating Digestion: Hormones as Chemical Messengers

A hormone is a chemical messenger produced and secreted from cells within endocrine glands or organs, into the blood. Hormones specifically influence the processes within certain cells of the body in order to regulate metabolism. Some hormones coordinate the digestive processes. For example, hormones like **gastrin, secretin,** and **cholecystokinin (CCK)** are involved in chemical digestion. Gastrin stimulates the secretion of gastric juice. Secretin is released from duodenal cells and signals the pancreas to secrete pancreatic juice (made of digestive enzymes) and sodium bicarbonate and the liver to produce bile. Cholecystokinin is released from duodenal cells in response to the presence of fat-rich chyme. Cholecystokinin signals pancreatic secretions and the gallbladder to release bile.[28]

Can you create an analogy for the job of a hormone in a practical everyday situation?

The Small Intestine The small intestine includes the **duodenum, jejunum,** and **ileum.** More digestion occurs in the small intestine, particularly in the duodenum, as compared to other segments of the intestine. Here, about 2 liters of intestinal juices per day—composed largely of mucus, enzymes, and hormones—are released to aid digestion. Some juices are released when the accessory organs—the liver, gallbladder, and pancreas—are signaled. After digestion is complete, absorption of many different nutrients, such as monosaccharides, amino acids, fatty acids, vitamins, and minerals, must occur in order to deliver these essential nutrients to body cells. All the while, peristalsis continues to take place and propels the chyme and sloughed-off intestinal cells into the large intestine.

Let's take a deeper look at the accessory organs, digestion, absorption, transportation, utilization, and storage of nutrients before venturing into the large intestine and discussing excretion.

Accessory Organs Involved in Aiding Digestion at the Duodenum Three accessory organs contribute to digestion: the liver, gallbladder, and pancreas (see Figure 3.5). The liver and the pancreas perform many vital functions for the entire body, some of which will be discussed later. For now we will focus on their roles in digestion. The liver produces approximately 0.5 liters of bile each day. Bile is composed largely of cholesterol-rich bile acids. The gallbladder stores the bile that is produced by the liver. The gallbladder's sole role in the body is to store bile acids, which are used to aid fat digestion. Bile is released from the gallbladder upon a signal by the hormone CCK, causing the gallbladder to contract. The purpose of bile is to emulsify fat in the small intestine. The pancreas produces approximately 1.5 liters of pancreatic juice each day and secretes the juice containing enzymes for carbohydrate, protein, and fat digestion, and sodium bicarbonate for the neutralization of acidic chyme into the small intestine.

Digestion in the Small Intestine

Fat digestion: Fat must be emulsified by bile that is produced by the liver and stored in the gallbladder (see Figure 3.6). When the gallbladder receives a signal to contract, bile is released into the duodenum. Bile allows fat to interact with fat-splitting lipase enzymes. Once fat is emulsified by bile and digested into glycerol, fatty acids, and/or monoglycerides by lipase, the tiny fat droplets are absorbed by the intestinal **villi** (**epithelial** cells) in the small intestine. Inside the cells, the fat droplets are packaged into **chylomicrons** (a temporary lipid-carrying protein). Chylomicrons are then released into the lymphatic system and later into the blood stream. The chylomicrons are taken up by the liver and repackaged into useable **lipoproteins**, which are released again into the blood stream for delivery to body cells.

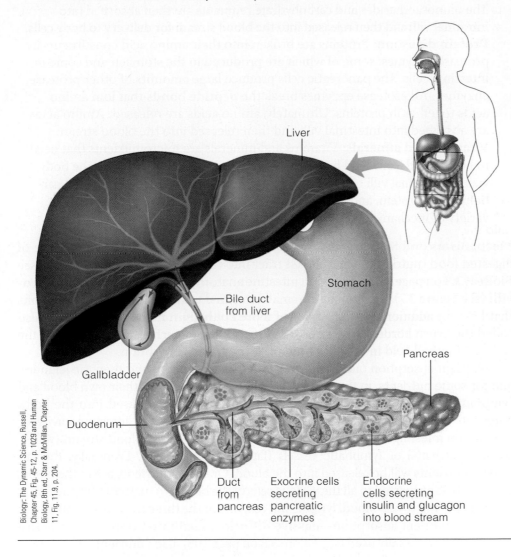

Biology: The Dynamic Science, Russell, Chapter 45, Fig. 45-12, p. 1029 and Human Biology, 8th ed. Starr & McMillan, Chapter 11, Fig. 11.9, p. 204.

Liver

Stomach

Pancreas

Bile duct from liver

Gallbladder

Duodenum

Duct from pancreas

Exocrine cells secreting pancreatic enzymes

Endocrine cells secreting insulin and glucagon into blood stream

FIGURE 3.5 The liver, gallbladder, and pancreas are the accessory organs that aid digestion.

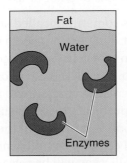

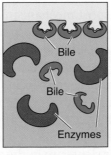

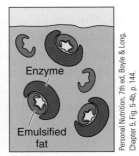

Personal Nutrition, 7th ed, Boyle & Long, Chapter 5, Fig. 5-4b, p. 144.

Fat

Water

Enzymes

Bile

Bile

Enzymes

Enzyme

Emulsified fat

A. Fats and water tend to separate; enzymes are in the water and can't get at the fat.

B. Bile (emulsifier) has affinity for fats and for water so it can bring them together.

C. Small droplets of emulsified fat. The enzymes now have access to the fat, which is mixed in the water solution.

FIGURE 3.6 The emulsification of fat by bile.

Carbohydrate digestion: Salivary and pancreatic amylase enzymes cleave the alpha bonds in starch to release glucose. In the small intestine, maltase, sucrase, and **lactase** break down maltose, sucrose, and lactose, respectively. The monosaccharides and carbohydrate remnants are then absorbed into intestinal villi and then released into the blood stream for delivery to body cells.

Protein digestion: Proteins are broken into their amino acid constituents by protease enzymes, some of which are produced in the stomach and some in intestinal cells. The pancreatic cells produce large amounts of other protease enzymes. All protease enzymes break the **peptide bonds** that join amino acids together in proteins. Ultimately, amino acids are released. Amino acids are absorbed into intestinal villi and then released into the blood stream.

Vitamins and minerals: Vitamins and minerals are micronutrients that need to be released from food by digestion so they can be absorbed into the body via the intestinal villi epithelial cells. They do not undergo digestion like carbohydrate, protein, and fat. Vitamins and minerals are absorbed through a variety of mechanisms.

Mechanisms and Sites of Absorption in the Small Intestine The passage of digested food (nutrients) from the GI tract into the body is called absorption (see BioBeat 3.3 on page 108). The small intestine anatomy includes the villi and **microvilli** (see Figure 3.7). The villi of the small intestinal wall are finger-like projections that have an additional surface layer of more epithelium made by microvilli (also called the brush border). The microvilli create a larger surface area (one-third the size of a football field in an adult human) for nutrient absorption.

Nutrient absorption takes place mostly in the duodenum and then the jejunum, and for some nutrients, in the ileum. The villi are supplied with their own blood and **lymphatic vessels**, so the nutrients can be absorbed and released into the body. Nutrients released by digestion are absorbed from the small intestine cavity or lumen into intestinal epithelial cells, then either passed to the blood stream (water-soluble nutrients) or lymphatic vessels (fat-soluble nutrients). Eventually, the fat-soluble nutrients will be released into the blood. Once the nutrients are in the blood, they can be transported to all the various cells of the body. No matter if the nutrient is entering the microvilli, blood, lymph, or a cell, one of the three most common mechanisms of nutrient absorption—**passive diffusion**, **facilitated diffusion**, or **active transport**—are usually used (see Figure 3.8 on page 108). Each nutrient, for the most part, specifically uses one of these **transport** mechanisms to enter or exit a cell:

Passive diffusion: This transport mechanism is sometimes called simple diffusion, passive transport, or simple transport. It allows nutrients to cross cell membranes freely from the intestinal lumen into cells by a concentration gradient created by osmotic pressure. This means that the higher concentration of nutrients on one side of the membrane creates a pressure to force the nutrients across the membrane to the lower concentration (see Figure 3.8). After food intake, there are higher concentrations of nutrients in the intestinal lumen compared to the inside of intestinal cells, so the nutrients are driven into villi. Once the nutrients are in the villi, they will be released into the blood or lymphatic system, depending on their solubility, for delivery and use by body cells.

Facilitated diffusion: This transport mechanism is sometimes called facilitated transport. The nutrient using this mechanism requires a specific carrier protein or receptor protein in order to cross the cell membrane. Therefore, it is a selective transport or absorption process. Nutrients that require facilitated diffusion bind to a specific receptor protein. The receptor protein is designed to selectively bind to the nutrient before the nutrient can gain

lactase An enzyme that breaks the bond between a glucose and a galactose molecule, or digests lactose.

peptide bonds The type of chemical bond that joins two amino acids together.

microvilli Tiny hair-like projections of cell membranes on epithelial cells of the small intestine that increase the absorptive surface area.

lymphatic vessels The tubular system of vessels where lymph fluid circulates.

passive diffusion Absorption mechanism in which nutrients cross the intestinal cell membrane freely by a concentration gradient that creates osmotic pressure from the forces of migration from an area of high concentration to a lower one.

facilitated diffusion A selective absorption mechanism in which nutrients cross cell membranes by binding to a specific carrier protein or receptor to cross the lipid bilayer; facilitated by a concentration gradient that creates osmotic pressure from the forces of migration from an area of high concentration to a lower one.

active transport A selective absorption mechanism in which nutrients cross cell membranes by binding to a specific carrier protein or receptor that requires ATP energy to cross the lipid bilayer.

transport The movement of molecules across a cell membrane boundary.

entry into the cell. Once the nutrient is assisted into the villi by the receptor protein, it will be released into the blood or lymphatic system, depending on its solubility, for delivery and use by body cells.

Active transport: This transport mechanism requires a selective carrier protein or receptor protein and adenosine triphosphate (ATP; energy) to

FIGURE 3.7 The anatomy of the small intestine.

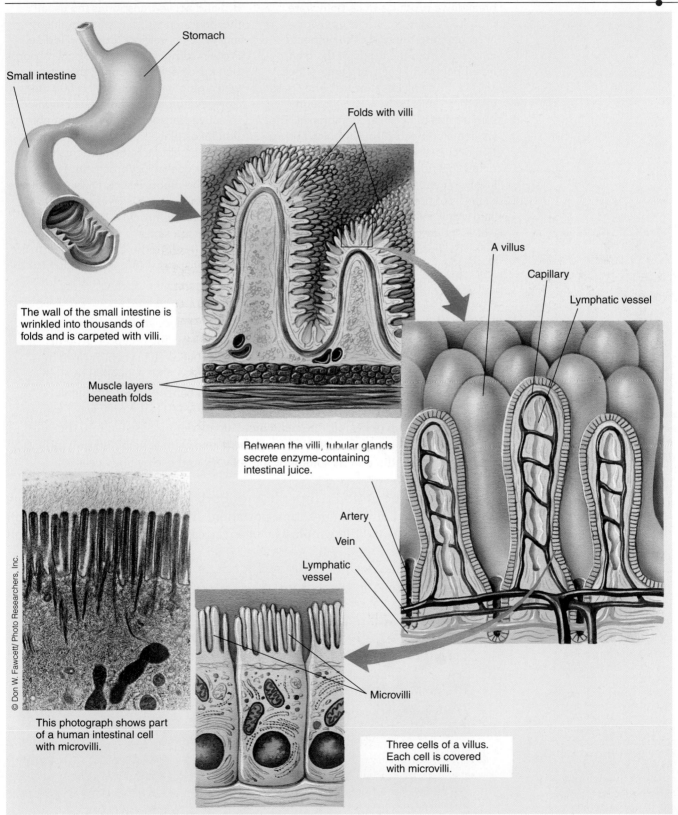

FIGURE 3.7 The anatomy of the small intestine.

Stomach

Small intestine

Folds with villi

The wall of the small intestine is wrinkled into thousands of folds and is carpeted with villi.

Muscle layers beneath folds

A villus

Capillary

Lymphatic vessel

Between the villi, tubular glands secrete enzyme-containing intestinal juice.

Artery

Vein

Lymphatic vessel

Microvilli

© Don W. Fawcett/ Photo Researchers, Inc.

This photograph shows part of a human intestinal cell with microvilli.

Three cells of a villus. Each cell is covered with microvilli.

Personal Nutrition, 7th ed, Boyle & Long, Chapter 3, Fig. 3-9, p. 87.

BioBeat 3.3

Nutrient Absorption in the Small Intestine

When nutrients (vitamins, minerals, and breakdown products of carbohydrate, protein, and fat from digestion) are absorbed into the body, the nutrients must pass from the inside of the **gastrointestinal tract** lumen into the epithelial cells in the small intestine villi. They pass into these cells by traveling through semipermeable cellular membranes. The three major mechanisms in which absorption occurs—passive diffusion, facilitated diffusion, and active transport—are involved in absorption of nutrients into the cells of the body or the transport across any cell membrane. For example, water and fatty acids are absorbed by passive diffusion, whereas facilitated diffusion is the usual mechanism needed to

absorb fructose, riboflavin, and vitamin B_{12}, and active transport is the usual mechanism needed to absorb glucose, amino acids, calcium, iron, and iodine. Protein carriers can speed up the absorption process. For example, the rate at which glucose enters a cell is 500 times slower with passive diffusion compared to active transport. After nutrient absorption takes place by the small intestinal cells, the cells pass the nutrients directly (or after packaging/processing) into either the blood vessels for water-soluble nutrients or lymphatic vessels for fat-soluble nutrients.

Can you relate the concept of diffusion and transport to your daily activity?

---

cross the cell membrane. Therefore, it is a highly selective absorption process. Nutrients that require active transport bind to a specific receptor protein and can only cross the cell membrane if ATP is used. The receptor protein is designed to selectively bind to the nutrient, and the ATP is needed to allow the nutrient to get into the cell while opposing a concentration gradient. The combination of the receptor protein and the ATP is the only way for the nutrient to gain entry into the cell. Once the nutrient is assisted into the villi by the receptor protein and ATP, it will be released into the blood or lymphatic system, depending on its solubility, for delivery and use by body cells.

Nutrient Transportation Through the Body The movement of nutrients throughout the body requires transportation. There are two transportation systems: the blood and the **lymph**. The blood stream provides the transportation

gastrointestinal tract The long, muscular tube extending from the mouth to the anus.

lymph A fluid containing lymphocytes that travels through the lymphatic vessels and is the first conduit for the transport of fat-soluble substances, packaged as chylomicrons, after digestion and intestinal cell absorption.

FIGURE 3.8 The fundamental mechanisms of nutrient absorption across a cell membrane.

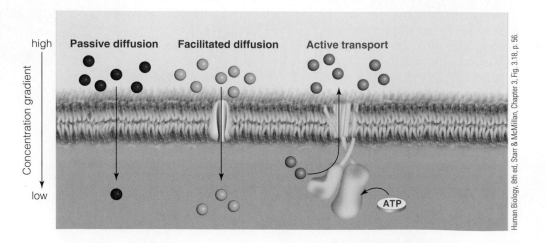

system for water-soluble nutrients, and the blood is the medium for transport. The lymph system provides the transportation for fat-soluble nutrients. Lymph fluid fills the lymphatic vessels. The lymph flows through the lymphatic vessels with the help of the pumping action of the heart and the pulses of the cardiovascular system. The lymph is filtered through lymph nodes and the spleen, but the fat-soluble nutrients are eventually released into the blood stream at the interface of a fast-flowing blood vessel. The lymphatic system is the initial transportation system for fat-soluble nutrients, and the lymph is the medium for transport.

Utilization and Storage Inside the Body After nutrients are digested, absorbed, and transported inside the body to cells, they are used in metabolism for ATP production, growth, repair, and storage (see BioBeat 3.4). The type of metabolic reactions that the nutrient participates in could be anabolic, catabolic, or transfer (neither anabolic nor catabolic). If the nutrient is not needed for one of these purposes, it can be stored. The storage of nutrients in the body may be short, intermediate, or long-term as follows:

Short-term storage: The substance is stored for seconds or minutes. An example is ATP in all cells.

Intermediate storage: The substance is stored for hours. An example is glycogen in the liver.

Long-term storage: The substance is stored for days, months, or years. An example is fat in adipose tissue; vitamin A and vitamin B_{12} in the liver; iron in the liver and bone marrow; protein in lean body mass; and calcium, phosphorus, and magnesium in bones.

BioBeat 3.4

Anabolic and Catabolic Reactions

Metabolism involves a constant process of chemical reactions, some of which are constructive in nature, whereas others are destructive. **Homeostasis** maintains a balanced state between building (**anabolism**) and destroying (**catabolism**) reactions. Anabolic, or building, processes are involved in growth, repair, synthesis, and repletion of tissues. A typical chemical reaction involved in anabolism is a **condensation reaction**, in which molecules are joined together as a chemical substance is built up (see Figure 3.9). In a condensation reaction, hydrogen and oxygen atoms are no longer needed, so they are condensed into water when the molecules are joined. Two examples of anabolic reactions are the combining of individual amino acids to make a protein and the combining of glucose units to make glycogen. On the other hand, catabolism is the breakdown of complex organic molecules into simpler components, accompanied by the release of energy (ATP). Catabolic processes occur in energy metabolism, stress, and tissue degradation. A typical chemical reaction involved in catabolism is a **hydrolysis** reaction: When the molecules of a chemical substance are broken apart, water is needed (see Figure 3.9). The water is split apart into atoms of hydrogen and oxygen, which are used to balance the chemical fragments resulting from the catabolic reaction. Two examples of catabolic reactions are breaking peptide bonds in a protein to release the individual amino acids and releasing glucose units from glycogen.

How does your body remain relatively the same in the midst of ongoing catabolism and anabolism in the dynamics of life?

glycogen A glucose-based storage molecule, sometimes called animal starch, that is synthesized and stored in the liver and muscle tissues.

homeostasis A balanced state (within a tolerable range) of the collection of anabolic biochemical reactions (synthesis) and catabolic biochemical reactions (degradation) occurring inside the body.

anabolism When small molecules are put together to build larger ones through metabolic condensation reactions (water is released).

catabolism When large molecules are broken apart to yield smaller ones through hydrolysis reactions, which require water.

condensation reaction An anabolic chemical reaction producing water and a larger molecule.

hydrolysis A chemical reaction that uses water to break the bonds in chemical compounds, in which water (H_2O) is split, generating a free hydrogen atom (H) and a hydroxyl group (OH), which are used to chemically balance the products of catabolism.

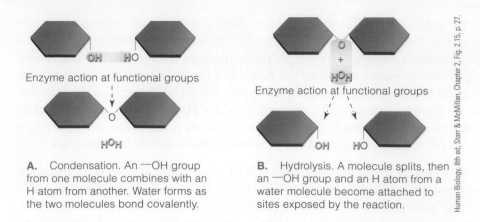

A. Condensation. An —OH group from one molecule combines with an H atom from another. Water forms as the two molecules bond covalently.

B. Hydrolysis. A molecule splits, then an —OH group and an H atom from a water molecule become attached to sites exposed by the reaction.

FIGURE 3.9 Condensation and hydrolysis reactions.

Human Biology, 8th ed, Starr & McMillan, Chapter 2, Fig. 2.15, p. 27.

The Large Intestine The journey through the GI tract ends with the passage of waste through the large intestine, found just after the ileum in the small intestine. Here the surface cellular anatomy changes from villi and microvilli to cells designed to absorb water, so the compaction of waste and absorption of water occurs in the large intestine as waste is prepared for excretion. The large intestine includes the cecum (the appendix, a small finger-like pouch, sticks out from this structure); colon (ascending, transverse, and descending); rectum; and anal canal (anus). The large intestine can also be called the bowel or colon.

Excretion The start of the journey through the GI tract was a conscious decision for food ingestion, and then the unconscious process of peristalsis took over to propel food through the tract. Now elimination occurs by conscious action through the circular anal sphincter. When the journey through the GI tract ends, a new one will surely begin as the food processor known as the gastrointestinal tract continues to operate to make the substances in food available for cellular metabolism (see Table 3.1 for a summary). Waste products and byproducts from cellular metabolism and the digestive process are eliminated from the body by excretion through one of four methods.

Feces Feces are excreted as the solid waste products of the GI system. Solid waste includes bacteria, fiber, sloughed-off intestinal epithelial cells, undigested food, and small amounts of GI tract juices. The speed at which food moves through the GI tract is based on meal size, dietary composition, and physiological factors and is referred to as **transit time**. Transit time from ingestion to excretion is typically 9 to 48 hours.

Urinary Output Urinary output is the volume of urine excreted in a 24-hour period. Urine is an aqueous waste solution excreted from the urinary system, or kidneys. Urine contains water, nitrogen waste (mostly **urea**), detoxified chemicals (including medications), mineral salts, and excess or degraded vitamins and minerals. The most efficient mechanism for disposing chemical metabolic waste is via the urine.

Respiration Respiration is excretion of gaseous metabolic waste products by the pulmonary system, or lungs. Respired air has high concentrations of carbon dioxide (CO_2) and water vapor.

Skin Secretions Aqueous and sebaceous waste are excreted as either sensible or insensible secretions from the integumentary system, or skin. Insensible excretion products include small amounts of water, mineral salts, lipids, and amino acids. Sensible secretion is noticeable sweating resulting from heat, exercise, fever, or hormonal imbalance.

transit time The amount of time it takes food to pass through the entire GI tract from mouth to anus.

urea A toxic nitrogen-containing organic compound that is a byproduct of amino acid catabolism and is excreted in the urine.

TABLE 3.1

Highlights of Digestion

Site	Mechanical Process	Chemical Process	Accessory Organs	Absorption
Along the GI tract	*Uses nerves and muscles*	*Uses hormones, enzymes, digestive juices, and acids*	*Aids digestion outside of the GI tract*	*Substance is taken up by the body*
Mouth	Mastication	Amylase (E)	N/A	N/A
Esophagus	Peristalsis	N/A	N/A	N/A
Stomach	Mixing, churning	HCl acid Pepsin (E) Gastrin (H)	N/A	Alcohol
Small intestine (duodenum, jejunum, ileum)	Peristalsis	Amylase (E) Protease (E) Lipase (E) Cholecystokinin (H) Secretin (H) Bile	*Liver*: Bile synthesis *Gallbladder*: Bile storage *Pancreas*: Enzymes and sodium bicarbonate	Carbo-hydrates, proteins, fats, vitamins, minerals
Large intestine	Peristalsis	N/A	N/A	Water
Rectum	Excretion	N/A	N/A	N/A

E (denotes enzyme)

H (denotes hormone)

© Cengage Learning 2013

Summary Points

- Digestion is the breakdown of food into absorbable molecular units.
- The GI tract and accessory organs participate in the mechanical and chemical digestion of food.
- Enzymes, hormones, acids, and bicarbonate are used in chemical digestion.
- Nerves and muscles are involved in mechanical digestion.
- The three major absorption mechanisms are passive diffusion, facilitated diffusion, and active transport.
- Once nutrients have been absorbed into intestinal cells, they are transported via the blood stream (water-soluble substances) or lymphatic vessels (fat-soluble substances) to cells.
- Nutrients are assimilated, utilized, stored, and/or metabolized by cells.
- Solid waste is excreted through the colon, and cellular metabolic waste is excreted via the kidneys, skin, and lungs.

Take Ten on Your Knowledge Know-How

1. What is digestion? What are some examples of chemical and mechanical digestion?
2. How do enzymes and hormones function in the body and in digestion?
3. How does food journey through your digestive tract? Which organs are involved, and what crucial role does each organ play in the process?
4. What are the accessory organs, and how do they function in aiding digestion?
5. How does the digestion of fat differ from that of carbohydrates and proteins?

6. How and where are most nutrients absorbed? What mechanisms does the body use to absorb nutrients? How does each absorptive mechanism work?
7. How are nutrients transported inside the body? How does the transportation of fat-soluble nutrients differ from that of water-soluble nutrients?
8. How are nutrients utilized by your body? What do *homeostasis*, *catabolism*, and *anabolism* mean? What are some examples of each?
9. How are nutrients stored in your body?
10. How is waste excreted from your body?

3.2 PROTEINS: FROM FOODS TO CELLS IN THE BODY

Introduction

Dietary protein has many critical functions in the body. It is important to understand how the human body uses the amino acid building blocks provided from dietary protein to synthesize its own proteins. Self-synthesized protein (endogenous protein synthesis) is needed for body tissue repair and maintenance, and it provides the basic structure of all cells in organs, muscles, the brain, nerves, skin, hair, and nails, as well as regulatory components such as enzymes, some hormones, and blood. The process of dietary protein **denaturation** and digestion is described below. Greater detail on protein is also provided with respect to **protein character**, types, deficiency, and excess.[57,74]

Protein Denaturation Versus Digestion

Protein denaturation is an important step in protein digestion. It causes the protein to change its three-dimensional structure as the coiled shape of the protein is loosened or modified (see Figure 3.10). Protein denaturation can be caused by heat. If you have ever cooked an egg, you have seen protein denaturation in action in both the white and the yolk. Denaturation also occurs with exposure of protein to acid or alkali (basic) treatments, as when lemon juice is added to milk: The milk proteins curdle. Acid denaturation occurs unseen when a food protein encounters stomach acid. Exposure of proteins to metals, such as by cooking foods in cast-iron pots, causes denaturation too. After denaturation, the protein is no longer functional, but enzymes can more easily access the various parts of the protein strand to digest (break apart) the amino acids in the protein; therefore, the process of denaturation is important for digestion, because it better prepares the protein for enzymatic digestion.

When a protein is digested, the protein strand is broken, and the amino acids are released. **Protein digestion** takes place in the stomach and small intestine through the action of protease enzymes. The amino acids that are released from digested proteins are absorbed into intestinal villi and transported to the body

denaturation The change in shape of proteins and loss of their function when exposed to heat, acids, bases, and/or heavy metals.

protein character The unique aspects of an individual protein, determined by the amino acid sequence and the folding and interacting of the protein strand that makes a three-dimensional functional molecular structure.

protein digestion When the peptide bonds between the amino acids in a protein strand are broken by a protease enzyme, causing the amino acids to be released for absorption, transportation, utilization, and excretion by body cells.

FIGURE 3.10 Protein denaturation and digestion.

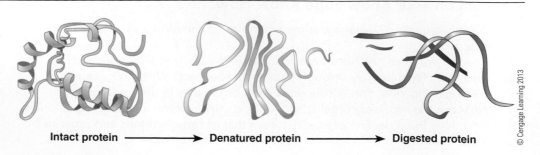

Intact protein ⟶ Denatured protein ⟶ Digested protein

© Cengage Learning 2013

cells via the blood stream. Inside body cells, the amino acids are ready to be incorporated into newly built or synthesized proteins based on the particular metabolic need of the cell at the time.

Protein Synthesis It is important for each cell in the body to make specific proteins in order to repair cell structures and carry out many cellular chemical processes. Protein is synthesized in the body from amino acids according to the genetic code, which is housed within the cell nucleus as part of a chromosome, in a process that begins with **gene expression**. Gene expression is regulated by cell signals, such as by hormonal messages, based on the cellular environment. Basically, when a cell needs a particular protein product to perform a particular task, the gene for that protein is expressed. After gene expression occurs, the **deoxyribonucleic acid (DNA)** gene code is copied to a ribonucleic acid (RNA) message, which is then translated to make an amino acid sequence that creates the unique protein strand. The formed protein strand folds and interacts with itself and sometimes with other protein molecules to make a unique three-dimensional shape and then performs its specific function. This process of making a functional protein, from copying DNA to RNA and then to creating a protein, is called **protein synthesis** (see BioBeat 3.5 on page 115 and Figures 3.11 and 3.12).

An example of hormonal control over gene expression and protein synthesis is the hormone testosterone's response to exercise. Testosterone is present in both men and women, although it exists in higher concentrations in men. In response to exercise stress, increased amounts of testosterone bind to cell receptors and signal an increase in protein synthesis, resulting in enhanced muscular strength and size. Thus, the genes that encode cellular proteins are turned on or expressed after the testosterone hormone signal is received.

gene expression The transcription of the nucleotide bases making up the genetic code of a gene into RNA and then the translation into a functional protein.

deoxyribonucleic acid (DNA) A linear, double-stranded polymer of the nucleic acids (nucleotides) adenine, guanine, cytosine, and thymine, found in the cell's nucleus, that makes up the cell's transmittable genetic information and encodes bioactive molecules.

protein synthesis A series of condensation reactions between amino acids that build a protein, directed by the genetic code, which occurs by transcription of DNA and translation of mRNA.

FIGURE 3.11 The cell houses the nucleus, where the chromosomes, with their DNA sequences, are located.

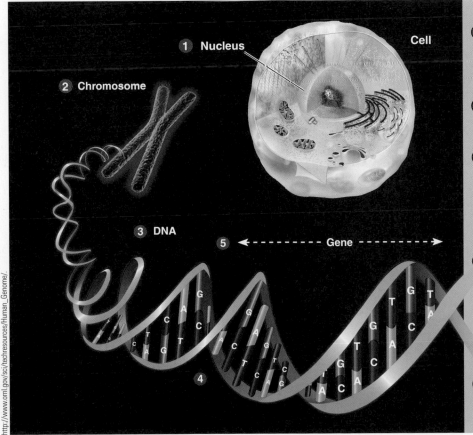

1 The human genome is a complete set of genetic material organized into 46 chromosomes, located within the nucleus of a cell.

2 A chromosome is made of DNA and associated proteins.

3 The double helical structure of a DNA molecule is made up of two long chains of nucleotides. Each nucleotide is composed of a phosphate group, a five-carbon sugar, and a base.

4 The sequence of nucleotide bases (C, G, A, T) determines the amino acid sequence of proteins. These bases are connected by hydrogen bonding to form base pairs—adenine (A) with thymine (T) and guanine (G) with cytosine (C).

5 A gene is a segment of DNA that includes the information needed to synthesize one or more proteins.

messenger ribonucleic acid (mRNA) A long polymer of the nucleic acids (nucleotides) adenine, cytosine, guanine, and uracil that are made inside the nucleus of the cell; upon the expression and transcription of deoxyribonucleic acid, mRNA then migrates to the ribosome in the cytosol of the cell, where it is translated for protein synthesis.

essential amino acids Amino acids that are needed by the body to build proteins but cannot be synthesized endogenously; thus they are needed from food sources.

nonessential amino acids Amino acids that are needed by the body to build proteins but are synthesized from nitrogen and carbohydrate intermediates and can also be provided by food sources.

FIGURE 3.12 Protein synthesis involves DNA transcription in the cell nucleus and then RNA translation with ribosomes and tRNA in the cell cytosol.

The expressed gene in the DNA template is transcribed or copied to **messenger ribonucleic acid (mRNA)**. This process is called *transcription*. Afterward, the mRNA leaves the cell nucleus and enters the cell cytosol. Here, the mRNA attaches to ribosomes (the site of protein synthesis), where the mRNA code is translated to an amino acid sequence. This process is called *translation* and is somewhat like code breaking. The RNA in the ribosome (ribosomal RNA or rRNA) reads the code and interprets the meaning to a specific amino acid that is needed in the protein sequence. Transfer RNA (tRNA) collects the needed amino acids and delivers them to the ribosome. Amino acids are then linked together in a protein strand. The bonds between the amino acids are called peptide bonds. The order of the amino acids in a protein is determined by transcription and translation of genetic material (see Figure 3.12 and BioBeat 3.5).

When the mRNA message is fully translated and the protein is synthesized, the mRNA is degraded, and the tRNA and ribosomes are freed to make other proteins. Further, the protein may only exist in the cell for minutes and then be degraded after it has done its job. The amino acids from the degraded protein reenter the cellular amino acid pool and are available to make other proteins.

Protein Character The human body uses a total of 20 amino acids. Nine are called **essential amino acids**. The remaining 11 are called **nonessential amino acids**, because the body can make them from carbohydrate intermediates and nitrogen molecules. The body can only acquire essential amino acids from dietary protein sources. The 20 different amino acids making up proteins have different properties and qualities that cause them to fold and interact with each other in unique ways to ultimately create a three-dimensional functional protein. This unique protein structure defines the protein character, which is

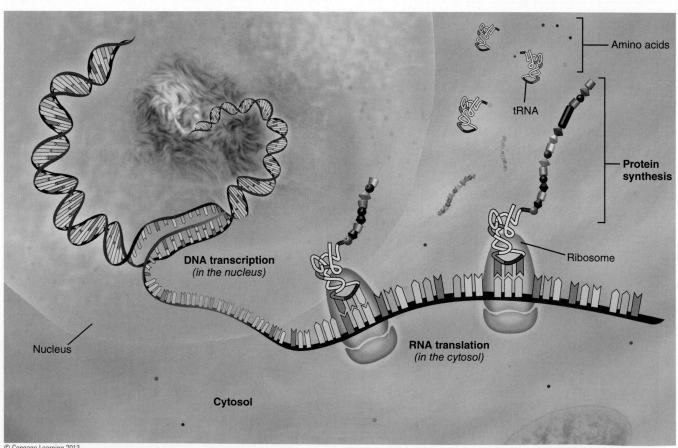

Amino acids

tRNA

Protein synthesis

Ribosome

DNA transcription
(in the nucleus)

RNA translation
(in the cytosol)

Nucleus

Cytosol

BioBeat 3.5

Protein Synthesis (DNA to RNA to Protein)

DNA and RNA are **nucleic acids.** DNA is made of four nucleotide bases: adenine (A), cytosine (C), guanine (G), and thymine (T). These bases are organic compounds containing carbon, hydrogen, oxygen, nitrogen, and phosphorus. They are made of a nitrogenous base, five-carbon sugar, and phosphate constituents. They form into a double-stranded DNA helix that has sugar and phosphate units attached. From the double strand of the DNA, one strand forms the "sense" sequence, and the other strand forms the "anti-sense" sequence. In the DNA double helix, A always pairs with T, and C always pairs with G. For example:

Sense	Antisense
A	T
T	A
G	C
C	G

DNA is converted to mRNA through a process called *transcription.* In mRNA, the nucleic acid uracil replaces thymine. Uracil pairs with adenine. mRNA is used to order the amino acids in the protein through a process called *translation.* Typically, three base pairs of RNA are read at a time to encode an amino acid. For example, AAG calls for lysine. As the mRNA is translated, the amino acids are linked together, forming a polypeptide and eventually a protein. Once the protein is formed, it can perform its function within the cell. DNA, RNA, and protein are considered **biomolecules.**[57]

What is the fate of the amino acids in the proteins you eat?

FIGURE 3.13 The primary, secondary, tertiary, and quaternary structure of the protein hemoglobin, found in red blood cells, which contains the mineral iron in the heme group to bind oxygen and deliver it to all body cells. Hemoglobin is a globular protein; it is part random and part alpha-helix.

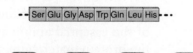

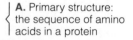
A. Primary structure: the sequence of amino acids in a protein

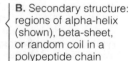
B. Secondary structure: regions of alpha-helix (shown), beta-sheet, or random coil in a polypeptide chain

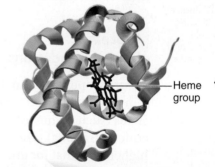
Heme group

C. Tertiary structure: overall three-dimensional folding of a polypeptide chain

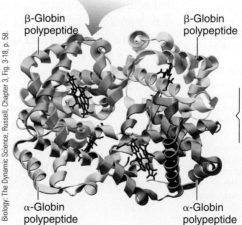
β-Globin polypeptide

β-Globin polypeptide

α-Globin polypeptide

α-Globin polypeptide

D. Quaternary structure: the arrangement of polypeptide chains in a protein that contains more than one chain

ultimately determined by the amino acid sequence in the protein. Proteins have many chemical functions as receptors, enzymes, and hormones. Their ability to participate in their unique role in the body depends on their three-dimensional structure. Proteins begin with a primary amino acid sequence that dictates each protein's three-dimensional secondary, tertiary, and even quaternary structure (see Figure 3.13).

Types of Protein

To better understand the types of proteins that make up lean body mass and function in the body, it is important to know about protein types. There are two basic types of protein: fibrous and globular. A **fibrous protein** is uniform in its structure. Its formation is either exclusively helical or in a pleated sheet. This means that the amino acids in the polypeptide strand fold and interact together to form a uniform three-dimensional structure. Fibrous protein three-dimensional structures can look like a slinky (alpha-helix) or accordion (beta-sheet). Some examples of fibrous proteins are the proteins found in hair and fingernails.

A **globular protein** has variation in its structure. It can be part helical, part sheet, part random, or completely random, meaning there are no consistent segments in the protein structures. Some examples of globular proteins in the body include proteins found in blood and mucus.

Protein Functions

Proteins and their amino acid building blocks are needed to build cellular proteins for many important body functions. The functions of protein include growth and maintenance of lean body mass (tissue turnover, maintenance, replacement, and deposition); creation of enzymes (chemical catalysts); creation of **antibodies** (immune agents); fluid and electrolyte balance (osmolarity); acid-base (pH) balance; creation of protein hormones like insulin and glucagon; nutrient transport in the blood and lymphatic vessels (lipoproteins); and release of energy (4 Calories per gram, which requires nitrogen removal or deamination [removal of an amine group from an amino acid] and increases the synthesis of urea, a nitrogenous waste product).

Each day, new protein needs to be consumed to replace the protein that was lost from the body in vital functions. Protein replacement is accomplished by the consumption of animal and plant proteins. Because most animal proteins have all the essential amino acids present in optimal amounts and ratios, they provide complete, high-quality, or high biological value proteins. Plant proteins, on the other hand, provide incomplete, low-quality, or low biological value proteins, because they are lacking one or more of the essential amino acids in their composition.

The quality of the protein is only as good as the most limiting essential amino acid. Egg white is the highest-quality food protein for humans, because the ratios of the essential amino acids in combination with all the amino acids present in the protein match human needs. When a protein is incomplete, it can be complemented with another incomplete protein to reach the equivalent of a complete protein. Complementary proteins occur by combining plant-based groups such as grains with legumes (see page 18 in Module 1).

Protein Recommendations

Both complete and incomplete proteins provide needed amino acids and other nutrients for the body. The Dietary Reference Intake (DRI) for protein for adults is 0.8 grams per kilogram of body weight. However, there are times when more protein may be needed. For example, the body uses more protein while in anabolic states (such as during childhood growth, pregnancy, or body building and rigorous athletic training). The DRI for protein for infants, children, and adolescents is set higher than that for grown adults (see Appendix A for values). For athletes needing muscle growth and/or tissue repair from bouts of exercise, approximately 1.2 to 1.6 grams per kilogram of body weight (high-quality dietary protein) can be consumed to support repair of lean body mass. Endurance athletes can safely consume up to 1.8 to 2 grams per kilogram of protein due to extended exercise stress on the body. The values given for adult athletes are not DRIs. For all adults, the Acceptable Macronutrient Distribution Range (AMDR) for protein is 10 to 35 percent of total Calories. However, when the dietary intake of protein exceeds twice the DRI amount of an individual, it is considered a high-protein diet and is not healthy.[136]

Protein Deficiency and Excess

When protein intake is below the DRI and energy intake is adequate, protein deficiency, or protein malnutrition, can occur. This state of protein deficiency is called **kwashiorkor**. Kwashiorkor, a term derived from Africa,

nucleic acids The building blocks of genetic material (DNA and RNA) used by all cells to divide, differentiate, or synthesize protein constituents; also called *nucleotides*.

biomolecule Organic, biological molecules produced by a living organism that make up cells, tissues, organs, and systems and include proteins, carbohydrates, lipids, and nucleic acids.

fibrous protein Proteins that are uniform in their structure; either in a helical structure or a pleated sheet.

globular protein Proteins with inconsistent and loosely interacting protein strands in their structure; the protein conformation can be part helical structure, part pleated sheet, part random, or completely random.

antibodies Proteins that are produced by the immune system in response to antigens, which are recognized as foreign substances that invade the body.

kwashiorkor Protein malnutrition resulting when a person consumes adequate Calories but an inadequate amount of protein, which is required to sustain growth and/or repair vital tissues.

AFP/Getty Images

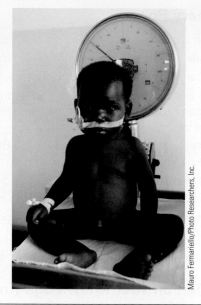

Mauro Fermariello/Photo Researchers, Inc.

FIGURE 3.14 Marasmus (left) and kwashiorkor (right).

means "the one who is displaced" and refers to the development of the condition after a sibling is born and the older child is weaned from the breast. If the older child then is placed on a diet that is inadequate in protein, the signs and symptoms of kwashiorkor (peripheral **edema**, protruding belly, decreased muscle mass, hair and skin changes, irritability, and lethargy) emerge (see Figure 3.14). Edema is a common sign of kwashiorkor. One of the functions of protein is to regulate fluid balance. With low protein status, fluid shifts from the blood stream to the extremities, causing peripheral edema.[10]

When both protein and energy (Calorie) intake are inadequate, **protein-energy malnutrition**, also called **marasmus**, can occur. Marasmus is characterized by total body wasting (see Figure 3.14). Both kwashiorkor and marasmus occur primarily in undeveloped countries. In developed countries, individuals who are on starvation diets, poor, abused, or in hypermetabolic states can experience marasmus. In severe cases, refeeding the individual is a delicate matter. The person cannot be immediately given a nutritionally balanced diet for protein. The unaccustomed ingestion of protein would cause fluid to shift too rapidly into the blood stream and lead to sudden death. Refeeding is gradual and progressive over several weeks to months until a healthy intake is established. Depending on the duration and severity of the deficiency, permanent damage may have occurred.

The intake of high-protein diets causing **protein excess** is most common in some athletes and fad dieters. Consuming excess dietary protein has negative health consequences. High-protein foods tend to be from animal sources that are also high in fat and cholesterol. Though high-protein diets have been touted to promote weight loss, the success of losing weight and keeping the lost weight off for more than two years from this type of dietary pattern is poor.[62,78,118,127,146] Further, this style of eating doesn't promote the adoption of a healthy diet, because it is limited in fruits, vegetables, whole grains, and low-fat dairy products or alternatives. Thus, there is a tendency in the long run for high-protein diets to promote obesity, heart disease, **cancer**, and osteoporosis.[30,32,95,120,127,147,156] Other health risks from consuming high amounts of protein in excess include dehydration, calcium and zinc loss, liver and spleen enlargement, and reduced liver and kidney function. Scientific studies also indicate that chronic protein overload accelerates kidney aging, and, when coupled with low carbohydrate intake, can cause metabolic acidosis. Interestingly, vitamin B_6 is needed to process protein, so high-protein diets may increase the chance for developing a vitamin B_6 deficiency.[17,20]

edema Fluid retention by body tissues as a result of protein deficiency (kwashiorkor) and other conditions that cause excessive amounts of fluid to be retained in interstitial (in between cells) spaces.

protein-energy malnutrition A condition known as *marasmus*, which results from dietary deficiencies of both protein and energy (Calories).

marasmus Protein-energy malnutrition resulting from both protein and energy (Calorie) deficiencies.

protein excess The intake of too much protein, causing dehydration, increased calcium and zinc excretion, liver and spleen enlargement, and long-term reduced liver and kidney function.

cancer A disease in which cells become engaged in uncontrolled cellular division and growth through the processes of initiation, promotion, and progression.

Summary Points

- Each day, amino acids need to be consumed from protein.
- There are high-quality (animal) and low-quality (plant) protein sources.
- Dietary protein is denatured and then digested.
- The amino acids from dietary intake are used by cells to make proteins by converting DNA to RNA in the processes of DNA transcription, RNA translation, and protein synthesis.
- Protein character is determined by the protein's amino acid sequence and three-dimensional structure.
- Proteins can be of two basic types: fibrous and globular.
- Proteins have many functions in the body.
- Protein deficiency is called kwashiorkor.
- Protein-energy malnutrition is called marasmus.
- Protein excess can led to negative health effects.

Take Ten on Your Knowledge Know-How

1. How are dietary proteins denatured and digested?
2. What is the role of DNA transcription and RNA translation in protein synthesis?
3. Explain how amino acids become integrated into your body proteins through protein synthesis.
4. How is protein character determined?
5. What are fibrous and globular proteins?
6. How do proteins function in your body?
7. How much protein should you consume in your diet, according to the DRI? What percent of Calories should come from protein in your diet?
8. What happens when your diet is deficient in protein? What is the name for this condition?
9. What health risks are associated with chronically consuming excess protein?
10. Are you consuming a healthy amount and type of protein? What can you do to improve your intake level and food choices if needed?

3.3 PHOTOSYNTHESIS AND FIBER

T-Talk 3.3

To hear Dr. Turley talk about photosynthesis and fiber, go to www.cengagebrain.com

photosynthesis The process by which green plants use the sun's energy to make carbohydrates from carbon dioxide and water.

chlorophyll Large, magnesium-containing molecules that can be seen as green pigments found in life forms capable of photosynthesis.

Introduction

In this section, we will discuss how carbohydrates are made (synthesized) through a process of carbon fixing in plants. We will look at the variety of carbohydrate molecules, some of which are digestible (starch and sugars) and some that are indigestible (cellulose and other fibers). We will see that consuming adequate amounts of the indigestible carbohydrate (fiber) has many health benefits.

Photosynthesis

Carbohydrates, including fiber, are found mostly in plant foods and are made by a process referred to as **photosynthesis** (see Figure 3.15 and BioBeat 3.6). Photosynthesis is the process of carbon fixing via this simple formula: Carbon dioxide plus water plus sunlight energy makes carbohydrate and water. This process takes place inside one of the plant cell organelles called the chloroplast and requires **chlorophyll**, the plant pigment that captures electromagnetic sunlight energy to make carbohydrate molecules, including sugars, starches, and fibers, from carbon dioxide and water.[29] (See Module 1, T-Talk 1.2, starting on page 11 for a review.)

Fiber

When plant foods are consumed, not only are a variety of nutrients consumed, but fiber is as well. Fiber consists of many glucose units linked together by beta chemical bonds. It is a type of complex carbohydrate. Because humans do not produce digestive enzymes to break the beta bonds in fiber, fiber is an indigestible, noncaloric (0 Calories per gram) polysaccharide.

It is important to know how much fiber a food item contains, so that the dietary intake can be appropriately planned for adequate intake. Presently, dietary and total fiber values are determined in foods. When either isolated or synthetic fiber is directly added to a processed food, the amount added is predetermined, and thus the quantity is known. Some fiber types have been termed **functional fiber** because of their beneficial physiological effects. Let's discuss dietary, functional, and total fiber in more detail.

Dietary Fiber Dietary fiber is the edible, indigestible components of carbohydrates and **lignin** naturally found in plant foods. The most common dietary fiber consumed is cellulose. Dietary fiber is measured as the residue remaining after digestion, absorption, and excretion has occurred *in vivo*. *In vivo* means "tested in a living system." In this case, an animal is fed a known quantity of a plant food, and the indigestible carbohydrate residue is measured after excretion. So the amount of dietary fiber present in food is the *in vivo* measurement of indigestible material. This measurement for fiber is used to label the grams of fiber per serving in human foods.[67,73]

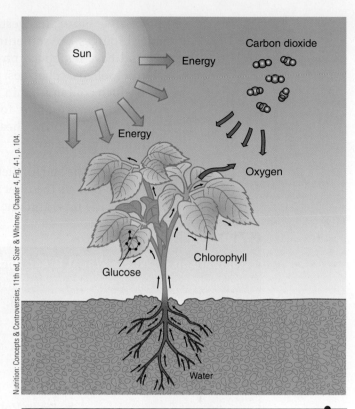

Nutrition: Concepts & Controversies, 11th ed, Sizer & Whitney, Chapter 4, Fig. 4-1, p. 104.

FIGURE 3.15
Photosynthesis is the process by which plants make carbohydrates.

BioBeat 3.6

Photosynthesis

Some organisms and green plants have the ability to convert light energy absorbed by chlorophyll into chemical energy through the process of photosynthesis. This process requires chlorophyll, carbon dioxide, water, and sunlight. Carbohydrate chemical structures are produced, and oxygen is released. Carbon fixing is an anabolic chemical reaction. Specifically, $6CO_2 + 12H_2O + $ sunlight energy $\rightarrow C_6H_{12}O_6 + 6H_2O + 6O_2$. The stored chemical energy in carbohydrate, when digested, is then available to the organism itself, as well as to other organisms that can degrade the organic material produced in the photosynthetic organism (e.g., a human eating spinach). Photosynthesis produces simple carbohydrates and complex carbohydrates, including fibers. An added bonus is that all animals and aerobic organisms can use the oxygen that plants generate while making carbohydrates.[29]

How does photosynthesis impact your well-being?

functional fiber Indigestible dietary components that have been isolated from natural sources or synthetically made, and have beneficial physiological effects in humans.

lignin An insoluble, indigestible organic molecule that holds cellulose fibers together in plants.

in vivo Refers to studies conducted, or any processes that occur, in living organisms.

Functional Fiber Indigestible carbohydrates, either synthetic or isolated from natural sources, are sometimes added as ingredients to processed foods (see Table 3.2) and have beneficial physiological effects in humans. Examples are cellulose gel, fructooligosaccharides, psyllium, inulin, xanthan gum, guar gum, or flax seed meal, which contains lignin. Adding functional fiber to the diet in processed foods provides a source of fiber to the individual as well as health benefits; for example, psyllium may improve heart health, and fructooligosaccharides may improve gastrointestinal health.[71,73]

Total Fiber The sum of dietary plus functional fiber in food is the total fiber value that is reflected as the fiber content on food package labels in the Nutrition Facts panel. It is the total amount of indigestible food present per serving.

TABLE 3.2
Comparison of Whole-Wheat Bread and High-Fiber White Bread
The nutrition facts are shown, and the functional fibers added to boost the fiber content are highlighted in the ingredient list.

Nutrition Facts: Whole-Wheat Bread		Nutrition Facts: High-Fiber White Bread	
Serving size: 1 slice		**Serving size:** 2 slices	
Servings per container: 20		**Servings per container:** 10	
Calories: 80 **Calories from fat:** 10		**Calories:** 108 **Calories from fat:** 4	
	% Daily Value		**% Daily Value**
Total fat: 1 g	2%	**Total fat:** 0.5 g	1%
Saturated fat: 0 g	0%	**Saturated fat:** 0 g	0%
Trans fat: 0 g		**Trans fat:** 0 g	
Cholesterol: 0 mg	0%	**Cholesterol:** 0 mg	0%
Sodium: 170 mg	7%	**Sodium:** 240 mg	10%
Total carbohydrate: 14 g	5%	**Total carbohydrate:** 18 g	6%
Dietary fiber: 2 g	8%	**Dietary fiber:** 5 g	20%
Sugars: 2 g		**Sugars:** 1 g	
Protein: 4 g		**Protein:** 8 g	
Vitamin A 0% **Calcium** 4%		**Vitamin A** 0% **Calcium** 10%	
Vitamin C 0% **Iron** 4%		**Vitamin C** 0% **Iron** 8%	
Simplified ingredients: Whole-Wheat Flour, Water, Wheat Gluten, High-Fructose Corn Syrup, Yeast, Molasses, Honey, Soybean Oil, Salt, Dough Conditioners, Calcium Sulfate, Vinegar, Yeast, Wheat Bran, Wheat Starch, Enzymes, Whey, Soy Flour, Calcium Propionate (to Retain Freshness), Soy Lecithin.		**Simplified ingredients:** Enriched Wheat Flour, Water, Wheat Gluten, High-Fructose Corn Syrup, Cracked Wheat, Cottonseed Fiber, Soy Fiber, Oat Fiber, Salt, Molasses, Dough Conditioners, Soy Flour, Yeast, Wheat Starch, Cornstarch, Cellulose Gel, Whey, Calcium Sulfate, Enzymes, Vinegar, Caramel Color, Calcium Propionate, Soy Lecithin.	
Contains wheat, milk, and soybeans		*Contains wheat, milk, and soybeans*	

© Cengage Learning 2013

soluble fibers Indigestible, mostly plant-derived dietary components that soften in water, including pectins, gums, and mucilages.

insoluble fibers Indigestible, mostly plant-derived dietary components that do not soften in water, including cellulose, hemicelluloses, and lignin.

Categories of Fiber There are many types of fiber, with varying chemical structures that affect the body in different ways (see the section called "Benefits and Actions of Fiber," page 122). Though there are different types of fibers found in whole plant foods, they have an overarching property that causes them to be categorized as soluble or insoluble. **Soluble fibers** soften and gel in water, whereas **insoluble fibers** do not soften in water but do attract water (see Table 3.3).[6,33]

TABLE 3.3

Some Dietary Fiber Facts

Category	Solubility	Fibers	Common Food Sources
Soluble	Soften and gel in water	Pectins, gums, mucilages	Fruits (like apple pectin), vegetables, legumes, and oats (e.g., oatmeal, cereals)
Insoluble	Do not soften or gel in water; do attract water	Cellulose[1], hemi-cellulose, lignins	Whole-grain foods, celery strings, apple peels

[1]Cellulose is the most common type of fiber.

© Cengage Learning 2013

Fiber Recommendations How much fiber is needed in the daily diet for optimal health and functioning? There are standard adult DRIs for fiber. The general DRI for total fiber intake is 38 grams for an adult male and 25 grams for an adult female. However, these values are not personalized. A more accurate calculation of an individual's fiber DRI is derived based on that person's daily Calorie intake. This personal DRI for fiber is 1.4 grams per 100 Calories consumed. It is important to understand that the fiber DRI is based on actual Calorie intake, not Calorie need.[67]

Let's put this concept into action by looking at an example. How much fiber does a person eating 4,200 Calories per day need? You can calculate the answer using this formula:

$$\text{Calories consumed} \div 100 \times 1.4$$

The answer would be:

$$4,200 \text{ Calories} \div 100 \times 1.4 = 58.8 \text{ g}$$

This gives approximately 59 grams as the personalized DRI for fiber for this Calorie intake level. If another person eats 1,430 Calories, then the fiber need would be:

$$1,430 \text{ Calories} \div 100 \times 1.4 = 20 \text{ g}$$

Data GENEie   Solah consumed 2,321 Calories and 17 grams of dietary fiber today. What is her DRI for fiber based on this Calorie intake? What percentage of her DRI did she consume for dietary fiber? What health implications can she expect to experience based on her dietary fiber intake?

To obtain the health benefits from consuming adequate fiber, high-fiber foods should be incorporated into the daily diet. A high-fiber food contains more than 2 grams of fiber per serving. In general, whole plant foods (legumes, grains, fruits, and vegetables) or products made with whole plant foods provide a good source of fiber. Examples of fiber-containing foods are provided in Table 3.4 and are shown by fiber category type in Figure 3.16. Adequate-fiber diets meet the fiber DRI of 1.4 grams per 100 Calories consumed, and high-fiber diets exceed the fiber DRI and provide at least 2 grams of fiber per 100 Calories.

TABLE 3.4

Where's the Fiber?

Food Group	Very High (> 4 g)	High (2–4 g)	Good (1–2 g)	Low (≤ 1 g)
Grain	½ cup bran flakes 1 cup shredded whole wheat or whole multi-grain cereals	1 cup oatmeal or puffed brown rice cereal 1 slice whole-wheat bread	1 slice rye bread ½ cup brown or wild rice 1 corn tortilla	1 cup corn flakes ½ cup white rice ½ cup pasta
Vegetable	½ cup legumes (dried beans)	½ cup broccoli, cauliflower, corn, beans, cabbage 1 ounce nuts/seeds	½ cup carrots, green pepper, celery, onion, lettuce	1 cup some vegetable juices
Fruit	N/A	1 apple, banana, orange, peach 1 cup berries 2 prunes	½ cup melon flesh	1 cup fruit juice

© Cengage Learning 2013

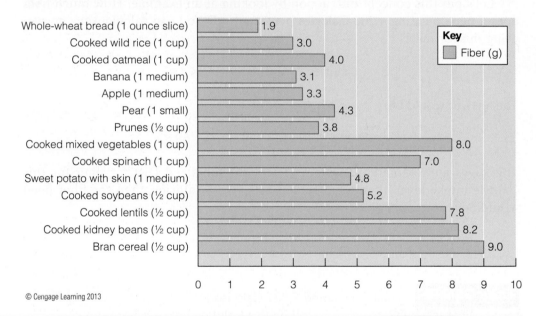

FIGURE 3.16 Food sources of dietary fiber.

© Cengage Learning 2013

There are many benefits from having an adequate fiber intake; however, as with many things, too much of a good thing can be bad. Though a Tolerable Upper Intake Level for fiber has not been established, there are negative effects associated with too much fiber in the diet. We will next discuss the benefits of adequate fiber and the negative effects of too much fiber.

Benefits and Actions of Fiber There are many benefits derived from consuming dietary fiber; several are highlighted here. Both soluble and insoluble fiber add *bulk* to the diet and increase stool volume. All types of fiber increase the volume of chyme through the GI tract. Also, high-fiber foods have *decreased caloric density*. In other words, you can eat larger amounts of food with fewer Calories absorbed. For example, having a slice of high-fiber bread with a glass of water or eating a salad with nonfat dressing to begin a meal will help reduce overall Calorie and food intake.

Dietary fiber is also a *stool softener*. The carbohydrate-related molecules that make up fiber associate strongly with water molecules, creating a softer stool that is easier to move through the GI tract, and is exceptional for relieving constipation (promoting laxation).[54] The abundant insoluble fiber cellulose is the main fiber that softens stool by attracting water to it, but all fiber types soften stool.

Fiber also *decreases transit time*. This means fiber decreases the amount of time food spends in the GI tract. Soft stool moves more quickly through both the small and large intestines. The exposure time of the large intestine cells to potential carcinogens in the feces is decreased, so the risk for colorectal cancer may be decreased. All types of fiber have this benefit.

A high-fiber diet also *improves the tone of GI tract muscles*. Peristalsis is the circular wave action that propels food through the GI tract. Because it retains water, fiber softens waste and increases the volume of chyme passing through the GI tract, which allows the muscles of the GI tract to exercise better, resulting in improved peristaltic action. Having strong muscles decreases the risk of having GI tract diseases like **diverticular disease**. All types of fiber have this benefit.

Unique to soluble fiber types is its molecular affinity to *bind fat and cholesterol* and thus promote heart health. The fiber in oat bran and legumes (soluble fiber) is particularly effective in binding the cholesterol-rich bile salts and also undigested breakdown products of fats that may have the potential for carcinogenesis. Besides the possibility of reducing the risk for colon cancer, soluble fiber reduces blood or **serum cholesterol** levels and can be beneficial in the prevention of heart disease.

Finally, soluble fiber *increases gastric emptying time*. This means fiber increases the time chyme spends in the stomach. Soluble fiber slows the rate of glucose absorption, thus the blood sugar level after eating is lower and may be more favorable for individuals with diabetes and **metabolic syndrome**.[6,7,32,39,52,73,84,100,101,113]

Negative Effects of Too Much Fiber Although fiber is beneficial in many ways, too much fiber has unpleasant effects; for example, an excess of fiber can cause *gas and bloating*. Gas and bloating are signs of the negative effects of eating a high-fiber diet suddenly. Bloating occurs because of gas produced when bacteria in the colon decompose or catabolize fiber molecules. To avoid this unpleasant side effect, a person should transition slowly to a high-fiber diet by gradually replacing low-fiber foods with higher-fiber foods.

A diet too high in fiber can cause *frequent bowel movements*. Having large, soft stools and frequent bowel movements is usually a benefit, but when fiber intake is too high, waste volume and frequency may become excessive and unhealthy. There is no real definition for normal bowel movements, as this is a personal thing—usually once or twice a day for most people—but anything from more than two per day and less than one per two days is probably approaching abnormal, causing the person to feel discomfort associated with the frequency.

Another drawback of consuming too much fiber is the *binding of positively charged minerals*. Certain fiber types can bind to and prevent the absorption of positively charged minerals like calcium, magnesium, zinc, copper, and iron. This effect has been shown in animals, but not humans. Also, certain fibers can *bind beta-carotene and other carotenoids*, blocking the absorption of beta-carotene (provitamin A).

The decreased Caloric value associated with high-fiber diets may cause *Calorie deficiency*. Fiber decreases energy intake because of the reduced caloric density. For an overweight adult, this characteristic of fiber is a benefit, but for an undernourished child, it is a drawback. For the very young and the very old, it may be difficult to achieve energy balance if dietary fiber intake is too high.

Lastly, a lack of sufficient water consumption with a high-fiber intake can cause *intestinal blockage*. This is especially true with encapsulated fiber supplements.

diverticular disease
Diverticulosis, a condition where diverticula (out-pocketing) are present in the colon as a result of weak colon wall structures and increased pressure against them; when the diverticula are inflamed, it is termed *diverticulitis*.

serum cholesterol The level of total cholesterol in the blood stream, including mostly LDL and HDL; it is at a desirable level when it is less than 200 mg/dL.

gastric emptying time The amount of time it takes chyme to pass through the stomach.

metabolic syndrome A lifestyle disease where the diagnosis is made by having three or more of the five following metabolic risk factors: central adiposity, hypertension, hyperglycemia, hypertriglyceridemia, or low HDL cholesterol.

FIGURE 3.17 Basic components of a wheat plant and wheat kernel.

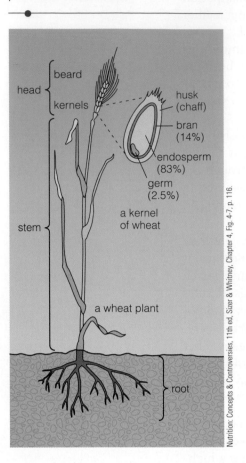

Nutrition: Concepts & Controversies, 11th ed, Sizer & Whitney, Chapter 4, Fig. 4-7, p. 116.

Encapsulated fiber dietary supplements intended to promote weight loss have been known to cause blockages of the gastrointestinal tract. Such blockages usually require medical intervention.

Effect of Whole-Grain Processing on Fiber Grains are the seeds of grass-type plants and are also consumed as foods or used to make food products like bread, cereal, and pasta. Some examples of whole grains that can be processed or eaten in their whole state include wheat, barley, rye, oats, corn, and rice. Some other grains eaten as staple foods in various parts of the world include amaranth, buckwheat, millet, quinoa, spelt, and teff. The kernel of grain is protected by a husk that is always removed before marketing. So a whole-grain or whole-grain–containing food product has the bran, germ, and endosperm of the grain (see Figure 3.17). When a whole grain is refined, the bran layer containing fiber and the germ layer containing vitamins, minerals, and fatty acids are removed. This leaves the endosperm for use in refined-grain food products. The endosperm contains starch and protein, which are both needed for baked-food product quality, but it has far less fiber, vitamins, and minerals.

When whole grains are refined, the nutrient content of the food product is diminished. Because of this loss of essential nutrients, a law, the Enrichment Act of 1942, was passed with the goal of improving the nutritional quality of refined grains. Nutrients that have been added back by law in the **enrichment** process include iron, thiamin, riboflavin, and niacin. As of January 1, 1998, folic acid has also been added back into refined bread and cereal products. These nutrients are added back to enriched products at levels close to those present in the whole grain.[96,114,121]

It is important to realize that whole-grain products are still more nutritious than enriched products, because not all of the nutrients lost in the grain refinement process are added back into processed foods; vitamin B_6, magnesium, and zinc are not included in the enrichment process. Further, fiber content is greatly diminished by refining whole grains. Thus, fewer nutrients are obtained from consuming the refined food product than from the whole, unrefined food. A refined product with functional fiber added to the processed food, such as soy fiber added as an ingredient to refined wheat bread or white bread that has added fiber and calcium, are examples of refined-grain products with fiber and nutrients added back (recall Table 3.2). To compare the nutritional differences between whole-grain bread and enriched white bread, see Figure 3.18.

Summary Points

- Plants make carbohydrates via photosynthesis.
- Carbon dioxide, water, and sunlight are required for photosynthesis.
- Fiber is noncaloric.
- Categories of fiber are soluble and insoluble.
- Total fiber includes functional and dietary fiber.
- The DRI for fiber is 1.4 grams per 100 Calories eaten.
- There are health benefits for adequate fiber intake.
- There are negative effects from eating too much fiber.
- Whole plant foods provide the best source of fiber and nutrients compared to refined food products.

enrichment The term associated with the process of adding iron, thiamin, riboflavin, niacin, and folic acid (but not vitamin B_6, magnesium, zinc, or fiber) back to processed grains at approximate levels of the original whole grain.

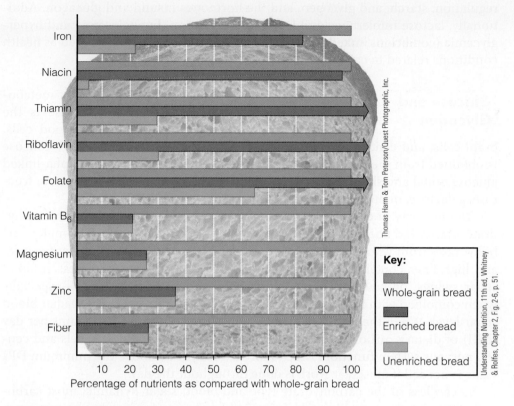

Key:

- Whole-grain bread
- Enriched bread
- Unenriched bread

Percentage of nutrients as compared with whole-grain bread

(Iron, Niacin, Thiamin, Riboflavin, Folate, Vitamin B$_6$, Magnesium, Zinc, Fiber)

Thomas Harm & Tom Peterson/Quest Photographic, Inc.

Understanding Nutrition, 11th ed, Whitney & Rolfes, Chapter 2, Fig. 2-6, p. 51.

FIGURE 3.18 Nutrients in whole-grain, enriched white, and unenriched white breads.

Take Ten on Your Knowledge Know-How

1. How does photosynthesis translate to edible carbohydrates that you consume?
2. What are the categories of fiber? What are the fiber types in each category?
3. What are food sources of the various fiber types?
4. What is your personalized DRI for dietary fiber? What is the calculated DRI for a person who consumes 1,500 Calories in one day?
5. What are the health benefits of eating adequate dietary fiber?
6. Do you think that your dietary fiber intake is helping you achieve health benefits? If so, why? If not, what food choice changes could you make to bring your fiber intake to a beneficial level?
7. What are some negative effects of consuming too much dietary fiber?
8. What happens to whole grains when they are refined?
9. What happens to the nutritional value of the food made with the refined processed grain? Which nutrients are required by law to be added back to refined grains?
10. Does your diet consist of consuming whole grains or refined enriched grains? What are the health benefits or health implications of your grain choices?

3.4 CARBOHYDRATE STORAGE AND DISORDERS

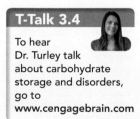

T-Talk 3.4

To hear Dr. Turley talk about carbohydrate storage and disorders, go to www.cengagebrain.com

Introduction Glucose is a very important carbohydrate molecule. The majority of six-carbon sugars consumed in the diet are glucose, and the concentrations of glucose in the blood are tightly regulated. In this section we will focus on blood sugar (also called blood glucose), blood sugar regulation, starch, and glycogen, and the hormones insulin and glucagon. Additionally, lactose intolerance and lactose maldigestion, hypoglycemia and hyperglycemia (conditions linked to diabetes), and diabetes will be discussed as health conditions related to carbohydrate intake.[33]

Glucose and Glycogen

The body prefers to use glucose for energy metabolism.[57] Under normal circumstances, glucose is the only energy-producing nutrient that red blood cells, brain cells, and central nervous system cells can use to generate ATP. Glucose is obtained from dietary carbohydrate sources, including starches (alpha-linked glucose units) and simple sugars (glucose and fructose, or glucose linked to fructose, galactose, or another glucose).

The majority of carbohydrates consumed should be from complex carbohydrate (starch and fiber) sources as opposed to simple sugar sources. Complex carbohydrates from whole foods are ideal, because they have higher nutrient density and high fiber content (when eaten as unrefined foods), and they lessen blood sugar swings, or highs and lows. Basically, complex carbohydrates—especially when combined with fiber, protein, and some fat—help maintain a steadier blood sugar level. When the body is at rest, a minimum amount of 130 grams per day (DRI) of dietary carbohydrate is needed to support blood sugar levels and central nervous system function. Much more carbohydrate above the minimum DRI value is needed if muscular work is done throughout the day.

Regardless of the carbohydrate type and blood sugar dynamic, most carbohydrate molecules are ultimately converted to glucose for energy production in the body. Because not all glucose obtained from dietary carbohydrate intake is immediately needed for energy production, some can be stored in the liver and muscle in the form of glycogen, usually for a few hours in the liver and a few days in the muscle.

Glycogen, known as animal starch, is a complex carbohydrate because it is made up of many glucose units. It differs from plant starch in that it is a highly branched molecule (see Figure 3.19). Glycogen is synthesized and stored in liver and muscle cells in animals. Liver glycogen is used primarily to regulate blood sugar levels that support the brain, central nervous system, and red blood cell functioning, whereas muscle glycogen is used to meet the energy needs of high-intensity muscular work. The liver can store a maximum of 100 grams (400 Calories) of glycogen. To achieve this level, adequate amounts of carbohydrate need to be consumed, preferably from healthy food sources, throughout the day.

Greater amounts of carbohydrate intake are required to support an active lifestyle, especially for the working muscle. The muscle has its own store of glycogen to draw on during high-intensity muscular work. Muscle glycogen storage levels are influenced by diet and exercise. Muscle glycogen levels can range from 1 to 4 grams of glycogen per 100 grams of muscle tissue, depending on the amount of dietary carbohydrate eaten after the amount of exercise. To achieve the higher end of muscle glycogen storage for better physical performance, the muscle needs to be stimulated with daily moderate exercise, and dietary carbohydrate intake needs to range between 60 and 65 percent of Calories consumed.

Recall that blood sugar levels are tightly regulated to ensure brain function and respiration. Because red blood cells and nervous tissues are using glucose constantly, glucose needs to be constantly put back into the blood stream. Liver glycogen is slowly degraded upon hormonal signaling in between meals to maintain normal blood sugar levels and normal functioning of the brain, the central nervous system, and red blood cells. The normal fasting blood sugar range is 70 to 99 milligrams of glucose per deciliter of blood (mg/dL). This level is regulated by hormonal control.

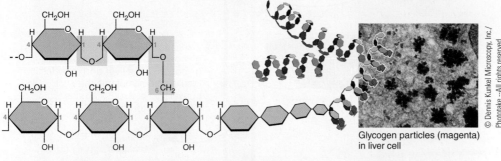

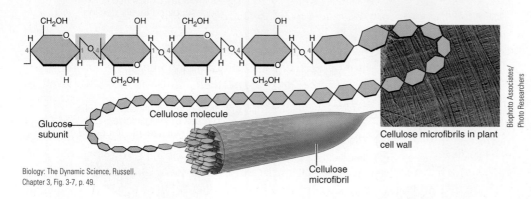

Amylose grains (purple) in plant root tissue

Glycogen particles (magenta) in liver cell

Glucose subunit

Cellulose molecule

Cellulose microfibril

Cellulose microfibrils in plant cell wall

Biology: The Dynamic Science, Russell. Chapter 3, Fig. 3-7, p. 49.

FIGURE 3.19 A comparison of the chemical structure of amylose (a plant starch), glycogen (animal starch), and cellulose (a plant fiber).

BioBeat 3.7

Blood Glucose Regulation by Insulin and Glucagon

Normally, human blood glucose (sugar) levels are maintained between 70 and 99 milligrams of glucose per 100 milliliters (one deciliter [dL]) of blood (70 to 99 mg/dL). Two regulatory hormones, insulin and glucagon, regulate blood glucose levels. Blood glucose levels rise rapidly after the ingestion, digestion, and absorption of carbohydrate. The rise in blood glucose causes insulin to be secreted. Insulin is a protein hormone produced by the beta cells of the pancreas and secreted into the blood stream in response to elevated blood sugar levels. Insulin binds to insulin receptors on liver and muscle cells. You can

think of insulin as the key and the insulin receptor as the lock. When the key goes into the lock, the permeability of the cell membrane changes to allow blood glucose to enter the cell. Cells use the glucose for immediate energy production or to replenish glycogen stores. Glycogen repletion or synthesis is an anabolic process. Insulin can cause blood glucose levels to drop very rapidly, while the target cells are being flooded with glucose. Conversely, when carbohydrate has not been eaten for an extended period and liver glycogen stores are becoming low, blood glucose levels drop slowly. In response to the decreasing

blood glucose levels, glucagon, a protein hormone produced by the alpha cells in the pancreas, is secreted into the blood. Glucagon sends a catabolic signal to liver cells, which stimulates the breakdown of glycogen stores and releases glucose back into the blood stream. The maintenance of normal blood glucose levels is critical for brain, central nervous system, and red blood cell functioning. Thus, blood glucose levels are tightly regulated by insulin and glucagon.[57,93]

Which of your feeding behaviors cause your blood glucose to fluctuate the most?

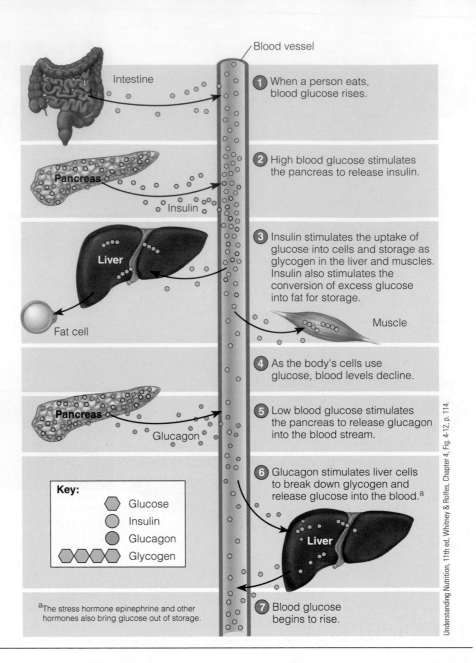

Blood vessel

Intestine

Pancreas

Insulin

Liver

Fat cell

Muscle

Pancreas

Glucagon

Key:

⬡ Glucose

⬤ Insulin

⬤ Glucagon

⬡⬡⬡⬡ Glycogen

Liver

(1) When a person eats, blood glucose rises.

(2) High blood glucose stimulates the pancreas to release insulin.

(3) Insulin stimulates the uptake of glucose into cells and storage as glycogen in the liver and muscles. Insulin also stimulates the conversion of excess glucose into fat for storage.

(4) As the body's cells use glucose, blood levels decline.

(5) Low blood glucose stimulates the pancreas to release glucagon into the blood stream.

(6) Glucagon stimulates liver cells to break down glycogen and release glucose into the blood.[a]

(7) Blood glucose begins to rise.

[a]The stress hormone epinephrine and other hormones also bring glucose out of storage.

Understanding Nutrition, 11th ed, Whitney & Rolfes, Chapter 4, Fig. 4-12, p. 114.

FIGURE 3.20 The regulation of blood glucose by the hormones insulin and glucagon.

insulin A protein hormone secreted by pancreatic beta cells in response to elevated blood glucose, which changes the cell membrane permeability to allow glucose to enter liver and muscle cells, thus lowering blood glucose levels.

glucagon A protein hormone that is secreted by pancreatic alpha cells in response to low blood glucose levels, stimulates glycogen breakdown in the liver, and causes increases in blood glucose levels from the release of glucose into the blood from the liver glycogen stores.

protein sparing When carbohydrates provide adequate energy so that the amino acids from protein can be used for tissue repair, maintenance, and growth purposes, rather than energy production.

The two main hormones involved in blood sugar (glucose) regulation are **insulin** and **glucagon**. They work in concert to ensure that blood sugar levels are optimal for body functioning, including having an energy supply for the brain, central nervous system, and red blood cells (see BioBeat 3.7 on page 127 and Figure 3.20).

When dietary carbohydrate intake is restricted, the fastest way to generate glucose internally is by degrading protein from amino acid pools, muscle tissue, and, if needed, from essential organs. The breakdown of amino acids occurs when glucose is not plentiful and fat breakdown cannot keep up with energy needs. Thus, carbohydrate can be said to be **protein sparing**, because when dietary carbohydrate intake is adequate it prevents the breakdown of protein and amino acids. Very simply put, the body can't work efficiently without carbohydrate available as the primary source of energy. Dietary carbohydrates are converted to glucose (and used for energy production to meet ATP needs), glycogen (for replenishing carbohydrate stores), and fat (given excess energy intake when glycogen stores are full).

Carbohydrate Disorders

There are many disorders related to carbohydrates. Some that will be explored include lactose intolerance, milk allergy, hypoglycemia, and diabetes.[21,93,107]

Lactose Intolerance and Lactose Maldigestion Even though drinking cow's milk throughout life is common among Americans, what species drink the mother's milk of other species? Are there any genetic signatures leading to natural selection from this behavior? Why do some people, especially those from certain cultural heritages, tolerate milk while others do not? The answers are tied to human biocultural **evolution** and how several alterations to the human genome can contribute to genetic variance at the population level (see BioBeat 3.8). The ability to digest and absorb milk is a necessity for the survival of any infant mammal, including humans. Historically, when milk consumption ends after weaning, the lactase enzyme gene is no longer required, so the gene is turned off. With this gene inactive, the consumption of milk or milk products leads to a severe gastrointestinal condition known as **lactose intolerance**.

BioBeat 3.8

The Evolution of Lactase Persistence

As late as 5000 BC, fossil evidence indicates that ancient adult Europeans were unable to digest milk. It has been estimated that 7,000 to 9,000 years ago, one of the strongest human genetic natural selection signatures took place by the origination of cattle ranching in the Middle East, making cow's milk readily available for humans to drink. **Natural selection** is the evolution of a species through genetic alterations in heritable traits that enable the species to adapt and survive in its environment. It works by having a new genetic trait that makes the species stronger or gives it a greater ability to endure. This trait is passed on to new offspring in successive generations. Being able to drink milk into and through adulthood (by keeping the lactase enzyme gene active) is advantageous because of the nutritional value and source of fluid for the body.

The Fulani and Tutsi are two cattle-ranching tribes of East Africa. They coevolved with European cattle ranchers. Yet, the two cultures have specific DNA mutations or molecular signatures enabling their lactase gene persistence and thus lactose tolerance. Several distinct single-gene mutations occurring at the population level have been tracked genetically through these and other cultures. The studies of inheritance patterns and evolution reveal varying degrees of lactase gene persistence and thus lactose tolerance. Three such common gene mutations are shown in Figure 3.21.[J,12,111,134]

How do genetics and evolution affect your tolerance of lactose?

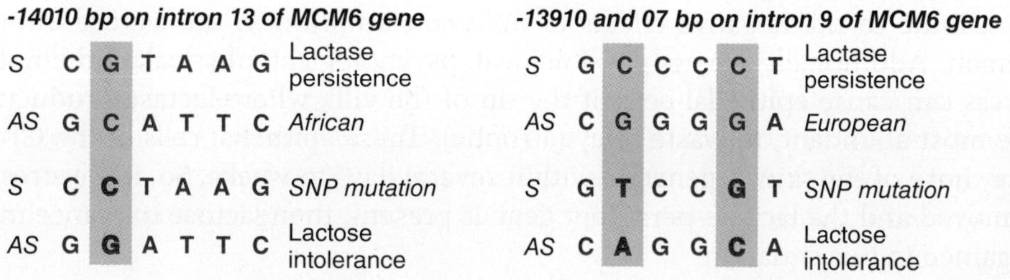

evolution The change in heritable genetic composition of a population, such as by gene mutation and as a result of natural selection, which leads to adaptation over successive generations.

lactose intolerance An inborn error of metabolism in which a person cannot make lactase and thus has no ability to digest the milk sugar lactose; signs include severe gas, bloating, and diarrhea after lactose ingestion.

natural selection The evolution of a species through genetic alterations in heritable traits that enable the species to adapt and survive in its environment over time.

FIGURE 3.21 The genetics and co-evolution of lactase persistence in East Africans and Northern Europeans, with mutations highlighted on the minichromosome maintenance 6 (MCM6) gene, which alters the sense (S) and antisense (AS) base pair (bp) sequence. When a gene sequence alteration occurs commonly in a population, it is called a single nucleotide polymorphism (SNP).

-14010 bp on intron 13 of MCM6 gene

S	C	G	T	A	A	G	Lactase persistence
AS	G	C	A	T	T	C	African
S	C	C	T	A	A	G	SNP mutation
AS	G	G	A	T	T	C	Lactose intolerance

-13910 and 07 bp on intron 9 of MCM6 gene

S	G	C	C	C	C	T	Lactase persistence
AS	C	G	G	G	G	A	European
S	G	T	C	C	G	T	SNP mutation
AS	C	A	G	G	C	A	Lactose intolerance

© Cengage Learning 2013

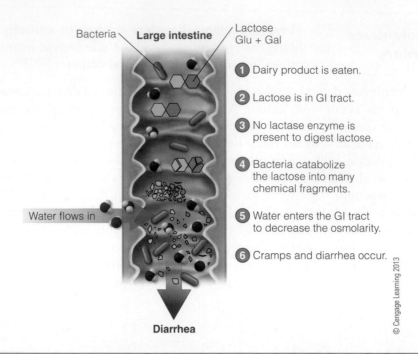

Bacteria **Large intestine** Lactose Glu + Gal

1. Dairy product is eaten.

2. Lactose is in GI tract.

3. No lactase enzyme is present to digest lactose.

4. Bacteria catabolize the lactose into many chemical fragments.

Water flows in

5. Water enters the GI tract to decrease the osmolarity.

6. Cramps and diarrhea occur.

© Cengage Learning 2013

Diarrhea

FIGURE 3.22 Osmotic diarrhea results from lactose intolerance.

lactose maldigestion A condition in which a person has a limited ability to make lactase and thus has a compromised ability to digest the milk sugar lactose, resulting in the signs of gas, bloating, and diarrhea after lactase ingestion.

gut-associated lymphoid tissue A significant portion of the immune system (40 percent) that is largely integrated into the intestines of the digestive system.

anaphylaxis A sudden, life-threatening, systemic or multisystem hypersensitivity allergic reaction.

allergen A substance that can cause allergy or an allergic response; also known as an *antigen*.

food intolerance An adverse reaction to a food or food component that is not mediated by the immune system.

antigen A substance that elicits the formation of antibodies by the immune system.

hypoglycemia A blood glucose concentration that is below normal (70 to 99 mg/dL) and becomes symptomatic at about 60 mg/dL.

In general, many individuals of American and Northern European descent tend to be more lactase persistent and lactose tolerant, whereas Native Americans and people from Asia and West Africa, whose ancestors are from non-cattle-ranching cultures, tend to be more lactase deficient and lactose intolerant. Given these findings, we can label the conditions of not being able to digest lactose from milk products as **lactose maldigestion** or lactose intolerance.[53,107]

Lactose intolerance occurs when an enzyme called lactase, which is needed to digest lactose (break the glucose-galactose bond), is missing. With lactose maldigestion, the lactase enzyme has reduced function. Normally, cells in the small intestine produce lactase, and the enzyme is present in pancreatic digestive secretions. When the lactase enzyme is missing or insufficient, and a person consumes lactose from dairy products, lactose digestion does not occur. Intact, undigested lactose is not absorbable, and the result is that high concentrations of lactose enter the colon and are metabolized by bacteria in the GI tract. The metabolic byproducts produced by the bacteria from the lactose cause fluid to shift into the GI tract, leading to osmotic diarrhea and the discomforts of gas, bloating, cramps, and diarrhea (see Figure 3.22).[50,111]

Individuals with lactose maldigestion may be able to tolerate a small amount of lactose from dairy products, especially those produced with bacteria (such as yogurt, aged cheese, and milk products with bacteria added, like acidophilus milk). Individuals with lactose intolerance may need to avoid the intake of dairy products altogether or use an over-the-counter product to supply the missing lactase enzyme upon ingestion of lactose from dairy products. Dairy products include milk, yogurt, cheese, and many puddings and frozen desserts like ice cream. These products are commonly made from cow's or goat's milk.

The conditions of lactose intolerance and lactose maldigestion are influenced ecologically by intestinal bacteria. These bacteria can metabolize lactose and thus contribute to the negative effects of milk consumption in the lactose-intolerant person. Additionally, excess physiological, psychological, physical, and emotional stress can cause epithelial cells at the tip of the villi, where lactase production is the most abundant, to waste away (atrophy). These epithelial cells of the GI tract, like those of the skin, regenerate within several days to weeks. So, if the stressor is removed and the lactase-persistent gene is present, then lactose tolerance may be regained in this situation.

Milk Allergy Lactose intolerance is different from milk allergy. Many people experience a significant histamine-mediated allergic response when dairy products are consumed. The histamine response causes mucus membranes to secrete fluid. This is why allergic responses usually involve skin, gut, and respiratory distress. With milk allergy, the culprit is not the lactose sugar, but rather the casein protein. The immune system of the person with a milk allergy reacts to the ingested casein protein as if it were a foreign substance invading the body (see BioBeat 3.9 and Figure 3.23). The GI tract is lined with an enormous number of immune cells, collectively called the **gut-associated lymphoid tissue**. These cells are meant to fight off foreign substances such as bacteria and viruses. However, some people have inherited allergies, and their immune cells react to foods, environmental agents (e.g., pollen, mold, pollution), and/or animal dander.[65]

Allergies involve the immune system, and when food is the cause of the allergy, the responses are usually immediate and can include gastrointestinal distress (nausea, vomiting, cramps, and diarrhea), skin reactions (urticaria or hives, rashes, and eczema), and respiratory problems (from mucus formation to narrowing of the airway or **anaphylaxis**). The common food **allergens** are specific proteins in milk, eggs, fish, crustacean shellfish, tree nuts, peanuts, wheat, and soybeans. Many other foods cause allergies, including oral allergy, which is characterized by tingling, itching, and/or swelling of the lips and mouth. An adverse allergy-like reaction to a food or food component that does not trigger an immune response is termed **food intolerance**. There is really no treatment for food allergy or intolerance besides eliminating the offending food from the person's diet.[85,152,159]

Hypoglycemia **Hypoglycemia** is a condition characterized by low blood sugar (glucose). Generally, a blood sugar value of less than 60 mg/dL causes symptoms of hypoglycemia. Low blood sugar levels can occur in some people as reactive, spontaneous, or drug-induced hypoglycemia. In this condition, the signs and symptoms are feeling nervous, light-headed, shaky, weak, sweaty, and/or faint, as well as having a headache, blurred vision, and/or hunger.[93]

Reactive Hypoglycemia   This is a condition in which blood sugar levels drop after eating a load of digestible carbohydrate. Reactive hypoglycemia occurs because too much insulin is secreted in response to a quick, sharp increase in blood sugar. Too much glucose (blood sugar) is removed too rapidly from the blood and enters the muscle and liver, resulting in low blood sugar.

Spontaneous Hypoglycemia   Spontaneous hypoglycemia occurs when liver stores of glycogen are depleted, diminishing the body's ability to maintain a normal blood sugar level. This condition happens to everyone between meals or when adequate amounts of carbohydrate have not been consumed.

Drug-Induced Hypoglycemia   Drug-induced hypoglycemia lowers blood sugar as a result of a drug reaction. Antiinflammatory and thyroid medications are known to cause hypoglycemia. In persons with diabetes, overadministration of insulin and/or excess exercise without proper dietary carbohydrate intake results in hypoglycemia.

BioBeat 3.9

Food Allergies

Food allergies occur when the protein components within foods are recognized by the immune system as foreign, causing the body to produce antibodies directed against them. Antibodies (also called immunoglobulins [Ig]) are large protein structures produced by B lymphocytes (immune cells) that circulate in the body to fight disease and infection. There are five classes of antibodies: IgA, IgD, IgE, IgG, and IgM. IgEs are involved most commonly in allergic reactions. The IgE is made by the immune system to specifically bind to a protein component of the offending food. When a protein in food is known to cause an allergic response in a person, the protein is termed an *allergen*. Normally, antibodies are directed against proteins in viruses or bacteria. In this case, the offending protein is called an **antigen**. Allergens and antigens are foreign substances that elicit an immune response. A classic allergic response that is IgE mediated is the type I immediate hypersensitivity reaction. This reaction occurs within seconds to minutes of allergen exposure. Allergic reactions may also be delayed. These reactions occur within hours to days of allergen exposure. Signs and symptoms associated with the allergic reaction include nausea, vomiting, diarrhea, skin reactions like itchy hives and dermatitis, or in some cases, increased secretions of the mucus membranes that are most noticeable in the upper respiratory tract or the GI tract. Asthma and anaphylactic shock are severe and even life-threatening responses to allergens.[21,24,42,60,61,79,102,124,133,152]

Is your immune system reactive to food protein antigens?

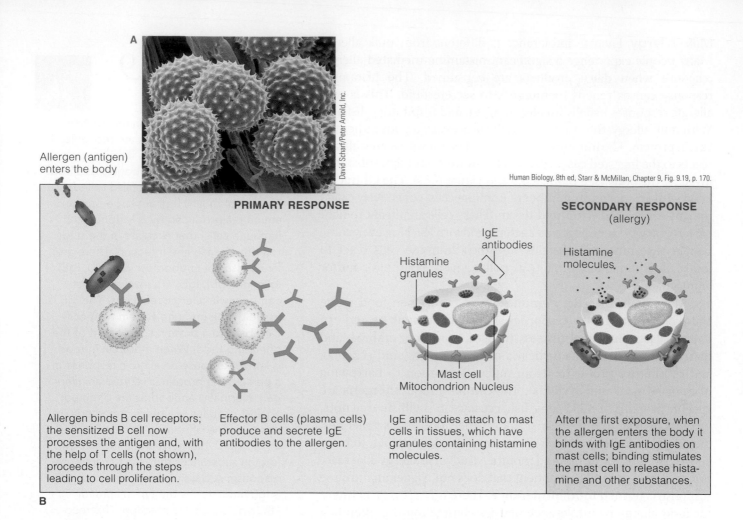

A

Allergen (antigen) enters the body

David Scharf/Peter Arnold, Inc.

PRIMARY RESPONSE

SECONDARY RESPONSE
(allergy)

IgE antibodies

Histamine granules

Histamine molecules

Mast cell
Mitochondrion Nucleus

Allergen binds B cell receptors; the sensitized B cell now processes the antigen and, with the help of T cells (not shown), proceeds through the steps leading to cell proliferation.

Effector B cells (plasma cells) produce and secrete IgE antibodies to the allergen.

IgE antibodies attach to mast cells in tissues, which have granules containing histamine molecules.

After the first exposure, when the allergen enters the body it binds with IgE antibodies on mast cells; binding stimulates the mast cell to release histamine and other substances.

B

FIGURE 3.23 Mast cells (A) in an allergic immune-mediated response (B).

hyperglycemia A blood glucose concentration that is greater than or equal to 126 mg/dL, which is both above normal (70 to 99 mg/dL) and prediabetes (100 to 125 mg/dL).

inflammation A cellular response to injury that is characterized by pain, swelling, redness, and heat.

type 2 diabetes The most prevalent type of diabetes, in which resistance to self-produced insulin occurs because of poor diet and lifestyle, obesity, and genetic factors.

Diabetes Diabetes is a chronic disease of the endocrine system resulting in the inability to reduce blood sugar (glucose). Thus, **hyperglycemia** (high levels of glucose sugar in the blood stream), or simply high blood sugar, occurs. Diabetes is categorized as type 1 or type 2. As discussed below, both types may be aggravated by increased stress, oxidative stress, and **inflammation**. The most common type of diabetes is type 2. This type of diabetes is strongly linked to obesity (although not all type 2 diabetics are obese) and living a physically inactive lifestyle. The incidence of **type 2 diabetes** has been rising dramatically, in concert with the obesity epidemic occurring in America. Currently, 26 million Americans (or more than 8 percent of the population) are estimated to have diabetes, compared to 5.76 million in 1980. Several million American people are undiagnosed or unaware of their disease state. The high level of sugar in the blood puts a diabetic person at increased risk for heart disease, stroke, kidney disease, retinopathy (loss of vision), neuropathy (loss of nerve function), limb atrophy leading to amputation (due to a combination of problems associated with damage to the circulatory system and nervous system), and decreased life expectancy. Knowing how to manage blood sugar levels in the type 1 or 2 diabetes disease state is critical in order to prevent the complications of hyperglycemia resulting from extensive cellular, tissue, and organ damage (see Table 3.5).[A,C,5,13,38,45,59,104,117,125,140,143]

TABLE 3.5

Common Complications of Unmanaged Diabetes by Body System

Organ/Site	Damage
Circulatory system	Damage to blood vessels, causing increased risk for atherosclerosis and high blood pressure as well as heart attack, stroke, and blindness (from damage to the retina of the eye, causing changes to lens shape and vision)
Digestive system	Gum disease, gastroparesis (nerve damage) causing delayed stomach emptying, heartburn, nausea, and vomiting
Endocrine system	Abnormal hormonal control of blood glucose, causing hyperglycemia
Integumentary system	Increased risk of skin infection, patchy skin discoloration, and skin thickening on the back of the hands
Peripheral nervous system	Nerve damage, impairing sensations of pain, poor circulation, foot ulcers, and possible foot/leg amputation from formation of necrotic (dead) tissue
Reproductive system	Possibility of gestational diabetes in pregnant women, leading to high birth weight, prenatal complications, and type 2 diabetes later in life
Urinary system	Damage to blood vessels of the kidney, leading to kidney damage and failure

© Cengage Learning 2013

Type 1 Diabetes This type of diabetes occurs in approximately 5 percent of diabetes cases. The onset of the disease is typically acute and occurs most often during the childhood, adolescent, or early adult years. Thus, it has been termed juvenile-onset diabetes. **Type 1 diabetes** is the severe form of diabetes and is difficult to control, because an autoimmune reaction against the pancreatic beta cells has destroyed or reduced the ability of the pancreas to synthesize adequate amounts of insulin to lower blood glucose levels. Because the pancreatic beta cells have been destroyed by the **autoimmune response,** the body loses its ability to produce insulin. Since blood sugar can only be lowered with insulin administration, insulin injections are essential to control blood sugar (bring it down to normal levels). Diabetics must eat amounts of carbohydrates compatible with the type and amount of insulin injected and measure their blood sugar levels frequently throughout the day to achieve normal blood sugar levels.

Insulin is usually injected into soft tissue just under the skin, although an inhaled form holds promise for those with diabetes.[c] It cannot be taken in pill form because insulin is a protein hormone. It would be denatured and digested like any other ingested protein if taken orally, thus rendering it nonfunctional in the blood stream.[93]

Type 2 Diabetes This type of diabetes occurs in approximately 95 percent of diabetes cases. Typically, the onset of the disease occurs during adulthood. However, with the rise in childhood obesity, children are being diagnosed with type 2 diabetes more frequently.

Developing type 2 diabetes is predisposed by obesity and **genetics** (see Figure 3.24). In type 2 diabetes, insulin resistance occurs, partly because of a decreased insulin receptor response in the liver and muscle cells. The "lock" has changed, so the "key" doesn't work. In other words, the cells of the muscle and liver that normally

type 1 diabetes Diabetes in which the pancreas fails to produce any insulin, usually because of autoimmune destruction of insulin-producing pancreatic beta cells.

autoimmune response When the immune system targets self and destroys the cells, tissues, organs, and systems within the body.

genetics The study of an organism's genetically transmitted heredity (familial or inherited) defined by deoxyribonucleic acid (DNA), the gene sequences, or genetic material.

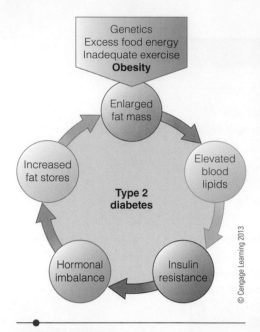

FIGURE 3.24 The type 2 diabetes implicating factors and cycle when unmanaged.

hyperinsulinemia High levels of insulin in the blood.

oral hypoglycemic agents Drugs used to treat type 2 diabetes that restore the functioning of self-produced insulin.

post-prandial In a fed state, or after eating.

prediabetes When fasting blood glucose levels are in the range of 100 to 125 mg/dL, which is higher than normal (70 to 99 mg/dL) but not yet high enough to be considered diabetic (greater than or equal to 126 mg/dL).

take up glucose in response to insulin binding to insulin receptors have become resistant to the insulin signal. Thus, even though insulin is present in the blood, blood sugar remains high. Hyperglycemia and **hyperinsulinemia** occur. The high blood sugar levels occurring in type 2 diabetes can be reduced with lifestyle changes (diet, exercise, weight control, and stress management) and **oral hypoglycemic agents**. Type 2 diabetes was initially called non-insulin-dependent diabetes mellitus. Now it is considered a lifestyle disease because blood sugar control is quite achievable with regular exercise and consuming a healthy diet that promotes weight loss when needed. Often, medication is prescribed to a person with type 2 diabetes. The oral hypoglycemic agent drug is taken in pill form to help augment the action of self-produced insulin.[46,69,83,110,112,140]

Diagnosing Diabetes Diabetes, whether type 1 or type 2, is characterized by hyperglycemia. Thus, a fasting blood sugar reading will indicate whether a person has normal glucose levels or hyperglycemia. A 12-hour fasting blood glucose level of 126 mg/dL or higher or a 2-hour **post-prandial** (fed) blood glucose level of 200 mg/dL or higher is considered hyperglycemia and indicates diabetes. A fasting blood glucose level of 100 to 125 mg/dL indicates **prediabetes**, and is one of the diagnostic criteria for metabolic syndrome (pages 141–143).

A person who is believed to have type 2 diabetes will typically also undergo a glucose tolerance test. This test evaluates the blood sugar response to a large glucose feeding and is used to diagnose diabetes. The procedure for the glucose tolerance test involves eating a normal diet for three days before the test. Then the person fasts for 12 hours (no food or drink besides water), and a baseline fasting blood sugar level is taken. Once again, blood sugar levels of 126 mg/dL or higher indicate diabetes. Next, a glucose load of 1 gram of carbohydrate per kilogram of body weight (or a maximum of 100 grams for adults and 1.75 grams of carbohydrate per kilogram of body weight for children) is consumed quickly. The person's blood sugar is monitored every half hour for up to 6 hours. Blood samples are taken and blood sugar levels are compared to standard, normal values. Blood sugar levels of 200 mg/dL or higher 2 hours after the glucose load indicate diabetes (see Figure 3.25).

FIGURE 3.25 The glucose tolerance test reveals normal and diabetic carbohydrate metabolism.

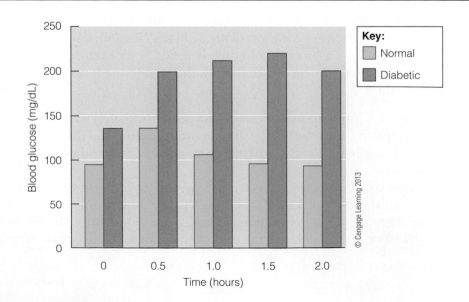

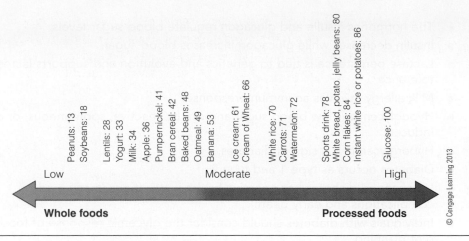

FIGURE 3.26 The approximate glycemic index values of various foods as low, moderate, and high.

A typical diabetic response involves hyperglycemia when fasting and an increase in blood sugar after the glucose load for a prolonged period. Basically, the glucose tolerance test is a physiological demonstration that the person's body has reduced or no ability to lower blood sugar on its own. The test may not differentiate type 1 diabetes from type 2, but usually type 1 blood sugar levels are much higher compared to type 2.

The Glycemic Response of Foods For individuals with diabetes, it is helpful to eat foods that have a lower glycemic response or have a slower entrance of glucose into the blood stream (the exception to this would be if the individual was in shock and needed a rapid increase of blood sugar). Figure 3.26 shows the **glycemic index** of various foods, and Table 3.6 shows the glycemic index of foods in a sample meal. The higher the glycemic index number, the faster and greater the rise in blood glucose levels will be after eating the food. Foods or meals with high protein, fat, and fiber content tend to lessen (blunt) the **glycemic response** and are therefore more beneficial in routinely controlling blood sugar levels.[43,45,52,82,91,101,119]

TABLE 3.6

The Glycemic Index Value of Food in a Higher versus Lower Sample Meal

Higher Glycemic Index Lunch	Lower Glycemic Index Lunch
2 slices white bread: 70	1 cup chili with beans: 27
3 ounces turkey: 22	2 bananas: 53
1 cup watermelon: 72	2 ounces potato chips: 54
3 ounces corn chips: 73	1 cup 1% milk: 33
1 cup cola drink: 71	

© Cengage Learning 2013

Summary Points

- Glucose is required by the brain, red blood cells, and central nervous system for energy (ATP).
- Carbohydrate is the preferred energy source of the body.
- All digestible carbohydrate is converted to glucose for energy.
- Excess carbohydrate is stored as glycogen in liver and muscle cells.
- If glycogen stores are full, excess carbohydrate is converted to and stored as fat.
- Liver glycogen maintains blood sugar for brain, red blood cell, and central nervous system function.
- Muscle glycogen is used by the working muscle in high-intensity exercise.

glycemic index A number assigned to a food based on the blood-glucose-raising potential of the given food compared to glucose; has a standard of 100.

glycemic response The rise and fall of blood glucose levels after the consumption of a food.

- The hormones insulin and glucagon regulate blood sugar levels.
- Insulin decreases while glucagon increases blood sugar.
- Lactase persistence is tied to genetics and evolution and supports lactose tolerance.
- Milk allergy involves an immune response.
- Hypoglycemia is low blood sugar and can be reactive, spontaneous, or drug induced.
- Hyperglycemia is a sign of diabetes.
- Diabetes occurs as type 1 and type 2.
- Unmanaged diabetes has negative health implications (many medical complications and early death).
- Individuals with diabetes should consider the glycemic response of foods for diet planning.
- Selecting foods with a low glycemic index is useful in controlling blood sugar spikes.

Take Ten on Your Knowledge Know-How

1. How do hormones regulate blood glucose (sugar)?
2. How and where is glucose stored inside your body? What is the purpose of this stored glucose?
3. What is lactose intolerance? How is it tied to genetics and evolution? How does lactose intolerance differ from a milk allergy?
4. What dietary changes would you need to make if you had lactose intolerance? Milk allergy?
5. What is a normal blood glucose (sugar) level? What would be considered too low or too high if the value was obtained in a state of fasting?
6. What is hypoglycemia and what are three common ways that hypoglycemia occurs?
7. How are type 1 and type 2 diabetes similar and yet different?
8. How is diabetes diagnosed?
9. What health complications are associated with unmanaged hyperglycemia?
10. How can the glycemic index be useful for a person with diabetes? How can this apply to your diet?

3.5 LIPIDS IN HEART DISEASE AND CANCER

T-Talk 3.5
To hear Dr. Turley talk about lipids in heart disease and cancer, go to www.cengagebrain.com

Introduction

In this section we will focus on the incidence, contributing factors, and prevention of heart disease and cancer. We will see that some dietary lipids promote good health, whereas others, when out of balance, have negative effects on health, such as promoting heart disease and cancer. For heart disease, blood lipids, dietary fat, oxidation, antioxidants, *trans* fatty acids, omega-3 fatty acids, and other factors will be discussed. Diet, lifestyle, and dietary lipids will be explored in the context of the cancer process.

Lipids in Heart Disease

Diseases of the heart are the leading cause of death in the United States. In this unit we will focus on coronary heart disease.

Heart Attack Coronary heart disease occurs when a blood vessel leading to the heart muscle is occluded (blocked), and oxygen delivery to the consistently contracting heart muscle is diminished or absent. Lack of oxygen eventually causes

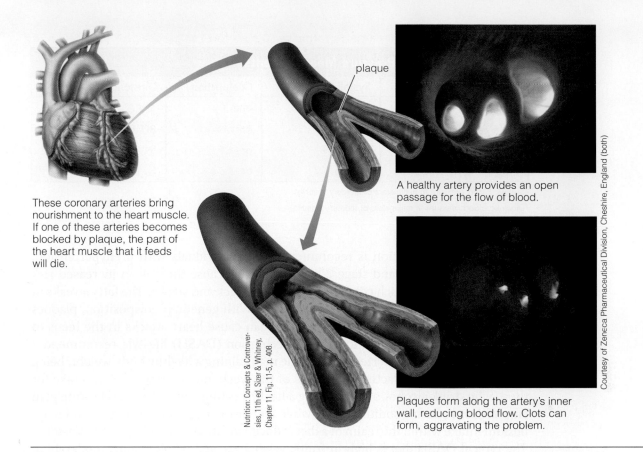

These coronary arteries bring nourishment to the heart muscle. If one of these arteries becomes blocked by plaque, the part of the heart muscle that it feeds will die.

Nutrition: Concepts & Controversies, 11th ed, Sizer & Whitney, Chapter 11, Fig. 11-5, p. 408.

A healthy artery provides an open passage for the flow of blood.

Plaques form along the artery's inner wall, reducing blood flow. Clots can form, aggravating the problem.

Courtesy of Zeneca Pharmaceutical Division, Cheshire, England (both)

FIGURE 3.27
Atherosclerosis in coronary arteries.

a heart attack, or *myocardial infarction*. During the heart attack, heart muscle cells die as a result of a lack of oxygen. A heart attack can be considered mild to massive depending on the degree of oxygen deprivation and the amount of heart muscle destroyed.

Atherosclerosis It is estimated that one-third of the individuals who perish from diseases of the heart die from atherosclerosis. Atherosclerosis in the blood vessels that lead to the brain also contributes to stroke. Atherosclerosis in the heart blood vessels is more commonly known as coronary artery disease (CAD), caused by clogging, narrowing, and/or hardening of the **arteries** (see Figure 3.27). This condition occurs because of plaque buildup. Plaques, also initially called fatty streaks, develop in the arterial walls, usually after arterial injury has occurred, causing the progression of plaque buildup. At the site of arterial damage, inflammation and plaque deposits occur. Plaque contains fatty material, including cholesterol, which is waxy and sticky. When blood levels of cholesterol increase, cholesterol-rich plaque buildup also increases, and as the plaque builds this is known as the progression of coronary artery disease.[H,76,150,154]

Injury to the arterial wall is commonly caused by **hypertension**, tobacco use, free radicals, hyperglycemia, and **hyperhomocystemia**. Hypertension is also known as high blood pressure and is one of the leading causes of arterial damage leading to atherosclerosis and heart disease.

Blood pressure is expressed as systolic pressure over diastolic pressure. Systolic pressure is arterial pressure caused by the contraction of the heart muscle. Diastolic pressure is arterial pressure when the heart is between contractions or beats. The measurements are given in millimeters of mercury (mmHg). Normally, resting blood pressure should be less than 120 mmHg systolic and less than 80 mmHg diastolic. Blood pressure values greater than these indicate either prehypertension or hypertension, which causes damage to every vital organ (see Table 3.7).[89,144]

atherosclerosis A disease of the arteries that is characterized by the accumulation of lipid-containing material called *plaque* on the inner walls of the arteries, particularly the coronary and cranial arteries.

arteries The largest blood vessels that carry blood from the heart to the tissues.

hypertension Elevated blood pressure above 120/80 millimeters of mercury; the prehypertension stage is established by the range 120–139/80–89, stage 1 of hypertension is given by the range 140–159/90–99, and stage 2 is greater than or equal to 160/100.

hyperhomocystemia High levels of homocysteine in the blood.

TABLE 3.7

Interpreting Blood Pressure Measurements			
Category	Systolic[1]	Conjunction	Diastolic[2]
Optimal	< 120	and	< 80
Prehypertension	120–139	or	80–89
Stage 1 hypertension	140–159	or	90–99
Stage 2 hypertension	≥ 160	or	≥ 100

[1]Systolic blood pressure in millimeters of mercury (mmHg)

[2]Diastolic blood pressure in millimeters of mercury (mmHg)

© Cengage Learning 2013

Lifestyle modification is recommended for individuals with prehypertension, stage 1 hypertension, and stage 2 hypertension because there is an increased risk for atherosclerosis, which contributes to heart attack and stroke. The fatty streaks in atherosclerosis can begin to develop at birth. With genetic predisposition, plaques may be present even at 5 years of age and can cause heart attacks in the teens to 20s. The Dietary Approach to Stop Hypertension (DASH) lifestyle recommendations for controlling hypertension include maintaining a healthy body weight, being moderately physically active most days of the week, moderating alcohol intake for those who drink alcohol, and following a healthy eating plan. The DASH eating plan emphasizes reducing sodium intake while increasing potassium, calcium, and magnesium intake within the realm of a diet reduced in fat, saturated fat, and cholesterol. The typical DASH diet is high in fruits, vegetables, and whole grains, moderate in low-fat dairy products, lean meats, and legumes, and low in sweets. Liquid plant oils, nuts, and seeds are included as healthy dietary fat sources.[F,15,16,19,26,31,44,103,135,139,142,145]

Heart disease is of particular concern for men. The rate of heart disease in men ages 25 to 34 is three times that of women of the same age. From ages 35 to 44, the rate of heart disease is twice as high in men as in women. From ages 55 to 75, women are catching up to men in terms of death from heart disease. During this time, women are losing their protective role of estrogen as they undergo menopause. From ages 75 to 80, the rates of deaths from heart disease in men and women are equal.[23,47,77]

There are many risk factors for heart disease, and having two or more risk factors greatly increases the chance of death from heart disease. Risk factors include elevated serum cholesterol, genetics, tobacco use, excessive alcohol consumption, hypertension, diabetes, obesity, cerebrovascular or peripheral vascular disease, sedentary lifestyle, stress, male gender, and consuming a low-fiber and high-fat diet. Many of these risk factors are controllable or modifiable. So, atherosclerosis and thus heart disease is preventable by lifestyle change, including diet and aerobic fitness. Aerobic fitness causes arteries to enlarge, blood pressure to go down, and **blood lipids** to be reduced. Practicing a healthy lifestyle during childhood and through all stages of life is important.[25,47,97,123,147,148,158]

Serum Triglycerides Fat and cholesterol are packaged in carrier proteins called lipoproteins (see Table 3.8 and Figure 3.28). The protein component helps them to travel throughout the circulatory system. The protein in lipoprotein also serves as a communication piece with cell receptors. There are a few common different lipoproteins made from triglycerides, phospholipids, cholesterol, and protein. Triglyceride-rich lipoprotein fractions include chylomicrons and **very-low-density lipoproteins (VLDLs)**. Chylomicrons are the lipoproteins made by intestinal cells after fat-soluble substances undergo digestion and absorption. VLDLs are produced by liver cells and are converted to LDL in the blood stream after the triglycerides present in them are delivered to cells.[58,72,157]

blood lipids A variety of carrier lipoproteins in blood, including high-density lipoproteins (HDLs), low-density lipoproteins (LDLs), very-low-density lipoproteins (VLDLs), and chylomicrons.

very-low-density lipoprotein (VLDL) The type of lipoprotein cholesterol made primarily by liver cells to transport triglycerides to body cells.

When serum triglyceride levels are elevated (i.e., high chylomicrons and VLDL), this is called **hypertriglyceridemia** (high blood triglycerides) and is associated with heart disease because **serum triglycerides** can thicken the blood. After consuming a low-fat diet for three days before testing, a fasting 12-hour blood test is needed to determine an accurate triglyceride level. A level of less than 150 mg/dL is considered normal and a level greater than 150 mg/dL is considered high. A level of 450 mg/dL is so high and abnormal that pumping the blood is like pumping ketchup.

With hypertriglyceridemia, dietary fat intake should be reduced to less than 20 percent of Calories consumed. Omega-3 fatty acid supplementation can reduce serum trigycerides. The intake of simple sugars and alcohol should be reduced to meet healthy dietary recommendations. Taking steps to normalize blood lipids can reduce the risk of heart disease and stroke as well as liver and pancreas problems.[58,76,141,147,149]

hypertriglyceridemia A fasting serum triglyceride level of 150 mg/dL or greater.

serum triglycerides The level of total triglycerides in the blood stream, including chylomicrons and VLDL; it is at a desirable level when it is less than 150 mg/dL.

TABLE 3.8

Lipoprotein Fractions

Serum Triglycerides	Serum Cholesterol
Very-low-density lipoprotein cholesterol (VLDL)	Low-density lipoprotein cholesterol (LDL)
Chylomicrons	High-density lipoprotein cholesterol (HDL)
	Intermediate-density lipoprotein cholesterol (IDL)
*These are triglyceride-rich lipoprotein fractions.	*These are cholesterol-rich lipoprotein fractions.

© Cengage Learning 2013

FIGURE 3.28 The lipoproteins: A typical lipoprotein (A), the variety of lipoproteins (B), and the composition of lipoproteins (C).

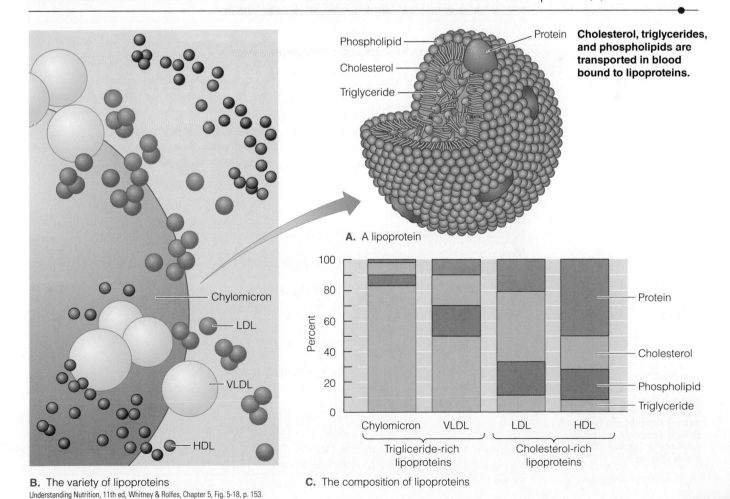

Phospholipid
Cholesterol
Triglyceride
Protein

Cholesterol, triglycerides, and phospholipids are transported in blood bound to lipoproteins.

A. A lipoprotein

Chylomicron
LDL
VLDL
HDL

Protein
Cholesterol
Phospholipid
Triglyceride

Chylomicron   VLDL   LDL   HDL

Triglyceride-rich lipoproteins   Cholesterol-rich lipoproteins

Percent: 100, 80, 60, 40, 20, 0

B. The variety of lipoproteins
Understanding Nutrition, 11th ed, Whitney & Rolfes, Chapter 5, Fig. 5-18, p. 153.

C. The composition of lipoproteins

Serum Cholesterol Most of the serum cholesterol packaged in the liver is in the form of low-density lipoprotein (LDL) and high-density lipoprotein (HDL). **Intermediate-density lipoprotein (IDL)** is present to a minor extent (see Table 3.9). These lipoproteins—whether HDL, LDL, or IDL—are carriers of a variety of lipids and protein. The varying amount of cholesterol, triglycerides, phospholipids, and protein identifies the lipoprotein. For example, HDLs contain more protein and less cholesterol than LDLs or IDLs. Further, the functions of the lipoproteins differ. HDL returns cholesterol to the liver for synthesis of myelin, bile, hormones, and vitamin D. LDL transports cholesterol to cells throughout the body for functions such as incorporation into cell membranes.

Elevated levels of total serum cholesterol are associated with atherosclerosis, especially if two other risk factors exist, such as male gender, family history, hypertension, high-fat diet, sedentary lifestyle, hyperhomocystemia, tobacco use, and excess alcohol intake. Having a level of LDL that is too high or of HDL that is too low increases a person's risk for heart attack and stroke. The highest levels of HDL cholesterol have been reported in premenopausal women who are aerobically fit. Because post-menopausal women have lost the protective effect of estrogen, their level of LDL may rise, thus increasing their risk for heart disease. Cholesterol levels increase with age, so it is important to know what your cholesterol levels are now and monitor them in the future in order to reduce the risk of heart disease (see Tables 3.9 and 3.10).[63,76,93]

TABLE 3.9

National Cholesterol Education Program Criteria

Total Cholesterol (mg/dL)	Recommendation
< 200 *Low risk*	Repeat test in five years (desirable).
200–239 *Borderline high risk*	Refer to medical doctor for follow-up if history of CAD, or if two or more CAD risk factors. If no reported history of CAD, or less than two risk factors, recheck blood level in one year. LDL > 130 mg/dL and/or HDL < 35mg/dL
≥ 240 *High risk*	Refer to medical doctor for intervention and follow-up. Recheck in about three months. Prescribe drugs. LDL > 160 mg/dL and/or HDL < 35mg/dL

© Cengage Learning 2013

TABLE 3.10

American Heart Association Serum Cholesterol Recommendations by Category

Total Serum Cholesterol (mg/dL)	LDL Serum Cholesterol (mg/dL)	HDL Serum Cholesterol (mg/dL)
< 200 *Desirable*	< 100 *Optimal*	< 40 *Low (indicates risk)*
200–239 *Borderline high*	100–129 *Near optimal*	> 60 *High*
≥ 240 *High*	130–159 *Borderline high*	
	160–189 *High*	
	≥ 190 *Very high*	

© Cengage Learning 2013

intermediate-density lipoprotein (IDL) Lipoproteins that are transiently formed by the degradation of very-low-density lipoproteins to low-density lipoproteins and do not promote or regress coronary artery disease.

Metabolic Syndrome Metabolic syndrome, also known as syndrome X or insulin resistance syndrome, is characterized by the presence of a set of metabolic risk factors that are associated with the development of heart disease and type 2 diabetes. Metabolic syndrome, as defined by the National Cholesterol Education Program Adult Treatment Panel III, includes five abnormalities: (1) **central adiposity**, as defined by a waist circumference greater than 40 inches in men and 35 inches in women, (2) fasting blood triglyceride level of 150 mg/dL or greater, (3) fasting blood HDL cholesterol level of less than 40 mg/dL in men and less than 50 mg/dL in women, (4) elevated blood pressure of 130 over 85 mmHg or greater, and (5) fasting blood glucose of 110 mg/dL or greater. Individuals having three or more of the five abnormalities are considered to have metabolic syndrome.[9,41,64,69,70,83,108]

The prevalence of metabolic syndrome is on the rise worldwide. It is estimated that nearly 47 million American adults have metabolic syndrome, as evidenced by a constellation of abnormalities including disturbed glucose metabolism, excess abdominal fat, **dyslipidemia**, and hypertension; sadly, it is becoming more common among young people and is linked to food processing techniques such as those employed to generate high-fructose corn syrup (see BioBeat 3.10).[129]

There is a theory, called the thrifty gene theory, which ties metabolic syndrome to the "survival of the fittest" concept but within a misfit genotype in the present-day environment. Normally, genotypic changes occur very slowly and over many generations. Metabolic syndrome has only become an emerging medical phenomenon for about the past 20 years. What has occurred quickly is a change in our way of living. Essentially, we could say that the genotype designed for hunter-gatherers is ill-suited or in disequilibrium with today's industrialized foods society and sedentary lifestyle.[12,92]

It is hypothesized that humans who could efficiently use food and store energy as body fat resulting in weight gain would be able to survive tough times during the hunter-gatherer period of human existence. But now, in modern-day society, where people eat more and exercise less, these naturally selected traits are undesirable and even detrimental. The best method of combating metabolic disturbances is lifestyle management, including slimming down when needed, increasing physical activity, increasing whole fruit and vegetable consumption, consuming a low to moderate fat intake level, and limiting the intake of added sugars and refined carbohydrates.[22,49,70,86,98,99,141,149]

Studies of master athletes show that those who participate in competitive sports in their middle to later years (women older than 35 years and men older than 40 years) have shown lipid profiles similar to those of young adults. Further, they have better glucose tolerance and waist circumferences compared to their sedentary counterparts; thus, they are at reduced risk for heart disease, type 2 diabetes, and metabolic syndrome. Other added benefits of having an active lifestyle include reduced risk for osteoporosis and the ability to eat more food energy and essential nutrients while maintaining a lower body weight than their sedentary counterparts. Exercise physiologists have referred to metabolic syndrome as sedentary death syndrome. So the concept of "eat more, weigh less, and be healthier" is true for those who engage in regular vigorous exercise at any age, starting at any point in life.[115,128]

Dyslipidemia This term describes the many abnormal circulating levels of blood lipids and/or lipoproteins that can occur. This could mean high (greater than or equal to 240 mg/dL) total serum cholesterol (**hypercholesterolemia**); high (greater than or equal to 160 mg/dL) LDL cholesterol; low (less than 40 mg/dL) HDL cholesterol; high (greater than or equal to 150 mg/dL) triglycerides (**hypertriglyceridemia**); or any combination of these abnormal circulating lipoprotein fractions.

central adiposity A waist circumference greater than 40 inches in men and greater than 35 inches in women.

dyslipidemia Any combination of abnormal circulating lipoprotein fractions of hypercholesterolemia, high LDL cholesterol, low HDL cholesterol, and hypertriglyceridemia.

hypercholesterolemia A fasting total blood cholesterol level of 240 mg/dL or greater.

BioBeat 3.10

High-Fructose Corn Sweetener: Not So Sweet for Metabolic Syndrome

High-fructose corn sweetener (HFCS) is made from the glucose syrup yield from corn starch. HFCS products were introduced into the food industry in the 1970s. A large percentage of the glucose molecules are chemically altered to become fructose. There are two common HFCSs used to sweeten processed foods. HFCS 55 is 55 percent fructose and 45 percent glucose, and HFCS 42 is 42 percent fructose and 58 percent glucose. HFCS 55 is of particular concern.

Many potentially negative health effects have been associated with the increased consumption of HFCS worldwide. The increased risk for developing metabolic syndrome is one of them. Metabolic syndrome is an epigenetic phenomenon, meaning that the expressions of the genes are influenced by the environment. This is clear by the way that the Therapeutic Lifestyle Changes diet and regular exercise can improve the health markers and control the progression of the syndrome (see Table 3.11). More than 38 genes have been identified and associated with the manifestations of metabolic syndrome, and scientists are beginning to understand how the expression of each gene is influenced. One of the most influential underlying conditions of metabolic syndrome is insulin resistance. Many dietary and exercise factors affect the production, secretion, and effectiveness of insulin.[3,41,64]

A diet containing HFCS has been shown to alter lipid and carbohydrate metabolism, notably lipid metabolism in the liver and in the insulin response. As a result, elevated serum triglyceride levels, an increase in nonalcoholic fatty liver accumulation, and several other lipid profile changes that are associated with increased risk of heart disease have been well documented. Because of the changes in liver metabolism, it is possible for the production of **uric acid** to increase. An increased level of uric acid increases insulin resistance. Finally, the intake of HFCS reduces the levels of three key hormones in weight control: Insulin levels are lower, which is good, but chronic intake of HFCS also reduces the leptin and ghrelin levels, which are bad for weight control. Leptin suppresses appetite and increases metabolic rate, thus favoring weight control. Ghrelin is also a hormone that suppresses appetite. It is hypothesized that these may be the mechanisms that promote weight gain.

Chronic intakes of HFCS (more than 50 grams or just over 3 tablespoons per day) alters liver and carbohydrate metabolism in unhealthy ways. There is supporting evidence that chronic, high intakes of HFCS and several potential mechanisms can explain the promotion of metabolic syndrome. More research is needed to prove the cause-and-effect relationship of HFCS and metabolic syndrome. The elaboration of the epigenetic mechanisms will fortify the understanding of this complex syndrome.[58,81]

Are you at risk for metabolic syndrome (genetically and environmentally) by your HFCS intake?

Dietary Fat and Lipoproteins The type of dietary fatty acids consumed influences the type of lipoproteins that the liver produces (see pages 21–22 of Module 1 for a review). Scientific studies have presented enough evidence to say that consuming a diet high in fat, and particularly in saturated fatty acids, increases total serum cholesterol and LDL cholesterol; polyunsaturated fatty acids decrease total serum cholesterol and LDL and HDL cholesterol; monounsaturated fatty acids decrease total serum cholesterol and LDL and increase HDL; and cholesterol can increase total cholesterol and LDL cholesterol.[4,76,80,109,130] Phospholipids have not been indicated in the disease process. Considering the profound effect that diet has on serum cholesterol, it is no surprise that the Adult Treatment Panel III has guidelines for cholesterol management. Their guidelines used to be called the step I and step II diets, but they are now called the Therapeutic Lifestyle Changes (TLC) diet. These new guidelines are summarized in Table 3.11 and are endorsed by the American Heart Association (AHA). A sample of a TLC diet plan is given in Table 3.12.

uric acid A nitrogen-containing waste product generated from purine catabolism that contributes to gout and other illnesses.

TABLE 3.11

The Therapeutic Lifestyle Changes (TLC) Diet

Summary of the TLC Diet	
Total fat	25–35% of Calories
Saturated fatty acids	< 7% of Calories
Polyunsaturated fatty acids	≤ 10% of Calories
Monounsaturated fatty acids	≤ 20% of Calories
Trans fatty acids	As low as possible
Carbohydrate	50–60% of Calories
Protein	~ 15% of Calories
Cholesterol	< 200 mg/day
Plant stanols/sterols	2 g/day
Soluble fiber	10–25 g/day
Total Calories	Balance energy intake and expenditure to maintain desirable body weight and prevent weight gain. Expend 200 Calories per day in moderate physical activity.

© Cengage Learning 2013

TABLE 3.12

Sample 2,200-Calorie TLC Diet and Menu

Examples of Food in a 2,200-Calorie, One-Day TLC Diet	
Grains	7 ounce equivalents with one-half in the form of whole grains
Vegetables	3 cup equivalents
Fruits	2 cup equivalents
Low-fat dairy	3 cup equivalents
Lean meat/fish/alternatives	6 ounce equivalents; soy protein may replace animal products
Eggs	< 2 yolks per week
Oils	6 teaspoon equivalents
Sample breakfast	1 cup cooked oatmeal
	1 cup skim milk
	¼ cup raisins
	1 cup orange juice
Sample lunch	3 ounces turkey meat
	2 slices whole-wheat bread
	1 tomato
	1 lettuce leaf
	1 tablespoon mustard
	1 cup bean salad with 1½ tablespoons Italian dressing
	1 cup skim milk
Sample dinner	1½ cups whole-wheat pasta
	1 cup spaghetti sauce
	3 meatballs made with extra lean hamburger
	1 slice French bread with 2 teaspoons of margarine

© Cengage Learning 2013

Demo GENEie The solid shortening demonstration: Bring in a small can of shortening, a teaspoon, and two paper plates. Look up the Nutrition Facts information for total fat grams for a fast-food meal (such as a double cheeseburger, fries, and a milkshake). For every 4 grams of fat, scoop out 1 teaspoon of shortening and put it on the plate. Now do the same thing for a healthy meal alternative (1 cup brown rice, 4 ounces of baked chicken, 1 cup of steamed broccoli, and 8 ounces of soy milk). Compare the piles of shortening on the two plates. Discuss the health implications of eating fast food too frequently.

Oxidation and Antioxidants The AHA recommends limiting saturated fatty acids to less than 7 percent of total Calories consumed in a day but also recommends not exceeding 10 percent of Calories from polyunsaturated fatty acids. The double bonds of polyunsaturated fatty acids are weaker bonds and are more vulnerable to **oxidation** from **reactive oxygen species** and radicals. Following oxidation the double bond breaks, generating electron-deficient lipid fragments that are chemically unstable and very sticky and damaging. LDL cholesterol commonly undergoes oxidation. Oxidized LDL cholesterol accelerates atherosclerotic plaque formation.

The function of an **antioxidant** is to prevent and terminate oxidation reactions (see BioBeat 3.11).[66,151] Antioxidants donate needed electrons to electron-deficient molecules, thus reducing the reactivity. This also protects other molecules in the body from damage and maintains normal cell structure and integrity. Several essential nutrients function as antioxidants, either in water-soluble cellular spaces or fat-soluble cellular spaces. Lipoproteins and vitamin E are both fat-soluble. LDL cholesterol is particularly susceptible to oxidative damage. Antioxidants have been implicated

oxidation The process of a substance combining with oxygen, resulting in the loss of an electron and the creation of a chemically unstable (more reactive) molecule.

reactive oxygen species Highly unstable and reactive molecules produced as a byproduct of oxygen metabolism, which damage cell structures unless stabilized by an antioxidant.

antioxidant A chemical compound that can donate an electron without becoming chemically reactive itself, and thus can inhibit oxidation and reduce the damage that electron-deficient chemicals cause.

cardiovascular disease A variety of diseases that affect the heart and blood vessels.

BioBeat 3.11

Antioxidants in Heart Disease and Cancer

Antioxidants have become extremely popular molecules during the last decade and should be consumed in a balanced, healthy diet. Antioxidants control the damage that radical molecules and reactive oxygen species do in the body. A radical is a chemical with an unpaired set of electrons, and a reactive oxygen species is an electron-deficient molecule. Both radical molecules and reactive oxygen species are chemically very reactive and damaging. An antioxidant supplies an electron to the radical molecules and reactive oxygen species, thus reducing the reactivity and damaging effects of the molecule (see Figure 3.29). Radical molecules and reactive oxygen species damage is associated with heart disease, cancer, and many other chronic diseases. The essential vitamins that function as antioxidants include vitamin E (tocopherols), vitamin C (L-ascorbic acid), and beta-carotene (provitamin A). Vitamin E is of interest in heart disease because it is fat-soluble and associates with lipids, including LDL cholesterol. In human clinical trials, daily vitamin E intake levels of 134 milligrams have been shown to reduce oxidation of LDL cholesterol, which in turn slows the progression of atherosclerosis. This level of vitamin E intake can only be achieved by supplementation in order to preserve a healthy, low-fat diet. Not all studies demonstrate that vitamin E supplementation, whether from synthetic or natural forms, reduces **cardiovascular disease** mortality, though no unfavorable effects are believed to occur from supplementation of 30 to 600 milligrams per day.[51,66,68,88,155]

Do the foods in your diet supply antioxidant nutrients to protect against heart disease and cancer?

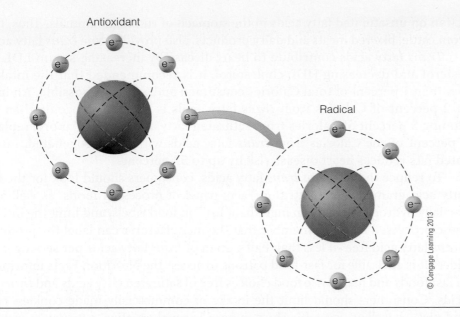

Antioxidant

Radical

© Cengage Learning 2013

FIGURE 3.29 An antioxidant donates an electron (e⁻) to a radical molecule.

in the prevention of heart disease through their action of reducing LDL oxidation. Because of its fat solubility, vitamin E can incorporate directly into LDL cholesterol packages. So, vitamin E is especially good at blocking LDL oxidation. Food sources of vitamin E include plant oils, wheat germ, whole grains, nuts, seeds, and, to a lesser extent, dark green leafy vegetables.[68] The negative side to good food sources of vitamin E is that most of them are high in fat. It's good to eat these foods as a healthy source of needed fat and vitamin E, but don't overdo their intake. Vitamin E supplementation may be recommended to achieve vitamin E's antioxidant protection.

Hydrogenation and Partial Hydrogenation The process of **hydrogenation** involves removing the double bonds from polyunsaturated fatty acids and monounsaturated fatty acids and adding hydrogen molecules to stabilize the carbon bonds, thus making the fats become more saturated. Carbon atoms in organic compounds are most stable, with four sites of bonding. In the hydrogenation process, the double bonds are removed, and hydrogen binding creates the stability at the four sites. Adding hydrogen to the fatty acid chain while removing the double bonds hardens the fat. The harder the fat, the more saturated it is. Conversely, the softer the fat, the more unsaturated it is. Stick margarine and shortening are highly hydrogenated and saturated. The problem with the process of hydrogenation and **partial hydrogenation** is that the unsaturated fatty acids become saturated, and *trans* fatty acids can be created. *Trans* fatty acids contribute to heart disease even more than saturated fatty acids do.

Cis and Trans Fatty Acids *Cis* **fatty acids** are the naturally occurring configuration of the two hydrogen atoms in a double bond found in unsaturated fatty acids. In the *cis* configuration, the hydrogen atoms are on the same side of the double bond in the fatty acid carbon chain.

In the *trans* fatty acid configuration, the hydrogen atoms are on the opposite side of the double bond in unsaturated fatty acids. The *trans* configuration is predominately a chemical fluke that occurs when polyunsaturated fatty acids and monounsaturated fatty acids are partially hydrogenated. So, processed foods containing partially hydrogenated oils provide most of the consumed *trans* fatty acids. Since the 2003 amendment to the Nutrition Labeling and Education Act regarding *trans* fatty acids, levels have reduced in the U.S. food supply (see page 47 in Module 2 for food label information). Additionally, *trans* fatty acids can be formed by bacterial

hydrogenation A food processing technique that chemically forces the addition of hydrogen atoms by saturating monounsaturated or polyunsaturated fatty acids (the double bonds are eliminated, making the fats saturated, solid, resistant to oxidation, shelf stable, and less healthy).

partial hydrogenation A food processing technique that chemically forces partial saturation (addition of hydrogen atoms and the removal of the double bonds) of monounsaturated or polyunsaturated fatty acids, which can generate *trans* fatty acid configurations.

cis **fatty acids** The natural configuration of unsaturated fatty acids, in which the hydrogen atoms are on the same side of the double bond in the carbon chain.

action on unsaturated fatty acids in the stomach of ruminant animals. Thus, foods from cattle, like red meats and dairy products, also provide some *trans* fatty acids.

Trans fatty acids contribute to heart disease by increasing serum LDL cholesterol and decreasing HDL cholesterol. It is recommended that the intake be less than 1 percent of total Calories consumed and as low as possible. An intake of 2 percent of Calories from *trans* fatty acids is more atherogenic than consuming 5 percent of Calories from saturated fatty acids. Furthermore, replacing 2 percent of the Calories from *trans* fatty acids with unhydrogenated, unsaturated fats reduces heart disease risk by up to 53 percent.[90,123]

To reduce the intake of *trans* fatty acids, consumers should look for the *trans* fatty acid grams on the Nutrition Facts panel of processed foods, as well as for partially hydrogenated oils in ingredient lists on food labels, and limit the intake of these processed foods. Remember that the manufacturer can label the product as "*trans* fatty acid free" if less than half a gram of *trans* fatty acids per serving is provided. It is advisable for fast-food patrons to access the Nutrition Facts information of fast foods and make wise food choices free of saturated fatty acids and *trans* fatty acids. Consumers should limit the intake of commercially-made cookies, cakes, and pies, as well as vegetable shortenings. As an alternative, consumers can bake with vegetable oils, use margarines that are soft, and/or choose margarines that are *trans* fatty acid free. In response to the numerous scientific studies demonstrating the detrimental effects of *trans* fatty acids on heart health, the food industry is formulating new methods of producing margarine so that *trans* fatty acid levels are minimized. Basically, more saturated fatty acids, such as from coconut and palm oil, are being formulated into margarines with liquid plant oil, water, and an **emulsifier** to replace partial hydrogenation and the generation of *trans* fatty acids. Though saturated fatty acids from coconut and palm oils are not health-promoting either, they are not believed to be as detrimental as *trans* fatty acids.[2,37,132]

DA+ GENEie   Go to Diet Analysis Plus and enter some foods like coconut oil, olive oil, flax seeds, fish, and processed foods that have crunchy or flaky aspects to them, such as pot pie, fast-food breakfast potatoes, crackers, and so forth. Create a source analysis spreadsheet report. Study the fatty acid content by food entered for sat-fat, mono-fat, poly-fat, *trans*-fat, omega-3, and omega-6. Note your observations for discussion.

Eicosanoides and Fatty Acids Inside the cells of the body, omega-3 fatty acids and **omega-6 fatty acids** are used to synthesize **eicosanoids**. Eicosanoids are organic chemical compounds that are at least 20 carbons long. Cells produce eicosanoids and further process them into a variety of hormone-like compounds that alter cellular processes. Some fatty acid–derived eicosanoids are called prostaglandins. Prostaglandins cause an array of biological activities, including inflammatory and antiinflammatory responses, as well as influences that increase the tendency of blood to clot or not and make respiratory tissues or blood vessels relax or constrict. The prostaglandin effect is determined by whether it is made from omega-3 fatty acids or omega-6 fatty acids.[126]

The terms *omega-3 fatty acids* and *omega-6 fatty acids* are defined by the chemical structure of the polyunsaturated fatty acids. Omega-3 fatty acids are termed this way because the first site of unsaturation (double bond) in the fatty acid chain is after the third carbon from the omega end of the molecule (see page 22 in Module 1 for a review). Omega-6 fatty acids have the first site of unsaturation (double bond) in the fatty acid chain after the sixth carbon from the omega end of the molecule.

emulsifier A substance with aqueous and lipid affinities (water-soluble and fat-soluble chemical attractions), such as bile and lecithin, that promotes the formation of a stable mixture, or emulsion, of oil and water.

omega-6 fatty acids Polyunsaturated fatty acids in which the first double bond is after the sixth carbon, counting from the methyl (CH_3) end of the carbon chain.

eicosanoids Biologically active compounds, such as prostaglandins, derived from long-chain polyunsaturated fatty acids that help to regulate blood pressure, blood clotting, blood lipids, inflammation, and other body functions.

The beneficial prostaglandins are made from omega-3 fatty acids and, more efficiently, from fish oils that are highly unsaturated fatty acids (HUFAs) and positively impact cardiovascular health. Some of the actions include reducing blood clotting, blood pressure, and blood lipids. Specifically, they increase HDL and decrease LDL cholesterol levels. Additionally, they have also been found to improve **immunity**, alleviate arthritis and asthma, and reduce inflammation. Popular dietary omega-3 fatty acids that are HUFAs include **eicosapentaenoic acid (EPA)** and **docosahexaenoic acid (DHA)**, both from fish, and the essential polyunsaturated fatty acid alpha-linolenic acid, from flax seeds, walnuts, and some plant oils. On the other hand, diets high in omega-6 fatty acids have been associated with inflammation and the growth of tumors. If too many of the prostaglandins made from omega-6 fatty acids are not balanced with those produced from the omega-3 fatty acids, wellness is definitely compromised.[11,31,34,94,105]

Recommended Intakes of Omega-3 and Omega-6   The AHA recommends consuming 0.5 to 1.8 grams of omega-3 fatty acids per day as fatty fish or supplements. Further, the AHA recommends an intake of 1.5 to 3 grams of alpha-linolenic acid (an omega-3 fatty acid) per day. The ratio of omega-3 fatty acids to omega-6 fatty acids consumed in the diet is a current area of investigation. The popular ratio advocated to promote optimal health is 1:4. An intake of 2 to 4 grams of a combination of EPA and DHA can improve the blood lipid profile. To achieve this intake level and ratio, very careful management of dietary fats is required. Interestingly, dietary sources of the various fatty acid types can be associated with geographic region (see Figure 3.30). The closer the fat-producing species is to the equator, the hotter the environmental temperature, and thus more solid and saturated fatty acids are present. The farther away the fat-producing species is from the equator, the colder the temperature and thus more fluid and unsaturated fatty acids are present.[D,137]

Other Factors Affecting Heart Health   There is more to the dietary impact on heart health and the promotion or prevention of atherosclerosis than fatty acids. Other noteworthy lifestyle factors include appropriate intakes of folate and niacin, use of margarine with plant stanols and omega-3 fatty acids, moderate intake of alcohol (especially red wine) for those who drink, and consuming soluble fiber and soy protein, along with a regular program of aerobic physical activity.

immunity The body's immune system defense mechanisms against invasion of foreign entities.

eicosapentaenoic acid (EPA) An omega-3 polyunsaturated fatty acid with 20 carbons and five double bonds with health-promoting properties, present in fish and synthesized in limited amounts in the body from the essential omega-3 fatty acid alpha-linolenic acid (18 carbons with three double bonds).

docosahexaenoic acid (DHA) An omega-3 polyunsaturated fatty acid with 22 carbons and six double bonds with health-promoting properties, present in fish and synthesized in limited amounts in the body from the essential omega-3 fatty acid alpha-linolenic acid (18 carbons with three double bonds).

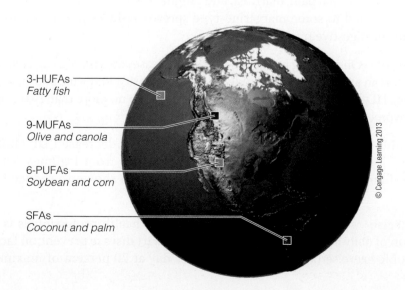

3-HUFAs
Fatty fish

9-MUFAs
Olive and canola

6-PUFAs
Soybean and corn

SFAs
Coconut and palm

© Cengage Learning 2013

FIGURE 3.30 The geography of fatty acids: Highly unsaturated fatty acids (HUFAs), monounsaturated fatty acids (MUFAs), polyunsaturated fatty acids (PUFAs), and saturated fatty acids (SFAs).

Recall the health claims that can be made on the food package label regarding soy protein, plant stanols and sterols, dietary fiber, and eating low-fat foods (see page 55 in Module 2 for a review).[48,56,116,131,153]

Role of Folate Evidence shows that elevated levels of **homocysteine** cause arterial wall damage; thus, it is established as an independent risk factor contributing to heart disease. Homocysteine is an intermediate amino acid in the conversion to methionine and cysteine. Elevated homocysteine is a condition known as hyperhomocystemia or homocystemia. The B vitamins—especially folate, B_6, and B_{12}—are required as enzyme cofactors to lower homocysteine levels in the blood. Most individuals eat plenty of animal-based foods providing enough vitamin B_6 and B_{12}, but not enough plant-based foods providing folate. Thus, many individuals need to optimize their intake of folate by consuming more dark green leafy vegetables. This can improve heart health and normalize homocysteine metabolism. Taking 400 micrograms of folic acid per day in supplement form is also safe and believed to be heart healthy. Some people with hyperhomocystemia can be successfully treated with folic acid supplementation. Also noteworthy is the scientific evidence showing that individuals with alcoholism often manifest folic acid deficiency and hyperhomocystemia and also benefit from taking a folic acid supplement.[48,114]

Role of Niacin High doses (1,000 to 4,000 milligrams per day) of niacin have been used to increase HDL and decrease LDL cholesterol. However, niacinamide is the usual form of vitamin B_3 in dietary supplements and fortified foods. Niacinamide will not alter lipoprotein fractions. The Tolerable Upper Intake Level for niacin is 35 milligrams per day for adults. Commonly, 50 milligrams will cause a flushing response (vasodilation akin to a hot flash) and a slight lowering of LDL cholesterol and elevation of HDL cholesterol.

Role of Statins Prescription medications known as statin-type cholesterol-lowering drugs reduce the synthesis of cholesterol in the liver by blocking the activity of the enzyme HMGCoA reductase. Because of the effect on the liver, liver function needs to be routinely monitored in statin drug users. Some people who take statin drugs experience negative side effects, such as muscle pain and fatigue, which, like abnormal liver function, can implicate serious ill effects of the drug. The dietary supplement red yeast rice also contains the statin compound and is effective for lowering blood cholesterol. Additionally, there are drugs that inhibit cholesterol absorption or block cholesterol and bile absorption in the small intestine, but they may cause abdominal pain, diarrhea, and fatigue. A natural plant stanol named benecol that is added to some margarine-type spreads reduces the absorption of cholesterol in the digestive tract.[27,35,40,106]

Role of Alcohol One serving per day of alcohol decreases the risk of a cardiovascular accident. It is an anticoagulant and reduces blood viscosity. Red wine and/or grape juice increases HDL cholesterol. Men should limit intake to no more than two drinks per day and women to no more than one drink per day.[31,105,122]

Role of Soluble Fiber Soluble fiber has been shown to decrease LDL cholesterol without lowering HDL. The consumption of fiber from legumes, whole oats, and psyllium-based natural fiber laxatives has been shown to reduce LDL cholesterol.

Role of Exercise Almost universally accepted by all health professionals is the incorporation of daily aerobic physical activity as a heart disease prevention factor. Regular aerobic exercise (at least 30 minutes per day at 70 percent of maximum

TS Photography/Getty Images

Red wine has heart health–promoting properties.

homocysteine An intermediary compound in the metabolism of the amino acids methionine and cysteine that can build up in the blood stream when the diet is inadequate in folic acid, vitamin B_6, and vitamin B_{12}, which are all needed for the metabolic conversions.

heart rate, most days of the week) has been shown to increase HDL cholesterol and decrease LDL cholesterol and blood pressure.[47,88]

| Lipids in Cancer | Cancer is a disease in which cells become abnormal and engage in uncontrolled cell growth. The cancerous cells take nourishment from the normal cells, and their growth impairs normal tissue and organ functioning nearby. When the cancer becomes advanced and possibly spreads throughout the body, the body systems can shut down, and life fails to be sustained.

Cancer is the second leading cause of death in American adults. The term *cancer* represents a group of cellular diseases from different origins, lineages, or tissues. The most prevalent types of adult cancers that are diet and lifestyle related include lung, breast, prostate, and colon cancers, followed by pancreas, esophagus, stomach, and liver cancers (see Figure 3.31). When the cancer cell is in a solid tissue such as lung, breast, or prostate, a tumor forms. When it is in the blood stream or derived from bone marrow, leukemia occurs.

One thing all cancer cells have in common is uncontrolled cell growth. Cancer cells become abnormally engaged in cellular division and lose the normal cell cycle controls through the processes of initiation, promotion, and progression (see BioBeat 3.12 and Figure 3.32). Collectively, this is called carcinogenesis. The carcinogenic initiating and promoting agents could come from the environment or diet, coupled with metabolic activity and genetic predisposition. Despite the complexity of cancer as a disease state, years of scientific study have demonstrated that diet can promote or reduce the risk for developing diet-related cancers.[B,I,36,55,75,87,138]

Exercise improves blood lipids, promotes heart health, and reduces the risk for cancer.

Nutrition and Cancer In general, scientific research demonstrates that individuals who consume Calories in excess of their need and/or consume a high total fat diet have increased incidence of cancer and heart disease (see Table 3.13 on page 151). When it comes to dietary fat, the type of fat makes a difference. The intake of too much total fat (more than 35 percent of total Calories) and saturated fatty

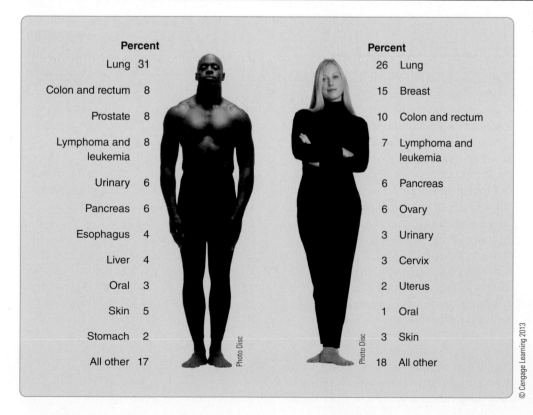

FIGURE 3.31 The leading types of cancer by gender.

Percent

Lung	31
Colon and rectum	8
Prostate	8
Lymphoma and leukemia	8
Urinary	6
Pancreas	6
Esophagus	4
Liver	4
Oral	3
Skin	5
Stomach	2
All other	17

Percent

26	Lung
15	Breast
10	Colon and rectum
7	Lymphoma and leukemia
6	Pancreas
6	Ovary
3	Urinary
3	Cervix
2	Uterus
1	Oral
3	Skin
18	All other

© Cengage Learning 2013

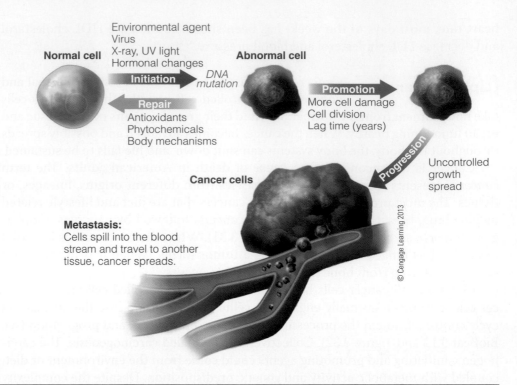

Environmental agent
Virus
X-ray, UV light
Hormonal changes

Normal cell

Initiation → *DNA mutation*

Abnormal cell

Repair
Antioxidants
Phytochemicals
Body mechanisms

Promotion
More cell damage
Cell division
Lag time (years)

Progression

Uncontrolled growth spread

Cancer cells

Metastasis:
Cells spill into the blood stream and travel to another tissue, cancer spreads.

© Cengage Learning 2013

FIGURE 3.32 The process of carcinogenesis.

acids (more than 7 percent of total Calories) promotes heart disease, and intake of too much total fat and polyunsaturated fatty acids promotes cancer. Assessing the healthiness of a person's fatty acid intake can be determined by calculating the poly-unsaturated fatty acid to saturated fatty acid ratio (P:S) (see Appendix B for details and a sample calculation). Because it is difficult to consume a palatable diet without any saturated fatty acids, and there are essential fatty acids that are polyunsaturated

BioBeat 3.12

Cancer and the Impact of Nutrition

Cancer occurs through the process known as carcinogenesis. The genetic material in a normal cell can be altered to become a cancerous cell by a variety of things. Ultraviolet light, radiation, chemicals in tobacco, and environmental pollutants can alter DNA and initiate cancer, or change a normal cell into an abnormal cell that engages in uncontrolled growth. There is also strong evidence linking dietary practices to cancer. Diets high in fat are known to promote cancer. This means the initiated cell is further enhanced to a cancerous state. High-fat diets especially promote cancer of the breast and prostate. Diets low in antioxidant nutrients are believed to contribute to stomach, esophageal, and lung cancers, while low-fiber diets

are linked to increased colorectal cancer. After a normal cell has undergone initiation and promotion, it becomes a cancer cell that engages in progression. At this stage, the cell divides uncontrollably. The cancer cells disrupt normal tissue function and can metastasize (move) to other sites in the body. At the point of progression, cancer treatment is needed and early detection is critical. Cancers that have metastasized are difficult to treat, so survival from the disease is reduced. However, some nutrients like retinoic acid (a form of vitamin A) have been used to treat precancerous lesions.[B,I,1,36,87]

Are there aspects of your diet that support carcinogenesis?

TABLE 3.13

Impact of Dietary Fats in Heart Disease and Cancer

Dietary Factor	Heart Disease	Cancer
Low fat intake (< 25% of total Calories)	Reduces disease risk	Reduces disease risk
Moderate fat intake (25–35% of total Calories)	Does not contribute to disease	Does not contribute to disease
High fat intake (> 35% of total Calories)	Increases disease risk	Increases disease risk
Fat distribution: • Excess saturated fatty acids • Low monounsaturated fatty acids • Inadequate polyunsaturated fatty acids	Increases total blood cholesterol, LDL (is atherogenic)	Is not ideal
Fat distribution: • Low saturated fatty acids • Low monounsaturated fatty acids • Excess polyunsaturated fatty acids	Decreases total blood cholesterol, HDL, and LDL (is not ideal)	Increases cancer risk (is tumorigenic)
Fat distribution: • Low saturated fatty acids • Majority monounsaturated fatty acids • Adequate polyunsaturated fatty acids	Decreases total blood cholesterol, LDL (is ideal)	Is not associated with cancer or heart disease risk (is not tumorigenic or atherogenic)

© Cengage Learning 2013

fatty acids, it is recommended to limit saturated fatty acid intake and strive to consume a P:S ratio of about 1:1. A diet high in fat with a P:S ratio value that is greater than or equal to 3:1 is **tumorigenic**. Further, when the diet is high in fat, especially polyunsaturated fatty acids, total blood cholesterol is lowered, but HDL is lowered along with LDL, so this is not ideal for heart health either. Having a P:S value that is less than or equal to 1:3 (or 0.33:1) is atherogenic (promotes atherosclerosis).

When the diet meets the AMDR, has 20 to 35 percent of total Calories from fat, limits saturated fatty acids and *trans* fatty acids, is balanced in providing enough essential polyunsaturated fatty acids, and emphasizes the consumption of more monounsaturated fatty acid, then health is optimized. Monounsaturated fatty acid intake can approach 15 percent of total Calories as recommended by the AHA. Additionally, the AMDRs for the essential fatty acids, linoleic acid (an omega-6 fatty acid), and alpha-linolenic acid (an omega-3 fatty acid) need to be met. Meeting the 5 to 10 percent of Calories AMDR for linoleic acid and 0.6 to 1.2 percent of Calories AMDR for alpha-linolenic acid reduces the risk of heart disease. However, to improve the lipid profile, an intake of 2 to 4 grams per day of the omega-3 fatty acids, DHA, and EPA from fish or fish oil is needed.

There is more to cancer than dietary fat. When a person's diet is low in fiber and phytochemicals from antioxidant-rich fruits and vegetables, the cancer risk may be increased. Eating plant foods in a rainbow of colors provides a safeguard from developing cancer. This is because of the augmentation of a variety of cancer-fighting

tumorigenic Cancer or a tumor caused by alterations in the DNA from many things, such as tobacco, drinking alcohol, being sedentary, eating high-fat animal foods, not eating a plant-based diet, and so forth.

mechanisms that occur as a result of the phytochemicals provided by the plant pigments (see Table 3.14). Of course, lifestyle factors and genetics are also involved in carcinogenesis. Individuals who use tobacco, drink alcohol, are sedentary, are obese, and are exposed to radiation (including ultraviolet sunlight such that redness and skin burning occurs) have an increased cancer risk. So, in general, eating a plant-based diet that is balanced for caloric intake, fat, and type of fat and practicing a healthy lifestyle helps to prevent cancer.[14,30,75,95]

TABLE 3.14

Jennifer Turley

© Cengage Learning 2013

The Color Code for Phytochemicals in Fruits and Vegetables		
Color	**Phytochemical**	**Fruit and Vegetable**
Red	Lycopene	Tomatoes, red peppers, raspberries, strawberries, watermelon
Yellow/green	Lutein, zeaxanthin	Honeydew melons, kiwis, leafy greens, avocados
Red/purple	Anthocyanins	Grapes, berries, wine, red apples, plums, prunes
Orange	Carotenoids	Carrots, mangos, papayas, apricots, pumpkins, yams
Orange/yellow	Flavonoids	Oranges, tangerines, lemons, peaches, cantaloupe
Green	Glucosinolates	Broccoli, Brussels sprouts, kale, cabbage
White/green	Allyl sulfides	Onions, leeks, garlic

Summary Points

- For heart health, it is important to limit saturated fatty acids to less than 7 percent of Calories and *trans* fatty acids to less than 1 percent of Calories and replace these fatty acids with polyunsaturated fatty acids up to 10 percent of Calories and monounsaturated fatty acids up to 15 percent of Calories while keeping total fat between 20 and 35 percent of Calories.

- Consume a 1:4 ratio of omega-3 fatty acids to omega-6 fatty acids, and for those with hypertriglyceridemia, take 2 to 4 grams of EPA and DHA per day.

- Consume an adequate amount of protein from plant sources, including soy products, meat alternatives, and legumes, and from the animal kingdom, low-fat dairy foods and lean meats with an emphasis on fish.

- High-fat diets with too many saturated fatty acids promote heart disease, and too many polyunsaturated fatty acids promote cancer.

- Heart health can be promoted by consuming legumes, whole grains, fruits, and vegetables to achieve 45 to 65 percent of Calories from carbohydrate, sufficient antioxidants, and adequate fiber intake, with an emphasis on soluble fiber.

- Folic acid prevents hyperhomocystemia, an independent risk factor for heart disease.

- Soluble fiber, soy, and plant stanols and sterols should be included in the diet for their heart health–promoting properties.

- Engage in exercise, especially aerobic physical activity, and avoid tobacco use to improve blood pressure and heart health and to reduce cancer risk.

- Choose to eat mostly plant foods for the prevention of cancer.

- Eat a variety of whole grains, fruits, and vegetables to support an antioxidant-rich diet, which is needed to prevent heart disease (by preventing the oxidation of polyunsaturated fatty acids, monounsaturated fatty acids, and LDL cholesterol) and cancer (by preventing the propagation of radical species to the point of DNA damage).

- If alcohol is consumed, limit intake and consider drinking red wine.
- Practice all aspects of a sound diet—Calorie control, adequacy, balance, moderation, and variety—for disease prevention.

Take Ten on Your Knowledge Know-How

1. What is atherosclerosis? What causes it? How does it contribute to heart disease?
2. What is a normal blood pressure reading? What are risks associated with having hypertension (high blood pressure)? What are some diet and lifestyle strategies to normalize blood pressure?
3. What are the various risk factors for heart disease? Which risk factors apply to you? Which risk factors are modifiable? What diet and lifestyle changes can you make to improve your heart health?
4. What are lipoproteins? What are the serum triglyceride and serum cholesterol carrying lipoproteins? What are their functions in the body? What are the risks associated with elevated levels or insufficient levels?
5. What is a normal (optimal or desirable) total cholesterol, LDL cholesterol, HDL cholesterol, and triglyceride serum level when taken in a fasting state? Are your personal levels healthy? If not, how can you improve them?
6. What is metabolic syndrome? What kind of TLC diet plan would you make based on your daily caloric need? What are the therapeutic attributes of the food choices in your diet plan?
7. What is dyslipidemia? How do the various dietary lipids, including omega-3 fatty acids, affect your serum lipid profile? What are other health benefits of consuming the correct ratio of omega-3 fatty acids?
8. Describe oxidation, antioxidants, hydrogenation, and partial hydrogenation in relation to heart disease and/or cancer. Which foods provide a good source of antioxidant nutrients?
9. What additional dietary interventions can you implement to improve your heart health?
10. What is carcinogenesis? Which diet and lifestyle factors reduce the risk of carcinogenesis?

 **SUMMARY**

CONTENT KNOWLEDGE

IN THIS MODULE YOU HAVE LEARNED:

- About the processes governing mechanical and chemical digestion, as well as the structure and function of the gastrointestinal tract.
- How plants make a variety of carbohydrates, some digestible and some indigestible, in a process called photosynthesis.
- How consuming enough fiber has many health benefits.
- Why carbohydrates are so important for central nervous system and red blood cell function, yet that very few are stored as glycogen in either the liver or the muscles.
- About common issues of carbohydrate metabolism, including lactose intolerance, hypoglycemia, and hyperglycemia.
- About the associations of dietary fat intake and the risks for developing heart disease and cancer.
- About the dangers of consuming too high or too low of a dietary protein intake.
- That the most common trigger of food allergies stems from the proteins in foods.

PERSONAL IMPROVEMENT GOALS

YOU WILL NOTE THAT IF YOU CONSUME:

- Your food when you are calm, you can promote digestion and absorption of nutrients more efficiently.
- A low glycemic index diet, healthy fats, moderate protein, and adequate fiber, you can reduce your risks of developing obesity, heart disease, and diet-related cancers.
- Mostly whole plant foods and limit your intake of processed foods, you can reduce your exposure to saturated fatty acids, *trans* fatty acids, and high-fructose corn syrup and provide your digestive tract the fiber it needs to maintain health.

Here is a tip for you: Have your nutritional risks assessed; know your blood pressure, blood cholesterol, and blood sugar. These and other nutritional risk factors are modifiable (can be improved) by making positive lifestyle changes. Thus, you can become a much healthier person and prevent a nutrition related medical condition in the future.

You can assess if you met the learning objectives for this module by successfully completing the Homework Assessment and the Total Recall activities (sample questions, case study with questions, and crossword puzzle).

Homework Assessment

50 questions

1. Which of the following statements is true regarding digestion?
   A. It occurs in the large intestine (also called the colon).
   B. It involves the immune system.
   C. It requires enzymes and hormones.
   D. It can occur in the kidney.
   E. All of the above.

2. Bile is required to emulsify:
   A. Starch
   B. Protein
   C. Alcohol
   D. Fat
   E. Fiber

3. Which statement below is false about enzymes and/or hormones?
   A. Enzymes are proteins that catalyze metabolic reactions.
   B. Hormones are chemical mediators that are produced at one site in the body, are released into the blood stream, and control the action of cells at a distal site in the body.
   C. Insulin and glucagon are hormones that regulate blood sugar.
   D. Cholecystokinin is an enzyme that digests starch.
   E. None of the above.

4. The type of enzyme that chemically digests fat is called:
   A. Bile
   B. Amylase
   C. Lipase
   D. Protease
   E. Cholecystokinin

5. An absorption mechanism that requires a specific carrier protein and adenosine triphosphate (ATP) energy is termed:
   A. Passive diffusion
   B. Facilitated diffusion
   C. Active transport
   D. Active facilitation
   E. Active diffusion

6. An anabolic reaction involves:
   A. Condensation
   B. Hydrolysis
   C. Digestion
   D. A and C
   E. B and C

7. An example of an anabolic reaction is:
   A. Glycogen production
   B. Emulsification of fat
   C. Digestion of protein
   D. A and B
   E. B and C

8. A catabolic reaction involves:
  A. Condensation
  B. Hydrolysis
  C. Synthesis
  D. A and C
  E. B and C

9. An example of a catabolic reaction is:
  A. Glycogen production
  B. Red blood cell formation
  C. Digestion of protein
  D. A and B
  E. B and C

10. Which of the following nutrients is required for brain, central nervous system, and red blood cell energy production and functioning?
  A. Amino acids
  B. Fatty acids
  C. Glucose
  D. Albumin
  E. Folate

11. Gary performed a dietary analysis and discovered that he consumed 3,900 Calories and 38 grams of dietary fiber per day. Which statement below is true regarding his fiber intake as it relates to the dietary recommendation of 1.4 grams of fiber per 100 Calories consumed?
  A. It is adequate, and he needs at least 20 to 30 grams of fiber to be adequate.
  B. It is adequate, and he needs at least 38 grams of fiber to be adequate.
  C. It is inadequate, and he needs at least 55 grams of fiber to be adequate.
  D. It is inadequate, and he needs at least 62 grams of fiber to be adequate.

12. Gary ate 40 grams of saturated fat. Given that he consumed 3,900 Calories, Gary is:
  A. Consuming a high saturated fat diet
  B. In compliance with the American Heart Association recommendation for saturated fat intake
  C. In need of reducing his saturated fat intake
  D. Not in compliance with the American Heart Association recommendation for saturated fat intake
  E. A, C, and D are true

13. Amanda is a 15-year-old female. Her fasting blood glucose level was determined to be 145 mg/dL. Therefore, she may be at risk for:
  A. Anemia
  B. Hypoglycemia
  C. Homocystemia
  D. Uremia
  E. Diabetes

14. Insulin is a hormone that:
  A. Is produced by cells in the pancreas in response to high blood glucose levels
  B. Catalyzes the catabolic reaction of converting glucose to adenosine triphosphate (ATP) in the liver cells
  C. Causes blood sugar levels to decline
  D. A and B
  E. A and C

15. Glucagon is:
  A. An enzyme that metabolizes glucose in the pancreas
  B. A hormone that increases glucose levels in the blood stream
  C. An antibody that reacts against glucose in the blood stream
  D. Another name for glycogen
  E. None of the above

16. Which of the following is true regarding lactose intolerance?
  A. It causes hyperactive peristalsis
  B. Sickness occurs after eating dairy products
  C. It is due to an inadequate lactase enzyme
  D. It is tied to genetics
  E. All of the above

17. Which of the following dietary factors have been shown to be protective against heart disease?
  A. *Trans* fatty acids (decrease HDL)
  B. Omega-3 fatty acids (increase HDL and decrease LDL)
  C. Linoleic acid (decreases LDL and HDL)
  D. The soluble fiber pectin (decreases LDL)
  E. B and D

18. Which of the following is a cancer protective agent found in the plant kingdom?
  A. Soluble fiber
  B. Insoluble fiber
  C. Water-soluble antioxidant vitamin C
  D. Fat-soluble antioxidant vitamin E
  E. All of the above

19. An antibody is produced by the immune cells in response to an antigen.
  A. True   B. False

20. An antigen is produced by the immune cells in response to an antibody.
  A. True   B. False

21. Protein excess is called:
  A. Marasmus
  B. Aproteinemia
  C. Kwashiorkor
  D. Kwashiorkor-marasmus
  E. None of the above

22. Amanda's blood pressure is 135 over 87 mmHg. Her blood pressure:
    A. Is optimal
    B. Indicates prehypertension
    C. Indicates stage 1 hypertension
    D. Indicates stage 2 hypertension

23. Gary has an LDL cholesterol reading of 171 mg/dL. This LDL cholesterol value is:
    A. Optimal
    B. Borderline high
    C. High
    D. Very high

24. Gary has an HDL cholesterol reading of 63 mg/dL. This HDL cholesterol value is:
    A. Low
    B. High

25. Antioxidants:
    A. Control oxidation
    B. Include vitamins C and E
    C. Play a role in cancer risk reduction
    D. Play a role in heart disease risk reduction
    E. All of the above

26. Janalee and Jacob are college students and athletes. Janalee plays basketball and Jacob runs track. They are both enrolled in a nutrition class. They have become friends and spend time studying together. They both need to improve their exam performances, so they began to review their notes for the upcoming test. Because they are both athletes, their liver and muscle glycogen stores are always depleted after practice. They begin to talk about what glycogen is. Janalee is correct when she states that glycogen is a complex carbohydrate made up of glucose molecules.
    A. True   B. False

27. Jacob replies, "Oh, yes. It is animal starch and my stores will allow me to run fast for a longer time." Janalee begins to complain about how sluggish her muscles felt at practice, so they begin to talk about what they should eat in order to replete their glycogen stores. Janalee is knowledgeable about nutrition when she suggests peanuts are a great food selection to replenish muscle glycogen stores due to the high glycemic index and glucose content.
    A. True   B. False

28. Jacob spouts out that he really likes the insulin effect. He has heard that muscle glycogen stores will increase after exercise in response to eating carbohydrate. He is correct when he tells Janalee that eating enough foods with high glycemic indexes will cause blood sugar to rise high enough

to stimulate insulin production and secretion from the beta cells in the pancreas, and insulin will allow glucose to enter muscle cells.
    A. True   B. False

29. Janalee asks Jacob if he knows his blood cholesterol level, but before he can respond, Janalee explains that her total blood cholesterol is in the borderline high category. Janalee should begin to consume more legumes in her diet for the additional soluble fiber source, which can have a blood cholesterol–lowering effect.
    A. True   B. False

30. Jacob cannot remember what the dominant lipoprotein fractions contain. Then, Janalee responds correctly, "All lipoprotein fractions, to some extent, are made up of phospholipids, cholesterol, protein, and triglycerides."
    A. True   B. False

31. Jacob then asks, "What lipoprotein fraction circulates in the lymph before entering the blood stream?" Janalee is correct when she states, "Chylomicrons."
    A. True   B. False

32. Jacob correctly states that the lipoprotein fractions that make up total blood cholesterol levels include high-density lipoprotein cholesterol (HDL) and very-low-density lipoprotein cholesterol (VLDL).
    A. True   B. False

33. Jacob has had enough chit-chat about blood cholesterol and lipoproteins. He asks Janalee to explain what antioxidants do to protect polyunsaturated fatty acids, and she is correct when she states, "Antioxidants help to oxidize the double bonds in the fatty acid chains."
    A. True   B. False

34. Jacob asks Janalee about what she knows regarding the absorptive mechanisms or how substances are transported across membranes, and she replies, "Not much." However, Janalee is correct when she tells Jacob that passive diffusion is a selective transport mechanism.
    A. True   B. False

35. Next, Janalee asks, "How about hormones, Jacob? What are they?" Jacob is correct when he explains that hormones are catalysts for chemical reactions.
    A. True   B. False

36. Jacob asks, "Do you remember what the accessory organs of digestion are?" Janalee answers correctly, "The duodenum, jejunum, and ileum."
    A. True   B. False

37. "What does the pancreas do?" asks Janalee. Jacob answers correctly when he says, "It produces bile and bicarbonate."

    A. True   B. False

38. "I understand now how we make energy, but how do plants make energy?" asks Janalee. Jacob answers correctly when he says, "Plants use photosynthesis, which requires hemoglobin, carbon, and water to make carbohydrate."

    A. True   B. False

39. Janalee is correct when she adds that photosynthesis is an anabolic process.

    A. True   B. False

40. Janalee further inquires about plants' abilities to make fiber. Jacob explains to Janalee correctly that the most abundant type of fiber consumed in the diet is cellulose, and cellulose is a complex carbohydrate, made from glucose molecules.

    A. True   B. False

41. "Wait, don't plants make protein, too?" Janalee asks. Jacob is correct when he says that the proteins generally made by plants are missing one of the essential amino acids; thus, plant proteins are usually considered sources of low-quality proteins.

    A. True   B. False

42. "Hey, but eating a plant-based diet promotes health. Even the MyPlate food guidance system recommendations support eating most of the foods from the plant kingdom," says Janalee. Jacob is correct when he says that consuming a diet rich in whole grains and colorful fruits and vegetables can reduce the risk of developing heart disease and cancer.

    A. True   B. False

43. "A diet rich in whole grains and colorful fruits and vegetables would support a high fiber intake, but those nutritionists are categorizing fanatics. What are the categories for the types of dietary fibers?" asks Jacob. Janalee is correct when she explains to Jacob that the two fiber categories are soluble and insoluble fiber.

    A. True   B. False

44. "Yes, but what does total fiber mean?" Jacob asks. Janalee is correct when she replies, "It is the total amount of indigestible carbohydrate provided from dietary fiber and any functional fiber that may have been consumed."

    A. True   B. False

45. "Oh my! My mind is getting foggy and I am having a hard time concentrating," Janalee professes. "It has been a while since I have eaten." Jacob thinks that Janalee should eat a cooked egg to help ease her foggy mind. He is correct when he tells Janalee that her mental capacity is diminished because her blood cholesterol levels drop between meals.

    A. True   B. False

46. Jacob thinks that when you cook an egg, the egg protein is digested. Janalee sets him straight when she explains to Jacob that proteins are denatured by heat, acid, alkali, and metals, while proteins are digested by enzymes.

    A. True   B. False

47. Janalee tells Jacob that she can't eat eggs. Since she was a young child, she has struggled with food allergies, especially to eggs, peanuts, and cow's milk. Which of the following physiological responses could Janalee experience after eating eggs?

    A. Gastrointestinal discomfort
    B. Dermatological disturbances
    C. Upper respiratory tract complications
    D. Anaphylactic shock
    E. All of the above

48. Jacob tells Janalee that her food allergies were inherited. Janalee says, "Now that you mention it, my mom has all sorts of funny food reactions. I used to think she was so weird!" Which of the following foods would probably not cause an adverse reaction if Janalee's mom ate it?

    A. Wheat
    B. Soybeans
    C. Rice
    D. Shellfish
    E. Tree nuts

49. Jacob tells Janalee not to feel bad, because his gene pool isn't so squeaky clean either. Jacob's dad found out that his total blood cholesterol level was high, and so was his LDL. Which of the following recommendations would you make to Jacob's father to lower his total blood cholesterol and his LDL?

    A. Reduce protein intake
    B. Increase sugar intake
    C. Reduce saturated fat intake
    D. Increase folic acid intake
    E. Maintain insoluble fiber intake

50. Jacob's mother is obese and was just diagnosed with type 2 diabetes. Which of the following recommendations would you make to Jacob's mom to restore blood sugar control?

    A. Reduce blood pressure
    B. Increase insulin intake
    C. Reduce body fat content
    D. Increase folic acid intake
    E. Maintain insoluble fiber intake

Total Recall

SAMPLE QUESTIONS

True/False Questions

1. The liver produces enzymes that aid in digestion.

2. Glucagon is an enzyme that decreases blood sugar levels.

3. Type 1 diabetes usually develops during adulthood.

4. HDL is the "good" kind of cholesterol.

5. The amino acids in dietary proteins are released when the protein is denatured.

Multiple Choice Questions: Choose the best answer.

6. Which type of fatty acid has the worst effect on blood cholesterol?
   A. Saturated
   B. Polyunsaturated
   C. Monounsaturated
   D. *Cis*
   E. Omega-3

7. Which of the following is true about fiber?
   A. It provides 4 Calories per gram.
   B. It decreases gastric emptying time.
   C. It decreases transit time.
   D. It decreases stool volume.
   E. All of the above.

8. Which food source below provides the most dietary fiber?
   A. Lettuce
   B. Orange juice
   C. Corn flakes
   D. Melon
   E. Legumes

9. Which tissue is dependent upon glucose as an energy source?
   A. Liver
   B. Pancreas
   C. Heart
   D. Brain
   E. Bone

10. Cancer is characterized by:
    A. Uncontrolled cell growth
    B. Autoimmunity
    C. Cell suicide
    D. Atherosclerosis
    E. Hypertension

CASE STUDY

Sasha and Toby have been married for two years. They are both musicians; Sasha plays the flute, and Toby plays the tuba. Sasha is 21 years old, is 5'1" tall, weighs 110 pounds, and is female. Toby is 25 years old, is 5'9" tall, weighs 207 pounds, and is male. They are both physically inactive (low active). Sasha has a normal body weight, while Toby is obese. Sasha and Toby's Estimated Energy Requirements are approximately 1,970 and 3,150 Calories per day, respectively. Let's explore their diet and disease risk and relationships based on the information in Table 3.15.

1. Based on the information provided for Sasha, she may have:
   A. Heart disease
   B. Cancer
   C. Diverticulosis
   D. Hypoglycemia
   E. Diabetes

2. Based on the information provided for Toby, he may be at an increased risk for:
   A. Heart disease
   B. Cancer
   C. Diverticulosis
   D. A and C
   E. A, B, and C

TABLE 3.15

The Laboratory, Dietary, and Clinical Case Study Data for Sasha and Toby

	Sasha	Toby
Fasting Blood Values	Total cholesterol: 147 mg/dL LDL: 95 mg/dL HDL: 52 mg/dL Triglycerides: 130 mg/dL Glucose: 58 g/dL	Total cholesterol: 218 mg/dL LDL: 183 mg/dL HDL: 35 mg/dL Triglycerides: 147 mg/dL Glucose: 105 g/dL
Blood Pressure	100/65 mmHg	131/92 mmHg
Typical Food Intake	Breakfast: yogurt and whole-grain cereal with fresh berries Lunch: salads (bean, green, fruit, pasta, etc.) Dinner: meat-based casseroles, spaghetti, pizza Snacks: licorice, jelly beans, hard candy, fruit, nuts Drinks: water, soda, herbal tea	Breakfast: scrambled eggs with ham or doughnut/pastry Lunch: fast-food sandwich or burger, with fries and soda or milkshake Dinner: meat-based casseroles, spaghetti, pizza Snacks: beef jerky, chips, cheese, crackers, pretzels Drinks: soda, whole milk, milkshakes, occasional water
Typical Selected Nutrient Profile	Total Calories: 1,800–2,000 Total fat: 25–30% Total carbohydrate: 60–65% Total protein: 10–15% Saturated fatty acid: 6–9% Cholesterol: 100–200 mg Fiber: 30–35 g Sugar: 25–30% Alcohol: 0 g Vitamin E: 90–110% of DRI Vitamin C: 100–120 mg Sodium: 1,200–1,800 mg	Total Calories: 3,300–3,500 Total fat: 35–40% Total carbohydrate: 30–45% Total protein: 35–40% Saturated fatty acid: 12–16% Cholesterol: 300–500 mg Fiber: 20–25 g Sugar: 15–20% Alcohol: 0 g Vitamin E: 40–50% of DRI Vitamin C: 50–60 mg Sodium: 3,500–4,000 mg
Symptoms	Frequent nausea, diarrhea, and skin itching. Felt weak, sweaty, and had a headache at time of fasting assessment.	Frequent constipation.

© Cengage Learning 2013

3. Based on the information provided for Sasha, she may have allergies.

   A. True   **B.** False

4. Which of the following factors could be contributing to Toby's elevated blood pressure?

   A. High sugar intake
   B. High sodium intake
   C. Lack of alcohol intake
   D. Lack of dietary fiber
   E. High cholesterol intake

5. The fasting blood sugar (glucose) values for both Sasha and Toby are normal.

   A. True   **B.** False

6. Toby has more than five risk factors for heart disease.

   A. True   **B.** False

7. Which statement is the most true regarding Sasha and Toby's protein intake?

   A. Sasha is at risk for kwashiorkor.
   B. Toby is at risk for kwashiorkor.
   C. Sasha is at risk for accelerated kidney aging.
   D. Toby is at risk for accelerated kidney aging.
   E. Both Sasha and Toby have a healthy and balanced protein intake.

8. Toby's dietary fiber intake should be ___ grams per day, according to his personalized DRI.

   A. 22 to 25
   B. 38 to 40
   C. 46 to 49
   D. 53 to 56
   E. 61 to 63

9. Based on Toby's usual Calorie intake, his body will be engaged in fat synthesis and storage. Metabolically, fat synthesis is _____ while fat storage is _____.
   A. Catabolic, anabolic
   B. Anabolic, catabolic
   C. Anabolic, neither catabolic nor anabolic
   D. Catabolic, neither catabolic nor anabolic
   E. Catabolic, both catabolic and anabolic

10. Sasha and Toby should absorb the vitamin E and C from their diet in their:
    A. Stomach
    B. Liver
    C. Small intestine
    D. Colon
    E. Mouth

FUN-DUH-MENTAL PUZZLE

ACROSS
4. A risk factor for heart disease.
5. Organ that makes bile and stores glycogen.
6. Disease caused by excess sugar in the blood stream.
9. Process causing protein shape to change in response to heat.
10. Food group that decreases cancer risk.
12. Clinical name for protein deficiency.
15. Type of food that may contain *trans* fatty acids.
17. Transportation vessels for water-soluble substances.
19. Anabolic process by which plant cells make carbohydrates.
20. Action that moves food along the gastrointestinal tract.
22. Fiber does what to transit time?
24. Energy source of the brain.
25. People with type 2 diabetes usually have insulin _____.

DOWN
1. Amino acid intermediate that causes arterial damage.
2. Enzyme that breaks down starch.
3. Digestion involving muscles and nerves.
7. Protein type that has a uniform alpha-helix or beta-sheet formation.
8. Process generating sticky lipid fragments that increase atherosclerosis.
11. An excellent food source of fiber.
13. Leading cause of arterial wall injury and atherosclerosis.
14. A soluble fiber found in apples.
16. Gallbladder secretion that emulsifies fat.
18. Monounsaturated fatty acids decrease this type of serum cholesterol.
21. Hormone made in the pancreas to reduce blood sugar levels when needed.
23. Common allergic food.

References

Web Resources

A. American Association of Diabetes Educators: www.diabeteseducator.org
B. American Cancer Society: www.cancer.org
C. American Diabetes Association: www.diabetes.org
D. American Heart Association: www.heart.org
E. Centers for Disease Control, National Center for Chronic Disease Prevention and Health Promotion: www.cdc.gov/chronicdisease
F. DASH Eating Plan: www.nhlbi.nih.gov/health/public/heart/hbp/dash/new_dash.pdf
G. Gastrointestinal System, Medical College of Wisconsin: chorus.rad.mcw.edu/index/4.html
H. Lipids Online: www.lipidsonline.org
I. National Cancer Institute: www.cancer.gov
J. Understanding Evolution: evolution.berkeley.edu/evolibrary/home.php

Works Cited

1. Afman, L., & Muller, M. (2006). Nutrigenomics: From molecular nutrition to prevention of disease. *Journal of the American Dietetic Association, 106*(4), 569–576.
2. Albers, M. J., Harnack, L. J., Steffen, L. M., & Jacobs, D. R. (2008). 2006 marketplace survey of *trans-*fatty acid content of margarines and butters, cookies and snack cakes, and savory snacks. *Journal of the American Dietetic Association, 108*(2), 367–370.
3. Allman-Farinelli, M. A. (2009). Do calorically sweetened soft drinks contribute to obesity and metabolic disease? *Nutrition Today, 44*(1), 17–20.
4. Allman-Farinelli, M. A., Gomes, K., Favaloro, E. J., & Petocz, K. (2005). A diet rich in high-oleic-acid sunflower oil favorably alters low-density lipoprotein cholesterol, triglycerides, and factor VII coagulant activity. *Journal of the American Dietetic Association, 105*(7), 1071–1079.
5. American Diabetes Association. (2006). Standards of medical care in diabetes, 2006. *Diabetes Care, 29*, S4–S42.
6. American Dietetic Association. (1993). Position of the American Dietetic Association: Health implications of dietary fiber. *Journal of the American Dietetic Association, 93*, 1446–1447.
7. American Dietetic Association. (2002). Health implications of dietary fiber. *Journal of the American Dietetic Association, 102*, 993–1000.
8. American Dietetic Association. (2006). Diseases and conditions. *Evidence Analysis Library*. Accessed at: http://adaevidencelibrary.com.
9. Anderson, J. J. B., Prytherch, S. A., Sparling, M., Barrett, C., & Guyton, J. R. (2006). The metabolic syndrome. *Nutrition Today, 41*(3), 115–122.
10. Archer, S. L. (2007). Staying focused on the undernourished child—India. *Journal of the American Dietetic Association, 107*(11), 1879–1881.
11. Baik, I., Abbott, R. D., Curb, J. D., & Shin, C. (2010). Intake of fish and omega-3 fatty acids and future risk of metabolic syndrome. *Journal of the American Dietetic Association, 110*(7), 1018–1026.
12. Barnicot, N. A. (2005). Human nutrition: Evolutionary perspectives. *Integrative Physiological & Behavioral Science, 40*, 114–117.
13. Bassuk, S. S., & Manson, J. E. (2008). Lifestyle and risk of cardiovascular disease and type 2 diabetes in women: A review of the epidemiologic evidence. *American Journal of Lifestyle Medicine, 2*(3), 191–213.
14. Bazzano, L. A. (2006). The high cost of not consuming fruits and vegetables. *Journal of the American Dietetic Association, 106*(9), 1364–1368.

15. Behall, K. M., Scholfield, D. J., & Hallfrisch, J. (2006). Whole-grain diets reduce blood pressure in mildly hypercholesterolemic men and women. *Journal of the American Dietetic Association, 106*(9), 1445–1449.

16. Berkow, S. E., & Barnard, N. D. (2005). Blood pressure regulation and vegetarian diets. *Nutrition Reviews, 63*(1), 1–8.

17. Bernstein, A. M., Treyzon, L., & Li, Z. (2007). Are high-protein, vegetable-based diets safe for kidney function? A review of the literature. *Journal of the American Dietetic Association, 107*(4), 644–650.

18. Bertrand, P. P., & Bertrand, R. L. (2007). Teaching basic gastrointestinal physiology using classic papers by Dr. Walter B. Cannon. *Advances in Physiology Education, 31*(2), 136–139.

19. Bibbins-Domingo, K., Chertow, G. M., Coxson, P. G., Moran, A., Lightwood, J. M., Pletcher, M. J., & Goldman, L. (2010). Projected effect of dietary salt reductions on future cardiovascular disease. *The New England Journal of Medicine, 362*(7), 590–598.

20. Bilsborough, S., & Mann, N. (2007). A review of issues of dietary protein intake in humans. *International Journal of Sport Nutrition and Exercise Metabolism, 16*(2), 129–152.

21. Boguniewicz, M., Moore, N., & Paranto, K. (2008). Allergic diseases, quality of life, and the role of the dietitian. *Nutrition Today, 43*(1), 6–10.

22. Burrowes, J. D. (2006). Metabolic syndrome: Controversy and consensus. *Nutrition Today, 41*(3), 131–137.

23. Burrowes, J. D. (2007, November/December). Preventing heart disease in women: What is new in diet and lifestyle recommendations? *Nutrition Today, 42*(6), 242–247.

24. Cappellano, K. L. (2008). Food allergy and intolerances. *Nutrition Today, 43*(1), 11–14.

25. Cefalu, W. (2005). Glycemic control and cardiovascular disease: Should we reassess clinical goals? *The New England Journal of Medicine, 353*, 2707–2709.

26. Champagne, C. M. (2006). Dietary interventions on blood pressure: The dietary approaches to stop hypertension (DASH) trials. *Nutrition Reviews, 64*(2), S53–S56.

27. Chan, Y.-M., Varady, K., Lin, Y., Trautwein, E., Mensink, R. P., Plat, J., & Jones, P. J. H. (2006). Plasma concentrations of plant sterols: Physiology and relationship with coronary heart disease. *Nutrition Reviews, 64*(9), 385–409.

28. Chandra, R., & Liddle, R. A. (2007). Cholecystokinin. *Current Opinion in Endocrinology, Diabetes and Obesity, 14*(1), 63–67.

29. Chow, W. S. (2003). Photosynthesis: From natural towards artificial. *Journal of Biological Physics, 29*(4), 447–459.

30. Cleary, M. P., & Grossmann, M. E. (2009). Obesity and breast cancer: The estrogen connection. *Endocrinology, 150*(6), 2537–2542.

31. Coleman, E. (2010). The Mediterranean diet: A proved CVD preventive. *Today's Dietitian, 12*(2), 48–53.

32. Coups, E. J., Manne, S. L., Meropol, N. J., & Weinberg, D. S. (2007). Multiple behavioral risk factors for colorectal cancer and colorectal cancer screening status. *Cancer Epidemiology Biomarkers and Prevention, 16*(3), 510–516.

33. Cummings, J. H., & Stephen, A. M. (2007). Carbohydrate terminology and classification. *European Journal of Clinical Nutrition, 61*, S5–S18.

34. Cupples, C. (2008). Functional foods: Omega-3 polyunsaturated fatty acids for heart health. *Nutrition in Complementary Care, 10*(3), 41, 48–51.

35. Devaraj, S., & Jialal, I. (2006). The role of dietary supplementation with plant sterols and stanols in the prevention of cardiovascular disease. *Nutrition Reviews, 64*(7), 348–354.

36. Divisi, D., Di Tommaso, S., Salvemini, S., Garramone, M., & Crisci, R. (2006). Diet and cancer. *Acta Biomedica, 77*(2), 118–123.

37. Eckel, R. H., Borra, S., Lichtenstein, A. H., & Yin-Piazza, S. Y. (2007). Understanding the complexity of *trans* fatty acid reduction in the American diet: American Heart Association *Trans* Fat Conference, 2006 report of the *Trans* Fat Conference Planning Group. *Circulation, 115*, 2231–2246.

38. Edelman, R. (2005, May/June). Obesity, type 2 diabetes, and cardiovascular disease. *Nutrition Today, 40*(3), 119–123.

39. Elia, M., & Cummings, J. H. (2007). Physiological aspects of energy metabolism and gastrointestinal effects of carbohydrates. *European Journal of Clinical Nutrition, 61*, S40–S74.

40. Ellegard, L. H., Andersson, S. W., Normen, A. L., & Andersson, H. A. (2007). Dietary plant sterols and cholesterol metabolism. *Nutrition Reviews, 65*(1), 39–45.

41. Fernandez, M. L. (2007). The metabolic syndrome. *Nutrition Reviews, 65*(6), S30–S34.

42. Ferreira, C. T., & Seidman, E. (2007). Food allergy: A practical update from the gastroenterological viewpoint. *Journal of Pediatrics, 83*(4), 381–382.

43. Finley, C. E., Barlow, C. E., Halton, T. L., & Haskel, W. L. (2010). Glycemic index, glycemic load, and prevalence of the metabolic syndrome in the Cooper Center Longitudinal Study. *Journal of the American Dietetic Association, 110*(12), 1820–1829.

44. Folsom, A. R., Parker, E. D., & Harnack, L. J. (2007). Degree of concordance with DASH diet guidelines and incidence of hypertension and fatal cardiovascular disease. *American Journal of Hypertension, 20*(3), 225–232.

45. Foster-Powell, K., Holt, S. H. A., & Brand-Miller, J. C. (2002). International table of glycemic index and glycemic load values: 2002. *American Journal of Clinical Nutrition, 76*, 5–56. Accessed at: http://www.ajcn.org/cgi/content/full/76/1/5.

46. Franz, M. J. (2007). Lifestyle interventions across the continuum of type 2 diabetes: Reducing the risks of diabetes. *American Journal of Lifestyle Medicine, 1*(7), 327–334.

47. Gabriel, K. K., & Ainsworth, B. E. (2009). Building Healthy Lifestyles Conference: Modifying lifestyles to enhance physical activity and diet and reduce cardiovascular disease. *American Journal of Lifestyle Medicine, 3*(S1), 6S–10S.

48. Ganji, V., & Kafai, M. R. (2009). Demographic, lifestyle, and health characteristics and serum B vitamin status are determinants of plasma total homocysteine concentration in the post-folic acid fortification period, 1999–2004. *Journal of Nutrition, 139*(2), 345–352.

49. Garza, C., & Pelletier, D. L. (2007). Dietary guidelines: Past, present, and future. In E. Kennedy & R. Deckelbaum (Eds.), *The nation's nutrition*. Washington, DC: International Life Sciences Institute.

50. Gaskin, D. J., & Ilich, J. Z. (2009). Lactose maldigestion revisited: Diagnosis, prevalence in ethnic minorities, and dietary recommendations to overcome it. *American Journal of Lifestyle Medicine, 3*(3), 212–218.

51. Getz, L. (2008). Making sense of antioxidants. *Today's Dietitian, 10*(9), 50–54.

52. Getz, L. (2010). Spill the beans: Tips to help clients get their fiber from the best sources. *Today's Dietitian, 12*(5), 20–24.

53. Gibbons, A. (2006). Human evolution: There's more than one way to have your milk and drink it, too. *Science, 314*, 2.

54. Grabitske, H. A., & Slavin, J. L. (2008). Laxation and the like. *Nutrition Today, 43*(5), 193–200.

55. Grant, B., & Evert, A. (2008). Managing patients diagnosed with cancer and diabetes. *Today's Dietitian, 10*(8), 8–16.

56. Greenstone, C. L. (2007) A commentary on lifestyle medicine strategies for risk factor reduction, prevention, and treatment of coronary artery disease. *American Journal of Lifestyle Management, 1*(2), 91–94.

57. Gropper, S. S., Smith, J. L., & Groff, J. L. (2009). *Advanced nutrition and human metabolism* (5th ed.). Belmont, CA: Wadsworth.

58. Havel, P. J. (2005). Dietary fructose: Implications for dysregulation of energy homeostasis and lipid/carbohydrate metabolism. *Nutrition Reviews, 63*(5), 133–157.

59. Hayes, C., & Kriska, A. (2008). Role of physical activity in diabetes management and prevention. *Supplement to the Journal of the American Dietetic Association, 108*(4), S19–S23.

60. Heffler, E., Guida, G., Badiu, I., Nebiolo, F., & Rolla, G. (2007). Anaphylaxis after eating Italian pizza containing buckwheat as the hidden food allergen. *Journal of Investigational Allergology and Clinical Immunology, 17*(4), 261–263.

61. Hegde, V. L., & Venkatesh, Y. P. (2007). Anaphylaxis following ingestion of mango fruit. *Journal of Investigational Allergology and Clinical Immunology, 17*(5), 341–344.

62. Herman, D. R., Vargas, M. A., & Lin, Y. (2008). Therapies: Strategies for successful weight management programs: What are the options? *Nutrition in Complementary Care, 10*(3), 41, 44–48.

63. Hodorowicz, M. A. (2006, April). Interventions to normalize blood lipid levels. *Today's Dietitian, 8*(4), 20–22.

64. Hollander, J. M., & Mechanick, J. I. (2008). Complementary and alternative medicine and the management of the metabolic syndrome. *Journal of the American Dietetic Association, 108*(3), 495–509.

65. Hwang, J. B., Lee, S. H., Kang, Y. N., Kim, S. P., Suh S. I., & Kam, S. (2007). Indexes of suspicion of typical cow's milk protein-induced enterocolitis. *Journal of Korean Medicinal Science, 22*(6), 993–997.

66. Institute of Medicine. (1998). *Dietary Reference Intakes: Proposed definition and plan for review of dietary antioxidants and related compounds.* Washington, DC: National Academies Press.

67. Institute of Medicine. (2001). *Dietary Reference Intakes: Proposed definition of dietary fiber.* Washington, DC: National Academies Press.

68. Jambazian, P. R., Haddad, E., Rajaram, S., Tanzman, J., & Sabate, J. (2005). Almonds in the diet simultaneously improve plasma α-tocopherol concentration and reduce plasma lipids. *Journal of the American Dietetic Association, 105*(3), 449–454.

69. Janiszewski, P. M., Janssen, I., & Ross, R. (2007). Does waist circumference predict diabetes and cardiovascular disease beyond commonly evaluated cardiometabolic risk factors? *Diabetes Care, 30*(12), 3105–3109.

70. Johnson, T., Avery, G., & Byham-Gray, L. (2009). Vitamin D and metabolic syndrome. *Topics in Clinical Nutrition, 24*(1), 47–54.

71. Johnston, C. (2009). Functional foods as modifiers of cardiovascular disease. *American Journal of Lifestyle Medicine, 3*(S1), 39S–43S.

72. Jones, D. (2005). *Textbook of functional medicine.* Gig Harbor, WA: Institute for Functional Medicine.

73. Jones, J. R., Lineback, D. M., & Levine, M. J. (2006). Dietary Reference Intakes: Implications for fiber labeling and consumption: A summary of the International Life Sciences Institute North America Fiber Workshop, June 1–2, 2004, Washington DC. *Nutrition Reviews, 64*(1), 31–38.

74. Keskin, O., Gursoy, A., Ma, B., & Nussinov, R. (2008). Principles of protein: Protein interactions: What are the preferred ways for proteins to interact? *Chemical Reviews, 108*(4), 1225–1244.

75. Krukowski, R. A., & West, D. S. (2010). Consideration of the food environment in cancer risk reduction. *Journal of the American Dietetic Association, 110*(6), 842–844.

76. Kuller, L. H. (2006). Nutrition, lipids, and cardiovascular disease. *Nutrition Reviews, 64*(2), S15–S26.

77. Lapointe, A., Balk, E. M., & Lichtenstein, A. H. (2006). Gender differences in plasma lipid response to dietary fat. *Nutrition Reviews, 64*(5), 234–249.

78. Larsen, T. M., Dalskov, S.-M., van Baak, M., Jebb, S. A., Papadaki, A., Pfeiffer, A. F. H., . . . & Astrup, A., for the Diet, Obesity, and Genes (Diogenes) Project. (2010). Diets with high or low protein content and glycemic index for weight-loss maintenance. *The New England Journal of Medicine, 363*, 2102–2113.

79. Li, H., Chehade, M., Liu, W., Xiong, H., Mayer, L., & Berin, M. C. (2007). Allergen-IgE complexes trigger CD23-dependent CCL20 release from human intestinal epithelial cells. *Gastroenterology, 133*(6), 1905–1915.

80. Lopes, C., Aro, A., Azevedo, A., Ramos, E., & Barros, H. (2007). Intake and adipose tissue composition of fatty acids and risk of myocardial infarction in a male Portuguese community sample. *Journal of the American Dietetic Association, 107*(2), 276–286.

81. Lustig, R. H. (2010). Fructose: Metabolic, hedonic, and societal parallels with ethanol. *Journal of the American Dietetic Association, 110*(9), 1307–1321.

82. Mangels, R., Lucas, D., Barnard, N., & Lacey, J. (2010). Vegetarian approaches to type 2 diabetes. *Vegetarian Nutrition, 18*(3), 1, 8–9.

83. Mann, J. I. (2006). Nutritional recommendations for the treatment and prevention of type 2 diabetes and the metabolic syndrome: An evidence-based review. *Nutrition Reviews, 64*(9), 422–427.

84. McIntosh, G. H. (2001). Cereal foods, fibres and the prevention of cancers. *Australian Journal of Nutrition and Dietetics, 58*, S35–S45.

85. Meerschaert, C. M. (2007). Food allergy: A look at traditional and complementary diagnosis and treatment. *Today's Dietitian, 9*(7), 40–43.

86. Meydani, M. (2005). A Mediterranean-style diet and metabolic syndrome. *Nutrition Reviews, 63*(8), 312–314.

87. Michels, K. B., Mohllajee, A. P., Roset-Bahmanyar, E., Beehler, G. P., & Moysich, K. B. (2007). Diet and breast cancer: A review of the prospective observational studies. *Cancer Supplement, 109*(12), 2712–2748.

88. Miner, M. (2007). Fitness, antioxidants, and moderate drinking: All to lower cardiovascular risk. *American Journal of Lifestyle Medicine, 1*(2), 110–112.

89. Moon, M. (2008). Supplements: The role of a high potassium diet in managing blood pressure. *Nutrition in Complementary Care, 10*(4), 72–76.

90. Mozaffarian, D., Katan, M. B., Ascheria, A., Stampfer, M. J., & Willet, W. C. (2006). Trans fatty acids and cardiovascular disease. *The New England Journal of Medicine, 354*(15), 1601–1613.

91. Murphy, M. M., Douglass, J. S., & Birkett, A. (2008). Resistant starch intakes in the United States. *Journal of the American Dietetic Association, 108*(1), 67–78.

92. National Academy of Sciences, Institute of Medicine. (2008). *Science, evolution, and creationism.* Washington, DC: National Academies Press.

93. Nelms, M., Sucher, K., & Long, S. (2007). *Nutrition therapy and pathophysiology.* Belmont, CA: Brooks/Cole.

94. Nettleton, J. A., & Katz, R. (2005). Omega-3 long-chain polyunsaturated fatty acids in type 2 diabetes: A review. *Journal of the American Dietetic Association, 105*(3), 428–440.

95. Norat, T., & Riboli, E. (2001, February). Meat consumption and colorectal cancer: A review of epidemiological evidence. *Nutrition Reviews, 59*, 37–47.

96. Olson, B. H., & Keast, D. R. (2004, July/August). Effectiveness and safety of folic acid fortification. *Nutrition Today, 39*(4), 169–175.

97. Palmer, S. (2009). Fighting heart disease: The Dean Ornish way. *Today's Dietitian, 11*(2), 48–52.

98. Pan, Y. (2008). Overall diet quality is key in preventing metabolic syndrome. *Journal of the American Dietetic Association, 108*(2), 286.

99. Panagiotakos, D. B., Pitsavos, C., Skoumas, Y., & Stefanadis, C. (2007). The association between food patterns and the metabolic syndrome using principal components analysis: The ATTICA study. *Journal of the American Dietetic Association, 107*(6), 979–987.

100. Park, Y., Hunter, D. J., Spiegelman, D., Bergkvist, L., Berrino, F., van den Brandt, P. A., . . . & McCullough, M. L. (2005). Dietary fiber intake and risk of colorectal cancer: A pooled analysis of prospective cohort studies. *Journal of the American Medical Association, 294*, 2849–2857.

101. Perrigue, M. M., Mosivais, P., & Drewnowski, A. (2009). Added soluble fiber enhances the satiating power of low-energy-density liquid yogurts. *Journal of the American Dietetic Association, 109*(11), 1862–1868.

102. Peterson, C. G., Hansson, T., Skott, A., Bengtsson, U., Ahlstedt, S., & Magnussons, J. (2007). Detection of local mast-cell activity in patients with food hypersensitivity. *Journal of Investigative Allergology and Clinical Immunology, 17*(5), 314–320.

103. Phumg, O. J., Makanji, S. S., White, C. M., & Colman, C. I. (2009). Almonds have a neutral effect on serum lipid profiles: A meta-analysis of randomized trials. *Journal of the American Dietetic Association, 109*(5), 865–873.

104. Pierce, D. (2008). Exercise for diabetes prevention and treatment. *Today's Dietitian, 10*(10), 8–16.

105. Pineo, C. E., & Anderson, J. J. B. (2008). Cardiovascular benefits of the Mediterranean diet. *Nutrition Today, 43*(3), 114–120.

106. Polagruto, J. A., Wang-Polagruto, J. F., Braun, M. M., Lee, L., Kwik-Uribe, K., & Keen, C. L. (2006). Cocoa flavanol-enriched snack bars containing phytosterols effectively lower total and low-density lipoprotein cholesterol levels. *Journal of the American Dietetic Association, 106*(11), 1804–1813.

107. Posthauer, M. E. (2007). Lactose intolerance: Testing and challenges. *Today's Dietitian, 9*(7), 24–26.

108. Procopiou, M., &, Philippe, J. (2005). The metabolic syndrome and type 2 diabetes: Epidemiological figures and country specificities. *Cerebrovascular Diseases, 20*, 2–8.

109. Rakhovskaya, M., Jonnalagadda, S. S., & Khosla, P. (2006). Dietary fat, post-prandial lipids, and atherosclerosis. *SCAN'S (A Publication for Sports, Cardiovascular, and Wellness Nutritionists) PULSE, 25*(4), 12–16.

110. Rigby, A. J. (2004, September). Insulin resistance: A weighty issue. *Today's Dietitian, 6*(9), 47–53.

111. Robayo-Torres, C. C., & Nichols, B. L. (2007). Molecular differentiation of congenital lactase deficiency from adult-type hypolactasia. *Nutrition Reviews, 65*(2), 95–98.

112. Romao, I., & Roth, J. (2008). Genetic and environmental interactions in obesity and type 2 diabetes. *Supplement to the Journal of the American Dietetic Association, 108*(4), S24–S28.

113. Rose, D. J., DeMeo, M. T., Keshavarzian, A., & Hamaker, B. R. (2007). Influence of dietary fiber on inflammatory bowel disease and colon cancer: Importance of fermentation pattern. *Nutrition Reviews, 65*(2), 51–62.

114. Rosenberg, I. H. (2007). Folic acid fortification. *Nutrition Reviews, 65*(11), 503.

115. Rosenbloom, C., & Bahns, M. (2005, November/December). What can we learn about diet and physical activity from master athletes? *Nutrition Today, 40*(6), 253–256.

116. Sacks, F. M., Lichtenstein, A., Van Horn, L., Harris, W., Kris-Etherton, P., & Winston, M., for the American Heart Association Nutrition Committee. (2006). Soy protein, isoflavones and cardiovascular health. *Circulation, 113*, 1034–1044.

117. Saydah, S. H., Fradkin, J., & Cowie, C. C. (2004). Poor control of risk factors for vascular disease among adults with previously diagnosed diabetes. *Journal of the American Medical Association, 291*, 335–342.

118. Simkin-Silverman, L. R., Conroy, M. B., & King, W. C. (2008). Treatment of overweight and obesity in primary care practice: Current evidence and future directions. *American Journal of Lifestyle Medicine, 2*(4), 296–304.

119. Small, P., & Brand-Miller, J. (2009). From complex carbohydrate to glycemic index. *Nutrition Today, 44*(6), 236–243.

120. Smith, S. R. (2009). A look at the low-carbohydrate diet. *The New England Journal of Medicine, 361*(23), 2286–2288.

121. Solomons, N. W. (2007). Food fortification with folic acid: Has the other shoe dropped? *Nutrition Reviews, 65*(11), 512–515.

122. Sparling, M. C., & Anderson, J. J. B. (2009). The Mediterranean diet and cardiovascular diseases: Translating research findings to clinical recommendations. *Nutrition Today, 44*(3), 124–135.

123. Stamler, J. (2008). Population-wide adverse dietary patterns: A pivotal cause of epidemic coronary heart disease/cardiovascular disease. *Journal of the American Dietetic Association, 108*(2), 228–232.

124. Stein, K. (2009). Are food allergies on the rise, or is it misdiagnosis? *Journal of the American Dietetic Association, 109*(11), 1832–1837.

125. Steinbrook, R. (2006). Facing the diabetes epidemic: Mandatory reporting of glycosylated hemoglobin values in New York City. *The New England Journal of Medicine, 354*, 545–548.

126. Stephensen, C. B. (2004). Fish oil inflammatory disease: Is asthma the next target for omega-3 fatty acid supplements? *Nutrition Reviews, 62*(12), 486–489.

127. St. Jeor, S. T., Howard, B. V., Prewitt, E., Bovee, V., Bazzarre, T., & Eckel, R. H. (2001). Dietary protein and weight reduction: A statement for healthcare professionals from the Nutrition Committee of the Council on Nutrition, Physical Activity, and Metabolism of the American Heart Association. *Circulation, 104*, 1869–1874.

128. Suh, S. H., Paik, I. Y., & Jacobs, K. (2007). Regulation of blood glucose homeostasis during prolonged exercise. *Molecules and Cells, 23*(3), 272–279.

129. Sullivan, V. K. (2006). Prevention and treatment of the metabolic syndrome with lifestyle intervention: Where do we start? *Journal of the American Dietetic Association, 106*(5), 668–671.

130. Swain, J. F., McCarron, P. B., Hamilton, E. F., Sacks, F. M., & Appel, L. J. (2008). Characteristics of the diet patterns tested in the optimal macronutrient intake trial to prevent heart disease (OmniHeart): Options for a heart-healthy diet. *Journal of the American Dietetic Association, 108*(2), 257–265.

131. Talati, R., Sobieraj, D. M., Makanji, S. S., Phung, O. J., & Coleman, C. I. (2010). The comparative efficacy of plant sterols and stanols on serum lipids: A systematic review and meta-analysis. *Journal of the American Dietetic Association, 110*(5), 719–726.

132. Tarrago-Trani, M. T., Phillips, K. M., Lemar, L. E., & Holden, J. M. (2006). New and existing oils and fats used in products with reduced *trans*-fatty acid content. *Journal of the American Dietetic Association, 106*(6), 867–880.

133. Teufel, M., Biedermann, T., Rapps, N., Hausteiner, C., Henningsen, P., Enck, P., & Zipfel, S. (2007). Psychological burden of food allergy. *World Journal of Gastroenterology, 13*(25), 3456–3465.

134. Tishkoff, S. A., Reed, F. A., Ranciaro, A., Voight, B. F., Babbitt, C. C., Silverman, J. S. . . . & Deloukas, P. (2007). Convergent adaptation of human lactase persistence in Africa and Europe. *Nature Genetics, 39*(1), 31–40.

135. Trumbo, P. R., & Ellwood, K. C. (2007). Supplemental calcium and risk reduction of hypertension, pregnancy-induced hypertension, and preeclampsia: An evidence-based review by the US Food and Drug Administration. *Nutrition Reviews, 65*(2), 78–87.

136. Trumbo, P., Schlicker, S., Yates, A. A., & Poos, M. (2002). Dietary Reference Intakes for energy, carbohydrates, fiber, fat, fatty acids, cholesterol, protein and amino acids. *Journal of the American Dietetic Association, 102*, 1621–1631.

137. Tufts University. (2007, July). Studies find new omega-3 benefits: But are you getting the right healthy fats? *Tufts University Health and Nutrition Letter, 25*(5), 4–5.

138. Tufts University. (2009). Dietary do's and don'ts from the latest research on cancer prevention. *Tufts University Health & Nutrition Letter, 27*(5), 1–2.

139. Tufts University. (2010). Shake the salt habit to reduce your risk of stroke and heart disease. *Tufts University Health & Nutrition Letter, 28*(1), 1–2.

140. Tufts University. (2010). Lifestyle changes produce lasting benefit against diabetes risk. *Tufts University Health & Nutrition Letter, 28*(2), 1–2.

141. Tufts University. (2010). Added sugars not so sweet for cholesterol levels. *Tufts University Health & Nutrition Letter, 28*(5), 1–2.

142. Tufts University. (2010). "Best evidence yet" for heart-health benefits of nuts. *Tufts University Health & Nutrition Letter, 28*(6), 1–2.

143. Turpyn, A., Rankin, J. W., & Davy, B. (2006, Summer). Diabetes: Something to stress about. *SCAN'S (A Publication for Sports, Cardiovascular, and Wellness Nutritionists) PULSE, 25*(3), 8–10.

144. U.S. Department of Health and Human Services. (2003). The seventh report of the Joint National Committee on Prevention, Detection, Evaluation, and Treatment of High Blood Pressure. *NIH Publication*, No. 03-5233.

145. U.S. Department of Health and Human Services, National Institutes of Health, & National Heart, Lung, and Blood Institute. (2006 revision). Your guide to lowering your blood pressure with DASH. *NIH Publication*, No. 06-4082. Accessed at: http://www.nhlbi.nih.gov/health/public/heart/hbp/dash/new_dash.pdf.

146. U.S. Department of Health and Human Services, National Institutes of Health, & National Institutes of Diabetes and Digestion and Kidney Diseases. (2009, March). Weight-loss and nutrition myths: How much do you really know? *NIH Publication*, No. 04-4561.

147. Van Horn, L. (2008). Diet and heart disease: Continuing contributions. *Journal of the American Dietetic Association, 108*(2), 203.

148. Van Horn, L., McCoin, M., Kris-Etherton, P. M., Burke, F., Carson, J. S., Champagne, C. M., Karmally, W., & Sikand, G. (2008). The evidence for dietary prevention and treatment of cardiovascular disease. *Journal of the American Dietetic Association, 108*(2), 287–331.

149. Van Horn, L., Johnson, R. K., Flickinger, B. D., Vafiadis, D. K., & Yin-Piazza, S. (2010). Translation and implementation of added sugars consumption recommendations: A conference report from the American Heart Association Added Sugars Conference 2010. *Circulation, 122*(23), 2470–2490.

150. Vargas, M. (2009). Atherosclerosis, functional foods, and nutritional genomics. *Dietitians in Integrative and Functional Medicine, 12*(2), 20, 23–30.

151. Verhagen, H., Buijsse, B., Jansen, E., & Bueno-de-Mesquita, B. (2006). The state of antioxidant affairs. *Nutrition Today, 41*(6), 244–250.

152. Vickerstaff, J. M. (2007). Food allergies: The immune response. *Today's Dietitian, 9*(7), 12–16.

153. Watkins, B. A., & Hutchins, H. (2010). Omega-3 fatty acids: Past, present, and future. *Dietitians in Integrative and Functional Medicine, 13*(2), 21, 24–32.

154. Watson, R. R., & Preddy, V. R. (2003). *Nutrition and heart disease: Causation and prevention.* New York: CRC Press.

155. Webb, A. L., & Villamor, E. (2007). Update: Effects of antioxidant and non-antioxidant vitamin supplementation on immune function. *Nutrition Reviews, 65*(5), 181–217.

156. Wood, R. J. (2006). Effect of dietary carbohydrate restriction with and without weight loss on atherogenic dyslipidemia. *Nutrition Reviews, 64*(12), 539–545.

157. Yuan, G., Al-Shali, K. Z., & Hegele, R. A. (2007). Hypertriglyceridemia: Its etiology, effects and treatment. *Canadian Medical Association Journal, 176*(8), 1113–1120.

158. Zimmet, P., & Thomas, C. R. (2003). Genotype, obesity and cardiovascular disease: Has technical and social advancement outstripped evolution? *Journal of Internal Medicine, 254*, 114–125.

159. Zinkernagel, R. M., & Hengartner, H. (2001). Regulation of the immune response. *Science, 293*, 251–253.

The Science of Nutrition in Energy Balance, Body Composition, Weight Control, and Fitness

Valentyn Volkov/Shutterstock.com

MODULE GOAL

To investigate energy balance, physical activity, and weight control to prevent obesity and achieve nutritional adequacy.

LEARNING OBJECTIVES

After completing this learning module, you will be able to explain:

- Several scientific study types used in nutritional science.
- The effects of positive energy balance on disease promotion.
- The epidemic of obesity in the United States.
- How dietary analysis can be used to evaluate nutritional adequacy.
- How physical fitness and physical activity principles are applied to achieve and maintain a healthy body weight and expend enough energy to reduce the risk of developing chronic disease.

PERSONAL IMPROVEMENT GOALS

After studying this module, you will be able to:

- Perform a diet analysis and an energy expenditure analysis.
- Structure a physical fitness program.
- Identify all of the dietary and lifestyle changes you need to make in order to reduce the risks for many health complications.
- Make and prioritize a lifestyle changes list, with realistic goals for each change you want to make.
- Create a plan for reaching each goal and positively reinforce yourself when you reach each goal.

As with any of the other life sciences, nutrition research employs a variety of scientific study types in order to identify, understand, and provide evidence to support knowledge. In this module, we discuss basic research methods to lay a foundation for understanding the scientific studies dealing with the recent worldwide epidemic of obesity. You will see how physical activity plays a role in energy expenditure and body composition so that you can achieve and maintain a healthy body weight. In this module, you will learn about scientific inquiry, obesity, energy balance, healthy body weights, and the fundamentals of physical fitness.

4.1 SCIENTIFIC INQUIRY TIED TO GENETICS, EVOLUTION, AND OBESITY

Introduction

Science is a way of knowing. Its purpose is to investigate and explain the mechanisms that govern nature and to identify ways in which all natural phenomena are interrelated. Science produces knowledge that is based on evidence from the natural world, and that knowledge is repeatedly tested against observations of nature. **Scientific knowledge** is enduring because it has been repeatedly tested. However, scientific knowledge changes because scientists continue to question and test results to refine their knowledge. The tentative nature of scientific knowledge is its greatest strength: Ideas and explanations that are consistent with evidence are refined or discarded and replaced by those that are more consistent.

Science provides personal fulfillment that comes from an understanding of the natural world. In addition, experience with the process of science develops skills that are increasingly important in the modern world. These include creativity, critical thinking, problem solving, and the communication of ideas. A person who is **scientifically literate** is able to evaluate and propose explanations based on evidence and is able to apply conclusions from such explanations appropriately. The scientifically literate individual can assess whether a claim is scientific and distinguish a scientific explanation from those that are not scientific.

The field of nutrition is based on scientific study that integrates biology, physiology, microbiology, immunology, botany, chemistry, genetics, and molecular biology. An example of nutrition as an integrated scientific field is **nutrigenomics** (see BioBeat 4.1). In order to understand how evidence is produced, it is crucial to understand the basics of scientific inquiry. In this section we will explore scientific inquiry and apply it to genetics, evolution, and obesity.

T-Talk 4.1

To hear Dr. Turley talk about scientific inquiry, go to www.cengagebrain.com

---

BioBeat 4.1

Nutrigenomics: A Rapidly Emerging Field in Nutrition

Nutrigenomics is a new area of investigation in the field of nutrition and medicine. One's genetic makeup influences disease susceptibilities, and now it is known that gene expression can be modified by environmental factors such as nutrition and other lifestyle choices. Nutritional genomics is the study of how nutrients and other biologically active food components affect gene expression. Also, mistakes in the gene base pairs are being defined by the resulting physiological dysfunction and disease. This evolving appreciation for the central role genetics plays in health and disease will have a significant impact on the way medicine will be practiced in the future. The role that nutrients and other food components play in the modulation of gene expression will be the future basis for nutritional recommendations. There are already more than 50 nutrition-related medical conditions that exist as a result of genetic aberrations. An effective and significant way to manage these nutrition-related medical conditions (although considered alternative medicine approaches) is through diet and lifestyle practices.[3,10,25,28,37,53,69,73,80,81]

In what ways can your diet affect gene expression?

science A way of knowing and explaining the phenomena and processes of the natural world.

scientific knowledge The enduring understanding of the natural world when tested repeatedly using the scientific method.

scientifically literate Capable of proposing and evaluating evidence based on scientific methods.

nutrigenomics The study of the relationships between dietary influences and components and gene expression.

Scientific Method

Scientific study follows a basic format known as the **scientific method**. It is **hypothesis**-driven. The scientist begins by asking a question, developing a hypothesis, and then designing an experiment that allows for the collection of data to answer the question and support or refute the hypothesis. The scientist then analyzes the data to produce results, makes interpretations and draws conclusions about the results, and communicates the evidence and ideas through publications and presentations. The nature of scientific inquiry and research is to continue to ask more questions and advance knowledge by understanding more and more details of the phenomena through experimentation (see Figure 4.1). The validation of a hypothesis by numerous studies done by different scientists, all reaching the same conclusion, may lead to the formation of a **theory**, such as the theory of evolution or the thrifty gene theory (see Table 4.1).

scientific method The scientific investigation process used by scientists to produce evidence for understanding the natural world; involves identifying a problem to solve, formulating a hypothesis, testing the hypothesis, collecting and analyzing the data, and making conclusions.

hypothesis A proposed scientific belief based on facts and previous observations that needs additional validation by the scientific method.

theory Phenomena believed to be true based on scientific evidence, but lacking conclusive evidence, thus not taken as fact.

case studies Noncontrolled medical investigation that may be published in medical literature to provide suggestive, scientific information.

clinical trials Controlled medical investigation where statistical analyses can be applied to determine the significance of the study results.

intervention trial A prospective statistical comparison between populations that compares the effect of performing an intervention program in one population against the other (similar) population that had no intervention.

epidemiology Retrospective statistical investigation of populations to provide supportive data and ideas for further research.

laboratory experiments Well-controlled, experimental conditions taking place in a laboratory setting, often using animals, tissues, cells, insects, microbes, or other living systems, where the results of the study produce strong, supportive scientific evidence.

control group A group in a study that is not treated or is given a placebo and is used as a comparison to the experimentally treated group.

placebo A blank treatment given to subjects in a control group, sometimes referred to as the "sugar pill."

TABLE 4.1

Examples of Theories in Science and Nutrition

Theory	Scientific Explanation
Cell	The cell is the most basic unit of life; organisms are made of one or more cells, and new cells arise from existing cells.
Evolution	The change in heritable genetic composition of a population, such as by gene mutation and as a result of natural selection.
Set point	Body weight seems to be relatively easy to maintain within a given range (different for each individual), and is genetically and physiologically controlled. The lower body weight range is more tightly defended to prevent starvation and preserve life.
Thrifty gene	Humans who efficiently stored food energy as body fat survived tough times during the hunter-gatherer period of human existence. Natural selection of these thrifty genetic traits through evolution may lead to present-day obesity when the person lives in a processed-food environment and leads a sedentary lifestyle.
Glucostasis	Blood glucose levels affect overall body weight; mild hypoglycemia is associated with elevated levels of insulin, which inhibits the satiety center and promotes positive energy balance. Normal glycemia results in satiety and better weight control.

© Cengage Learning 2013

Types of Scientific Studies The design of the experiment may take several forms. In the field of nutrition, these study types can include **case studies, clinical trials, intervention trials, epidemiology,** and **laboratory experiments.** Recall the nature of scientific inquiry, where the scientist creates a hypothesis, designs an experiment to produce the data, and conducts an analysis to prove the hypothesis true or false. A scientist's reputation is built on reporting the findings.

We will now describe the different study types:

Case studies are commonly used in medical literature, where they may be presented as an intervention protocol to an individual. Even though many parameters are measured objectively, there is no control in the experimental design. Case study reports can serve as suggestive evidence. They may perpetuate more case studies and possibly lead to a clinical trial.

Clinical trials are conducted with basic experimental design, meaning that there is a **control group (placebo)** and an experimental group tested and compared in the study. The data collected from each group are statistically compared. This statistical analysis is used to determine the significance of the results. In clinical trials, the participants are free living and only report back to the clinic for evaluation and compliance.

Intervention trials apply an intervention protocol to one group of people or population and give no treatment to another population that has similar characteristics. Usually, intervention protocols attempt to improve health with the implementation of the strategies under investigation in the study and are evaluated prospectively, or looking forward in time. Data from these trials offer strong evidence, though no single study is viewed as conclusive.

Epidemiological studies are studies of populations. Data are collected retrospectively (looking back in time at events that have already taken place) within populations. Because many lifestyle, genetic, and environmental factors could have affected the population up to the beginning of the study, the results of epidemiological studies provide supportive evidence for a possible relationship of the factors tested or the relationships shown in the results.

Laboratory experiments are the studies with the greatest control. There is always a basic experimental design, and studies can use any life forms (e.g., microorganisms, viruses, animals, plants, cell lines) to produce scientific evidence and observation of confined life forms. Because of the cost of doing human research, the scarcity of facilities, and the difficulty of recruiting subjects, much of the nutritional scientific inquiry is initiated using experimental animal models.

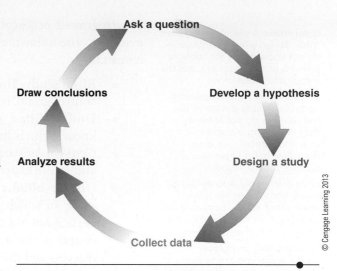

FIGURE 4.1 The scientific method.

© Cengage Learning 2013

Markers of Good Science Things to consider when judging the quality of the science include many aspects of the experimental design. Ultimately, to provide strong evidence in human nutrition, human nutrition experiments need to be conducted, with the participants confined during the study. Most human studies that are conducted are clinical trials; in other words, the participants are free living. They are checked for compliance with the study protocol and are infrequently evaluated. Many lifestyle factors could also influence the study results. Designing an experiment to produce high-quality evidence requires much consideration. The following are examples of some considerations that must be made in experimental design:[64,70]

- *Number of subjects:* Are there enough subjects for statistical significance? The more participants who are involved in the study, the better the statistical power is in the data analysis.
- *Matching groups:* Are the study groups matched according to characteristics such as ethnicity, gender, age, disease state, and lifestyle habits like smoking or alcohol use? The experimental and control groups need to be matched according to the purpose of the study so that any comparisons made in the study results are valid.
- *Control groups:* Does the study include both an experimental and a placebo (control) group to validate that the results are not caused by the placebo effect?
- *Duration of the study:* Is the duration of the study sufficient to show full effects so correct conclusions can be made?
- *Reproducible results:* Can the experiment be repeated with the same results?

FIGURE 4.2 The regulation of the number and size of fat cells.

The treatment protocol can make a huge difference in the quality of the scientific evidence. The following are some of the types of designs commonly used in human research:

- **Blind study**: The subject does not know if he or she is in the experimental or control group, but the researchers do know who is receiving the placebo.
- **Double-blind study**: Neither the subject nor the primary investigator knows who is in the experimental or control group.
- **Cross-over study**: Each group is evaluated for a period in the placebo group and in the experimental group. The design may or may not be blinded.
- **Double-blind, cross-over study**: Neither the researcher nor the participants know in which group (experimental or control) they are placed, and each participant is a member of both experimental and control groups during the course of the study. In other words, each participant is tested while using a placebo and the substance being investigated. In this way, data can be compared both within groups and among individuals. This is a highly reputable study design.

Consistent results from multiple credible studies published over time are used to make dietary recommendations. A whole body of evidence, collected over many years of experimentation, is interpreted by the scientific community to create public recommendations regarding diet and health (disease prevention). Keep in mind that testimonials, which are commonly used to sell nutritional products, do not have to be part of any sort of study and should not be considered as scientific statements. The many research methods that are directed at understanding, preventing, or treating obesity have brought us to our current understanding of this medical condition. Studies have shown that about 70 percent of adult Americans are overweight or obese and are either suffering from or at risk of developing the medical conditions that contribute to the current healthcare crisis. The obesity problem is far from being solved, but here is what research has contributed to our understanding.

Obesity

Obesity is considered by the scientific community as body fatness significantly in excess of the level that is consistent with optimal health. Excess body fat is stored inside fat cells. With body fat **weight gain**, the mass or size of fat cells increases. This is referred to as **hypertrophy**. Once a person's fat cells are filled up, then the fat cells can divide or increase in cell number. This is referred to as **hyperplasia**. Once a fat cell is created, there is no biological method of eliminating the cell. However, the amount of fat inside the cell can be reduced with **weight loss** (see Figure 4.2). Obesity is determined by **body mass index (BMI)** (discussed later in this module). A BMI score of greater than or equal to 30 is **obese**, while a BMI score of 25 to 29.9 indicates **overweight**.

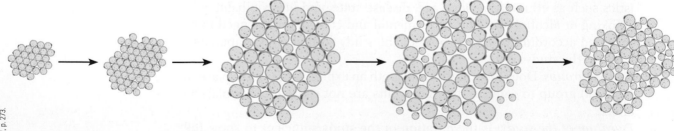

During growth, fat cells increase in number.

When energy intake exceeds expenditure, fat cells increase in size.

When fat cells have enlarged and energy intake continues to exceed energy expenditure, fat cells increase in number again.

With fat loss, the size of the fat cells shrinks (reduce cell volume) but not the number.

Understanding Nutrition, 12th ed, Whitney & Rolfes, Chapter 9, Fig. 9-2, p. 273.

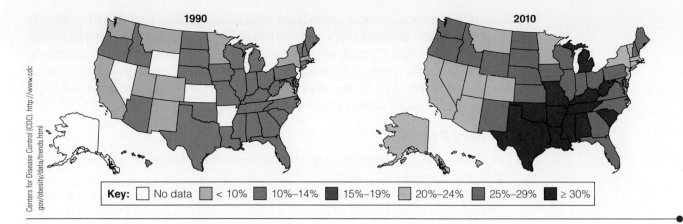

1990   2010

Key: ☐ No data ■ < 10% ■ 10%–14% ■ 15%–19% ■ 20%–24% ■ 25%–29% ■ ≥ 30%

Centers for Disease Control (CDC). http://www.cdc.gov/obesity/data/trends.html

Currently, the U.S. population is experiencing an **obesity epidemic** (see Figure 4.3 and BioBeat 4.2). When you think of an epidemic, usually an outbreak of an infectious disease comes to mind. However, even the Centers for Disease Control, which functions as the National Center for Disease Prevention and Health Promotion, has used the word *epidemic* to describe the uncontrolled rise in obesity in the United States. The prevalence of obesity among the entire U.S. adult population has increased from 12 percent in 1991 to 33.8 percent in 2010. More than 72 million adult Americans were considered to be obese. More than 17 percent of children and adolescents ages 2 to 19 years of age are obese. This reflects a dramatic rise in obesity since 1991. The incidence of obesity during the last 20 years has doubled in adults and tripled in children. The prevalence of overweight and obesity is greater in blacks (51 percent higher) and Hispanics (21 percent higher) compared to whites.[A,13,14,40,54] Overall, the incidence of obesity continues to grow.[B,C,D,14]

The reason for concern about obesity is simple: Obesity is scientifically linked to increased risk for chronic diseases and medical conditions. Health problems associated with obesity include increased incidence of heart disease, cancer, type 2 diabetes, arthritis, liver or gallbladder disease, high-risk pregnancy, hernia, varicose veins,

FIGURE 4.3 Percent of the U.S. population considered obese in 1990 and 2010, by state.

BioBeat 4.2

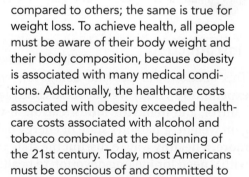

Size-Wise: The Epidemic of Obesity

Obesity is a public health problem in many industrialized nations worldwide. It is evident that a genetic predisposition leads to obesity, but strong environmental factors (such as less physical activity and consuming more processed food) can also play a role. We know this from identical twin studies. Investigators overfed sets of identical twins by 1,000 Calories per day, and then starved them by 1,000 Calories per day. Within the sets of twins, weight was gained and lost similarly, but among all the sets of twins, there was great variability regarding the amount of weight gained and lost from the 1,000-Calorie difference. Some individuals have a much easier time getting fat from overfeeding compared to others; the same is true for weight loss. To achieve health, all people must be aware of their body weight and their body composition, because obesity is associated with many medical conditions. Additionally, the healthcare costs associated with obesity exceeded healthcare costs associated with alcohol and tobacco combined at the beginning of the 21st century. Today, most Americans must be conscious of and committed to weight control, because more than two-thirds of adult Americans are overweight or obese.[10,14,32,41,43,48,54,63,66,67,69,80]

What can be done about the obesity epidemic?

obesity epidemic A rapidly growing number of people within a population with excesses of stored body fat (BMI of 30 or greater) that leads to increased morbidity and mortality.

4.1 SCIENTIFIC INQUIRY TIED TO GENETICS, EVOLUTION, AND OBESITY   **175**

accidents, depression, and low self-esteem. Medical conditions resulting from obesity include high blood pressure (hypertension), high blood cholesterol and triglycerides, and high blood sugar. Overall, obesity leads to decreased longevity and decreased quality of life. Furthermore, it is contributing to the crippling of the healthcare system in the United States. We will now explore how to measure obesity and control the epidemic through healthy dietary practices and physical activity.[F,4,7,9,12,22,38,42,45,58,59,76,78,88,89]

Summary Points

- Nutrition relies on scientific study to understand the relationships between diet and disease.
- Data synthesis from multiple study types (laboratory experiments, case studies, intervention trials, clinical trials, and epidemiology) is used to establish the nutrition science literature.
- Data synthesis across numerous disciplines (physiology, anatomy, chemistry, immunology, virology, etc.) over time is needed to study nutrition.
- Consistent results pertaining to human health from human trials are used to make dietary recommendations.
- Epidemiological study results have verified the obesity epidemic.
- Obesity increases the risk for many chronic diseases and medical conditions.

Take Ten on Your Knowledge Know-How

1. What is scientific inquiry?
2. What is a theory?
3. What are some of the various study types used in scientific inquiry?
4. What is the best treatment protocol to have in a human trial?
5. Imagine that you are a nutritional scientist. What scientific study would you want to conduct? What is your hypothesis? How would you set up your experiment?
6. How has the field of nutritional science impacted your life?
7. What is nutrigenomics?
8. What causes obesity?
9. What BMI number is associated with obesity?
10. Is the obesity epidemic of concern to you? Why or why not?

4.2 ENERGY BALANCE

T-Talk 4.2
To hear Dr. Turley talk about energy balance, go to www.cengagebrain.com

energy balance The consumption of the same number of Calories as expended; Calories eaten equals Calories burned; isocaloric; body weight is maintained.

behavior modification Techniques used to promote desired behavior changes, where thoughts and activities can be changed to achieve personal goals.

Introduction

Understanding **energy balance** can lead to an improved ability to achieve and maintain a healthy body weight. This section will provide details on energy balance equations (intake versus expenditure), body weight versus body fat, body composition assessment, weight control, and **behavior modification**. It will become clear that in order to lose or gain weight, a person's state of energy balance will need to be disrupted. Furthermore, a healthy way to disrupt the state of energy balance will be explained, along with a need for lifelong devotion to healthy eating and activity practices.[1,38,47,56,72,84]

States of Energy Balance and Energy Balance Equations

Energy balance is the relationship between the number of Calories consumed and the number burned or expended. When the energy balance equation is used to predict body weight changes, it is the change in fat weight that can be best represented, because fat is the energy storage form in the body. There are 3,500 Calories stored in 1 pound of body fat. Body **fat mass** change can be calculated with these formulas:

$$\text{Calorie intake} - \text{Calorie expenditure} = \text{Calorie difference}$$

$$\text{Calorie difference} \div 3{,}500 \text{ Calories per pound of fat} = \text{Change in fat mass}$$

If a person consumes 2,500 Calories in a day and only expends 2,000 Calories in that same day, then that person's change in fat mass is:

$$2{,}500 \text{ intake} - 2{,}000 \text{ expenditure} = +500 \text{ Calories} \div 3{,}500 \text{ Calories per}$$
$$\text{pound} = 0.14 \text{ pounds of fat mass or fat weight change}$$

Because the intake exceeded the expenditure in this example, the person has an excess of energy and would store that energy inside fat cells and thus accumulate excess body fat (see Appendix B for additional sample calculations). Changes in **lean body mass** can occur, but it is very difficult to relate the Calorie deficit or excess to an amount of change in lean body mass. When increasing lean body mass, one must implement a rigorous strength training program and consume additional protein to achieve the lean body mass gain. The loss of lean body mass is usually a result of severe Caloric deficit resulting from starving or disease, and both fat mass and lean body mass are lost.[79,85,86]

Demo GENEie Show a model of 5 pounds of fat. Calculate how many Calories are present in the 5 pounds. Calculate how Calorie-deficient a person would need to be over time to lose 5 pounds of fat.

There are three states of energy balance: positive, negative, and balanced (see Figure 4.4). If the number of Calories consumed equals the number of Calories expended, and the person is consuming an **isocaloric** diet, then the state of energy balance occurs. When in the state of energy balance, body weight is typically maintained. The state of **positive energy balance** occurs when the Calories consumed are greater than the Calories expended. In the state of positive energy balance, body weight is typically gained, and for every additional 3,500 Calories consumed, 1 pound of fat is gained. If Calories consumed are less than the Calories expended, then body weight is typically lost, and the state of **negative energy balance** occurs. Energy is furnished in the diet by the Calories consumed from the energy-producing nutrients: carbohydrate, protein, and fat, as well as alcohol. Recall that each of the different energy-producing dietary components provides a specific number of Calories per gram consumed: 4, 4, 9, and 7 Calories per gram, respectively.[47,72,87]

fat mass The pounds of fat estimated from the percentage of total body weight from fat; includes essential fat mass and stored body fat.

lean body mass Animal tissue, including mostly vital organs and muscle mass, but also connective tissue.

isocaloric When energy intake is equal to energy expenditure; the Calorie amount consumed meets the energy need.

positive energy balance The consumption of more Calories than expended; Calories eaten are greater than the Calories used; body weight increases.

negative energy balance The consumption of fewer Calories than expended; Calories eaten are less than the Calories used; body weight decreases.

FIGURE 4.4 States of energy balance.

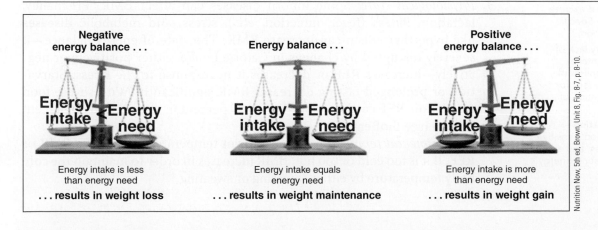

Negative energy balance . . .
Energy intake < Energy need
Energy intake is less than energy need
. . . results in weight loss

Energy balance . . .
Energy intake = Energy need
Energy intake equals energy need
. . . results in weight maintenance

Positive energy balance . . .
Energy intake > Energy need
Energy intake is more than energy need
. . . results in weight gain

Nutrition Now, 5th ed, Brown, Unit 8, Fig. 8-7, p. 8-10.

Energy Expenditure

There are three categories within **energy expenditure**: (1) Calories required for **Basal Metabolic Rate (BMR)**, also called **Resting Energy Expenditure (REE)**; (2) voluntary muscle movement from physical activity; and (3) the **Specific Dynamic Action of food**, also termed the **thermic effect of food** (see Figure 4.5). It is the sum of the Calories expended in each of the categories—BMR or REE, physical activity, and Specific Dynamic Action of food—that makes up total energy expenditure. Each of the three categories is discussed below.[72]

Basal Metabolic Rate (BMR) BMR is the energy required by the body to sustain life or minimally function (the heart to beat, the lungs to breathe, etc.) during a day. It accounts for approximately 60 percent of the total daily energy expenditure in a typical person's life. BMR is determined in a **fasting** state (12 hours), when the body is at complete rest, and it accounts for the majority of the ways that energy is expended in a sedentary person (determined by the MyPlate activity levels). REE is similar to BMR in that it considers the amount of energy used by a person at rest for one day (a 24-hour time period) (see BioBeat 4.3). REE can be calculated using the **Mifflin-St. Jeor equation**. This equation considers gender, height in centimeters, weight in kilograms, and age in years.[18,26,72]

To calculate REE using the Mifflin-St. Jeor equation, select the gender-appropriate equation. Plug the personal information on height, weight, and age into the equation, and solve. The answer is the estimated number of Calories needed to sustain life (vital function) in a 24-hour period (see Appendix B for a calculated example).[2]

For men: REE (Calories/day) = (10 × Weight in kilograms) + (6.25 × Height in centimeters) − (5 × Age in years) + 5

For women: REE (Calories/day) = (10 × Weight in kilograms) + (6.25 × Height in centimeters) − (5 × Age in years) − 161

There are four major parameters that affect BMR, or the number of Calories used in vital processes:

1. *Age*: As one ages, the metabolic rate slows.
2. *Gender differences*: The metabolic rate is influenced by the percentage of lean body mass. Males in general have more lean body mass and lower fat mass than females. Lean body mass requires more energy to sustain compared to fat mass. The more lean body mass that is present, the greater the BMR; thus males usually have a higher BMR than females. If BMR is adjusted to fat-free body mass, then there are no gender differences.
3. *Physiological state*: Some inborn diseases can affect BMR. Pregnancy, lactation, illness (fever, infection, etc.), stress, and metabolic diseases like hyperthyroidism can increase BMR. The state of energy balance—if severely disrupted by changes in Calorie intake, either positively or negatively—increases REE or decreases it in response to the stress. Starvation or prolonged fasting decreases BMR significantly. With severe food restriction, REE can be reduced up to 80 percent in women and 60 percent in men (see BioBeat 4.3).
4. *Environmental temperature*: The ambient temperature influences BMR or REE. If it is too cold or too hot, BMR increases in order to maintain the core body temperature by either shivering or sweating.[26]

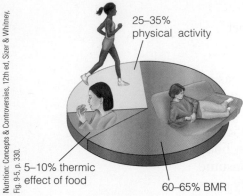

25–35% physical activity

5–10% thermic effect of food

60–65% BMR

FIGURE 4.5 Components of energy expenditure.

energy expenditure The number of Calories used by the body in a 24-hour period in the categories of Basal Metabolic Rate, Specific Dynamic Action of food, and physical activity.

Basal Metabolic Rate (BMR) The number of Calories expended in a 24-hour period to sustain vital function under basal conditions.

Resting Energy Expenditure (REE) An estimated number of Calories needed to sustain vital functions under resting conditions in a 24-hour period.

Specific Dynamic Action of food The energy expended to ingest, digest, assimilate, and excrete food; also called the *thermic effect of food*.

thermic effect of food The energy expended to ingest, digest, assimilate, and excrete food; also called the *Specific Dynamic Action of food*.

fasting The voluntary lack of food and beverage intake, promoting the catabolism of lean and fat body masses for energy and the depletion of stored nutrients.

Mifflin-St. Jeor equation Mathematical equation, sensitive to gender, body mass, and age, used to estimate Resting Energy Expenditure.

BioBeat 4.3

Estimating and Altering Resting Energy Expenditure

The "metabolic cart" is a machine used to measure oxygen consumption and carbon dioxide production, along with a wide variety of other measurements associated with human performance. From the number of milliliters of oxygen consumed per kilogram of body weight per minute, one can estimate Calorie burn or Calorie expenditure. This method of measuring energy expenditure is called **indirect calorimetry**. Resting Energy Expenditure (REE) is measured while an individual is lying at rest, and the amount of oxygen consumed or Calories measured reflects the amount of Calories required just to stay alive or support vital function. The resulting value is used to estimate the number of Calories a person requires in a 24-hour period (day) just to sustain life or vital function. You can calculate this number of Calories by using the Mifflin-St. Jeor equation. Additionally, Calorie burn for various physical activities (voluntary muscle movements) can be calculated from REE.

REE can be altered. It is increased by increasing the amount of metabolically active muscle tissue in the body, or overeating. It is decreased by the loss of lean body mass and by starvation and semistarvation. Women who starve can suppress REE by 80 percent, and men can do so by 60 percent. For example, let's say that Jane's normal REE is 1,295. Tuesday, she went to aerobics for an hour and burned 270 Calories. She did not eat the rest of Tuesday, and she went to aerobics Wednesday morning before breakfast. Her REE theoretically could be suppressed 80 percent. For the same hour of aerobics on Wednesday, she burned 54 Calories.[6,18,24]

What is wrong with starving yourself for even one day?

Physical Activity or Voluntary Muscle Movement The energy expended to perform physical activity is dependent on the individual's body weight and the intensity of the exercise. It accounts for approximately 30 percent of total energy expenditure. BMR or REE can be used to determine the energy spent or Calorie burn in physical activity. You will note a direct relationship between Calorie burn, body weight, and the intensity of the exercise. As body weight increases, so does Calorie expenditure, and as the intensity of the exercise increases, so does Calorie burn. When calculating Calorie burn, the level or intensity of physical activity is assigned a certain factor. As the level of the activity rises (intensity increases), the value of the factor increases too. When the person is at complete rest, the only Calorie requirement is to sustain vital function, thus the factor for the resting level is 1, and essentially amounts to the Calories expended as REE. To calculate Calorie burn for other levels of activity, REE is then multiplied by the factor associated with the level of activity and the amount of time (in hours) spent doing the activity. This value is then divided by 24 hours, and the basis for determining the Calories expended for the hour can be calculated. The simple formula to determine Calories burned in an activity is as follows:

$$\text{Calories burned} = \text{Hours spent in activity} \times \text{REE} \times \text{Activity factor} \div 24 \text{ hours}$$

In general, the following activity factors associated with five levels of activities are used to calculate the Calorie expenditure (burn).

- The level with no activity is the *Rest* or *Sleeping/Reclining* level: It has a factor of 1, because all activities in this category are at rest.
- The second category is the *Sedentary* or *Very Light Activity* level: Activities that are mostly sitting or standing in a small space are in this category. The REE factor of 1.5 is used as the multiplier to determine Calorie burn.

indirect calorimetry A measurement of Resting Energy Expenditure using oxygen consumption, carbon dioxide production, and mathematical conversions to determine the Calories used by a person.

- The third category is the *Easy* or *Low Activity* level: Activities that demand whole-body movement but do not promote sweating fit into this category. The REE factor of 2.5 is used as the multiplier when calculating Calorie burn.
- The next level of activity is the *Rigorous* or *Moderate Activity* level: Activities that demand whole-body movement and promote profuse sweating fit in this category. The REE factor of 5 is used as the multiplier.
- *Strenuous Activity* is the most extreme category of exercise. Only all-out efforts that cannot be sustained for more than a few minutes without exhaustion fit into this category. The REE factor of 7 is the multiplier for this level (see Appendix B for activity factors, sample calculations, and sample activities in these categories).

Specific Dynamic Action of Food   The Specific Dynamic Action of food can be thought of as the "food processing charge." Another technical term used for Specific Dynamic Action is the *thermic effect of food*. The Specific Dynamic Action requires approximately 10 percent of the total number of Calories consumed. The Specific Dynamic Action accounts for all of the work or energy used to ingest, digest, absorb, and process food. Specific Dynamic Action is the increased rate of metabolism measured during the post-prandial (after feeding) period.[70,86]

Summary Points

- Energy balance is the state determined by energy intake versus energy expenditure.
- Intake occurs with the consumption of the energy-producing nutrients: carbohydrate, protein, and fat, as well as alcohol.
- Expenditure is the sum of Calories burned by BMR, physical activity, and Specific Dynamic Action.
- Weight gain and body fat accumulation occurs with positive energy balance.
- Weight loss and fat mass reduction occurs with negative energy balance.
- When Calorie intake equals Calorie expenditure, energy balance occurs.
- Starving can reduce REE and thus reduce Calorie burn in every level of activity.

Take Ten on Your Knowledge Know-How

1. How do the states of energy balance relate to weight control?
2. What state of energy balance is one in when eating an isocaloric diet?
3. How is energy gained and expended by the body?
4. What are the various aspects of energy expenditure?
5. Calculate your REE using the Mifflin-St. Jeor equation. What does this number mean?
6. How can the REE value change?
7. How many Calories would you expend in a single activity in one day based on your calculated REE?
8. What is your estimated total energy expenditure and total energy intake for one day? Based on this data, what is your state of energy balance?
9. What is your estimate for pounds of body fat gained or lost in one week's time based on your energy intake and expenditure data?
10. It is very common for people to underreport food intake and overestimate energy expenditure. How can these errors be minimized when you are recording your intake and expenditure data?

4.3 BODY COMPOSITION AND WEIGHT CONTROL

Introduction

In determining a healthy weight, one must consider what types of tissues are contributing to the total amount of body weight. In this unit, we will look at each type of tissue.

Types of Body Mass

Your weight is the sum of a variety of tissue types. The main components contributing to your body weight or body mass include lean body mass, fat mass, mineral mass, and water.[87]

Lean Body Mass Lean body mass consists largely of the vital organs and muscle. In the average young adult, lean body mass constitutes about 55 percent of total body weight. This tissue is about 70 percent water by weight and provides about 4 Calories per gram when catabolized to energy with insufficient dietary intake.

Fat Mass There are two parts of fat mass. One part of fat mass is **essential fat mass** (or essential body fat). The percent of essential fat mass differs between genders. In males, essential fat mass contributes about 3 percent of total body weight, and in females, essential fat mass contributes about 12 percent of total body weight. The higher body fat in women is required for a normally functioning reproductive cycle. When a woman's body fat drops below the essential fat mass level, she experiences disruptions in the menstrual cycle, notably **amenorrhea**, a condition characterized by absence of the normal menstrual cycle.[46]

The other part of fat mass is **stored body fat**. Stored body fat is 15 percent water by weight and provides 9 Calories per gram when catabolized. Stored body fat is variable depending on the category of obesity; however, in a normal range of healthy body weights, stored body fat ranges between 10 and 15 percent of total body weight.[46]

Mineral Mass The third category of tissue type that contributes to total body weight is **mineral mass**, basically the body's bones. Mineral mass is present fairly consistently across body weights. Large body weights usually have larger bone mass compared to thinner individuals. Mineral mass constitutes about 4 percent of total body weight. Most minerals inside the body localize in the skeletal system. However, minerals do have other structural roles in some proteins and hormones, and nonstructural roles, such as in muscle contraction, nerve impulse transmission, fluid and pH balance, blood clotting, and signal transduction.

Water It should be noted that, in an average young adult, water weight—or the percent of total body weight contributed by water—is about 60 percent. Recall the high percentage of water in lean body mass and the much lower percentage in fat weight. Additionally, the fluid in the blood stream adds to the total percent of body weight that is water weight.

Body Weight Versus Body Fat

Body composition, a term commonly heard in nutrition and exercise contexts, refers to the types of tissues contributing to total body weight. Unhealthy body weights or body compositions are caused by a high percentage of body fat. Body composition assessments can be used to measure the percent of total body weight

T-Talk 4.3
To hear Dr. Turley talk about body composition and weight control, go to www.cengagebrain.com

essential fat mass The percentage of total body weight from fat that is minimally required for structure and function, also known as essential body fat.

amenorrhea Absence of the normal menstrual cycle for three months, or no period within the last six months.

stored body fat The difference between total body fat and essential fat mass.

mineral mass The noncombustible, inorganic substances incorporated in animal bodies; the ash after cremation.

body composition The percentages of fat mass, mineral mass, and lean body mass contributing to the total body weight of a person.

that is body fat. Body composition is a critical factor of body weight with respect to maintaining good health, personal appearance, and/or performance.[23,77]

The best way to determine whether a person is overweight or obese is to determine the percentage of fat that constitutes an individual's body composition. A male is considered too fat if he has more than 20 **percent body fat**. When a male exceeds 20 percent body fat, functional capacity is diminished, and the ability to exercise is compromised. A woman is considered too fat if she has more than 26 percent body fat, again for the same reasons. Table 4.2 further categorizes body fat measures in terms of very lean, lean, physically fit, average, fat, and obese. In some instances, individuals with a high amount of lean body mass have a high calculated BMI, but their body composition analysis shows that they are lean. The body composition assessment should be made on individuals with high BMI values to confirm obesity.[86]

TABLE 4.2		
Interpreting Percent Body Fat		
Body Fat Categories	**% Body Fat (Male)**	**% Body Fat (Female)**
Essential fat	*3*	*12*
Very lean	≤ 10	≤ 13
Lean	11–15	14–19
Physically fit**	12–15	18–22
Not fat (average)	16–19	20–25
Fat*	20–27	26–32
Obese	≥ 28	≥ 33

*Strong recommendations for fat cell reduction are made when males exceed 20 percent body fat and females exceed 26 percent body fat.

**If you are a male with less than 12 percent body fat or a female with less than 18 percent body fat, you may also interpret the result as being physically fit.

© Cengage Learning 2013

percent body fat The percent of total body weight from fat, which can be estimated using many body composition evaluation techniques.

FIGURE 4.6 Commonly used body fat assessment techniques: hydrostatic weighing (upper left), Bod Pod (lower left), bioelectrical impedance (upper right), and skin-fold calipers (lower right).

Jennifer Turley

Determining Percent Body Fat Professional evaluators use a variety of techniques for determining an individual's percentage of body fat. The most commonly used techniques are shown in Figure 4.6 and discussed in the following text.

Hydrostatic Weighing Technique This is one of the most reliable ways to measure body composition. Underwater weighing is used to compare a person's weight on land versus in the water. A person with more fat mass weighs less in water than a person with more lean body mass. This is a very accurate test for measuring body fat if done correctly. The accuracy depends on the person's comfort in the water and the ability to blow all the air out of their lungs while underwater.

Air Displacement Techniques A machine called the Bod Pod® uses air displacement, which is the volume of air in the pod that has been displaced by the body, to assess body fat. The technique is accurate when the person tested is wearing tight-fitting spandex shorts or swimwear plus a spandex cap, with normal breathing and when one remains still in the pod.

Skin-Fold Measurement   The most practical technique to measure body composition is measuring skin folds. Skin-fold caliper instruments are used to measure skin-fold thickness. Most commonly, the millimeters of fat folds from three body sites are summed to determine body composition. Most elaborately, the seven possible measurement sites are the midaxillary, chest, abdominal, tricep, subscapular (below the shoulder blade), suprailiac (above the hip bone), and thigh. The values from the seven sites are put into a gender specific equation to interpret the body composition. However, the three-site skin-fold measurement technique can be reliable for young adults when an experienced evaluator is performing the measurements with a high-quality pair of calipers. The more sites measured by a skilled fitness professional, the better the body fat estimate.

Bioelectrical Impedance   Recently popularized because of its ease of assessment, bioelectrical impedance uses an undetectably low voltage of electric current that is sent through the body. The current is impeded (slowed down) with high fat mass. This technique can be accurate if the person is hydrated, has not exercised in 12 hours, has not consumed any food for 4 hours, and has abstained from caffeinated and alcoholic beverages for 12 hours.

Light Absorption Technique   This technique uses a modification to the machine developed by the USDA to grade beef (called the Futrex 5000®). This method is based on the principles of light absorption, reflectance, and near-infrared spectroscopy. An infrared monitor is placed on the midpoint of the bicep to make the assessment. If a person is too lean or too fat, the reading is inaccurate. If the person is within a normal body fat range, the technique can be accurate.

Other Research Techniques   A few more techniques have been used in a research setting. These research methods include dual energy x-ray absorptiometry (DEXA), total body potassium, substance dilution, and magnetic resonance imaging (MRI). They are very expensive but accurate techniques that are usually reserved for the acquisition of scientific data for research purposes.[86]

Central Obesity or Central Adiposity
Individual control over body weight and composition is internally regulated (see BioBeat 4.4). Furthermore, the location of fat on the body has an effect on a person's health and is controlled by genetic and lifestyle signals. Central adiposity, or excess fat in the abdominal region of the body, is associated with increased health risks for type 2 diabetes, hypertension, dyslipidemia, and heart disease.[57,62] Central adiposity is determined by girth (waist) measurement. A waist circumference measurement of greater than 40 inches for men or greater than 35 inches for women indicates central adiposity and obesity (a BMI of 30 or more). An error can occur in using the girth measurement to determine central adiposity when a person is very short. With central adiposity, **visceral fat** accumulates deep inside the body's abdominal cavity (see Figure

visceral fat Abdominal fat mass that indicates unhealthy accumulated fat around vital organs.

FIGURE 4.7 Visceral and subcutaneous abdominal fat in a normal (left) and overweight (right) person.

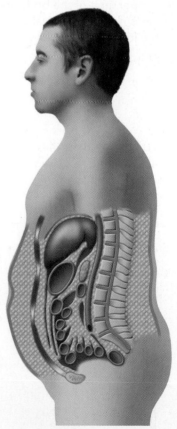

In healthy weight people, some fat is stored around the organs of the abdomen.

In overweight people, excess abdominal fat increases the risks of diseases.

Understanding Nutrition, 12th ed, Whitney & Rolfes, Chapter 8, Fig. 8-8, p. 255.

subcutaneous fat Fat mass under the skin, which can be used to estimate total body fat.

set point theory The concert of psychological and physiological regulatory mechanisms that maintain a comfortable body weight range.

4.7) and creates interference with optimal organ function. In many men, extra fat deposits in the abdominal cavity cause an apple body shape.

Another type of obesity is gynoid, which causes a pear body shape. Gynoid obesity is characterized by **subcutaneous fat**. Subcutaneous fat is not as detrimental to heart health. In many women, subcutaneous fat forms around the hips and thighs, and causes the pear body shape.

BioBeat 4.4

The Set Point Theory and the Weighty Issues of Today

The **set point theory** of body weight regulation is based on the idea that there is a range of body weight that is comfortable to maintain within the living environment. From an evolutionary and adaptation perspective, an explanation for why weight gain is a worldwide problem stems from the genetic influence for survival (preventing death from starvation), which continues to enable the species to propagate. To consume a few more Calories than are needed is not threatening to the survival of the species. Thus, the theory is that the upper limit of the body weight range is not as biologically defended as the lower limit of the body weight range. The defense of the lower

range of body weight prevents starving by promoting the intake of food. The motivation to consume food is great when one feels hunger, and the pleasure felt by feeding provides the positive reinforcement needed to associate food with comfort.

Appetite is the psychological influence that affects food intake, which in turn affects body weight. Appetite is regulated by many external cues and internal thoughts. Appetite can be controlled by paying attention to the internal cues given to the brain to start and/or stop eating (hunger and satiety). These cues are considered internal in nature, as the body and mind know when it is necessary to eat. Many people inadver-

tently use external cues for appetite control and eating. And in our media-intensive world, there are many misleading cues promoting the intake of food for pleasure. Seeing food and wanting to eat it, or the cultural influence of needing to leave a clean plate after a meal, promote positive energy balance. Strengthening the use of internal signals or physiological cues for hunger and satiety, in addition to exercising daily and suppressing the psychological influences, is best for achieving and maintaining a healthy body weight.[32,52,70,78,82,86,88]

What can you do to maintain your set point in a healthy body weight range?

Body Mass Index The determination of a healthy body weight has been evolving. For many decades, weight-for-height tables were used, but starting in the 1990s, it became more common to use body mass index (BMI). BMI is a clinical tool that is commonly used to define a healthy body weight. The BMI-calculated number reflects body leanness more effectively than weight-for-height tables, even though it also uses weight and height measurements. Currently, body weight assessments should be made using BMI and confirmation of obesity should be made using body composition techniques.[23,44]

FIGURE 4.8 Body mass index (BMI) and increased mortality when underweight and obese.

BMI is used to define the healthiness of the body weight in relation to height. BMI can be determined using a nomogram (see Table 4.3) or by calculation. To calculate BMI, take the body weight (in kilograms) and divide it by height (in meters squared). The body weight in pounds can be converted to kilograms by dividing by 2.2, and the height in meters can be determined from inches by dividing by 39.37. The calculated BMI can then be interpreted using Table 4.3 (see Appendix B for facts, formulas, conversions, and a sample calculation). BMI is a screening tool used to determine whether a person is **underweight** (see BioBeat 4.5) or overweight and at risk for a variety of nutrition related medical conditions. However, BMI is not an accurate measurement of percent body fat; rather, BMI indicates an increased risk for morbidity (or illness), and increased mortality risk (see Figure 4.8). If the BMI is 25 or above and the waist circumference is greater than 40 inches for a male or 35 inches for a female, then the individual is at high risk for developing heart disease, hypertension, dyslipidemia, and type 2 diabetes (see page 141).

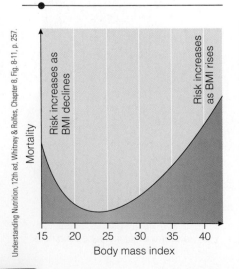

BioBeat 4.5

Reasons and Ways to Increase Body Weight When Underweight

Even though very little morbidity is associated with being underweight, being underweight is associated with a more rapid decline in health and increased risk for mortality than being a little overweight. There are many reasons that a person may be underweight. It could be due to voluntary food restriction, extreme exercise behaviors, genetics, and health conditions such as hyperthyroidism. Health risks associated with having a BMI below 18.5 are complicated by severe medical conditions, such as eating disorders, cystic fibrosis, and cancer, or undernutrition associated with poverty and food shortage. A clinical sign that occurs in underweight women is amenorrhea, or cessation of a woman's menstrual cycle, which causes infertility, or the inability to conceive and reproduce.

Gaining body weight can be as challenging as losing body weight. Individuals in need of weight gain could have a high metabolism genetically and/or hormonally, an active lifestyle, extensive food aversions, intolerances, allergies, or sickness or disease. The goal for weight gain for healthy individuals should be to add lean body mass, primarily muscle mass. This is done by combining a positive energy balance (200 Calories per day), an adequate protein dietary regime (1.6 grams of protein per kilogram of body weight per day), and a hypertrophy-promoting strength training program. When weight gain is needed, the individual can facilitate positive energy balance by choosing Calorie-dense foods, eating them regularly throughout the day, and curtailing weight-bearing aerobic activity to reduce energy expenditure.[85,86]

What can you do to increase body weight and add lean body mass if you are underweight?

underweight Having a lower body weight in relation to height than what is considered normal; a BMI less than 18.5.

TABLE 4.3

Interpreting Body Mass Index (BMI)[E]

BMI	19	20	21	22	23	24	25	26	27	28	29	30	31	32	33	34	35
Height (inches)	Body weight (pounds)																
58	91	96	100	105	110	115	119	124	129	134	138	143	148	153	158	162	167
59	94	99	104	109	114	119	124	128	133	138	143	148	153	158	163	168	173
60	97	102	107	112	118	123	128	133	138	143	148	153	158	163	168	174	179
61	100	106	111	116	122	127	132	137	143	148	153	158	164	169	174	180	185
62	104	109	115	120	126	131	136	142	147	153	158	164	169	175	180	186	191
63	107	113	118	124	130	135	141	146	152	158	163	169	175	180	186	191	197
64	110	116	122	128	134	140	145	151	157	163	169	174	180	186	192	197	204
65	114	120	126	132	138	144	150	156	162	168	174	180	186	192	198	204	210
66	118	124	130	136	142	148	155	161	167	173	179	186	192	198	204	210	216
67	121	127	134	140	146	153	159	166	172	178	185	191	198	204	211	217	223
68	125	131	138	144	151	158	164	171	177	184	190	197	203	210	216	223	230
69	128	135	142	149	155	162	169	176	182	189	196	203	209	216	223	230	236
70	132	139	146	153	160	167	174	181	188	195	202	209	216	222	229	236	243
71	136	143	150	157	165	172	179	186	193	200	208	215	222	229	236	243	250
72	140	147	154	162	169	177	184	191	199	206	213	221	228	235	242	250	258
73	144	151	159	166	174	182	189	197	204	212	219	227	235	242	250	257	265
74	148	155	163	171	179	186	194	202	210	218	225	233	241	249	256	264	272
75	152	160	168	176	184	192	200	208	216	224	232	240	248	256	264	272	279
76	156	164	172	180	189	197	205	213	221	230	238	246	254	263	271	279	287

A BMI of < 18.5 is underweight, 18.5–24.9 is normal/healthy, 25–29.9 is overweight, 30–34.9 is class I obesity, 35–39.9 is class II obesity, and ≥ 40 is class III obesity.
Understanding Nutrition, 12th ed, Whitney & Rolfes, Chapter 8, Table 8.4, p. 252.

Combating Obesity

Individuals who are overweight or obese often seek nondietary solutions to combat their body weight issues. These may involve surgery, the use of drugs, or fad diets. Common surgical procedures to combat obesity include liposuction, intestinal resection, gastric bypass, and the gastric band. Drugs to combat obesity may be purchased over the counter or prescribed. Low-carbohydrate and very-low-Calorie diets (less than 800 Calories per day) may also be employed to combat obesity. There is a 6 percent success rate for reducing weight and maintaining the weight loss by semistarvation dieting. During prolonged fasting (or low-Calorie, low-carbohydrate diets), the BMR declines, and lean body mass is used to make glucose for brain, central nervous system, and red blood cell functioning. In general, these methods for weight loss can be health threatening and usually do not result in long-term sustained weight loss or a healthy lifestyle. A good measure of a successful weight loss program is if the person has kept at least two-thirds of the total weight lost off for at least two years.[49,60,66,74]

The body metabolism changes under conditions of **feasting** and fasting. Figure 4.9 depicts the fate of energy intake and stored energy during periods of feasting and fasting. First, you will see that when a person overeats, excess Calories from carbohydrate, protein, fat, and alcohol are stored in fat cells as fat. Second, you will see that when a person fasts or eats a very-low-Calorie diet, the body exhausts glycogen stores within hours and burns fat, but also burns body protein or lean body mass. Losing lean body mass or muscle mass is not a desirable type of weight loss, because it causes metabolism and vital function to decline (see BioBeat 4.6). If fat weight loss is desired while lean body mass is preserved, the rate of weight loss must be slow. Thus, the recommendation to lose 1 pound of fat weight per week results in stored body fat loss and lean body mass preservation. A weight loss program that goes at a faster rate will force the body to use lean body mass (muscle protein) to meet the glucose energy needs of the body.[47,76]

feasting The continuation of food intake beyond satiety, promoting fat synthesis.

hydration Of or pertaining to water; the status of body water content assessed by fluid and electrolyte balance in the intracellular and extracellular fluids.

BioBeat 4.6

Body Weight and Preserving the Rate of Metabolism

The state of energy balance and that of **hydration** determine the changes in weight a person experiences on a daily basis. Remember that 1 cup of water equals 8 ounces, so if a person dehydrates 2 cups of water, that will amount to 1 pound of weight loss. However, weight changes resulting from tissue changes are influenced by the state of energy balance. Recall that there are 3,500 Calories per pound of stored body fat. In order to lose primarily fat weight and maximize weight loss, 1 pound of fat weight loss is recommended per week. This means that a caloric deficit of 500 Calories per day is the goal. Research has shown that the negative energy balance created by reducing Calorie intake by about 200 Calories per day and increasing energy expenditure by 300 Calories per day will maximize fat weight loss, preserve lean body mass, and maximize total weight loss. If one creates too great an energy deficit, or starves cells for energy, the metabolic response is to decrease Resting Energy Expenditure. This means that every minute through every level of activity, the individual will burn fewer Calories overall (see Appendix B). REE needs to be maintained in order to maximize Calorie burn. This is why when one is practicing weight loss, daily devotion and patience is required. Let's say that normally Jane burns an average of 2,687 Calories per day. She has been starving herself on an 800-Calorie daily diet, and now her REE may theoretically be suppressed by 80 percent. Now, her average Calorie burn in a day is 537 Calories.[18,87]

Are you or have you been metabolically suppressed?

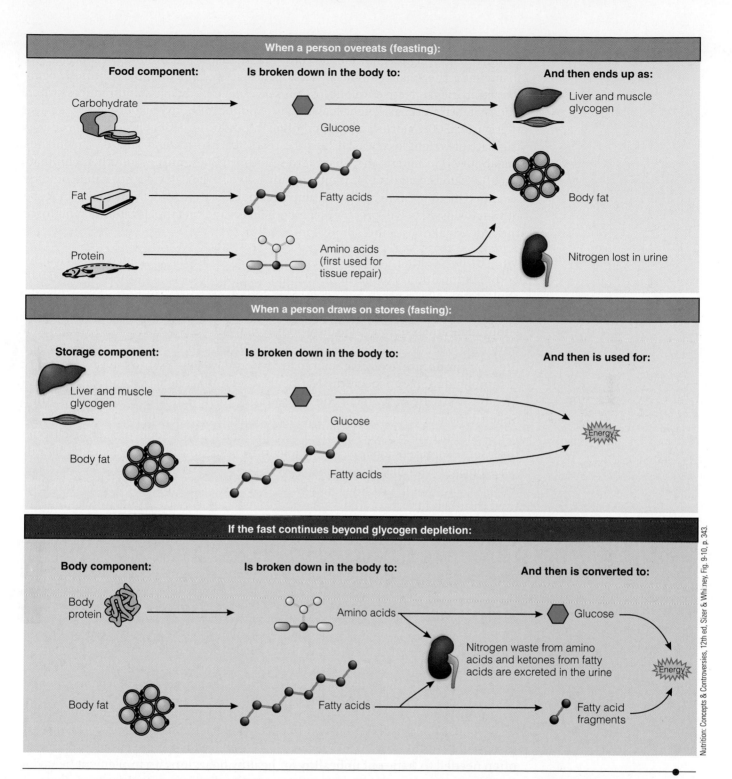

When a person overeats (feasting):

Food component:	Is broken down in the body to:	And then ends up as:
Carbohydrate	Glucose	Liver and muscle glycogen
Fat	Fatty acids	Body fat
Protein	Amino acids (first used for tissue repair)	Nitrogen lost in urine

When a person draws on stores (fasting):

Storage component:	Is broken down in the body to:	And then is used for:
Liver and muscle glycogen	Glucose	Energy
Body fat	Fatty acids	

If the fast continues beyond glycogen depletion:

Body component:	Is broken down in the body to:	And then is converted to:
Body protein	Amino acids	Glucose
		Nitrogen waste from amino acids and ketones from fatty acids are excreted in the urine
Body fat	Fatty acids	Fatty acid fragments → Energy

Nutrition: Concepts & Controversies, 12th ed, Sizer & Whitney, Fig. 9-10, p. 343.

Weight Control and Weight Change Weight is maintained when energy balance occurs, and that is when energy intake equals energy expenditure. The body continually strives for metabolic equilibrium. Metabolic rate is the greatest within a narrow range of energy balance. Metabolism is more easily maintained when foods are provided every 4 to 6 hours when a person is active and every 12 to 14 hours within a sleep period. If the body is not fed, it is pushed metabolically to produce energy from alternative sources (stored proteins and fats), which is metabolically stressful. When carbohydrates are limited and the body relies more on amino acids and fatty acids for ATP energy production, then metabolic byproducts such as ketones accumulate in the blood; these byproducts are indicators of malnutrition.

FIGURE 4.9 Feasting and two phases of fasting.

Weight control involves manipulating the energy balance equations so that a healthy body weight can be achieved or maintained. If you want to lose weight by losing fat rather than lean body mass or muscle, you must employ sound weight-loss strategies. Effective weight loss preserves lean body mass and reduces stored body fat. An optimal weight-loss rate is 1 pound of body fat per week. Because there are 3,500 Calories in 1 pound of fat, a 500-Calorie deficit per day produces this rate of weight loss. It is best to create a 500-Calorie deficit per day by including about 40 minutes of aerobic exercise to burn 300 Calories and then eliminating 200 Calories of dietary fat intake daily as well. The caloric restriction could be accomplished by eliminating 22 grams of hidden fat in the diet each day. This type of weight-loss rate will preserve lean body mass and metabolic rate, and will force the individual to use stored body fat.[8,9,11,35,39]

The Role of Exercise Exercise is critical in the maintenance of a healthy body composition. Furthermore, it improves cardiovascular fitness, raises HDL cholesterol, slows heart rate, decreases blood pressure, supports brain health, maintains lean body mass and bone density, increases flexibility, strength, and endurance, and promotes a healthy body weight. Exercise can also improve mental outlook and increase metabolic rate. The DRI for physical activity is 60 minutes of **moderate exercise** cumulative per day. This same physical activity recommendation is made by the National Academy of Sciences. Moderate exercise includes activities in which body parts are moving and the body is sweating. Scientific studies have shown that being physically active, so that 2,000 to 3,000 Calories are burned per week through exercise, results in reduced risk for chronic diseases. Exercises such as walking, jogging, and running result in about 100 Calories burned per mile.[7,27,29,42,46,51,55,72,83,89]

Spot reduction, or reducing fat mass in one part of the body, is a weight-loss myth. Fat-cell fill or stored body fat increases and decreases uniformly among all fat cells depending on the state of energy balance. Fat cells are filled wherever they are. However, there is some gender or genetic predisposition for where the fat cells tend to exist. Thus, there is no chance for spot reduction to occur from extensively exercising one group of muscles in the body.[85]

DA+ GENEie Go to Diet Analysis Plus and build your profile. Build a profile report to see your Estimated Energy Requirement (DRI for Calories). Now change your activity level and rebuild the profile. Do this with several activity levels. Change your age, height, and/or weight and notice the change in Calories. Discuss the effect of height, weight, activity, and age on the amount of Calories recommended.

Changing One's Behavior Weight control is easier if a person develops long-term, consistent, healthy lifestyle behaviors. Because many lifestyle habits are essentially learned or acquired behaviors, a behavior modification program is often needed to trade out unhealthy for healthy behaviors. To implement behavior modification, a person needs to first *set a realistic and healthy goal*. Second, one needs to *identify current behaviors that need to change to achieve the goal and identify new behaviors* that will help achieve the goal. Now the person can *create a behavior modification plan* and follow the plan. Finally, there is a need to constantly *evaluate progress and modify the weight control plan* as needed to promote continued success.[5,19,33]

The characteristics of lifelong diet habits that have greater long-term weight control success rates include incorporating *daily, hour-long exercise*; using *behavior modification* to permanently incorporate desired health behaviors;

weight control Diet and lifestyle measures taken to achieve or maintain a healthy body weight.

moderate exercise An intensity of physical activity that causes the body to sweat.

and *eating a well-balanced diet* to promote nutritional adequacy from a variety of foods.[21,30,31,35,45]

Optimal Dietary Planning for Adequacy

Optimal dietary planning involves ensuring that the diet meets standards for promoting health and **nutrient adequacy**. To do so, one must limit saturated and *trans* fatty acid, cholesterol, simple sugar, and sodium intakes; make sure the percentages of Calories from carbohydrate, protein, and fat are within the Acceptable Macronutrient Distribution Ranges (AMDRs); ensure fiber intake is adequate; ensure DRIs are met for the essential nutrients; and make sure total Calories consumed are controlled. Furthermore, the diet, lifestyle, and behavior may need to be optimally planned and matched to a person's genetic profile to achieve optimal nutrition and health (see BioBeat 4.1 on page 171).[8,36]

When a person is planning to maintain, lose, or gain weight, nutritional adequacy needs to be considered. This implies that the diet provides essential nutrients, fiber, and energy to maintain health. Dietary assessment tools need to be incorporated into the weight control plan. This may include using a variety of tools, such as diet analysis software programs, food composition tables, and food guidance systems like MyPlate and the Exchange Lists, to monitor dietary intake.[8,34,50]

Calorie-controlled diets need to be planned so that there are no nutrient inadequacies (nutrients less than 100 percent of the DRI) and certainly no nutrient deficiencies (nutrients less than 66 percent of the DRI). Consuming diets with caloric levels of 1,200 Calories or less makes it difficult to achieve nutritional adequacy. In such cases, a multivitamin and mineral supplement or the use of heavily fortified foods may be needed to meet the DRIs. Furthermore, Tolerable Upper Intake Levels (ULs) should be considered when selecting nutritional supplements and highly fortified foods to prevent nutrient toxicities or excesses. Each individual should monitor his or her nutrient intake to promote nutrient adequacy and avoid **nutrient inadequacy**, **nutrient deficiency**, or **nutrient excess** (see Table 4.4). In Module 5, vitamin and mineral deficiency and toxicity signs and symptoms will be explored.[36] These principles—including behavior modification, energy balance, weight control, body composition assessment, and nutritional adequacy—should be applied in establishing lifelong weight control and health.[9]

TABLE 4.4

Dietary Interpretation of Vitamin and Mineral Intake	
Vitamin and Mineral Dietary Interpretation	**% Intake**
Excessive	Above the UL
Adequate	100% of the DRI up to the UL
Inadequate	66–99% of the DRI
Deficient	Less than 66% of the DRI

© Cengage Learning 2013

Summary Points

- Individuals need to know their body composition (lean body mass versus fat mass) in the context of their BMI.
- There are different ways to determine body composition.
- The distribution of body fat affects health risks (visceral versus subcutaneous fat).
- BMI is a clinical tool to determine appropriate body weight for height.

nutrient adequacy The dietary intake of the needed amount of energy, nutrients, and fiber to support optimal health and function; numerically defined as 100 percent of the DRI up to the Tolerable Upper Intake Level.

nutrient inadequacy The dietary intake of less than the needed amount of energy, nutrients, and fiber to support optimal health and function; numerically defined as 66 to 99 percent of the DRI.

nutrient deficiency The dietary intake of much less than the needed amount of energy, nutrients, and fiber to support optimal health and function; numerically defined as less than 66 percent of the DRI.

nutrient excess The dietary intake of much more than the needed amount of energy, nutrients, and fiber to support optimal health and function; numerically defined as above the Tolerable Upper Intake Level for vitamins and minerals.

- A slow rate of weight loss is suggested to preserve lean body mass and metabolism while reducing fat mass.
- An optimal rate of weight loss is 1 pound per week.
- A behavior modification program is useful in weight control.
- Lifelong diet and exercise habits should be practiced for optimal body weight, health, and nutrient adequacy.

Take Ten on Your Knowledge Know-How

1. What does body composition mean?
2. What are some accurate methods for determining percent body fat?
3. What is your interpretation of your percent body fat for your gender?
4. How and why is having central adiposity detrimental to health?
5. What is the BMI? What are the strengths and weaknesses of using the BMI to determine obesity?
6. What is your calculated BMI? What does this number mean to you?
7. What is a safe and effective weight-loss protocol that will preserve a person's lean body mass and rate of metabolism? What are the risks of eating too few Calories? What are the benefits of daily exercise?
8. What is behavior modification? How does behavior modification fit into weight management, health, and disease?
9. Do you have a need to modify behaviors related to your diet, exercise, and lifestyle? If yes, why? If not, why not?
10. What are the key principles of optimal dietary planning for nutritional adequacy?

4.4 PRINCIPLES OF FITNESS FOR HEALTH

T-Talk 4.4
To hear Dr. Turley talk about principles of fitness for health, go to www.cengagebrain.com

Introduction

There is no question that the sports-specific training programs that competitive athletes use to advance their athletic performances impact human physiology. However, the training program may not produce a high rating on overall physical fitness testing or prevent injury. As understanding is gained about assessing **physical fitness** and creating a total fitness program, it may be realized that the physical activity programs that some athletes use to prepare for competition may not support wellness or healthy aging. Personal fitness goals, health, safety, and pleasure of exercising must be considered in the development of a personal fitness program. We will explore total fitness, aerobic and resistive exercise, the FIT classification system, and logging activity for weight control and health in this section.[46,61]

Total Fitness for Health

A total fitness program is an exercise program designed to prepare an individual to rank high on physical fitness testing procedures that include evaluation of aerobic capacity, muscular endurance, muscular strength, flexibility, and body composition. Over the years, a paradigm shift has occurred for the goal of a fitness program. The purpose of exercise has changed from developing the skills to do well in physical fitness testing to deriving health from exercise (see Table 4.5). As a result of the total fitness approach to exercise, exercise programs have become more diverse, and the health values of exercise are clearer. The elements of a total fitness program include a warm-up, an aerobic component, a resistive component, a flexibility component, and a cool-down (see BioBeat 4.7 for information on the warm-up and cool-down).

physical fitness Having good cardiovascular, respiratory, and muscular capacity assessed by body composition, aerobic activity, and muscular performance and flexibility.

TABLE 4.5

The Values of Exercise

Exercise Value	Specific Areas of Benefit
General health	Improved mental health Increased feeling of vigor Improved sleep pattern Increased bone density Decreased risk of chronic diseases such as obesity, heart disease, vascular disease, cancer, diabetes, arthritis, and osteoporosis
Body composition and weight control	Decreased percent body fat Increased lean body mass Stabilized body weight Increased metabolic rate Improved muscle tone
Cardiorespiratory health and fitness	Improved circulation (pumping capacity of the heart, stroke volume, oxygen delivery capacity to the tissues, promotion of wound healing, decreased tendency for infections) Increased oxygen uptake Improved lipid profile, decreased total cholesterol (increased HDL, decreased LDL), decreased triglycerides Decreased blood pressure Decreased heart rate Increased efficiency of metabolic processes (fat metabolism and carbohydrate metabolism)
Athletic performance	Improved competitiveness in sporting events Body sculpting in fitness competitions

© Cengage Learning 2013

splenic shunt An increase in blood flow to the working muscle caused by the increased oxygen demands of exercise, thus diverting the blood supply away from the vital organs to the skeletal muscles.

BioBeat 4.7

The Warm-Up and Cool-Down

Many physiological adjustments are made in the cardiorespiratory system and in cellular metabolism in response to exercise stress. A warm-up is important for minimizing the stress response to exercise and reducing the production of metabolic cellular waste products, generated due to a lack of oxygen, that contribute to pain and fatigue. The objectives of a warm-up are to (1) slowly increase the pulse rate to an aerobic level, (2) facilitate vascular changes (**splenic shunt**) to accommodate aerobic exercise, (3) gear up the energy metabolism (i.e., increase the delivery of oxygen and nutrients to the working muscle, and raise the core body temperature to an optimal level for energy production), (4) increase the elasticity of the muscle fibers (which helps prevent muscle, tendon, ligament, and joint injuries, as well as

to improve muscle contraction), and (5) preserve muscle glycogen and decrease lactic acid formation.

To warm up most effectively, you should begin with a very-low-level activity for a 2- to 5-minute period. As you progress through the warm-up, the intensity of the activity can gradually increase. Acceptable types of warm-up activities include fast walking, slow jogging, or light calisthenics (trunk circling, jumping jacks, arm circling). However, you can also accomplish many purposes of the warm-up by passively heating the body up in a whirlpool or sauna. Once the body has warmed up, the oxygen delivery to the working muscles is more efficient, and thus the physiological benefits of the exercise session may be greater.

The cool-down portion of an exercise session promotes recovery from exercise stress. It is a pro-

cess that is targeted to reestablish homeostasis in the body. Many physiological adjustments are made to adapt back to a comfortable level of metabolism. The objectives of the cool-down are to (1) slowly decrease pulse rate, (2) decrease core body temperature, (3) slowly constrict blood vessels, (4) continue the adequate supply of blood to the brain, and (5) facilitate removal of metabolic waste products from the muscles. To accomplish an effective cool-down, it is important to continue to move for about a 5-minute period after the aerobic session. This period can be followed with a series of flexibility exercises for about another 5 minutes.

What do you do to warm up for an exercise session, and what do you do to cool down?

Fitness Testing Comprehensive physical fitness testing evaluates four aspects of functional capacity: flexibility (see BioBeat 4.8), muscular strength (weight training), muscle endurance (repetitive muscular contractions), and **aerobic capacity** (cardiovascular endurance). There are national standards for performance and percentile rankings by age and gender for each aspect of fitness testing. Typical tests to assess aerobic capacity may include the Three-Minute Step Test, the 1-mile walk, the 1.5-mile run, or the YMCA bike test. Typical strength tests may include the bench press, pushups, abdominal crunches (situps), grip dynomometer, and leg press. Flexibility may be assessed using the sit and reach, modified sit and reach, or a shoulder rotation test. Body composition may be assessed many different ways, including hydrostatic weighing, air displacement, infrared spectrophotometry, bioelectrical impedance, or skin-fold calipers (refer to page 182). The person's overall fitness rating is influenced by his or her performance in each of the areas.

BioBeat 4.8

Stretching for Flexibility

The purposes of stretching exercises are to maintain muscle fiber and connective tissue elasticity, reduce the risk for injury, facilitate muscle recovery from exercise, and preserve the range of muscle movement around a joint. Stretching exercises should be included in the daily exercise routine for all of the major muscle groups. Stretching improves flexibility, which is the range of motion possible around a specific joint. A person may not be able to move normally if the normal range of motion around the joint is compromised. A reduced range of motion decreases the efficiency of biomechanical movement. The ability to move a joint through an adequate range of movement is important for daily activities and sports performance. For example, a sprinter may be handicapped by tight, inelastic hamstring muscles. Because the ability

to flex the hip joint will be limited as a result of the tight muscles, the sprinter's stride will be shortened. Activities such as gymnastics, ballet, diving, karate, and yoga require improved flexibility or even the ability to hyperextend some joints for superior performance.

Adequate ranges of motion for many articulations may be more important for long-term injury prevention than just the rating for flexibility that one could get in physical fitness testing. Individuals involved with physical activity who have poor flexibility (in specific body areas or generally) have an increased risk of injury. Once flexibility is assessed and flexibility insufficiencies are identified, a stretching program can be customized, emphasizing those areas in need of improvement.[65,85,86]

What is your level of flexibility?

aerobic capacity The ability of the respiratory and cardiovascular systems to deliver oxygen to working muscles.

aerobic exercise Physical activity where the heart rate can be elevated and maintained steadily for at least 10 minutes and there is increased oxygen uptake and delivery to the body and muscles.

resistive exercise A form of physical activity done to increase the size and strength of the skeletal muscles.

Aerobic and Resistive Exercise

There are two basic types of exercise: aerobic and resistive. **Aerobic exercise** places physiological demand on the cardiorespiratory systems, and **resistive exercise** places physiological demand on the skeletal-muscular system.

Aerobic Exercise Aerobic exercise includes activities in which an elevated heart rate can be maintained steadily for at least a 10-minute period. Examples of aerobic exercise include walking, jogging, running, jumping rope, rowing, bicycling, swimming, stair stepping, side sliding, cross-country skiing, skating, and hiking. Tennis, basketball, soccer, bowling, volleyball, and weight lifting are not considered aerobic activities, because the elevated heart rate is not maintained steadily; however, these types of activities do provide exercise and health values.

The purpose of aerobic exercise is to condition the **cardiorespiratory systems**. Oxygen uptake and oxygen delivery are key to aerobic exercise and one's ability to perform aerobically. In order to have a conditioning effect on the **cardiovascular system**, it is necessary to elevate the heart rate to at least 70 percent of the **maximum heart rate (MHR)** (see Figure 4.10 and BioBeat 4.9). The percent of one's MHR that is sustained is referred to as the intensity of the exercise.

An evaluation tool of aerobic exercise stress is **perceived exertion**. It is used in conjunction with the working heart rate (WHR) to monitor aerobic fitness tests. Perceived exertion is also a great personal tool to use to gauge the intensity of the aerobic exercise if a more precise way to monitor the WHR (such as a heart rate monitor device) is unavailable. Working at 60 percent of MHR feels very comfortable. At this intensity level, it is possible to carry on a conversation and work for long periods. The perceived exertion would be "somewhat hard." On the other hand, when working at 80 percent, the exercise is not comfortable; talking is difficult, breathing is deep, and muscle fatigue can be felt within a short period. The perceived exertion at this intensity is "very hard." In general, the goal is to sustain an increased heart rate within a range of at least 70 to 80 percent of the MHR; the perceived exertion rating would be "hard."

The FIT Classification System   Exercise physiologists have been able to set classifications for aerobic fitness regarding cardiovascular exercise behaviors. The three parameters used to influence the exercise effects on the cardiovascular system include the frequency of the exercise session (referred to in terms of the number of days per week that one exercises), the intensity of the exercise (based on the percentage of MHR while exercising), and the time or duration that the elevation of heart rate has been maintained (in continuous minutes). With

Mike Watson Images/Jupiter Images

Aerobic exercise.

cardiorespiratory systems Systems of the body, including the heart and lungs and associated blood vessels.

cardiovascular system Of or pertaining to the heart and blood vessels or the circulatory system of the body.

BioBeat 4.9

Maximum Heart Rate, Working Heart Rate, and VO$_2$ Max

A person's maximum heart rate (MHR) is strongly dependent on the person's age. The MHR (given in units of beats per minute) for humans peaks at 20 years old at a level of approximately 200 beats per minute. Thus, after the age of 20, you will notice an age-associated decline in the MHR value of 1 beat per minute per year. If you are 20 years old or older, your MHR can be easily calculated as 220 – Age in years = MHR. This formula works for about 70 percent of the population. About 15 percent of adults have a significantly faster-beating heart, and about 15 percent have a significantly slower-beating heart than the average adult. Thus, if the outliers use the sample formula, it is erroneous.

Working heart rate (WHR) is the number of heartbeats per minute that are measured during exercise. The WHR formula can be used to determine the workout intensity that you want to achieve. For example, if you are interested in fat cell reduction, you will want to engage in low-intensity aerobic exercise, which maintains your heart rate within 60 to 70 percent of your MHR to maximize the body's ability to use fat as a fuel. On the other hand, if you are interested in high-level fitness, you will want to engage in high-intensity aerobic exercise, which maintains your heart rate within 75 to 85 percent of your MHR. To calculate the intensity of the exercise, follow the formula

WHR ÷ MHR × 100 = % MHR. To determine one's true MHR, a maximum oxygen uptake measurement (VO$_2$ max) test is performed. **VO$_2$ max** is the maximum amount of oxygen (O$_2$) that can be taken up by the body, and is given in units of milliliters of oxygen per kilogram of body weight per minute. A graded exercise protocol is used to increase the workload while evaluating the elevation in heart rate. When the workload is increased, the point at which the heart rate does not increase is the VO$_2$ max. At VO$_2$ max, the true MHR is determined.

What percentage of your MHR should you exercise at to meet your goals?

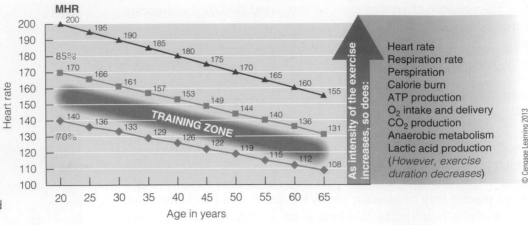

FIGURE 4.10 Heart rate, exercise intensity, and physiological responses.

As intensity of the exercise increases, so does:
Heart rate
Respiration rate
Perspiration
Calorie burn
ATP production
O_2 intake and delivery
CO_2 production
Anaerobic metabolism
Lactic acid production
(*However, exercise duration decreases*)

© Cengage Learning 2013

maximum heart rate (MHR) The highest number of times the heart can beat competently or contract per minute, given in beats per minute.

perceived exertion Standard words that are used to describe how hard you feel you are working during exercise.

VO_2 max The maximum amount of oxygen that can be taken up by the body, determined under maximum exercise stress and given in milliliters of oxygen per kilogram of body weight per minute.

eccentric When the muscle is lengthening while it is contracting, and when it is opposing resistance; it is known as a negative lift.

these three parameters, exercise physiologists have created a "frequency, intensity, time" (FIT) format that is used to define and communicate a person's cardiovascular fitness based on his or her aerobic exercise habits (see Table 4.6). By evaluating how many times a person engages in aerobic activity per week and the intensity and duration of the exercise, the aerobic fitness rating can be determined.

TABLE 4.6

The FIT Classification System for Aerobic Activity

Factor	Low	Average	High
F = frequency (days per week)	3	4	5+
I = intensity (% MHR)	60–69 (low)	70–79 (moderate)	80–90 (high)
T = time (continuous minutes)	10–20	15–45	30–60

© Cengage Learning 2013

Resistive exercise.

Glowimages/Corbis

Resistive Exercise for Strength Weight training is the most popular resistive exercise and incorporates **eccentric** and **concentric** muscle contractions (see Table 4.7). Muscles move the bones. The movement of the bones occurs as a result of the shortening and elongating of the muscles that are located adjacent to joints in the body. There are two sets of muscles around each joint. One set of muscles decreases the angle of the joint when it contracts, and the other increases the angle of the joint when it contracts. These sets of muscles are called **antagonistic muscles**. When muscle strengthening occurs, the contractility of the muscle fibers has increased. There are three parameters that influence muscle contractility: (1) the amount of resistance (weight) applied during contraction, (2) the number of repetitions (consecutive contractions of the antagonistic muscle) within a set and the number of sets performed, and (3) the speed of the muscle contractions. Depending on how these parameters are manipulated, the muscles respond (adapt) in different ways. For example, if the resistance during contraction is provided by a light weight, and the number of contractions is at least 16, the muscles will gain primarily endurance with little increase in muscle mass. On the other hand, if a heavy weight is applied during contraction so that only 8 to 12 repetitions are possible, the muscle gains primarily strength and mass.

To develop an effective strength-training program, one needs to evaluate the "one repetition maximum" (one-rep-max) for each type of lifting exercise. To do this on a machine, a light weight is selected. One repetition of the exercise is done, then the next brick or increment of weight is added, and another repetition is done. The person keeps working through this graded exercise process until a weight is added that makes it impossible to complete the full range of motion, or one rep. The last weight level at which one rep can be completed is the one-rep-max.

TABLE 4.7		
The Muscle Contractions, Training Types, and Balanced Lifts		
Muscle contractions	*Eccentric muscle contraction* Muscle is elongating while contracting or generating tension *Concentric muscle contraction* Muscle is shortening while contracting or generating tension *Isometric muscle contraction* Muscle length is static while the muscle generates tension; this type of contraction is used when joints are injured to attempt to maintain some muscle strength	
Training types	***Isometric** training* Program involves a series of isometric muscle contractions—muscles contracting against a fixed amount of force ***Isokinetic** training* The use of special equipment to control the speed of the muscle contractions *Circuit training* Combines aerobic activity with resistive activity for the purposes of increasing Calorie burn, reducing boredom, and improving fitness *Weight training* Program involves lifting weights	
Balance lifts	Bench press/seated row Dumbbell bench press/bent-over row Shoulder press/lateral pulls Seated leg press/truck flex Pectoral fly/bent-over fly Upright row/dips Lateral raises/medial pulls Leg extension/leg curls Bicep curl/tricep extension Abdominals/back extension	

© Cengage Learning 2013

An all-around muscle strengthening program should include exercises that involve all of the major muscle groups in balanced lift action format (see Table 4.7). A general guide for beginners to improve strength is to train at a level of resistance that is about 80 percent of their one-rep-max. Somewhere between 8 and 12 repetitions should be achievable. It usually takes three to four sets before the muscles are tired. At all times, it is important to balance the strength exercises so the antagonistic muscles develop fairly equitably in order to minimize the risk for joint injury. About every 8 to 12 weeks, a strength cycle should be followed by an endurance cycle so that connective tissue can be developed to stabilize the muscle fibers. An endurance lift cycle is established by lifting at 60 percent of the one-rep-max and attempting to complete 15 to 20 repetitions per set. Again, it takes three to four sets before the muscles are tired. Depending on personal fitness preferences, any of the types of muscular efforts noted in Table 4.7 can be used to maintain or gain strength. To maintain strength, it is recommended to engage in

concentric When the muscle shortens and it is contracting, usually when it is opposing resistance.

antagonistic muscles A pair of opposing muscles positioned adjacent to an articulating joint where, when one set of muscles contracts, the angle of the joint decreases, and when the other set contracts, the angle of the joint increases.

isometric When the muscle remains the same length while contracting or the muscle is static when tension occurs.

isokinetic The muscle force exerted or muscle resistance applied during limb movement is at a fixed speed or constant velocity.

three to four sets of 8 to 12 repetitions at 80 percent of the one-rep-max of resistive exercise twice per week. To gain strength, it is recommended to engage in three to four sets of 8 to 12 repetitions at 80 percent of the one-rep-max of resistive exercise at least three times per week.

The Overload Principle

To improve fitness or functional capacity, physiological adaptations must be made to a higher demand than what is comfortable to maintain. In order for the body to increase functional capacity, the **overload principle** is applied progressively until the fitness and/or strength goals are achieved. The overload principle can be applied to aerobic and resistive exercise parameters to promote functional capacity. The aerobic parameters applied to the overload principle include exercise frequency, duration, and intensity. The resistive exercise parameters applied to the overload principle include resistance, repetitions, and sets, and the speed of the muscle contractions. It takes time, persistence, consistency, and devotion to gain functional capacity. However, reversing or losing functional capacity as a result of the absence of exercise (detraining) unfortunately occurs fairly rapidly.

Logging Activity for Weight Control and Health

All physical activity done in a day counts. Whether you want to subscribe to a total fitness program, enjoy your physical activity, achieve weight loss, burn enough Calories for health, or do well in physical fitness testing, all physical activity helps promote health. The common denominator for exercise is Calorie burn. No matter what type of physical activity you engage in, energy expenditure reflects the impact of the exercise value on health, and logging your exercise patterns helps you plan for and achieve your fitness and health goals.[7,55,75]

Activity for Weight Control Calorie restriction with physical activity is the best approach for promoting weight loss. Research has shown that moderate-intensity physical activity for 30 to 45 minutes, three to five days per week, promotes weight control. To lose weight, to keep the weight off (control body weight), and to improve functional capacity, 60 minutes of rigorous activity is needed daily. People who are new to daily exercise and want to start exercising should choose an aerobic medium and the duration of exercise that is comfortable for them to do. They should be encouraged to exercise every day and to extend the duration of each aerobic exercise session during the week by 2 minutes. As each week begins, another 2 minutes is added to each aerobic exercise session for the week. This schedule could accumulate 14 additional minutes of physical activity weekly for them. The duration of the activity should continue to be increased until the desired energy expenditure is achieved each day.[7,55,75]

Logging Activity for Health To reduce the risk of developing chronic disease, it is important to expend between 2,000 and 3,000 Calories per week in structured physical activity. Adults burn an average of about 100 Calories per mile. This means that people who walk, jog, or run need to put in the equivalent of 20 to 30 miles per week. Diet Analysis Plus can be used to track activity and estimate energy expenditure, or energy expenditure can be calculated as described in the sections on physical activity or voluntary muscle movement on pages 178–179. In either case, keeping an activity log is beneficial in evaluating your fitness program with respect to the total fitness format, the rigor of your activity, and your weekly Calorie burn or energy expenditure with respect to health.

You can use the activity log in Figure 4.11 to record all of your structured physical activity. From this log, you will evaluate your warm-up and cool-down pro-

overload principle A greater than normal workload demand placed on the cardiorespiratory or skeletal-muscular systems that leads to increased functional capacity.

Name: Date: Time of day:		
Warm-up activity:	Duration:	Flexibility:
Cardiovascular aerobic activity:	Duration:	WHR:
		% MHR:
Cool-down activity:	Duration:	Flexibility:

Muscular strength and endurance Resistive exercise	Amount of weight	# of reps	# of sets
1			
2			
3			
4			
5			
6			
7			
8			
9			
10			

Other activity:	Duration:	

Comments:		

© Cengage Learning 2013

FIGURE 4.11 Exercise activity log.

cesses (see BioBeat 4.7 on page 191), flexibility (see BioBeat 4.8 on page 192), aerobic conditioning exercises, muscular strengthening exercises, muscular endurance exercises, and the amount of time you engage in a variety of physical activities that will count for Calorie burn. From this log you can determine your daily Calorie expenditure and your fulfillment of your total fitness program. You should keep comments regarding what inspired you to exercise and what your reward was for exercising. On days that you do not engage in physical activity, it is valuable to note the barriers to exercise on those days of the week. On days that you do engage in exercise, make note of the promoters of exercise. Remember, you should be striving for 60 minutes per day or at least 7 hours per week.

Summary Points

- The common denominator for exercise is Calorie burn; a minimum amount of Calorie burn to yield health benefits is between 2,000 and 3,000 Calories per week.
- An optimal amount of exercise to positively influence weight control and functional capacity is 60 minutes of rigorous activity per day.

- The parameters evaluated in physical fitness testing include body composition, flexibility, aerobic capacity, muscular strength, and muscular endurance.
- A total fitness program can support a good rating on physical fitness testing because it includes warm-up and cool-down activities, flexibility exercises, aerobic activities, and resistive muscular work.
- The three parameters to increase for improving aerobic capacity are frequency, intensity, and time.
- The three parameters to increase for improving muscular strength are the amount of resistance, the number of repetitions and sets, and the speed of the muscle contractions.

Take Ten on Your Knowledge Know-How

1. How can you evaluate your physical activity to understand your total fitness program and your Calorie burn?
2. If you only had three days per week to exercise for an hour each time, how could you still gain aerobic fitness?
3. You want to get stronger. What percentage of your one-rep-max would you want to lift?
4. What is the common denominator for evaluating the value of a variety of physical activities on health?
5. What physiological changes occur in a warm-up? What physiological changes occur in a cool-down?
6. What is aerobic exercise? Give some examples of aerobic exercise.
7. How many types of muscle contractions are there? What types of muscle contractions are typically used when lifting weights?
8. What are the parameters of muscular endurance exercises?
9. If you want to lose weight, how much time should you allot for exercise? How about to maintain weight loss?
10. How many miles could you run in an hour? How many Calories would you burn?

4.5 FUNDAMENTALS OF EXERCISE FOR NUTRITION

electrolyte balance The equilibrium or normal ratio of charged particles in a solution that function as electrolytes (such as sodium, potassium, and chloride) in the body.

Introduction Exercise requires energy, and the diet provides energy from the three types of dietary fuels: carbohydrates, proteins, and fats. Carbohydrates are known as the high-performance fuel because they produce ATP quickly (see page 200). You can think of digestible carbohydrates as the kindling in a fire. Fats are low-level performance fuels because they are slow to produce ATP, but the level of production is very steady. You can think of fats as the logs on the fire, giving off a steady source of heat but not as rapid, intense, and short-lived as the kindling on the fire. Proteins can be used to produce ATP, but their first purpose in the body is to supply the building blocks (amino acids) required by the body for protein synthesis (used for tissue growth and repair, both of which are anabolic processes). Protein is used as a fuel if an excess amount of protein has been consumed in the diet, a person is under intense stress, or the person is starving; however, in the late stages of intense endurance exercise, which is a very stressful physiological condition, protein becomes an important muscle fuel and can supply up to 15 percent of the ATP produced.

Exercise also requires optimal fluid and **electrolyte balance**, as well as a healthy diet adequate in the essential vitamins and minerals. We will explore fuel utilization and the dietary fuels needed before, during, and after exercise, as well as thermal regulation and hydration, in more detail.[15,16,17,20,71,85,86]

Fuel Utilization During Exercise

During exercise, ATP is largely produced from a combination of carbohydrates and fats. The percentage from each fuel depends primarily on the intensity and duration of the activity. In general, carbohydrate use increases with increasing intensity and falls with increasing duration of physical activity. However, the absolute amount of carbohydrate and fat used by muscles can be shifted, depending on fuel availability and fitness. The availability of fatty acids increases fatty acid oxidation (fat breakdown for ATP production), and when more carbohydrate is available, more carbohydrate is metabolized for energy. This reciprocal interplay between fat and carbohydrate use should be carefully considered when deciding on food consumption for athletic performance and competition. Nutrition plays its chemical roles in energy metabolism by providing the substrates (the fuels, glucose, amino acids, and fatty acids) and essential structural components (vitamins and minerals) of enzymes needed to drive the chemical reactions to generate ATP (energy) in **glycolysis**, the **citric acid cycle**, and the **electron transport system** (see Figure 4.12). As you can see in Figure 4.12, adequate nutrition is vital to supply the energy-producing nutrients (glucose, fatty acids, and amino acids) into the energy pathways. Additionally, by looking closely at Figure 4.12, you can see that every water-soluble essential vitamin except for vitamin C is required for the variety of enzymes to drive the chemical reactions needed to produce ATP.

The goals of dietary intervention for the athlete are to fill carbohydrate (glycogen) stores in the muscles and liver and to make both carbohydrate and fat readily available in the muscle, and glucose in the blood stream for use by the central nervous system. Carbohydrate as a fuel can support higher-intensity exercise than

glycolysis The energy pathway where glucose is either broken down anaerobically (without oxygen) to pyruvate to release ATP for the body, or aerobically (with oxygen) to acetyl CoA and to enter the citric acid cycle.

citric acid cycle A series of oxygen-dependent enzymatic reactions taking place inside mitochondria, in which covalent bonds are broken and the energy is captured to produce ATP; also known as the Krebs cycle or the tricarboxylic acid cycle.

electron transport system A membrane-bound system of proteins in the mitochondria of cells that couples an electron donor and an electron acceptor to create a membrane potential that generates ATP energy.

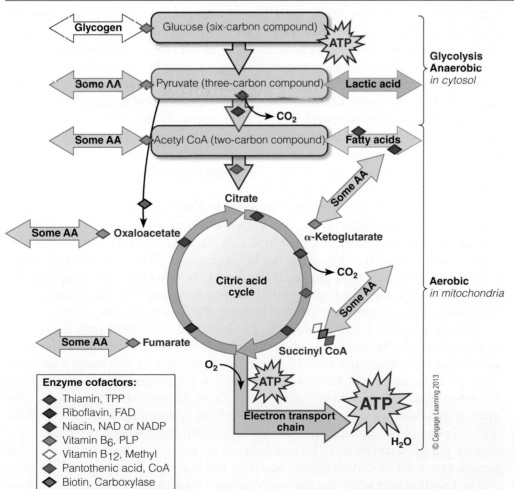

FIGURE 4.12 ATP energy production through glycolysis, the citric acid cycle, and the electron transport chain from glucose, fatty acids, and amino acids (AA). The role of B vitamin cofactors are illustrated for thiamin pyrophosphate (TPP), flavin adenine dinucleotide (FAD), nicotinamide adenine dinucleotide (NAD), nicotinamide adenine dinucleotide phosphate (NADP), pyridoxal phosphate (PLP), Coenzyme A (CoA), and methylation (methyl) and carboxylation reactions.

fat, but far less carbohydrate than fat is stored in the body. The metabolic challenge is to ensure the carbohydrate supply in the muscles and slow down its depletion by relying more on fat as a fuel and supplying carbohydrate during the exercise. Dietary carbohydrate plays a key role in fuel partitioning because glucose availability from absorbed carbohydrates tends to preserve glycogen levels and improve fatty acid availability. Supplementing small amounts of carbohydrates during the exercise improves the ability to exercise.[17]

Performance during exercise depends, in part, on the provision of adequate fuel to working muscles. Therefore, athletes often ingest carbohydrate during intense endurance exercise to support plasma glucose concentrations and spare muscle glycogen oxidation. Many studies have investigated the **ergogenic** value of consuming carbohydrate before, during, or after an exercise bout. There is overwhelming evidence that carbohydrate consumption before and during prolonged exercise can enhance endurance performance. Additionally, there is a 2-hour window after exercise where enzymes in the muscles are up-regulated to promote glycogen repletion. To take advantage of this chemistry, all one needs to do is eat a recovery meal. See the section "Recovery Nutrition after Exercise" later in this module.[15,17]

Carbohydrates

There is a limited supply of carbohydrates available in the body and a need for carbohydrates to support ATP production in the central nervous system, red blood cells, and high-intensity exercise. The enhancement of performance resulting from carbohydrate supplementation is especially noted in strenuous-to-exhaustive periods of exercise. Exhaustive exercise is high-intensity exercise (above 75 percent VO_2 max) that lasts more than 90 minutes.

Carbohydrates in the Body and Diet

Only about 0.5 percent of a person's total body weight is carbohydrate. This amounts to about three-quarters of a pound. Carbohydrate is available in the body from the blood stream as blood glucose and from the liver and muscles as glycogen. A normal fasting blood glucose level ranges from 70 milligrams of glucose per deciliter of blood to 99 milligrams of glucose per deciliter of blood (see page 127). There are about 5 liters of blood in an adult circulatory system, which amounts to roughly 4.5 grams of glucose (almost 1 teaspoon of sugar) or about 18 Calories in the blood stream in normal, fasting conditions. If blood sugar becomes low, there are serious detriments to thought processes, perception, central functioning, and exercise performance. The upper brain and red blood cells depend solely on blood glucose for ATP production.

In a healthy adult liver, there are roughly 75 to 100 grams of liver glycogen stored in a maximally replenished organ. This amount could provide 300 to 400 Calories to the central nervous system and red blood cells. Liver glycogen is used to maintain a steady concentration of blood glucose between meals. Most liver glycogen stores can maintain normal blood glucose levels for 4 to 6 hours between meals, and for about 10 hours during the overnight fast. However, during exercise, blood glucose requirements are dramatically elevated because of the increased energy demands.

Muscle glycogen levels are variable in adults. The amount of muscle glycogen is dependent on the amount of dietary carbohydrate consumed and the amount of exercise the muscle has had. Muscle glycogen is used by the muscle for high-intensity muscular work. An untrained individual consuming a moderate amount of carbohydrate (50 percent of Calories) has about 12 grams of glycogen per kilogram of muscle, and a well-trained individual athlete consuming a high-carbohydrate diet (65 percent of Calories) will have 20 to 30 grams of glycogen per kilogram of muscle. If a person exercises a lot and does not eat enough carbohydrates, or if that person eats a lot of carbohydrates but does not exercise, muscle glycogen levels will be low. It is only when the person exercises a lot and eats a high-carbohydrate

ergogenic Any agent that, when applied, improves the capacity to exercise when tested against a placebo.

diet (toward the upper realm of the AMDR for carbohydrate) that the muscle glycogen levels are high.

When evaluating dietary carbohydrate, sedentary humans need at least 130 grams per day of carbohydrate (the DRI) in order to support brain, central nervous system, and red blood cell function (about 2 grams per kilogram of body weight for a 150-pound person). An active person needs at least 6 grams of dietary carbohydrate per kilogram of body weight for optimal muscle glycogen levels. For endurance athletes, 7 to 10 grams of dietary carbohydrate per kilogram of body weight are needed to replenish muscle glycogen stores.

Carbohydrate Management for Enhanced Exercise Performance The ergogenic effects of carbohydrate management have been best displayed during prolonged exercise, because work volume can be consistently controlled and measured. Dietary carbohydrate and carbohydrate supplementation become a significant endurance factor in events that last more than 90 minutes. Optimal management of carbohydrate includes regulating the intake of the total amount of carbohydrate during the last large meal before the exercise or competition, just before the exercise or competition, during the exercise or competition, and after the exercise or competition (see Table 4.8).

If the last large meal is consumed 4 hours before the exercise or event, it should contain 4 grams of carbohydrate per kilogram of body weight. This amounts to a large meal. A modified plan (instead of the 4-hour prior large meal) that has shown ergogenicity is to provide 2 grams of dietary carbohydrate per kilogram of body weight 2 hours before the exercise or competition—a moderate-size meal (see Table 4.8). Keep in mind that these meals should be low in fat (less than 25 percent of Calories from fat). Just before the exercise or competition (5 to 10 minutes), 0.5 to 1 gram of carbohydrate per kilogram of body weight can be comfortably consumed. Again, the food choices should be low in fat. Consuming this small amount of carbohydrate can delay the onset of fatigue and increase the potential for the total amount of work to be done in the exercise bout.

During exercise, the body can absorb about 60 grams of carbohydrate per hour (1 gram per minute). It is easy to provide the carbohydrate in a fluid replacement beverage. Adjustment of the carbohydrate concentration can be made based on how much water the athlete is willing to drink. Runners more commonly use glucose tabs and drink water or a sports drink, or a glucose electrolyte replacement solution (GES). GESs are made or defined in terms of the percentage of glucose in the solution. A well-tolerated concentration is a 6 percent solution, which means that there are 60 grams of sugars (sucrose, glucose, dextrose, etc.) in 1 liter of fluid. Six-percent solutions provide about 15 grams of carbohydrate per cup of fluid, which can be consumed about every 15 minutes.[15,17,20]

Recovery Nutrition after Exercise After the exercise, there is a 6-hour time period when the metabolism of muscle cell recovery is elevated. From the time the exercise ceases, the level of muscle metabolism begins to decrease and continues to lower during the post-exercise 6-hour period. If nutrients are made available to the muscle by consuming proper nutrition, the rate of muscle glycogen replenishment can be maximized. The fastest rate of glycogen replenishment occurs during the 2-hour window following exercise. Carbohydrate should be consumed along with protein in a 3:1 ratio immediately after the exercise and again 2 hours later. This exercise nutrition strategy can maximize muscle glycogen repletion after exhaustive exercise. The individualized prescription would be 1.65 grams of carbohydrate per kilogram of body weight and 0.55 grams of protein per kilogram of body weight. For proper nutrition, about 600 Calories per recovery meal for a 150-pound person would be needed. The two recovery meals roughly amount to 1,200 Calories in the

TABLE 4.8

The Timing of Carbohydrate Intake for Intense Athletic Exercise for a 150-Pound Person

Timing and Amount of Intake	Exchange List Pattern	Food and Beverage Options for Exchange List Patterns
4 hours prior *Carbs: 4 g/kg or 1.82 g/lb* OR	10 starches 5 fruits 3 other carbohydrate exchanges	2 cups pasta (90 g) 1 bagel (60 g) 1 cup juice (45 g) 1 cup fat-free frozen yogurt (45 g) 1 large banana (30 g) *(270 g total)*
2 hours prior *Carbs: 2 g/kg or 0.91 g/lb*	6 starches 3 fruit exchanges	1.33 cups pasta (60 g) 2 slices of bread (30 g) 1 cup juice (45 g) *(135 g total)*
5–10 minutes prior *Carbs: 0.5–1 g/kg or 0.23–0.45 g/lb*	2–4 other carbohydrate exchanges	1–2 low-fat sport bars (2-ounce size) or the equivalent of 2–4 tablespoons of sugar *(34–68 g total)*
During *Carbs: 60 g/hr*	2–4 other carbohydrate exchanges/hour	A 6% glucose electrolyte replacement solution provides about 15 g of carbohydrate per cup of fluid *(1 tablespoon of sugar and ¼ teaspoon of salt per 1 cup water; drink 2–4 cups per hour)*
Immediately after *Carbs: 1.65 g/kg or 0.75 g/lb* *Protein: 0.55 g/kg or 0.25 g/lb*	7 starches, 3 meats or 2 milks, 2 fruits, 4 starches, 2 meats or 1 milk, 2 fruits, 4 starches, 3 meats exchanges	2 cups low-fat milk (24 g carbs, 16 g protein) 2 cups ready-to-eat cereal (60 g carbs, 12 g protein) 4 tablespoons raisins (30 g carbs) 2 egg whites (14 g protein) *(114 g of carbohydrate and 42 g of protein total)*
2–6 hours after *Carbs: 1.65 g/kg or 0.75 g/lb* *Protein: 0.55 g/kg or 0.25 g/lb*	5 starches, 2.5 vegetables, 3 meats, 2 fruits exchanges (similar to immediately after meal)	3 ounces chicken (21 g protein) 1 large baked potato (60 g carbs, 12 g protein) 1 ¼ cups cooked broccoli (12 g carbs, 5 g protein) 1 small roll (15 g carbs) 1 large pear (30 g carbs) *(117 g of carbohydrate and 38 g of protein total)*

© Cengage Learning 2013

post-exercise time to achieve optimal replenishment. This is the amount of energy expended to run 12 miles.

If exhaustive exercise was not achieved, then estimate the Calorie expenditure and create a recovery meal that would supply a 3:1 ratio of carbohydrate to protein for the Calories used. It is important to gain an understanding of the number of Calories you use in exercise so that you do not eat too many Calories and gain fat weight. Remember, the goal of recovery nutrition is to replenish glycogen stores and provide the building blocks for lean body mass repair. You can estimate your Calorie expenditure using the information learned on page 179 or by using the "track activity" option in the Diet Analysis Plus software. If estimating your Calorie expenditure is not always possible, then you should consume the first recovery meal as prescribed, and for the second recovery meal, you should eat to satiety (until you are full) with a carbohydrate-to-protein ratio of 3:1.[17,71]

Data GENEie Solah weighs 138 pounds. How many grams of carbohydrate does she need to eat 2 hours after exercise if she does not eat 4 hours before exercise?

Proteins

Protein utilization is influenced by dietary and physiological carbohydrate conditions. Protein needs are typically met by consuming 100 to 200 percent of the DRI through proper nutrition and dietary planning. Protein management for enhanced exercise performance is tied to maintaining nitrogen balance, supporting lean body mass gains, and consuming a post-exercise meal.

Proteins in the Body and Diet The amino acid pool in the body comprises about 300 grams of protein. The amino acids that make up this pool are found throughout the **intracellular** and **extracellular** fluids in the body. To stay in protein balance, the typical person loses about 0.6 grams per kilogram of body weight protein per day. This is measured as nitrogen loss. Protein is lost in sweat (1 gram of protein per liter of sweat) and urine (about 3 grams per day is not reabsorbed by the kidneys), and the greatest amount is lost in a combination of shedded skin, finger and toe nails, hair, gastrointestinal tract fluids, sloughed-off intestinal cells, and degraded hormones, enzymes, and tissues. Consuming adequate protein to replace the loss and provide for tissue repair and maintenance is referred to as **nitrogen balance**.

If carbohydrate is adequate, then protein metabolism is very efficient. If carbohydrate is limited, then protein is used inefficiently by body cells. However, if the person is in negative energy balance or muscle glycogen stores are depleted, significantly more protein is wasted because it is used to produce energy. Normally, protein may contribute 5 percent of total fuel during exercise. In the later stages of endurance exercise, or under stress, protein catabolism contributes up to 15 percent of the fuel. When protein is used for energy, it comes predominantly from using the branched-chained amino acids isoleucine, leucine, and valine.

Protein Management for Enhanced Exercise Performance Dietary protein needs depend on the sport or activity being performed. Most adult athletes will stay in nitrogen balance at the DRI for protein of 0.8 grams of protein per kilogram of body weight, or by consuming the lower range of the AMDR for protein (approximately 12 to 17 percent of Calories). Competitive, strenuously-training athletes require 1.2 to 1.6 grams of protein per kilogram of body weight per day. For heavily-training athletes, whether in endurance or resistive activities, 1.2 to 1.8 grams of protein per kilogram of body weight per day may be required to maintain nitrogen balance. For strength training for lean body mass gain, 1.6 grams of protein per kilogram of body weight per day, plus an additional 200 Calories per day to promote positive energy balance, is a sound dietary approach. Endurance and ultra-endurance athletes may have a protein need of 1.8 to 2.0 grams of protein per kilogram of body weight per day. In general, approximately two-thirds of the protein consumed should be either high-quality protein or the protein complement equivalent of all the athletic protein requirements mentioned.

More attention needs to be paid to protein in the recovery from exercise phase. The combination of carbohydrate and protein is needed to stimulate the insulin response and accomplish the influx of amino acids into the muscle cells. There is no question

intracellular Inside cells.

extracellular Outside cells.

nitrogen balance Nitrogen loss, compared to nitrogen intake from protein sources.

that there is a significant improvement in achieving muscle glycogen recovery with the addition of protein to the carbohydrate post-exercise meal (see Table 4.8).[71]

Fats

Dietary fat can impact the health status of any person, even when the person is a highly trained athlete. The type of fat consumed is of great importance for health promotion purposes. Planning the amount and type of fat to consume for exercise nutrition usually comes after planning for carbohydrate and protein.

Fats in the Body and Diet Once carbohydrate and protein needs are met, dietary fat should comprise the remaining energy needed by the athlete, usually the lower end of the AMDR for fat (20 to 25 percent of Calories). The type of fat that is consumed makes a significant difference in health and performance. Remember all the health effects of the various types of dietary fats from Modules 1, 2, and 3 (see pages 23, 60, and 147). The serious athlete should focus on consuming adequate essential fatty acids (linoleic and alpha-linolenic acid), omega-3 fatty acids (up to 2 percent of total Calories), and monounsaturated fatty acids. It is also wise to avoid an excess of unhealthy saturated fatty acids (less than 7 percent of Calories) and *trans* fatty acids (less than 1 percent of Calories).

Fat Management for Enhanced Exercise Performance Ingestion of dietary fat is not a useful approach for providing fuel during exercise, because it may take several hours for the long-chain fatty acids to actually reach the muscle cells and thus become an available fuel source. Dietary triglycerides are emptied slowly from the stomach, packaged into chylomicrons in the small intestine, and secreted into the lymphatic system before entering the blood stream. Only a portion of the triglycerides present in circulating chylomicrons ultimately provide fatty acids to muscle.

In contrast, **medium-chain triglycerides** have been touted as a potential ergogenic fuel during exercise and are currently present in some commercially prepared sport bars. Medium-chain triglycerides are emptied rapidly from the stomach, rapidly absorbed and hydrolyzed by the small intestine cells, and secreted directly into the blood stream. Furthermore, medium-chain triglycerides are easily taken up by cells to be used for energy production. Thus, medium-chain triglycerides are readily available for oxidation. However, several studies have demonstrated that oral supplementation with medium-chain triglycerides is unlikely to improve performance during endurance exercise. The amount of medium-chain triglycerides that can be given orally is limited to approximately 25 to 30 grams, because diarrhea and other gastrointestinal side effects are common with higher doses. Furthermore, although orally administered medium-chain triglycerides are readily oxidized, they do not spare muscle glycogen during either moderate or high-intensity exercise.

Thermal Regulation and Hydration

Humans are warm-blooded. Heat comes from the incomplete capture of energy in the formation of ATP from carbohydrates, proteins, and fats. About 20 percent of the energy that exists within the macromolecules of carbohydrate, fat, and protein is actually captured as ATP. The rest of the energy is lost or released as heat. As the rate of metabolism increases, more and more heat is generated inside the body. This heat must be lost from the body, or the **core body temperature** will increase. When the core body temperature rises, the body's mechanisms, such as sweating, are employed to reduce the core body temperature. Core body temperature is measured orally or rectally. A normal core body temperature is 98.6 degrees Fahrenheit. The functional body temperature range is 97 to 104 degrees Fahrenheit. Once core body temperature rises above 104 degrees Fahrenheit, the function of enzymes is impaired, because proteins,

medium-chain triglycerides Fatty acids that have 6 to 12 carbons in the chain and are esterified to glycerol.

core body temperature The temperature of the body at the deep central part.

from which enzymes are made, begin to denature (change their conformation) at high temperatures. When the enzymes do not work, the normal biochemical reactions cannot occur, so the physiology fails and the diagnostic signs and symptoms of **thermal injury** exist.

Understanding and Controlling Body Temperature The signs and symptoms of elevated core body temperature, overheating, or getting too hot (thermal injury) include weakness, chills, goose flesh appearing on the upper arms and chest, headache, nausea, faintness, muscle cramps, hypotension (low blood pressure), increased heart rate, lack of sweating, shock, and death. Sweating helps cool the body and prevent thermal injury. Sweat is 99 percent water, with **salt** (sodium chloride) and other minerals composing the remaining 1 percent. On average, there are 2.6 grams of sodium chloride per liter of sweat, 1 gram of potassium per liter, and trace amounts of iron, magnesium, copper, zinc, and amino acids. A high sweat rate is 2 to 3 liters per hour, while an average sweat rate range when noticeably sweating is 1 to 1.5 liters per hour.

On average, there is a 2.6-gram (2,600-milligram) loss of salt per liter of sweat. Because salt is 40 percent sodium, there are 1,040 milligrams of sodium lost per liter of sweat. There are about 5 grams of salt per teaspoon. So it takes about one-half a teaspoon of salt to replace the amount of salt lost during an hour of hard exercise. The amount of sodium contained in 6 grams of salt (the Daily Reference Value) is 2.4 grams or 2,400 milligrams. This amount of sodium per day will cover the sodium needs of the body with 1 hour of moderate exercise. If you are looking for sodium content in packaged foods and beverages, it is found on the Nutrition Facts panel of the food package label and is listed in milligrams of sodium.

Dehydration to Hydration **Thirst** mechanisms are not stimulated in the body until the body is 2 percent dehydrated. Cardiovascular and aerobic performance is significantly reduced when 2 percent of body water has been lost. This amount of water loss means a person weighing 200 pounds would lose 4 pounds of body weight and now weigh 196 pounds. With water losses of 5 percent—the equivalent loss of 10 pounds for a 200-pound person—there is risk for serious thermal injury (see Figure 4.13).

Once you feel thirsty, your body's ability to perform exercise at a high level has already declined (see Table 4.9). It is helpful to know your hydrated (juiced-up) body weight. If you are sweating hard during exercise, you can expect to lose 1 liter of fluid per hour. Two cups of water weigh 1 pound, and 1 liter weighs about 4 pounds. It is

thermal injury Damage to the body as a result of elevated temperature above 104 degrees Fahrenheit, causing reduced functional capacity and performance.

salt A crystalline mineral compound of sodium and chloride.

thirst A craving for water when there is a need to rehydrate tissues.

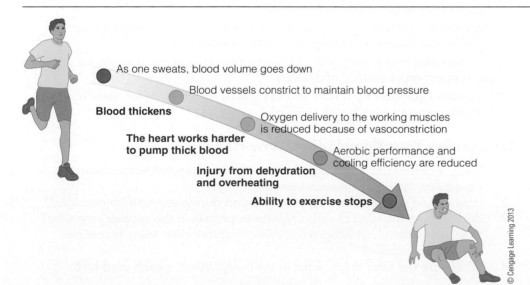

As one sweats, blood volume goes down

Blood vessels constrict to maintain blood pressure

Blood thickens

Oxygen delivery to the working muscles is reduced because of vasoconstriction

The heart works harder to pump thick blood

Aerobic performance and cooling efficiency are reduced

Injury from dehydration and overheating

Ability to exercise stops

© Cengage Learning 2013

FIGURE 4.13 Dehydration causes aerobic performance to decline and may cause injury.

a good practice to weigh in before strenuous exercise and weigh out after the bout of exercise. For every pound lost, you should drink 2 cups of water.

The human body can absorb about 1 liter of fluid per hour, maximally. Cold fluids are absorbed faster than room-temperature or warm ones. Dilute fluids are absorbed faster than concentrated ones. Small volumes of fluid are absorbed faster than large ones. Thus, small, frequent amounts of cold fluids (or diluted GES drinks) should be consumed during exercise. A good principle to follow is to consume 1 cup of fluid, at a temperature of 41 degrees Fahrenheit, every 10 to 15 minutes during exercise. Exercisers must learn to drink before they feel thirsty and be conscientious about the temperature, volume, and **osmolarity** of the fluids consumed.[15,20,68]

TABLE 4.9

Hydration Happy and Dehydration Demise

Importance of Hydration During Exercise	Cardiovascular Demise During Dehydration
• Optimize muscle strength	• Plasma volume decreases
• Optimize aerobic capacity	• Osmolarity of the blood increases
• Prevent thermal injury, which includes dizziness, cramping, fainting, heat exhaustion, and heat stroke	• Blood pressure goes down
	• Blood vessels are constricted
• Maintain a pool of fluid to draw upon for sweat loss	• Heart rate increases
	• Heart filling decreases
• Avoid stress to the cardiovascular system, which is affected dramatically by fluid losses	• Cardiac output decreases
	• Sweating decreases
	• Core body temperature increases

© Cengage Learning 2013

Summary Points

- As the intensity of the exercise increases, so does the percentage of carbohydrate used to produce ATP.
- Carbohydrate stores in the body are limited.
- Dietary carbohydrate intake can increase muscle glycogen levels and can enhance exercise performance.
- A combination of carbohydrate and protein consumed after the exercise can support muscle glycogen replenishment.
- The protein needs of most competitive athletes range between 1.2 to 1.6 grams of protein per kilogram of body weight per day.
- A low-fat diet that provides all of the essential fatty acids in adequate amounts can support most athletic endeavors.
- Adequate hydration from fluid and salt intake is important for aerobic performance and avoiding thermal injury.

Take Ten on Your Knowledge Know-How

1. What recommendations for carbohydrate management before exercise have been shown to enhance exercise ability?
2. How should carbohydrate supplementation during exercise be managed?
3. What should the ratio of carbohydrate to protein in the recovery meal be? How many grams per kilogram of carbohydrate? How many grams per kilogram of protein?
4. Where are the sites of glycogen in the body? What is each used for?

osmolarity A measure of the concentration of particles in a solution.

5. What is the range of protein intake in terms of grams of protein per day for a 150-pound competitive athlete?
6. What nutrient spares protein? When are amino acids used to produce ATP?
7. What do medium-chain triglycerides provide? What may be the usefulness of medium-chain triglycerides in sport performance?
8. What types of fatty acids are beneficial for health and performance?
9. What percentage of sweat is water? How much salt is lost in 1 liter of sweat?
10. What are some examples of thermal injury? At what percentage of total body weight loss from fluids do the effects of thermal injury become more likely to occur?

 **SUMMARY**

CONTENT KNOWLEDGE

AFTER READING THIS MODULE, YOU SHOULD BE ABLE TO:

- Explain how the field of nutrition has soundly and scientifically demonstrated the role of diet and lifestyle in health and disease.
- Describe the importance of identifying unhealthy nutritional and lifestyle practices, setting reachable goals, and establishing a plan to modify your behavior for each of the changes that you want to make, so that you can live healthier each day of your life.
- Demonstrate the knowledge, skills, and competencies needed to perform a diet and energy expenditure analysis.
- Identify some areas of your diet and lifestyle that may need to change in order to achieve optimal health.
- Describe basic principles of exercise and design a fitness program that promotes health.
- Apply your understanding of sport nutrition basics before, during, and after exercise, and how to enhance the ability to exercise.

PERSONAL IMPROVEMENT GOALS

NOW THAT YOU HAVE DECIDED THAT YOU WILL CHANGE, IT IS TIME TO SET YOUR GOALS AND TAKE ACTION. YOU CAN ENHANCE YOUR QUALITY OF LIFE BY:

- Making a list of lifestyle changes you have become aware of that you need to make and addressing them one at a time.
- Realizing that by the time you state that you will change, you have progressed through a few stages of thought: "I can choose to change," "I know how to change," "I want to change," "I have decided to change," "I will change," and "I am changing."
- Implementing a behavior modification program and rewarding yourself for reaching the action stage of change ("I am changing") and meeting your goals.

Here is a tip for you: Realize that today is the first day of the rest of your life. Find the time and energy required to live a healthy lifestyle every day. Making small changes is an easy thing to do and very satisfying.

You can assess if you met the learning objectives for this module by successfully completing the Homework Assessment and the Total Recall activities (sample questions, case study with questions, and crossword puzzle).

Homework Assessment

50 questions

Project Overview

1. Use the information from Module 4 to answer questions 1–20, beginning on page 211, related to scientific inquiry, energy balance, body composition, and weight control.

2. Use the case study below to prepare to answer questions 21–50, beginning on page 212. Analyze the two days of eating and activity in the case study using the Diet Analysis Plus (DA+) online software. Track diet and activities according to the exact descriptions provided in this assessment.

3. Evaluate the case study diet record patterning according to the MyPlate model. Use the MyPlate report from DA+ to evaluate goals versus intakes for all food groups except oils. For oils, you will have to use Appendix C to pattern the teaspoons of oil for each day, and then calculate the average over the two days.

4. Follow the instructions provided in this assessment for creating DA+ reports. Create the two-day average reports for Macronutrient Ranges, Fat Breakdown, Intake vs. Goals, MyPlate Analysis, and Energy Balance. Use these reports to answer the questions related to the case study.

5. Submit your answers to the 50 questions that follow the case study.

Use Diet Analysis Plus: Analyze the diet and activity given in the case study below using the DA+ software.

Build the Profile: Enter the following information in DA+. Elliot is a 23-year-old male. He is 6 feet 2 inches tall and weighs 215 pounds. Elliot is physically "active" (exercises 60 minutes or more each day).

Track Diet for Day 1: To enter Elliot's diet for day 1 (see Table 4.10), click on Track Diet.

1. Select the day of the week you want for day 1.
2. To enter the foods and select the food items exactly, do the following.
3. Type "English muffin" in the search box.
4. Now, click to select "English Muffin, Plain, Enriched, Toasted."
5. The serving size box for the food entry will appear.
6. Select the unit (item, tablespoon, ounces, etc.) using the dropdown menu.
7. Enter the amount of the food listed in the diet record.
8. Assign the food to Breakfast, Lunch, Dinner, or Snack using the dropdown menu.
9. Click Save to enter the food.
10. Continue to enter all of the items Elliot ate on day 1 using the following table of Elliot's day 1 food intake.

Elliot's Day 1 Food Intake

Meal	Type In:	Select	Amount
Breakfast	English muffin	English Muffin, Plain, Enriched, Toasted	2 items
	Butter	Butter	4 tablespoons
	Jelly	Jelly (Includes Grape)	4 tablespoons
	Coffee	Coffee, Brewed	16 fluid ounces
	Coffee creamer	Coffeemate Non Dairy Creamer, Original, Liquid	4 tablespoons
Lunch	Hot dog	Hot Dog Weiner or Frankfurter, on Bun	3 items
	Chips, salted	Chips, Potato, Salted	2 ounces
	Vanilla Coke	Coca-Cola Vanilla Coke Soda	12 fluid ounces
Dinner	Sausage	Sausage, Pork, Polish Cooked	4 ounces
	Fries, medium	Burger King French Fries, Medium, Salted	1 serving
	Beer	Beer	24 fluid ounces
	Ice cream	Ice Cream, Chocolate	½ cup
Snack	Doughnut	Doughnut, Glazed	2 items
	Tea	Tea, Herbal, Prepared	16 fluid ounces

© Cengage Learning 2013

Track Activity for Day 1: To enter Elliot's activity for day 1 (see Table 4.11), click on Track Activity.

1. Select the day of the week you want for day 1 and type "sleeping" in the search box.
2. Select "Sleeping." In the activity duration box, select 6 hours and click Save.
3. Repeat the process for each of Elliot's activities on day 1 using the following table of Elliot's day 1 activity record.

Elliot's Day 1 Activity Record

Type In:	Select	Hours
Sleeping	Sleeping	6
Showering	Showering, toweling off, standing	1
Dressing	Dressing, undressing	1
Eating	Eating, sitting	1
Driving	Driving, automobile or light truck	2
Sitting	Sitting quietly	6
Cleaning	Cleaning, light (dusting, vacuuming, emptying trash)	1
Standing	Playing with children, light (standing)	3
Walking mph	Walking, slow, 2.0 mph, level	2
Walking mph	Walking, 3.0 mph, moderate, level	1

© Cengage Learning 2013

Track Diet for Day 2: To enter Elliot's diet for day 2 (see Table 4.12), click on Track Diet.

1. Select the day of the week you want for day 2.
2. Follow the step-by-step instructions to enter foods given for day 1.
3. Enter all of the items for day 2 using the following table of Elliot's day 2 food intake.

TABLE 4.12

Elliot's Day 2 Food Intake

Meal	Type In:	Select	Amount
Breakfast	Milk	Milk, Non fat, Skim or Fat Free	8 fluid ounces
	Total	General Mills Total Raisin Bran Cereal	2 cups
	Strawberries	Strawberries	1 cup
	Orange juice	Juice, Orange	8 fluid ounces
	Flax seed	Seeds, Flax or Linseed, Ground	2 tablespoons
Lunch	Turkey	Turkey, Light Meat, Meat Only, Roasted	4 ounces
	Broccoli	Broccoli	1 cup
	Carrots	Carrots, Sliced, Boiled, Drained	1 cup
Dinner	Brown long rice	Rice, Brown, Long Grain, Cooked	1 cup
	Nonfat milk	Milk, Non Fat Skim or Fat Free	8 fluid ounces
	Broiled salmon	Salmon, with Butter, Broiled or Baked	6 ounces
	Mashed potatoes	Potatoes, Mashed	1 cup
	Spinach	Spinach, Chopped, Boiled, Drained	1 cup
	Tap water	Water, Tap	16 fluid ounces
Snack	Nonfat fruit yogurt	Yogurt, Fruit, Non Fat, Sweetened with Low Calorie Sweetener	1 cup
	Almonds, dry	Almonds, Dry Roasted, Salted	¼ cup
	Mixed bread	Bread, Mixed Grain	2 slices
	Peanut butter	Peanut Butter, Smooth	2 tablespoons

© Cengage Learning 2013

Track Activity for Day 2: To enter Elliot's activity for day 2 (see Table 4.13), click on Track Activity.

1. Select the day of the week you want for day 2.
2. Follow the step-by-step instructions to enter activities given for day 1.
3. Enter all of the day 2 activities using the following table of Elliot's day 2 activity.

TABLE 4.13

Elliot's Day 2 Activity Record

Type In:	Select	Hours
Sleeping	Sleeping	8
Showering	Showering, toweling off, standing	1
Dressing	Dressing, undressing	1
Driving	Driving, automobile or light truck	2
Sitting	Sitting quietly	7
Eating	Eating, sitting	1
Standing	Playing with children, light (standing)	2
Walking mph	Walking, slow, 2.0 mph, level	1
Walking mph	Walking, 3.0 mph, moderate, level	1

© Cengage Learning 2013

Build Reports: After all your day 1 and day 2 diet and activity data is entered and tracked in DA+, it is time to build reports for the average of the two days. To view reports, select the Reports tab and then the report type (Macronutrient Ranges, Fat Breakdown, Intake vs. Goals, MyPlate Analysis, or Energy Balance) under Nutrients. Use the following instructions to print reports.

1. Select the Reports tab.
2. Select the Advanced Report Option and then Custom Averages.
3. Choose the correct start and end dates for a two-day average report.
4. Uncheck the box next to these report types: Profile DRI Goals, Source Analysis, Intake Spreadsheet, Exchanges Spreadsheet, and Activities Spreadsheet.
5. Leave the box checked next to these report types to generate the following:
   a. Macronutrient Ranges
   b. Fat Breakdown
   c. Intake vs. Goals
   d. MyPlate Analysis
   e. Energy Balance
6. Select Print Custom Report for a PDF file to be built.

Save Your Work: Your work is automatically saved on the date you entered the diet and activity information. You will NOT turn in your DA+ data entry results or printouts. You will use your data, whether onscreen or as a printout from your computer, to answer questions 21–50 below.

Answer the following 20 questions using your nutrition knowledge from Module 4.

1. The scientific method is driven by proven facts.
   A. True   B. False

2. A retrospective epidemiological study is used to track cohorts into the future.
   A. True   B. False

3. The duration of a study should be considered when evaluating the quality of experimental design.
   A. True   B. False

4. A study utilizing a double-blind, cross-over, placebo-controlled design has a better scientific design than a single-blind clinical trial.
   A. True   B. False

5. Which of the following best describes an intervention trial?
   A. A description of a population status associated with what has happened in the past.
   B. The description of the outcomes an individual displays as a result of a treatment protocol.
   C. A basic experimental design with an experimental group.
   D. A study of the outcomes of a prospective protocol applied to one population compared to another population that remained untreated.
   E. A treatment of a small group where all of the participants are treated the same way.

6. The body mass index (BMI) can be calculated to determine if a person has a normal body weight for their height.
   A. True   B. False

7. Which of the following is a health risk associated with obesity?
   A. Arthritis
   B. Gallbladder disease
   C. Hernias
   D. Varicose veins
   E. All of the above

8. Which of the following states of positive energy balance would promote a fat weight gain of approximately 10 pounds in one year?
   A. The overconsumption of 35 Calories per day
   B. The overconsumption of 59 Calories per day
   C. The overconsumption of 96 Calories per day
   D. The overconsumption of 121 Calories per day
   E. The overconsumption of 211 Calories per day

9. Which of the following states of negative energy balance would promote a fat weight loss of approximately 3 pounds in 30 days?
   A. The underconsumption of 35 Calories per day
   B. The underconsumption of 70 Calories per day
   C. The underconsumption of 140 Calories per day
   D. The underconsumption of 350 Calories per day
   E. The underconsumption of 500 Calories per day

10. If a female has 23 percent body fat, what percentage of stored body fat would she have?
    A. 3
    B. 7
    C. 11
    D. 17
    E. 20

11. A percent body fat value of 19 percent for an adult female would be considered:
    A. Very lean
    B. Average
    C. Fat
    D. Obese
    E. Physically fit

12. A percent body fat value of 12 percent for an adult male would be considered:
    A. Very lean
    B. Average
    C. Fat
    D. Obese
    E. Physically fit

13. An adult body mass index (BMI) value of 31 would be considered:
    A. Underweight
    B. Normal
    C. Overweight
    D. Obese (class I)
    E. Obese (class II)

14. Running a 200-meter race would fall into which activity level? (Use Appendix B for help.)
    A. Resting (*Reclined activity*)
    B. Sedentary (*Very light activity*)
    C. Easy (*Light activity*)
    D. Rigorous (*Moderate activity*)
    E. Strenuous (*Extreme activity*)

15. Which of the following is an evaluation tool of aerobic exercise?
    A. Overload principle
    B. Perceived exertion
    C. Citric acid cycle
    D. One repetition maximum
    E. Electrolyte balance

16. The intake of which nutrient does not need to be managed during exercise?
    A. Carbohydrate
    B. Electrolytes
    C. Protein
    D. Water
    E. None of the above

17. Flexibility is one of the four aspects of functional capacity evaluated in a comprehensive physical fitness test.
    A. True   B. False

18. During exercise, ATP production is mainly from carbohydrates and fats.
    A. True   B. False

19. Elliot is a male, age 23, who weighs 215 pounds and is 6 feet 2 inches tall. What is his approximate Resting Energy Expenditure (REE), using the Mifflin-St. Jeor equation? (Use Appendix B for help.)
    A. 2,490
    B. 1,876
    C. 2,042
    D. 2,324
    E. 2,510

20. If Elliot washed windows at home for 2 hours without resting, approximately how many Calories would he expend in this time? (Use Appendix B for help.)
    A. 519
    B. 425
    C. 391
    D. 255
    E. 235

Answer the following 30 questions using the results of your analysis of Elliot's diet and activity for the average of the two days. Also use your nutrition knowledge from Modules 1, 2, 3, and 4, as well as Appendix A for DRI tables, Appendix B for facts and formulas, and Appendix C for MyPlate information. Choose the best answer.

21. On average over the two days, what state of energy balance was Elliot in? (Use the energy balance report and compare Calories consumed to Calories expended.)
    A. Positive
    B. Negative
    C. Balanced

22. On average over the two days, Elliot should have lost approximately 0.05 pounds per day. (Use the energy balance report and compare Calories consumed to Calories expended.)
    A. True   B. False

23. On average over the two days, Elliot is losing weight. (Use the energy balance report and compare Calories consumed to Calories expended.)
    A. True   B. False

24. On average over the two days, Elliot's diet met the Average Macronutrient Distribution Ranges (AMDRs) for carbohydrate.
    A. True   B. False

25. On average over the two days, Elliot's diet met the Average Macronutrient Distribution Ranges (AMDRs) for protein.
    A. True   B. False

26. On average over the two days, Elliot's diet met the Average Macronutrient Distribution Ranges (AMDRs) for fat.

    A. True   B. False

27. On average over the two days, Elliot's diet met the DRI for protein (at least 100 percent of his DRI).

    A. True   B. False

28. On average over the two days, Elliot's diet met or exceeded the 130 gram DRI for carbohydrate.

    A. True   B. False

29. On average over the two days, Elliot's diet met or exceeded the DRI for the omega-6 essential fatty acid linoleic acid.

    A. True   B. False

30. On average over the two days, Elliot's diet met or exceeded the DRI for the omega-3 essential fatty acid alpha-linolenic acid.

    A. True   B. False

31. On average over the two days, the primary type of fatty acid that Elliot consumed was:

    A. Saturated
    B. Monounsaturated
    C. Polyunsaturated

32. On average over the two days, Elliot's diet met the American Heart Association guidelines for percent of Calories from saturated fat.

    A. True   B. False

33. On average over the two days, Elliot's diet met the 2010 Dietary Guidelines for cholesterol.

    A. True   B. False

34. On average over the two days, Elliot's calculated DRI for fiber was approximately 38 grams (plus or minus 1 gram). (Using the Intake vs. Goals report for the two-day average, calculate his DRI for fiber using this formula: Fiber DRI = Calories consumed ÷ 100 × 1.4.)

    A. True   B. False

35. On average over the two days, Elliot's diet was adequate in fiber. (Using the Intake vs. Goals report for the two-day average, calculate his percent DRI using this formula: % DRI = Fiber intake ÷ Calculated fiber DRI need × 100.)

    A. True   B. False

36. On average over the two days, Elliot's diet was deficient in the following vitamins:

    A. Vitamin D
    B. All of the B vitamins
    C. Vitamin C
    D. Vitamin D and vitamin E
    E. None of the above

37. On average over the two days, Elliot's diet was inadequate in the following vitamins:

    A. Vitamin D
    B. All of the B vitamins
    C. Vitamin C
    D. Vitamin D and vitamin E
    E. None of the above

38. On average over the two days, Elliot's diet was excessive or potentially toxic in the following vitamins:

    A. Niacin
    B. All of the vitamins
    C. Folate
    D. Both niacin and folate
    E. None of the above

39. On average over the two days, Elliot's diet was deficient in the following minerals:

    A. Calcium and potassium
    B. Iron and potassium
    C. Potassium
    D. Sodium
    E. None of the above

40. On average over the two days, Elliot's diet was inadequate in the following minerals:

    A. Calcium
    B. Iron
    C. Potassium
    D. Sodium
    E. None of the above

41. On average over the two days, Elliot's diet was excessive or potentially toxic in the following minerals:

    A. Sodium
    B. Zinc
    C. Potassium
    D. Iron
    E. None of the above

42. On average over the two days, Elliot's diet met the MyPlate grains group recommendation. (Refer also to Appendix C.)

    A. True   B. False

43. On average over the two days, Elliot's diet met the MyPlate vegetable group recommendation.

    A. True   B. False

44. On average over the two days, Elliot's diet met the MyPlate fruit group recommendation.

    A. True   B. False

45. On average over the two days, Elliot's diet met the MyPlate dairy group recommendation.

    A. True   B. False

46. On average over the two days, Elliot's diet exceeded the MyPlate protein foods group recommendation.
   A. True   B. False

47. On average over the two days, Elliot's diet exceeded the MyPlate empty Calorie limitation.
   A. True   B. False

48. On average over the two days, Elliot's diet met and exceeded the MyPlate oils recommendation. (Pattern his intake of oils using the food records and Appendix C.)
   A. True   B. False

49. On average over the two days, Elliot's diet met the Average Macronutrient Distribution Range (AMDR) for sugars. (Using the Intake vs. Goals report for the two-day average, calculate his percent of Calories from sugar as follows: Grams of sugar intake × 4 Calories/gram = Calories from sugar. Calories from sugar ÷ Total Calories × 100 = % of Calories from sugar.)
   A. True   B. False

50. On average over the two days, Elliot's diet was adequate in water and fluid intake.
   A. True   B. False

Total Recall

SAMPLE QUESTIONS

True/False Questions

1. Diet and disease relationships are identified through the process of scientific inquiry.

2. Excess body fat is associated with increased risk for cancer.

3. The BMI indicates the healthiness of a person's body weight and composition.

4. An optimal ratio of carbohydrate to protein in a recovery meal is 1:4.

5. Practicing nutritional adequacy is important during weight loss but not weight gain.

Multiple Choice Questions: Choose the best answer.

6. Weight loss occurs when:
   A. Calorie intake exceeds expenditure
   B. Calorie expenditure exceeds intake
   C. Calorie intake matches expenditure
   D. Calorie balance occurs
   E. Calorie monitoring occurs

7. A person expends most of his or her energy each day in:
   A. Physical activity
   B. Basal Metabolic Rate
   C. Hormonal gender controls
   D. Voluntary muscle contractions
   E. Eating

8. Which of the following exercises is considered aerobic?
   A. Basketball
   B. Tennis
   C. Hockey
   D. Jogging
   E. Weight lifting

9. Fat located around the _____ is associated with increased risk for disease.
   A. Thighs
   B. Buttocks
   C. Waist
   D. Shoulders
   E. Extremities

10. Which of the following is an important factor in achieving and maintaining a healthy body weight?
   A. Using drugs for appetite regulation
   B. Incorporating behavior modification
   C. Undergoing cosmetic surgery
   D. Practicing junk food denial
   E. Eliminating dietary fat

CASE STUDY

Andrew is a 22-year-old graduating college senior. During his college years, he gradually and continuously put on extra body weight. He is 5 feet 10 inches tall. When he first started college four years ago, he had a normal and healthy body weight of 165 pounds. Today he weighs 190 pounds and has a waist circumference measurement of 41 inches. Now he wants to adopt a total fitness program that will promote weight loss.

1. What is Andrew's BMI today, and how would you interpret Andrew's BMI?
   A. 19.5, underweight
   B. 23.7, normal weight
   C. 27.3, overweight
   D. 32.1, obese
   E. 48.7, extremely obese

2. Andrew's body weight was normal when he started college four years ago.
   A. True   B. False

3. Which of the following is not a component of a total fitness program?
   A. Meditation
   B. Warm-up and cool-down
   C. Muscular strength and endurance exercises
   D. Aerobic exercise
   E. Flexibility

4. What is the best way for Andrew to determine if he has excess body fat?
   A. Calculating his BMR
   B. Obtaining a Bod Pod measurement
   C. Using a bioelectrical impedance scale
   D. Estimating his body water content
   E. Assessing glycogen storage

5. How much physical activity does Andrew need to do to promote weight loss each week?
   A. 60 minutes of walking once a week
   B. 45 minutes of jogging five times a week
   C. 30 minutes of hiking twice a week
   D. 40 minutes of weight lifting twice a week
   E. 5 to 10 minutes of stretching seven times a week

6. What is the shortest amount of time for Andrew to lose the extra body weight in a healthy way?
   A. 2–3 months
   B. 6–7 months
   C. 9–10 months
   D. 12–13 months
   E. 16–17 months

7. What is Andrew's approximate Resting Energy Expenditure (REE) as calculated by the Mifflin-St. Jeor equation?
   A. 1,441
   B. 1,706
   C. 1,872
   D. 2,985
   E. 3,366

8. Andrew's Estimated Energy Requirement (EER) at the beginning of his weight loss is approximately 3,366 Calories per day. Approximately what caloric intake level would you recommend for him to achieve weight loss of 1 pound per week? (Consider reduced food intake and increased energy expenditure with a moderately physically active lifestyle including 40 minutes of aerobic exercise per day.)
   A. 1,706
   B. 1,872
   C. 3,566
   D. 2,966
   E. 3,166

9. Which of the following answers provides the soundest advice for Andrew concerning weight loss and maintenance of his weight loss afterward?
   A. Use over-the-counter weight loss pills to suppress appetite and increase catabolic hormone secretions.
   B. Fast for five to seven days to cleanse the body, and then adopt a diet of animal products.
   C. Implement behavior modification techniques to develop a well-balanced, nutritionally adequate diet, along with an exercise program.
   D. Focus on spot reduction to reduce waist measurement.
   E. Either eat what you want and exercise, or limit your diet so you don't need to exercise.

10. What are the health consequences of Andrew's body weight and waist circumference today?
    A. Hypertension
    B. Heart disease
    C. Type 2 diabetes
    D. A and B
    E. A, B, and C

FUN-DUH-MENTAL PUZZLE

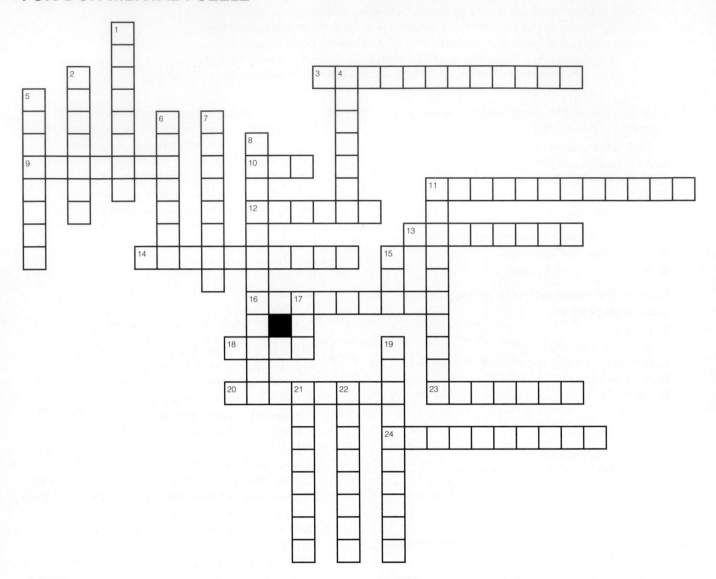

ACROSS
3. Data collection across a population.
9. A female with 21 percent body fat would be in what body fat category?
10. Factor that causes BMR to slow down.
11. A person with a pear shape has excess _____ fat.
12. Body fat assessment method that uses air displacement.
13. Mineral excreted in urine when protein catabolism occurs.
14. Drives a scientific study.
16. A diet is _____ in a nutrient when less than 66 percent of the DRI is provided.
18. Mineral compound lost in sweat.
20. Type of nutrition needed after exercise.
23. Type of adiposity associated with type 2 diabetes.
24. Interpretation of a BMI of 28.

DOWN
1. Type of fat that accumulates inside the abdominal cavity.
2. Can be converted to glucose when needed.
4. Another name for a control group.
5. Type of energy balance resulting in weight loss.
6. Occurring at epidemic proportions.
7. May need modification to achieve a healthy lifestyle.
8. Energy-producing nutrient that is very limited in the body.
11. The field of nutrition is based on what kind of study?
15. Effective rate of weight loss is _____ pounds per week.
17. 3,500 Calories are in 1 pound of ____.
19. Metabolic pathway that produces ATP from glucose anaerobically.
21. Principle that must be applied for the body to increase fitness.
22. Necessary to maintain a healthy body weight.

References

Web Resources

A. Behavioral Risk Factor Surveillance System: www.cdc.gov/brfss/index.htm and apps.nccd.cdc.gov/gisbrfss/default.aspx
B. Centers for Disease Control: www.cdc.gov
C. Centers for Disease Control Obesity Trends: www.cdc.gov/obesity/data/trends.html
D. Centers for Disease Control, National Center for Chronic Disease Prevention and Health Promotion: www.cdc.gov/chronicdisease/index.htm
E. National Heart Lung and Blood Institute: www.nhlbi.nih.gov/guidelines/obesity/bmi_tbl.htm
F. National Institutes of Health: www.nih.gov

Works Cited

1. Abbot, J. M., Thomson, C. A., Ranger-Moore, J., Teixeira, P. J., Lohman, T. G., Taren, D. L., . . . & Houtkooper, L. B. (2008). Psychosocial and behavioral profile and predictors of self-reported energy underreporting in obese middle-aged women. *Journal of the American Dietetic Association, 108*(1), 114–119.
2. American Dietetic Association. (2006). How do you determine accurate resting metabolic rate? *Evidence Analysis Library*. Accessed at: http://adaevidencelibrary.com.
3. Anderson, P., & Milner, J. (2005, July/August). Highlights of the ILSI functional foods meeting: Reports from the special conference on functional foods for health promotion: Implications for reducing obesity, part 2. *Nutrition Today, 40*(4), 165–172.
4. Angulo, P. (2007). Obesity and nonalcoholic fatty liver disease. *Nutrition Reviews, 65*(6), S57–S63.
5. Baranowski, T. (2006). Advances in basic behavioral research will make the most important contributions to effective dietary change programs at this time. *Journal of the American Dietetic Association, 106,* 808–811.
6. Bauer, J., Reeves, M. M., & Capra, S. (2004). The agreement between measured and predicted Resting Energy Expenditure with pancreatic cancer: A pilot study. *Journal of the Pancreas, 5,* 32–40.
7. Beals, K. A. (2009). Physical activity: How much is enough? *SCAN'S (A Publication for Sports, Cardiovascular, and Wellness Nutritionists) PULSE, 28*(3), 10–13.
8. Benezra, L. M., Nieman, D. C., Nieman, C. M., Melby, C., Cureton, K., Schmidt, D., . . . & Osterberg, K. (2001). Intakes of most nutrients remain at acceptable levels during a weight management program using the food exchange system. *Journal of the American Dietetic Association, 101,* 554–558, 561.
9. Berentzen, T., & Sorensen, T. I. A. (2006). Effects of intended weight loss on morbidity and mortality: Possible explanation of controversial results. *Nutrition Reviews, 64*(11), 502–507.
10. Bifulco, M., & Caruso, M. G. (2007). From the gastronomic revolution to the new globesity epidemic. *Journal of the American Dietetic Association, 107*(12), 2058–2060.
11. Bray, G. A. (2007). The missing link: Lose weight, live longer. *The New England Journal of Medicine, 357*(8), 818–820.
12. Calle, E. E., Rodriguez, C., Walker-Thurmond, K., & Thun, M. J. (2003). Overweight, obesity, and mortality from cancer in a prospectively studied cohort of U.S. adults. *New England Journal of Medicine, 348,* 1625–1638.
13. Centers for Disease Control and Prevention. (2004). Surveillance for certain health behaviors among selected local areas: United States, Behavioral Risk Factor Surveillance System, 2002. *Morbidity and Mortality Weekly Report, 53*(SS-5), 1–100.

14. Centers for Disease Control and Prevention. (2009). Obesity trends: 1985–2009 prevalence of obesity among U.S. adults, by characteristics. *National Center for Chronic Disease Prevention and Health Promotion*. Accessed at: http://www.cdc.gov/nccdphp/dnpa/obesity/trend/maps.

15. Coleman, E. (2007). Fueling during exercise. *Today's Dietitian, 9*(3), 12–20.

16. Coleman, E. (2008). Fluid replacement guidelines for exercise. *Today's Dietitian, 10*(3), 10–14.

17. Coleman, E. (2010). Reconsider athletes' carbohydrate needs. *Today's Dietitian, 12*(3), 46–51.

18. Compher, C., Frankenfield, D., Keim, N., & Roth-Yousey, L. (2006). Best practice methods to apply to measurement of resting metabolic rate in adults: A systematic review. *Journal of the American Dietetic Association, 106*(6), 881–903.

19. Cullen, K. W., Baranowski, T., & Smith, S. P. (2001). Using goal setting as a strategy for dietary behavior change. *Journal of the American Dietetic Association, 101*, 562–566.

20. Dada, J. H. (2010). Marathon fueling: Runners need proper nutrition and hydration for the 26.2-mile stretch. *Today's Dietitian, 12*(3), 36–39.

21. Davis, E. M., Cullen, K. W., Watson, K. B., Konarik, M., & Radcliffe, J. (2009). A fresh fruit and vegetable program improves high school students' consumption of fresh produce. *Journal of the American Dietetic Association, 109*(7), 1227–1231.

22. Dietrich, M., & Jiala, I. (2005). The effect of weight-loss on a stable biomarker of inflammation, C-reactive protein. *Nutrition Reviews, 63*(1), 22–23.

23. Di Renzo, L., & De Lorenzo, A. (2007). Normal weight obese syndrome: How fat is obese? *SCAN'S (A Publication for Sports, Cardiovascular, and Wellness Nutritionists) PULSE, 26*(4), 9–13.

24. Doyle-Lucas, A., & Davy, B. M. (2009). A guide to indirect calorimetry. *Today's Dietitian, 11*(4), 50–55.

25. Escott-Stump, S. (2009). A perspective on nutritional genomics. *Topics in Clinical Nutrition, 24*(2), 92–113.

26. Frankenfield, D., Roth-Yousey, L., & Compher, C. (2005). Comparison of predictive equations for resting metabolic rate in healthy nonobese and obese adults: A systematic review. *Journal of the American Dietetic Association, 105*(5), 775–789.

27. Gabriel, K. K., & Ainsworth, B. E. (2009). Building Healthy Lifestyles Conference: Modifying lifestyles to enhance physical activity and diet and reduce cardiovascular disease. *American Journal of Lifestyle Medicine, 3*(S1), 6S–10S.

28. Gillies, P. J. (2003). Nutrigenomics: The rubicon of molecular nutrition. *Journal of the American Dietetic Association, 103*, S50–S55.

29. Going, S. B., & Laudermilk, M. (2009). Osteoporosis and strength training. *American Journal of Lifestyle Medicine, 3*(4), 310–319.

30. Goodyear, L. J. (2008). The exercise pill: Too good to be true? *The New England Journal of Medicine, 359*(17), 1842–1844.

31. Grimm, K. A., Blanck, M. H., Scanlon, K. S., Moore, L. V., Grummer-Strawn, L. M., & Foltz, J. L. (2010). State-specific trends in fruit and vegetable consumption among adults: United States, 2000–2009. *Morbidity and Mortality Weekly Report, 59*(35), 1125–1130.

32. Harris, J. E. (2008). The need for a concerted effort to address global obesity. *Topics in Clinical Nutrition, 23*(3), 216–228.

33. Harris, M. A., & Flomo, D. (2007). Changing and adhering to lifestyle changes: What are the keys? *American Journal of Lifestyle Medicine, 1*(3), 214–219.

34. Heber, D., & Bowerman, S. (2001). Applying science to changing dietary patterns. *Journal of Nutrition, 131*, 3078S–3081S.

35. Herman, D. R., Vargas, M. A., & Lin, Y. (2008). Therapies: Strategies for successful weight management programs: What are the options? *Nutrition in Complementary Care, 10*(3), 41, 44–48.

36. Institute of Medicine. (2003). *Dietary Reference Intakes: Applications in dietary planning.* Washington, DC: National Academies Press.

37. Jackson, K. (2004, September). Pioneering the frontier of nutrigenomics. *Today's Dietitian, 6*(9), 34–37.

38. Jakicic, J. M., & Otto, A. D. (2006). Treatment and prevention of obesity: What is the role of exercise? *Nutrition Reviews, 64*(2), S57–S61.

39. Janiszewski, P. M., Saunders, T. J., & Ross, R. (2008). Lifestyle treatment of metabolic syndrome. *American Journal of Lifestyle Medicine, 2*(2), 99–108.

40. Jayachandran, J., Banez, L. L., Aronson, W. J., Terris, M. K., Presti, J. C. Jr., Amling, C. L., Kane, C. J., & Freedland, S. J. (2009). Obesity as a predicator of adverse outcome across black and white race: Results from the Shared Equal Access Regional Cancer Hospital (SEARCH) database. *Cancer, 115*(22), 5263–5271.

41. Johnson, D. B., Gerstein, D. E., Evans, A. E., & Woolward-Lopez, G. (2006). Preventing obesity: A life cycle perspective. *Journal of the American Dietetic Association, 106*(1), 97–102.

42. Journal of the American Dietetic Association. (2005). *Obesity: Etiology, treatment, prevention, and application in practice*, supplement 1.

43. Khan, L. K., Sobush, K., Keener, D., Goodman, K., Lowry, A., Kakietek, J., & Zaro, S. (2009). Recommended community strategies and measurements to prevent obesity in the United States. *Morbidity and Mortality Weekly Report, 58*(RR-7), 1–26.

44. Lin, D. C. (2007). Actual measurements of body weight and height are essential: Most self-reported weights and heights are unreliable. *Nutrition Today, 42*(6), 263–266.

45. Liou, T., Pi-Sunyer, F. X., & Laferrere, B. (2005). Physical disability and obesity. *Nutrition Reviews, 63*(10), 321–331.

46. Maud, P. J., & Foster, C. (2006). *Physiological assessment of human fitness* (2nd ed.). Champaign, IL: Human Kinetics.

47. Melanson, K. J. (2007). Dietary considerations of obesity treatment. *American Journal of Lifestyle Medicine, 1*(6), 433–436.

48. Melanson, K. J., Angelopoulos, T. J., Nguyen, V. T., Martini, M., Zukley, L., Lowndes, J., . . . & Rippe, J. M. (2006). Consumption of whole-grain cereals during weight loss: Effects on dietary quality, dietary fiber, magnesium, vitamin B-6, and obesity. *Journal of the American Dietetic Association, 106*(9), 1380–1388.

49. Merchant, A. T., Vantaparast, H., Barlas, S., Dehghan, M., Ali Shah, S. M., De Koning, L., & Steck, S. E. (2009). Carbohydrate intake and overweight and obesity among healthy adults. *Journal of the American Dietetic Association, 109*(7), 1165–1172.

50. Mettler, S., & Meyer, N. L. (2010). Food pyramids in sports nutrition. *SCAN'S (A Publication for Sports, Cardiovascular, and Wellness Nutritionists) PULSE, 29*(1), 12–18.

51. Miner, M. (2007). Fitness, antioxidants, and moderate drinking: All to lower cardiovascular risk. *American Journal of Lifestyle Medicine, 1*(2), 110–112.

52. Morrison, C. D., & Berthoud, H. R. (2007). Neurobiology of nutrition and obesity. *Nutrition Reviews, 65*(12), 517–534.

53. Müller, M., & Kersten, S. (2003). Nutrigenomics: Goals and strategies. *Nature Reviews, 4*, 315–322.

54. The National Center for Health Statistics. (2010). Prevalence of obesity (class I, II, and III) among adults aged ≥ 20 years, by age group and sex: National Health and Nutrition Examination Survey, United States, 2007–2008. *Morbidity and Mortality Weekly Report, 59*(17), 527.

55. Nelson, M. E., & Kennedy, M. A. (2009). The steps toward developing the physical activity guidelines for Americans. *Nutrition Today, 44*(3), 98–103.

56. O'Connell, B. S. (2001, November/December). Weight management: What works? *Dietitians Edge*, 18–19.

57. Odegaar, A. O., & Pereira, M. A. (2006). Trans fatty acids, insulin resistance, and type 2 diabetes. *Nutrition Reviews, 64*(8), 364–372.

58. Ozier, A. D., Kendrick, O. W., Leeper, J. D., Knol, L. L., Perko, M., & Burnham, J. (2008). Overweight and obesity are associated with emotion- and stress-related eating as measured by the eating and appraisal due to emotions and stress questionnaire. *Journal of the American Dietetic Association, 108*(1), 49–56.

59. Palmer, S. (2010). A unique perspective: Dr. David Katz's take on reversing obesity in America. *Today's Dietitian, 12*(1), 28–31.

60. Pastors, J. G. (2010). Bariatric surgery for type 2 diabetes. *Today's Dietitian, 12*(9), 44–49.
61. Pedersen, B. K. (2007). Health benefits related to exercise in patients with chronic low-grade systemic inflammation. *American Journal of Lifestyle Medicine, 1*(4), 289–298.
62. Pi-Sunyer, F. X. (2004). The epidemiology of central fat distribution in relation to disease. *Nutrition Reviews, 62*(7), S120–S126.
63. Popkin, B. M. (2004). The nutrition transition: An overview of world patterns of change. *Nutrition Reviews, 62*(7), S140–S143.
64. Rowe, S., Alexander, N., Clydesdale, F. M., Applebaum, R. S., Atkinson, S., Black, R. M., . . . & Wedral, E. (2009). Funding food science and nutrition research: Financial conflicts and scientific integrity. *Nutrition Today, 44*(3), 112–113.
65. Schroeder, J. (2010). Stretching: What is the research telling us? *American Fitness, 28*(3), 24–30.
66. Schryver, T. (2001, November/December). Weight loss and obesity, NHLBI style. *Dietitians Edge,* 30–41.
67. Schwartz, J., & Byrd-Bredbenner, C. (2006). Portion distortion: Typical portion sizes selected by young adults. *Journal of the American Dietetic Association, 106*(9), 1412–1418.
68. Shirreffs, S. M. (2008). Assessment of fluid and electrolyte status. *SCAN'S (A Publication for Sports, Cardiovascular, and Wellness Nutritionists) PULSE, 27*(1), 1–4.
69. Simopoulos, A. P. (2002). Genetic variation and dietary response: Nutrigenetics/nutrigenomics. *Asia Pacific Journal of Clinical Nutrition, 11,* S117–S128.
70. Smolin, L. A., & Grosvenor, M. B. (2003). *Nutrition science and applications* (4th ed.). Hoboken, NJ: Wiley.
71. Spano, M., & Kerksick, C. M. (2007). Speeding recovery: Nutrition & supplementation for exercise. *Today's Dietitian, 9*(3), 32–35.
72. Stigler, P., & Cunliffe, A. (2006). The role of diet and exercise for the maintenance of fat-free mass and resting metabolic rate during weight loss. *Sports Medicine, 36,* 239–262.
73. Stover, P. J., & Caudill, M. A. (2008). Genetic and epigenetic contributions to human nutrition and health: Managing genome–diet interactions. *Journal of the American Dietetic Association, 108*(9), 1480–1487.
74. Strychar, I. (2006). Diet in the management of weight loss. *Canadian Medical Association Journal, 174,* 56–63.
75. Terre, L. (2009). Promoting physical activity in minority populations. *American Journal of Lifestyle Medicine, 3*(3), 195–197.
76. Tillotson, J. E. (2005, May/June). 10 things Congress needs to know about obesity. *Nutrition Today, 40*(3), 126–129.
77. Tremblay, A. (2004). Dietary fat and body weight set point. *Nutrition Reviews, 62*(7), S75–S77.
78. Tufts University. (2009). To live to a biblical old age, stay physically active. *Tufts University Health & Nutrition Letter, 27*(10), 1–2.
79. Tyler, C., Johnston, C. A., & Foreyt, J. P. (2007). Lifestyle management of obesity. *American Journal of Lifestyle Medicine, 1*(6), 423–429.
80. Vakili, S., & Caudill, M. A. (2007). Personalized nutrition: Nutritional genomics as a potential tool for targeted medical nutrition therapy. *Nutrition Reviews, 65*(7), 301–315.
81. Vargas, M. (2009). Atherosclerosis, functional foods, and nutritional genomics. *Dietitians in Integrative and Functional Medicine, 12*(2), 20, 23–30.
82. Weber, J. A. (2006). Talking about hunger in a land of plenty. *Journal of the American Dietetic Association, 106*(6), 804–807.
83. Welland, D. (2010). Think exercise for brain health. *Today's Dietitian, 12*(3), 24–27.
84. Westerterp, K. R. (1993). Food quotient, respiratory quotient, and energy balance. *American Journal of Clinical Nutrition, 57,* 759S–765S.

85. Wildman, R., & Miller, B. (2004). *Sports and fitness nutrition*. Belmont, CA: Wadsworth.
86. Williams, M. H. (2005). *Nutrition for health, fitness, and sport* (7th ed.). New York: McGraw-Hill.
87. Wolfe, R. R. (2006). The underappreciated role of muscle in health and disease. *American Journal of Clinical Nutrition, 84*(3), 475–482.
88. Yanovski, A. Z., & Yanovski, J. A. (2002). Obesity. *New England Journal of Medicine, 346*, 591–602.
89. Zoeller, R. F. (2009). Physical activity and fitness in African Americans: Implications for cardiovascular health. *American Journal of Lifestyle Medicine, 3*(3), 188–194.

The Vitamins and Minerals

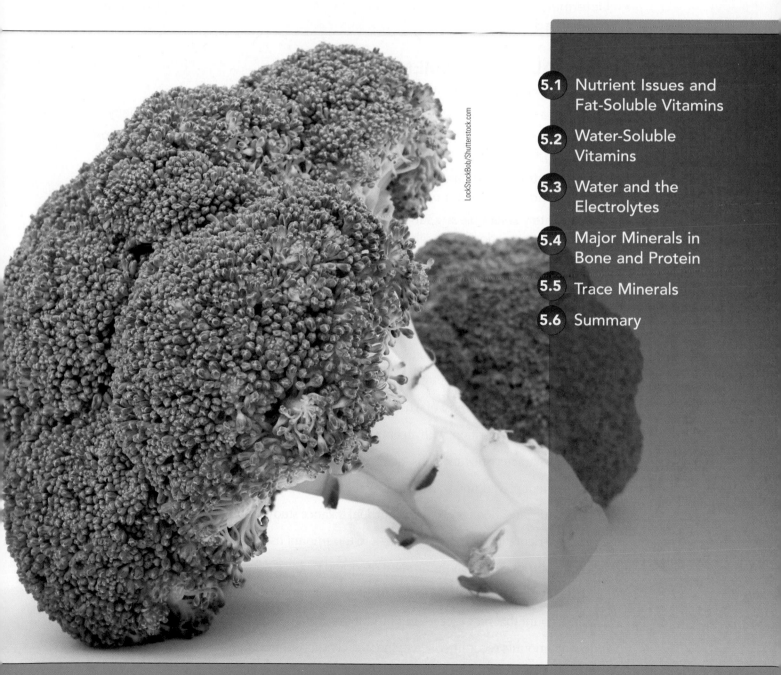

LockStockBob/Shutterstock.com

MODULE GOAL

To understand the appropriate intake of vitamins and minerals to regulate metabolism and maintain health.

LEARNING OBJECTIVES

After completing this learning module, you will be able to:

- Determine the recommended amount of each essential nutrient for healthy persons.
- Determine the deficiency and toxicity signs and symptoms for each essential nutrient in human nutrition.
- Explain the function(s) of nutrients in metabolism.
- Identify good food sources for each essential nutrient.
- Identify any significant health-promoting or therapeutic uses for some essential nutrients.
- Determine the Tolerable Upper Intake Levels (ULs) for the nutrients that have been established.

PERSONAL IMPROVEMENT GOALS

When you complete this learning module, you will know how to:

- Use the results of dietary analysis to evaluate vitamin and mineral adequacy.
- Incorporate your skills, knowledge, and competencies into daily dietary action regarding food choice.
- Correct nutrient imbalances with appropriate nutrient food sources and recommendations based on individual age and gender.
- Appropriately select dietary supplements when supplementation is needed, such as with food intake restrictions from allergy, intolerance, or vegetarian eating.
- Ensure adequate nutrient needs daily to promote optimal health and body functioning.

Nutrition is a relatively young science. The body of scientific knowledge continues to grow rapidly as researchers seek better understanding of the multiple functions of nutrients in the body. As recently as 30 years ago, most of what was taught in a basic nutrition course involved protein, vitamin, and mineral nutrition, with an emphasis placed on the essential amino acids, vitamins, and minerals, the characteristics and consequences of deficiencies and toxicities, the role in metabolism that each nutrient played, and good food sources of the nutrient. Up until that same time, scientists determined nutrient adequacy by using nutrient balance studies, a type of study that measures the amount of nutrient taken in and compares it to the amount used by the body. Nutrient balance occurs when the intake equals the amount used, and that was considered adequate nutrition. Today, the goal for adequate nutrition is to provide the amount of the nutrient that optimizes function and health, an amount that may exceed the amount that achieves nutrient balance. The standards for the nutrient amounts recommended in the 21st century support optimal function and health.

5.1 NUTRIENT ISSUES AND FAT-SOLUBLE VITAMINS

Introduction

Vitamins and minerals play integral roles in cell metabolism as cofactors for enzymes (see BioBeat 1.6 on page 30). In the body, vitamins and minerals are required at specific levels for normal function. Essential nutrient intake level recommendations are currently established to support optimal function and health (see BioBeat 5.1). Too much or too little of a nutrient taken into the body usually results in serious health consequences. Specific deficiency and toxicity symptoms are associated with inappropriate intakes of the essential vitamins and minerals.[25,69]

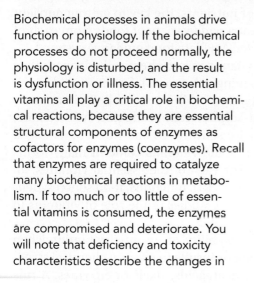

BioBeat 5.1

Nutrients, Biochemistry, Physiology, and Health

Biochemical processes in animals drive function or physiology. If the biochemical processes do not proceed normally, the physiology is disturbed, and the result is dysfunction or illness. The essential vitamins all play a critical role in biochemical reactions, because they are essential structural components of enzymes as cofactors for enzymes (coenzymes). Recall that enzymes are required to catalyze many biochemical reactions in metabolism. If too much or too little of essential vitamins is consumed, the enzymes are compromised and deteriorate. You will note that deficiency and toxicity characteristics describe the changes in physiology. Many of the essential minerals also play biochemical roles, but many are required for the structure of tissue as well. The iron molecule required in **hemoglobin** is a good example of the structural role a mineral plays in a biological molecule. When iron is deficient, the blood has too little hemoglobin. Without sufficient hemoglobin in red blood cells, there is not enough oxygen-carrying capacity in the red blood cells, and the individual experiences iron-deficiency **anemia**.[24,25,75]

Have you consumed the nutrients you need to optimize your structure and function?

DA+ GENEie Go to Diet Analysis Plus and enter 2 cups of Total® breakfast cereal, 2 cups of puffed rice cereal, and two skinless, boneless grilled chicken breasts. Create a Source Analysis spreadsheet report. Study the vitamin and mineral content. Which of these food selections provides the most nutrients?

Vitamins in the Body

Vitamins are organic compounds needed in tiny amounts in the body. They are categorized by their solubility in water or oil (fat). The fat-soluble vitamins (A, D, E, and K) are soluble in oil. Fat-soluble vitamins accumulate primarily in the fatty tissue in the body and to a small extent can be excreted through bile. Because there is a longer storage reserve of fat-soluble vitamins in the body, it takes a longer time for either toxicity or deficiency symptoms to occur. Furthermore, because of the large reserves of **fat-soluble nutrients**, toxicities are long lived. The **water-soluble nutrients** include the water-soluble B vitamins—**thiamin** (B$_1$), **riboflavin** (B$_2$), **niacin** (B$_3$), **vitamin B$_6$**, **vitamin B$_{12}$**, **biotin**, **pantothenic acid**, **folate**,

T-Talk 5.1
T-Talk 5.1
To hear Dr. Turley talk about nutrient issues and fat-soluble vitamins, go to www.cengagebrain.com

hemoglobin A large, protein-containing iron that is essential for oxygen transport and carbon dioxide removal by red blood cells.

anemia An inability for the red blood cells to deliver oxygen to body cells, causing the individual to feel tired, weak, breathless, apathetic, and often experience headache; presents in many types, such as macrocytic, microcytic, and hemolytic.

fat-soluble nutrients The nutrients that are soluble in oil.

water-soluble nutrients Nutrients that dissolve in an aqueous solution or water.

thiamin A water-soluble vitamin needed for the enzyme thiamin pyrophosphate (TPP), which is needed for ATP production from carbohydrate, normal appetite, and nervous system functioning; the deficiency causes beriberi, and toxicity is poorly documented in humans.

riboflavin A water-soluble vitamin that plays its role as a coenzyme for FMN and FAD in fatty acid energy metabolism; deficiency causes ariboflavinosis.

niacin A water-soluble vitamin that plays a role as a coenzyme in NAD, NADP, energy metabolism, and steroid synthesis; deficiency causes pellegra.

vitamin B$_6$ Various forms of this water-soluble B vitamin (PN, PL, PM, PNP, PLP, and PMP) function as coenzymes in the metabolism of amino acids, glycogen, and some lipids.

vitamin B$_{12}$ A B vitamin called cyanocobalamin that functions in nucleic acid synthesis; deficiency results in pernicious anemia.

biotin A water-soluble essential vitamin that functions in carboxylation reactions, and is present in food in a free form or bound to protein.

pantothenic acid A water-soluble vitamin that is a component of coenzyme A and widespread in foods; deficiency is poorly documented in humans.

and **choline**—and water-soluble **vitamin C**; all dissolve in water. Because of their solubility in water, they can be excreted readily via urine. In general, the stores of water-soluble vitamins are not extensive. Therefore, water-soluble vitamins can rapidly be depleted from the body. With fast depletion, water-soluble vitamins have a shorter time to manifest toxicity symptoms as well. Furthermore, because replenishment of water-soluble vitamins is so fast when given high levels, deficiencies, like toxicities, are typically short lived.[29,31,32,33]

The term *vitamins* represents a group of organic compounds essential for normal metabolism, growth, development, and cellular function. Each vitamin has specific functions, and most work together with enzymes as cofactors or in other ways necessary to sustain normal metabolism and health. Each essential vitamin needs to be consumed in the diet at the DRI level based on age, gender, and conditions of pregnancy and lactation. DRI values are expressed as Recommended Dietary Allowances (RDAs) or Adequate Intakes (AIs) for the essential vitamins (see Appendix A). The quantity values for the essential vitamins are very small (milligrams or micrograms), depending on the nutrient. The intake values reflected on food package labels are Reference Daily Intakes (RDIs). The RDIs are based on the highest 1968 RDA adult values, so they are a set of single values that is not age-, gender-, or condition-specific and a "one-size-fits-all" general standard.[76]

Essential vitamins are vital to life and vital for body functioning at specific levels. Because foods have differing levels of nutrients, the key to getting nutrient needs met and having a healthy diet is to eat a wide variety of foods from each of the food groups, by applying the personalized MyPlate food guidance system.

In the early days of nutrition science, vitamins were chemically defined; they were thought to be amines (nitrogen-containing substances). So these vital nutrients, thought to be amines, were called vitamins.[43,44] Water-soluble B vitamins do contain nitrogen, but vitamin C, and the fat-soluble vitamins do not. Furthermore, it has been discovered that some vitamins are sensitive to and thus are damaged by many chemical stressors (see Table 5.1). Some vitamins come in many forms or exist as a family of related compounds. For each of the essential vitamins, an exploration of the nutrient forms, dietary need (DRI), functions, food sources, deficiencies, and toxicities will be addressed. Recent studies of the role of vitamins in promoting good health will be covered in BioBeats.

Many people have questions about the stability of essential vitamins when cooking, processing, or storing foods. The major insults to vitamin chemical structure include exposure to oxygen, light, cold, heat, acids, alkali, or enzymes. A rule of thumb is that the damage to a vitamin's effectiveness is directly related to the length of exposure to one or more of these factors. Natural exposures to these insults from food handling for human consumption do not necessarily result in complete vitamin destruction (see Table 5.1).

folate A generic term used to describe a collection of chemical compounds found in food that can be converted to folic acid in the body and function in the formation of nucleic acids; deficiency results in macrocytic normochromic anemia.

choline A water-soluble vitamin that is used to make the neurotransmitter acetylcholine.

vitamin C A water-soluble vitamin that functions as an antioxidant, in collagen synthesis, and in the maintenance of connective tissue; deficiency results in scurvy; also known as L-ascorbic acid.

TABLE 5.1

The Susceptibility and Reduced Stability of Vitamins to Various Factors

Factor	Damaged Vitamin
Oxygen	Vitamins A, E, and C
Light	Vitamins A, B_6, C, D, E, riboflavin, and folate (folic acid)
Cold	Vitamin E and pantothenic acid
Heat	Vitamin C, vitamin K, thiamin, folate, pantothenic acid, and vitamin B_{12}
Acid	Pantothenic acid and vitamin B_6
Alkali	Thiamin, vitamin C, pantothenic acid, and vitamin B_6
Enzyme	Thiamin is destroyed by the enzyme thiaminase, which is abundant in raw herring, shrimp, clams, and carp

© Cengage Learning 2013

Diagnosing Nutritional Deficiency and Toxicity

If a nutrient deficiency or toxicity is suspected, the problem should be properly diagnosed by evaluating the **signs** and **symptoms** in context of the diet and physiological state of an individual (see BioBeat 5.2). Evaluating the cause of the signs and symptoms involves a four-step process that includes looking at the person's dietary records, assessing clinical symptoms, studying biochemical values, and then adjusting the nutrient intake to an appropriate level.

Unfortunately, diagnosing a subclinical nutrient deficiency has not been adequately developed. Subclinical deficiency, which occurs before the onset of specific nutritional deficiency signs and symptoms, results in suboptimal functioning of physiological systems. An example of suboptimal functioning is the decreased cognitive function in children who are iron depleted, but whose blood chemistry does not yet indicate iron deficiency. Individuals with subclinical deficiency are often the marketing target of the dietary supplement industry. The process of diagnosing a nutrient deficiency or toxicity is outlined below.[3,17,39,53,76]

- *Dietary records:* To identify low intake of a nutrient, dietary records can be used. Additionally, a medical history can confirm a metabolic or physiological problem that creates an increased need for a nutrient. For toxicity, tracking the use of supplements can reveal a high intake.

- *Clinical deficiency or toxicity symptoms:* The symptoms should be compatible with the low (deficiency) or high (toxicity) dietary intake or increased need (see the individual nutrient deficiencies and toxicities associated with each nutrient described throughout this module). Self-diagnosing a nutrient deficiency or clinical evaluation is not reliable. Abnormal body nutrient levels must also support the clinical manifestation. Individuals who are more likely to be undernourished include those who are poverty stricken, young children, and women who bear children frequently. Individuals who are more likely to exceed nutrient recommendations are those who take single nutrient supplements.

- *Biochemical tests:* The appropriate tests may include assessing blood levels, tissue levels, urine levels, and the like. They can demonstrate low (deficiency) or high (toxicity) levels of the nutrient. For most nutrients, reliable biochemical tests confirm the nutrient malnutrition.

- *Nutrient supplementation or withdrawal:* For deficiency, nutrient supplementation should correct the deficiency signs and symptoms. For toxicity, nutrient withdrawal to normal physiological levels should correct the toxicity signs and symptoms. This is the biological evidence (correcting the signs and symptoms) that a person was truly deficient or toxic in a nutrient. There are times, depending on the length and severity of the deficiency or toxicity, that permanent damage occurs.[7,17,77]

Fat-Soluble Vitamins

The fat-soluble vitamins include **vitamin A**, **vitamin D**, **vitamin E**, and **vitamin K** (see Table 5.2 on pages 236–237 for a summary and Appendix A for chemical structures).

Vitamin A[G,I,16,30,32,33,43,54,65,66,83] (See Table 5.2 on page 236 for a summary and Appendix A for vitamin A chemistry.)

Vitamin A Forms   Retinol, retinal, and retinoic acid are active vitamin A forms from animal sources; these forms are also known as previtamin A. Previtamin A forms can be somewhat interconverted (see Figure 5.1). The family of carotenoids can be made into vitamin A. Beta-carotene, which provides the bright orange pigment in plant sources, is the best-known carotenoid, also known as **provitamin A**. In human metabolism, beta-carotene can be converted to vitamin A. Twelve

signs Measurable markers or marked visual changes that indicate a medical or nutritional ailment.

symptoms A subjective complaint about the state of wellness.

vitamin A A fat-soluble vitamin in the active forms retinol, retinal, and retinoic acid that functions in vision, epitheial cell maintenance, and gene expression; deficiency causes blindness.

vitamin D A fat-soluble vitamin that is synthesized in the body from cholesterol and functions in the mineralization of bone by regulating calcium and phosphorus; deficiency causes rickets and osteomalacia.

vitamin E The family of fat-soluble vitamins that functions in antioxidant activity and is also known as tocopherols and tocotrienols.

vitamin K A fat-soluble vitamin that functions in blood clotting; most of the dietary source is produced by the bacteria in the gut.

provitamin A Beta-carotene and other antioxidant carotenoids that can be converted to vitamin A inside the body and are found in yellow/orange or dark green fruits and vegetables.

BioBeat 5.2

Signs and Symptoms of Nutrient Deficiencies or Toxicities

Adequate nutrient recommendations for the intake of essential nutrients help us to avoid deficiency and toxicity. Regarding most of the essential nutrients, if too much or too little of the nutrients are consumed, physiology fails, and distinct signs and symptoms manifest depending on the roles the nutrients play in metabolism. A *sign* is a recognizable physical change, or a measurable abnormal level of a biological marker specific to the nutrient. You will notice that many of the signs are visible or tangible, such as brittle nails, hair loss, skin dryness, paleness, and so on. A *symptom*, on the other hand, is a subjective complaint about one's state of wellness that cannot be measured comparatively among individuals, such as pain, swelling, headache, dry mouth, joint soreness, and so on. Each of the essential micronutrients has its own set of signs and symptoms for deficiency and toxicity.[52,53]

Have you ever experienced a vitamin or mineral deficiency or toxicity?

FIGURE 5.1 The interconversion and primary functions of the various forms of vitamin A.

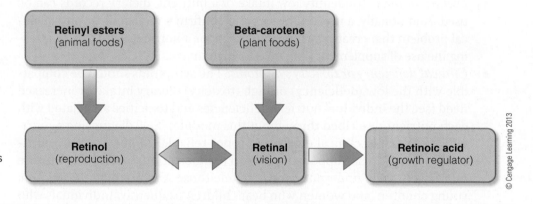

© Cengage Learning 2013

microgrobams of beta-carotene are equivalent to one microgram of retinol, which is equivalent to one **Retinol Activity Equivalent (RAE)**. You can see some inefficiency in the conversion from beta-carotene to vitamin A, and this is why beta-carotene is far less toxic to the human body than previtamin A forms.

Functions of Vitamin A   Vitamin A functions in vision (including light and dark adaptation in the form of retinal; see Figure 5.2) and the maintenance of epithelial cells for internal and external surface linings (skin, mucus membranes, eyes, respiratory tract, urinary tract, and gastrointestinal tract). Vitamin A is also required for growth, reproduction, embryonic development, gene expression, and immune function. Provitamin A forms have some antioxidant properties (see BioBeat 5.4 on page 234). Beta-carotene has been shown to be beneficial in reducing the damage caused by overexposure to the sun (sunburn), slowing the aging process, and reducing the damage from radicals in the body.

Vitamin A's DRI and UL   The DRI level is based on nutritional balance, as well as maintaining total body stores. The DRI for vitamin A, expressed in micrograms of retinol (the most active form of vitamin A), is 900 micrograms per day for an adult male 19 to 30 years old; for an adult female 19 to 30 years old, it is 700 micrograms per day. The UL for an adult male or adult female 19 to 70 years old is 3,000 micrograms of retinol per day. The UL for carotenoids is not determined. Refer to Appendix A for all life span and gender-specific values.

Some Specifics on the Units of Vitamin A for the RDI   On food labels, vitamin A is often given in **International Units (IU)**. However, the percent of RDIs can also be converted to **Retinol Equivalents (REs)** of vitamin A. This conversion of

Retinol Activity Equivalent (RAE) The conversion of vitamin A activity from provitamin A and previtamin A, also known as Retinol Equivalents (REs).

International Units (IU) A standardized unit of measure used on food package labels for vitamins A, D, and E.

Retinol Equivalent (RE) The conversion of vitamin A activity from provitamin A and previtamin A, also known as Retinol Activity Equivalents (RAEs).

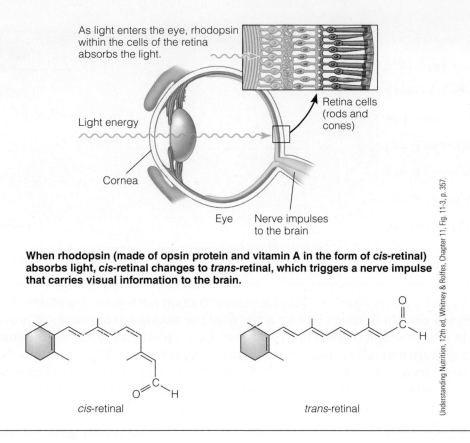

As light enters the eye, rhodopsin within the cells of the retina absorbs the light.

Retina cells (rods and cones)

Light energy

Cornea

Eye Nerve impulses to the brain

When rhodopsin (made of opsin protein and vitamin A in the form of *cis*-retinal) absorbs light, *cis*-retinal changes to *trans*-retinal, which triggers a nerve impulse that carries visual information to the brain.

cis-retinal

trans-retinal

Understanding Nutrition, 12th ed, Whitney & Rolfes, Chapter 11, Fig. 11-3, p. 357.

FIGURE 5.2 The role of retinal (a form of vitamin A) in vision.

units is needed to compare the nutrient amounts listed on food and supplement labels to the amounts of nutrients that are recommended for individuals in the DRI tables. In food composition tables and databases, the vitamin A content of food is given in REs or RAEs. This conversion allows you to understand how to convert food package label units into those of the DRI.

Vitamin A Deficiency (Hypovitaminosis A)   This is the leading cause of blindness worldwide. In blindness caused by a vitamin A deficiency, called **xerophthalmia**, the cornea of the eye thickens until it becomes a "blinder." Blindness and the thickening of the cornea are not reversible. Other deficiency signs include the decreased ability for the eye to adapt to light and dark environments (night blindness), hyperkeratosis or thickened skin, abnormal bone growth, diarrhea, and depressed immunity.

Vitamin A Toxicity   Toxicities of previtamin A are well known and include the signs and symptoms of nausea and vomiting, headache, vertigo (impaired balance), and incoordination. Chronic, long-term toxicity can lead to loss in bone mineral density, liver abnormalities, and birth defects. **Hypercarotenemia**, when beta-carotene is high in the blood stream and is deposited in the skin and the subcutaneous fat, can cause the skin to appear orange. However, it is not associated with adverse health effects when the cause is eating an abundance of foods high in beta-carotene, as opposed to taking supplements.

Acutane, isotretinoin, or 13-*cis* retinoic acid has been used to treat deep cystic acne. Because acutane is a derivative of vitamin A, and vitamin A can be extremely toxic, clinical signs and symptoms must be monitored carefully in the person taking acutane. Females of reproductive age are generally not prescribed acutane because of the risk of serious birth defects. Acutane also raises LDL cholesterol levels significantly, so blood values should be monitored with an acutane prescription.

hypovitaminosis A A condition of vitamin A deficiency characterized by impaired growth, night blindness, xerophthalmia, diarrhea, depressed immunity, and hyperkeratosis.

xerophthalmia The name of the eye condition causing blindness resulting from a vitamin A deficiency.

hypercarotenemia A state of the blood that is characterized by high blood levels of beta-carotene.

acutane A prescription drug made from a form of vitamin A (retinoic acid) that is used to treat acne.

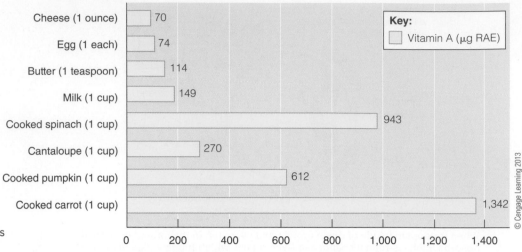

FIGURE 5.3 Food sources of vitamin A.

Vitamin A Food Sources   Animal sources contain microgram quantities of preformed vitamin A, which is more active than the microgram quantities of provitamin A that the body derives from plant sources (such as beta-carotene). The animal sources are butter, egg yolk, fortified milk, liver, and cheese (see Figure 5.3). Plant sources include yellow/orange or dark green fruits and vegetables (see Figure 5.3). Beta-carotene is a bright orange pigment. Examples of these food sources are apricot, cantaloupe, squash, carrot, pumpkin, and sweet potato.

Vitamin D[G,I,4,9,13,14,20,22,27,29,30,38,40,45,48,50,53,78,79]   (See Table 5.2 on page 237 for a summary and Appendix A for vitamin D chemistry.)

Vitamin D Forms   There are two forms of vitamin D that you should know about: **cholecalciferol** (vitamin D_3) and **1,25-dihydroxycholecalciferol**, or calcitriol. Vitamin D_3 is the precursor form of the vitamin that is fortified in milk, and it can be made in the body. However, calcitriol is the active form of vitamin D in the body. Synthesis of calcitriol depends on adequate exposure of the skin to sunlight and healthy liver and kidney function (see Figure 5.4).

Vitamin D Synthesis   Vitamin D is synthesized in the body from cholesterol when the skin is exposed to ultraviolet B (UVB) light given off by the sun (see BioBeat 5.3 on page 232 and Figure 5.4). If the exposure to sunlight is limited, the body cannot synthesize enough vitamin D; thus, it is considered an essential nutrient. In classical vitamin D synthesis, cholesterol in the skin undergoes a chemical transformation upon UVB light exposure. This new chemical structure, 7-dehydrocholesterol, is converted to cholecalciferol (vitamin D_3) in the skin and released into the blood. It is then **hydroxylated** in the liver at the 25th position (the site of the chemical change in the molecule). This chemical structure, $25(OH)D_3$, is released back into the blood stream and then hydroxylated again in the kidney at the first position (another site of chemical change in the molecule). After these reactions, the vitamin D molecule is active and is chemically 1,25-dihydroxyvitamin D_3, also called 1,25-dihydroxycholecalciferol (calcitriol), or simply vitamin D.

Vitamin D Functions   Vitamin D functions in the body as a regulator of calcium and phosphorus balance in the blood. It has steroid-like chemistry and hormone-like activity. Its primary effect is to increase the absorption of calcium by the small intestine, but it is also involved with increasing phosphorus excretion and calcium reabsorption by the kidney. Vitamin D functions in cell metabolism to reduce cell division (antiproliferation) and promote cell maturation or specialization (prodifferentiation).

cholecalciferol Vitamin D_3, which is the form of vitamin D that is used to fortify food and is added to dietary supplements.

1,25-dihydroxycholecalciferol The active form of vitamin D, also called 1,25-dihydroxyvitamin D_3 and calcitriol.

hydroxylated The addition of a hydroxyl (OH) group to a compound, which often results in an electron-deficient chemical.

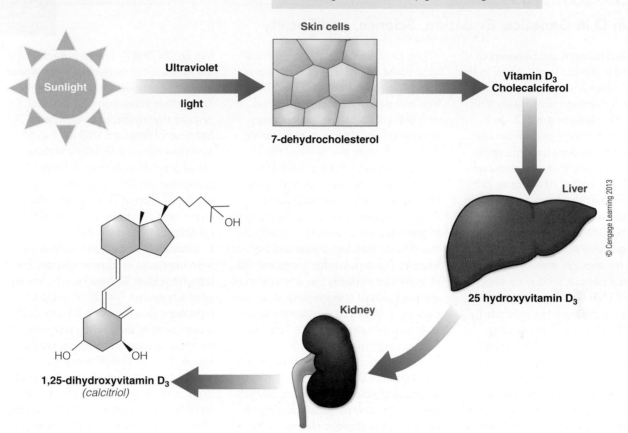

Skin tone evolution variations exist due to a "G" or "A" allele change in the melanin pigmentation gene

Skin cells

Sunlight

Ultraviolet

light

7-dehydrocholesterol

Vitamin D₃
Cholecalciferol

Liver

© Cengage Learning 2013

25 hydroxyvitamin D₃

Kidney

OH

HO   OH

1,25-dihydroxyvitamin D₃
(calcitriol)

FIGURE 5.4 The synthesis and chemical structure of vitamin D, with mention of skin tone evolution variations.

Vitamin D's DRI and UL   The DRI, expressed in micrograms of cholicalciferol, is based on an inadequate exposure to sunlight. The DRI for an adult male or an adult female 19 to 30 years old is 5 micrograms per day. The UL for an adult male or an adult female 19 to 70 years old is 50 micrograms per day. Refer to Appendix A for all life span and gender-specific values.

Some Specifics on the Units of Vitamin D for the RDI   On food labels, vitamin D is given in IU. However, the percent of RDIs can also be converted to micrograms of vitamin D₃ (cholecalciferol). One microgram of vitamin D₃ is equal to 40 IU. Thus, the DRI of 5 micrograms of vitamin D is equal to 200 IU, and the UL of 50 micrograms of vitamin D is equal to 2,000 IU. This conversion allows you to understand how to convert food package label units into those of the DRI. The level of vitamin D fortification in 1 quart of milk is 400 IU.

Vitamin D Deficiency   The deficiency states of vitamin D are **rickets** and **osteomalacia**. Rickets occurs in children and is characterized by soft bones, bowed legs, pigeon chest, poor growth, and knocked knees. Osteomalacia occurs in adults and is characterized by porous bones and an increased risk for osteoporosis. This condition is caused by the poor regulation of calcium and phosphorus metabolism. Vitamin D deficiency causes malformed teeth, increased circulating levels of parathyroid hormone (PTH) and alkaline phosphatase (AlkP), and decreased circulating levels of serum phosphorus.

rickets The deficiency condition of vitamin D as it manifests in children; seen by bowed legs and a pigeon chest.
osteomalacia A softening of the bone caused by a vitamin D deficiency in adults.

BioBeat 5.3

Vitamin D in Genetics, Evolution, Science, and Society

The earliest humans are believed to have lived in Africa. There, close to the equator, a darker skin tone provided a survival advantage. Eventually, the original "G" allele of the SCL gene (involved in melanin, or dark brown pigment, production) became dominant. This genetic trait, still dominant in 93 to 100 percent of Africans, produced a darker skin tone, which provided a natural selection advantage in preventing DNA damage and death from aggressive skin cancer. It also allowed for enough vitamin D synthesis in these people, who were exposed to a lot of UVB light because of their proximity to the equator, outdoor living, and lack of need for clothing.

Sometime between 55,000 and 85,000 years ago, humans migrated from Africa to higher latitudes in Europe. The environmental change caused a strong and rapid natural selection pressure, resulting in a point mutation in the SCL gene to the "A" allele, which is present in all people of European descent. This mutation inhibits melanin production; the resulting lighter skin tone allows for more efficient vitamin D synthesis with less sunlight exposure. Individuals at higher latitudes with the primary "G" allele would develop

rickets because of the decreased ability to make vitamin D_3. Rickets causes skeletal abnormalities, including bowed legs and pelvic deformities. Women with osteomalacia (the adult version of rickets) have narrowed birth canals, which may cause the death of the woman and baby during childbirth.

The food supply was and still is relatively poor in vitamin D sources. Thus, adequate solar fuel is needed to trigger vitamin D chemical activity in the skin for human survival and health. Vitamin D is a powerful, hormone-like vitamin that controls the absorption of dietary calcium and incorporation of calcium in bones. It is essential to survival of the human species. Thus, conflicting selective forces of protection from UV sunlight damage and synthesis of vitamin D were resolved in evolution. Archeologists, epidemiologists, geneticists, and molecular biologists have tied the genetic change primarily to a single base-pair substitution in the SCL gene that encodes for melanin pigment by geographical latitude and genetic heritage.

In the present day, poor vitamin D status in adults is occurring in epidemic proportions as measured by the blood levels of the intermediate form of vita-

min D, $25(OH)D_3$. The geography of the epidemic is evident in areas where vitamin D production is insufficient. The farther away from the equator, the greater the insufficiency of vitamin D because of reduced UVB light exposure (see Figure 5.5). Many factors are at play: air pollution, the use of sunscreen (an SPF factor of 8 or above blocks all vitamin D synthesis), cloud cover, and skin tone. Because of the number of factors involved, it is difficult to estimate a population-wide sunlight exposure recommendation. For latitudes closer to the equator, somewhere between 5 to 30 minutes of sunlight exposure between 10 a.m. and 3 p.m. several times per week may be needed to support sufficient vitamin D status. Vitamin D deficiency is associated with many chronic medical conditions, including osteoporosis, fractures from falls, psychiatric problems, colon, breast, and prostate cancers, pain, hypertension, cardiovascular disease, and increased susceptibility to autoimmune inflammatory diseases such as type 1 diabetes, multiple sclerosis, and rheumatoid arthritis.[J,4,8,27,38,40,48,50,78,79]

How do genetics, evolution, and society affect your vitamin D status?

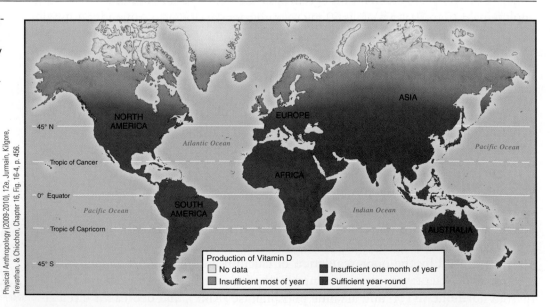

FIGURE 5.5 The geography of vitamin D production. The farther away a person lives from the equator, the greater their insufficiency of vitamin D because of reduced sunlight exposure.

Physical Anthropology (2009-2010), 12e, Jurmain, Kilgore, Trevathan, & Chiochon, Chapter 16, Fig. 16-4, p. 456.

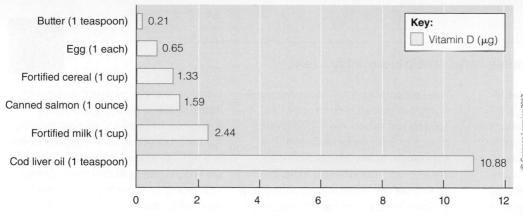

Food	Vitamin D (µg)
Butter (1 teaspoon)	0.21
Egg (1 each)	0.65
Fortified cereal (1 cup)	1.33
Canned salmon (1 ounce)	1.59
Fortified milk (1 cup)	2.44
Cod liver oil (1 teaspoon)	10.88

Key: Vitamin D (µg)

© Cengage Learning 2013

FIGURE 5.6 Food sources of vitamin D.

Vitamin D Toxicity   Vitamin D synthesis signaled from sunlight exposure is not believed to cause toxicity because of feedback mechanisms. Toxicity, usually from excess supplementation, is called hypervitaminosis D and is characterized by excess serum $25(OH)D_3$. Hypervitaminosis D poses the risk of **hypercalcemia** (high levels of calcium). Hypercalcemia is associated with calcium deposits in soft tissue and organs (including kidneys, causing kidney stones and reduced renal function). Signs and symptoms include nausea, vomiting, weakness, anorexia, excessive urination and thirst, and excess calcium in the urine.

Vitamin D Food Sources   In general, vitamin D is not abundantly present in the food supply. Some sources of vitamin D are cod liver oil, butter, egg yolk, fatty fish, and fortified products like milk, margarine, and some cereals (see Figure 5.6).

Vitamin E[G,I,5,30,32,49,53,63]   (See Table 5.2 on page 237 for a summary and Appendix A for vitamin E chemistry.)

Vitamin E Forms   Vitamin E refers to a family of compounds called **tocopherols** and **tocotrienols** that are found naturally in foods. The common biological form is **alpha-tocopherol**, which is added to foods and supplements. Other members of the vitamin E family include beta, gamma, and delta tocopherol and alpha, beta, gamma, and delta tocotrienol, which are found naturally in foods. The activity of these members can be compared to and expressed as Tocopherol Equivalents (TE).

Vitamin E Functions   Vitamin E's major function is that of an antioxidant (see Figure 5.7 and BioBeat 5.4). It protects cells, especially red blood cells, against the effects of electron-deficient chemicals such as oxygen radicals and other radical species, which are potentially damaging byproducts of the body's metabolism. It is fat-soluble, so it protects unsaturated lipid substances, including cell membranes, from **oxidative stress**. Vitamin E also functions at the molecular level, affecting cell proliferation and differentiation. It may also reduce blood clotting and improve vasodilation (the dilation of blood vessels).

Vitamin E's DRI and UL   The DRI level is established for vitamin E based on its antioxidant effect. The DRI for vitamin E, expressed in milligrams of alpha-tocopherol, is 15 milligrams per day for an adult male or an adult female 19 to 30 years old. The UL for an adult male or an adult female 19 to 70 years old is 1,000 milligrams per day. Refer to Appendix A for all life span and gender-specific values.

Some Specifics on the Units of Vitamin E for the RDI   On food labels, vitamin E is given in IU. However, the percent of RDIs can also be converted to milligrams of vitamin E (alpha-tocopherol). One milligram of d-alpha-tocopherol from food, fortified foods, and supplements is equal to 0.67 IU. This conversion allows you to understand how to convert food package label units into those of the DRI.

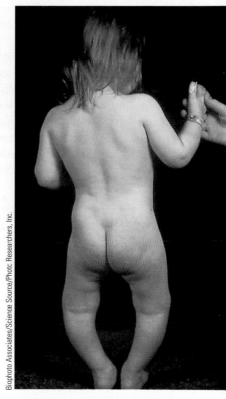

Biophoto Associates/Science Source/Photo Researchers, Inc.

A child with vitamin D deficiency rickets.

hypercalcemia A state of the blood that is characterized by high blood levels of calcium.

tocopherols A class of chemical compounds, along with tocotrienols, that make up the vitamin E family.

tocotrienols A class of chemical compounds, along with tocopherols, that make up the vitamin E family.

alpha-tocopherol A form of vitamin E most commonly used in dietary supplements and food fortification.

oxidative stress A measurable shift to a more electron-deficient state in a biological system that results in injury.

BioBeat 5.4

Oxidative Stress, Disease, and Antioxidant Nutrients

Oxidative stress is a measurable shift in the chemical state in a biological system where more electron-deficient, or oxidized, chemicals exist. This chemical state increases the potential for biological injury. Biological injury occurs when the viability or essential function of a cell, tissue, or organism is reduced. Electron-deficient chemicals cause injury. Many diseases have been attributed to oxidative stress. Some examples of these diseases include heart disease, cancer, Parkinson's disease, asthma, and cataracts. An antioxidant can donate an **electron** to an electron-deficient chemical without becoming reactive itself, and thus prevent damage from oxidation. An antioxidant is any substance that, when present at low concentrations, compared with those of an **oxidizable substrate**, significantly delays or inhibits oxidation of that substance. Many phytochemicals are electron donors and have shown heart-healthy effects, such as the allyl sulfides from onions and garlic, and anticancer effects from the indoles, isothiocynates, and sulforaphane from cruciferous (cabbage-family) vegetables. Vitamin C is a potent antioxidant in the water-soluble spaces, whereas beta-carotene and vitamin E are effective antioxidants in fat-soluble spaces in the body.

Vitamin E in various forms, through high levels of supplementation, has been shown in some studies to prevent the oxidation of LDL cholesterol, thereby protecting against heart disease, slowing the aging response, improving aerobic training for athletes (especially athletes training at high altitudes), reducing cancer risk, slowing Alzheimer's disease, and possibly reducing pain associated with **fibrocystic breast disease**.

You can eat an antioxidant-rich diet by eating a plant-based diet and choosing fruits like citrus fruits, tropical fruits, melons, berries, kiwis, and apricots, and choosing vegetables from the cabbage family, dark leafy greens, carrots, sweet potatoes, winter squashes, and appropriate amounts of nuts, seeds, and plant oils.[1,19,63,67]

Do you eat an antioxidant-rich diet?

FIGURE 5.7 The role of vitamin E as a membrane antioxidant and electron donor, which reduces damage to the cell membrane lipid bilayer.

Lipid bilayer

Vitamin E

Free radicals

© Cengage Learning 2013

electron A negatively-charged particle in a molecular structure.

oxidizable substrate Any chemical compound capable of losing an electron.

fibrocystic breast disease The presence of lumpy cysts in the breast that may be treated with vitamin E.

hemolytic anemia A type of anemia resulting from red blood cell destruction, which can be caused by insufficient vitamin E levels in prematurely born infants or by vitamin E toxicity.

peripheral neuropathy Decreased sensation of the nerves that go to the limbs.

hemolysis When red blood cells break.

Vitamin E Deficiency   An adult human deficiency of vitamin E is extremely rare and not well characterized; however, in premature infants, **hemolytic anemia** results. Vitamin E deficiency is better characterized in individuals with a rare genetic abnormality. In this case, **peripheral neuropathy**, skeletal weakening, and red blood cell **hemolysis** occur. Vitamin E deficiency is well characterized in animal models, but extrapolations to humans are not clear.

Vitamin E Toxicity   A variety of consequences have been delineated regarding toxicities of vitamin E in humans, but only when taken in very high levels as a dietary

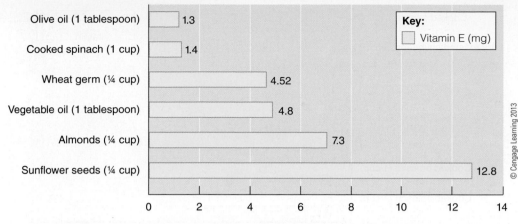

Olive oil (1 tablespoon) — 1.3
Cooked spinach (1 cup) — 1.4
Wheat germ (¼ cup) — 4.52
Vegetable oil (1 tablespoon) — 4.8
Almonds (¼ cup) — 7.3
Sunflower seeds (¼ cup) — 12.8

Key: Vitamin E (mg)

0 2 4 6 8 10 12 14

© Cengage Learning 2013

FIGURE 5.8 Food sources of vitamin E.

supplement. High intakes of vitamin E (levels above the UL) can potentially interfere with vitamin K's role in blood clotting and act as a pro-oxidant, increasing red blood cell hemolysis.

Vitamin E Food Sources The main dietary sources of vitamin E are vegetable oils, nuts, seeds, and green leafy vegetables. Other sources of vitamin E include fortified cereals, wheat germ, and whole grains (see Figure 5.8).

Vitamin K[G,I,10,33,35,36] (See Table 5.2 on page 237 for a summary and Appendix A for vitamin K chemistry.)

Vitamin K Forms The biologically active forms of vitamin K are in families called **phylloquinones** (vitamin K_1) and **naphthoquinones** (vitamin K_2, menaquinone, is an active member of this family). A variety of quinones occur naturally in foods and can be used to function as vitamin K in the body.

Vitamin K Functions Vitamin K functions in blood clotting (see Figure 5.9); thus, it prevents **hemorrhage**. It is also involved in bone metabolism.

Vitamin K's DRI, UL, and RDI The units of the DRI are in micrograms. The DRI for an adult male 19 to 30 years old is 120 micrograms per day; for an adult female 19 to 30 years old, it is 90 micrograms per day. The UL for an adult male or an adult female 19 to 70 years old is not determined. The RDI is 80 micrograms of vitamin K per day. Refer to Appendix A for all life span and gender-specific values.

Vitamin K Deficiency Though natural deficiency is rare, lack of vitamin K increases blood-clotting time and causes hemorrhaging if the person is cut or injured. Vitamin K deficiency occurs naturally in newborns and from certain medical measures. Newborns have limited gastrointestinal (GI) tract bacteria, which synthesize vitamin K. Thus, vitamin K shots are necessary for newborns. Medical treatments that destroy the GI tract bacteria that synthesize vitamin K include

phylloquinones A family of vitamin K chemical compounds known as vitamin K_1, which are in dark green and cabbage-family vegetables.

naphthoquinones A family of vitamin K chemical compounds known as vitamin K_2, which are made by the bacteria in the gut but are present in small amounts in dairy and meat products.

hemorrhage Uncontrolled, excessive bleeding that can also occur from a reduced ability for the blood to clot.

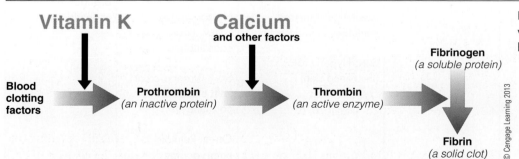

FIGURE 5.9 The role of vitamin K and calcium in blood clotting.

Vitamin K → Blood clotting factors → **Prothrombin** (an inactive protein) → **Calcium** and other factors → **Thrombin** (an active enzyme) → **Fibrinogen** (a soluble protein) → **Fibrin** (a solid clot)

© Cengage Learning 2013

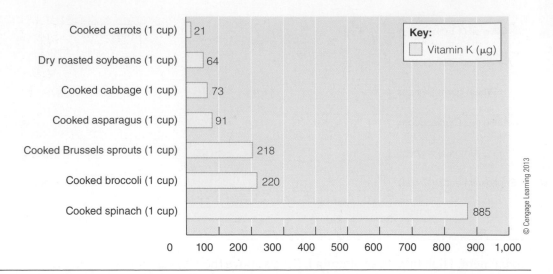

FIGURE 5.10 Food sources of vitamin K.

antibiotics, chemotherapy, and abdominal radiation. Cardiac patients on anticlotting drugs like Coumadin have an increased risk for bleeding and hemorrhaging because these drugs are vitamin K **antagonists**.

Vitamin K Toxicity Toxicity of vitamin K in adults is poorly delineated. In infants, high levels of vitamin K supplementation (with menadione, a synthetic version of vitamin K) induces jaundice and liver damage.

Vitamin K Food Sources Bacteria in the GI tract make 300 to 500 micrograms of vitamin K in the form of menaquinone, which is then absorbed in the jejunum and the ileum of the small intestine. Menaquione is found in smaller amounts in animal meats and dairy products. Good food sources of vitamin K as phylloquione are dark green leafy vegetables, cruciferous (cabbage-family) vegetables (cabbage, bok choy, collards, broccoli, Brussels sprouts, kohlrabi, kale, mustard greens, turnip greens, and cauliflower), other vegetables, soybeans, and some plant oils (see Figure 5.10).

antagonists A chemical substance that counteracts the effects of another chemical substance.

TABLE 5.2

The Fat-Soluble Vitamins: Functions, Deficiencies, Toxicities, and Food Sources

Nutrient and Other Names	Functions in the Human Body	Deficiency Name, Signs and Symptoms	Toxicity Signs and Symptoms	Sources (Food and Other)
Vitamin A PREVITAMIN A: retinol retinal retinoic acid 1 μg of retinol = 1 RAE PROVITAMIN A: carotenoids (beta-carotene) 12 μg beta-carotene = 1 μg of retinol or 1 RAE	Vision, gene expression, reproduction, embryonic development, epithelial cell maintenance and growth, and immune function	*Hypovitaminosis A* BONE & TOOTH: Impaired growth CENTRAL NERVOUS SYSTEM: Night blindness, complete blindness (xerophthalmia) GI SYSTEM: Diarrhea IMMUNITY: Depressed immunity, more infections SKIN: Hyperkeratosis (thickened skin)	BONE & TOOTH: Decreased bone mineral density CENTRAL NERVOUS SYSTEM: Headache, vertigo GI SYSTEM: Nausea and vomiting, liver abnormalities NEUROMUSCULAR SYSTEM: Poor coordination SKIN: Orange color with excess beta-carotene OTHER: Retinoid embryopathy	RETINOL: Fortified milk, cheese, butter, margarine, eggs, liver BETA-CAROTENE: Dark green leafy vegetables, broccoli, deep orange fruits and vegetables

Nutrient and Other Names	Functions in the Human Body	Deficiency Name, Signs and Symptoms	Toxicity Signs and Symptoms	Sources (Food and Other)
Vitamin D PROVITAMIN D: cholecalciferol (vitamin D$_3$) ACTIVE VITAMIN D: 1,25-dihydroxyvitamin D$_3$, also called 1,25-dihydroxycholecalciferol (calcitriol) 1 μg cholecalciferol = 40 IU	*Hormone-like* Mineralization of bone (raises blood calcium levels by increasing absorption of calcium by the small intestine and increasing calcium retention by the kidneys; reduces blood phosphorus levels by increasing kidney excretion), cell metabolism (antiproliferation and prodifferentiation)	*Rickets (children) Osteomalacia (adults)* BONE & TOOTH: Poor growth, bowed legs, soft bones, pigeon chest, knocked knees, and malformed teeth in children; porous bones in adults CARDIOVASCULAR SYSTEM: Increased circulating levels of PTH and AlkP, decreased circulating levels of serum phosphorus GI SYSTEM: Decreased calcium absorption	*Unknown from self-synthesis Hypervitaminosis D (characterized by high levels of 25(OH)D$_3$ from supplementation)* CARDIOVASCULAR SYSTEM: High blood calcium CENTRAL NERVOUS SYSTEM: Weakness GI SYSTEM: Nausea, vomiting, anorexia OTHER: Kidney stones, increased thirst, increased urination, increased urinary calcium	Fortified milk, margarine, and some cereals, eggs, fatty fish Self-synthesis with unprotected sunlight exposure
Vitamin E TOCOPHEROLS TOCOTRIENOLS Tocopherol equivalents (TEs) are based on alpha-tocopherol 1 mg of d-alpha-tocopherol from food, fortified foods, and supplements = 0.67 IU	*Functions at the molecular level* Antioxidant, protector of unsaturated lipids, cell membrane stabilizer	*No disease name, rare, poorly characterized in adults, common in premature infants and a rare genetic condition* CARDIOVASCULAR SYSTEM: Hemolytic anemia in infants CENTRAL NERVOUS SYSTEM: Neuropathy in the genetic disorder NEUROMUSCULAR SYSTEM: Skeletal weakening in the genetic disorder	*Only described when taken in very high levels as supplements* CARDIOVASCULAR SYSTEM: Interferes with vitamin K's role in blood clotting, augmentation of anti-blood-clotting medication, increased hemolysis	Nuts, seeds, plant oils, wheat germ, fortified cereals, green leafy vegetables, other vegetables
Vitamin K VITAMIN K$_1$: phylloquinones VITAMIN K$_2$: naphthaquinones	Blood clotting, bone metabolism	*No disease name, rare except with newborns and following certain medical measures* CARDIOVASCULAR SYSTEM: Increased clotting time, hemorrhaging with a cut or injury	*Poorly described in adults* GI SYSTEM: Jaundice and liver damage in infants (caused by high levels of vitamin K from supplemented menadione) CARDIOVASCULAR SYSTEM: Interferes with anti-blood-clotting medication	Bacterial synthesis in the GI tract, green leafy and cruciferous vegetables, soybeans, some plant oils

© Cengage Learning 2013

Summary Points

- Diagnosing and confirming a nutrient deficiency or toxicity requires analysis of the diet, clinical evaluation, biochemical analysis, and evaluation of the response to corrected intake levels.
- Toxicities and deficiencies take longer to develop for fat-soluble vitamins compared to water-soluble vitamins.
- The fat-soluble vitamins are grouped by their solubility in oil.
- The fat-soluble vitamins have specific chemical forms and functions in the body.
- DRIs exist for essential fat-soluble vitamins: A, D, E, and K.
- Deficiency and toxicity signs and symptoms are unique for each nutrient, as are food sources.

Take Ten on Your Knowledge Know-How

1. How would you determine if you are at risk for nutritional deficiency or toxicity?
2. Why is it important to eat a varied diet?
3. What are the nutrient forms, adult DRI, functions, food sources, and deficiency and toxicity signs and symptoms for vitamin A?
4. What are the nutrient forms, adult DRI, functions, food sources, and deficiency and toxicity signs and symptoms for vitamin D? What is the role of vitamin D in genetics, evolution, and society?
5. What are the nutrient forms, adult DRI, functions, food sources, and deficiency and toxicity signs and symptoms for vitamin E?
6. What are the nutrient forms, adult DRI, functions, food sources, and deficiency and toxicity signs and symptoms for vitamin K?
7. What is an antioxidant? Which nutrients function as antioxidants? How do they promote health? What are good food sources of antioxidant nutrients?
8. Which fat-soluble vitamins can be synthesized in the body? What is the precursor or substance required for their synthesis?
9. Which fat-soluble vitamin is produced by bacteria in the GI tract?
10. What are the differences between fat-soluble vitamins and water-soluble vitamins regarding absorption, deficiencies, and toxicities?

(5.2) WATER-SOLUBLE VITAMINS

Introduction The water-soluble vitamins include thiamin (B_1), riboflavin (B_2), niacin (B_3), vitamin B_6, vitamin B_{12}, folate, vitamin C, pantothenic acid, biotin, and choline (see Table 5.3 for a summary and the chemical structures in Appendix A). Many water-soluble vitamins function as cofactors in metabolism (see BioBeat 5.5).

Thiamin: Vitamin B_1[D,G,I,24,31,44,53,75,76,84] (See Table 5.3 on page 249 for a summary and Appendix A for thiamin chemistry.)

B_1 Forms First known as vitamin B_1 and now as thiamin, this vitamin has a single chemical structure technically known as aneurin.

B_1 Functions B_1 functions as the active group of the coenzyme thiamin pyrophosphate (TPP) and is an enzyme cofactor in energy metabolism (see Figure 5.11). It is specifically used in the metabolism of carbohydrates and branch-chain amino acids (BCAA). It also supports normal appetite and nervous system function.

BioBeat 5.5

Energy Production and So Many Water-Soluble Vitamins

Energy production requires a concert of biochemical reactions tightly regulated by many different enzymes and cofactors (see Figure 5.11). There are three interacting biochemical pathways that produce the majority of adenosine triphosphate (ATP) aerobically. The three pathways include glycolysis, the citric acid cycle, and the electron transport chain. All of the water-soluble vitamins except for vitamin C and choline are utilized in some fashion in enzymes as cofactors (coenzymes) in the regulation of energy production from glucose, fatty acid, and amino acid substrates. It is interesting that the majority of **substrate** used for ATP generation comes from fatty acids and

glucose. Furthermore, as the intensity of exercise increases, more glucose is used in order to generate a faster rate of ATP production. It is only under severe stress that the body utilizes amino acids for ATP production. In the late stages of endurance exercise, only about 15 percent of all ATP produced comes from amino acids. By eating a balanced diet according to MyPlate, one can ensure that the substrates and the regulators of metabolism will be provided at adequate levels.[75]

Does your diet follow your personal MyPlate food intake plan to provide the appropriate amount of nutrient enzyme cofactors to support energy production?

substrate A chemical compound that is acted on by enzymes that support life.

FIGURE 5.11 Vitamins as enzyme cofactors in energy metabolism.

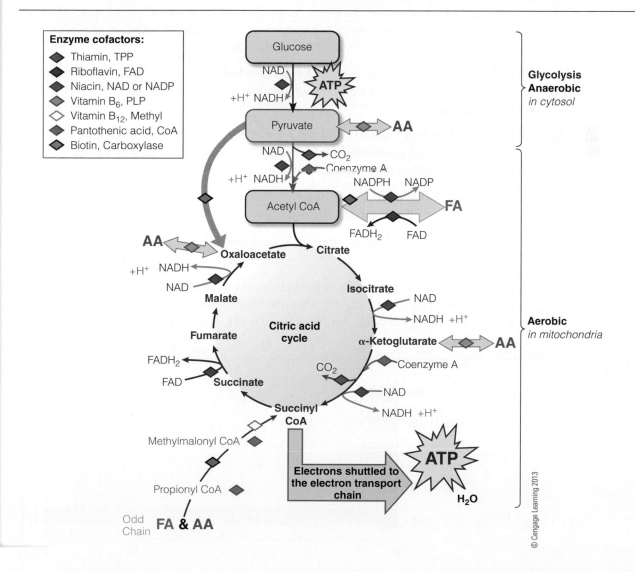

© Cengage Learning 2013

B₁'s DRI, UL, and RDI The DRI for thiamin (B_1), expressed in milligrams, is based on 0.5 milligrams of B_1 per 1,000 Calories. The DRI for an adult male 19 to 30 years old is 1.2 milligrams per day; for an adult female 19 to 30 years old, it is 1.1 milligrams per day. The UL for an adult male or an adult female 19 to 70 years old is not determined. The RDI is 1.5 milligrams per day. Refer to Appendix A for all life span and gender-specific values.

B₁ Deficiency General symptoms of B_1 deficiency include anorexia, weight loss, mental changes, muscle weakness, and cardiovascular effects. A deficiency of B_1 results in a disease called **beriberi**, which presents itself in two forms: dry beriberi and wet beriberi.

Dry beriberi is characterized by nausea, exhaustion, and tingling hands and feet. Continuous deprivation of the nutrient progresses to numb hands and feet and eventually to loss of limb control (peripheral paralysis). Also, the deficient person experiences heart rhythm changes.

Wet beriberi is common in alcoholics who are consuming very little food and need increased levels of B_1 to metabolize alcohol, a carbohydrate-related structure. This form of beriberi is characterized by edema, decreased ability to coordinate motor movement, peripheral paralysis, confusion, and heart rhythm changes consistent with congestive heart failure. This syndrome is termed **Wernicke-Korsakoff syndrome**. The Wernicke's syndrome (encephalopathy including disturbed gait, strange eye movement, and confusion) is immediately responsive to thiamin administration; however, regarding the Korsakoff psychosis (amnesia and limitations in short-term memory processes), the role of thiamin's effects on correcting this part of the condition is unclear.

B₁ Toxicity B_1 toxicity has not been reported in humans from food or supplements because there is a rapid decline in absorption and increase in excretion with excess intake.

B₁ Food Sources Good food sources of B_1 are pork, whole grains, enriched bread, and cereal products (see Figure 5.12). Under the Enrichment Act of 1942, vitamin B_1 is added to refined grain products.

Riboflavin: Vitamin B₂[D,G,I,24,31,44,53,75,76] (See Table 5.3 on page 249 for a summary and Appendix A for riboflavin chemistry.)

B₂ Forms First known as vitamin B_2 and now as riboflavin, this vitamin has a single chemical structure.

B₂ Functions B_2 functions as the active group of two coenzymes, **flavin mononucleotide (FMN)** and **flavin adenine dinucleotide (FAD)**, and is an enzyme cofactor in energy production (see Figure 5.11). It is specifically used in fatty acid metabolism.

---

beriberi The condition caused by a deficiency of the water-soluble vitamin thiamin; can present in a dry or wet form.

Wernicke-Korsakoff syndrome The name of the syndrome associated with a thiamin deficiency resulting in wet beriberi, caused by excessive alcohol intake, manifesting with encephalopathy and psychosis.

flavin mononucleotide (FMN) A coenzyme made from riboflavin that carries one electron.

flavin adenine dinucleotide (FAD) A coenzyme made from riboflavin that carries two electrons.

FIGURE 5.12 Food sources of thiamin.

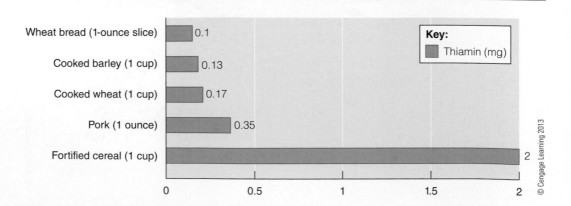

Key:
■ Thiamin (mg)

Food	Thiamin (mg)
Wheat bread (1-ounce slice)	0.1
Cooked barley (1 cup)	0.13
Cooked wheat (1 cup)	0.17
Pork (1 ounce)	0.35
Fortified cereal (1 cup)	2

© Cengage Learning 2013

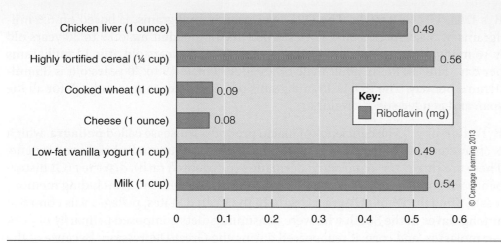

FIGURE 5.13 Food sources of riboflavin.

B$_2$'s DRI, UL, and RDI   The DRI for B$_2$, expressed in milligrams, is based on 0.6 milligrams of B$_2$ per 1,000 Calories. The DRI for an adult male 19 to 30 years old is 1.3 milligrams per day; for an adult female 19 to 30 years old, it is 1.1 milligrams per day. The UL for an adult male or an adult female 19 to 70 years old is not determined. The RDI is 1.7 milligrams per day. Refer to Appendix A for all life span and gender-specific values.

B$_2$ Deficiency   The B$_2$ deficiency syndrome is known as **ariboflavinosis**, but rarely occurs as a single deficiency. Deficiency halts growth and causes sore and swollen throat, swollen, magenta-colored tongue, cracking at the corners of the mouth, and dandruff.

B$_2$ Toxicity   No clear symptoms of B$_2$ toxicity have been identified. It has limited absorption and is rapidly excreted in urine.

B$_2$ Food Sources   Good sources of riboflavin are milk products, organ meats, and whole and enriched grains and products such as bread, cereal, and pasta (see Figure 5.13). Under the Enrichment Act of 1942, vitamin B$_2$ is added to refined grain products.

Niacin: Vitamin B$_3$ [D,G,I,24,30,31,44,53,75,76]   (See Table 5.3 on page 249 for a summary and Appendix A for niacin chemistry.)

B$_3$ Forms   The term *niacin* refers to **nicotinic acid** and **nicotinamide** (also called niacinamide), which are active in the body's chemistry. Because the two compounds do not act the same way in the body, the units of the DRI have been adjusted for the variance in biological activity. To convert the forms of niacin derivatives present in the diet, niacin equivalents (NEs) are used. Niacin equivalents are based on the conversion of the compound to nicotinic acid. One NE equals 1 milligram of nicotinic acid.

B$_3$ Functions   Vitamin B$_3$ functions as a coenzyme used in energy metabolism. It plays its role in energy metabolism as a part of two coenzymes, **nicotinamide adenine dinucleotide (NAD)** and **nicotinamide adenine dinucleotide phosphate (NADP)**. These are involved in generating ATP from amino acids, fatty acids, and glucose. Thus, it supports energy production (see Figure 5.11 on page 239). It also supports steroid synthesis and health of the skin, nervous system, and digestive system.

B$_3$ Synthesis   The essential amino acid **tryptophan** can be converted to niacin. Sixty milligrams of tryptophan can produce one NE. Thus, if tryptophan is plentiful in the diet while niacin is limited, niacin can be produced from tryptophan and the niacin requirement can be met.

ariboflavinosis The condition caused by a deficiency of the water-soluble vitamin riboflavin.

nicotinic acid The chemical form of niacin that is used to reduce total blood cholesterol but causes flushing if an aspirin has not been taken before the supplementing of this water-soluble vitamin.

nicotinamide The chemical form of niacin that is fortified in foods and used for most dietary supplements; also known as niacinamide.

nicotinamide adenine dinucleotide (NAD) A coenzyme made from niacin.

nicotinamide adenine dinucleotide phosphate (NADP) A phosphorylated coenzyme made from niacin.

tryptophan An essential amino acid that can be used to synthesize niacin in the body.

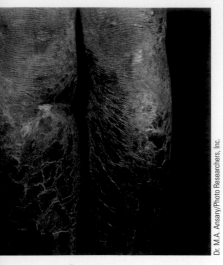

Pellagra from niacin deficiency causes severe dermatitis.

pellagra The deficiency disease of the water-soluble vitamin niacin, characterized by dermatitis, diarrhea, and dementia.

pyridoxine (PN) The name for the coenzyme made from vitamin B_6 that carries a pyrimidine and is found in plant foods.

pyridoxal (PL) The name for the coenzyme made from vitamin B_6 that carries the aldehyde group and is found in animal foods.

pyridoxamine (PM) The name for the coenzyme made from vitamin B_6 that carries an amine group and is found in animal foods.

B_3's DRI, UL, and RDI The DRI, expressed in milligrams, is based on 6.6 milligrams of niacin per 1,000 Calories. The DRI for an adult male 19 to 30 years old is 16 milligrams per day; for an adult female 19 to 30 years old, it is 14 milligrams per day. The UL for an adult male or an adult female 19 to 70 years old is 35 milligrams per day. The RDI is 20 milligrams per day. Refer to Appendix A for all life span and gender-specific values.

B_3 Deficiency The deficiency of niacin produces a disease called **pellagra**, which is characterized by depression, apathy, headache, fatigue, and a bright red tongue. There are three "Ds" of pellagra: dermatitis (pigmented rash), diarrhea (GI disturbances of vomiting, diarrhea, and constipation), and dementia (including memory loss). A fourth "D" of pellagra is death. In the United States, pellagra was common among slaves in the South who were consuming diets composed primarily of pork fat, molasses, and corn; it reappeared during the Great Depression. Because of the American history of niacin deficiency, it was one of the original nutrients included in the Enrichment Act of 1942.

B_3 Toxicity Naturally occurring niacin in food is not associated with causing adverse effects. However, nicotinic acid supplementation induces a violent flushing reaction (the first symptom of excess nicotinic acid) because of the intense blood vessel dilation that is produced when megadoses are consumed. Also, nausea, vomiting, liver toxicity, blurred vision, and impaired glucose tolerance can occur. High doses of nicotinic acid have been used in the treatment of hypercholesterolemia. An increased tendency to utilize muscle glycogen and decreased ability to mobilize fat from adipose tissue may facilitate the effect of increased amounts of HDL cholesterol and decreased amounts of LDL cholesterol in the blood. No more than 500 milligrams of nicotinic acid should be taken per day, and liver function tests should be run periodically because of the liver injury that may occur with chronic, high-dose intake of niacin. Nicotinamide is not associated with causing the flushing reaction, altering LDL cholesterol levels, or causing liver injury.

B_3 Food Sources Good sources of niacin are meats, fish, poultry, whole and enriched grains, and products such as bread, cereal, and pasta (see Figure 5.14). Under the Enrichment Act of 1942, vitamin B_3 is added to refined grain products. Dairy products and legumes are good sources of tryptophan.

Vitamin B_6[D,G,I,24,30,31,44,53,75,76] (See Table 5.3 on page 250 for a summary and Appendix A for vitamin B_6 chemistry.)

B_6 Forms There are six forms of vitamin B_6 that have biological activity in the body: **pyridoxine (PN)**, **pyridoxal (PL)**, and **pyridoxamine (PM)**, and their

FIGURE 5.14 Food sources of niacin.

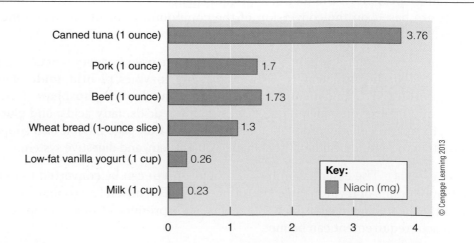

respective 5'phosphates, **pyridoxine phosphate (PNP)**, **pyridoxal phosphate (PLP)**, and **pyridoxamine phosphate (PMP)**. The food source, whether from plant or animal, determines the form of vitamin B_6 that is being consumed. PLP is the major metabolically active coenzyme form in the body.

B_6 Functions   The various B_6 forms function as coenzymes in the metabolism of amino acids, glycogen, and some lipids (see Figure 5.11 on page 239). B_6 is needed for **heme** synthesis and the interconversion of amino acids to carbohydrate intermediates and lipids. That is, B_6 is involved in the removal of nitrogen molecules off amino acids (**deamination**, or **transamination** if the amine group is attached to a carbohydrate intermediate to build a nonessential amino acid). Thus, B_6 is involved in many aspects of protein metabolism. B_6 also helps to convert tryptophan to niacin.

B_6's DRI, UL, and RDI   The DRI for B_6, expressed in milligrams, is based on 0.016 milligrams of B_6 per gram of protein consumed in the diet. The DRI level for vitamin B_6 is based on the typical protein intake of Americans, as opposed to the DRI for protein, because Americans typically eat more protein than the body needs. The DRI for B_6 for an adult male or an adult female 19 to 30 years old is 1.3 milligrams per day. The UL for an adult male or an adult female 19 to 70 years old is 100 milligrams per day. The RDI is 2 milligrams per day. Refer to Appendix A for all life span and gender-specific values.

B_6 Deficiency   A deficiency of vitamin B_6 rarely occurs alone. Usually a B_6 deficiency is seen with the lack of several water-soluble vitamins. The deficiency symptoms of vitamin B_6 are anemia (small cell type or microcytic caused by decreased heme synthesis), confusion, depression, epileptic convulsions (altered brain wave patterns), and seborrheic (greasy) dermatitis. People taking certain medications require more B_6.

B_6 Toxicity   Toxicity of B_6 in huge doses (2,000 milligrams per day) leads to sensory neuropathy, a type of nerve damage, and skin lesions. As a result, the toxic person may experience tingly to numb hands and feet (which is reversible to a certain extent when the toxicity is corrected).

B_6 Food Sources   PLP and PMP are found in animal foods, whereas PN and PNP are found in plant foods. Good sources of B_6 include all meats, fish, poultry, fortified cereals, legumes, nonstarchy vegetables, and noncitrus fruits (see Figure 5.15).

pyridoxine phosphate (PNP) The name for the coenzyme made from vitamin B_6 that is phosphorylated and carries a pyrimidine, and is found in plant foods.

pyridoxal phosphate (PLP) The name for the most important form of the coenzyme made from vitamin B_6 that is phosphorylated and carries the aldehyde group, and is found in animal foods.

pyridoxamine phosphate (PMP) The name for the coenzyme made from vitamin B_6 that is phosphorylated and carries an amine group, and is found in animal foods.

heme A portion of hemoglobin and myoglobin proteins; a portoporphrine, which is a chemical ring structure that holds iron.

deamination The removal of a nitrogen group from an amino acid by a vitamin B_6 coenzyme.

transamination The transfer of a nitrogen group from an amino acid by a vitamin B_6 coenzyme.

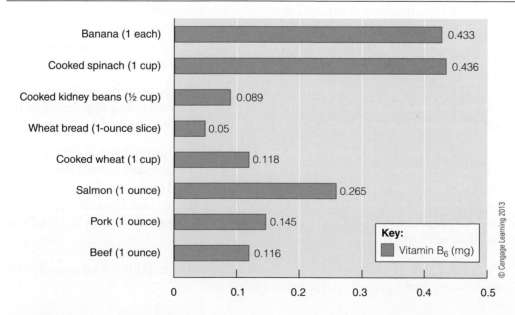

FIGURE 5.15 Food sources of vitamin B_6.

© Cengage Learning 2013

Vitamin B$_{12}$[D,G,I,24,31,37,44,53,59,60,75,76] (See Table 5.3 on page 250 for a summary and Appendix A for vitamin B$_{12}$ chemistry.)

B$_{12}$ Forms Vitamin B$_{12}$ is known as cyanocobalamin (**cobalamin**), a single chemical structure. The centerpiece of the molecule is the trace mineral cobalt.

B$_{12}$ Functions B$_{12}$ is involved in nucleic acid synthesis (used to make DNA and RNA), red blood cell synthesis, the synthesis of other new cells, and nerve cell maintenance. B$_{12}$ functions as a cofactor in energy production (especially from fatty acids and amino acids; see Figure 5.11 on page 239) and is critical for the conversion of homocysteine to methionine, and thus prevents hyperhomocystemia. B$_{12}$ synergizes with the vitamin **folic acid**.

B$_{12}$ Absorption and Storage B$_{12}$ needs the protein **intrinsic factor** and gastric acid in order to be bound and absorbed. Intrinsic factor is produced by the cells in the stomach lining; thus, if the stomach is damaged in any way, B$_{12}$ deficiency can occur, because in order for B$_{12}$ to be absorbed, the vitamin must be bound to intrinsic factor. If gastric acid production is decreased, such as in the elderly, then B$_{12}$ is poorly absorbed. Also, the complex of B$_{12}$ and intrinsic factor is absorbed in the last part of the small intestine called the ileum. If one has a damaged stomach or ileum, then intramuscular injections of B$_{12}$ must be taken every couple of months. Nasal sprays are also emerging as a form of B$_{12}$ supplementation. B$_{12}$ should be supplemented in the diets of pure vegans because of the complete exclusion of animal products in their diet. Unlike any other B vitamin, B$_{12}$ is stored in the liver; thus, it takes a long time for a B$_{12}$ deficiency to develop in humans, possibly three to five years without an intake.

B$_{12}$'s DRI, UL, and RDI The DRI for vitamin B$_{12}$, expressed in micrograms, is established based on stores and daily needs for the nutrient. The DRI for an adult male or an adult female 19 to 30 years old is 2.4 micrograms per day. The dietary need for cobalt is met by meeting the DRI for vitamin B$_{12}$. The UL for an adult male or an adult female 19 to 70 years old is not determined. The RDI is 6 micrograms per day. Refer to Appendix A for all life span and gender-specific values.

B$_{12}$ Deficiency Deficiency signs are **macrocytic normochromic anemia**, compounded with nerve damage that can progress to paralysis. The combination of these signs in vitamin B$_{12}$ deficiency is termed **pernicious anemia**. Other signs and symptoms include cognitive disturbances, such as loss of concentration, memory loss, disorientation, and dementia. Furthermore, vision and gait disturbances, tingling and numbness in the hands and feet (sensory neuropathy), insomnia, impotency, impaired bowel (including increased gas production and constipation), reduced bladder control, sore tongue, and loss of appetite can occur.

B$_{12}$ Toxicity Toxicity symptoms are poorly delineated, and documented cases are few; however, caution should still be taken when consuming high levels of B$_{12}$ as a supplement, even though absorption may decrease as intake levels increase.

B$_{12}$ Food Sources Vitamin B$_{12}$ is naturally found in animal foods. Thus, good sources of vitamin B$_{12}$ include organ and muscle meat from any animal (beef, poultry, fish, etc.), dairy products, and eggs (see Figure 5.16). In addition, very small amounts of B$_{12}$ may be provided by bacteria, yeast, fungi, and algae. If a processed food package label from a plant food shows B$_{12}$ as being present, then the food was fortified with this nutrient.

Folate (Folic Acid)[D,G,I,18,21,24,30,31,37,41,44,53,68,72,74,75,76] (See Table 5.3 on page 250 for a summary and Appendix A for folate chemistry.)

---

cobalamin The chemical form of vitamin B$_{12}$, which contains cobalt.

folic acid The form of a water-soluble vitamin (pteroylmonoglutamic acid) that synergizes with vitamin B$_{12}$ and has been associated with reducing the risk of spina bifida.

intrinsic factor A vitamin B$_{12}$ binding protein that is needed for absorption.

macrocytic normochromic anemia An anemia that can be used to describe the red blood cells as large and normal color.

pernicious anemia The deficiency of vitamin B$_{12}$, characterized by macrocytic normochromic anemia and central nervous system impairment.

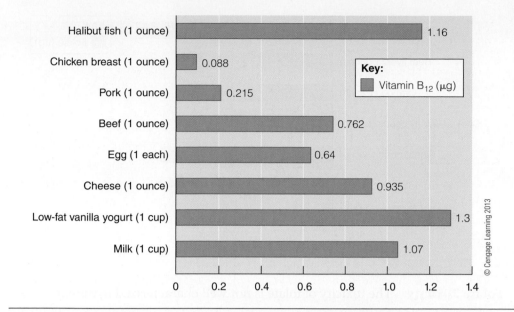

Halibut fish (1 ounce) — 1.16
Chicken breast (1 ounce) — 0.088
Pork (1 ounce) — 0.215
Beef (1 ounce) — 0.762
Egg (1 each) — 0.64
Cheese (1 ounce) — 0.935
Low-fat vanilla yogurt (1 cup) — 1.3
Milk (1 cup) — 1.07

Key:
Vitamin B$_{12}$ (μg)

© Cengage Learning 2013

FIGURE 5.16 Food sources of vitamin B$_{12}$.

Folate Forms   Folate is a generic term referring to plant food folate (pteroylpolyglutamates, a precursor molecule to folic acid in the body) and active folic acid (pteroylmonoglutamic acid, formed in the body from the precursor or added in active form in supplements and fortified foods). Folic acid is the most absorbable (bioavailable) form. In the past, folate has also been known as vitamin B$_9$, factor U, factor R, vitamin M, and vitamin B-C. Some people cannot adequately convert plant food folate to the active form of folic acid in the body. When folic acid is added to foods or provided in dietary supplements, it is in the active form.

Folate Functions   Folate functions in the formation of nucleic acids (DNA and RNA), which play a role in cell division; thus, deficiency signs can be noted in cells that are rapidly turning over, such as red blood cells. Folate functions as a cofactor in amino acid metabolism and is critical for the conversion of homocysteine to methionine, thus preventing hyperhomocystemia. Folate synergizes with vitamin B$_{12}$.

Folate's DRI, UL, and RDI   The RDI for folate, expressed in micrograms, is based on Dietary Folate Equivalents. One microgram of precursor food folate equals 0.6 micrograms of folic acid (from supplements taken with food or from fortified foods) or 0.5 micrograms of folic acid (from a supplement taken on an empty stomach). The DRI level is established based on stores, daily needs for the nutrient, and catabolism of homocysteine. The DRI for an adult male or an adult female 19 to 30 years old is 400 micrograms per day. The DRI is 600 micrograms during pregnancy. The UL for an adult male or an adult female 19 to 70 years old is 1,000 micrograms per day. This level is set to prevent the masking of the neuropathy caused by vitamin B$_{12}$ deficiency. The RDI is 400 micrograms per day. Refer to Appendix A for all life span and gender-specific values.

Folate Deficiency   The folate deficiency sign is a macrocytic (large cell) normochromic anemia, thus the symptoms are those of an anemia: weakness, fatigue, decreased concentration, irritability, headache, heart palpitations, shortness of breath, and red tongue. Deficiency can occur through inadequate dietary intake and also from skin exposure to UV sunlight, which destroys folic acid. Folate is required for fetal neural tube development. This is another example of a nutrient tied to genetics and evolution. Early humans living in Africa are believed to have developed dark skin from greater melanin pigmentation to protect against the destruction of folic acid and thus preserve life. During embryonic development, maternal folate deficiency can cause **spina bifida**, a type of severe, life-threatening **neural tube defect**.

spina bifida A neural tube defect present at birth that can be caused by poor preconceptual folic acid status.

neural tube defect The malformations of the brain and/or spinal cord occurring during embryonic development that can be caused by insufficient folate status.

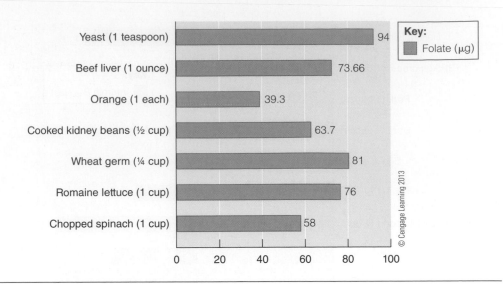

FIGURE 5.17 Food sources of folate.

Folate Toxicity   The toxicity of folate is not well characterized in humans.

Folate Food Sources   Good sources of folate are raw, leafy green vegetables (like spinach and turnip greens), green vegetables (like asparagus and broccoli), legumes (especially pinto beans, navy beans, and lentils), some fruit juices (like orange juice), liver, brewer's yeast, wheat germ, and some fortified cereals (see Figure 5.17).

Vitamin C[D,E,G,I,11,24,30,32,49,53,56,57,75,76]   (See Table 5.3 on page 250 for a summary and Appendix A for vitamin C chemistry.)

Vitamin C Forms   The chemical structure of **L-ascorbic acid** is the most active form of vitamin C. However, a less active form, known as dehydroascorbic acid, is present in food. Both forms are active in the body.

Vitamin C Functions   Vitamin C functions as a cofactor for the enzymes needed for the synthesis of collagen, a component of connective tissues such as skin, bones, teeth, blood vessels, epithelial tissue, tendons, and ligaments. Vitamin C is also a cofactor for enzymes needed for the synthesis of **carnitine** and **neurotransmitters**. Vitamin C is a well-known water-soluble antioxidant. It is converted to the oxidized form of vitamin C, **dehydroascorbic acid**, after donating electrons (see BioBeat 5.4 on page 234).

Vitamin C's DRI, UL, and RDI   The DRI for vitamin C, expressed in milligrams, is established at a level that maintains a body pool of vitamin C (ascorbic acid) that is about one-half of the maximum and can prevent deficiency symptoms from occurring for about a six-week period with no dietary source. Vitamin C is the most widely supplemented vitamin in the United States. It is commonly used in megadoses or levels that exceed the DRI by 10 times. When it is used at these levels, it far exceeds the nutritional application, but rather exerts a pharmacological effect with no clear benefits. The DRI for an adult male 19 to 30 years old is 90 milligrams per day; for an adult female 19 to 30 years old, it is 75 milligrams per day. It is recommended that cigarette smokers consume an additional 35 milligrams of vitamin C per day above their DRI to compensate for the vitamin C losses associated with the increased stress on the body resulting from smoking. The UL for an adult male or an adult female 19 to 70 years old is 2,000 milligrams per day. The RDI is 60 milligrams per day. Refer to Appendix A for all life span and gender-specific values.

Vitamin C Deficiency   A deficiency of vitamin C leads to a disease called **scurvy** and is characterized by defects in connective tissue. Signs and symptoms of scurvy include weakness, fatigue, follicular hyperkeratosis (or thicker skin, especially

L-ascorbic acid The chemical name for vitamin C.

carnitine A vitamin-like compound that helps long-chain fatty acids get into the mitochondria for the generation of ATP energy.

neurotransmitter An active chemical substance that allows transmission of messages in the nervous system.

dehydroascorbic acid A reduced form of vitamin C, or L-ascorbic acid, that is commonly caused by oxidation but can be restored by gluthione peroxidase.

scurvy The deficiency condition of vitamin C (or L-ascorbic acid) that has signs of swollen, red, bleeding gums, petechia, and follicular hyperkerotosis.

around the hair follicle), **petechia** (pinpoint subcutaneous capillary bleeding), swollen, red, bleeding gums, impaired wound healing, easy bruising, joint pain, coiled hairs, depression, and edema. Ten milligrams of vitamin C per day will prevent scurvy from occurring; however, it takes about 150 to 200 milligrams per day to saturate the body tissues of vitamin C.

Vitamin C Toxicity   Toxicity of vitamin C causes gastrointestinal disturbances, especially diarrhea, and is only caused by supplementing large doses above the UL. Some people exhibit an increased risk for kidney stones and iron toxicity.

Vitamin C Food Sources   Good sources of vitamin C are citrus fruits (oranges, grapefruits, lemons, and limes), berries, kiwis, melons, cabbage family vegetables (cauliflower, broccoli, cabbage, Brussels sprouts), peppers, and vitamin C–fortified foods (see Figure 5.18).

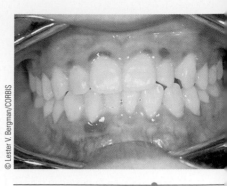

Early signs of vitamin C deficiency scurvy.

Pantothenic Acid[D,G,I,24,31,44,53,75,76]   (See Table 5.3 on page 251 for a summary and Appendix A for pantothenic acid chemistry.)

Pantothenic Acid Forms   The chemical structure of pantothenic acid is unique. This single molecular structure is used for all of its roles in metabolism.

Pantothenic Acid Functions   Pantothenic acid functions as an essential component in coenzyme A (see Figure 5.11 on page 239). Coenzyme A is involved in a wide variety of metabolic pathways for fatty acids, cholesterol, steroid hormones, vitamins A and D, neurotransmitters, amino acids, and porphyrin and corrin rings (large organic compounds that provide structure or function).

Pantothenic Acid's DRI, UL, and RDI   The DRI, expressed in milligrams, for an adult male or an adult female 19 to 30 years old is 5 milligrams per day. The UL for an adult male or an adult female 19 to 70 years old is not determined. The RDI is 10 milligrams per day. Refer to Appendix A for all life span and gender-specific values.

Pantothenic Acid Deficiency   Burning feet, fatigue, apathy, irritability, restlessness, sleep disturbances, nausea, vomiting, abdominal cramps, muscle cramps, impaired gait, hypoglycemia, and increased insulin sensitivity are characteristic of pantothenic acid deficiency.

Pantothenic Acid Toxicity   Toxicity of pantothenic acid is poorly delineated in humans. No adverse affects are associated with high intakes.

Pantothenic Acid Food Sources   Pantothenic acid is found in many foods. Major sources include high-protein animal products, yeast, legumes, and whole grains (see Figure 5.19).

petechia Pinpoint hemorrhaging (bleeding) under the skin.

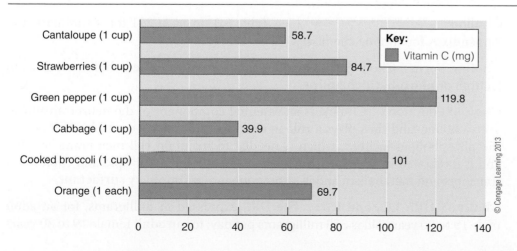

FIGURE 5.18 Food sources of vitamin C.

Food	Vitamin C (mg)
Cantaloupe (1 cup)	58.7
Strawberries (1 cup)	84.7
Green pepper (1 cup)	119.8
Cabbage (1 cup)	39.9
Cooked broccoli (1 cup)	101
Orange (1 each)	69.7

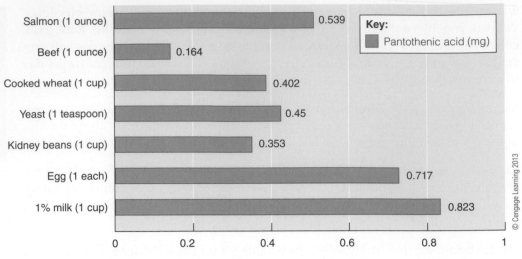

FIGURE 5.19 Food sources of pantothenic acid.

Biotin[D,G,I,24,31,44,53,75,76,86] (See Table 5.3 on page 251 for a summary and Appendix A for biotin chemistry.)

Biotin Forms Biotin exists as free biotin and protein-bound forms in food.

Biotin Functions Biotin functions as a coenzyme in carboxylation reactions involved in energy metabolism (see Figure 5.11 on page 239). It further participates in cell proliferation, gene silencing, and DNA repair.

Biotin's DRI, UL, and RDI The DRI for biotin, expressed in micrograms, is 30 micrograms per day for an adult male or an adult female 19 to 30 years old. The UL for an adult male or an adult female 19 to 70 years old is not determined. The RDI is 300 micrograms per day. Refer to Appendix A for all life span and gender-specific values.

Biotin Deficiency Dietary biotin deficiency is rare and seen only in severe undernutrition. Lethargy, depression, hallucinations, muscle pain, weakness, fatigue, scaly red rash around the eyes, nose, and mouth, conjunctivitis, and hair loss occur with biotin deficiency. Biotin is known as the anti–raw egg white factor because the protein avidin, present in raw eggs, binds to biotin and makes it unavailable for the body.

Biotin Toxicity Toxicity of biotin is poorly delineated in humans. No adverse affects are associated with high intakes.

Biotin Food Sources Biotin is widely found in food. Major food sources include animal products, yeast, legumes, nuts, whole grains, and chocolate. Small amounts of biotin are made by intestinal bacteria.

Choline[D,G,I,12,24,30,31,44,53,70,75,76,85] (See Table 5.3 on page 251 for a summary and Appendix A for choline chemistry.)

Choline Form Choline exists as free choline and esterified choline, such as in lecithin or **phosphatidylcholine**.

Choline Functions Choline is an essential component of the neurotransmitter acetylcholine, and thus plays a role in memory and muscle control. Choline is in the phospholipid lecithin, which is needed to maintain cell membrane stability and is integrated in bile acids. Furthermore, it is needed for lipid and cholesterol transport and metabolism and is a component of **pulmonary surfactant**.

Choline's DRI, UL, and RDI The DRI, expressed in milligrams, for an adult male 19 to 30 years old is 550 milligrams per day; for an adult female 19 to 30 years

phosphatidylcholine The chemical term for the phospholipid lecithin, which is made out of choline.

pulmonary surfactant A chemical substance produced by lung cells that is critical for the structure and function of the very small lung sacs.

old, it is 425 milligrams per day. The UL for an adult male or an adult female 19 to 70 years old is 3,500 milligrams per day. There is no RDI for choline. Refer to Appendix A for all life span and gender-specific values.

Choline Deficiency Fatty liver and liver damage occur with choline deficiency.

Choline Toxicity Hypotension (low blood pressure), altered liver function, increased salivation, and a fishy body odor result from choline toxicity.

Choline Food Sources Egg yolks, soybeans, peanuts, and high-protein animal products provide choline.

TABLE 5.3

The Water-Soluble Vitamins: Functions, Deficiencies, Toxicities, and Food Sources

Nutrient and Other Names	Functions in the Human Body	Deficiency Name, Signs and Symptoms	Toxicity Signs and Symptoms	Sources (Food and Other)
Thiamin VITAMIN B$_1$: aneurin	Coenzyme TPP in energy (carbohydrate and BCAA) metabolism, regulates appetite and the nervous system	*Beriberi* CARDIOVASCULAR SYSTEM: Heart rhythm changes CENTRAL NERVOUS SYSTEM: Mental changes, confusion, amnesia, disturbed eye movement GI SYSTEM: Nausea, anorexia NEUROMUSCULAR SYSTEM: Muscle weakness, tingling hands and feet, peripheral paralysis, decreased ability to coordinate motor movement, disturbed gait OTHER: Edema, exhaustion, weight loss	*Poorly documented in humans*	Pork, whole grains, enriched bread, cereal products
Riboflavin VITAMIN B$_2$	Coenzyme FMN and FAD in energy (fatty acid) metabolism	*Ariboflavinosis* BONE & TOOTH: Halts growth GI SYSTEM: Sore and swollen throat, swollen, magenta-colored tongue, cracking at the corners of the mouth SKIN: Dandruff	*Poorly documented in humans*	Milk products, organ meats, whole and enriched grains, products such as bread, cereal, and pasta
Niacin VITAMIN B$_3$: nicotinic acid nicotinamide 1 mg nicotinic acid = 1 NE 60 mg tryptophan can = 1 mg niacin	Coenzymes NAD and NADPH in general energy metabolism and steroid synthesis	*Pellegra* CENTRAL NERVOUS SYSTEM: Depression, apathy, headache, dementia (including memory loss) GI SYSTEM: Bright red tongue, diarrhea (GI disturbances of vomiting, diarrhea, and constipation) SKIN: Dermatitis OTHER: Fatigue	*Not from food but rather nicotinic acid supplements* CARDIOVASCULAR SYSTEM: Violent flushing reaction from vasodilation CENTRAL NERVOUS SYSTEM: Blurred vision GI SYSTEM: Nausea, vomiting, liver toxicity OTHER: Impaired glucose tolerance	Meats, fish, poultry, whole and enriched grains, products such as bread, cereal, and pasta Dairy products and legumes are good sources of tryptophan

continued

The Water-Soluble Vitamins: Functions, Deficiencies, Toxicities, and Food Sources (*continued*)

Nutrient and Other Names	Functions in the Human Body	Deficiency Name, Signs and Symptoms	Toxicity Signs and Symptoms	Sources (Food and Other)
Vitamin B$_6$ PN, PL, PM, PNP, PLP, PMP	Coenzyme (especially PLP) in metabolism of amino acids, glycogen, and some lipids; needed for heme synthesis, and helps convert tryptophan to niacin	*No disease name* Cardiovascular system: Microcytic anemia Central nervous system: Confusion, depression, epileptic convulsions Skin: Seborrheic (greasy) dermatitis	*Not from food but from large supplement doses* Neuromuscular system: Nerve damage, sensory neuropathy, tingly to numb hands and feet Skin: Dermatological lesions	Meats, fish, poultry, fortified cereals, legumes, nonstarchy vegetables, noncitrus fruits
Vitamin B$_{12}$ Cobalamin	Synthesis of nucleic acids and new cells, nerve cell maintenance, cofactor in energy production (fatty acids and amino acids), and the conversion of homocysteine to methionine; synergizes with folate	*Pernicious anemia* Cardiovascular system: Macrocytic (large cell) normochromic anemia Central nervous system: Loss of concentration, memory loss, disorientation, dementia, visual disturbances GI system: Impaired bowel function (including increased gas production and constipation), sore tongue, loss of appetite Neuromuscular system: Nerve damage, gait disturbances, tingling and numbness in the hands and feet (sensory neuropathy) Other: Insomnia, impotency, reduced bladder control	*Poorly documented in humans*	Meat, fish, poultry, milk, cheese, eggs
Folate Food folate: pteroylpolyglutamates Folic acid: pteroylmonoglutamic acid	Nucleic acid synthesis, needed to make new cells, cofactor in amino acid metabolism, conversion of homocysteine to methionine; synergizes with B$_{12}$	*No disease name* Cardiovascular system: Macrocytic normochromic anemia, heart palpitations, shortness of breath Central nervous system: Weakness, fatigue, decreased concentration, irritability, headache GI system: Red tongue	*Poorly documented in humans*	Raw, leafy green vegetables, other green vegetables, legumes, oranges, liver, brewer's yeast, wheat germ, some fortified cereals
Vitamin C L-ascorbic acid	Cofactor for the enzymes needed for the synthesis of collagen (needed for connective tissues), carnitine, and neurotransmitters; water-soluble antioxidant	*Scurvy* Central nervous system: Weakness, fatigue, depression GI system: Swollen, red, bleeding gums Skin: Follicular hyperkeratosis, petechia, easy bruising, impaired wound healing Other: Joint pain, edema, coiled hairs	GI system: Gastrointestinal disturbances, especially diarrhea Other: Increased risk for kidney stones and iron toxicity in certain people	Citrus fruits, berries, kiwis, melons, cabbage-family vegetables, peppers, vitamin C–fortified foods

Nutrient and Other Names	Functions in the Human Body	Deficiency Name, Signs and Symptoms	Toxicity Signs and Symptoms	Sources (Food and Other)
Pantothenic Acid	Essential component in coenzyme A (required in the metabolism of fatty acids, cholesterol, steroid hormones, vitamins A and D, neurotransmitters, amino acids, and porphyrin and corrin rings)	*No disease name* CENTRAL NERVOUS SYSTEM: Fatigue, apathy, irritability, restlessness GI SYSTEM: Nausea, vomiting, abdominal cramps NEUROMUSCULAR SYSTEM: Muscle cramps, impaired gait OTHER: Burning feet, sleep disturbances, hypoglycemia, increased insulin sensitivity	*Poorly documented in humans*	Widespread in foods, though highest in high-protein animal products, yeast, legumes, whole grains
Biotin FREE AND PROTEIN BOUND	Coenzyme in carboxylation reactions involved in energy metabolism, cell proliferation, gene silencing, and DNA repair	*No disease name* CENTRAL NERVOUS SYSTEM: Lethargy, depression, hallucinations, weakness, fatigue NEUROMUSCULAR SYSTEM: Muscle pain SKIN: Scaly red rash around the eyes, nose, and mouth OTHER: Conjunctivitis, hair loss	*Poorly documented in humans*	Widespread in foods, though highest in high-protein animal products, yeast, legumes, nuts, whole grains, chocolate Small amounts are made by intestinal bacteria
Choline FREE AND ESTERIFIED: lecithin phosphatidylcholine	Essential component of the neurotransmitter acetylcholine (plays a role in memory and muscle control), maintains cell membrane stability, integrated in bile acids, needed for lipid and cholesterol transport and metabolism, component of pulmonary surfactant	*No disease name* GI SYSTEM: Fatty liver, liver damage	CARDIOVASCULAR SYSTEM: Hypotension GI SYSTEM: Altered liver function, increased salivation OTHER: Fishy body odor	Egg yolks, soybeans, peanuts, high-protein animal products

© Cengage Learning 2013

Summary Points

- The water-soluble vitamins have specific functions.
- DRIs are established for all essential water-soluble vitamins: B_1, B_2, B_3, B_6, B_{12}, folate, vitamin C, pantothenic acid, biotin, and choline.
- Deficiency and toxicity signs and symptoms are established for most of the water-soluble vitamins.

Take Ten on Your Knowledge Know-How

1. Which vitamins function in energy production, and in which type of energy metabolism: carbohydrate, protein, and/or fat?
2. What is the DRI for the nutrients that are based on the amount of Calories or protein consumed in the diet?

3. What are the functions, food sources, and deficiency and toxicity signs and symptoms of thiamin (B$_1$), riboflavin (B$_2$), and niacin (B$_3$)?
4. What are the functions, food sources, and deficiency and toxicity signs and symptoms of vitamins B$_6$, B$_{12}$, and folate?
5. What are the functions, food sources, and deficiency and toxicity signs and symptoms of vitamin C?
6. What are the functions, food sources, and deficiency and toxicity signs and symptoms of pantothenic acid, biotin, and choline?
7. Deficiencies of which water-soluble vitamins cause anemia?
8. What biochemical pathways generate most of the ATP in the body?
9. How many chemical forms does vitamin B$_{12}$ have? Which mineral is present in the chemical structure of vitamin B$_{12}$?
10. Which water-soluble vitamins do not participate in energy production?

5.3 WATER AND THE ELECTROLYTES

T-Talk 5.3

To hear Dr. Turley talk about water and the electrolytes, go to www.cengagebrain.com

calcium Ninety-nine percent of this major mineral is found in bone and teeth, but it is also important for blood clotting, muscle contraction, and nerve conduction.

phosphorus Eighty-five percent of this major mineral is in bone and teeth; if too much is consumed, bones demineralize to increase the calcium levels in the blood.

magnesium A major mineral, abundant in plant foods, needed to mineralize bones and teeth, as well as to participate in muscle contractions, nerve conduction, and blood clotting.

sodium A major mineral that maintains electrolyte balance in the extracellular fluid and makes up 60 percent of salt.

chloride A major mineral that is part of salt, but also used to make HCl (hydrochloric acid).

potassium The major mineral in intracellular fluid that maintains cell volume.

sulfur A major mineral that helps stabilize protein structures; deficiency is poorly documented in humans; the need is met by consuming a protein-adequate diet.

iron A trace mineral that is a component of several functional proteins, including hemoglobin and myoglobin.

copper A trace mineral; Menkes disease is associated with a deficiency and Wilson's disease is associated with a toxicity.

iodine A trace mineral that is essential for the structure of thyroid hormone.

Introduction

Minerals cannot be made by living organisms, but they can become a part of living organisms. Plants obtain minerals directly from soil. Animals obtain minerals indirectly from eating plants or other animals and small amounts from water. The amount of minerals in water varies greatly by geographic location. There are numerous minerals in earth, but only some of them are essential for normal human metabolism, growth, development, and cellular function. Each essential mineral has specific functions. Some work as enzyme cofactors, some have structural roles, some regulate water balance, and some maintain normal acid–base (pH) balance. The essential minerals will be explored in the body and in the food supply.

Minerals in the Body

Minerals are inorganic elements that are naturally found in earth. Like vitamins, they are needed in tiny amounts in the body. Minerals are categorized as major and trace based on their levels in the body. Major minerals are found in quantities greater than 5 grams in an adult reference body, and they have a DRI level greater than 100 milligrams per day. The major minerals are **calcium, phosphorus, magnesium, sodium, chloride, potassium**, and **sulfur**. The key major minerals that maintain normal fluid balance and acid–base balance are called electrolytes and include sodium, potassium, and chloride. The major minerals that contribute to bone structure are calcium, phosphorus, and magnesium. Sulfur is a major mineral associated with protein. Trace minerals are found in quantities of 5 grams or less in an adult reference body, and they have a DRI level less than 100 milligrams per day. The trace minerals are **iron, copper, iodine, manganese, fluoride, chromium, molybdenum, selenium, zinc**, and **cobalt**.

Demo GENEie Showcase the quantity difference between the amounts of major minerals found in the body. Place 3 pounds of flour in a clear plastic bag to represent the amount of calcium a reference adult has in the body. Contrast this amount to 5 grams of trace mineral, which would be ¼ ounce of flour.

Each essential mineral, like each essential vitamin, should be consumed in the diet at a level appropriate for healthy individuals based on age, gender, and conditions of pregnancy and lactation. DRI values are expressed as RDAs or AIs for the essential minerals (see Appendix A). The quantity values for the essential minerals are very small, milligrams or micrograms, depending on the nutrient. The intake values reflected on food package labels are RDIs. The RDIs are based on the highest adult 1968 RDA values. The RDIs include a single value for each nutrient that is not age, gender, or condition specific.

When the dietary intake of essential minerals is inadequate, a deficiency state results over time, and specific deficiency signs and symptoms occur. When dietary intake levels are too high, toxicity signs and symptoms often result. Some foods or food groups provide a good source of some minerals and not others. The key to getting mineral needs met and for having a healthy diet is to eat a wide variety of foods from each of the food groups, such as by applying the personalized MyPlate food guidance system. For each of the essential minerals, the exploration of the DRI, functions, food sources, deficiencies, and toxicities will be addressed. Water, also an inorganic compound, will also be addressed, because it is needed for maintaining hydration, converting food into energy, carrying nutrients through the body, and removing waste.[17,29,34]

Because water is such an important nutrient, it will be the first topic we discuss.

Water[15,34,42,51,62,71] (See Table 5.4 on page 258 for a summary of water.)

Form   Water is an inorganic compound composed of two hydrogen atoms and one oxygen atom. It is abbreviated chemically as H_2O.

Function   Water has many essential functions in the body. It functions as the medium for nutrient transport and chemical reactions. It is also needed for the protection of body contents, lubrication, and body temperature regulation. Approximately 60 percent of body weight is water, and the body tries to maintain that percentage through a water balance process that involves the brain and kidneys working in concert (see BioBeat 5.6). Sixty to 65 percent of total body water is **intracellular fluid (ICF)** (inside the cells), and 35 to 40 percent is **extracellular fluid (ECF)**. Extracellular fluid is fluid outside the cells or between cells (interstitial) and includes tears, saliva, GI tract juices, and **plasma** volume (fluid in the blood stream).

manganese A trace mineral, different from the major mineral magnesium, that plays a role in bone formation and the metabolism of amino acids, carbohydrates, and cholesterol.

fluoride A trace mineral that functions to strengthen tooth enamel, resist dental caries, and stabilize bone.

chromium A trace mineral that potentiates the action of insulin and may improve the glucose tolerance test.

molybdenum A trace mineral that is a coenzyme for molybdoenzymes.

selenium A trace mineral that functions through selenoproteins and is an antioxidant.

zinc A trace mineral that is a structural component in more than 100 metalloenzymes; deficiency is caused by a genetic disorder resulting in acrodermatitis enteropathica.

cobalt A trace mineral that is essential for vitamin B_{12} structure; cobalamin.

intracellular fluid (ICF) The fluid inside of cells.

extracellular fluid (ECF) The fluid that is outside of cells.

plasma The largest of the extracellular fluid volumes; provides a medium for the red and white blood cells to be transported through the circulatory system.

BioBeat 5.6

Body Fluid: Intracellular Fluid, Extracellular Fluid, Electrolytes, and Osmolarity

Water is the medium for metabolism and nutrient transport. It is the fluid of life. Without an intake of this nutrient, dysfunction or even death will occur faster than with the limitation of any other of the essential nutrients in human nutrition. Fluid is compartmentalized in two different spaces in the body. Intracellular fluid, or the fluid inside of cells, comprises 60 to 65 percent of total body water. Extracellular fluid is the fluid outside of cells. Extracellular fluid comprises 35 to 40 percent of total body water. There are three major compartments of extracellular fluid: vascular fluid (plasma), **interstitial fluid** (fluid between cells), and **cerebral spinal fluid**. Intracellular fluid and extracellular fluid balance is primarily regulated by **osmotic pressure**. Osmotic

pressure is created by the concentration of particles in the fluid on two sides of semipermeable membranes. Osmotic pressure is maintained between the intracellular fluid and the extracellular fluid primarily by sodium (Na), potassium (K), and chloride (Cl). Na and Cl are minerals in extracellular fluid. K is the major mineral of intracellular fluid (see Figure 5.20). NaCl (salt) and K are collectively known as the *electrolytes*. Electrolytes are molecules that, when dissolved in a solution, become ionized by either losing an electron or gaining an electron. Fluids are balanced within animal bodies through these chemical and physical properties.[71]

What can you do to improve your personal hydration?

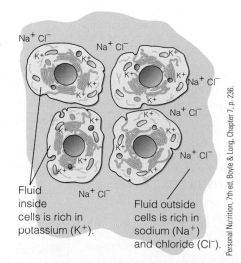

FIGURE 5.20 The major electrolytes and their primary cellular location.

Fluid inside cells is rich in potassium (K^+).

Fluid outside cells is rich in sodium (Na^+) and chloride (Cl^-).

Personal Nutrition, 7th ed, Boyle & Long, Chapter 7, p. 236.

DRI A DRI for water has been established for adults to encourage appropriate fluid intake. The DRI for water is set at 2.7 liters (11 cups) per day for adult women and 3.7 liters (15 cups) per day for adult men (see Appendix A). Think of the value of a 2-liter soda bottle as a reference to envision the daily volume of fluid. These levels may meet the needs of many women and men but may not be optimal fluid intakes. Furthermore, it is well known that individuals who engage in strenuous physical activity or experience heat stress have greatly increased fluid needs. Basically, the fluid lost in strenuous team or individual endurance sports, approximately 1 to 2 liters per hour, should be replaced to prevent dehydration. Fluid intake should be large enough to produce the need to void every 1 to 2 hours (see page 205 in Module 4). Currently, there is no UL for water because water toxicity is not a risk under normal circumstances and in individuals with healthy kidneys.[34]

Deficiency The body is constantly losing fluids. Water is lost through conscious and unconscious (obligatory) means. The major ways that body water or fluid is lost include urine as liquid waste, respiration as vapor in breath, feces as solid and somewhat moist waste, and sweat as liquid waste. The organs directly involved in water loss are the kidneys for urine production, the lungs for respiration, the large intestine (colon) for fecal waste, and the skin for sweat (see Figure 5.21). Water loss via the renal system has the most variability. If water is plentiful, the urine is dilute and produced at a faster rate, and if fluid intake is insufficient, the urine is concentrated and produced at a slower rate.

Without water, a person dies in a few days; thus, one may conclude that water is the most essential nutrient for the body. However, possibly more real than dying of thirst is suffering from impaired daily mental and physical performance and long-term chronic disease, all of which may be prevented by proper hydration. When water needs are not met, many body systems are negatively affected. Mild **dehydration** has been shown to reduce brain function, as evidenced by compromised short-term memory, ability to concentrate, and alertness. Some of the consequences of inadequate water intake include impaired mental function and motor control, reduced aerobic and endurance exercise performance, increased core body temperature during exercise (thermal injury), increased fever response, increased resting heart rate, impaired blood pressure regulation, decreased cardiac output, and fainting in susceptible people. Individuals who are heavily exercising and losing water and electrolytes need to be conscientious about replacing both water and electrolytes.[34,42,47,71]

Chronic diseases are multifactorial, meaning that many things contribute to their onset and progression. Lack of water is believed to be one component associated with increased risk of urinary tract infections, kidney stones, constipation, hypertension, heart disease, kidney dysfunction, and hyperglycemia in individuals with diabetes. Furthermore, some studies suggest that lack of good hydration possibly increases colon and bladder cancer and disorders of the lungs and airway (bronchopulmonary).[42,62,71]

interstitial fluid One of the extracellular fluid compartments, which refers to fluid between cells.

cerebral spinal fluid Extracellular fluid that circulates around the spinal cord.

osmotic pressure The pressure exerted against a semipermeable membrane that is created by the concentration of particles in a solution.

dehydration A reduction of total body water when thermal regulation, cardiovascular function, and electrolyte balance are all impaired.

FIGURE 5.21 Fluid sources and losses from the body in liters.

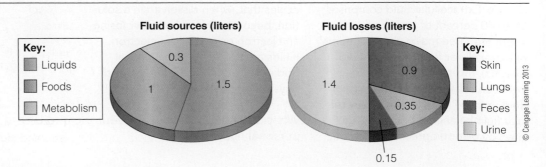

Toxicity   Excess water intake causes **water intoxication**. This condition is rare, usually occurring with forced water intake, but it has serious consequences. Potential life threats from water intoxication include low blood levels of sodium (**hyponatremia**), which is associated with central nervous system edema, lung congestion, and muscle weakness. The maximum renal clearance (urinary output) ranges from 700 milliliters (24 ounces) up to 1,000 milliliters per hour.

Sources   Most people think that water itself is the only source of fluid or water for the body. In reality, fluids such as milk, juice, and noncaffeinated beverages, foods such as whole fruits and vegetables, and even metabolic reactions that produce H_2O as a byproduct (metabolic water) inside the body all provide a source of water (see Figure 5.21).

Some drinking water is a source of nutrients. Although some bottled water is marketed as nutritious, it is important to check the nutrient density value on the food package label. Hard water contains very small amounts of calcium and magnesium, while manmade soft water contains sodium from the process of extracting calcium and magnesium. "Natural" soft water contains few minerals, whereas processed and marketed water may have minerals added—just read the food label to find out about the mineral quality of the water.

Besides the presence or absence of minerals in water, other water quality issues are of utmost importance. These include contamination by microorganisms, organic compounds, or heavy metals. Public agencies evaluate water conservancies regularly, and private well water sources should be regularly tested for contaminants by the users. Water with poor quality contains amounts of one or more of these substances above the standard. Water with good quality does not contain (or has low levels of) these substances. Consider drinking only bottled or boiled water when traveling abroad, depending on the water quality in the region.

Sodium[1,24,30,34,53] (See Table 5.4 on page 258 for a summary of sodium, molecularly abbreviated Na, and shown in the periodic chart in Appendix A.)

Sodium Functions   Sodium is a major cation (positively charged) electrolyte in extracellular fluid (see Figure 5.20 on page 253). Sodium's critical functions include affecting extracellular fluid balance and volume, plasma osmolarity (the concentration of active particles in solution in plasma) and volume, pH balance, and the membrane potential of cells (such as for muscle contraction and active transport of nutrients).

Sodium's DRI, UL, and DRV   The DRI, expressed in milligrams, for sodium for males and females ages 19 to 30 is 1,500 milligrams per day for adults. The UL is 2,300 milligrams per day, and the Daily Reference Value (DRV) is 2,400 milligrams per day. Refer to Appendix A for all life span and gender-specific values.

Sodium Deficiency   Too little sodium leads to hyponatremia, or low blood levels of sodium, which causes mental apathy, weakness, muscle cramping, skeletal muscle cell injury, renal and cardiac failure, and loss of appetite.

Sodium Toxicity   Too much dietary sodium, usually provided by too much salt and salted or salt-added processed foods, can increase blood pressure and cause fluid retention (edema). In less than 10 percent of hypertensive patients, **sodium sensitivity** is the cause of a marked elevation in blood pressure in response to consuming sodium, which increases the risk for heart attack and stroke.

Sodium Food Sources   Sodium is found mostly in salt (sodium chloride, or NaCl), including sea salt, soy sauce, and processed foods (especially canned soups, salt-cured meats, pickles, salted pretzels, olives, and fast foods) (see Figure 5.22). Approximately 10 percent of the sodium consumed in the typical American diet comes from natural sources, 10 percent from using the salt shaker, and 80 percent from consuming processed and fast foods.

water intoxication The result of drinking too much water.

hyponatremia A state of the blood that is characterized by low blood levels of sodium.

sodium sensitivity A person who has a significant elevation in blood pressure associated with sodium intakes above the Tolerable Upper Intake Level (UL).

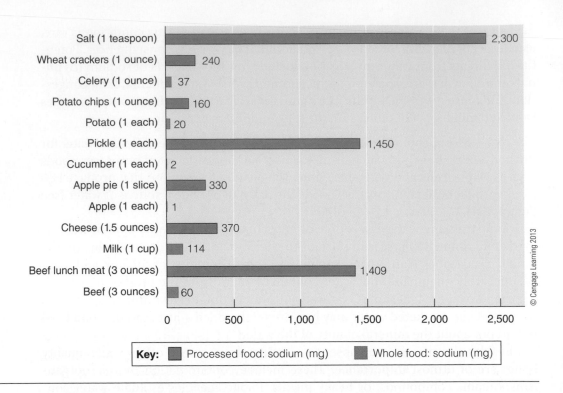

FIGURE 5.22 Food sources of sodium.

Key: ▮ Processed food: sodium (mg)   ▮ Whole food: sodium (mg)

Chloride[24,30,34,53] (See Table 5.4 on page 258 for a summary of chloride, molecularly abbreviated Cl, and shown in the periodic chart in Appendix A.)

Chloride Functions Chloride is a major anion (negatively charged) electrolyte in extracellular fluid (see Figure 5.20 on page 253). Chloride's critical functions include affecting extracellular fluid balance and volume and plasma osmolarity and volume. Additionally, chloride is in hydrochloric acid, which is needed for food digestion.

Chloride's DRI, UL, and RDI The DRI, expressed in milligrams, is set for sodium chloride (salt) at 3,800 milligrams per day for adults (males and females 19 years and older). Because the DRI for sodium is 1,500 milligrams and salt is 40 percent sodium and 60 percent chloride, the DRI for chloride is 2,300 milligrams per day. This amount of chloride can maintain electrolyte balance when the body is at rest. The UL is 3,600 milligrams per day, and the RDI is 3,400 milligrams per day. Refer to Appendix A for all life span and gender-specific values.

Chloride Deficiency Common excessive electrolyte losses are from vomiting and diarrhea. Minor electrolyte losses occur during exercise (1,500 milligrams of sodium chloride per liter of sweat). This level of loss is small in comparison to vomiting. Deficiency of chloride can cause growth failure in children, muscle cramps, mental apathy, and loss of appetite.

Chloride Toxicity Elevated levels of chloride are normally harmless, typically occurring from chlorine gas, which is poisonous but evaporates from water. However, in combination with sodium, it increases blood pressure, which increases the risk for heart attack and stroke.

Chloride Food Sources Salt and sea salt (sodium chloride, or NaCl), salt substitute (potassium chloride, or KCl), lite salt (a combination of NaCl and KCl), soy sauce, and processed foods all provide chloride. Some vegetables such as seaweed, rye, tomatoes, lettuce, celery, and olives provide some chloride. Because salt and salt substitute are a major source of chloride, symptoms of too much salt can be difficult to identify.

Potassium[1,24,34,47,53,55] (See Table 5.4 for a summary of potassium, molecularly abbreviated K, and shown in the periodic chart in Appendix A.)

Potassium Functions Potassium is the primary cation electrolyte for maintaining osmolarity in the intracellular fluid (see Figure 5.20 on page 253). Potassium has critical functions in maintaining fluid volume in cells and thus is important for normal cellular function. Potassium affects nerve transmission, muscle contraction, and vascular tone. Furthermore, it blunts the rise in blood pressure from excess sodium intake and decreases kidney stone reoccurrence.

Potassium's DRI, UL, and DRV The DRI, expressed in milligrams, is 4,700 milligrams per day for all adults. This amount of potassium can maintain electrolyte balance at rest. The UL is not established. The DRV is 3,500 milligrams per day. Refer to Appendix A for all life span and gender-specific values.

Potassium Deficiency Severe potassium deficiency usually accompanies dehydration and is characterized by **hypokalemia** (low blood levels of potassium). Adverse affects of hypokalemia include cardiac arrhythmia, muscle weakness, and glucose intolerance. Certain medications, especially the thiazide diuretics given as a first-line drug in the treatment of high blood pressure, can induce hypokalemia. Moderate potassium deficiency can occur without hypokalemia and causes increased blood pressure, salt sensitivity, risk of kidney stones, bone turnover, and possibly heart attack and stroke.

Potassium Toxicity It is very difficult to reach toxic levels of potassium in the body if the kidneys are healthy. Toxicity is usually only a natural threat in individuals with renal (kidney) failure. However, a person with kidney failure can easily suffer from potassium toxicity. Too much potassium causes **hyperkalemia**, which induces muscle **tetany** and, if toxicity levels are high enough, cardiac arrest. This is common in people with chronic renal failure.

Potassium Food Sources Food sources of potassium include all whole foods like meats, milk, fruits, vegetables, grains, and legumes. Potassium as potassium chloride is in salt substitutes (see Figure 5.23).

hypokalemia A state of the blood that is characterized by low blood levels of potassium.

hyperkalemia A state of the blood that is characterized by high blood levels of potassium.

tetany Involuntary muscle contraction without relaxation.

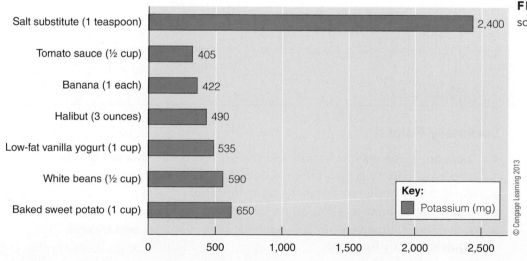

FIGURE 5.23 Food sources of potassium.

© Cengage Learning 2013

TABLE 5.4

Water and Electrolytes: Functions, Deficiencies, Toxicities, and Food Sources

Nutrient and Abbreviation	Functions in the Human Body	Deficiency Name, Signs and Symptoms	Toxicity Signs and Symptoms	Sources (Food and Other)
Water H_2O	Medium for metabolism, thermoregulation, and cardiovascular function	CARDIOVASCULAR SYSTEM: Increased resting heart rate, decreased cardiac output and blood pressure CENTRAL NERVOUS SYSTEM: Impaired mental function NEUROMUSCULAR SYSTEM: Impaired motor control OTHER: Thermal injury during exercise, increased fever response	CARDIOVASCULAR SYSTEM: Hyponatremia CENTRAL NERVOUS SYSTEM: Edema NEUROMUSCULAR SYSTEM: Muscle weakness OTHER: Lung congestion	Water, fluid-based beverages, fluid in foods (fruits and vegetables)
Sodium Na	Extracellular fluid balance and volume, plasma osmolarity and volume, pH balance, and the membrane potential of cells	CARDIOVASCULAR SYSTEM: Hyponatremia, cardiac failure CENTRAL NERVOUS SYSTEM: Mental apathy, weakness GI SYSTEM: Loss of appetite NEUROMUSCULAR SYSTEM: Muscle cramping, skeletal muscle cell injury OTHER: Renal failure	CARDIOVASCULAR SYSTEM: Increased blood pressure with increased risk for heart attack and stroke OTHER: Fluid retention (edema)	Salt, sea salt, soy sauce, processed foods
Chloride Cl	Extracellular fluid balance and volume, plasma osmolarity and volume, and hydrochloric acid production	BONE & TOOTH: Growth failure in children CENTRAL NERVOUS SYSTEM: Mental apathy GI SYSTEM: Loss of appetite NEUROMUSCULAR SYSTEM: Muscle cramps	CARDIOVASCULAR SYSTEM: In combination with sodium, increased blood pressure with increased risk for heart attack and stroke	Salt, sea salt, salt substitute, soy sauce, processed foods, seaweed, rye, tomatoes, lettuce, celery, olives
Potassium K	Intracellular fluid volume and function, nerve transmission, muscle contraction, vascular tone, blunts the rise in blood pressure from excess sodium, and decreases kidney stone reoccurrence	BONE & TOOTH: Bone turnover CARDIOASCVULAR SYSTEM: Hypokalemia, cardiac arrhythmia, increased blood pressure with increased risk for heart attack and stroke NEUROMUSCULAR SYSTEM: Muscle weakness OTHER: Glucose intolerance, risk of kidney stones	*In renal (kidney) failure only* CARDIOVASCULAR SYSTEM: Hyperkalemia, cardiac arrest NEUROMUSCULAR SYSTEM: Muscle tetany	All whole foods like meats, milk, fruits, vegetables, grains, and legumes, as well as salt substitutes

© Cengage Learning 2013

Summary Points

- There are seven major minerals and ten trace minerals that need to be consumed daily to support normal body functioning in humans.
- Water is very important for many body functions.
- Adequate fluid intake is important for health and optimal functioning.
- Dehydration negatively affects cognitive, cardiovascular, and thermal regulatory body functions.
- Fluid intake should support production of a clear urine every 2 hours while awake; water intoxication results from deliberate excessive intake.

- The electrolytes function to regulate fluid and acid–base balance.
- The major extracellular minerals are sodium and chloride.
- The major intracellular mineral is potassium.

Take Ten on Your Knowledge Know-How

1. What percentage of your body weight is water? Why is water so essential for your life?
2. How much water or fluid should you consume each day? Do you normally consume this amount? What are sources of water and fluid for your body?
3. How are fluids lost from the body?
4. What happens when one becomes dehydrated?
5. What is an electrolyte, and which nutrients function as electrolytes? How do electrolytes maintain osmotic pressure?
6. What are the functions, good food sources, and deficiency and toxicity signs and symptoms of sodium?
7. What are the functions, good food sources, and deficiency and toxicity signs and symptoms of potassium?
8. What are the functions, good food sources, and deficiency and toxicity signs and symptoms of chloride?
9. What type of disease usually causes hyperkalemia?
10. What is hyponatremia? How can it be caused?

5.4 MAJOR MINERALS IN BONE AND PROTEIN

Introduction   In this section the functions, deficiencies, toxicities, and food sources of the major minerals in bone (calcium, phosphorus, and magnesium) will be explored, along with osteoporosis types and risk factors. Furthermore, the functions, deficiencies, toxicities, and food sources of sulfur, a major mineral provided in sulfur-containing amino acids (which are richly incorporated in protein in the body), are discussed.

T-Talk 5.4

To hear Dr. Turley talk about major minerals in bone and protein, go to www.cengagebrain.com

Calcium[A,F,G,H,I,6,8,14,20,24,26,29,30,45,53,64,80,81,82]   (See Table 5.5 on page 264 for a summary of calcium, molecularly abbreviated Ca, and shown in the periodic chart in Appendix A.)

Calcium Functions   Calcium is a principle mineral in bones (see BioBeat 5.7 on page 261) and teeth. Although about 99 percent of calcium functions to mineralize bone and teeth, the remaining 1 percent functions in blood clotting, muscle contraction, and nerve conduction.

Calcium's DRI, UL, and RDI   The DRI, expressed in milligrams, for calcium for a male or a female 19 to 30 years old is 1,000 milligrams per day. The UL for calcium from childhood through adulthood is 2,500 milligrams per day. The RDI for calcium is 1,000 milligrams per day. Refer to Appendix A for all life span and gender-specific values.

Calcium Deficiency   Calcium deficiency has been described as causing stunted growth in children, low bone density (**osteopenia**), and accelerated bone loss in adults, leading to osteoporosis. This condition increases the risk for bone fractures.
　　Osteoporosis, which usually occurs later in life, results from a low calcium intake, among other factors, earlier in life. Osteoporosis is a condition typified by fragile bones that fracture easily, and loss of stature. Osteoporosis occurs as a result of bone

osteopenia Low bone density that can be caused by not consuming enough calcium in the diet or a toxicity of vitamin A.

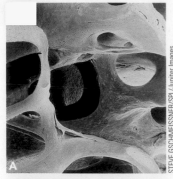

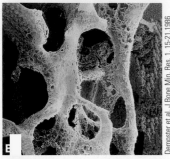

STEVE GSCHMEISSNER/SPL/Jupiter Images

Dempster et al., J Bone Min. Res. 1, 15-21,1986

Normal (A) and osteoporotic (B) trabecular bone.

mineral loss. Several things can demineralize bone or cause mineral (especially calcium) loss from the bone: a high-protein/phosphorus diet, lack of exercise, tobacco use, alcohol consumption, low calcium intake in the early years of life, and lack of estrogen after menopause. Women are four times more likely to become osteoporotic than men because a woman's rate of bone loss is accelerated by the loss of estrogen protection during menopause (see BioBeat 5.7). A woman can dramatically decrease her risk of severe consequences from osteoporosis just by ensuring optimum bone density during her entire life. In other words, it is advantageous to build your calcium bone bank or maximize your bone mineral density early in life. The ways to promote bone health include eating a calcium-rich diet, exercising (minimally meet the DRI for physical activity), not using tobacco, not drinking alcohol excessively, and not consuming a high-protein/phosphorus diet. Additionally, if bone mass is too low, physicians do prescribe medication to help slow down bone loss.

Calcium Toxicity   Calcium toxicity leads to hypercalcemia (high blood levels of calcium). Hypercalcemia can occur from hormonal imbalance and milk alkali syndrome. This can occur when a person consumes a lot of chewable calcium antacids throughout the day to control heartburn, while consuming large amounts of milk products. It is associated with increased risk for kidney stones, calcium deposits in soft tissues, constipation, and reduced absorption of iron, magnesium, zinc, and phosphorus.

Calcium Food Sources   Good sources of calcium are dairy products such as milk, yogurt, and cheese (see Figure 5.24). Nondairy sources include fortified nondairy milk alternatives such as soy, rice, and almond milks, fortified cereals, other fortified foods, canned fish with the bones included, spinach, turnip greens, tofu, broccoli, and kidney beans (see Figure 5.24). The bioavailability of calcium is highest (about 50 percent) from cruciferous vegetables; moderate (about 30 percent) from dairy products and calcium-fortified foods; low (about 20 percent) from beans, nuts, and seeds; and lowest (less than 5 percent) from spinach. Many people also choose to supplement their calcium intake with a variety of candy-like products (such as calcium gummies and chews) or with supplement pills composed of calcium carbonate or calcium citrate.

FIGURE 5.24 Food sources of calcium.

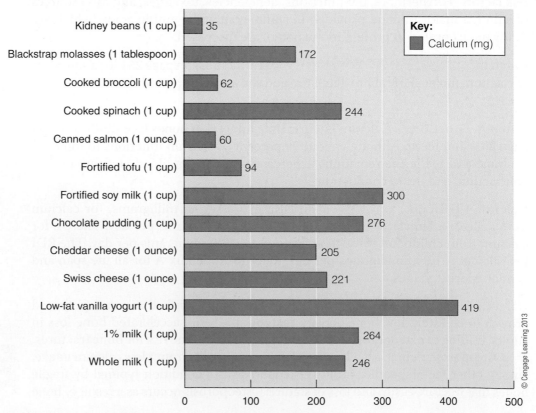

Key:
█ Calcium (mg)

Food	Calcium (mg)
Kidney beans (1 cup)	35
Blackstrap molasses (1 tablespoon)	172
Cooked broccoli (1 cup)	62
Cooked spinach (1 cup)	244
Canned salmon (1 ounce)	60
Fortified tofu (1 cup)	94
Fortified soy milk (1 cup)	300
Chocolate pudding (1 cup)	276
Cheddar cheese (1 ounce)	205
Swiss cheese (1 ounce)	221
Low-fat vanilla yogurt (1 cup)	419
1% milk (1 cup)	264
Whole milk (1 cup)	246

© Cengage Learning 2013

BioBeat 5.7

Oh, the Bones! Take Care of Them Early in the Life Span

Bone is living tissue. There are two dominant cell types that constantly remodel bone. They are **osteoblasts** (bone mineralizers) and **osteoclasts** (bone demineralizers). During growth, the osteoblasts dominate, and bone density continues to increase as bones grow. **Peak bone mass** for the average adult female occurs at 25 to 35 years old and for a male at 25 to 55 years old (see Figure 5.25). During this period there is a balance between the osteoblasts and osteoclasts. After the age of 35 years for a female, bone begins to demineralize

because the activity of the osteoclasts begins to dominate, and after menopause female bones demineralize at a faster rate due to the loss of estrogen protection. For males, bone begins to demineralize at about 55 years old. The greater the amount of bone mineral density an adult has at peak bone mass, the longer that individual will have stronger bones. Adequate dietary calcium, weight-bearing physical activity, a healthy dietary calcium-to-phosphorus ratio, and avoiding tobacco use and alcohol abuse are recommended

to promote bone mineral density. Other useful nutrients that support **bone mineralization** include vitamins D and K, magnesium, boron, and fluoride. Vitamin K appears to play a role in the carboxylation of a protein called osteocalcin that is involved in bone formation and thus supports high-quality bone mineral density.[H,20,23,24,40,61]

Which of your behaviors are contributing to or taking away from your lifelong bone mass regeneration and health?

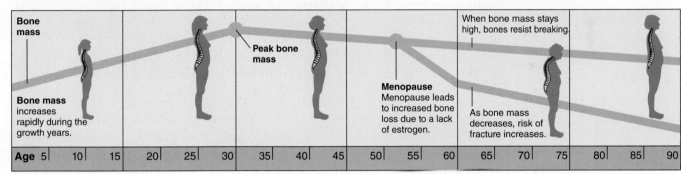

Adapted from The Stay Well Company, Krames StayWell, 780 Township Line Road, Yardley, PA 19067. (267) 685-2500.

FIGURE 5.25 Bone mass across the life span.

Data GENEie Solah is a 22-year-old female who is lactose intolerant and thus typically consumes only 500 milligrams of calcium each day from plant sources. What is her DRI for calcium? What percent of her DRI for calcium does she typically consume? What are several health implications of her intake level? What are some dietary recommendations for her?

Phosphorus[I,24,29,30] (See Table 5.5 on page 264 for a summary of phosphorus, molecularly abbreviated P, and shown in the periodic chart in Appendix A.)

Phosphorus Functions   The majority of phosphorus (85 percent) is found in bones and teeth. Phosphorus is important for the cell's genetic material (DNA and RNA), phospholipids in cell membranes, **phosphorylation** reactions, energy transfer (ATP), and buffering systems (maintaining pH).

Phosphorus's DRI, UL, and RDI   The DRI for phosphorus, expressed in milligrams, for a male or a female 19 to 30 years old is 700 milligrams per day. The UL for phosphorus ranges from 3,000 to 4,000 milligrams per day from childhood through adulthood. The RDI for phosphorus is 1,000 milligrams per day. Refer to Appendix A for all life span and gender-specific values.

Phosphorus Deficiency   Phosphorus deficiency is rare but can occur in starvation, diabetic ketoacidosis (a type of **metabolic acidosis**), and the aftermath of

osteoblasts The type of bone cell that builds or mineralizes bone.

osteoclasts The type of bone cell that breaks down or demineralizes bone.

peak bone mass The highest attainable bone density for an individual that is developed typically up to approximately ages 25 to 35.

bone mineralization The process in which minerals such as calcium, phosphorus, magnesium, and fluoride crystallize on the collagen matrix of bone.

phosphorylation When a phosphate group is added to a molecule, such as in the chemical reaction converting adenosine diphosphate (ADP) to adenosine triphosphate (ATP).

metabolic acidosis A blood pH less than 7.35 that results from too many hydrogen ions in the blood.

an alcoholic binge. **Hypophosphatemia** (low blood phosphorus) causes anorexia, muscle weakness, bone pain, and general debility.

Phosphorus Toxicity Excess phosphorus causes calcium excretion and bone demineralization. The body strives for a calcium-to-phosphorus ratio in the blood stream of 1:1. If dietary phosphorus is high, the body will break down bone to release calcium and return the ratio to 1:1. **Hyperphosphatemia** (high blood phosphorus) occurs in end-stage renal disease and with vitamin D excess. It causes reduced calcium absorption and the calcification of nonskeletal tissues (metastatic calcification), especially the kidneys.

Phosphorus Food Sources Good sources of phosphorus are animal proteins and processed foods (see Figure 5.26). Phosphorus-containing chemicals are added in generous quantities during food processing for preservation. A person who consumes a lot of processed foods and does not have a calcium-adequate diet may experience accelerated bone demineralization.

Magnesium[1,24,29,30,58] (See Table 5.5 on page 264 for a summary of magnesium, molecularly abbreviated Mg, and shown in the periodic chart in Appendix A.)

Magnesium Functions Magnesium is needed to build bones, teeth, and proteins. About 60 percent of the magnesium in the body is stored in the bones and teeth, with the remainder used metabolically in muscle and soft tissue. Magnesium is involved in more than 300 different enzymatic processes in the body. Magnesium also functions in muscle contraction, blood clotting, and nerve impulse transmission.

Magnesium's DRI, UL, and RDI The DRI for magnesium, expressed in milligrams, is 400 milligrams per day for an adult male 19 to 30 years old; for an adult female 19 to 30 years old, it is 310 milligrams per day. The UL for an adult male or an adult female 19 to 70 years old is 350 milligrams per day. The UL for this mineral is uniquely established. Specifically, it is for magnesium obtained from supplements and laxatives. It does not apply to magnesium levels from foods. The RDI is 400 milligrams per day. Refer to Appendix A for all life span and gender-specific values.

Magnesium Deficiency Magnesium depletion can occur in people with cardiovascular and neuromuscular diseases, malabsorption syndrome or chronic diarrhea, diabetes, renal wasting syndrome, osteoporosis, and chronic alcoholism. Magnesium deficiency is associated with hypocalcemia, muscle cramping, vitamin D

hypophosphatemia A state of the blood that is characterized by low blood levels of phosphorus and causes general debility.

hyperphosphatemia A state of the blood that is characterized by high blood levels of phosphorus that leads to metastatic calcification.

FIGURE 5.26 Food sources of phosphorus.

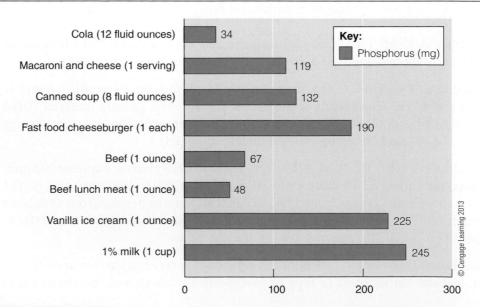

Key:
Phosphorus (mg)

Food	Phosphorus (mg)
Cola (12 fluid ounces)	34
Macaroni and cheese (1 serving)	119
Canned soup (8 fluid ounces)	132
Fast food cheeseburger (1 each)	190
Beef (1 ounce)	67
Beef lunch meat (1 ounce)	48
Vanilla ice cream (1 ounce)	225
1% milk (1 cup)	245

© Cengage Learning 2013

metabolism interference, neuromuscular hyperexcitability (when the muscles contract easily), and seizures.

Magnesium Toxicity Magnesium toxicity is not caused by food sources but rather by magnesium-containing supplements and laxatives. Most laxatives, such as milk of magnesia, contain magnesium to stimulate solid waste excretion. Also, some antacids, such as magnesium hydrochloride, are sources of magnesium. Excess magnesium causes diarrhea, nausea, and abdominal cramps. Larger pharmacological doses of magnesium can cause **metabolic alkalosis**, hypokalemia, and paralytic ileus (paralyzed ileum).

Magnesium Food Sources Magnesium is found mostly in plant foods, including nuts, legumes, whole grains, dark green leafy vegetables, chocolate, and cocoa (see Figure 5.27). It can also be found in some seafood and tap water. Up to 80 percent of magnesium can be lost in processed grains, and the Enrichment Act of 1942 does not mandate that magnesium be added back to processed foods.

Sulfur[24,34] (See Table 5.5 for a summary of sulfur, molecularly abbreviated S, and shown in the periodic chart in Appendix A.)

Sulfur Functions This mineral is needed for the biosynthesis of sulfur- and sulfate-containing compounds, which have numerous functions in the body. Furthermore, sulfur is a component of the organic compounds biotin, thiamin, cysteine, methionine, glutathione, taurine, and insulin. Sulfur helps stabilize protein shape and structure by forming disulfide bridges.

Sulfur's DRI, UL, and RDI Sulfur is an inorganic element naturally found in the earth. Sulfur is present in a wide variety of molecules. It is present in organic or inorganic sulfur- or sulfate-containing compounds. Sulfate is an inorganic compound made of sulfur plus four molecules of oxygen (SO_4). Sulfate is commonly attached to biological molecules, but unbound sulfate can exist in water. There is no DRI, UL, or RDI for sulfur or sulfate because animals, including humans, can derive the sulfur they need from the sulfur-containing amino acids methionine (essential) and cysteine (conditionally essential). The need for these amino acids and thus sulfur is satisfied by meeting the DRI for protein.

Sulfur Deficiency Sulfur has no known deficiency state in humans. Protein deficiency would appear first.

Sulfur Toxicity Excess sulfur intake can occur from consuming drinking water containing high inorganic sulfate levels. Such intake is associated with osmotic diarrhea. Also, sulfate and undigested sulfur-containing compounds are converted to damaging molecules in the colon by sulfate-reducing bacteria. Thus, excess sulfur and sulfate may contribute to inflammatory bowel diseases such as ulcerative colitis.

metabolic alkalosis A blood pH greater than 7.45 that results from not enough hydrogen ions in the blood; leads to paralytic ileus and can be caused from magnesium toxicity.

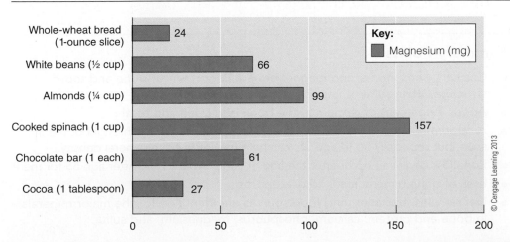

FIGURE 5.27 Food sources of magnesium.

Sulfur Food Sources Inorganic sulfate is present in foods and in tap water, which is consumed and used in a variety of beverages. Roughly 19 percent of dietary sulfate comes from dried fruits, commercial breads, soy, and sausages. Roughly 17 percent of sulfate intake comes from beverages such as beers, wines, ciders, and some juices. The other 64 percent comes from organic compounds in protein-containing foods.

TABLE 5.5

Major Minerals in Bone and Protein: Functions, Deficiencies, Toxicities, and Food Sources

Nutrient and Abbreviation	Functions in the Human Body	Deficiency Name, Signs and Symptoms	Toxicity Signs and Symptoms	Sources (Food and Other)
Calcium Ca	Bone and tooth structure, muscle contraction, nerve conduction, and blood clotting	*Osteopenia* BONE & TOOTH: Stunted growth in children, low bone density, osteoporosis	*Hypercalcemia* GI SYSTEM: Constipation, reduced absorption of iron, magnesium, zinc, and phosphorus OTHER: Kidney stones, calcium deposits in soft tissues	Dairy products, calcium-fortified nondairy milk alternatives, fortified foods, canned fish with the bones included, tofu
Phosphorus P	Bones and teeth, DNA and RNA, phospholipids, phosphorylation reactions, energy transfer (ATP), and buffering systems	*Hypophosphatemia* BONE & TOOTH: Bone pain GI SYSTEM: Anorexia NEUROMUSCULAR SYSTEM: Muscle weakness OTHER: General debility	*Hyperphosphatemia* GI SYSTEM: Reduced calcium absorption OTHER: Calcification of nonskeletal tissues	Animal proteins, processed foods
Magnesium Mg	Needed to build bone, teeth, and proteins; used in enzyme action, muscle contraction, blood clotting, and nerve impulse transmission	*Rare in healthy people Associated with hypocalcemia* CENTRAL NERVOUS SYSTEM: Seizures NEUROMUSCULAR SYSTEM: Muscle cramping, hyperexcitability OTHER: Vitamin D metabolism interference	*From supplements and laxatives* GI SYSTEM: Diarrhea, nausea, abdominal cramps, paralytic ileus OTHER: Metabolic alkalosis, hypokalemia	Nuts, legumes, whole grains, dark green leafy vegetables, chocolate, cocoa
Sulfur S ORGANIC OR INORGANIC SULFUR- OR SULFATE-CONTAINING COMPOUNDS	Biosynthesis of sulfur- and sulfate-containing compounds, stabilize protein shape via disulfide bridges	*No deficiency state known* Protein deficiency would appear first	GI SYSTEM: Osmotic diarrhea, may contribute to inflammatory bowel diseases such as ulcerative colitis	Dried fruits, commercial breads, soy, sausages, tap water, some beverages, protein-containing foods

© Cengage Learning 2013

Summary Points

- Calcium, phosphorus, and magnesium are important for bone and tooth structure, while sulfur is important in protein structure.
- About 99 percent of calcium in the body is in bone and teeth.
- A calcium-adequate diet and weight-bearing physical activity throughout the life span, but especially up to age 25, can help maximize bone mineral density.
- Bone loss begins to occur after the age of 35 for women and after age 55 for men.
- Loss of bone mineral leads to osteoporosis.
- Deficiencies, toxicities, and food sources are identified for the major minerals in bone and protein: calcium, phosphorus, magnesium, and sulfur.

Take Ten on Your Knowledge Know-How

1. Which major minerals are important for bone health? Which are important for protein shape?
2. What are the functions, the deficiency and toxicity signs and symptoms, and good food sources of calcium?
3. What are the functions, the deficiency and toxicity signs and symptoms, and good food sources of phosphorus?
4. What are the functions, the deficiency and toxicity signs and symptoms, and good food sources of magnesium?
5. What are the functions, the deficiency and toxicity signs and symptoms, and good food sources of sulfur?
6. What is osteoporosis, and how can you reduce the risk of developing it?
7. How would you know if your sulfur intake is adequate?
8. What happens if your intake of phosphorus is greater than your calcium intake?
9. What percentage of total body magnesium is in the bones and teeth? What about phosphorus? What about calcium?
10. What are the major sources of inorganic sulfate? What type of gastrointestinal condition may result from a high sulfate intake?

5.5 TRACE MINERALS

Introduction   The trace minerals are required in and exist in small amounts in the body. They include iron, iodine, zinc, fluoride, selenium, manganese, molybdenum, chromium, and copper. The role of trace elements in the body is for the structure of chemicals needed for tissue integrity or regulatory function.

Iron[A,I,28,30,33,46,73,74]   (See Table 5.6 on page 275 for a summary of iron, molecularly abbreviated Fe, and shown in the periodic chart in Appendix A.)

Iron Functions   Iron serves as a component of several functional proteins, including hemoglobin, **myoglobin**, cytochromes (such as those required for electron transport in ATP production as well as in liver detoxification), enzymes (flavoproteins), and the iron storage proteins (**transferrin**, **lactoferrin**, and **ferritin**). Hemoglobin is the protein that carries oxygen in the blood and accounts for about 67 percent of the body's iron. Myoglobin is the protein that makes oxygen available for the muscle and accounts for about 15 percent of the body's iron.

Iron's DRI, UL, and RDI   The DRI for iron, expressed in milligrams, is based on maintaining adequate iron stores in the body. The DRI for an adult male 19 to 30 years old is 8 milligrams per day; for an adult female 19 to 30 years old, it is 18 milligrams per day, more than 50 percent higher than that for a man, to provide more iron to cover iron losses during the menstrual cycle. Pregnant women have the highest iron need, at 27 milligrams per day. A pregnant woman needs more iron to support the increase in blood for the baby and herself. Lactating women have a DRI of 9 or 10 milligrams per day, depending on age. For individuals consuming a vegetarian diet, the iron DRI may be up to 80 percent greater (1.8 times) than the DRI value. The UL for an adult male or an adult female 19 to 70 years old is 45 milligrams per day. The RDI is 18 milligrams per day. Refer to Appendix A for all life span and gender-specific values.

Iron Deficiency   Iron-deficiency anemia is the most common type of anemia found in the United States and the world. Iron status can be viewed as normal, depleted, or deficient. There are two distinct phases in the development of iron-deficiency

T-Talk 5.5
To hear Dr. Turley talk about trace minerals, go to www.cengagebrain.com

myoglobin An iron-containing protein of the muscle cells that is needed for oxygen uptake.

transferrin An iron-containing transport protein in the body.

lactoferrin An iron-containing protein that is produced in mother's milk and is well absorbed.

ferritin An iron-containing protein that is used to evaluate total body iron stores.

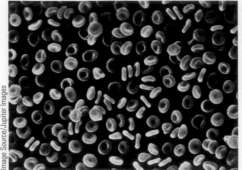

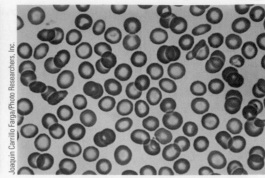

Normal red blood cells (left) and iron-deficient microcytic hypochromic anemic red blood cells (right), shown at a higher magnification.

anemia. In the first stage, iron depletion occurs without apparent signs or symptoms. In the second stage, moderate to severe iron deficiency occurs, with biochemical and physiological manifestations. Severe deficiency is clinically diagnosed and characterized by the sign of **microcytic hypochromic anemia**. The diagnostic symptoms are those related to anemia in general, such as fatigue and apathy.

Women of childbearing age (premenopausal), adolescents, children, infants, and athletes are at greater risk for having iron-deficiency anemia. Because of the iron losses incurred during menses, women are at much greater risk for developing iron-deficiency anemia than men. Women of childbearing age who are reproducing have the highest iron needs. Thus, pregnant women are routinely given iron supplements to meet the increased production of both maternal and fetal blood and the recovery of iron lost during delivery. It is estimated that 3 to 5 percent of women and less than 1 percent of men have iron-deficiency anemia in the United States. The statistics are much higher for children and adolescents, especially those involved in sports. In America, 25 to 35 percent of adolescent girls and 11 to 15 percent of adolescent boys have anemia. During infancy, childhood, and adolescence, there is an increased demand for blood formation, and thus iron, because of rapid growth and development. Iron deficiency in infants causes decreased psychomotor development, and in children it causes impaired cognitive function.

With sport activity, there is a higher rate of red blood cell destruction (especially for athletes in weight-bearing sports) and an increased demand for aerobic (oxygen-dependent) respiration. Remember that iron is required for delivering oxygen to all of the body cells (bound to hemoglobin), including the working muscles (bound to myoglobin).

All types of anemia (see BioBeat 5.8), including iron-deficiency anemia, are characterized by an inability to deliver oxygen to body cells and tissues. Iron deficiency with anemia is most commonly assessed by low hemoglobin blood values.

microcytic hypochromic anemia An anemia that can be used to describe the red blood cells as small and low in color.

nutritional anemias The types of anemia caused by nutrient deficiencies.

BioBeat 5.8

Nutritional Anemias

Several essential nutrients are involved in the differentiation of bone marrow stem cells into blood cells, and several are involved in the maturation of blood cells. If one or more of these nutrients are limited, the differentiation of blood cells or the maturation of the blood cells is altered, and dysfunctional blood cells develop. Dysfunctional red blood cells may have macrocytic, microcytic, or hemolytic characteristics. Dysfunctional red blood cells prevent adequate delivery of oxygen to cells, and the symptoms of anemia manifest. **Nutritional anemias** most frequently result from inadequacy of vitamins E (in prematurely born infants), B_6, B_{12}, and folic acid, and the minerals iron and copper. No matter what essential nutrient is inadequate, the individual feels tired (mentally and physically), weak, apathetic, cold, dizzy, irritable, and often headachy.[C,E,F,39]

Are you at risk for a nutrition-related anemia?

Other laboratory tests to evaluate a person's iron status include total iron-binding capacity, serum ferritin, transferrin saturation, and other complete blood count markers. Anemia, in general, results in the common signs and symptoms of weakness, mental apathy, headache, and pallor (paleness).

Iron deficiency without anemia can have symptoms of muscle fatigue and reduced capacity to work. These symptoms are believed to occur even with normal hemoglobin levels, because reduced iron, measured as ferritin, results in lower activity of iron-dependent enzymes needed for oxidative metabolism. Moderate iron deficiency without anemia is determined by having normal hemoglobin levels but low serum ferritin levels and/or a low transferrin receptor/ferritin index. Iron deficiency without anemia occurs in 12 to 16 percent of American women and 2 percent of American men. Studies have shown that an even higher percentage of adult athletes have iron deficiency without microcytic hypochromic anemia: 36 percent of trained adult female athletes and 6 percent of trained adult male athletes.

Iron Toxicity   Iron toxicity (overload) from food intake is traditionally unlikely, though possibly more likely with the intake of a large amount of fortified foods. Supplement users can develop toxicity. However, the primary form of toxicity is called **hemochromatosis** and is a genetic disorder in which iron absorption is upregulated, leading to toxicity (see BioBeat 5.9). Excess iron storage is known as **hemosiderosis**. This condition can be life threatening and occurs more frequently in men (because of the menstrual cycle in women). Gastrointestinal acute effects from high-dose iron supplementation include constipation, nausea, vomiting, and diarrhea. Chronic iron toxicity damages the central nervous, cardiovascular, and renal systems, and the liver. Skin color may be bronze or gray. Treatment for iron overload includes low iron consumption and regularly scheduled phlebotomy (blood removal).

hemochromatosis A genetic disease that causes high absorption rates of iron, resulting in iron toxicity; classically diagnosed by the triad of bronze-colored skin, diabetes (hyperglycemia), and liver disease.

hemosiderosis An iron toxicity condition, characterized by excess iron storage, that is commonly genetically linked and is more prevalent in men and often tied to hemochromatosis.

BioBeat 5.9

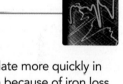

Iron Overload and Hereditary Hemochromatosis

Nearly all life forms require iron. The role of iron ecologically can be seen in oceans, on land, and in microbes, plants, and animals. In humans, iron is vital for oxygen delivery to cells and functions in metabolism. A delicate balance exists for regulating iron status in the body, because too much iron can be just as damaging as not enough iron. Hereditary hemochromatosis (HHC) is common in people of Western European genetic heritage. It is a result of a combination of genetic and environmental factors that lead to excess absorption of iron. HHC is a genetic (autosomal recessive) disorder caused by a mutation of the HFE gene that causes excessive intestinal absorption of iron. Evidence suggests natural selection played a part in the preservation of this genetic trait. Theories include evolution of the trait from undernourished populations, from warrior cultures such as the Vikings suffering from frequent

blood loss and thus iron loss, and from women carrying the trait having a reproductive advantage because of iron loss during menstruation. HHC is also tied historically to the bubonic plague that traveled across Europe in the early 1300s. People with HHC have excess iron, which would normally increase the risk for microbial infection and would kill them. However, with HHC, macrophages (specialized immune cells) are low in iron. The low iron distribution to these cells is believed to provide a selective advantage, because macrophages engulf and destroy the bacteria that cause the plague and lock them out of an iron source for their survival.

HHC presents clinically as the classic triad of bronze skin color, diabetes (hyperglycemia), and liver cirrhosis; it usually occurs in adults in their 40s and 50s and can be diagnosed earlier in life, especially if there is a family history. Over time, total body iron con-

tinues to accumulate more quickly in men than women because of iron loss through menstruation and child bearing. When vital organ levels of iron get too high, normal function is impeded. Liver function is compromised because of cirrhosis from iron toxicity, and there is a 200 percent increased risk for liver cancer. Damage to the pancreas results in decreased insulin production and sensitivity, and in the heart, arrhythmias and congestive heart failure are common. In addition, a common symptom is bone changes that resemble osteoarthritis, and several tissues in the endocrine system will also become damaged if the HHC is not managed. The best modern way to treat HHC is for the person to have blood removed regularly. In early history, bloodletting would have been an effective method to reduce total body iron.[B,E,F]

What is your genetic heritage, and does it predispose you to HHC?

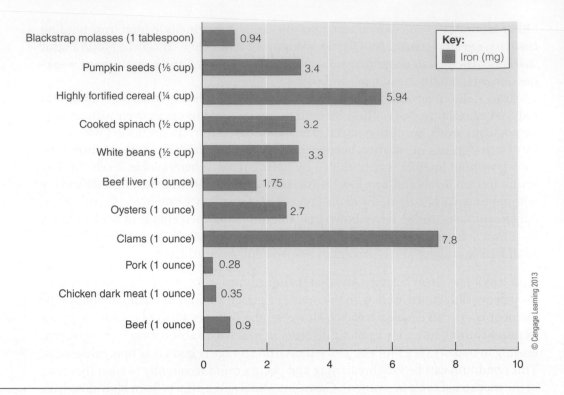

FIGURE 5.28 Food sources of iron.

Iron Food Sources   There are two sources of iron in the diet: plant and animal. Plant sources of iron provide *non-heme* iron, which is poorly absorbed (2 to 5 percent). However, the absorption of non-heme iron is increased by jointly consuming a food source rich in vitamin C or mixing vegetable intake with meat at mealtime. Another cooking technique to add non-heme iron is to use cast-iron cookware. The iron can be leached from the pot (usually by cooking acidic foods like tomato sauce) into the food, thus increasing the available iron content of the food. Foods providing non-heme iron include iron-fortified cereals and some legumes (dried beans), whole grains, vegetables, fruits, nuts, and seeds. The best non-heme sources include highly fortified cereals, soybeans, white beans, pumpkin seeds, blackstrap molasses, and spinach (see Figure 5.28). *Heme* iron is from animal meat sources and is well absorbed (10 to 20 percent) compared to non-heme iron. Foods providing a good source of heme iron include red meats (especially beef), poultry (especially dark meat), liver, and shellfish. The best heme iron sources include clams, oysters, organ meats, and red muscle meats. Lower amounts are found in fish, lamb, and pork (see Figure 5.28).

Iodine[24,30,33,48,75,76]   (See Table 5.6 on page 275 for a summary of iodine, molecularly abbreviated I, and shown in the periodic chart in Appendix A.)

Iodine Functions   Iodine functions as an essential component of the thyroid hormones **thyroxine** (T4) and triiodothyronine (T3). T3, the active form, regulates metabolic rate and influences the function of the brain, muscles, heart, pituitary gland, and kidneys.

Iodine's DRI, UL, and RDI   The DRI for iodine, expressed in micrograms, is 150 micrograms per day for an adult male or an adult female 19 to 30 years old. The UL for an adult male or an adult female 19 to 70 years old is 1,100 micrograms. The RDI is 150 micrograms per day. Refer to Appendix A for all life span and gender-specific values.

Iodine Deficiency   If maternal iodine intake is low, there is a deficiency of iodine in utero. If maternal iodine intake is corrected early in the pregnancy, then the risk of developmental defects and mental retardation in the offspring may be reduced or prevented. Deficiency in utero and in children causes **cretinism**. This condition

thyroxine A hormone produced by the thyroid gland that regulates the metabolic rate and requires iodine in its chemical structure.

cretinism Mental retardation at birth, which can be caused by an iodine deficiency.

is characterized by growth and developmental abnormalities and mental retardation. At any life span stage, iodine deficiency signs and symptoms are associated with **goiter** (enlarged thyroid gland), **hypothyroidism** (which is associated with elevated thyroid stimulation hormone (TSH) levels), and reduced metabolic rate. Goiter results when elevated TSH levels cause the thyroid gland to be overstimulated. Because iodine is lacking, thyroxine cannot be made properly by the thyroid gland, and the gland grows in size.

Iodine Toxicity   Excess intake of iodine from food is tolerated by most people. Adverse effects from chronic excess intake from food, water, and supplements include goiter, hypothyroidism (with elevated TSH levels), thyroiditis, sensitivity reactions, and thyroid cancer. Acute toxicity signs include burning of the mouth, throat, and stomach, abdominal pain, fever, nausea, vomiting, diarrhea, weak pulse, cardiac irritability (causing a weak heartbeat), coma, and cyanosis (blue skin).

Iodine Food Sources   Iodine content in food is relatively low and is affected by iodine levels in the soil, which in turn is affected by farming practices. Land that was once covered by sea water has iodine. In areas that do not have iodine in the soil, the development of iodine-deficiency-related goiter has been very common. The advent of iodized salt has successfully controlled iodine deficiency in developed nations. Seafood and iodized salt are rich in iodine (depending on the salt manufacturer, 200 to 400 micrograms of iodine are added per teaspoon of salt; see Figure 5.29) and should be included in the diet in order to ensure adequate iodine intake. Even though sea salt sounds like it contains iodine, it does not provide significant amounts that are bioavailable.

Zinc[24,30,33,53,74,75,76]   (See Table 5.6 on page 275 for a summary of zinc, molecularly abbreviated Zn, and shown in the periodic chart in Appendix A.)

Zinc Functions   Zinc has catalytic, structural, and regulatory functions and supports growth and development. It is an essential component of more than 100 different enzymes that are involved in a plethora of chemical reactions. Zinc is needed to make DNA and proteins and for immune reactions, for vitamin A transport, for taste perception, for wound healing, for spermatogenesis, for gene expression, and for fetal development. Zinc is also required for insulin synthesis, storage, and release.

Zinc's DRI, UL, and RDI   The DRI for zinc, expressed in milligrams, is 11 milligrams per day for an adult male 19 to 30 years old; for an adult female 19 to 30 years old, it is 8 milligrams per day. For individuals consuming a vegetarian diet, the actual zinc DRI may be up to 50 percent (1.5 times) greater than the DRI value. The UL for an adult male or an adult female 19 to 70 years old is 40 milligrams per day. The RDI is 15 milligrams per day. Refer to Appendix A for all life span and gender-specific values.

Zinc Deficiency   Zinc deficiency was characterized in humans in the Middle East during the 1960s and is called **acrodermatitis enteropathica**. This form of zinc deficiency is caused by an **inborn error of metabolism**. Aside from this genetic condition, human deficiency of zinc is rare, even though zinc deficiency can occur with only modest degrees of zinc restriction. Furthermore, chronic malabsorption syndromes and alcoholism can result in poor zinc status. The signs and symptoms of zinc deficiency include retarded growth and delayed sexual maturation, impotence, decreased taste acuity, impaired appetite, diarrhea, poor wound healing, hair loss, and eye and skin lesions.

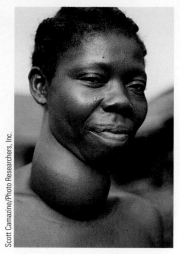

A woman with an enlarged thyroid gland (goiter) caused by iodine deficiency.

goiter An enlargement of the thyroid gland caused by an iodine deficiency or toxicity, malfunction of the gland, or overconsumption of goitrogens.

hypothyroidism Low blood levels of thyroid hormone fractions

acrodermatitis enteropathica An inborn error in metabolism that results in zinc deficiency.

inborn error of metabolism A disease caused by an inherited genetic defect causing a specific abnormality in metabolism.

FIGURE 5.29 Food sources of iodine.

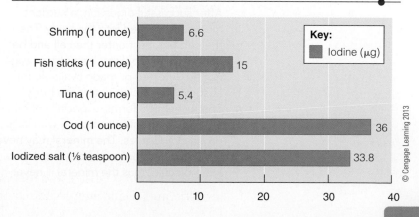

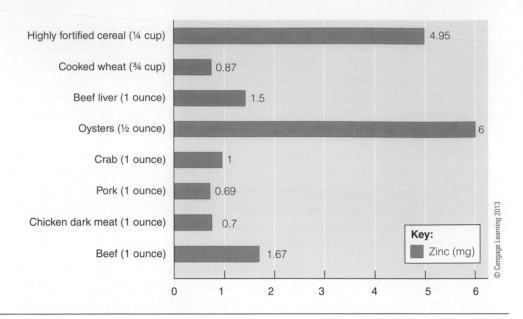

Food	Zinc (mg)
Highly fortified cereal (¼ cup)	4.95
Cooked wheat (¾ cup)	0.87
Beef liver (1 ounce)	1.5
Oysters (½ ounce)	6
Crab (1 ounce)	1
Pork (1 ounce)	0.69
Chicken dark meat (1 ounce)	0.7
Beef (1 ounce)	1.67

Key: Zinc (mg)

© Cengage Learning 2013

FIGURE 5.30 Food sources of zinc.

Zinc Toxicity   Excess zinc intake from natural food sources has not been reported to cause toxicity. Chronic high zinc intakes from supplements can result in altered and suppressed immunity, decreased HDL cholesterol, and reduced copper status. The signs and symptoms of acute zinc toxicity include loss of appetite, upper gastrointestinal pain, nausea, vomiting, diarrhea, abdominal cramps, and headache.

Zinc Food Sources   Good food sources of zinc are shellfish (especially oysters; however, not all seafood provides a good source of zinc), meats (especially red meats), and organ meats (see Figure 5.30). Zinc can also be found in whole grains and some fortified cereals. Plant sources of zinc are not as bioavailable for the body (see BioBeat 5.10). Refinement of grains causes up to 80 percent loss of zinc, and zinc is not one of the nutrients required by the Enrichment Act of 1942 to be added to processed foods.

BioBeat 5.10

The Biological Availability of Nutrients from Food or Dietary Supplements

How effectively a nutrient is utilized depends on the chemical form of the nutrient. The nutritional value of a nutrient cannot be realized until it enters the cell and becomes involved in metabolism. After successful digestion, a nutrient must be absorbed into the body. Then the nutrient must enter the cell and be utilized in metabolism. If a nutrient cannot be digested or made available for absorption, it never enters the body. Some chemical forms, especially of minerals, can be chelated or bound with large, organic molecules. The mineral may never be dissociated or freed from the chelating molecule, thus the mineral is never biologically available for absorption. The chemical analysis of the nutrient content of food often appears adequate, but the biological utilization of the nutrient from the food may show deficiency. This has been documented in the medical literature regarding chemical analysis of the iron and zinc content in unleavened whole-wheat bread causing, respectively, iron-deficiency anemia and delayed sexual maturation in Middle Eastern children.

Have you ever wondered about what forms a nutrient is in, such as in fortified cereal, and whether it is optimal for the body to use?

Fluoride[D,G,2,6,7,24,29,30,75,76] (See Table 5.6 on page 276 for a summary of fluoride, molecularly abbreviated F, and shown in the periodic chart in Appendix A.)

Fluoride Functions Fluoride is vital for bone and tooth calcification. Fluoride stimulates new bone formation and protects against dental caries (cavities). In teeth, fluoride replaces the hydroxy portion of **hydroxyapatite**, causing the formation of more decay-resistant **fluorhydroxyapatite**. This compound hardens tooth enamel and stabilizes bone mineral structure.

Fluoride's DRI, UL, and RDI The DRI for fluoride, expressed in milligrams, is 4 milligrams per day for an adult male 19 to 30 years old; for an adult female 19 to 30 years old, it is 3 milligrams per day. The UL for an adult male or an adult female 19 to 70 years old is 10 milligrams per day. There is no RDI for fluoride. Refer to Appendix A for all life span and gender-specific values.

Fluoride Deficiency A lack of fluoride causes dental caries. Optimal fluoride levels are typically achieved by supplementation of fluoridated water and/or supplementation through the guidance of dental health professionals. Low fluoride intake is also associated with vulnerable bones and teeth.

Fluoride Toxicity If too much fluoride is ingested during the period of tooth development (even pre-eruptive), tooth mottling, or **fluorosis** (discoloring of the teeth), will occur. Children chronically ingesting large amounts of fluoridated toothpaste and mouth rinses may develop fluorosis. Acute fluoride toxicity (1 to 5 milligrams per kilogram of body weight per day) signs and symptoms include nausea, vomiting, diarrhea, abdominal pain, and excessive salivation. Seizures, cardiac arrhythmias, and coma have also been reported. Chronic high intake (at least 10 milligrams per day for 10 years or longer) can lead to skeletal fluorosis, which starts with stiffness or pain in the joints accompanied by osteosclerosis (abnormal hardening of the bone), and progresses to hypercalcification, muscle wasting, and neurological defects.

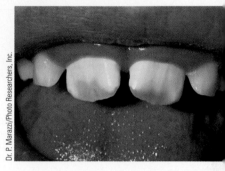

Fluorosis (discoloration of the teeth) caused by excess fluoride consumption during tooth formation.

Fluoride Food Sources Sources of fluoride include fluoridated water (1 part per million provides 1 milligram of fluoride per liter of water), some marine fish (like sardines when the bones are consumed), and tea.

Selenium[D,G,I,24,30,32,39,53,75,76] (See Table 5.6 on page 276 for a summary of selenium, molecularly abbreviated Se, and shown in the periodic chart in Appendix A.)

Selenium Functions Selenium functions through selenoproteins. Selenium is a cofactor for the antioxidant enzyme glutathione peroxidase and has a role in thyroid hormone actions. Furthermore, selenium functions in the reduction-oxidation (redox) status of vitamin C and other molecules. Thus, selenium is an antioxidant nutrient.

Selenium's DRI, UL, and RDI The DRI for selenium, expressed in micrograms, for an adult male or female age 19 to 30 is 55 micrograms per day. The UL for an adult male or female 19 to 70 years old is 400 micrograms per day. The RDI is 70 micrograms per day. Refer to Appendix A for all life span and gender-specific values.

Selenium Deficiency **Keshan disease** has been documented as a result of a deficiency of selenium. Keshan disease involves **cardiomyopathy** (weak heart muscle) and is documented in selenium-deficient children who have other stressors, like infection or chemical exposure.

Selenium Toxicity Inorganic and organic (such as selenomethionine) selenium can cause toxicity (**selenosis**), although signs and symptoms may be more rapid and occur at lower doses from inorganic selenium. Toxicity signs and symptoms

hydroxyapatite The mineralized protein matrix of bone and teeth that is rich in calcium, phosphorus, and magnesium.

fluorhydroxyapatite The mineralized protein matrix of bone where the phosphorous has been replaced because of the presence of fluoride.

fluorosis Tooth discoloration caused by too much exposure to fluoride.

Keshan disease The deficiency disease for selenium that causes cardiomyopathy.

cardiomyopathy A medical condition where the contractility of the heart muscle is weakened, and thus the pumping action of the heart muscle is inadequate to supply oxygen to the cells.

selenosis The toxicity condition of selenium commonly caused by selenomethionine that causes changes in connective and nervous tissue, garlic breath, and GI distress.

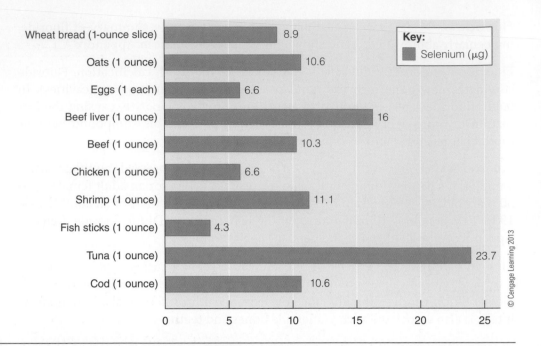

FIGURE 5.31 Food sources of selenium.

Key:
- Selenium (μg)

Food	Selenium (μg)
Wheat bread (1-ounce slice)	8.9
Oats (1 ounce)	10.6
Eggs (1 each)	6.6
Beef liver (1 ounce)	16
Beef (1 ounce)	10.3
Chicken (1 ounce)	6.6
Shrimp (1 ounce)	11.1
Fish sticks (1 ounce)	4.3
Tuna (1 ounce)	23.7
Cod (1 ounce)	10.6

© Cengage Learning 2013

include hair and nail brittleness and loss, gastrointestinal distress, a garlic breath odor, skin rash, fatigue, irritability, and nervous system abnormalities.

Selenium Food Sources   Selenium is found in meats, seafood, grains, dairy products, fruits, and vegetables (see Figure 5.31). The amount of selenium in plants is dependent on the selenium content of the soil, which varies greatly; thus, meat and seafood may provide the most reliable dietary sources.

Manganese[24,30,33] (See Table 5.6 on page 276 for a summary of manganese, molecularly abbreviated Mn, and shown in the periodic chart in Appendix A.)

Manganese Functions   Manganese functions in bone formation and in amino acid, cholesterol, and carbohydrate metabolism. Manganese is required for, and/ or activates, several **metalloenzymes.**

Manganese's DRI, UL, and RDI   The DRI for manganese, expressed in milligrams, is 2.3 milligrams per day for an adult male age 19 to 30. The DRI for an adult female age 19 to 30 is 1.8 milligrams per day. The UL for an adult male or female 19 to 70 years old is 11 milligrams per day. The RDI is 2 milligrams per day. Refer to Appendix A for all life span and gender-specific values.

Manganese Deficiency   There is limited data on manganese deficiency in humans; however, deficiency has been found to cause scaly, red dermatitis and reduced blood cholesterol levels. Furthermore, in reproductively capable women, low manganese status is associated with altered mood and increased pain.

Manganese Toxicity   Manganese toxicity occurs most frequently from airborne dust in industrial work settings. Toxicity causes central nervous system effects similar to Parkinson's disease (tremors, rigidity of limbs and trunk, and slow gait and lack of coordination). People with chronic liver disease seem to be particularly sensitive to manganese toxicity.

Manganese Food Sources   Manganese is found mostly in grains and grain products (37 percent), tea (20 percent), and vegetables (18 percent) (see Figure 5.32).

Molybdenum[24,30,33] (See Table 5.6 on page 276 for a summary of molybdenum, molecularly abbreviated Mo, and shown in the periodic chart in Appendix A.)

metalloenzymes Enzymes containing minerals that are metals.

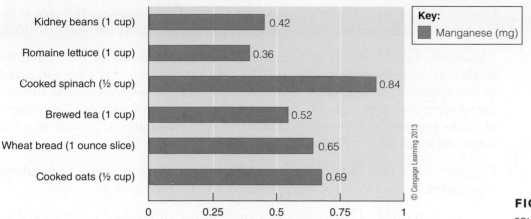

FIGURE 5.32 Food sources of manganese.

Molybdenum Functions   Molybdenum functions as a cofactor for enzymes called molybdoenzymes that are needed for the catabolism of sulfur-containing amino acids and some RNA compounds.

Molybdenum's DRI, UL, and RDI   The DRI for molybdenum, expressed in micrograms, is 45 micrograms per day for an adult male or female age 19 to 30. The UL for an adult male or female 19 to 70 years old is 2,000 micrograms per day. The RDI is 75 micrograms per day. Refer to Appendix A for all life span and gender-specific values.

Molybdenum Deficiency   Molybdenum deficiency has not been observed in healthy people but is well documented in people with a specific genetic defect affecting molybdoenzymes. A case study report of a man with Crohn's disease (a separate preexisting medical condition not caused from molybdenum deficiency) who had a dietary deficiency of molybdenum reported symptoms including tachycardia, headache, and night blindness, which where all corrected after molybdenum administration.[33]

Molybdenum Toxicity   Molybdenum toxicity has not been fully defined in humans, possibly because high intakes are rapidly excreted in urine. Individuals with copper deficiency and/or dysfunction in copper metabolism may be at increased risk of molybdenum toxicity.

Molybdenum Food Sources   The amount of molybdenum in foods depends on the soil in which the food is grown. Legumes, grain products, and nuts provide the greatest source of this trace mineral.

Chromium[24,33,48]   (See Table 5.6 on page 277 for a summary of chromium, molecularly abbreviated Cr, and shown in the periodic chart in Appendix A.)

Chromium Functions   Chromium plays a chemical role in the body largely to enhance the action of insulin and thus appears to improve glucose tolerance.

Chromium's DRI, UL, and RDI   The DRI for chromium, expressed in micrograms, is 35 micrograms per day for an adult male age 19 to 30. The DRI for an adult female age 19 to 30 is 25 micrograms per day. The UL for chromium has not been determined. The RDI is 120 micrograms per day. Refer to Appendix A for all life span and gender-specific values.

Chromium Deficiency   The effect of chromium deficiency in humans is poorly documented. Three patients who lacked chromium in the formula given to them through a feeding line into their blood stream experienced weight loss, peripheral neuropathy, and increased free fatty acid oxidation as a result of chromium deficiency. These symptoms were corrected by the administration of chromium.[33]

Chromium Toxicity   Chromium is poorly absorbed, and thus toxicity is rare. Individuals with kidney and liver disease may be more susceptible to chromium toxicity.

Chromium Food Sources   Chromium is widely distributed in foods, and its levels may be increased or decreased by food processing. Levels are increased in acidic foods that are processed in stainless steel cookware. The best sources are whole grains, especially the bran component of the grain kernel. Refinement of grains generally reduces chromium levels. Chromium is also present in some beers and wines (especially French red wines).

Copper[24,30,33,39]   (See Table 5.6 on page 277 for a summary of copper, molecularly abbreviated Cu, and shown in the periodic chart in Appendix A.)

Copper Functions   Copper acts as a cofactor for antioxidant enzymes and in the electron transport chain. It is important for melanin, collagen, and elastin (connective tissue proteins) biosynthesis, and is a component of the enzyme ceruloplasmin, which is instrumental in iron oxidation and binding to transferrin. Copper helps maintain neurochemical balance and inactivates histamine release during allergic reactions.

Copper's DRI, UL, and RDI   The DRI for copper, expressed in micrograms, is 900 micrograms per day for an adult male or female age 19 to 30. The UL for an adult male or female 19 to 70 years old is 10,000 micrograms per day. The RDI is 2,000 micrograms per day. Refer to Appendix A for all life span and gender-specific values.

Copper Deficiency   Copper deficiency is rare in adults but has occurred in prematurely born infants. The deficiency signs and symptoms include **normocytic hypochromic anemia**, **leukopenia** (lowered white blood cell count), **neutropenia** (lowered **neutrophil** cell count), and osteoporosis (in copper-deficient infants and children). There is an inherited syndrome called **Menkes disease** (also known as Menkes kinky hair, steely hair disease, and kinky hair disease). This relatively rare and life-threatening disorder affects primarily males and causes reduced copper absorption and transport to peripheral tissues.

Copper Toxicity   Toxicity from copper is unlikely but may occur from supplements and some beverages and drinking water. Signs and symptoms of copper toxicity are primarily GI tract–related and include abdominal pain, cramps, nausea, diarrhea, and vomiting. A genetic disease called **Wilson's disease** leads to copper toxicity from an impaired ability to excrete copper in bile. In Wilson's disease, copper toxicity causes liver damage. Low dietary intake of copper is a treatment component for Wilson's disease.

Copper Food Sources   Copper is found in many foods but is most concentrated in organ meats, seafood, nuts, seeds, whole grains, wheat bran, and cocoa products (see Figure 5.33).

Cobalt   Cobalt (molecularly abbreviated Co, and shown in the periodic chart in Appendix A) is the central component in vitamin B_{12} (cobalamin). See the section on vitamin B_{12} for its forms, functions, food sources, deficiency information, and toxicity information.

normocytic hypochromic anemia The characterization of the red blood cells that are normal in size but low in color.

leukopenia A condition of the blood that is characterized by lowered white blood cell count.

neutropenia A condition of the blood that is characterized by lowered neutrophil cell count.

neutrophil An abundant type of white blood cell (leukocyte) that is an important phagocyte (engulfing and destroying cell).

Menkes disease An inborn error in metabolism that causes a copper deficiency, which is characterized by osteoporosis in infants and children, normocytic hypochromic anemia, and a low white blood cell count.

Wilson's disease The name of the disease caused by a genetic disorder resulting in copper overload, largely causing GI tract distress and liver damage.

FIGURE 5.33 Food sources of copper.

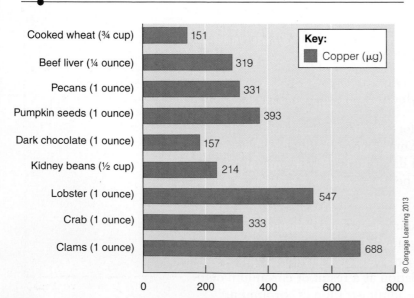

Food	Copper (µg)
Cooked wheat (¾ cup)	151
Beef liver (¼ ounce)	319
Pecans (1 ounce)	331
Pumpkin seeds (1 ounce)	393
Dark chocolate (1 ounce)	157
Kidney beans (½ cup)	214
Lobster (1 ounce)	547
Crab (1 ounce)	333
Clams (1 ounce)	688

Key:
■ Copper (µg)

© Cengage Learning 2013

TABLE 5.6

The Trace Minerals: Functions, Deficiencies, Toxicities, and Food Sources

Nutrient and Abbreviation	Functions in the Human Body	Deficiency Name, Signs and Symptoms	Toxicity Signs and Symptoms	Sources (Food and Other)
Iron Fe	Component of hemoglobin (for oxygen transport), myoglobin, cytochromes, enzymes, and iron storage proteins	*Microcytic hypochromic anemia* CENTRAL NERVOUS SYSTEM: Mental apathy, cold feeling, dizziness, irritability, headache NEUROMUSCULAR SYSTEM: Weakness, fatigue SKIN: Pallor (paleness)	*Unlikely from food* *Most likely from supplements or a genetic disorder* CARDIOVASCULAR SYSTEM: Damages the cardiovascular system CENTRAL NERVOUS SYSTEM: Damages the central nervous system GI SYSTEM: Constipation, nausea, vomiting, diarrhea SKIN: Bronze or gray skin color OTHER: Damages the renal system and the liver	HEME: Clams, oysters, organ meats, red muscle meats NON-HEME: Highly fortified cereals, soybeans, white beans, pumpkin seeds, blackstrap molasses, spinach
Iodine I	Essential component of the thyroid hormones thyroxine (T4) and triiodothyronine (T3), regulates the metabolic rate and influences the function of the brain, muscles, heart, pituitary gland, and kidneys	*Hypothyroidism* OTHER: Cretinism (growth and developmental abnormalities and mental retardation *in utero* and in children), goiter, reduced metabolic rate	*Hyperthyroidism* CARDIOVASCULAR SYSTEM: Weak pulse, cardiac irritability (with acute toxicity) CENTRAL NERVOUS SYSTEM: Coma (with acute toxicity) GI SYSTEM: Burning of the mouth, throat, and stomach, abdominal pain, fever, nausea, vomiting, diarrhea (with acute toxicity) SKIN: Cyanosis (with acute toxicity) OTHER: Goiter, hypothyroidism, thyroiditis, sensitivity reactions, thyroid cancer (with chronic toxicity)	Iodized salt, seafood
Zinc Zn	Catalytic, structural, and regulatory functions to support growth, development, and the function of more than 100 enzymes; required for insulin synthesis, storage, and release	GI SYSTEM: Decreased taste acuity, impaired appetite, diarrhea SKIN: Acrodermatitis enteropathica, poor wound healing, eye and skin lesions OTHER: Retarded growth and delayed sexual maturation, impotence, hair loss	CARDIOVASCULAR SYSTEM: Decreased HDL cholesterol CENTRAL NERVOUS SYSTEM: Headache GI SYSTEM: Loss of appetite, upper gastrointestinal pain, nausea, vomiting, diarrhea, abdominal cramps OTHER: Altered and suppressed immunity, reduced copper status	Shellfish, red and organ meats, whole grains, some fortified cereals

continued

The Trace Minerals: Functions, Deficiencies, Toxicities, and Food Sources (*continued*)

Nutrient and Abbreviation	Functions in the Human Body	Deficiency Name, Signs and Symptoms	Toxicity Signs and Symptoms	Sources (Food and Other)
Fluoride F	Bone and tooth health, and formation of decay-resistant fluorhydroxyapatite	BONE & TOOTH: Tooth decay	BONE & TOOTH: Fluorosis (discoloring of the teeth, tooth mottling), skeletal fluorosis, stiffness or pain in the joints, osteosclerosis, hypercalcification (with chronic toxicity) CARDIOVASCULAR SYSTEM: Cardiac arrhythmias (with acute toxicity) CENTRAL NERVOUS SYSTEM: Seizures, coma (with acute toxicity) GI SYSTEM: Nausea, vomiting, diarrhea, abdominal pain, excessive salivation (with acute toxicity) NEUROMUSCULAR SYSTEM: Muscle wasting, neurological defects (with chronic toxicity)	Fluoridated drinking water, marine fish containing bones, tea
Selenium Se	Selenoproteins, antioxidant, vitamin C redox	*Keshan disease* CARDIOVASCULAR SYSTEM: Cardiomyopathy	*Selenosis* CENTRAL NERVOUS SYSTEM: Fatigue, irritability, nervous system abnormalities GI SYSTEM: Gastrointestinal distress, garlic breath odor SKIN: Skin rash OTHER: Hair and nail brittleness and loss	Meats and seafood (most dependable sources), grains, dairy products, fruits, vegetables
Manganese Mn	Bone formation, amino acid, cholesterol, and carbohydrate metabolism	CARDIOVASCULAR SYSTEM: Reduced blood cholesterol levels CENTRAL NERVOUS SYSTEM: Altered mood, increased pain (in women of childbearing age) SKIN: Scaly, red dermatitis	CENTRAL NERVOUS SYSTEM: Effects similar to Parkinson's disease NEUROMUSCULAR SYSTEM: Tremors, rigidity of limbs and trunk, slow gait, lack of coordination	Grains, tea, vegetables
Molybdenum Mo	Cofactor for molybdoenzymes needed for the catabolism of sulfur-containing amino acids and some RNA compounds	*Poorly documented in humans*	*Poorly documented in humans*	Legumes, grain products, nuts

The Trace Minerals: Functions, Deficiencies, Toxicities, and Food Sources (*continued*)

Nutrient and Abbreviation	Functions in the Human Body	Deficiency Name, Signs and Symptoms	Toxicity Signs and Symptoms	Sources (Food and Other)
Chromium Cr	Increases the action of insulin	*Poorly documented in humans*	*Poorly documented in humans*	Whole grains, bran
Copper Cu	Cofactor for antioxidant enzymes and in the electron transport chain, important for connective tissue, iron oxidation, neurochemical balance, and histamine inactivation	BONE & TOOTH: Osteoporosis in infants and children CARDIOVASCULAR SYSTEM: Normocytic hypochromic anemia OTHER: Leukopenia, neutropenia	GI SYSTEM: Abdominal pain, cramps, nausea, diarrhea, vomiting	Organ meats, seafood, nuts, seeds, whole grains, wheat bran, cocoa products
Cobalt Co	A central component of vitamin B_{12} (cobalamin)	*See section on vitamin B_{12}*	*See section on vitamin B_{12}*	*See section on vitamin B_{12}*

© Cengage Learning 2013

Summary Points

- Iron-deficiency anemia is the most common nutritional anemia in the United States and the world.
- Iodine is essential to thyroid hormones.
- Zinc is important for protein metabolism and the structure of insulin.
- Fluoride is critical for dental health.
- Selenium is important for structures of selenoproteins that are involved in a wide variety of metabolic reactions.
- Many metalloenzymes are dependent on manganese and are involved in diverse roles in metabolism.
- Molybdenum is required for the structure of molybdoenzymes, which are needed for the catabolism of RNA and sulfur-containing amino acids.
- Chromium augments the action of insulin.
- Copper is involved in a wide variety of biochemical actions.
- Cobalt is required in the structure of vitamin B_{12}.

Take Ten on Your Knowledge Know-How

1. What is anemia? Nutritional deficiencies of which nutrients cause anemia? What type of anemia is caused by the nutritional deficiency? How can anemia be treated through dietary management?
2. What factors affect the bioavailability of nutrients?
3. What are some pros and cons of fluoridating public drinking water?
4. When should vitamin and mineral supplements be used? Do you think that you should use them? Why or why not?
5. What are the functions, deficiency and toxicity signs and symptoms, and good food sources of iron and iodine?
6. What are the functions, deficiency and toxicity signs and symptoms, and good food sources of zinc and manganese?
7. What are the functions, deficiency and toxicity signs and symptoms, and good food sources of fluoride and selenium?
8. What are the functions, deficiency and toxicity signs and symptoms, and good food sources of molybdenum and chromium?
9. What are the functions, deficiency and toxicity signs and symptoms, and good food sources of copper and cobalt?
10. What trace minerals affect bone?

5.6 SUMMARY

CONTENT KNOWLEDGE

IN THIS MODULE, YOU HAVE LEARNED:

- Foundational information on the fat-soluble and water-soluble vitamins, water, and the major and trace minerals.
- The signs and symptoms of deficiency and toxicity of many essential nutrients.
- Good food sources of each essential nutrient.
- The importance of achieving adequate nutrient intakes according to the DRIs and evaluating nutrient intake level by performing diet analysis.

PERSONAL IMPROVEMENT GOALS

WHEN YOU ARE EVALUATING THE DIETARY INTAKE OF NUTRIENTS, KEEP IN MIND THAT:

- The more days you can include in your dietary analysis, the better your understanding will be about the nutritional status of your body.
- If your intake of a nutrient is marginal, deficiency signs and symptoms may not be evident, but your body may not function optimally.
- It is usually safe to take a multivitamin and mineral complex daily, without exceeding the UL, and doing so ensures that you will meet the recommendation for your nutrient intakes.
- When a UL is established for a nutrient, this level can be consumed daily with little risk for toxicity. However, as intake continues to exceed the UL by a greater margin, the risk for toxicity increases.
- If you consume several fortified foods (such as ready-to-eat breakfast cereals and sport bars), you may exceed the UL for some vitamins and minerals.

Here is a tip for you: Be proactive in achieving adequate nutrient intake. Be aware that values on food labels are RDIs and that you should compare these to your age- and gender-specific DRI. Pay special attention to achieving adequate calcium intake. Remember that calcium deficiency is a silent disease, and you can actually increase bone density until you reach 25 years of age if your calcium intake is adequate.

You can assess if you met the learning objectives for this module by successfully completing the Homework Assessment and the Total Recall activities (sample questions, case study with questions, and crossword puzzle).

Homework Assessment

50 matching questions

You may use some answers more than once or not at all.

A nutrient required for:

1. Cell membranes
2. Bones and teeth
3. Nerve cell maintenance
4. Gastric secretions
5. Oxygen transport

   A. Iron
   B. Vitamin B_{12}
   C. Vitamin E
   D. Chloride
   E. Fluoride

A good food source providing:

6. Zinc
7. Vitamin D
8. Vitamin E
9. Calcium
10. Beta-carotene (provitamin A)

   A. Fortified milk
   B. Nuts
   C. Oysters
   D. Cantaloupe
   E. Pasta

A name or symptom for a nutrient deficiency of:

11. Vitamin D in adults
12. Vitamin B_{12}
13. Selenium
14. Iron
15. Riboflavin

   A. Microcytic anemia
   B. Ariboflavinosis
   C. Keshan disease
   D. Pernicious anemia
   E. Osteomalacia

A nutrient important for:

16. Appetite regulation
17. Connective tissue
18. Osmolarity

   A. Biotin
   B. Thiamin
   C. Vitamin C
   D. Sodium
   E. Riboflavin

An increased need for _____ would occur from _____.

19. Water
20. Vitamin B_6
21. Vitamin C
22. Vitamin K
23. Iron

   A. Antibiotic use
   B. Tobacco use
   C. Sweating
   D. Blood loss
   E. High dietary protein intake

A nutrient that functions as/in:

24. Coenzyme
25. A component of thyroxine
26. An antioxidant
27. Fluid balance
28. A component of hemoglobin

   A. Iodine
   B. Iron
   C. Vitamin C
   D. Potassium
   E. Niacin

An excess of _____ causes _____.

29. Niacin
30. Vitamin D
31. Fluoride
32. Sodium
33. Magnesium

   A. Diarrhea
   B. Edema
   C. Flushing reaction
   D. Discolored teeth
   E. Calcium deposits in soft tissue

A good food source providing:

34. Chloride
35. Riboflavin
36. Chromium
37. Manganese
38. Sulfur

   A. Milk products
   B. Protein-containing foods
   C. Bran
   D. Tea
   E. Salt

An appropriate dietary intake quantity for:

39. Protein
40. Iodine
41. Iron
42. Vitamin B_{12}

   A. Grams
   B. Milligrams
   C. Micrograms
   D. Nanograms
   E. Kilograms

A nutrient synthesized in good amounts by/from:

43. Sunlight
44. Friendly bacteria in the GI tract
45. The amino acid tryptophan
46. Cholesterol
47. Animals mostly

   A. Niacin
   B. Phosphorus
   C. Vitamin K
   D. Vitamin D
   E. Vitamin B_{12}

Miscellaneous matching:

48. Vitamin D
49. Antioxidants
50. Regulators of energy metabolism

   A. Most water-soluble vitamins
   B. Genetics and evolution
   C. Hormones
   D. Electron donors

Total Recall

SAMPLE QUESTIONS

True/False Questions

1. Meat provides an excellent source of all essential vitamins and minerals.

2. Fruit provides an excellence source of fat-soluble vitamins.

3. Protein-containing plant foods provide a good source of bioavailable iron.

4. A deficiency of vitamin C causes beriberi.

5. Deficiency of selenium may contribute to abnormal glucose metabolism.

Multiple Choice Questions: Choose the best answer.

6. Vitamin K is a fat-soluble vitamin important for:
   A. Vitamin E synthesis
   B. Energy metabolism
   C. Collagen synthesis
   D. Lipid antioxidation
   E. Blood clotting

7. A deficiency of vitamin B_{12} causes:
   A. Microcytic anemia
   B. Pernicious anemia
   C. Hemolytic anemia
   D. Sickle cell anemia
   E. None of the above

8. A recent role of vitamin B_6, vitamin B_{12}, and folate in health promotion is in the prevention of:
   A. Bone loss
   B. Depressed immunity
   C. Night blindness
   D. Hyperhomocystemia
   E. Allergy

9. Toxicities of the following nutrients are poorly characterized in human beings:
   A. Vitamin C and thiamin
   B. Folate and vitamin B_{12}
   C. Vitamin A and pantothenic acid
   D. Niacin and zinc
   E. Iodine and potassium

10. The following major minerals are involved in water balance:
   A. Sodium, potassium, and chloride
   B. Zinc, copper, and chromium
   C. Calcium, phosphorus, and magnesium
   D. Selenium, fluoride, and iodine
   E. A and C

CASE STUDY

Gertrude is six months pregnant. She loves hamburgers and milkshakes. She is a woman of European descent who has anemia, but she did not respond to iron supplements.

Henry turned vegan three years ago after his father was diagnosed with cancer. He eats a diet of whole foods and does not use vitamin and mineral supplements. He also read a book by John Robbins about reclaiming health, which pointed out the environmental and resource inefficiencies of raising animal protein, the negative health data generated by the overuse of animal products, and the health-promoting characteristics of vegetarian diets. He was recently diagnosed with anemia and complained of numb, cold-feeling fingers and toes.

1. What could be Gertrude's problem?
   A. She could have iron toxicity, as this also causes anemia.
   B. She could have another type of anemia besides iron-deficiency anemia.
   C. She could have calcium toxicity from the milkshakes.

2. Gertrude's European ancestry could increase her risk of having low folate status if she exposed her skin to excess sunlight (UV radiation).
   A. True   B. False

3. Gertrude's anemia is probably not due to a vitamin B_{12} deficiency, because vitamin B_{12} is found in animal food sources and Gertrude eats plenty of these.
   A. True   B. False

4. What is most likely causing Gertrude's anemia? A deficiency of:
   A. Iron
   B. Folate
   C. Vitamin B_{12}
   D. Vitamin B_6
   E. Vitamin E

5. What foods in Gertrude's diet lack the nutrient(s) she needs to correct her anemia?
   A. Starchy vegetables
   B. Melons and berries
   C. Dark green vegetables
   D. Fats and oils
   E. Red and orange vegetables

6. How could you confirm the cause of Gertrude's nutritional deficiency?
   A. Do a complete dietary analysis to verify low dietary intake of the suspected nutrient.
   B. Verify clinical symptoms of anemia (weakness, pallor, fatigue, etc.).
   C. Perform biochemical tests to show a low level of the nutrient in her body.
   D. Introduce the nutrient that is believed to cause her anemia back into her diet and consider supplementation.
   E. All of the above

7. Gertrude's unborn baby is at risk for spina bifida.
   A. True   B. False

8. Henry could have a vitamin B_{12} deficiency because he doesn't consume animal products and doesn't take a vitamin B_{12} supplement.
   A. True   B. False

9. Henry can correct his deficiency and maintain his vegan lifestyle by:
   A. Taking a vitamin B_{12} supplement
   B. Eating foods that are designed for vegans and have vitamin B_{12} added to them
   C. Eating more broccoli and cauliflower
   D. A and B
   E. A, B, and C

10. Why did it take three years before Henry's deficiency showed up?
    A. His folate intake masked his vitamin B_{12} deficiency.
    B. Vitamin B_{12} can be stored long term in the liver.
    C. His stomach had a vitamin B_{12} storage reservoir.
    D. He had extra intrinsic factor to increase absorption of his low vitamin B_{12} intake.

FUN-DUH-MENTAL PUZZLE

ACROSS

3. Deficiency of this mineral may affect blood glucose levels.
8. Substance that can be converted to vitamin D upon UVB light exposure.
10. B vitamin functioning in fatty acid metabolism.
12. Form of vitamin A required for vision.
14. Too much of this nutrient causes bone demineralization.
15. Deficiency of _____ delays sexual maturation.
18. Disease name for thiamin deficiency.
19. A good food source for tocopherols.
21. The component of hemoglobin that binds oxygen.
23. Deficiency of _____ contributes to liver damage.
24. Type of test used to determine a nutritional deficiency or toxicity.
25. Absorption site for vitamin K.

DOWN

1. Vitamin used as a drug to improve blood cholesterol levels.
2. Mineral put in laxatives to stimulate solid waste excretion.
4. The most essential nutrient for the body.
5. Toxicity of this vitamin form leads to nerve damage.
6. Mineral used to make acid for food digestion.
7. Iodine deficiency causes this.
9. Long-term, excessive intake of niacin may damage this organ.
11. Mineral concentrated in intracellular fluid.
13. Adequate intake of this vitamin acid prevents scurvy.
16. This vitamin is widely distributed in food.
17. Trace mineral found in bones and teeth.
20. Condition in which the body is unable to deliver oxygen to cells.
22. Adequate intake of this prevents hyperhomocystemia.

References

Web Resources

A. Centers for Disease Control: www.cdc.gov
B. Fighting Celtic Curse: celticcurse.org
C. Food and Nutrition Information Center: fnic.nal.usda.gov
D. International Bibliographic Information on Dietary Supplements (Office of Dietary Supplements): ods.od.nih.gov/health_information/ibids.aspx
E. Linus Pauling Institute: lpi.oregonstate.edu
F. MedlinePlus Health Information, a service of the National Library of Medicine: www.nlm.nih.gov/medlineplus
G. National Institutes of Health: www.nih.gov
H. National Osteoporosis Foundation: www.nof.org
I. Nutrient Data Laboratory, United States Department of Agriculture: www.nal.usda.gov/fnic/foodcomp/search
J. Understanding Evolution: evolution.berkeley.edu/evolibrary/home.php

Works Cited

1. Amara, A. A. (2010). The philosophy behind exo/endo/existing antioxidants and our built-in oxidant and antioxidant system. *Pharmazie, 65*(10), 711–719.
2. American Dietetic Association. (2006). Position of the American Dietetic Association: The impact of fluoride on health. *Journal of the American Dietetic Association, 105*(10), 1620–1628.
3. American Dietetic Association. (2011). Nutrition care process. *Evidence Analysis Library*. Accessed at: http://adaevidencelibrary.com.
4. Barrack, M. (2009). Beneficial effects of vitamin D on immune function, cancer risk, neuromuscular function, and cardiovascular health. *SCAN'S (A Publication for Sports, Cardiovascular, and Wellness Nutritionists) PULSE, 28*(2), 11–14.
5. Basu, A., & Imrhan, V. (2005). Vitamin E and prostate cancer: Is vitamin E succinate a superior chemoprotective agent? *Nutrition Reviews, 63*(7), 247–251.
6. Bounds, W., Skinner, J., Carruth, B. R., & Ziegler, P. (2005). The relationship of dietary and lifestyle factors to bone mineral indexes in children. *Journal of the American Dietetic Association, 105*, 735–741.
7. Brownie, S., & Myers, S. (2004). Wading through the quagmire: Making sense of dietary supplement utilization. *Nutrition Reviews, 62*(7), 276–282.
8. Brunner, R. L., Cochrane, B., Jackson, R. D., Larson, J., Lewis, C., Limacher, M., . . . & Wallace, R. (2008). Calcium, vitamin D supplementation, and physical function in the Women's Health Initiative. *Journal of the American Dietetic Association, 108*(9), 1472–1479.
9. Calvo, M. S. (2003). Prevalence of vitamin D insufficiency in Canada and the United States: Important to health status and efficacy of current food fortification and dietary supplement use. *Nutrition Reviews, 61*, 107–113.
10. Cashman, K. D. (2005). Vitamin K status may be an important determinant of childhood bone health. *Nutrition Reviews, 63*(8), 284–289.
11. Catani, M. V., Savini, I., Rossi, A., Melina, G., & Avigliano, L. (2005). Biological role of vitamin C in keratinocytes. *Nutrition Reviews, 63*(3), 81–90.
12. Caudill, M. A. (2010). Pre- and postnatal health: Evidence of increased choline needs. *Journal of the American Dietetic Association, 110*(8), 1198–1206.

13. Chan, J. (2009). Vitamin D update for nutrition professionals. *Vegetarian Nutrition, 18*(1–2), 1, 10–14, 17.

14. Cooper, C. C. (2009). Nutrients for whole-body health: Spotlight on omega-3s, vitamin D, and calcium. *Today's Dietitian, 11*(9), 26–32.

15. Daniels, M. C., & Popkin, B. M. (2010). Impact of water intake on energy intake and weight status: A systematic review. *Nutrition Reviews, 68*(9), 505–521.

16. de Souza Genaro, P., & Martini, L. A. (2004). Vitamin A supplementation and risk of skeletal fracture. *Nutrition Reviews, 62*(2), 65–67.

17. Diet Analysis Plus. (2011). Cengage Learning version 10.0 [software]. Accessed at: http://login.cengage.com/sso.

18. Eichholzer, M., Tönz, O., & Zimmermann, R. (2006). Folic acid: A public-health challenge. *Lancet, 367*, 1352–1361.

19. Farbstein, D., Kozak-Blickstein, A., & Levy, A. P. (2010). Antioxidant vitamins and their use in preventing cardiovascular disease. *Molecules, 15*(11), 8098–8110.

20. Finkelstein, J. S. (2006). Calcium plus vitamin D for postmenopausal women: Bone appetite? *The New England Journal of Medicine, 354*(7), 750–752.

21. French, M. R., Barr, S. I., & Levy-Mine, R. (2003). Folate intakes and awareness of folate to prevent neural tube defects: A survey of women living in Vancouver Canada. *Journal of the American Dietetic Association, 103*, 181–185.

22. Garland, C. F., Grant, W. B., Mohr, S. B., Gorham, E. D., & Garland, F. C. (2007). What is the dose-response relationship between vitamin D and cancer risk? *Nutrition Reviews, 68*(8), S91–S95.

23. Going, S. B., & Laudermilk, M. (2009). Osteoporosis and strength training. *American Journal of Lifestyle Medicine, 3*(4), 310–319.

24. Gropper, S. S., Smith, J. L., & Groff, J. L. (2009). *Advanced nutrition and human metabolism* (5th ed.). Belmont, CA: Wadsworth.

25. Hathcock, J. N. (1997). Vitamins and minerals: Efficacy and safety. *American Journal of Clinical Nutrition, 66*, 427–437.

26. Heaney, R. P., Rafferty, K., Dowell, M. S., & Bierman, J. (2005). Calcium fortification systems differ in bioavailability. *Journal of the American Dietetic Association, 105*, 807–809.

27. Holick, M. F. (2007). Vitamin D deficiency. *The New England Journal of Medicine, 357*(3), 266–281.

28. Hurrell, R. (2002). How to ensure adequate iron absorption from iron-fortified food. *Nutrition Reviews, 60*, S7–S14.

29. Institute of Medicine. (1997). *Dietary Reference Intakes for calcium, phosphorus, magnesium, vitamin D, and fluoride.* Washington, DC: National Academies Press.

30. Institute of Medicine. (1998). *Dietary Reference Intakes: A risk assessment model for establishing upper intake levels for nutrients.* Washington, DC: National Academies Press.

31. Institute of Medicine. (2000). *Dietary Reference Intakes for thiamin, riboflavin, niacin, vitamin B_6, folate, vitamin B_{12}, pantothenic acid, biotin, and choline.* Washington, DC: National Academies Press.

32. Institute of Medicine. (2000). *Dietary Reference Intakes for vitamin C, vitamin E, selenium, and carotenoids.* Washington, DC: National Academies Press.

33. Institute of Medicine. (2001). *Dietary Reference Intakes for vitamin A, vitamin K, arsenic, boron, chromium, copper, iodine, iron, manganese, molybdenum, nickel, silicon, vanadium, and zinc.* Washington, DC: National Academies Press.

34. Institute of Medicine. (2004). *Dietary Reference Intakes: Water, potassium, sodium, chloride, and sulfate.* Washington, DC: National Academies Press.

35. Iwamoto, J., Takeda, T., & Sato, Y. (2006). Menatetrenone (vitamin K) and bone quality in the treatment of postmenopausal osteoporosis. *Nutrition Reviews, 64*(12), 509–517.

36. Johnson, M. A. (2005). Influence of vitamin K on anticoagulant therapy depends on vitamin K status and the source of chemical forms of vitamin K. *Nutrition Reviews, 63*(7), 91–97.

37. Johnson, M. A. (2007). If high folic acid aggravates vitamin B_{12} deficiency what should be done about it? *Nutrition Reviews, 65*(10), 451–458.

38. Johnson, T., Avery, G., & Byham-Gray, L. (2009). Vitamin D and metabolic syndrome. *Topics in Clinical Nutrition, 24*(1), 47–54.

39. Jones, D. (2005). *Textbook of functional medicine*. Gig Harbor, WA: Institute for Functional Medicine.

40. Jonnalagadda, S. S., Culp, J., Sharma, B., & Campbell, J. (2010). Impact of vitamin D and calcium on health outcomes: Reviewing the evidence. *SCAN'S (A Publication for Sports, Cardiovascular, and Wellness Nutritionists) PULSE, 29*(2), 11–13.

41. Kaluski, D. N., Amitai, Y., Haviv, A., Goldsmith, R., & Leventhal, A. (2002). Dietary folate and the incidence and prevention of neural tube defects: A proposed triple intervention approach in Israel. *Nutrition Reviews, 60*, 303–307.

42. Kleiner, S. M. (1999). Water: An essential but overlooked nutrient. *Journal of the American Dietetic Association, 99*, 200–206.

43. Landska, D. J. (2010). Chapter 29: Historical aspects of the major neurological vitamin deficiency disorders: Overview and fat-soluble vitamin A. *Handbook of Clinical Neurology, 95*, 435–444.

44. Landska, D. J. (2010). Chapter 30: Historical aspects of the major neurological vitamin deficiency disorders: The water-soluble B vitamins. *Handbook of Clinical Neurology, 95*, 445–476.

45. Lee, C., & Majka, D. S. (2006). Is calcium and vitamin D supplementation overrated? *Journal of the American Dietetic Association, 106*(7), 1032–1034.

46. Lynch, S. (2002). Food iron absorption and its importance for the design of food fortification strategies. *Nutrition Reviews, 60*, S3–S6.

47. Manz, F., & Wentz, A. (2005). The importance of good hydration for the prevention of chronic disease. *Nutrition Reviews, 63*(6), S2–S5.

48. Mark, B. L., & Carson, J. S. (2006). Vitamin D and autoimmune disease: Implications for practice from the multiple sclerosis literature. *Journal of the American Dietetic Association, 106*(3), 418–424.

49. Martin, A., Youdim, K., Szprengiel, A., Shukitt-Hale, B., & Joseph, J. (2002). Roles of vitamin E and C on neurodegenerative diseases and cognitive performance. *Nutrition Reviews, 60*, 308–334.

50. Martini, L. A., & Wood, R. J. (2006). Vitamin D status and the metabolic syndrome. *Nutrition Reviews, 64*(11), 479–486.

51. Maughan, R. J., & Shirreffs, S. M. (2010). Dehydration and rehydration in competitive sport. *Scandinavian Journal of Medicine and Science in Sports, 20*(S3), 40–47.

52. McLaren, D. S. (1981). *A colour atlas and text of nutritional disorders*. London, England: Wolfe Medical Publications.

53. McLaren, D. S. (1992). *A colour atlas and text of diet-related disorders* (2nd ed.). Aylesbury, England: Mosby-Year Book Europe.

54. Michaelsson, K., Litchell, H., Vessby, B., & Melhus, H. (2003). Serum retinal levels and risk of fracture. *New England Journal of Medicine, 348*, 287–294.

55. Moon, M. (2008). Supplements: The role of a high potassium diet in managing blood pressure. *Nutrition in Complementary Care, 10*(4), 72–76.

56. Moyad, M., & Kondracki, N. L. (2009). The efficacy of vitamin C supplementation for the common cold. *Nutrition in Complementary Care, 11*(3), 41, 44–46.

57. Moyad, M., & Kondracki, N. (2009). The year-round benefits of vitamin C supplementation. *Nutrition in Complementary Care, 11*(4), 64–66.

58. Musso, C. G. (2009). Magnesium metabolism in health and disease. *International Urology and Nephrology, 41*(2), 357–362.

59. Norris, J. (2007). Update on vitamin B_{12}. *Vegetarian Nutrition Update, XVI*(II), 1–5.

60. Park, S., & Johnson, M. A. (2006). What is an adequate dose of oral vitamin B_{12} in older people with poor vitamin B_{12} status? *Nutrition Reviews, 64*(8), 373–378.

61. Piehowski, K. E., & Nickols-Richardson, S. M. (2009). Osteoporosis and obesity: Inflammation as an emerging link. *SCAN'S (A Publication for Sports, Cardiovascular, and Wellness Nutritionists) PULSE, 28*(3), 1–6.

62. Popkin, B. M., D'Anci, K. E., & Rosenberg, I. H. (2010). Water, hydration, and health. *Nutrition Reviews, 68*(8), 439–458.

63. Prior, R. L. (2004). Biochemical markers of antioxidant status. *Topics in Clinical Nutrition, 19*(3), 226–238.

64. Radak, T. L. (2004). Caloric restriction and calcium's effect on bone metabolism and body composition in overweight and obese premenopausal women. *Nutrition Reviews, 62*(12), 468–481.

65. Rando, R. R. (2001). The biochemistry of the visual cycle. *Chemical Reviews, 101*, 1881–1896.

66. Ribaya-Mercado, J. D., & Blumberg, J. B. (2007). Vitamin A: Is it a risk factor for osteoporosis and bone fracture? *Nutrition Reviews, 65*(10), 425–438.

67. Romano, A. D., Serviddio, G., de Matthaeis, A., Bellanti, F., & Vendemiale, G. (2010). Oxidative stress and aging. *Journal of Nephrology, 23*(S15), S29–S36.

68. Rosenberg, I. H. (2007). Folic acid fortification. *Nutrition Reviews, 65*(11), 503.

69. Russel, R. M. (2001). New micronutrient Dietary Reference Intakes from the National Academy of Sciences. *Nutrition Today, 36*, 163–171.

70. Sanders, L. M., & Zeisel, S. H. (2007, July/August). Choline: Dietary requirements and role in brain development. *Nutrition Today, 42*(4), 181–186.

71. Sawka, M. N., Cheuvfont, S. N., & Carter, R., III. (2005). Human water needs. *Nutrition Reviews, 63*(6), S30–S39.

72. Shuaibi, A. M., House, J. D., & Sevenhuysen, G. P. (2008). Folate status of young Canadian women after folic acid fortification of grain products. *Journal of the American Dietetic Association, 108*(12), 2090–2094.

73. Sinclair, L. M., & Hinton, P. S. (2005). Prevalence of iron deficiency with and without anemia in recreationally active men and women. *Journal of the American Dietetic Association, 105*(6), 975–978.

74. Stang, J., Brown, J., & Jacob, D. (2002). Effect of iron and folic acid supplements on serum zinc levels among a cohort of pregnant women. *Topics in Clinical Nutrition, 17*, 15–26.

75. Stipanuk, M. H. (2000). *Biochemical and physiological aspects of human nutrition*. Philadelphia: Saunders.

76. Thompson, J., & Manore, M. (2006). *Nutrition: An applied approach*. San Francisco: Pearson Benjamin Cummings.

77. Timbo, B. B., Ross, M. P., McCarthy, P. V., & Lin, C. J. (2006). Dietary supplements in a national survey: Prevalence of use and reports of adverse events. *Journal of the American Dietetic Association, 106*(12), 1966–1974.

78. Toner, C. D., Davis, C. D., & Milner, J. A. (2010). The vitamin D and cancer conundrum: Aiming at a moving target. *Journal of the American Dietetic Association, 110*(10), 1492–1500.

79. Tufts University. (2009). Aging brains may benefit from vitamin D. *Tufts University Health & Nutrition Letter, 27*(2), 1–2.

80. Vatanparast, H., & Whiting, S. J. (2006). Calcium supplementation trials and bone mass development in children, adolescents, and young adults. *Nutrition Reviews, 64*(4), 204–209.

81. Weaver, C. M. (2009). Closing the gap between calcium intake and requirements. *Journal of the American Dietetic Association, 109*(5), 812–813.

82. Welch, J. M., & Weaver, C. M. (2005). Calcium and exercise affect the growing skeleton. *Nutrition Reviews, 63*(11), 361–373.

83. Wolf, G. (2007). Identification of a membrane receptor for retinol-binding protein functioning in the cellular uptake of retinol. *Nutrition Reviews, 65*(8), 385–388.

84. Wooley, J. A. (2008). Characteristics of thiamin and its relevance to the management of heart failure. *Nutrition in Clinical Practice, 23*(5), 487–493.

85. Zeisel, S. H. (2010). Choline: An essential nutrient for public health. *SCAN'S (A Publication for Sports, Cardiovascular, and Wellness Nutritionists) PULSE, 29*(3), 4–7.

86. Zempleni, J., Wijeratne, S. S., & Hassan, Y. I. (2009). Biotin. *Biofactors, 35*(1), 36–46.

Nutrition Information and the Food Industry

Fedorov Oleksiy/Shutterstock.com

To recognize scientifically based nutrition information and to understand the food industry, food safety, food processing, and food production.

LEARNING OBJECTIVES

After completing this learning module, you will be able to:

- Recognize the major markers of information quality to determine reliable nutrition sources.
- State and describe the responsibilities of the governmental agencies that regulate food safety and the laws the food industry abides by.
- Explain safe handling of food and the use of food additives.
- Describe the various types of microorganisms causing foodborne illness, microbes that commonly contaminate foods, and their food poisoning signs and symptoms.
- Evaluate your food handling habits and practices to minimize the incidence of foodborne illness and adverse food reactions.
- Describe energy and nutrient cycling within an ecosystem.
- Relate levels of organization to food production.
- Distinguish between conventional and organic food production.
- Articulate each person's part in reducing demand for and consumption of processed food to support sustainable food systems.

PERSONAL IMPROVEMENT GOALS

When you complete this learning module, you will know how to:

- Critically evaluate any written nutrition information using the markers of reliability before you accept the information.
- Use your fact-versus-fallacy skills to seek out reliable sources of nutrition information.
- Avoid potentially harmful eating situations, and practice safe handling of food in your own household.
- Use the information about food additives in packaged foods to become more knowledgeable about processed foods.
- Reduce your carbon footprint by eating seasonal, local foods grown using organic, sustainable methods, fewer meat products, and fewer processed, fast, and restaurant foods.

The nutrition industry in the United States is a multimillion-dollar business. It can be difficult for the average consumer to distinguish between reliable and unreliable nutrition information; thus, one sign of an educated consumer can be his or her ability to identify sound nutrition information.

Governmental agencies created by a series of laws monitor the food supply from supplements to food safety, chemical additives, food processing techniques, and food production methods. There are many issues surrounding food safety, including the various microorganisms implicated in foodborne illness, chemical contaminants, and natural toxicants. Conventional food production systems can also generate health and safety concerns for consumers and the environment because of pesticides, hormones, antibiotics, genetically modified organisms, and nitrogen-based fertilizers. Organic food production systems provide an alternative, with the goal to provide healthy food using methods that are sustainable for future generations.

6.1 CREDIBILITY OF NUTRITION INFORMATION

Introduction

Herbal and nutritional **supplements**, diets, and even foods can cause serious harm and add to healthcare costs (up to $30 billion annually in the U.S. healthcare system). Although there are many benefits to taking dietary supplements, sometimes harm can come from direct effects of some products; at other times, it comes from indirect effects of products, such as not seeking medical attention when needed. Before deciding to accept a statement or purchase a product, it is wise to put the information through the quality checks of credentials, affiliation, sources, and purpose described in this section; if it fails several of them, then you should look for more credible information sources to either support or discredit the unreliable source that caught your interest. The Office of Dietary Supplements within the National Institutes of Health provides public information about dietary supplements, as does the National Center for Complementary and Alternative Medicine.[I,K,L,N,Q,37]

The purpose of this section is to explain who is credible in the field of nutrition and which sources of published information are reliable. To determine credibility and reliability, one must consider scientific methodology (see Module 4 on page 172) and evaluate the author's **credentials** and **affiliation**, the source of the information, the referencing of the information, and the purpose of the information (see BioBeat 6.1 on page 291). After taking these things into consideration, you can better determine the overall reliability of the nutrition information.[3]

Author Credentials

Deciphering a person's knowledge and expertise in the field of nutrition is not always easy. Individuals with an advanced degree, such as a master of science (M.S.) or preferably a doctorate (Ph.D.), in nutrition from an accredited institution who are working in the **public sector** are credible sources of nutrition information. **Registered Dietitians** (R.D.s) can also provide reliable nutrition information, although, unless otherwise specified, they will hold (minimally) a bachelor of science degree (B.S.). If you are reading a source of nutrition information and you see that the author has the credentials of Ph.D., R.D., then you know that this person has extensive knowledge in nutrition. However, if the author's credentials are simply Ph.D., then you need to learn more about what that person's degree area was to determine if he or she is qualified to give accurate nutrition recommendations and information.

T-Talk 6.1
To hear Dr. Turley talk about credibility of nutrition information, go to www.cengagebrain.com

supplements Products, such as vitamins, minerals, amino acids, glandular extracts, and herbs, that are taken by mouth in addition to a normal diet.

credentials The qualification or competence issued to an individual by a third party, such as a degree title earned by a graduate of a university or a national credentialing association.

affiliation A partnership between two parties, such as an employee and employer; can also be the professional body of which you are a member.

public sector The part of the economy that is owned and controlled by the public through the government and is not-for-profit and "by the people, for the people."

Registered Dietitian A person who has studied diet and nutrition at an Academy of Nutrition and Dietetics (formerly American Dietetic Association)–approved university program and has passed a standardized exam.

Many medical doctors (M.D.s) can provide sound nutrition information; however, not all can. Although a person with an M.D. has the research and science skills to become nutritionally competent, doing so is not professionally mandated by the American Medical Association.

A common title used in the field of nutrition is **nutritionist**. This title is meaningless because it has no formal ties to a person's knowledge, skills, or competencies. Anyone can call him or herself a nutritionist, so this term does not denote a qualified nutrition expert. Keeping all of that in mind, we can say that, in general, a credible nutrition author should have the credentials of M.D., Ph.D., M.S., and/or R.D.[3]

Affiliation

We have said that it is important to know an author's credentials and whether the author has a degree in nutrition. It is also important to know if the degree was obtained from an accredited institution and if the individual is affiliated with a public or **private sector** institution. **Professional affiliation** in this context refers to an author's business relationships. You could think of it simply as "for whom does this person work"? It is important to know a nutrition author's affiliation, because it can reveal his or her motives for publishing. So learn to decipher whether the author of any nutrition information you may be reading is affiliated with an accredited institution, in the private sector or public sector. The overall reliability of the information is increased when the author is affiliated with an accredited educational institution and/or public sector organization.

Accredited Educational Institution

An accredited educational institution is one that has been successfully evaluated by a qualified association or governing body, such as the American Council on Education, to ensure that it meets national academic requirements and education standards as a whole and for the quality of its educational programs. So ask yourself if the author's credentials were obtained from an accredited institution and if the author is employed by an accredited institution. If the answer is yes, then his or her credibility increases. The state Board of Education can identify accredited institutions in your state.

Public Sector Organizations

Public sector refers to all aspects of government: national, state, and local, as well as government programs and state institutions that are funded by tax dollars and are subject to government control. They are agencies and organizations "for the people, by the people," and provide credible, comprehensive information. Examples include the National Institutes of Health, state-funded universities, the Centers for Disease Control, the Food and Drug Administration, and others shown in Appendix E.[B,H,N,T]

Private Sector Organizations

Private sector refers to businesses, societies, associations, and households. They are self-funded and controlled **not-for-profit** (nonprofit) organizations and **for-profit** private industry.

Not-for-Profit Organization

A not-for-profit organization is one that is not out to make a financial gain. Not-for-profit organizations that are not funded by tax dollars are private sector. Regarding nutrition information, a not-for-profit organization—such as the American Cancer Society, the American Heart Association, and others found in Appendix E—publishes nutrition information from the organization as a whole (i.e., without a specific author), or from an author affiliated with the organization for the purpose of raising health awareness to support the organization's mission and purpose. When reviewing nutrition information, ask yourself if a not-for-profit organization employs the author. If the answer is yes, then his or her credibility increases. Keep in mind that the information will be focused on the purpose of the organization, such as heart health from the American Heart

nutritionist Term for a person who advises people on dietary matters, with or without appropriate credentials.

private sector The part of the economy that is self controlled and can be for-profit or not-for-profit.

professional affiliation The public, private for-profit, or private not-for-profit agency a professional represents.

not-for-profit An agency or organization that exists to serve the public good and not for the purpose of making money; also called *nonprofit*.

for-profit An organization in business to make money.

BioBeat 6.1

Markers of Fallacy: Testimonials versus Scientific Methods

"Fat farms" (weight control programs or products) heavily advertise using **testimonials** as the evidence that their programs are effective. Sport performance products use the same strategies to promote sales of their products. Although they are very appealing to consumers, testimonials are unscientific, unsubstantiated evidence.

Evidence that the scientific community would recognize is based on research published in **peer-reviewed** journals. Findings reported in these journals result from studies of several types: epidemiological studies, intervention trials, clinical trials, and case studies; however, the experimental protocol that provides the strongest evidence is the well-designed, double-blind, cross-over format with repeated measures.

The reliability of the reported nutrition information increases when the intent of the information is to inform, when it is written by individuals who have earned terminal degrees in the field of nutrition and are affiliated with an institution of higher education or a **governmental agency**, and when the evidence used to support the information is from peer-reviewed sources like professional journals.[Q]

How can you find reliable nutrition information?

Association. Also ask yourself, does the nutrition information come from a not-for-profit organization? If the answer is yes, the credibility increases.

For-Profit Private Industry   A business operating in private industry is seeking to make financial profit from the nutrition information or product it provides. When reviewing nutrition information, ask yourself if the author is employed in private industry (for-profit). If the answer is yes, then credibility of the information declines.

The Information Source

One of the markers of the quality of nutrition information is its source. Various information sources are discussed as follows. See Appendix E for contact information (addresses, phone numbers, Web sites, and/or e-mail addresses) of reliable resources for nutrition information.

Types of Information Sources   There are many different sources of nutrition information. Some common sources include refereed journals, newsletters, lay magazines, books, newspapers and news reports, private not-for-profit professional service organizations, public not-for-profit government agencies and funded institutions, consumer advocacy groups, and Web sites. Each of these sources will be explored in terms of nutrition information credibility.

Refereed Journals   *Peer-reviewed* and *refereed* mean the same thing: The information that has been published in the journal has been critiqued by several experts or peers in the field. Furthermore, the peers who review the information proposed for publication do so as a free service to their field. They verify the scientific methods used and look to see that the information is appropriate subject matter for the journal, is current, and presents a balanced view (does not skew data and discusses and cites appropriate studies as references). Additionally, the author who succeeds in publishing in the journal does not profit from it, but rather pays the page charges! Refereed journals like *Journal of the American Dietetic Association, Journal of Nutrition, Nutrition Reviews, Nutrition Today, American Journal of Clinical Nutrition,* and *Topics in Clinical Nutrition* are examples of credible sources. See Appendix E for some more examples of refereed journals.

testimonials Nonscientifically-based, personal testaments that something is effective.

peer-reviewed When colleagues professionally evaluate the work of other colleagues; also called *refereed*; required for publication of scholarly work in prestigious journals.

governmental agency A not-for-profit, public sector agency that is supported by tax dollars—"by the people, for the people."

Various nutrition information sources.

Jennifer Turley

Newsletters Newsletters published from accredited institutions of higher education with excellent nutrition programs, such as Tufts University's *Health & Nutrition Letter* and Berkeley's *Wellness Letter*, are credible. But realize that information is often shortened to fit into the space available in a newsletter, and the information may not be fully explained or supported.

Lay Magazines Lay (general public) magazines like *Cosmopolitan* and *Bon Appétit* may not publish accurate nutrition information and tend to be sensational. The authors of the articles often do not have degrees in nutrition and attempt to come across as experts in the latest nutrition craze.

Books You should use the markers of quality (author's credentials, author's affiliation, references to support the information, and purpose of the information) to evaluate a book's credibility. Books that are well supported by scientific information and are written by credentialed individuals are more reliable compared to another book with more sensational information.

Newspapers and News Reports Information from newspapers and television is not always credible and tends to be sensational. Because freedom of the press exists, information from these sources can be based on the findings from a single study and may be skewed by a reporter's interpretation of scientific findings. Reporters and journalists can raise the reliability of their story by quoting credible sources. Most reporters and journalists, because they are not scientists or experts in the field on which they are reporting (i.e., they lack a degree in the area of expertise), may unintentionally misinterpret, misquote, or misrepresent the quoted source and may be pressed for time to get a story together.

Private Not-for-Profit Professional Service Organizations National health interest groups, such as the Academy of Nutrition and Dietetics, American Medical Association, American Diabetes Association, American Health Foundation, and the Society of Nutrition Education, provide credible nutrition information but may be focused on specific health issues. Groups such as these may not, however, address all of the important information to help you live healthy.

Public Not-for-Profit Government Agencies and Funded Institutions Agencies such as the U.S. Department of Agriculture (USDA), Food and Drug Administration, Department of Health and Human Services, National Institutes of Health, and Environmental Protection Agency publish credible information for the American public. Further, state government–funded hospitals and universities also provide reliable information, as discussed on page 290.

Consumer Advocacy Groups Consumer advocacy groups may publish credible information, but the information is biased toward a political cause. Examples of consumer advocacy groups include Community Nutrition Groups, Nutrition Legislation News, The Consumer Information Center, and the Center for Science in the Public Interest.[c]

Web Sites The Web is an incredible tool and information resource. However, it is a self-publishing medium that anyone can use. So, when surfing the Web, you must use the markers of quality to be discerning about information found on Web pages (see Figure 6.1). Look to see who the author is, what his or her credentials and affiliations are, what the purpose of the information is, what references are used to support the information, when the information was published, and so forth. Perhaps even before looking at these things, look at the Web site's Uniform Resource Locator (URL) address that begins with http://. If it ends in .com or .net, then the reliability decreases. If it ends in .edu, .gov, or .org, then the reliability increases.

Web sites ending in .com indicate the information is for commercial purposes. Web sites ending in .net indicate the information can be from any kind of institution. Web sites ending in .edu indicate the information is from an educational institution. Web sites ending in .gov indicate the information is from the government and is for public health purposes. Web sites ending in .org indicate the information is from a not-for-profit organization. Not all Web sites ending in .org provide accurate nutrition information. See Appendix E for some examples of reliable Web sites.

Editorial Board When reviewing print information sources, be sure to look in the front section of the journal or magazine for the **editorial board** of the publication. Make sure the individuals on the editorial board have good credentials (e.g., M.D., Ph.D., M.S., or R.D.). When the information source lacks a credible editorial board, then the credibility of the information published in the source declines.

References

In any of the sources discussed previously, it is important to see if the information contains **references** and if the references are current. When looking at references, consider that they typically do not contain the author's credentials (see Figure 6.2). If a reference is cited and it is from a credible journal, book, newsletter, organization, or government agency, this adds credibility to the information source. In general, literature published in refereed journals provides references and is credible.

The Information Purpose

In general, there are two basic purposes of nutrition information: (1) to sell products, such as foods, supplements, and even information sources (books, magazines, newspapers, etc.), or (2) to inform the public for health awareness purposes. Regarding advertising for product sales through private industry, evaluate

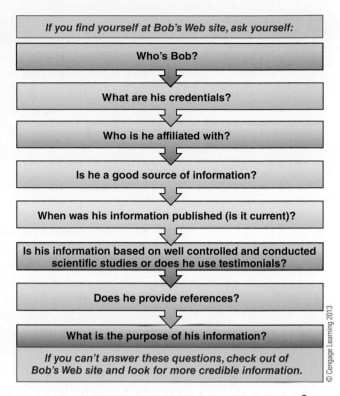

If you find yourself at Bob's Web site, ask yourself:

Who's Bob?

What are his credentials?

Who is he affiliated with?

Is he a good source of information?

When was his information published (is it current)?

Is his information based on well controlled and conducted scientific studies or does he use testimonials?

Does he provide references?

What is the purpose of his information?

If you can't answer these questions, check out of Bob's Web site and look for more credible information.

© Cengage Learning 2013

FIGURE 6.1 Deciphering the credibility of a Web site.

editorial board A group of individuals who are responsible for the quality and content of a frequent publication.

references Another term for citations, which are published sources recognizing the source of information or quotes used to support the author's work.

FIGURE 6.2 Decoding a reference and sample references.

Decoding a reference:

Author's last name, initials. (Year). Title. *Journal, Volume*(issue), page numbers and/or Web address.

Sample references:

Ball, K. (2009). Vegetarian diets and buying locally: Ways to increase environmental sustainability, support the local economy and contribute to individual health. *Vegetarian Nutrition Update, 17*(3,4), 1, 14–16, 20.

Dona, A., & Arvanitoyannis, I. S. (2009). Health risks of genetically modified foods. *Critical Reviews in Food Science and Nutrition, 49,* 164–175.

Monterey Bay Aquarium. (2008). Sustainable seafood guide. Available at: http://www.montereybayaquarium.org/cr/seafoodwatch.aspx.

Pimentel, D., Williamson, S., Alexander, C., Gonzales-Pagan, E., & Kontack, C. (2008). Reducing energy inputs in the US food system. *Human Ecology, 36,* 459–471.

Weber, C. L., & Matthews, H. S. (2008). Food-miles and the relative climate impacts of food choices in the United States. *Environmental Science & Technology, 42*(10), 3508–3513.

© Cengage Learning 2013

the information by asking the following questions: Does a person's involvement lead to personal gain with the profits from a sale? If yes, the credibility decreases. Is the material sensational, company propaganda, or used to sell a magazine? If yes, the information is not very credible. Remember that companies are bound by law to tell the truth about their products, but information may be incomplete or biased or supported by poorly designed scientific studies. Credibility goes down when information is provided for the promotion of a sale.[3]

Public Recommendations

Consumers often want to make drastic dietary changes as a result of reading the latest newspaper article or watching a news report on a recent study. Keep in mind that public recommendations are based on the interpretation of multiple studies with consistent results. In general, a whole body of evidence, collected from multiple study types (clinical, epidemiological, animal, cell culture, etc.) of credible design over many years of experimentation, needs to be interpreted by the scientific community before a public recommendation will be made. When a public recommendation is made, it is by a credible professional service organization or government agency. Furthermore, no single study, no matter how well designed or conducted, and no public recommendation based on the evaluation of years of scientific data, provides conclusive evidence. *Conclusive* implies the end of the story and no need for further study. This term should be reserved for laws in science and nature, like gravity and thermodynamics.[T]

Summary Points

- Think before you believe.
- Just because the information is published doesn't mean that it is accurate or true.
- Use your fact-versus-fallacy tools to decipher the overall credibility of nutrition information.
- In looking for credible nutrition information, consider the author (credentials and affiliation), source, references used to support the information, purpose, and scientific methodology involved in the research.
- Apply nutrition information credibility tools to all aspects of nutritional science, including foods, diets, and supplements.
- Seek credible information to affirm or discredit information from a less credible source.

Take Ten on Your Knowledge Know-How

1. What factors should be considered when determining nutrition information credibility?
2. Give an example of how you or someone you know has been potentially harmed by less-than-credible nutrition information.
3. What is a public recommendation?
4. When are public recommendations made?
5. What are some public recommendations that have impacted you?
6. What information is available in a professional reference (refereed) article?
7. Who is the most credible source from the field of nutrition for diet information?
8. Web sites with what URL endings are more likely to supply reliable nutrition information?
9. What is a private not-for-profit organization?
10. What is a public sector organization? Why are they referred to as "by the people, for the people" in the United States?

6.2 FOOD, DRUGS, AND SUPPLEMENTS

Introduction

Because food is necessary to sustain life, its safety is of utmost importance. The safety of food has been regulated by the U.S. government for more than a century. The safe handling of food by all segments of the food industry, especially with regard to prevention of **foodborne illness**, is regulated by several governmental agencies and laws. However, supplements and herbs are not as closely regulated as food products. This section will look at the food industry and how it is regulated. It will also summarize legislation related to food and supplements.

T-Talk 6.2

To hear Dr. Turley talk about food, drugs, and supplements, go to www.cengagebrain.com

Food Safety Legislation and Governmental Agencies

The role of the U.S. government in ensuring food safety extends all the way back to the early 1900s, during the time of increasing industrialization of foods, drugs, and cosmetics. The government wanted to ensure the safety of these types of products and began to create and pass laws to protect consumers.

Food Safety Legislation The following chronology of legislation highlights milestones in the history of U.S. food safety regulation.[1]

The Federal Food and Drugs Act of 1906 This is the first law regarding food and labeling, known as the "pure food law." Specifically, this act prevents the manufacture, sale, or transportation of adulterated, misbranded, poisonous, or deleterious foods, drugs, medicines, and liquors, and regulates their distribution.

The 1938 Food, Drug and Cosmetic (FD&C) Act This legislation is the basis of modern food law; it gives the Food and Drug Administration (FDA) authority over food and food ingredients and defines requirements for truthful labeling of ingredients. The FD&C Act mandated quality and identity standards for foods, prohibited false therapeutic claims for drugs, cosmetics, and medical devices, clarified the FDA's right to conduct factory inspections, and provided guidelines for control of product advertising. Stipulations for food were that food must be pure, safe to eat, prepared under sanitary conditions, and honestly packaged and labeled. Legally enforceable food standards were mandated to correct abuses in food packaging and quality, and the government was given authority to remove unsafe food from retail shelves.

The 1954 Miller Pesticide Act This legislation incorporated the Environmental Protection Agency (EPA) into the regulation of **pesticide** use and the evaluation of pesticide residues that remain on the foods when they are brought to market. The use of pesticides is strictly regulated, however, not without flaws.

The 1958 Food Additives Amendment to the FD&C Act This legislation requires FDA approval for the use of a food additive before its inclusion in food. Manufacturers must prove an additive's safety by conducting research and presenting the scientific safety data to the FDA. However, additives that were used before this act were not tested.

The 1960 Color Additives Amendment to the FD&C Act This law stipulates that dyes used in foods, drugs, cosmetics, and certain medical devices be FDA-approved. The FDA was given regulatory rights over the use of additives in foods: which additives were used, for what purpose, in what foods, and at what levels. The burden of safety also shifted. Instead of the government having to prove a food color additive is unsafe, the manufacturers must prove to the FDA, per safety criteria,

foodborne illness Sickness caused by the consumption of food contaminated by pathogenic microorganisms.

pesticide A heterogeneous category of chemical compounds (including fungicides, herbicides, insecticides, and rodenticides) that can be applied to conventionally produced agriculture to exterminate or ward off insect, microbial, plant, and animal pests; may also bioaccumulate in conventional animal foods; many are endocrine disruptors.

that the substance is safe to use. Two hundred colorants that were in use at the time of the amendment were allowed provisional use and required to undergo further testing to confirm their safety. Of those 200, 90 were later deemed safe by the FDA.

The Delaney Clause   The Food Additives and Color Additives Amendments include a provision called the Delaney Clause; thus, this clause is part of the FD&C Act. The Delaney Clause states that any additive substance known to cause cancer in humans or any animal at any level is not recognized as safe, and thus cannot be approved for use in food. There may be adverse effects to certain people caused by certain additives, but it is up to the consumer to be aware.

The GRAS List   The GRAS list includes the **food additives** that are Generally Recognized As Safe (GRAS) and can be added by the manufacturer in food processing. The inclusion of an additive on the GRAS list depends on its safety in a population. Each of the substances has a margin of safety associated with its use that is 1/100th of the level determined to be safe to use by the majority of the population. Certain individuals may display reactions to certain additives, but in a general population it is rare to experience an adverse reaction. Also, all of the food additives included on the GRAS list abide by the Delaney Clause. An exception to this clause is the use of saccharin. Though it is known to cause cancer at high doses in animal systems, in diabetic humans there was a need for a sugar substitute and no evidence for increased cancer risk in humans, so it has continued to be used as a food additive. Also, the American people didn't want saccharin removed from food, because at the time it was the only sugar substitute available for use in diet soda and other processed foods and products like toothpaste.[c]

Some substances were exempted from the food additive regulation process. Exempted substances included all substances that the FDA or USDA had determined were safe for food use before the 1958 Food Additives Amendment. These were prior-sanctioned substances. Sodium nitrite and potassium nitrite, which are used to preserve processed meats, are examples of prior-sanctioned substances.

GRAS substances were also exempted. The safety of GRAS substances is determined by either their extensive food use history before 1958 or credible published scientific evidence. There are hundreds of GRAS substances, including salt, sugar, spices, vitamins, and monosodium glutamate (MSG). Substances can be added to the GRAS list upon a manufacturer's request and FDA review. Substances can also be removed from the GRAS list as the FDA and USDA continue to monitor GRAS substances in relation to new scientific information. (For more information on food additives, see page 304.)[H]

The 1966 Fair Packaging and Labeling Act   This legislation was essentially the first food package label law. The law required all consumer products in interstate commerce to be honestly and informatively labeled. The FDA enforced the legal requirements on foods, drugs, cosmetics, and medical devices. However, most food products were not mandated to provide nutrition information to the consumer.

The 1968 Animal Drug Amendment to the FD&C Act   This amendment addresses residues of animal drugs in meat, eggs, and milk, and places all regulation of new animal drugs under the FD&C Act for efficient control of animal drugs and medicated feeds.

The 1976 Vitamins and Minerals Amendment   This amendment stops the FDA from establishing standards that would limit the potency of vitamins and minerals in food supplements. Furthermore, the FDA is not allowed to regulate vitamin and mineral supplements as drugs based on their potency. The quantity of sodium per serving was also included in the nutrition information on food package labels at this time.

food additives Chemicals that the Food and Drug Administration has approved to be added to processed foods from the following categories: antimicrobial agents, antioxidants, nutrients, stabilizers and thickening agents, artificial colors and flavors, bleaching agents, emulsifiers, or chelating agents.

The Nutrition Education and Labeling Act of 1990 This legislation is reflected on current food package labels' "Nutrition Facts" panel. The law requires that almost all packaged foods display nutrition labeling. Under the provisions of the law, health claims for foods are regulated and must be consistent with terms defined by the Secretary of Health and Human Services. Standards are created for the food ingredient panel, serving sizes, and terminology.

The Dietary Supplement Health and Education Act of 1994 This act establishes specific labeling requirements for supplements' "Supplement Facts" panel, defines **dietary supplements** and dietary ingredients, and classifies them as food. Furthermore, a supplement regulatory framework was built, a commission to recommend how to regulate claims was established, and the FDA was authorized to promote good manufacturing practices for dietary supplements.

The Food Allergy Labeling and Consumer Protection Act of 2004 This act requires that any food that contains a protein derived from peanuts, soybeans, cow's milk, eggs, fish, crustacean shellfish, tree nuts, or wheat be plainly labeled on the food package. These food proteins cause the majority of food allergies.

The Food Safety Modernization Act of 2011 When this bill is funded, it will direct the FDA to build a new food safety system overnight. Many of the gaps in monitoring food sources, the blocking of unsafe food, and the training of food safety personnel at every level will be improved.[58]

Governmental Agencies In the United States, monitoring and maintaining food safety for consumers and citizens is a team effort. Federal, state, and local governmental agencies work together and operate within the guidelines of the agencies' specific responsibilities, described as follows.

Local Health Departments Each state has a department of health that functions to promote health and wellness for the people living within counties in the state. One of the functions of each local health department is to monitor foodborne illness in the community. Citizens are encouraged to report diseases, including those resulting from food poisoning. Regarding food safety, local health departments issue food handlers permits to run food-providing establishments and conduct annual facility inspections for any public establishment that prepares food.

U.S. Department of Agriculture (USDA) The USDA oversees food wholesomeness of meat, poultry, and egg quality, in addition to conducting nutrition research and providing public education. The USDA grades all meat, poultry, and egg products. Qualities such as tenderness, juiciness, flavor, fullness, freedom from defects, and cleanliness are aspects of the grading. The USDA developed and released the MyPlate food guidance system and took responsibility for labeling USDA-certified organic foods.[T]

The Centers for Disease Control (CDC) The Centers for Disease Control and Prevention (CDC) is a unit of the Department of Health and Human Services (DHHS) in the U.S. government. The DHHS seeks to protect the health and safety of all Americans by providing essential human services. The CDC contributes to this mission by monitoring foodborne diseases as well as all diseases in the United States. Furthermore, the CDC tracks behavioral and nutritional risks of Americans, such as the incidence of obesity.[B]

The National Marine Fisheries Service (NMF) The NMF is a federal agency division of the Department of Commerce and is responsible for the stewardship of the nation's living marine resources and their habitat. The link between seafood

dietary supplement Processed food component(s) from any of the following categories: essential or nonessential mineral, essential or nonessential vitamin, herbal, glandular, enzyme, nutritional substance, or fiber.

and health is regulated by the NMF. Additionally, the NMF provides information about seafood products to the general public and supports the FDA in food safety and food health issues pertaining to seafood. This includes inspecting fish, boats, processing plants, brokers, and retail operators.[P]

The Environmental Protection Agency (EPA)   The EPA has a mission to protect human health and the environment. Part of this mission is supported by its regulation and enforcement of legislation pertaining to pesticide use and water quality.[D]

The World Health Organization (WHO)   This international agency focuses on the attainment of the highest possible level of health of all people on earth. The United States provides strong representation in the WHO. Regarding food, the agency oversees topics such as bioterrorism, mad cow disease, management of alcohol abuse, health care for chronic conditions (in an observer role), programs for micronutrient deficiency diseases, guidelines on drinking water quality, global food trade, and reports and contacts regarding food safety and additives.[U]

The Food and Drug Administration (FDA)   The FDA regulates all drugs rigorously, and regarding cosmetics, medical devices, and food, it implements varying degrees of control and intervention. For food, the FDA oversees general food safety issues in the food industry, with the exception of meat, poultry, eggs, and seafood. The FDA focuses primarily on approving food additives and biotechnology, overseeing food labeling issues, providing food safety information, and implementing Hazard Analysis and Critical Control Points (HACCP) (see page 317).[H,49]

The Division of Responsibility for the FDA in Drugs and Foods

The FDA is the governmental agency that regulates most of the food and all of the pharmaceutical (drug) industries in the United States. Pharmaceutical companies must conduct a substantial amount of research before the FDA approves a new drug, whether it is a prescription or an over-the-counter medication. Foods, however, are more loosely regulated by the FDA. Dietary supplements of all kinds are classed as food, and therefore their regulation is less strict than that of drugs.

Food and Dietary Supplements   According to the FDA, the term *food* includes food, nutritional substances, herbs, enzymes, glandular extracts, and fiber. In other words, the FDA considers all nutritional supplements to be food. Because food cannot be patented, except for genetically modified or cloned life forms, the rigorous testing for safety usually demanded by the FDA for drugs is not required and often isn't done. Adequate testing of the effectiveness of supplements would require studies done with humans. These tests are expensive and take a long time to do, so they are rarely done.[24]

The marketing of supplements and fortified food products is extensive. Unfortunately, many consumers are unaware that herbs and other supplements can have powerful negative effects on some people. If you plan to take a supplement, it is important to gather complete information about that supplement in order to make a wise decision about whether to take it and how much to take. Large amounts of certain nutrients can cause drug-like effects (such as significantly changing metabolism) rather than a nutritional effect and may not always promote health.

Product Assurance   Buyer, beware: Words like *natural, organic, high potency, stress formula, advanced formula,* and *tonic* have ambiguous definitions; these terms create connotations of great stuff, but the product may not be great. Furthermore, supplement costs vary dramatically. Fancy labels, processing techniques, and extensive advertising increase the cost of the products. The quality or the cost of the content is not necessarily the basis for the increased cost of the product. Generic brands of the supplements can be good. In any case, the consumer is looking for a guaranteed potency product. Consumers should look for specific **certification insignias** such as USP, USP-NF, BioFit, and PharmaPrint to determine quality.

- *USP and USP-NF:* The United States Pharmacopeial Convention (USP) has rigorous standards that are recognized in U.S. legislation, including the Dietary Supplement Health and Education Act of 1994. Manufacturers may use the letters "USP" in conjunction with the monograph name of the product as a way of proclaiming that they have tested their product and found it to comply with USP standards. Of more significance are dietary supplement products that have been verified by the USP Dietary Supplements Verification Program to have met USP standards. Products bearing the "USP Verified" mark have been voluntarily submitted by the manufacturer and independently verified by the USP to meet the purity, potency, and quality standards in the United States Pharmacopeia National Formulary (USP-NF). The "USP Verified" mark signifies that the dietary supplement contains the ingredients listed and declared on the label, is free of harmful, unacceptable levels of specified contaminants, will dissolve and release dietary ingredients into the body within a specified time frame, and has been made according to the FDA's current Good Manufacturing Practices. Additionally, USP products are labeled with an expiration date, supported by stability data.[59]
- *BioFit:* This insignia is a trademark owned by Paracelsian. This insignia implies that the products have passed a test for consistency among batches and for consistent biofunction associated with the health claims stated on the dietary supplement.
- *PharmaPrint:* This insignia is a trademark owned by PharmaPrint, Inc. The trademark is licensed to Whitehall-Robins Healthcare, a division of American Home Products, which markets the Centrum™ nutritional supplement line. Products using this trademark undergo a process that standardizes the active ingredients within the plant source and the herbal product, based on the predetermined quantity and mix of the active ingredients, to ensure batch-to-batch and pill-to-pill consistency.

The United States Pharmacopeia (USP) Dietary Supplements Verification Program product insignia.

Supplement Considerations When considering vitamin and mineral supplements, consumers should pay attention to the units of the Dietary Reference Intakes (DRIs) and Reference Daily Intakes, basis or criteria of the DRIs, Tolerable Upper Intake Level for the nutrient, function of the nutrient in the body, deficiency symptoms, toxicity symptoms, stability of the nutrient, and food sources of the nutrient. When considering a supplement, realize that nutrients can be used in medical nutrition therapy in high doses as drugs (see BioBeat 6.2) and that herbs have obvious drug effects and are used for medicinal purposes (see BioBeat 6.3).[1,17,51,68]

certification insignia A marker of quality for displaying product assurance or guaranteed potency.

Herbal Supplements Herbal preparations are also recognized as food by the FDA, even though many are marketed for their drug effects. Many people falsely believe that herbs are free of adverse reactions because they are often touted as all-natural. There are several concerns with the use of many herbs, including the lack of disclosure of the amount of the drug or pharmacologically active ingredients present in the preparation, the possibility that too much of the pharmacologically active ingredient is present, and contaminants that may have toxic effects may be present in the preparations. If one chooses to use herbal products, research by the consumer must be a prerequisite in addition

There are many consumer choices of nutrient and herbal supplements.

BioBeat 6.2

Nutrients Used in Nutritional Therapy

Medical nutrition therapy, recognized in managed care medicine, has expanded the use of nutrients in medical management. Under normal circumstances and in perceived healthy people, nutrient intake is at a level that maintains homeostasis and restores metabolic pools of nutrients, thus preventing deficiency. Vitamins and minerals are often consumed at higher levels than the DRI for a nutritional therapy purpose. Some nutrients, when taken in large doses, change cellular chemistry and act like a drug. When some nutrients are used in very large amounts, they exert completely different effects on the body's metabolism than when taken at the DRI level. When a nutrient is used at a high level to achieve the desired effect, then the nutrient is used for nutritional therapy. It is important to understand that the higher-than-normal nutrient intake does not mean that the person was deficient in the nutrient. And, if a person is deficient in a vitamin or mineral, supplementation will only correct abnormalities (signs and symptoms) that are caused by the deficiency of the particular nutrient, given that the deficiency was not so severe and long lasting that permanent, irreversible damage occurred. Contrary to popular belief, supplements are not the cure-all that they have often been hyped to be, and they will not improve the general (nonspecific) symptoms of not feeling good, being depressed, suffering from headaches, or not having enough energy if the body's metabolic pools are sufficient.[N,Q,17,26,31,51,66,68]

Examples of nutrients used in nutritional therapy at high doses as drugs include vitamin E, which is taken for the treatment of polycystic fibrous breast disease and for its antioxidant effects; niacin, which is used to decrease LDL and increase HDL cholesterol levels; vitamin A, which is used in the form of 13-*cis* retinoic acid for the treatment of acne; zinc, which is used in the form of lozenges to reduce the severity and duration of a common cold; and vitamin C, which is touted for the treatment of colds and for its antioxidant effects.

Do you use nutrients for nutritional therapy?

Germany's Commission E The governmental regulatory agency in Europe that controls the production, distribution, and use of herbal medicines.

product assurance standards Standards used to ensure batch-to-batch consistency, potency, dissolution, and disintegration of dietary supplements or drugs.

to the identification of product potency. Consumers may not be aware that ingredients may vary in herbal preparations, especially when they are sold as extracts without reference to the concentration (milligram or microgram value) of active ingredients, or that some herbal preparations have nutritional qualities that are often masked or associated with the herb's drug effect. If you use herbs, understand in detail the characteristics and effects of the herbs. Know the amount of the drug present in the preparation and the adverse side effects that can be experienced (see BioBeat 6.3).[29,33,64,69]

BioBeat 6.3

Herbal Remedies: Food or Medicine?

Many herbal remedies have been used since antiquity to manage illness and wellness. Germany has been the worldwide leader in herbal medicines. **Germany's Commission E** is a governmental agency similar to the FDA, but regulates the production, distribution, and use of herbal medicines. This agency sets the **product assurance standards** for the herbal medicines available in most of Europe; herbal products that meet Commission E's standards have guaranteed potency.

In 1998, the first Physicians' Desk Reference (PDR) for Herbal Medicines was published; it was based on Commission E monographs. The PDR describes the pharmacology, efficacy, and contraindications of about 700 herbal medicines. It includes, when available, information on herbal and prescription drug interactions.

Even though the FDA recognizes herbals as dietary supplements (or as food), herbal products do contain pharmacologically active ingredients and are used for their pharmaceutical effects. Herbal products marketed in the United States are not tightly regulated by the FDA; therefore, guaranteed potency is not a requirement of U.S. herbal products. Consumers must take exceptional care in evaluating the quality of herbal products, as well as the safety and the efficacy of each of the products. The U.S. National Library of Medicine and National Institutes of Health MedlinePlus herbal medicine Web site (www.nlm.nih.gov/medlineplus/herbalmedicine.html) is a good resource to use for understanding herbal and botanical remedies.[O,R,10,29,33]

Do you use herbal medicines?

Coenzyme Q₁₀ **Supplement Facts**		
Serving size: 1 softgel		
Amount per serving		
	% Daily Value	% DV
Vitamin E (as d-alpha-tocopherol acetate) 150 IU	150 %	
Coenzyme Q₁₀ (ubiquinone) 100 mg	*	Symbols
* Daily Value not established		
Other Ingredients: Gelatin, sorbitol, glycerin and purified water, polysorbate 80, hydroxylated soy lecithin, medium-chain triglycerides, annatto seed extract, and soybean oil.		

Like Nutrition Facts panel — Standard nutrients — Other nutrients — Ingredients

© Cengage Learning 2013

FIGURE 6.3 A sample Supplement Facts panel.

The Dietary Supplement Health and Education Act (DSHEA) of 1994

Being able to sell and purchase foods, supplements, and herbal products is a liberty and a burden under the FDA regulatory guidelines for food compared to drugs. Our cultural belief in "a pill for every ill" creates a high level of acceptance for dietary supplements as effective alternatives for managing many ills. It is a liberty to buy dietary supplements of all sorts because consumers are free to make their own choices and purchases. It is a burden because consumers need to educate themselves with sound information and make wise choices so as to not cause personal injury and financial distress. In an effort to guide the supplement industry, the Dietary Supplement Health and Education Act (DSHEA) of 1994 was passed. This is legislation pertaining to the sale, marketing, labeling, safety, and support statements made on nutritional supplements.[24]

In the DSHEA, the FDA defined nutritional supplements as any of the following: herbs, essential and nonessential vitamins, essential and nonessential minerals, nutritional substances, glandulars, fiber, and enzymes. All supplements must be labeled according to the DSHEA. Dietary supplement labels must include the name and quantity of the nutritional supplement per unit, a Supplement Facts panel (see Figure 6.3), and a disclaimer that the FDA has not evaluated the product. Products may include structure and function claims, so long as the product does not claim to cure or prevent diseases. The FDA-approved health claims that exist for food labels (see Module 2, page 55) can be used for supplements as long as the supplement meets the same criteria as food. Careful wording of health claims and structure and function claims is required. Table 6.1 gives some examples of prohibited and permissible health claims pertaining to the supplement industry.

TABLE 6.1

Examples of Prohibited Versus Permissible Health Claims

Prohibited	Permissible
Reduces pain and stiffness associated with arthritis	Promotes relaxation
Laxative	Promotes regularity
Prevents Alzheimer's disease	Reduces absentmindedness
Antiviral	Supports the immune system
Antidepressant	Reduces frustration or rejuvenates

© Cengage Learning 2013

Herbs Herbal dietary supplements are ground-up plant material or extracts that can be encapsulated or not. Additionally, **tinctures** and teas are common forms of the herbal products. The 16 most common herbal products sold in the United States are described in Table 6.2.

TABLE 6.2

Top Herbs and Their Uses

Herbal Supplement	Commonly Used or Touted To:
Ginkgo biloba	Increase brain blood flow, prevent dementia, improve memory and libido, and act as an antioxidant
St. John's wort	Manage mild to moderate depression
Ginseng	Boost energy and correct imbalances
Garlic	Reduce blood cholesterol, lower blood pressure, and act as an anti-cancer agent
Echinacea	Stimulate immunity that can combat colds, flus, and respiratory infections
Saw palmetto	Reduce the symptoms of an enlarged prostate
Kava kava	Calm anxiety and reduce the stress response
Soy	Reduce the symptoms of menopause, improve heart health, and act as a gender-specific anticancer agent
Valerian	Induce sleep and reduce anxiety
Evening primrose	Reduce inflammation, immune disorders, and the discomforts of pre-menstrual syndrome (PMS) and menopause
Grape seed extract	Reduce allergic responses and promote heart health with its antioxidant properties
Milk thistle	Improve liver disease by regenerating cells and protecting the liver from injury
Bilberry	Improve eye disorders
Black cohosh	Reduce the discomforts of menopause, PMS, and other menstrual disorders
Pycnogenol	Promote heart and connective tissue health with its antioxidant properties
Ginger	Reduce the nausea of pregnancy, chemotherapy, and motion sickness, improve digestion, and reduce inflammation

© Cengage Learning 2013

Essential and Nonessential Vitamins Recall that vitamins are organic chemical compounds that are integral structural components of enzymes or coenzymes in the body. A standard vitamin pill contains only the essential vitamins (covered in Module 5, beginning on page 235). Nonessential vitamin compounds are made in the body, and healthy people do not benefit from taking additional nonessential vitamins, but some individuals may improve function by supplementing with additional nonessential vitamins (see Table 6.3).[46,65]

TABLE 6.3

Nonessential Vitamins and Their Uses

Nonessential Vitamin	Commonly Supplemented and Touted to Improve:
L-carnitine	Fatty acid oxidation, vigor in end-stage renal patients, and contractility of the heart muscle in cardiomyopathy
Coenzyme Q$_{10}$	Contractility of the heart muscle in cardiomyopathy; it also reduces periodontal disease and protects mitochondrial membranes from oxidation
Lipoic acid	Liver function; it protects against further damage by providing additional antioxidant support

© Cengage Learning 2013

tinctures Alcohol extracts of biologically active plant compounds that are concentrated and used in a liquid form to remedy illness.

Essential and Nonessential Minerals Recall that minerals are inorganic elements that are naturally found on earth. Each mineral is defined by a unique set of physical and chemical properties. Standard vitamin/mineral supplements usually only contain the essential minerals (covered in Module 5, beginning on page 252). Some nonessential minerals have been popularly marketed to improve health or function. Some of these nonessential minerals include boron and vanadium. Boron is commonly sold in 3-milligram doses to improve calcium metabolism in bone mineralization. Vanadium in the form of vanadyl sulfate (100 milligrams per day) has been used to normalize blood sugar levels in people with type 2 diabetes.

Nutritional Substances Nutritional substances are biologically active chemical compounds that have been extracted from food sources, concentrated, and commonly encapsulated, and then marketed to have a variety of effects, from improving many discomforts of medical conditions to enhancing sport performance (see Table 6.4).

TABLE 6.4

Popular Nutritional Substances and Their Uses

Dietary Supplement	Commonly Used or Touted To:
Glucosamine	Improve joint health and increase joint comfort
Chondroitin	Improve joint health and increase joint comfort
Beta-alanine	Increase speed and strength
Creatine	Increase speed and strength
HMB (beta-hydroxy, beta-methyl butyrate)	Increase strength and lean body mass, reduce muscle soreness, and increase aerobic capacity
Fish oil	Improve heart health, reduce inflammation, reduce muscle soreness, and increase aerobic capacity
Conjugated linoleic acid	Reduce fat mass and reduce the risk of cancer

© Cengage Learning 2013

Glandulars A glandular is a dietary supplement that contains hormones. Very few glandulars can be sold over the counter. The two most popular include melatonin and dehydroepiandrosterone (DHEA). Melatonin is a protein hormone made from the essential amino acid tryptophan in the pineal gland in response to dark. It is the natural hormone that induces sleep and has the greatest influence on **circadian rhythm** in the human body. DHEA is a steroid hormone. The largest amount is produced in the adrenal glands in the body. It is known as the anti-aging hormone, because the peak production in the body is at age 20, and by 40 years old, half of the peak amount is produced.

Fiber Fiber is an eclectic term used for indigestible molecules normally consumed in foods. There is a subcategory for fiber called functional fiber, which provides a beneficial effect in addition to its role in health (see page 120). Some of the functional fibers include fructooligosaccharides, which may be beneficial for those who suffer from inflammatory bowel disease; flax seed lignins, which may be beneficial in the modulation of estrogens in the body; and psyllium, which may be beneficial for altering blood lipids to improve heart health. Because fibers bind strongly with water, good hydration is important to maintain, especially when supplementing functional fiber.

Enzymes A variety of digestive enzymes (see page 111) and other digestive aid products are available over the counter. They may improve many complications of poor digestion seen commonly in the elderly and in stressed individuals. Common signs and symptoms of poor digestion include gas, bloating, malnutrition, and a wide variety of other gastrointestinal discomforts resulting from indigestion.

circadian rhythm A daily cycle of biochemical, physiological, and/or behavioral processes of living organisms, greatly influenced by the exposure to 24-hour cycles of light and dark.

Betaine hydrochloric is a common ingredient in enzyme products. It increases stomach acid production. Pepsin, bromelain, and papain, also common ingredients, are proteases that digest protein. Bile extract may also be added to emulsify fat and facilitate the absorption of vitamins A, D, E, and K. Another common ingredient is pancreatin, which provides the combination of amylase, lipase, and protease that digests proteins, carbohydrates, and fats. Furthermore, additional amounts of amylase, lipase, and protease can be included in supplements to increase the digestion of proteins, carbohydrates, and fats.[30,38,43,63,78]

Summary Points

- Many governmental agencies are responsible for the safety of food.
- Governmental food safety agencies enforce the laws that pertain to food safety.
- Take supplements only purposefully and knowledgeably.
- Remember, rigorous scientific testing is not FDA-mandated on nutritional supplements.
- Become familiar with product certification insignias.
- Read the Supplement Facts panel.
- Look for credible publications supporting the safety and efficacy of the supplement.

Take Ten on Your Knowledge Know-How

1. What is the role of the FDA in the food and drug industries? What are your feelings on this topic?
2. What piece of legislation is enforced by the EPA?
3. What is the GRAS list? Have all of the chemicals on the list been thoroughly tested?
4. What federal agency oversees meat, eggs, and poultry?
5. What types of foods are overseen by the NMF?
6. What is product assurance? What are some high-quality certification insignias?
7. What essential nutrients have been used therapeutically?
8. What are a few references or resources to use for herbal information?
9. What is regulated by the Dietary Supplement Health and Education Act of 1994? Are all dietary supplements safe?
10. Give two examples of dietary supplements from each of the supplement categories.

6.3 FOOD ADDITIVES

T-Talk 6.3
To hear Dr. Turley talk about food additives, go to www.cengagebrain.com

Introduction Food additives are used to enhance nutritional value, prevent disease, keep food from spoiling, add flavor, retain color, and/or prolong the shelf life of processed food. The FDA approves the use of food additives in the United States. Since 1960, approved food additives must be extensively tested. The food additives that were sunseted onto the GRAS list in 1958 did not undergo extensive safety testing. Today, food additives must be proven effective, safe, and detectable in the final food product before they are approved for use.[H,8]

Food additives can fall into one of three categories: those added intentionally, incidentally, or indirectly. An intentional additive is a substance used for restoration, enrichment, fortification, preservation, or enhancement (flavor, texture, appearance, and consumer appeal) of the product to which it is added. An incidental additive is a chemical compound or material present in processed foods that was unintentionally included in the food, such as a substance that comes in contact with food during growth or processing (e.g., insects, detergents, etc.). An indirect additive is a substance present in the food package, which may then be present in the food. In the following sections, intentional food additives will be explored in more detail.

Categories of Intentional Food Additives

The thousands of intentional food additives that can be added to processed foods fit into eight different categories. All food additives have their own set of regulations, meaning that the FDA has defined acceptable amounts and situations for use (see Table 6.5 and BioBeat 6.4).[C,E,F]

TABLE 6.5

Categories of Food Additives

Food Additive	Function
Antimicrobial agents	Preservatives that prevent spoilage by reducing the growth of mold or bacteria. Examples include acetic acid (vinegar), sodium chloride (salt), benzoic, proprionic, and sorbic acids, nitrates, nitrites, and sulfur dioxide.
Antioxidants	Preservatives that prevent foods containing fat from going rancid. Examples include vitamin E, butylated hydroxyanisole (BHA), and butylated hydroxytoluene (BHT).
Artificial colors and flavors	FDA-certified chemicals added to food to enhance appearance are artificial colors. Examples include vegetable dyes and synthetic dyes such as FD&C yellow #5. Chemicals added to food to enhance flavor or mimic flavor are artificial flavors or flavor enhancers. An example is monosodium glutamate (MSG).
Bleaching agents	Substances used to whiten foods like flour. An example is peroxide.
Chelating agents	Molecules that bind other molecules and prevent discoloration, flavor changes, and rancidity of food. Examples include citric acid, malic acid, and tartaric acid.
Emulsifying agents	Chemicals that keep water and oil in solution. Examples include soy and egg lecithin and mono- and diglycerides.
Nutrient additives	Vitamins and minerals added to food for nutrient quality.
Stabilizing and thickening agents	Ingredients that maintain emulsions, foams, or suspensions or lend a desirable, thick consistency to foods. Examples include dextrins, starch, pectin, and gums.

© Cengage Learning 2013

Antimicrobial Agents These include a variety of chemicals that prevent **microbial growth**. A substance as simple as salt (sodium chloride) added to food is effective as an antimicrobial agent. Other compounds reduce pH to control microbial growth, while sorbic acid is added to cheese wrappers and other processed foods to prevent molding.

Antioxidants These substances retard the oxidation of unsaturated fats in food and include butylated hydroxyanisol (BHA), butylated hydroxytoluene (BHT), and vitamin E. These are commonly added to unsaturated fat food sources like oils. BHA and BHT are also added to processed foods that are high in fat, such as potato chips and baked foods; they are even added to some cereals.

Nutrients Essential vitamins and minerals are added to many restored, enriched, or fortified foods, such as breakfast bars, cereals, refined grains, and meal replacers. Many nonessential nutrients are added to **functional foods**, such as omega-3 fatty acids and taurine. Examples of essential nutrients added to food include ascorbic acid (vitamin C) and thiamin mononitrate (thiamin).[4,34,74]

microbial growth The many ways microorganisms perpetuate themselves, which are affected by temperature, protein, pH, and moisture.

functional foods Foods that may have additional health-promoting properties beyond providing basic nutrition needs (Calories and essential nutrients); may also be called medicinal foods.

DA+ GENEie Go to Diet Analysis Plus and enter vitamin water and regular water, Ovaltine and Hershey's chocolate drinks, and Cheerios and regular oatmeal cooked in water. Use the intake spreadsheet to compare the nutrient analysis results. Draw conclusions about the nutrient additives in these foods.

Stabilizers and Thickening Agents Starches, gums, and gels add form and structure to food. They can improve the mouth feel and texture of processed foods. A few examples of foods that may contain these agents are regular and low-fat mayonnaise, dressings, ice cream, yogurt, and cream cheese.

Artificial Colors and Flavors (Flavor Enhancers) These substances are added to improve a food's visual appearance or to accentuate a food's natural taste. Examples include FD&C yellow #5 (tartrazine), a food dye, and monosodium glutamate (MSG), a flavor enhancer. MSG is added to many processed foods, including soups, frozen entrées, chips, and many other snack foods. Dyes such as FD&C colorants are added to many processed foods to make them look more appealing. Examples include popsicles, punch, candy, flavored yogurt, and the like. Drugs and cosmetics commonly have added colorants. Just read the ingredient list and look for the presence of these chemicals.

Bleaching Agents These chemicals make foods look whiter. Examples of bleaching agents include benzoyl peroxide and azodicarbonamide. These are used to whiten foods such as refined flours and baked goods.

Chelaters These chemicals stabilize chemical reactivity by binding metal ions that cause foods to change color and/or go rancid. Examples include citric acid, malic acid, and tartaric acid (cream of tartar). Some foods containing chelating agents are canned goods, beverages, dressings, and sauces. The iron that is added to bleached flour is chelated.

BioBeat 6.4

Food Chemicals: Authentic or Artificial Substances?

Food, like every other substance around us, is made up of chemicals. As you have learned, some chemicals—such as vitamins, minerals, amino acids, and fatty acids—are essential for human nutrition. Some chemicals that are produced in life forms play a role in metabolism. These are known as biological molecules and are produced naturally. Examples are adenosine triphosphate (ATP), glucose, galactose, stanols, and cholesterol. Some chemicals are produced totally synthetically in a food science laboratory and used as food additives. Aspartame, an example of an artificial sweetener, is synthesized from two amino acids: phenylalanine and aspartic acid. On the other hand, salt or sodium chloride is a natural compound and has been used as a preservative since antiquity.

The chemicals that are not of natural origin are completely synthetic, manmade, artificial, or simply not biological. Some examples of these compounds include olestra, tartrazine (FD&C yellow #5), butylated hydroxytoluene, and sucralose. Although these artificial substances are FDA-approved and Generally Recognized As Safe (GRAS), some are not free of adverse reactions. Other substances like nitrates and nitrites are used as preservatives in many processed foods, especially meats. Although these fake chemicals are GRAS, they can cause adverse reactions, such as headache and hives, in some people. Additionally, nitrates and nitrites can combine with amines, which are present in foods and in the gastrointestinal tract, to make nitrosamines. Nitrosamines are cancer-causing chemicals, and their presence increases cancer risk.[M,36]

As you make food choices, decipher whether the food is more authentic or artificial and additive laden. Read the ingredients list on food package labels. If you are shopping for grape juice, take the time to find out if the juice is made from real grapes or artificial flavor plus artificial colorants. You will be healthier if you consume authentic foods and minimize the intake of artificial additives.[A,C]

Do you experience adverse reactions to chemicals in processed foods?

Emulsifers These chemicals keep water and oil mixtures combined in solution. Examples of emulsifiers include soy and egg lecithin and mono- and diglycerides (see page 25 in Module 1).

Fat Substitutes and Indigestible Fat

Fat substitutes use food chemistry to provide a good "mouth feel" with fewer Calories by replacing 9-Calorie-per-gram nutrients with 0- or 4-Calorie-per-gram nutrients. These substances replace or substitute fat (triglycerides) in the food by acting as stabilizing agents. A stabilizing agent gives food a pleasing form, appearance, and/or texture. Fat substitutes that act as stabilizing agents are used in many food products and are relied on heavily when the food product is reduced Calorie, reduced or low fat, or fat free. There is only one **fake fat**: olestra, which is fake because it is an artificial, manmade compound that is indigestible. It is used in traditionally high-fat snack foods like chips and crackers. Foods processed with olestra are typically fat free. It is important to understand that foods made with fat substitutes and fake fats are not Calorie free.

Fat Substitutes That Are Stabilizing Agents These chemicals are added to thicken, add texture, or improve the mouth feel of processed foods; some may be low fat or fat free, but others may not be. These agents are usually carbohydrates, proteins, or soluble fiber types.

Carrageenan, Guar, and Xanthum Gum Gels (Fiber-Based Products) These products are used to retain moisture because the molecules have a great affinity for water, so they thicken or have a gelling effect. Furthermore, they are indigestible and thus virtually noncaloric. Carrageenan is made from seaweed.

Starches These are complex carbohydrates that act as fillers and are digestible, thus caloric. Starches hold water and impart a creamy texture. They are often labeled as modified food starch and are usually wheat or corn based. These are found in many processed food products, including fast-food hamburgers, ice cream, some brands of fat-free milk, salad dressings, desserts, sauces, yogurt, and more. Starches are used to replace fats with digestible carbohydrates.

Simplesse This is a fat substitute made from a mixture of food proteins (egg white, whey, casein), cooked and blended to form tiny particles that trap water. Simplesse is perceived as fat in the mouth and adds flavor to the food product. Because simplesse is made from protein, it is a protein-stabilizing agent, it is caloric, and it can be denatured. Heat causes it to gel. Simplesse is found in cheese, ice cream, other frozen desserts, mayonnaise, and so on. The protein is used to replace the fat in the processed food.

Fake Fat Olestra is a manmade chemical designed to mimic fat. It provides all of the properties of added fat, like those found in baked goods, but because it is indigestible, products that use it can be labeled fat free.

The trade name of olestra is Olean. This fake fat has a complicated chemical structure of eight fatty acids linked to a sucrose molecule. The fatty acids are linked in such a way that the body does not digest olestra; therefore, it is noncaloric. Olestra does interfere with the absorption of fat-soluble nutrients (including vitamins A, D, E, and K) and can cause gastrointestinal distress. Products that use it must add additional amounts of the fat-soluble vitamins.[1]

fat substitutes A wide variety of chemical compounds (such as starches, gums, gels, and proteins) used in processed food to reduce the amount of fat and Calories in the food.

fake fat An artificial chemical compound (such as olestra) used in processed food that is indigestible and noncaloric, but provides the properties of fat.

Sugar and Artificial Sweeteners

Many food consumers are interested in reducing their intake of table sugar (sucrose) and thus turn to sugar substitutes. Despite these efforts, the average American diet has 25 percent of Calories derived from sugar, which adds up to approximately 70 pounds per year per person. Many consumers eat honey because they believe it to be healthier, or use artificial sweeteners to control body weight. Artificial sweeteners provide intense sweetness, so the small quantities used in foods can reduce the Calories from normal sugar. It is important to understand that foods made with **sugar substitutes** or **artificial sweeteners** are not always Calorie free. Let's explore these substances in more detail.[2,18,70]

Honey All sugars provide empty Calories (and are low in nutrients). Honey is not nutritious because of the very tiny amounts of nutrients present in the sugar Calories provided. The quantities of nutrients in honey have very little significance in relation to one's diet. However, honey is sweeter than sucrose, and thus less sugar can be used to achieve the desired sweetness.

Sugar Free Sugar free indicates that sucrose (table sugar) is not present in the food product. Other sugar substitute sweeteners such as sorbitol, a sugar alcohol, provide Calories but do not promote tooth decay and provide a sweet flavor.

Artificial Sweeteners Aspartame, saccharin, sucralose, and Aceulfame-K are all artificial sweeteners that have been approved by the FDA and provide few Calories because of their intense sweetness or chemical structure. Aspartame is packaged in blue, saccharin in pink, and sucralose in yellow, and all are usually offered in restaurants.[2]

Aspartame (NutraSweet, Equal) Aspartame is a popular artificial sweetener made up of two amino acids: phenylalanine and aspartic acid. It is 160 to 220 times sweeter than sugar and is found in more than 1,500 food products in the United States, such as diet drinks, sweetened cereals, and baked goods. The safe consumption level set by the FDA is 50 milligrams per kilogram of body weight per day. This sweetener is GRAS; however, some side effects have been reported in association with its use, but only by a few people when evaluated on a population basis. The side effects have also been variable in nature but generally include headaches, seizures, and behavioral changes. Scientific studies have not confirmed these associations with aspartame. The one exception to the safety of aspartame is that individuals with a rare genetic abnormality called phenylketonuria cannot use it. These individuals cannot metabolize the amino acid phenylalanine and therefore should avoid using aspartame (see page 365 in Module 7). Again, the blue packets of artificial sweetener at most restaurants contain aspartame.

Saccharin (Sweet'N Low, Sugar Twin) This is a controversial artificial sweetener. Huge amounts have been shown to cause cancer in rats, but not in humans. It has remained approved for use because of consumer demand (see page 296). It is 200 to 700 times sweeter than sucrose. Again, the pink packets of artificial sweetener at most restaurants contain saccharin.

Sucralose (Splenda) This sweetener is made from a chemical modification of real sugar and is used in products such as diet soft drinks and desserts. It is also available in a cup-for-cup sugar substitute product that can be used in cooking. It is 600 times sweeter than sucrose and may have the fewest safety concerns. Again, the yellow packets of artificial sweetener at most restaurants contain sucralose.

Jennifer Turley

Equal, Sweet'N Low, and Splenda are artificial sweeteners.

sugar substitutes A wide variety of chemical compounds that are perceived as sweet to the human tastebuds and are added to processed foods to reduce the sucrose content.

artificial sweeteners Artificial chemicals such as aspartame, saccharin, sucralose, and Aceulfame-K that have been approved by the Food and Drug Administration to replace sugar in processed foods.

Demo GENEie Bring in labels of processed foods, over-the-counter medications, and cosmetics (lotions, toothpaste, etc.). Read the ingredients list and take note of any additives present. Explain and discuss the purpose of some of the key additives that you find in the items studied.

Aceulfame-K (Sunette) This is an artificial sweetener, 200 times sweeter than sucrose, that is used in diet soft drinks and desserts. It can be combined with aspartame to make an even sweeter taste.

Summary Points

- Chemicals are added to food intentionally, incidentally, or indirectly.
- Food additives are used for many reasons in the food industry and can fit into eight different categories.
- Chemicals in food are FDA approved and GRAS, although not free of adverse reactions in some individuals.
- Intentional additives are used in processed foods for their nutritional value, preservative qualities, or aesthetic effects (appearance-, flavor-, and/or texture-enhancing properties).
- Fat substitutes, fake fats, and artificial sweeteners are commonly used in processed foods to reduce fat or sucrose content.

Take Ten on Your Knowledge Know-How

1. Do some food additives have adverse effects? Do food additives cause cancer?
2. What are the basic types of foods additives? Give an example of the presence of each type of food additive in the foods that you typically consume.
3. What are the basic categories of food additives? Give an example of chemicals and their purpose in each category. What are some foods that you commonly consume that contain these chemicals?
4. If you wanted to consume a diet with fewer intentionally added chemicals, what would you do? Try to plan a nutritionally adequate day's worth of eating using this approach.
5. How does the food industry reduce fat in processed foods while maintaining the sensation of fat in the mouth upon consumption? Can your mouth tell the difference between real fat and fat substitutes or fake fats?
6. Which fat substitutes, fake fats, and/or artificial sugars do you regularly consume? Do you plan on altering your intake? If yes, why? If not, why not?
7. What are the differences between an authentic and an artificial food additive?
8. What types of additives are intended to reduce sugar in the diet?
9. Does sugar free mean Calorie free?
10. What are incidental additives?

6.4 FOOD SAFETY: MICROBIAL GROWTH

Introduction The most prevalent food safety issue today is foodborne illness related to food handling. The CDC estimates there are 76 million foodborne illness cases annually in the United States. Mishandling of food is by far the most frequent cause of people getting sick from food. In this section, we will look at select **microorganisms** that cause food poisoning, the ways food can be contaminated, and ways of preventing contamination. This section will explain the importance of safe handling of food by consumers and food industry personnel. Microbial growth and the common microorganisms that threaten the safety of food will be discussed. These will be contrasted to beneficial **microbes** supporting health.

Microorganisms and Microbial Growth So far we have learned that there are laws and regulatory agencies to govern the food industry. One main purpose is to control microbial growth to prevent food poisoning. In

T-Talk 6.4
To hear Dr. Turley talk about microbial growth, go to www.cengagebrain.com

microorganisms A wide variety of life forms, also called microbes, that are so small that a microscope is needed to see them.

microbes A general term for microorganisms such as bacteria, protozoa, parasites, fungi, and algae.

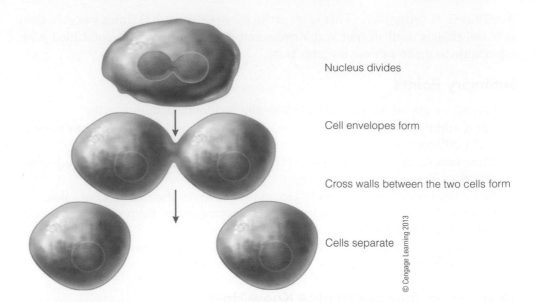

FIGURE 6.4 Microbial cell division by fission.

Nucleus divides

Cell envelopes form

Cross walls between the two cells form

Cells separate

© Cengage Learning 2013

this section, we will see where nutrition and microbiology intersect, because microbes are abundant in the food and water supply. Each year in the United States, hundreds of thousands of people are hospitalized and thousands of deaths occur from outbreaks of food poisoning. Thus, it is critical to gain an understanding of the types of microbes and their growth processes so that you can minimize their presence and harmful effects. Fortunately, foodborne illness can be prevented, so let's learn how to keep food safe and recognize unsafe foods and eating situations.[13,14,48,49]

Microbial Growth Microorganisms may divide by replicating their components and separating into two cells. In this process of replication, the nucleus divides, a cell envelope forms, a cross wall forms between the two cells, and finally the cells separate (see Figure 6.4). When one cell divides into two, the two into four, the four into eight, and so on, **clonal expansion** occurs; cell number increases exponentially. This type of cell division is called **fission**. Most microorganisms divide rapidly. A single cell can grow into a clonal population in a matter of hours (see Table 6.6). Depending on the microorganism, reproduction can also occur by forming endospores or by budding. Viruses divide by a different mechanism that will be explained later.[48]

TABLE 6.6		
Clonal Expansion of Microorganisms		
Time	Minutes at 97°F	Theoretical Number of Organisms
8:00	0	10
8:23	23	20
8:46	46	40
9:09	69	80
9:32	92	160
9:55	115	320
10:18	138	640
10:41	161	1,289
11:04	184	2,560
11:27	207	5,120
11:50	230	10,240

© Cengage Learning 2013

Owen Franken/CORBIS

Eating buffet- and potluck-style can give rise to a potentially unsafe eating situation.

Data GENEie Pretend you are at a potluck dinner where a large bowl of potato salad made with eggs is available to eat on the dinner table. It has been handled by 20 people at room temperature over the course of 3 hours. One of the first people to dish from the salad bowl sneezed on the front side of the salad and contaminated it with 200 bacteria. People eating the salad were scooping from the back side of the bowl. The microbes divided every 20 minutes. How many microbes are in the salad from the sneeze after 3 hours?

Classification of Microorganisms Microorganisms vary in their shape, cell arrangement (chains versus clusters), inclusions, spores and/or capsules, staining, motility, and nutrient and oxygen requirements (see photographs on page 311). A microorganism that grows in the presence of oxygen is an **aerobe**, whereas

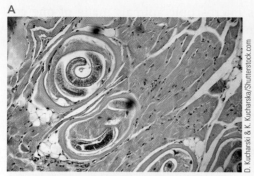
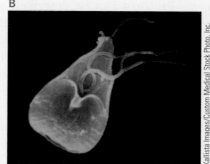
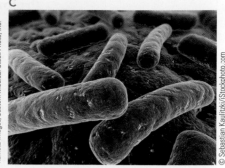

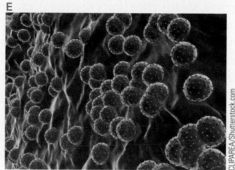
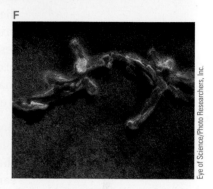

one that grows in the absence of oxygen is an **anaerobe**. Microorganisms can be encountered through air, water, food, improper or lack of hand washing (fecal contamination), soil, and person-to-person interactions. Types of microorganisms include **parasites**, **protozoa**, **bacteria**, **fungi**, **viruses**, and **algae**. All of these microbes can be implicated in foodborne illness, except freshwater algae.

Common Microorganisms Causing Food Poisoning Microorganisms commonly infect food, reproduce, are metabolically active, and cause sickness when ingested. Foodborne illness or food poisoning can have mild to severe and life-threatening flulike symptoms and, in some cases, can cause neurological and organ damage to those who are afflicted. Contaminated food can cause severe gastrointestinal tract upset (cramps, diarrhea, nausea, and vomiting). Acid–base imbalance and/or fluid and electrolyte loss can result from severe diarrhea and vomiting. Other organs may be damaged depending on the microorganism causing the foodborne illness.[25,77]

Table 6.7 on pages 314–315 identifies many common microorganisms and infectious agents that cause disease and classifies the microorganism or agent as a parasite, protozoan, bacterium or bacterial toxin, or virus. It also shows which microorganisms frequently infect foods, the signs and symptoms of the sickness, the time of onset, and methods for disease/sickness prevention. Several infectious agent types—parasites, protozoa, bacteria, bacterial toxins, fungi, viruses, and **prions**—are discussed briefly with an example.[G,82]

Parasites   A parasite is a living organism that lives in or on another living organism, the host. Parasites take their nourishment from the host without benefiting the host. *Trichinella spiralis* is a nematode parasite that causes trichinosis. The larvae of the *Trichinella spiralis* worm can reside in the muscle meat of infected animals. Eating undercooked meat from the contaminated animal, usually undercooked pork, transmits the disease to the human consumer. Achieving the recommended cooking temperatures for each type of meat can prevent illness. When consumed, the stomach acid dissolves the larvae cyst and releases the worms. In the small intestine, the worms replicate and enter the body to target muscles.

Various microorganisms:
A: *Trichinella spiralis* parasite,
B: *Giardia lamblia* protozoan,
C: *Escherichia coli* bacterium,
D: *Asperigillus flavus* mycotoxin producer,
E: Hepatitis A virus, and
F: a prion protein.

clonal expansion Asexual reproduction of one organism, producing a population of genetically equivalent organisms.

fission The splitting of one unicellular organism to become two.

aerobe A living organism that requires oxygen.

anaerobe A living organism that does not require oxygen.

parasites Living organisms that live in or on another living organism (the host), and take their nourishment from the host without benefiting the host.

protozoa Microscopic animals made of one cell or a group of like cells that live predominately in water; some are human pathogens.

bacteria A large group of single-celled organisms of various shapes, lacking a nucleus, that are everywhere on earth, important for nutrient cycling, and abundant on and in the human body.

Protozoa   Protozoa can be described as microscopic animals made of one cell or a group of like cells that live predominantly in water. Some protozoa, such as *Giardia lamblia*, are human pathogens. *Giardia lamblia* infection is known as the "backpacking bug." When water containing this protozoan is consumed, the small pear-shaped **trophozoites** begin to live in the duodenum. The trophozoites become infective cysts as they pass through the lower small intestine. A useful test to diagnose *Giardia lamblia* infection is stool examination for the organism. This parasitic protozoan can be identified under the microscope by its distinctive flagellate trophozoite with two nuclei and an adhesive disk. *Giardia lamblia* reproduces by binary fission of trophozoites in the small intestine. *Giardia lamblia* infection may be asymptomatic or may cause a variety of intestinal symptoms, including chronic diarrhea, steatorrhea (fatty diarrhea), cramps, bloating, fatigue, and weight loss. The presence of intestinal trophozoites causes intestinal epithelial cells to die early; the result is a reduced ability to digest and absorb fats and fat-soluble vitamins, compromising the host's nutritional status. Human immune system mechanisms are inadequate to control *Giardia lamblia* infection. In the United States, infections are treated with drug therapy because *Giardia lamblia* infection is not self-limiting.[G]

Bacteria   *Bacterium* is the singular form of the word *bacteria*. Bacteria represent a large group of single-celled organisms lacking a nucleus. They can have various shapes (such as rod, spherical, or spiral) and are ubiquitous, found everywhere on earth. They are abundant on most foods and in the human body. A single human may compete with hundreds of thousands of bacterial species. Bacteria are also important for nutrient cycling, especially in soil.

Escherichia coli, abbreviated *E. coli*, is a group of bacteria that commonly reside in the small intestine. Some strains of *E. coli* are **pathogenic** and implicated in foodborne illness. *E. coli* 0157:H7 is one of the emerging foodborne pathogens. Because bacteria replicate so quickly and frequently, they evolve rapidly. *E. coli* 0157:H7 is an *E. coli* bacterium strain that evolved to become highly pathogenic. The nature and extent of its rapid evolution from a nonpathogenic strain has revealed highly divergent genes and implicated **recombination** with natural selection. The *E. coli* 0157:H7 bacterium is a worldwide public health threat because it contaminates the food supply (originally ground beef products) and causes **enterohemorrhagic colitis** and bloody diarrhea.[7,35]

Bacterial Toxins   Some microbes directly cause foodborne illness or produce **enterotoxins**, which act on the mucus membranes and cells in the digestive tract, or **neurotoxins**, which act on the nervous system. For example, the neurotoxin produced by the bacterium *Clostridium botulinum* causes botulism. It causes symptoms such as double vision, inability to swallow, and speech difficulty; these symptoms progress to respiratory paralysis and death if they are not treated quickly.

Fungi   The term *fungi* can be defined as a taxonomic kingdom that includes yeast, molds, mushrooms, and other species similar to plants but lacking chlorophyll. Mycotoxins are produced from fungal infections of plant crops and pose a biohazard risk. The common mold species *Asperigillus flavus*, which produces aflatoxin, can grow on fruits, vegetables, grains, nuts, and seeds, especially those produced and/or stored in warm and humid climates. Aflatoxin is a potent liver cancer-causing agent that is monitored carefully by the FDA.

Viruses   Viruses originate as a **virion**, which is encapsulated DNA or RNA, depending on the virus. DNA viruses directly integrate into the host's nuclear DNA. RNA viruses (also called retroviruses) are reverse transcribed to DNA, which then integrate into the host's nuclear DNA. Once it is a part of the host's DNA, the virus uses the host's metabolism to replicate and make capsids. The viral capsids are then released from the cell to infect more cells. Sometimes, the cell becomes so full of

fungi A taxonomic kingdom that includes yeast, molds, mushrooms, and other species similar to plants but lacking chlorophyll.

viruses Encapsulated genetic material that require a host in order to perpetuate.

algae A large and varied group of chlorophyll-containing photosynthetic organisms that are primarily aquatic and include seaweed and kelp.

prion An infectious protein believed to cause mad cow disease and be transmittable to humans.

trophozoites Protozoa that are in a growing phase rather than a reproductive or resting phase.

pathogenic Disease-causing.

recombination The breaking down of DNA (usually), which then rejoins or crosses over to a different DNA molecule and begins a generational transfer of the genetic change.

enterohemorrhagic colitis Intestinal bleeding causing bloody diarrhea.

enterotoxin A protein toxin that can be released by a microorganism and affects the intestine.

neurotoxin A toxin that can be produced by bacteria and targets nerve cells.

virion A single, complete virus particle.

virions that it lyses (breaks apart), increasing the viral infection. Unfortunately, food can be a carrier of pathogenic viruses such as norovirus, rotavirus, and Hepatitis A.

For a viral example, let's explore Hepatitis A virus (HAV) in more detail. Hepatitis means "inflamed liver." Primates are the only natural host in the animal kingdom for HAV. The route of infection is fecal, which can be transmitted orally or through contaminated water. Once the primate has been infected with HAV, the animal is a lifelong carrier of the virus. Even though HAV enters the body through the gastrointestinal tract, the virus reproduces in the liver and is excreted from the liver through the bile acids. Large concentrations of the virus can be detected in the stool of the host. The time from infection to onset of illness from HAV infection ranges from 15 to 50 days. The median day of onset is 28. Seventy-six to 97 percent of adults who have been infected become symptomatic, and 40 to 70 percent develop yellowing of the skin, called jaundice, because of the liver injury. To make a definitive diagnosis of HAV infection, a blood test for the IgM antibody to HAV is performed. The antibodies can be detected 10 days before the onset of symptoms and do not disappear until six months after symptoms have subsided.

There is a vaccine for HAV and rotavirus. These are the only vaccines available to prevent foodborne illness in the United States. The HAV vaccine is recommended for adult drug abusers, homosexual males, or international travelers who plan to go to poorly sanitized geographic locations. HAV and rotavirus vaccines are also recommended in pediatric immunizations.[G]

Prions A prion (proteinaceous infectious particle) is an infectious protein believed to cause mad cow disease (bovine spongiform encephalopathy) and is transmittable to humans. The prion protein has been found to survive in soil for 29 months and maintain its pathogenicity. Soil can become contaminated from excrement of infected animals. Human and animal outbreaks occur from consuming contaminated feed. Creutzfeldt-Jakob disease is a neurological and fatal human form of the disease caused by prions.

Common Microorganisms Supporting Health We have seen that certain microorganisms can cause serious illness or even be fatal if contaminated foods are eaten. But not all microorganisms are bad; some microorganisms in food are intentional additives and have beneficial effects (see BioBeat 6.5 on page 315). Intentionally added microbes are in the ingredients list on the food package label.[F] In traditional fermented yogurt, the two main bacteria used are *Lactobacillus bulgaricus* and *Streptococcus thermophilus*. In yogurt containing **probiotics**, many different bacteria may be intentionally added, but the two main bacteria added for health are *Lactobacillus acidophilus* and *Bifidobacterium bifidus*. Probiotics provide one or more health benefits to the consumer via the many live microorganisms that will survive in the gastrointestinal tract. The word *probiotics* was combined from *pro* and *biota*, which means "for life." Probiotic supplements have been shown to be beneficial in preventing or managing diarrhea, lactose intolerance, yeast infections, inflammatory bowel disease, and allergies. Because supplements vary and require proper storage, it is difficult to know about the true viability of these supplements.

If the probiotic is for maintenance or preventive purposes, a reasonable dose would be between one billion (10^9) organisms per gram two or three times weekly (achieved with supplements) and four billion organisms per gram daily. For therapeutic purposes, such as fighting an intestinal infection, coping with an illness, or after a course of **antibiotics**, an effective dose would be five to ten billion organisms per gram, two or three times daily. It should be noted that once yogurt is prepared, it contains 10^9 organisms per gram, but once it is stored at 5 degrees Celsius for 60 days, the bacteria concentration gradually decreases to 10^6 organisms per gram. Still, live-cultured yogurt is a good source of probiotics.

probiotics Live microorganisms that, when administered in adequate amounts, confer a health benefit on the host.

antibiotics A chemical substance that is administered as a drug to kill bacteria.

TABLE 6.7

Some Common Culprits Causing Foodborne Illnesses

Organism and Disease Caused	Food Attributes and Sources	Onset and Symptoms	Prevention Methods
Campylobacter jejuni bacterium **Campylobacteriosis**	Raw poultry, beef, lamb, unpasteurized milk	Onset: 2–5 days Symptoms: Diarrhea, nausea, vomiting, cramps, fever, bloody stools	Cook food thoroughly. Use sanitary food-handling methods. Use pasteurized milk.
Listeria monocytogenes bacterium **Listeriosis**	Raw meat, seafood, milk, soft cheeses	Onset: 3–70 days Symptoms: Flulike symptoms, blood poisoning, meningitis (stiff neck), pregnancy complications, headache, fever	Cook food thoroughly. Use sanitary food-handling methods. Use pasteurized milk.
Clostridium perfringins bacterium **Perfringins**	Improperly stored meats and meat products	Onset: 8–12 hours Symptoms: Diarrhea, nausea, vomiting	Cook food thoroughly. Use sanitary food-handling methods. Refrigerate food promptly and properly.
Salmonella bacterium **Salmonellosis**	Raw or undercooked eggs, meats, poultry, dairy products, shrimp, frog legs, yeast, coconuts, pasta, chocolate	Onset: 6–48 hours Symptoms: Nausea, fever, chills, vomiting, cramps, diarrhea, headache. Can be fatal.	Cook food thoroughly. Use sanitary food-handling methods. Use pasteurized milk. Refrigerate food promptly and properly.
Shigella bacterium **Shigellosis**	Moist foods and liquids undercooked and handled by an infected person	Onset: 18–36 hours Symptoms: Abdominal cramps, diarrhea, vomiting, fever, chills, malaise, nausea, headache	Cook food thoroughly. Use sanitary food-handling methods.
Escherichia coli bacterium **Traveler's diarrhea**	Contaminated water, undercooked ground beef, raw foods, unpasteurized soft cheeses	Onset: 12–18 hours Symptoms: Loose and watery stools, nausea, bloating, cramps. *E. coli* 0157:H7 causes bloody diarrhea and can be fatal.	Cook food thoroughly. Use safe water and pasteurized milk. Wash fruits and vegetables.
Vibrio parahaemolyticus, *Vibrio vulnificus,* *Vibrio cholerae,* and *Vibrio fluvialis* bacterium **Vibrio infection**	Undercooked seafood, especially oysters	Onset: 18–36 hours Symptoms: Abdominal cramps, diarrhea, vomiting, fever, chills, malaise, nausea, headache, bloody or mucoid diarrhea, skin lesions	Keep food preparation areas clean. Cook and store foods at the appropriate temperature.
Staphylococcus aureus bacterium toxin **Staphylococcal poisoning**	Meats, poultry, egg products, tuna, potato and pasta salads, cream-filled pastries	Onset: 0.5–8 hours Symptoms: Diarrhea, nausea, vomiting, cramps, fatigue. Mimics the flu.	Cook food thoroughly. Use sanitary food-handling methods. Refrigerate food promptly and properly.
Clostridium botulinum bacterium toxin **Botulism**	Anaerobic environment of low acidity (canned corn, peppers, green beans, soups, beets, mushrooms, tuna, chicken, ham, sausages, etc.)	Onset: 4–6 hours Symptoms: Nervous system symptoms like double vision, inability to swallow, speech difficulty. Progresses to respiratory paralysis. Is often fatal. Can leave prolonged symptoms in survivors.	Use proper canning methods. Avoid commercially prepared canned products that are bulging, bent, rusted, or broken.
Hepatitis A virus **Hepatitis**	Contaminated shellfish, foods handled by unsanitary food handler who is a carrier	Onset: 15–50 days Symptoms: Liver inflammation, tiredness, nausea, jaundice, muscle pain	Cook food thoroughly.

Some Common Culprits Causing Foodborne Illnesses (*continued*)

Organism and Disease Caused	Food Attributes and Sources	Onset and Symptoms	Prevention Methods
Norovirus and Norwalk-like virus **Gastroenteritis**	Raw shellfish from polluted water, ready-to-eat foods, salad ingredients handled by infected persons	Onset: 24–48 hours Symptoms: Nausea, vomiting, diarrhea, abdominal pain lasting 24–60 hours. Headache and low-grade fever may also occur.	Cook food thoroughly. Use sanitary food-handling methods.
Rotavirus **Gastroenteritis**	Raw foods handled by infected persons, oral-fecal route, person-to-person contact, contaminated water. Can be carried by asymptomatic people.	Onset: 1–3 days Symptoms: Vomiting and diarrhea lasting 4–8 days, temporary lactose intolerance in some cases, low-grade fever	Use sanitary food-handling methods.
Giardia lamblia parasitic protozoan **Giardiasis**	Contaminated water, uncooked foods	Onset: 1–6 weeks Symptoms: Diarrhea (or constipation), nausea, vomiting, abdominal pain and distention, gas	Use sanitary food-handling methods. Avoid raw fruits and vegetables where protozoa are endemic. Dispose of sewage properly.
Trichinella spiralis parasite **Trichinosis**	Raw or undercooked pork or wild game	Onset: 24 hours Symptoms: Abdominal pain, nausea, vomiting, diarrhea, fever. 1–2 weeks later, muscle pain, low-grade fever, edema, skin eruptions, pain when breathing, loss of appetite. Drug therapy can kill the worms.	Cook food thoroughly.
Toxoplasma gondii parasite **Toxoplasmosis**	Raw or undercooked meats, mishandled infected cat feces, unwashed fruits and vegetables in certain conditions	Onset: 10–13 days Symptoms: Fever, headache, rash, sore muscles. Life-threatening to unborn fetuses.	Cook food thoroughly. Use sanitary food-handling methods.

© Cengage Learning 2013

BioBeat 6.5

Beneficial Flora in the Gastrointestinal Tract Support Health

Most bacterial infections that affect the gastrointestinal (GI) tract are self-limiting, meaning that the host's immune system can combat the infection successfully in a few days. The combination of humoral immunity (IgA antibodies secreted from the GI tract cells) and a healthy complement of friendly bacteria that control the growth of pathogenic organisms limits the risk of foodborne illness.

It has been documented for years in the biological literature that a variety of *Lactobacillus* bacteria prevent or reduce the infection caused by pathogenic microorganisms. Antibiotic use, chemotherapy, and abdominal radiation are notorious for altering GI flora. To reestablish a **healthy flora** in the GI tract, the intake of live cultures of a variety of species and strains of *Lactobacillus* and *Bifidobacterium*

bacteria is helpful. These can be consumed from some types of live-cultured yogurts or from supplements. Read the food labels for the microorganisms present. Additionally, the inclusion of fructooligosaccharides, or a specific type of functional dietary fiber, supports the growth of the friendly bacteria in the colon.[23,28,39,61]

Do you have healthy GI flora?

flora The collective bacteria and other microorganisms in an ecosystem, in a host, or the gastrointestinal tract of the host.

healthy flora The collective bacteria (such as *lactobacillus* and *bifidobacterium*) that colonize the gastrointestinal tract and prevent or reduce infection caused by pathogenic microorganisms.

Summary Points

- Bacteria can multiply quickly and exponentially.
- Many types of microorganisms can cause foodborne illness.
- To identify the cause of foodborne illness, the first information to consider is the time of onset of the illness, the signs and symptoms of the illness, and the type of food consumed.
- Some microorganisms are beneficial for health.

Take Ten on Your Knowledge Know-How

1. What does rapid clonal expansion mean? Does cell division happen exponentially?
2. Give one example of a parasite that causes foodborne illness. Describe the time of onset and the signs and symptoms of the illness.
3. Give one example of a protozoan that causes foodborne illness. Describe the time of onset and the signs and symptoms of the illness.
4. Give one example of a bacterium that causes foodborne illness. Describe the time of onset and the signs and symptoms of the illness.
5. Give one example of a fungus that causes foodborne illness. Describe the time of onset and the signs and symptoms of the illness.
6. Give one example of a virus that causes foodborne illness. Describe the time of onset and the signs and symptoms of the illness.
7. What is a probiotic?
8. Give two examples of probiotics. What conditions may improve by supplementing with them?
9. What type of functional fiber supports the growth of probiotics?
10. Describe a couple of bacterial toxins. What actions do each perform?

6.5 FOOD ISSUES: CONSUMER AWARENESS

T-Talk 6.5

To hear Dr. Turley talk about consumer awareness, go to www.cengagebrain.com

Introduction

Fortunately, we know a great deal about procedures to control microbial growth, oxidative changes, and enzymatic destruction so that food is kept safe and desirable. These same measures also extend shelf life of food products. Killing or minimizing microorganisms can be achieved by **temperature adjustment**, dehydrating, chemical processing, **modified atmosphere packaging**, and **irradiation**.

Consumers need to be aware of other potential food safety issues, such as chemical contaminants, **natural toxicants**, antibiotics, hormones, pesticides, cloning, and genetic engineering. All of these factors are part of modern food production systems and affect food sustainability. It is important to be knowledgeable about energy systems, levels of organization, and nutrient cycles as they pertain to food production. **Organic food production** will be explored in detail because it has been shown to be a more environmentally responsible way to produce food and is **sustainable** for future generations.

temperature adjustment The process of chilling, freezing, and heating foods to minimize microbial growth or kill microbes.

modified atmosphere packaging A food preservation technique that replaces oxygen in packaging with carbon dioxide or nitrogen to extend shelf life.

irradiation The exposure of food to high-energy particle bombardment, which destroys microbial and insect contamination.

natural toxicants Chemical compounds naturally occurring in foods that can cause illness or death.

Food Handling

There are some effective measures for the prevention of foodborne illness. The food industry employs many methods to make food safe to eat and follows the Hazard Analysis Critical Control Points (HACCP) system (see BioBeat 6.6). Once food is in our homes, we need to exercise safe food-handling practices, too. This means cooking and storing foods at proper temperatures, handling raw and cooked foods properly, and recognizing troublesome foods and eating situations (such as picnics or buffets) that require extra efforts to keep the food safe (see Table 6.8).

BioBeat 6.6

Preventing Foodborne Illness Through HACCP

HACCP is an acronym for a system called Hazard Analysis Critical Control Points. This is a straightforward, logical system of food safety management and training that is focused on the prevention of foodborne illness. HACCP was originally created as a part of space age technology, when planners were concerned with providing safe food for astronauts. HACCP principles have since been applied to the larger food industry. The system looks at the food processing system from start to finish and identifies potential hazards, then puts in place controls to monitor practices at each stage. Furthermore, records are kept to document the controls and monitors, from personnel training to proper functioning of food processing, storage, and cooking equipment. Examples of hazards that are controlled include microbes, chemicals, and physical substances. Examples of critical control points that are identified are cooking, chilling, handling, cleaning, and storing. Critical limits for each control point include minimum cooking temperature and time. Examples of procedures include determining how and by whom cooking time and temperature are monitored and establishing corrective actions to be taken when problems occur, such as reprocessing or disposal of questionably safe foods.[47,48,49]

What are the benefits of HACCP?

TABLE 6.8

Basic Safe Food-Handling Instructions

Store Foods Properly	Cook Foods Properly	Handle Raw Foods Properly	Handle Cooked Foods Properly
Refrigerate or freeze. Thaw in the refrigerator or microwave. Keep cold foods cold: refrigerator temperature no warmer than 40°F and freezer temperature of 0°F.	Cook thoroughly. Don't consume raw or under-cooked meat or seafood. Cook to a high enough core temperature to kill pathogenic microorganisms.	Don't cross-contaminate raw fruits and vegetables with raw meats. Wash hands and cutting surfaces. Minimize food infection.	Keep hot foods hot. Refrigerate leftovers immediately or discard. Minimize the time foods are held in the danger zone of 40°F–140°F.

© Cengage Learning 2013

Factors Affecting Microbial Growth

Consumers need to understand that foods are infected with pathogenic microorganisms and viruses. Some microorganisms such as bacteria multiply rapidly and exponentially given optimal growth conditions. The ideal environment is warm, moist, protein-rich, and neutral in pH. The most common types of bacteria that thrive in this type of environment and cause foodborne illness are *salmonella*, *clostridium*, *listeria*, and *staphylococcus*. *Salmonella* and *clostridium* are naturally present in food, and foodborne illness results from mishandling the food. An infected person can introduce microbes such as *staphylococcus* to food. If the food is mishandled, the *staphylococcus* microorganisms can clonally expand and cause severe illness in those who consume the food. Temperature, water content, protein levels, and acidity all affect the rate of microbial growth potential.[F,11,16,21,44]

Temperature Adjustment Chilling, freezing, and high temperature heating (cooking, pasteurizing, canning, sterilizing) slow microbial growth or kill microbes. The temperature range in which microbes divide the fastest is between 40 degrees and 140 degrees Fahrenheit. This temperature range is called the **danger zone** (see Figure 6.5). Foods that are moist and protein rich should not be stored for long periods when held in the danger zone. The optimal growth temperature is at the midpoint of the range.

organic food production A food production system that follows specific standards that do not permit the use of pesticides, synthetic fertilizers, antibiotics, hormones, and/or genetically modified organisms and also strives to enhance the environment and conserve earth's resources.

sustainable Able to meet current needs without compromising the ability to meet future needs.

danger zone A range of food storage temperature from 40 to 140 degrees Fahrenheit where microorganisms can replicate quickly enough to cause foodborne illness.

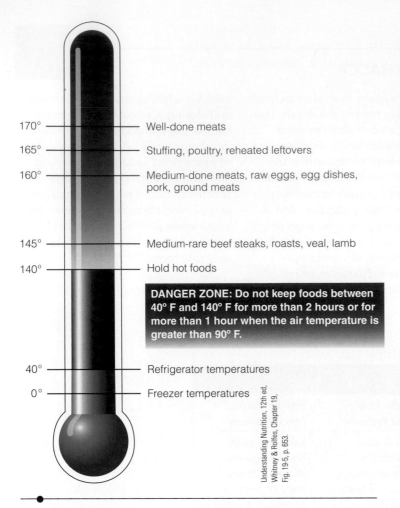

170° ———— Well-done meats

165° ———— Stuffing, poultry, reheated leftovers

160° ———— Medium-done meats, raw eggs, egg dishes,
pork, ground meats

145° ———— Medium-rare beef steaks, roasts, veal, lamb

140° ———— Hold hot foods

DANGER ZONE: Do not keep foods between 40° F and 140° F for more than 2 hours or for more than 1 hour when the air temperature is greater than 90° F.

40° ———— Refrigerator temperatures

0° ———— Freezer temperatures

Understanding Nutrition, 12th ed.,
Whitney & Rolfes, Chapter 19,
Fig. 19-5, p. 653.

FIGURE 6.5 A thermometer showing freezer, refrigerator, and cooking temperatures, as well as the danger zone.

Cooking temperatures to kill microorganisms are specific for the type of food being cooked. Minimally, seafood, beef, lamb, and pork should be cooked to 145 degrees Fahrenheit, medium-done meats, raw eggs, egg-containing dishes, pork, and ground meats to 160 degrees Fahrenheit, poultry, stuffing, and leftovers to 165 degrees Fahrenheit, and well-done meats to 170 degrees Fahrenheit (see Figure 6.5). For storing and serving foods, the temperature should be held below 40 degrees Fahrenheit or, if held hot in a steam tray, the food should be held above 140 degrees Fahrenheit. Food handlers should refrigerate foods as soon as possible after the meal service is complete; they should not allow foods to cool on the counter for several hours. The longer the food is held at the optimal growth temperature, the faster the bacteria grow in numbers, and the more potential there is for food poisoning.

Conversely, cold foods need to stay cold. Think of salad bars with the large, deep bowls sitting in ice. The ice slows bacterial growth, but often only the bottom part of the bowl is chilled enough for safety. When you scoop from the top of the bowl, you have taken food from a high-risk infection area. Furthermore, in buffet-style eating, more and more people have come in contact with the food, so the risk for cross-contamination increases.[47]

Water Water plays a big role in microbial growth. If the water is bound tightly within the food (such as in jam), or if there is little moisture available, bacteria cannot grow well. Removing water from food by dehydration greatly reduces microbial growth. This technique, along with adding salt or sugar to change the water properties of foods, has been used since antiquity to preserve foods.

Protein Protein-rich foods and food mixtures containing vegetables, eggs, milk, or meats are high-risk foods. These foods are common carriers of pathogenic bacteria. Bacteria thrive in a moist, high-protein medium, with neutral pH and a warm environment. Microbiologists grow bacteria on an agar plate, which is essentially a protein-rich, environmentally favorable bacterial diet!

pH The acidity level of the food also makes a big difference in microbial growth. Acidic and alkaline foods are less likely to support bacterial growth than are foods with a more neutral pH. More commonly, foods are preserved using acids because they are more pleasing to human taste. Alkaline foods are not palatable to humans.

Manufacturing Methods for Extending Shelf Life

There are many manufacturing methods in place to extend the shelf life of commercially prepared and/or processed foods. Dehydration was discussed previously. Additional methods to be discussed include chemical processing, irradiation, and modified atmosphere packaging, which can be used on fresh or dried foods.

Chemical Processing Sodium benzoate in margarine, calcium propionate in bread, sorbic acid in cheese wrappers, salt in cured meats, and sugar in syrups, jams, and jellies are common additives that minimize microbial growth or prevent degradation of the food product.

Irradiation The technology of irradiation involves exposing food to ionizing radiation by gamma rays or electron beams in a shielded facility. Irradiation controls insects and microbes in food. With gamma rays, food is placed near (within several feet) a radiation source such as cobalt[60] or cesium[137]. The food does not become radioactive; rather, the gamma rays penetrate fresh and frozen foods. For electron beams, an electrically powered electron accelerator projects a beam of electrons toward thinly packaged foods and liquids or free-flowing grains. Currently, the FDA has approved the irradiation of foods at a specific radiation level, as shown in Table 6.9.[54,71,79]

United States Department of Agriculture

The Radura symbol of irradiation sanctioned by the USDA, indicating a food product has been irradiated.

TABLE 6.9

Selected Food Irradiation Specifications

Food	Amount of Radiation (kilograys*)	Food Safety Effect
Wheat powder	0.2–0.5	Kills insects
White potatoes	0.05–0.15	Extends shelf life
Spices	30	Kills insects
Fresh fruits	1.0	Delays maturation
Pork	0.3–1.0	Controls microbial growth
Poultry	3.0	Controls microbial growth
Raw beef	4.5	Controls microbial growth
Frozen beef	7.0	Controls microbial growth

*One kilogray will increase the temperature of the product by 0.43 degrees Fahrenheit.
© Cengage Learning 2013

Modified Atmosphere Packaging Reducing the amount of oxygen in the environment prevents microbial growth and enzymatic destruction. The enzyme that causes plant deterioration is an oxidative enzyme. Without oxygen, the polyphenol oxidase enzyme cannot work. In modified atmosphere packaging, oxygen is usually replaced with carbon dioxide or nitrogen. This technique is used to bag fresh produce to extend the shelf life significantly.

Other Food Safety Concerns

The safety of food can be compromised by more than the presence of microorganisms. Natural toxicants, poisonous chemicals, pesticides, antibiotics, and hormones can have detrimental health effects. Evidence is also emerging that **genetically modified organisms (GMOs)** may also pose food safety concerns. Some of these factors are naturally occurring, whereas others are the result of human effort.

Natural Toxicants Certain molds produce toxicants called aflatoxins, which are deadly mycotoxins. Aflatoxin contamination is common in grains and nuts. Aflatoxin is **mutagenic** (causes cells to change), especially to liver cells, and thus is a potent liver carcinogen. Other natural toxicants include oxalic acid in rhubarb leaves, solanine, found mostly in the green part of potato skins, goitrogens in

genetically modified organisms (GMOs) Portions of DNA from one living organism are introduced into another living organism (transgenic), or the expression of a gene is blocked (antisense); also called genetic engineering or biotechnology.

mutagenic A physical or chemical agent that causes a mutation or a physical change to the genetic material.

Foods like green potatoes, cruciferous vegetables, raw lima beans, and apricot pits contain natural toxicants.

© Science Photo Library/Alamy · © LHB Photo/Alamy · © Stepan Popov/iStockPhoto.com · Valentyn Volkov/Shutterstock.com

cabbage-family vegetables or cruciferous vegetables, cyanogens in raw lima beans and apricot pits, red tide toxin in blooming sea algae, and other toxins that are present in certain herbs such as belladonna, hemlock, and sassafras.

Chemical Poisoning Examples of common **chemical poisons** include **lead** (when food is exposed to lead-containing air, dust, paint, and ceramic glazes), zinc (when acidic foods like lemonade are stored in galvanized cans), pesticides (as residues on plant foods like fruits and vegetables and in fat portions of meat and milk), **mercury** (found in fish and aquatics), and other substances like **cadmium** and **polychlorinated biphenyls** (substances in industrial waste that get into water and soil) (see Table 6.10).[15,40,53]

TABLE 6.10

Some Examples of Chemical Contaminants

Name	Sources	Toxic Effects	Route to Food
Cadmium (heavy metal)	Plastics, batteries, alloys, pigments, smelters, burning fuels, electroplates, cigarette and volcanic smoke, ash	Slow and irreversible damage to kidneys and liver	Enters air from smokestack emissions, settles on the ground, is absorbed into plants, is consumed by farm animals and people
Lead (heavy metal)	Lead crystal glassware, painted china, old house paint, batteries, pesticides, leaded gasoline	Displaces calcium, iron, zinc, and other minerals from their action in the bones, kidneys, and liver, resulting in organ damage	Originates from industrial waste; pollutes air, water, soil
Mercury (heavy metal)	Earth gases, industrial waste, electrical equipment, paints, agri- and aquaculture products, and some vaccines	Poisons the nervous system, especially in fetuses	Industries release into water; acid rain is converted to methylmercury by bacteria and ingested by fish
Polychlorinated biphenyls (organic compounds)	Produced for use in electrical equipment (transformers, capacitors)	Long-lasting skin eruptions, eye irritation, growth retardation, anorexia, fatigue	Comes from discarded electrical equipment and accidental industrial leakage

© Cengage Learning 2013

Chemicals are often released into the atmosphere through industrialization. The chemicals then return to earth, contaminate waterways, and work their way into the human food supply by accumulating in animal tissues. One such example is the neurotoxin methylmercury, which can enter the human body by eating contaminated fish. Mercury that has been emitted from power generation and industrial waste moves through the atmosphere, land, and water. In low-oxygen environments, such as the bottom of lakes and oceans, mercury biomethylation transformation occurs by certain bacteria. This methylmercury is then easily absorbed into the living tissue of aquatic organisms and their predators (see Figure 6.6). Furthermore, it is not easy to eliminate or excrete methylmercury (or any of the heavy metals, such as cadmium and lead) from animal bodies, thus it can **bioaccumulate**. As smaller life forms are consumed by larger carnivorous life forms, heavy metal amounts continue to increase as life goes on. To reduce exposure to methylmercury, it is recommended to eat fish that live closer to the surface of the water and that have a shorter

chemical poisons Toxic substances like lead, mercury, and pesticides that have the potential to kill.

lead A heavy metal that displaces calcium, iron, zinc, and other minerals from their action in the bones, kidneys, and liver, resulting in organ damage.

mercury A heavy metal that poisons the nervous system, especially in fetuses.

cadmium A heavy metal causing slow and irreversible damage to the kidneys and liver.

polychlorinated biphenyls Organic compounds causing long-lasting skin eruptions, eye irritation, growth retardation, anorexia, and fatigue.

bioaccumulate The buildup (increased concentration) of substances in living tissues that accumulate up food chains and within food webs.

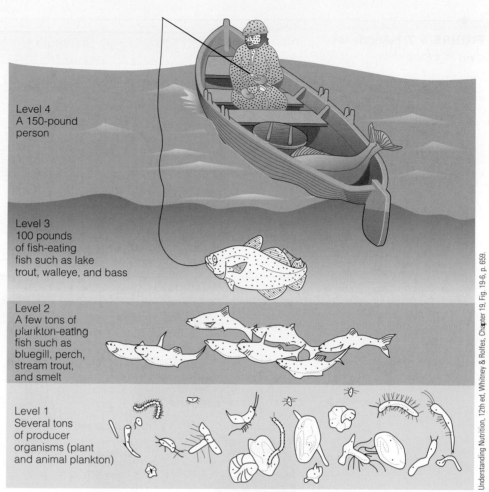

Key:
□ Toxic chemicals

4 If none of the chemicals are lost along the way, people ultimately receive all of the toxic chemicals that were present in the original plants and plankton.

Level 4
A 150-pound person

3 Contaminants become further concentrated in larger fish that eat the small fish from the lower part of the food chain.

Level 3
100 pounds of fish-eating fish such as lake trout, walleye, and bass

2 Contaminants become more concentrated in small fish that eat the plants and plankton.

Level 2
A few tons of plankton-eating fish such as bluegill, perch, stream trout, and smelt

1 Plants and plankton at the bottom of the food chain become contaminated with toxic chemicals, such as methylmercury (shown as red dots).

Level 1
Several tons of producer organisms (plant and animal plankton)

Understanding Nutrition, 12th ed, Whitney & Rolfes, Chapter 19, Fig. 19-6, p. 659.

FIGURE 6.6 Toxic chemical exposure through a marine food web.

life span. Fish with the highest methylmercury levels include king mackerel, shark, swordfish, and tilefish. Chilean bass, grouper, marlin, orange roughy, and tuna can also be relatively high in methylmercury.

Pesticides Pesticides are commonly used in **conventional food production** systems. They are of concern because they threaten **biodiversity** and the lives of humans and farmed, wild, and aquatic species. They persist in nature and can bioaccumulate in **food webs** and **food chains**, and are stored in fatty portions of meat and milk (see Figure 6.7). There are residues of pesticides on conventionally grown fruits and vegetables that also can accumulate in processed foods.

Pesticides are sources of **endocrine disruptors** and cancer promoters, and they have many other negative health effects. When pesticide residues enter the human body, they add to the pool of free estrogens in the body. When these compounds are oxidized, they become cancer-causing substances. The EPA enforces the Miller Pesticide Act; however, many scientists are producing evidence for a variety of safety concerns, including negative genetic effects associated with certain pesticides. Children are especially vulnerable to the harmful effects of pesticides. Ways to reduce pesticide intake include trimming fat off of meats, varying the types of meats eaten, washing fresh produce (scrubbing and rinsing), discarding outer leaves of leafy vegetables, and peeling waxed fruits and vegetables. Choosing to buy and consume organic foods can virtually eliminate foodborne pesticide exposure.[72]

Antibiotics and Hormones In conventional food production systems, antibiotics and hormones are commonly used when raising animals. Their purpose is to reduce animal sickness while accelerating their growth to meet food production

conventional food production A food production system used in many developed nations that employs the use of pesticides, synthetic fertilizers, antibiotics, hormones, and/or genetically modified organisms—all used to increase yields.

biodiversity The genetic variation among all life forms on earth.

food webs A network of interacting food chains within an ecosystem.

food chains The linking of plants and animals by their food relationships and transfer of food energy.

endocrine disruptors Certain exogenous, organic chemical compounds that, when taken into the body, alter or disrupt the hormonal balance or homeostasis.

FIGURE 6.7 Pesticide residues in a food chain.

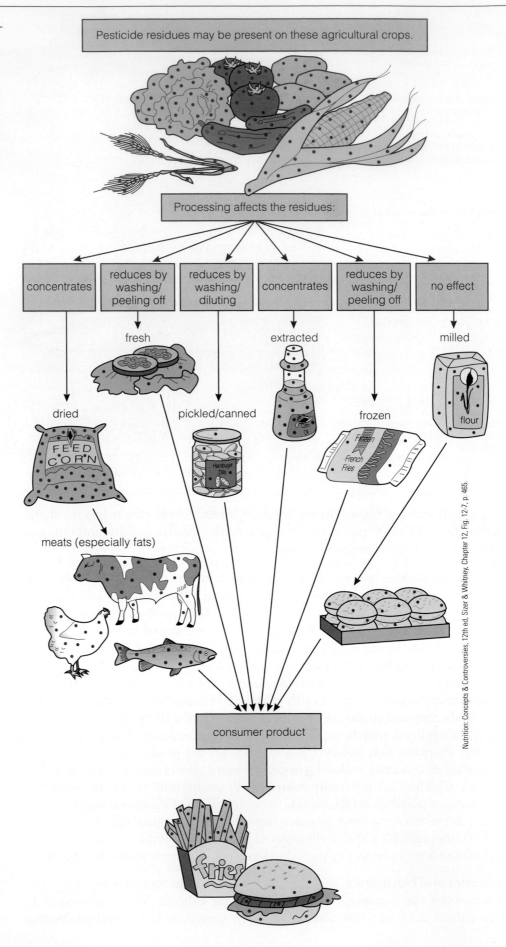

Pesticide residues may be present on these agricultural crops.

Processing affects the residues:

| concentrates | reduces by washing/peeling off | reduces by washing/diluting | concentrates | reduces by washing/peeling off | no effect |

fresh

extracted

milled

dried

pickled/canned

frozen

flour

FEED CORN

Hamburger Dills

Corn Oil

Frozen French Fries

meats (especially fats)

consumer product

fries

Nutrition: Concepts & Controversies, 12th ed, Sizer & Whitney, Chapter 12, Fig. 12-7, p. 465.

demands. Antibiotic and hormone use is widespread in the United States, and the health consequences are controversial for some people. The issue with the use of antibiotics is the serious risk of antibiotic resistance and multiple drug resistance of organisms that infect animals and humans. The issue with hormones is their potential for causing endocrine disruption, potentially leading to fertility problems and cancer. Steroid hormones are included in animal foods as feed efficiency enhancers, and increased residues are present in the food products. You can reduce your exposure to these substances by purchasing organic animal products.[52,57,72]

Genetically Modified Organisms (GMOs) When a food or organism is genetically modified (GM), a special set of technologies has been used to alter the genetic makeup of the species. Even though the terms *biotechnology* and *genetic modification* are used interchangeably, biotechnology is a more general term that refers to using organisms or their components, such as enzymes, to make products that include wine, cheese, beer, and yogurt. GMOs are also known as genetically engineered foods and are altered at the gene level by having a new gene inserted (transgenic) or the expression of an innate gene (antisense) or mRNA (RNA interference) blocked. GM products can include anything from foods to food ingredients, animal feeds, or fibers.[5,41,72]

GM is done to achieve desired characteristics, such as improving taste, increasing resistance to pests, pesticides, and diseases, shortening the growing period, delaying ripening, or instilling plant sterility (terminator gene) and/or promote antibiotic resistance. Conventional food production farmers are common users of GMO plant species. An example of genetic engineering of a tomato is shown in Figure 6.8.

Several concerns surface with GMOs, such as unexpected changes from gene activation or repression, increased allergic responses, the development of food sensitivities from increased plant toxicant levels, altering the food nutritional profile, reduction of biodiversity, legal issues, and more. Multiple negative health effects have been observed in numerous animal research models, affecting many different cells, organs, and organ systems. Some forms of GMOs, such as round-up-ready soybeans, actually enable farmers to use more pesticides on the GM crop, as well as herbicides around the plants. When GMO plant pollens are released into the wild outdoors, **cross-hybridization** can occur, which may threaten native species. Thus, all of these attributes could be detrimental for consumers and for the environment.[20,42,45,67,72]

cross-hybridization Genetically mixing different species to produce hybrids; cross-breeding.

FIGURE 6.8 Genetic engineering is a fast and deliberate way to change genetics, as shown here in this delayed ripening tomato.

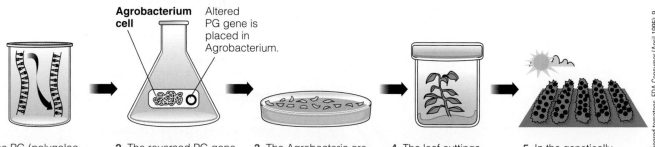

1. The PG (polygalacturonase) gene, which causes ripe tomatoes to soften and rot, is isolated and cloned. The sequence of PG gene is reversed so that the gene is backward (in what scientists call the "antisense" orientation).

2. The reversed PG gene is put into Agrobacterium, which is a bacterium that infects plants and is commonly used by genetic engineers to insert modified or foreign genes into other target cells.

3. The Agrobacteria are then placed in a petri dish with pieces of leaf from a tomato plant. The cut edges of the leaves absorb the Agrobacteria and antisense PG gene. Antisense gene thus becomes part of the genetic material of the tomato plant cells.

4. The leaf cuttings regenerate into tomato plants that contain the reversed PG gene. The new plants sprout roots, are transplanted to soil, and grow to mature tomato plants. The seeds collected from these greenhouse tomatoes are planted outdoors for field trials and more seed production.

5. In the genetically engineered tomato, the natural PG gene's production of the fruit-rotting PG enzyme is repressed by the reversed gene. This gives the commercial tomato extended shelf life, allowing it to ripen more fully on the vine and still have time to get to market before it spoils.

Adapted from Calgene's recipe for genetically engineered tomatoes, FDA Consumer (April 1995): 9.

There are numerous, important controversies regarding the safety, ethics, ownership, and labeling of GMOs. Some safety concerns boil down to the unknown, long-term effects of consuming GMOs, including the possibility of increased chronic disease and allergies. Also, the potential environmental impact of the unintended transfer of transgenes through cross-pollination, and the impact on other organisms, such as soil microbes, has environmentalists very concerned about the **ecology** of GMOs. This is compounded with the loss of flora and **fauna** biodiversity. These issues alter the natural evolution of organisms by tampering with nature when splicing genes among species. There are fierce objections by some about combining animal genes in plants and vice versa.

Furthermore, company ownership of GM products may result in domination of world food production by a few companies. This would increase dependence on industrialized nations by developing countries and could stimulate biopiracy and foreign exploitation of natural resources. Another issue is the labeling of GM products. The mixing of GM crops with non-GM products confounds labeling attempts and creates an information gap for consumers. Presently, the only way for a consumer to buy non-GMO foods is to purchase certified organic food or to read packages carefully to see if the manufacturer states that their product is non-GMO.[45,72,75]

Food Systems

Food systems are important to understand because they dictate energy and nutrient cycling and thus sustenance of life. Additionally, your impact on the environment can result from the consumer choices that you make.

The sun emits energy, some of which reaches the earth to support life within an energy pyramid. Energy flows through an **ecosystem** from producers to primary, secondary, and tertiary consumers (see Figure 6.9). An example of energy flow applied to a food web could begin with soil, made nutrient-rich by decomposers (bacteria and fungi that feed on small organic matter from excretion, death, and decomposition), which supports the growth of plant forms of food, such as grains. A small animal, such as a rodent, eats the plant foods, such as kernels of the grain; a larger

ecology The interrelationship of living organisms and the environment.

fauna All of the animal life in a region, time period, or environment.

ecosystem The interaction of living organisms with each other and their environmental surroundings.

FIGURE 6.9 The energy pyramid.

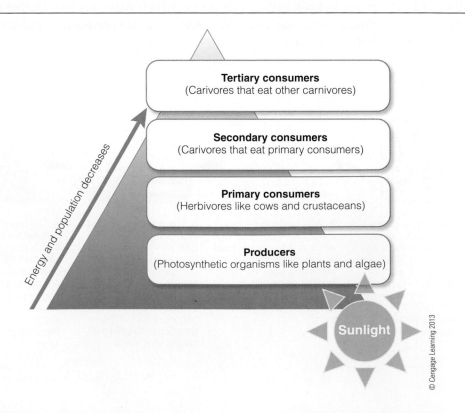

animal, such as a bird of prey, eats the rodent; then a larger animal, such as a fox, eats the bird of prey; and lastly, a hunting human eats the fox. When eating from a food production system, humans tend to be primary and secondary consumers. When leading a hunting-gathering lifestyle, human consumers could be viewed as primary, secondary, and, in some cases, tertiary consumers. It should be noted that usually humans don't eat carnivores, but in the wild, energy flow is better represented.

Food Production

The food system can be manipulated so that food is mass produced and readily available for the modern-day person living in an industrialized country. The consumer promotes massively produced ready-to-eat food, which drives the food production systems. The basics of a food production system are simple: Food of all types, both plant and animal, is produced on a farm. After it is harvested, food "in the raw" either goes directly to market or is processed, packaged, and then marketed. The food is then shipped to warehouses, and from there to local markets, where it is sold to the consumer. After the food is eaten, a demand for more food is created (see Figure 6.10). The demand for more processed ready-to-eat food has occurred with a growing population and with busier lifestyles. This system requires excessive natural resources.

Levels of Organization Within Food Production With mass food production and food processing, concerns about the imbalanced ecology of conventional U.S. food production have arisen. Ecology is the interrelationship of living organisms and the environment. Ecology and the levels of organization can easily be seen within a farming food production system (see Table 6.11). This concept is important to grasp so that the interconnected nature of the most basic life form, the cell, to all living things interacting with nonliving things in the natural world, is realized. The impact on one area can positively or negatively affect many others. This concept rings true with the harmony that is supported by organic food production and the disharmony from conventional farming practices.

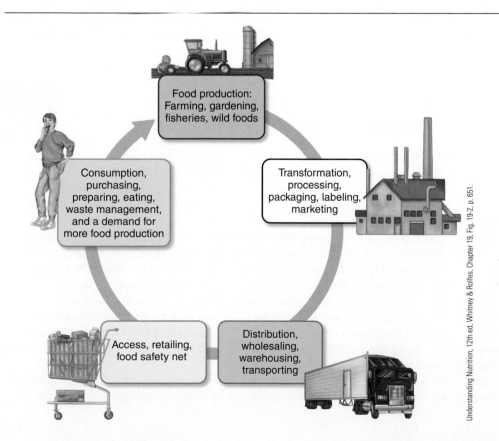

FIGURE 6.10 A food production system

Understanding Nutrition, 12th ed, Whitney & Rolfes, Chapter 19, Fig. 19-2, p. 651.

TABLE 6.11

Levels of Organization Within a Food Production System

Level of Organization	Broad Description	Food Production System Example
Biosphere	The earth's ecosystems	All farms on earth
Ecosystem	A community and its nonliving natural surroundings	The animals, plants, microbes, soil, rocks, water, air, and sunlight on the farm
Community	A population living in a defined area together	A farm with cattle, chickens, turkeys, sheep, and goats
Population	A group of the same organisms living in the same geographic area	The cattle on a farm
Organism	An individual living thing	One cow
Group of cells	Tissues, organs, and organ systems	The brain and central nervous system of a cow
Cell	The smallest functional unit of life	A single cow brain cell
Molecule	A group of atoms making a chemical compound	A prion protein made of carbon, oxygen, hydrogen, and nitrogen
Atom	The smallest component of an element	Hydrogen

© Cengage Learning 2013

Conventional Food Production Systems The techniques used in conventional food production and the amount of resources required to process, package, and transport the food can have a detrimental impact ecologically (see BioBeat 6.7). Conventional farming uses GMOs, antibiotics, hormones, pesticides, nitrogen-based fertilizers, sewage sludge, and heavy equipment to mass-produce food. The process depletes soil and pollutes water, including ground water, and contributes to global warming. Many conventionally produced foods are mass produced. Mass animal and food production facilities are often located in remote places because

BioBeat 6.7

The Emerging Issues of U.S. Food Production and Sustainability

There are many other concerns about food safety aside from foodborne illness. The concerns about food processing techniques range from food irradiation to the chemical additives in processed foods. The consumption of pesticide residues on plant foods and the residues that can concentrate in animal fat from bioaccumulation, in addition to antibiotic and hormone use, may have detrimental health consequences. Fish farming and more extreme food alterations resulting from genetic engineering of plant foods and the cloning of animals have raised serious food safety controversies. For the most part, these methods are employed to generate food products that will be more profitable to businesses and possibly more appealing to consumers. But consumers may not know what they are eating and may be unaware of the risks associated with current food production, whether their appetites are putting too heavy a burden on limited natural resources, and whether the current food system is sustainable.

The plain truth is that the U.S. food system is currently detrimental to the environment, human health, communities, and future food sustainability. The formula of eating too much animal meat and byproducts, less home cooking from whole, locally produced food sources, and the consumption of more processed, fast, and restaurant food equals an unsustainable food system and a global burden. Why? These practices deplete energy, water, soil, forests, and wild foods and compromise air and water quality, while generating solid waste and hazardous materials.[5,1,12,19,22,27,50,55,56,60,62,72,76,81]

How can you change your eating habits to be more sustainable?

of their unpleasant nature. Mass plant foods are produced as single crops, which results in a lack of genetic variety. Certain conventional foods are permitted to undergo irradiation and be processed with numerous chemical food additives to extend their shelf life.[72,76]

Organic Food Production Systems Organic food is produced by farmers who use renewable resources to protect the soil, water, and health of the environment now and for future generations. The organic food production system does not permit the use of pesticides, synthetic fertilizers, antibiotics, hormones, or GMOs. Consequently, organic farming generally improves soil fertility, maintains ecological harmony and biodiversity, reduces greenhouse gas emissions and thus global warming, enhances the environment, and conserves earth's resources. Food production per acre of organic production is also competitive with conventional food production.

The USDA-certified organic food label symbol.

At the beginning of the 21st century, the USDA set national standards for food labeled USDA Organic. Whether the food is grown in the United States or imported from other countries, the term *organic* has a specific meaning, but when used with other phrases can have a few different meanings (see Table 6.12). Before a product can be labeled organic, a government agent must inspect the farm where the food is produced to make sure the farmer is meeting all USDA organic standards. It can take a farm five years to become a USDA-certified organic farm. Companies that handle or process organic food before it gets to local supermarkets or restaurants must be certified as well. The USDA insignia of quality is labeled on the food if the food meets the standards for certified organic by the USDA.[T]

TABLE 6.12

The Many Meanings of the Term *Organic*

Term	Definition
100% organic	The food is made with all USDA-certified organic ingredients
Organic	More than 95% of the food ingredients are USDA-certified organic
Made with organic ingredient	More than 70% of the food ingredients are USDA-certified organic
No use of the term *organic*	Less than 70% of the food ingredients are USDA-certified organic

© Cengage Learning 2013

In essence, all aspects of conventional food production are prohibited in organic farming. Farmers who produce organic meats, poultry, eggs, or dairy products do not give medications, immunizations, feed-efficiency enhancers, or hormones to the animals during production. The animals are less confined than in many conventional, concentrated animal feedlots. Organically produced plant foods are grown without using standard pesticides, genetic engineering, sewage sludge, or synthetic fertilizers. Nitrogen in soil is maintained by composting and through the actions of decomposers and crop rotation (see BioBeat 6.8 and Figure 6.11). Organic farms are often local and multidimensional, growing many crop types and raising various animals, which enable them to produce food more locally and sustainably.[50,56,72,73,80]

Food System Sustainability The food production system has a large impact on the environment. Conventional systems contribute significantly to global greenhouse gas emissions and thus global warming. Using pesticides, fertilizers, and heavy equipment, along with excess tilling, irrigation, and overgrazing, depletes soil and pollutes water. Some solutions may begin with being knowledgeable and choosing to eat foods produced in a sustainable way.[22,56,73]

BioBeat 6.8

How Nitrogen Cycles Naturally

Nitrogen cycles through the air, soil, living matter, and dead matter continuously by complex chemical and biological processes (see Figure 6.11). Food webs on land contribute nitrogen to the soil through excretion, death, and decay of plants and animals. The conversion of ammonia (NH_3) to ammonium (NH_4) salts (**ammonification**) and the conversion of ammonium to nitrite and nitrate (**nitrification**) take place by soil microbial decomposers such as bacteria and fungi. The processed forms of nitrogen are taken up by organisms that produce their own food using light or heat energy (**autotrophs**), such as green plants, and supports soil quality. Bacteria can also further process nitrate to nitrogen gas, which is emitted into the atmosphere. Atmospheric nitrogen can reenter the soil naturally and be fixed by bacteria for plant use. Or, nitrogen can be fixed artificially by agricultural industrial processes used to make nitrogen-based fertilizer, which can be applied to soil in conventional farming.

A delicate nitrogen balance needs to be maintained ecologically. Nitrogen, as ammonia, ammonium, nitrate, and nitrite, can leach out of soil into water sources as a result of rain and irrigation. Thus, excess nitrogen in soil, such as from fertilizer application to crops in conventional systems, has the potential to be unhealthy for humans and harmful to the environment. Furthermore, nitrogen products are produced through a petroleum-based process and thus deplete fossil fuels.

Organic food production relies on the natural processes of ammonification and nitrification rather than the application of petroleum-based fertilizers. Thus, organic food production naturally supports the nitrogen cycle needed for life in terrestrial ecosystems.

How do your life and food choices affect the nitrogen cycle?

FIGURE 6.11 The nitrogen nutrient cycle, from soil decomposers to industrial agriculture and atmospheric gas.

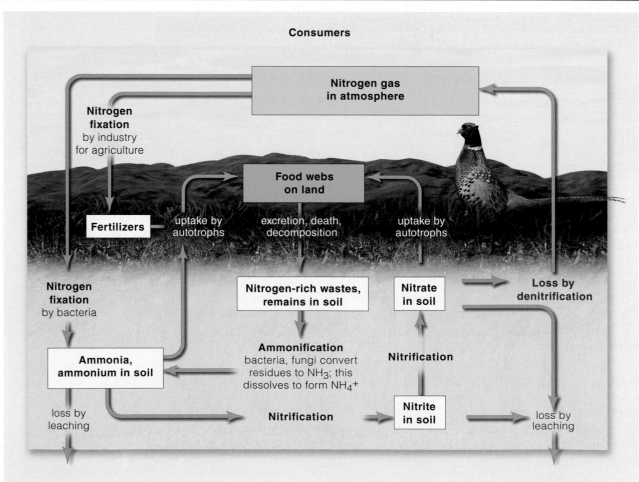

Fast foods and processed foods are not made using sustainable methods. Chemical fertilizers and pesticides are nonrenewable resources, thus conventional crops produced using them are not sustainable. Food and beverage processing, packaging, and transporting contribute greatly to the depletion of fossil fuels. The consumption of more conventionally produced food as well as processed, fast, and restaurant food in general contributes to an unsustainable food system and a global burden.

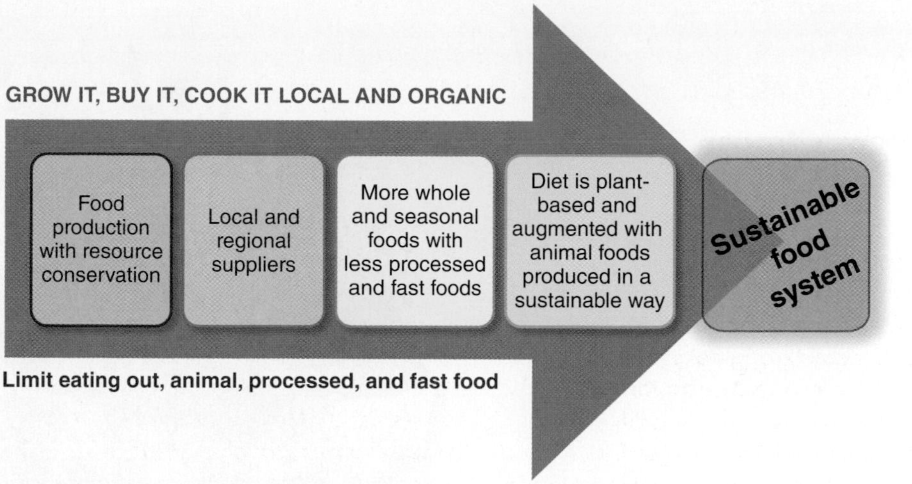

GROW IT, BUY IT, COOK IT LOCAL AND ORGANIC

| Food production with resource conservation | Local and regional suppliers | More whole and seasonal foods with less processed and fast foods | Diet is plant-based and augmented with animal foods produced in a sustainable way | Sustainable food system |

Limit eating out, animal, processed, and fast food

© Cengage Learning 2013

FIGURE 6.12 A sustainable food system.

Current aquatic techniques to harvest seafood and fish farming methods generally damage the environment and ecosystems. Gillnetting, long-lining, dredging, trawling, and dragging are irresponsible fishing methods. Furthermore, farming carnivorous fish such as salmon and tuna increases the burden on wild resources, because these fish are fed fish. To note, farming omnivorous tilapia fish is sustainable.

Eating Sustainably and Living Well Purchasing food from local and regional suppliers (such as from farmers' markets), choosing certified organic foods, preparing and eating more whole and seasonal foods, and eating a plant-based diet (augmented with animal foods produced in a sustainable way) are responsible consumer choices (see Figure 6.12). By choosing a local, organic, and mostly plant-based diet and lifestyle, you can support a sustainable food system that promotes ecological, human, and global health. The Holistic Health and Food Sustainability Model shown in Figure 6.13 was created to help food consumers achieve these goals. The model depicts basic food groups with applicable whole food choices and describes the need to eat organic, local, and mostly plant foods. Plant foods are emphasized, and more-sustainable animal choices are placed before less-sustainable choices. For example, beef is unsustainable, whereas chicken and eggs are more sustainable animal food sources. The model incorporates the need for water from free-flowing sources (unpackaged) and ties in important qualities of human life such as love, family, helping others, literacy, arts, music, rest, physical activity, and nourishing the body.[S,1,6,9,12,32,55,72,81]

Summary Points

- Food should be handled properly to avoid foodborne illness and the severe gastrointestinal distress, pH and electrolyte imbalances, and potentially life-threatening illnesses that result from it.
- Most microbes multiply rapidly in a warm environment that is rich in water and protein with a neutral pH.
- Chemical additives, dehydration, irradiation, and modified atmosphere packaging are employed by the food industry to control microbial growth, while HACCP is employed to reduce the risk of foodborne illness.
- Natural toxicants and heavy metals can contaminate food and water and cause serious negative health effects.
- Energy flows through an ecosystem from producers to primary, secondary, and tertiary consumers.

ammonification Conversion of ammonia to ammonium salts.

nitrification Conversion of ammonium to nitrite and nitrate.

autotrophs Organisms that produce their own food using light or heat energy.

FIGURE 6.13 Holistic health and food sustainability model.[73]

- The levels of organization in a food production system are atom, molecule, cell, group of cells (tissue, organ, organ system), organism, population, community, ecosystem, and biosphere.
- Conventional food production systems can have a detrimental ecological impact.
- Organic food production systems are more sustainable, health promoting, and ecologically friendly.
- A sustainable food system is one that is local, organic, and mostly plant based and requires minimal processing, packaging, and transportation.

Take Ten on Your Knowledge Know-How

1. Describe a high-risk food poisoning scenario. What steps could you take to convert this to a low-risk scenario?
2. How do temperature, moisture, oxygen, and pH impact food safety and microbial growth?
3. Besides microbes, what other factors can cause adverse food reactions leading to compromised health?
4. How does methylmercury bioaccumulate? What are the risks of consuming foods contaminated with methylmercury?
5. How can you reduce your exposure to pesticides?
6. What are the purposes, benefits, and risks of GMOs?
7. What are the purposes and risks associated with antibiotic and hormone use in animal food production?
8. What are the levels of organization and how would you relate them to a food production system?
9. How does nitrogen cycle through an organic food production system? How does a conventional food production system differ?
10. What is a sustainable food system?

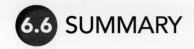

6.6 SUMMARY

CONTENT KNOWLEDGE

IN THIS MODULE, YOU WERE:

- Introduced to some markers of quality that will allow you to put nutrition information into context and gain an understanding of its reliability, so that you can make sound nutrition and supplement decisions.
- Informed about the legal and governmental history of food safety and information, so you can understand how the food industry operates. You now know that government efforts are made, whether with food processing techniques or chemical additives, to prevent foodborne illness and permit manufacturing liberties for the development and sales of consumable food products.
- Acquainted with many chemicals added to food to enhance flavor, texture, and appearance, and sometimes to reduce Calories (such as from fat or sugar) and to improve nutritional status. These chemicals are GRAS based on an entire population of people; however, some of the chemicals have known adverse reactions that some consumers may experience.
- Made aware of the many aspects of conventional food production that are not sustainable, and organic food production methods that are more environmentally friendly and sustainable; you, as a consumer, can support sustainable food systems.

PERSONAL IMPROVEMENT GOALS

IN THIS MODULE, YOU HAVE LEARNED:

- A marker of an educated person is that he or she can read and use information to improve well-being. You can use your markers of reliability to judge nutrition information and apply it for your own betterment.
- That every step along the way of storing and preparing a food for consumption requires attention to the food's safety. Ultimately, the consumer needs to be aware of food safety risks and take measures to reduce risk.
- That you can now more effectively read processed food package labels to make wiser choices to suit your individual diet and avoid adverse reactions to food additives.
- That doing your part in the food system includes consuming fewer shipped, processed, and away-from-home foods; purchasing food from local and regional suppliers such as farmers' markets, choosing certified organic foods, preparing and eating more whole and seasonal foods, and eating a plant-based diet augmented with animal foods produced in a sustainable way are responsible choices. By choosing this lifestyle, you can support a sustainable food system that promotes ecological, human, and community health now and in the future.

Here is a tip for you: Do most of your shopping along the periphery of the grocery store, in bulk, and/or at farmers' markets. Stock up on nonperishable items to save transportation costs and fossil fuel resources used on your trips to the market. Always strive to purchase and eat more whole, fresh, local, organic, and mostly plant foods.

You can assess if you met the learning objectives for this module by successfully completing the Homework Assessment and the Total Recall activities (sample questions, case study with questions, and crossword puzzle).

Homework Assessment

50 questions

1. The use of chemical fertilizers and pesticides to produce food is sustainable.
   A. True   **B.** False

2. Genetically modified organisms have no known health risks.
   A. True   **B.** False

3. When an author is affiliated with the public sector, the reliability of the information increases.
   A. True   **B.** False

4. A person with the credentials of R.D. would make a credible author for an article providing nutrition information.
   A. True   **B.** False

5. When the results of an American scientific study are favorable, the government will form a dietary recommendation for all Americans to follow.
   A. True   **B.** False

6. When an author of an article on dietary supplements has the credentials of Ph.D., it is important to know if his or her degree is in the area of nutrition.
   A. True   **B.** False

7. When an author cites published journal articles regarding laboratory, animal, and human studies as references in his or her own article, the overall reliability of the article decreases.
   A. True   **B.** False

8. When the author of an article on diet aids is affiliated with private industry, credibility of the information increases.
   A. True   **B.** False

9. When surfing the Internet, a Web site ending in .gov implies that the site is governmental and public domain.
   A. True   **B.** False

10. When reading an article on weight loss, you notice that the author lacks credentials but interviews and quotes individuals who have the credentials of M.D., Ph.D., and R.D. This would increase the overall reliability of the article.
    A. True   **B.** False

11. A celebrity providing a personal testimony about a product or diet provides evidence that the product is safe and effective.
    A. True   **B.** False

12. When the author of an article on diet aids uses before and after pictures of a single person as research evidence, credibility of the information increases.
    A. True   **B.** False

13. The term *nutritionist* has no legal definition; therefore, a nutritionist is not always a credible person in regard to nutrition information.
    A. True   **B.** False

14. The American Heart Association is a reliable source for nutrition information about heart disease.
    A. True   **B.** False

15. The reliability of a health information publication is reduced when the individuals on the editorial board lack the credentials of M.D., Ph.D., M.S., or R.D.
    A. True   **B.** False

16. When information is used to increase health awareness, the reliability is reduced compared to information provided to promote the sales of a product.
    A. True   **B.** False

17. The cattle on a farm would represent a population in terms of levels of organization in a food production system.
    A. True   **B.** False

18. In the nitrogen cycle, excretion, death, and decomposition contribute nitrogen to the soil, which supports plant growth.
    A. True   **B.** False

19. Hepatitis A is a virus that infects primates.
    A. True   **B.** False

20. Which of the following criteria can be useful in identifying an infection of *Giardia lamblia*?
    A. A stool specimen
    B. The appearance of flagellate trophozoites under the microscope
    C. An adhesive disk
    D. Binary fission
    E. All of the above

21. Which of the following microorganisms is useful to reestablish healthy flora in the gastrointestinal tract after antibiotic use?
    A. *Giardia*
    B. *Lactobacillus*
    C. *Clostridium*
    D. *Bifidobacterium*
    E. B and D

22. Which of the following processes is used to kill or minimize microorganisms?
    A. Dehydration
    B. Heat
    C. Sugar and salt
    D. Irradiation
    E. All of the above

23. The FDA is responsible for labeling a farm-fresh food with the insignia of organic.
    A. True   B. False

24. Olestra and sucralose are examples of synthetic food additives.
    A. True   B. False

25. Food additives that are on the GRAS list pose no adverse threats to people when consumed at levels allowed in the manufacturing process.
    A. True   B. False

26. Which of the following agencies oversees the safety of meat, poultry, and eggs?
    A. FDA
    B. USDA
    C. EPA
    D. NMF
    E. None of the above

27. Which of the following agencies oversees the safety of food additives in the United States?
    A. FDA
    B. USDA
    C. EPA
    D. NMF
    E. None of the above

28. Which of the following agencies monitors foodborne illness in the community?
    A. FDA
    B. USDA
    C. EPA
    D. NMF
    E. None of the above

29. Which of the following agencies regulates pesticide use in the United States?
    A. FDA
    B. USDA
    C. EPA
    D. NMF
    E. None of the above

30. Which of the following agencies oversees the safety of seafood in the United States?
    A. FDA
    B. USDA
    C. EPA
    D. NMF
    E. None of the above

31. Which of the following industries requires substantial research and is tightly regulated by the FDA?
    A. Food other than meat, poultry, eggs, and seafood
    B. Prescription drugs
    C. Dietary supplements
    D. Herbal supplements
    E. None of the above

32. Which of the following laws regulates the safety of dietary supplements?
    A. DSHEA
    B. Miller Pesticide Act
    C. GRAS list
    D. All of the above
    E. None of the above

33. Which of the following laws governs the safety of food additives?
    A. DSHEA
    B. Miller Pesticide Act
    C. The Delaney Clause (GRAS list)
    D. All of the above
    E. None of the above

34. The most common microorganisms that cause foodborne illness divide:
    A. Occasionally
    B. Periodically
    C. Rapidly
    D. Slowly
    E. None of the above

35. Which of the following classes of microorganisms is the least likely to cause foodborne illness?
    A. Bacteria
    B. Protozoa
    C. Fungi
    D. Freshwater algae
    E. None of the above

36. A common symptom of foodborne illness is:
    A. Fluid retention
    B. Nausea
    C. Heart palpitations
    D. Hearing loss
    E. None of the above

37. A bacterium that causes foodborne illness is:
    A. Hepatitis A
    B. *Clostridium perfringens*
    C. *Giardia lamblia*
    D. *Trichinella spiralis*
    E. None of the above

38. A virus that causes foodborne illness is:
    A. Hepatitis A
    B. *Clostridium perfringens*
    C. *Giardia lamblia*
    D. *Trichinella spiralis*
    E. None of the above

39. A protozoan that causes foodborne illness is:
    A. Hepatitis A
    B. *Clostridium perfringins*
    C. *Giardia lamblia*
    D. *Trichinella spiralis*
    E. None of the above

40. A parasite that causes foodborne illness is:
    A. Hepatitis A
    B. *Clostridium perfringins*
    C. *Giardia lamblia*
    D. *Trichinella spiralis*
    E. None of the above

41. A microorganism that lives in an anaerobic environment is:
    A. *Clostridium botulinum*
    B. *Staphylococcus aureus*
    C. *Campylobacter jejuni*
    D. Listeriosis
    E. None of the above

42. The best way to prevent foodborne illness is to:
    A. Store foods at room temperature
    B. Rinse vegetables before cooking
    C. Wash hands before handling food
    D. Say prayers before eating a meal
    E. Salt the food before serving

43. A food additive that enhances flavor is:
    A. Vitamin E
    B. Monosodium glutamate
    C. Pectin
    D. Acetic acid
    E. Tartrazine

44. A food additive that prevents microbial growth is:
    A. Vitamin E
    B. Monosodium glutamate
    C. Pectin
    D. Acetic acid
    E. Tartrazine

45. A food additive that thickens food is:
    A. Vitamin E
    B. Monosodium glutamate
    C. Pectin
    D. Acetic acid
    E. Tartrazine

46. A food additive that prevents fats from going rancid is:
    A. Vitamin E
    B. Monosodium glutamate
    C. Pectin
    D. Acetic acid
    E. Tartrazine

47. Which of the following substances is a fake fat?
    A. Olestra
    B. Carrageenan
    C. Aspartame
    D. Sorbic acid
    E. BHT

48. Which of the following substances is a stabilizing agent?
    A. Olestra
    B. Carrageenan
    C. Aspartame
    D. Sorbic acid
    E. BHT

49. Which of the following heavy metals causes liver damage when present at a high level in food?
    A. Cadmium
    B. Lead
    C. Mercury
    D. A and B
    E. A, B, and C

50. Which of the following heavy metals damages the nervous system when present at a high level in food?
    A. Cadmium
    B. Lead
    C. Mercury
    D. A and B
    E. A, B, and C

Total Recall

SAMPLE QUESTIONS

True/False Questions

1. Private sector nutrition information is more valid than public sector nutrition information.

2. Dietary supplements are approved and regulated by the Food and Drug Administration.

3. Viruses are not associated with foodborne illness.

4. Nitrates are added to food because they enhance the food's flavor.

5. An organic food production system is a detriment to the nitrogen cycle.

Multiple Choice Questions: Choose the best answer.

6. The DRIs are essentially public recommendations for nutrient intake levels. Given this, which answer most likely represents how DRI levels are established? Interpretation of:
   A. Animal studies by scientists
   B. Internet Web sites by consumers
   C. A whole body of credible scientific evidence by scientists
   D. Newspaper articles by journalists
   E. Public recommendations of other developed nations

7. Which of the following is a bacterium that evolved to a strain that causes potentially fatal hemorrhagic colitis?
   A. Hepatitis A
   B. *Giardia lamblia*
   C. *Bifidobacterium*
   D. Prions
   E. *Escherichia coli* 0157:H7

8. Which of the following is a disease caused by a pathogenic parasite?
   A. Trichinosis
   B. Shigellosis
   C. Traveler's diarrhea
   D. Botulism
   E. Listeriosis

9. Which of the following answers has the least food safety risk?
   A. Chemical poisoning
   B. Microorganisms
   C. Improper food handling
   D. Nutrient additives
   E. Natural toxicants

10. Which of the following is an artificial sweetener with few safety concerns?
    A. Sunnette
    B. Sucralose
    C. Starches
    D. Saccharin
    E. Simplesse

CASE STUDY

Megan and her mother, Martha, went to an all-you-can-eat buffet for lunch. Megan ate from the salad bar, while her mother Martha sipped all-natural herbal slimming tea she brought with her and read the newspaper and chatted about it with her daughter. In the newspaper, Martha was surprised to read a health story stating that a new published study showed that fiber supplements did not reduce colon cancer risk in elderly men with a history of polyps. Megan wasn't very interested in the article, so she took a trip to the salad bar to get a food source of dietary fiber. Megan selected the following conventionally produced food and beverage items to consume: mixed lettuce greens, diced tomatoes, sliced cucumbers, canned beets, shredded cheese, grilled chicken strips, croutons (from white bread), regular creamy ranch dressing, potato salad, side of fried potato skins, and a diet soda. About 40 minutes after eating, Megan and her mother Martha found themselves in the restaurant's public bathroom. Megan was experiencing flulike symptoms, while Martha was feeling the effects of her herbal slimming tea.

1. Which of Megan's foods listed below was most likely heat-treated to reduce food safety concerns?
   A. Croutons
   B. Tomatoes
   C. Lettuce greens
   D. Cheese
   E. Diet soda

2. Which of Megan's foods listed below was most likely modified atmosphere packaged?
   A. Croutons
   B. Canned beets
   C. Lettuce greens
   D. Cheese
   E. Diet soda

3. Which of Megan's foods listed below poses the greatest risk for ingestion of the natural toxicant solanine?
   A. Potato skins
   B. Canned beets
   C. Cucumbers
   D. Cheese
   E. Diet soda

4. Which of Megan's foods listed below poses the greatest risk for botulism if improperly processed?
   A. Potato salad
   B. Canned beets
   C. Cucumbers
   D. Cheese
   E. Diet soda

5. Which of Megan's foods listed below most likely caused her to experience flulike symptoms?
   A. Potato salad
   B. Canned beets
   C. Cucumbers
   D. Cheese
   E. Diet soda

6. Which microbe most likely caused Megan's food poisoning and flulike symptoms?
   A. *Staphylococcus aureus*
   B. *Trichinella spiralis*
   C. *Clostridium perfringins*
   D. *Campylobacter jejuni*
   E. *Salmonella*

7. Which statement below would be false regarding possible chemical additives in the foods that Megan ate?
   A. The ranch dressing could contain starches, gums, and/or gels.
   B. The diet soda could contain sucralose or aspartame.
   C. The cheese could have been packaged in a wrapper treated with sorbic acid.
   D. The bread used to make the croutons was probably vitamin and mineral enriched.
   E. The canned beets could have been treated with peroxide.

8. The findings published in the newspaper regarding fiber and cancer should lead scientists to come up with a lower DRI for dietary fiber.
   A. True   B. False

9. Megan chose a sustainable eating practice on this day.
   A. True   B. False

10. Martha's herbal slimming tea, because it is all natural, has no safety concerns.
    A. True   B. False

FUN-DUH-MENTAL PUZZLE

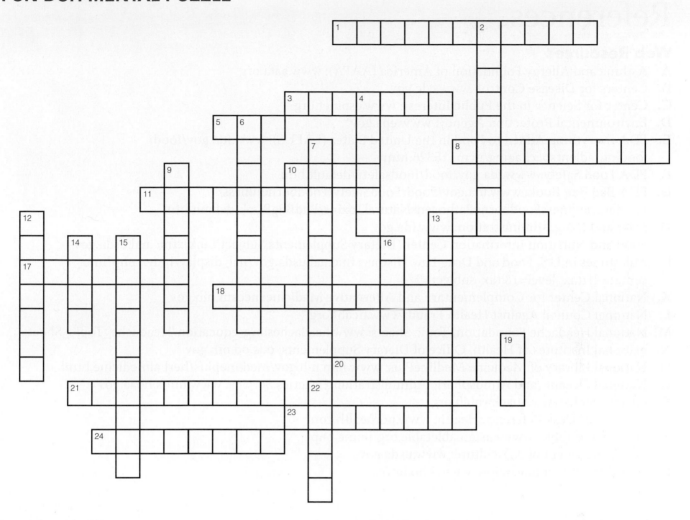

ACROSS

1. Process of removing water from food to reduce microbial growth.
3. Industrial heavy metal waste contaminating fish.
5. Acronym for agency that monitors foodborne diseases in the United States.
8. 1938 legislation links food to what?
10. Common gum added to processed food for moisture retention.
11. Type of FDA-approved claim used on qualifying supplement labels.
13. Washing _____ is important for preventing foodborne illness.
17. Foods rich in _____ are common carriers of pathogenic bacteria.
18. Common bacterium causing foodborne illness.
20. Consumers should know a supplement's deficiency and toxicity _____.
21. Another name for an organization not interested in financial gain.
23. Disease caused by the toxin of an anaerobic microbe.
24. Another name for peer-reviewed published information.

DOWN

2. Mold-produced carcinogenic toxin in improperly stored nuts.
4. Type of acid added to foods for its chelating properties.
6. Title for a person with a registered nutrition education.
7. Artificial sweetener with the fewest safety concerns.
9. Clause that prevents the addition of cancerous agents to processed foods.
12. Given optimal conditions, microbes grow/divide _____.
13. Disease caused by a virus that targets liver cells.
14. The FDA regulates nutritional supplements the same as _____.
15. Supporting information sources listed in more credible publications.
16. Common chemical added to processed food to reduce microbial growth.
19. Acronym for a common flavor enhancer added to processed foods.
22. Type of effect that can be experienced from herbal supplements.

References

Web Resources

A. Asthma and Allergy Foundation of America (AAFA): www.aafa.org
B. Centers for Disease Control: www.cdc.gov
C. Center for Science in the Public Interest: www.cspinet.org
D. Environmental Protection Agency: www.epa.gov
E. FDA Everything Added to Food in the United States (EAFUS): www.fda.gov/food/foodingredientspackaging/ucm115326.htm
F. FDA Food Safety: www.fda.gov/food/foodsafety/default.htm
G. FDA Bad Bug Book: www.fda.gov/Food/FoodSafety/FoodborneIllness/FoodborneIllnessFoodbornePathogensNaturalToxins/BadBugBook/default.htm
H. Food and Drug Administration: www.fda.gov
I. Food and Nutrition Information Center (Dietary Supplements Subject Link): fnic.nal.usda.gov
J. Milestones in U.S. Food and Drug Law History: fnic.nal.usda.gov/nal_display/index.php?info_center=4&tax_level=1&tax_subject=274
K. National Center for Complementary and Alternative Medicine: nccam.nih.gov
L. National Council Against Health Fraud: www.ncahf.org
M. National Headache Foundation, Topic Sheets: www.headaches.org/education/Headache_Topic_Sheets
N. National Institutes of Health, Office of Dietary Supplements: ods.od.nih.gov
O. National Library of Medicine MedlinePlus: www.nlm.nih.gov/medlineplus/herbalmedicine.html
P. National Oceanic and Atmospheric Administration Fisheries Service: www.nmfs.noaa.gov
Q. Quackwatch: www.quackwatch.org
R. Physicians' Desk Reference Health: www.pdrhealth.com
S. Sustainable Table: www.sustainabletable.org/home.php
T. U.S. Department of Agriculture: www.usda.gov
U. World Health Organization: www.who.int/en

Works Cited

1. American Dietetic Association. (1998). Position of the American Dietetic Association: Fat replacers. *Journal of the American Dietetic Association, 98,* 463–468.
2. American Dietetic Association. (1998). Position of the American Dietetic Association: Use of nutritive and nonnutritive sweeteners. *Journal of the American Dietetic Association, 98,* 580–587.
3. American Dietetic Association. (2002). Position of the American Dietetic Association: Food and nutrition misinformation. *Journal of the American Dietetic Association, 102,* 260–266.
4. American Dietetic Association. (2005). Position of the American Dietetic Association: Fortification and nutritional supplements. *Journal of the American Dietetic Association, 105*(8), 1300–1311.
5. American Dietetic Association. (2006). Position of the American Dietetic Association: Agricultural and food biotechnology. *Journal of the American Dietetic Association, 106*(2), 285–293.
6. American Dietetic Association. (2007). Position of the American Dietetic Association: Food and nutrition professionals can implement practices to conserve natural resources and support ecological sustainability. *Journal of the American Dietetic Association, 107*(6), 979–987.
7. Armelagos, G. J. (2009). The viral superhighway. *Physical anthropology* (Annual Editions, E. Angeloni, Ed.), Article 37, 170–174. New York: McGraw-Hill.
8. Asthma and Allergy Foundation of America. (2005). Food additives. Accessed at: http://www.aafa.org/display.cfm?id=9&sub=20&cont=285.

9. Ball, K. (2009). Vegetarian diets and buying locally: Ways to increase environmental sustainability, support the local economy and contribute to individual health. *Vegetarian Nutrition, 3/4*, 1, 14–16, 20.

10. Blumenthal, M., Buss, W. R., Goldberg, A., Gruenwald, J., Hall, T., Klein, S., Riggins, C. W., & Rister, R. S. (1998). *The complete German Commission E monographs: Therapeutic guide to herbal medicines.* Austin, TX: American Botanical Council.

11. Bondarianzadeh, D. (2007). Food risk to babies: Listeriosis. *Nutrition Today, 42*(6), 236–239.

12. Brannon, C. A. (2008). Organics: Separating science fiction from fact. *Today's Dietitian, 10*(4), 8–14.

13. Centers for Disease Control and Prevention. (2004). Preliminary FoodNet data on the incidence of infection with pathogens transmitted commonly through food-selected sites: United States, 2003. *Morbidity and Mortality Weekly Report, 53*(16), 338–342.

14. Centers for Disease Control and Prevention. (2010). Surveillance for foodborne disease outbreaks: United States, 2007. *Morbidity and Mortality Weekly Report, 59*(31), 973–979.

15. Chan, H. M., & Egeland, G. M. (2004). Fish consumption, mercury exposure, and heart disease. *Nutrition Reviews, 62*(2), 68–72.

16. Cifelli, C. J., Maples, I. S., & Miller, G. D. (2010). Pasteurization: Implications for food safety and nutrition. *Nutrition Today, 45*(5), 207–215.

17. Coleman, E. (2009). Dietary supplement quality and safety. *Today's Dietitian, 11*(10), 56–61.

18. Coulston, A. M., & Johnson, R. K. (2002). Sugar and sugars: Myths and realities. *Journal of the American Dietetic Association, 102*, 351–353.

19. De Lorenzo, A., Noce, A., Bigioni, M., Calabrese, V., Della Rocca, D. G., Di Daniele, N., Tozzo, C., & Di Renzo, L. (2010). The effect of Italian Mediterranean organic diet (IMOD) on health status. *Current Pharmaceutical Design, 16*(7), 814–824.

20. Dona, A., & Arvanitoyannis, I. S. (2009). Health risks of genetically modified foods. *Critical Reviews in Food Science and Nutrition, 49*, 164–175.

21. Dorner, B. (2007–2008, Winter). Clostridium-difficile and antibiotic associated diarrhea. *Healthy Aging Newsletter*, 8–9.

22. Eshel, G., & Martin, P. (2006, January). Diet, energy, and global warming. *Earth Interactions, 10*(1), 1–17.

23. Ezendam, J., & van Loveren, H. (2006). Probiotics: Immunomodulation and evaluation of safety and efficacy. *Nutrition Reviews, 64*(1), 1–14.

24. FDA Talk Paper. (1997, September 23). *FDA publishes final dietary supplement rules.* Rockville, MD: Food and Drug Administration, U.S. Department of Health and Human Services.

25. Foodborne Illness Primer Work Group. (2004, October/December). Foodborne illness primer for physicians and other health care professionals. *Nutrition in Clinical Care, 7*(4), 134–140.

26. Fragakis, A. S. (2007). *The health professional's guide to popular dietary supplements* (3rd ed.). Chicago: American Dietetic Association.

27. Getz, L. (2008). Beef beware? *Today's Dietitian, 10*(6), 36–39.

28. Grieger, L. (2007). Prebiotics and probiotics: The bacterial good guys. *Today's Diet & Nutrition, 3*(2), 54–57.

29. Gruenwals, J., Brendler, T., & Jaenicke, C. (Eds.). (2007). *PDR for herbal medicines* (4th ed.). Montvale, NJ: Thomson.

30. Hammer, H. F. (2010). Pancreatic exocrine insufficiency: Diagnostic evaluation and replacement therapy with pancreatic enzymes. *Digestive Diseases, 28*(2), 339–343.

31. Hendler, S. S., & Rorvik, D. (Eds.). (2001). *PDR for nutritional supplements.* Montvale, NJ: Thomson.

32. Holben, D. H. (2010). Farmers' markets: Fertile ground for optimizing health. *Journal of the American Dietetic Association, 110*(3), 364–365.

33. Jellin, J. M. (2010). *Natural medicines comprehensive database* (12th ed.). Stockton, CA: Therapeutic Research Faculty.

34. Johnston, C. (2009). Functional foods as modifiers of cardiovascular disease. *American Journal of Lifestyle Medicine, 3*(S1), 39S–43S.

35. Jurmain, R., Kilgore, L., Trevathan, W., & Ciochon, R. L. (2012). *Introduction to physical anthropology* (13th ed.). Belmont, CA: Wadsworth.

36. Knekt, P., Jarvinen, R., Dich, J., & Hakulinen, T. (1999). Risk of colorectal and other gastro-intestinal cancers after exposure to nitrate, nitrite and N-nitroso compounds: A follow-up study. *International Journal of Cancer, 80*, 852–856.

37. Kosta-Rokosz, M. D., Dvorkin, L., Vibbard, K. J., & Couris, R. R. (2005, January/February). Selected herbal therapies: A review of safety. *Nutrition Today, 40*(1), 17–28.

38. Lenard, L. (2009). Optimizing digestive health. *Life Extension, 15*(1), 48–55.

39. Lenoir-Wijnkoop, I., Saunders, M. E., Cabana, M. D., Caglar, E., Corthier, G., Rayes, N., . . . & Wolvers, D. A. W. (2007). Probiotic and prebiotic influence beyond the intestinal tract. *Nutrition Reviews, 65*(11), 469–489.

40. Levenson, C. W., & Axelrad, D. M. (2006). Too much of a good thing? Update on fish consumption and mercury exposure. *Nutrition Reviews, 64*(3), 139–145.

41. Lilyquist, K. (2010a). GM high-oleic soy: Despite concerns, it may soon be available for human consumption. *Today's Dietitian, 12*(12), 42–47.

42. Lilyquist, K. (2010b). Are genetically modified foods safe? *Today's Dietitian, 12*(5), 42–48.

43. Luteyn, J. (2006). Vital functions of digestive enzymes. *Total Health, 28*(2), 60–63.

44. Lynch, M. F., Tauxe, R. V., & Hedberg, C. W. (2009). The growing burden of foodborne outbreaks due to contaminated fresh produce: Risks and opportunities. *Epidemiology and Infection, 137*(3), 307–315.

45. Magaña-Gómez, J., & Calderón de la Barca, A. (2009, January). Risk assessment of genetically modified crops for nutrition and health. *Nutrition Reviews, 67*(1), 1–16.

46. Maleskey, G. (2005, Summer). Therapies: Statins, muscle damage and coenzyme Q10. *Nutrition in Complementary Care, 8*(1), 1, 6–8.

47. Medeiros, L. C., Hillers, V. N., Kendall, P. A., & Mason, A. (2001). Food safety education: What should we be teaching to consumers? *Journal of Nutrition Education, 33*, 108–113.

48. Medeiros, L. C., Kendall, P., Hillers, P., & Mascola, S. (2001). Identification and classification of consumer food-handling behaviors, for food safety education. *Journal of the American Dietetic Association, 101*, 1326–1332, 1337–1339.

49. Mortimore, S., & Wallace, C. (1998). *HACCP: A practical approach* (2nd ed.). New York: CRC Press.

50. Palmer, S. (2007). Organic, local & beyond. *Today's Dietitian, 9*(10), 45–46, 48, 50.

51. Palmer, S. (2009). Sorting out the science on multivitamins and minerals. *Today's Dietitian, 11*(8), 38–42.

52. Palmer, S. (2010). Antibiotics in animal agriculture. *Today's Dietitian, 12*(6), 32–35.

53. Park, S., & Johnson, M. A. (2006). Awareness of fish advisories and mercury exposure in women of childbearing age. *Nutrition Reviews, 64*(5), 250–256.

54. Parnes, R. B., & Lichtenstein, A. H. (2004, October/December). Food irradiation: A safe and useful technology. *Nutrition in Clinical Care, 7*(4), 149–155.

55. Pelletier, D. L. (2005). Science, law, and politics in the Food and Drug Administration's genetically engineered foods policy: FDA 1992 policy statement. *Nutrition Reviews, 63*(5), 171–181.

56. Pimentel, D., Williamson, S., Alexander, C. E., Gonzales-Pagan, O., & Kontack, C. (2008). Reducing energy inputs in the U.S. food system. *Human Ecology, 36*, 459–471.

57. Refsdal, A. O. (2000). To treat or not to treat: A proper use of hormones and antibiotics. *Animal Reproductive Science, 60–61*, 109–119.

58. Rodriguez, J. (2011). Food Safety Modernization Act. *ADA Times, 8*(2), 17.

59. Saldanha, L. G. (2007). The dietary supplement marketplace: Constantly evolving. *Nutrition Today, 42*(2), 52–54.

60. Santerre, C. R. (2008). Wild versus farm-raised salmon: Health benefits and risks. *SCAN'S (A Publication for Sports, Cardiovascular, and Wellness Nutritionists) PULSE, 27*(4), 1–5.

61. Santosa, S., Farnworth, E., & Jones, P. J. H. (2006). Probiotics and their potential health claims. *Nutrition Reviews, 64*(6), 265–274.

62. Scherr, S. J., Sthapit, S., & Mastny, L. (2009). *Mitigating climate change through food and land use.* Washington, DC: Worldwatch Institute and Ecoagriculture Partners.

63. Selimoğlu, M. A., & Karabiber, H. (2010). Celiac disease: Prevention and treatment. *Journal of Clinical Gastroenterology, 44*(1), 4–8.

64. Silvers, K. M. (2005). Herbal foods: Are they efficacious and safe? *Nutrition Today, 40*(1), 13–16.

65. Singh, U., Devaraj, S., & Jialal, I. (2007). Coenzyme Q10 supplementation and heart failure. *Nutrition Reviews, 65*(6), 286–293.

66. Sloan, E. (2007). Why people use vitamin and mineral supplements. *Nutrition Today, 42*(2), 55–61.

67. Stein, K. (2009). Are food allergies on the rise, or is it misdiagnosis? *Journal of the American Dietetic Association, 109*(11), 1832–1837.

68. Taylor, C. L. (2004). Regulatory frameworks for functional foods and dietary supplements. *Nutrition Reviews, 62*(2), 55–59.

69. Thomas, P. R. (2005, January/February). Dietary supplements for weight loss? *Nutrition Today, 40*(1), 6–12.

70. Thompson, F. E., McNeel, T. S., Dowling, E. C., Midthune, D., Morrissette, M., & Zeruto, C. A. (2009). Interrelationships of added sugars intake, socioeconomic status, and race/ethnicity in adults in the United States: National Health Interview Survey, 2005. *Journal of the American Dietetic Association, 109*(8), 1376–1383.

71. Tufts University. (2003, February). Supermarkets introduce irradiated ground beef—again. *Tufts University Health & Nutrition Letter, 12*, 3.

72. Turley, J. M., & Jackson, S. A. (2010). The personal and environmental health benefits of eating organic foods. *The International Journal of Environmental, Cultural, Economic, and Social Sustainability, 6*(1), 203–226.

73. Turley, J. M., & Thompson, J. (2010). A new holistic model portrays health and food sustainability. *The International Journal of Environmental, Cultural, Economic, and Social Sustainability, 6*(1), 345–354.

74. Vargas, M. (2009). Atherosclerosis, functional foods, and nutritional genomics. *Dietitians in Integrative and Functional Medicine, 12*(2), 20, 23–30.

75. Velimirov, A., Huber, M., Lauridsen, C., Rembialkowska, E., Seidel, K., & Bügel, S. (2010). Feeding trials in organic food quality and health research. *Journal of Food Agriculture, 90*(2), 175–182.

76. Weber, C. L., & Matthews, H. S. (2008). Food-miles and the relative climate impacts of food choices in the United States. *Environmental Science & Technology, 42*(10), 3508–3513.

77. Wilkins, J. (2007). Microorganisms that make us worry. *Today's Dietitian, 9*(6), 10–14.

78. Winstead, N. S., & Wilcox, C. M. (2009). Clinical trials of pancreatic enzyme replacement for painful chronic pancreatitis: A review. *Pancreatology, 9*(4), 344–350.

79. Wu, V. C. (2008). A review of microbial injury and recovery methods in food. *Food Microbiology, 25*(6), 735–744.

80. Yeager, D. (2008). Got organic? *Today's Dietitian, 10*(10), 60–64.

81. Yeager, V. (2007). Cloned or noncloned meat: The choice may not be ours. *Today's Dietitian, 9*(7), 4.

82. Zeind, C. S., & Couris, R. R. (2006, March/April). Prevention of food and waterborne diseases while traveling. *Nutrition Today, 41*(2), 78–86.

Nutrition Through the Life Span

Valentyn Volkov/
Shutterstock.com

MODULE GOAL

To showcase the benefits of a healthy diet and lifestyle during pre-pregnancy, pregnancy, lactation, infancy, childhood, adolescence, and older adulthood.

LEARNING OBJECTIVES

When you complete this learning module, you will be able to:

- Outline the maternal nutrient and physiological demands that come with reproduction.

- Describe to another person how establishing a healthy lifestyle and healthy eating practices supports normal growth and development of the child and provides the foundation for wellness during the life span.

- Explain how the processes of normal growth and development are greatly influenced by good nutrition.

- Argue that successful aging strategies during the later stages of the life span slows the rate of age-related physiological decline, thus preserving function and wellness in old age.

PERSONAL IMPROVEMENT GOALS

By adopting and implementing healthy lifestyle practices in your family, you can:

- Provide support and positive influences for each individual in your family throughout his or her life span.

- Cope with the fast pace of modern society and maintain a healthy lifestyle and eating practices.

- Likely prevent or delay the development of chronic diseases.

Living a healthy lifestyle is imperative for maximizing lifelong wellness. In this module, we will first highlight the influence that nutritional status has on birth outcome, including fertility, reproductive fitness, and pregnancy. We will explore nutrition as applied to growth and development for infants, children, adolescents, and older adults. Feeding challenges and good nutritional practices will be emphasized to promote optimal growth, development, and health throughout the **life span**.[XX,35]

Every person needs vitamins, minerals, amino acids, fatty acids, glucose, Calories/energy, fiber, and water, but the amount of each needed varies at different stages of life. In this module, we will look at how nutrient needs change during the life cycle; we will also address some inborn errors of metabolism that are screened at birth and some inherited genetic conditions that seriously affect lifelong nutritional status.[YY]

Associations between feeding and the environment are established very early in human development, and continue to influence feeding behaviors throughout life. Table 7.1 describes the stages of human development.[OO,16,139] Establishing healthy eating and dietary practices is critical for proper growth, development, and health throughout the life span. Food and taste preferences vary throughout the life span, but regardless of age, eating brings pleasure, comfort, and satiety. We will see how some foods, particularly sweet foods, intensify the perception of pleasure through an increased synthesis of **neurochemicals** in the brain. Chemical changes such as these have a great influence on food preference, food intake, and thus energy balance. This "food and mood" connection as it pertains to the intensity of food affection will also be addressed.[W]

TABLE 7.1	
Stages of Human Development	
Stage	**Description**
Prenatal	
Zygote	A single fertilized cell
Morula	A ball of cells
Blastocyst	A ball of cells with a surface layer and inner cell mass
Embryo	The developmental stages from two weeks to the end of eight weeks after fertilization
Fetus	The developmental stages from nine weeks until birth, usually about 38 weeks after fertilization
Postnatal	
Newborn	The neonatal period of two weeks after birth
Infancy	The period of time between neonate and childhood, usually two weeks of age until about one year of age
Childhood	The period of 12 to 13 years after infancy to puberty
Pubescence	The period of time (usually between ages 10–16 in girls and ages 13–16 in boys) when sexual traits develop; puberty
Adolescence	The period of time between childhood and adulthood when a child is transformed into an adult
Adulthood	Growth and development are complete, and body changes typically occur slowly compared to earlier life stages
Old age	General decrease in functional capacity of body systems occurs with advancing age

© Cengage Learning 2013

life span The period of time between birth and death; the maximum years alive attained by a member of a species.

neurochemicals A variety of organic molecules that activate functions in the central nervous system.

fertility The quality and ability of the natural capability to produce offspring.

pregnancy The period from conception to birth when a woman houses a developing embryo and then fetus in her womb.

7.1 REPRODUCTIVE FITNESS AND PRENATAL NUTRITION

Introduction A woman who is capable of reproduction lives with variable biology. During the menstrual cycle, her hormone levels vary from day to day. Food intake is influenced by the vacillation of hormone shifts during the menstrual cycle. Nutrition and nutritional status influence **fertility**, embryonic development, fetal development, and birth outcome. The goal of preconception fitness is to acquire the nutrient stores and the physiological preparedness to enjoy a **pregnancy** that will have no complications

T-Talk 7.1
To hear Dr. Turley talk about reproductive fitness and prenatal nutrition, go to www.cengagebrain.com

and a successful birth outcome. The maternal nutritional support required and the physiological stresses placed on the body during pregnancy are tremendous. Thus, physical fitness and excellent nutritional status support favorable reproduction events.[2,31,66]

Preconception Nutrition

From decades of studies of animal reproduction, we know that nutritional status affects fertility, **conception**, implantation, and the development of the **embryo**. Negative effects on fertility can come from being under- or overweight, having too many reactive oxygen species, or having folic acid, iron, zinc, magnesium, vitamin B_{12}, or vitamin B_6 deficiencies. On the other hand, consuming an antioxidant-rich diet that promotes a healthy body weight and provides all of the essential nutrients supports fertility.

One nutrient that needs to be supplemented before conception is folic acid. There is a mounting body of evidence suggesting that a mutation of a particular gene (5-methyltetrahydrofolate-homocysteine methyltransferase reductase [MTRR] or methylenetetrahydrofolate reductase [MTHFR]) prevents food folate from being converted to folic acid in the body and reduces fertility. These altered genes increase the concentrations of homocysteine in the body, which negatively affects fertility.

Acceptable iron stores are critical as well because of the anticipated maternal and fetal blood volume expansion that occurs during the third **trimester** of the pregnancy and the anticipated blood losses during delivery and the first month postpartum. Increasing body iron stores if they are low preconceptually may improve birth outcome.[4,8,66,138,142]

Nutrition and Fertility Fertility of both sexes may also be negatively affected by techniques used to produce food conventionally or by chemicals that may be added to processed food, in addition to nutrient adequacy and attributes of the diet. Human research is extremely difficult to do, and the evidence that has been reported in the literature is limited. Again, animal research provides the majority of the details that are associated with the evidence collected on fertility.[116]

Male and Female Fertility In females, low body fat levels inhibit **ovulation**, whereas too much body fat inhibits conception. A fertility diet includes foods low in *trans* fatty acids and saturated fatty acids, with higher amounts of monounsaturated fatty acids, vegetable proteins, fiber, and low-glycemic-index carbohydrate sources, along with adequate amounts of all of the essential vitamins and minerals. Exposure to pesticides has negatively affected men's fertility, so choosing organically produced foods may improve fertility as well.[4,31,66,116]

Celiac Disease and Infertility **Celiac disease** is an autoimmune genetic disease that damages the villi in the small intestine, decreasing absorption of nutrients from food. The autoimmune response is initiated when gluten, a protein complex in wheat, rye, and barley, is consumed. Without healthy villi, nutrient absorption is diminished and a person becomes malnourished, no matter how much food he or she eats. Celiac disease affects people differently. Three factors influence when and how severe the signs and symptoms of celiac are: (1) the length of time an **infant** was breastfed, (2) the age of the child when gluten-containing foods were first eaten, and (3) the amount of gluten-containing foods eaten. The longer the infant was breastfed, the later the symptoms of celiac disease appear.

Reproductive problems, such as delayed **menarche**, amenorrhea, early **menopause**, **infertility**, **hypogonadism**, recurrent abortions, and low-birth-weight or preterm deliveries, are now known to be among the symptoms of females who have celiac disease, while **impotence** occurs in males with celiac disease. The

---

conception The moment in time when an egg (ova) is fertilized by a sperm.

embryo The product or stage after the fertilization of the egg when rapid cellular differentiation occurs; precedes the fetal stage of development.

trimester One of the three three-month periods into which human pregnancy is divided.

ovulation The moment in time when egg(s) are released from the ovaries.

celiac disease An inheritable autoimmune disease that damages the villi in the small intestine when the person eats wheat, barley, or rye.

infant A term used to refer to the very young offspring of humans; usually includes the time from birth to at least age 1.

menarche The central event of female puberty, signaled by the first menstrual cycle.

menopause The time of a woman's life when her menstrual cycle ceases permanently; 12 months of amenorrhea (no menstrual cycle).

infertility The state of being unable to produce offspring; usually in a woman, it is an inability to conceive; in a man, it is an inability to impregnate.

hypogonadism A condition in which decreased production of gonadal hormones leads to below-normal function of the gonads and retardation of sexual growth and development.

impotence A male sexual dysfunction characterized by the inability to develop or maintain an erection of the penis sufficient for satisfactory sexual performance.

pathogenesis of reproductive disorders is unclear, but autoimmunity and macro- and/or micronutrient deficiencies may play a role.[E,V,KK,4,45,66,85,105,116]

Polycystic Ovarian Syndrome   Polycystic ovarian syndrome (PCOS) is the most common female **endocrine** or hormonal disorder and affects approximately 5 to 10 percent of all females. PCOS is believed to be caused by insulin resistance. The combination of high blood levels of insulin, elevated blood sugar, and food affection (the excessive pleasuring response that some people feel when they eat) results in 60 percent of women with PCOS having weight management issues. Acne and excess hair growth on the face and body with thinning scalp hair are also signs of PCOS. An unhealthy lifestyle leads to an increased risk of diabetes and gender-specific cancers. There is no cure for PCOS, but it can be successfully managed through diet, exercise, and in some cases medical intervention. Women with PCOS often have great difficulty getting pregnant because they have irregular or no menstrual periods or irregular ovulation, with or without monthly bleeding. The most painful aspect of the disorder is the accumulation of unruptured follicles on the periphery of the **ovaries**. These follicles are often mislabeled as cysts but are another physical manifestation of the syndrome.[Y,16,60,71]

Building the Blue Ribbon Baby   A "blue ribbon baby" is a healthy baby born at term from an uncomplicated pregnancy, within the ideal range for weight and length: 7 pounds, 8 ounces, and 20 inches long. This is a successful parameter of a birth outcome. Genetic testing may be employed to determine the risks for or evidence of disease that may affect the offspring's quality of life. Several issues regarding the intakes of essential nutrients have been identified as causing complications of pregnancy or negatively affecting embryonic and fetal development. Sound nutritional practices, controlling behavioral risks, and **prenatal** monitoring are all associated with a successful birth outcome.[2,38,39,48]

There are many notable features shown in Figure 7.1 that display the variable biology that a woman of reproductive capability cycles through roughly every four weeks. Four key hormones predominantly influence the menstrual cycle: **estrogen, progesterone, luteinizing hormone (LH),** and **follicle-stimulating hormone (FSH).** Hormones are potent biological molecules that cause cell metabolism to change. The four female gender tissue-targeting hormones are illustrated in the menstrual cycle in Figure 7.1. Vacillations in FSH and LH control ovum (**egg**) development and ovulation (release), while estrogen and progesterone increase the lining of the uterine wall and the sloughing of the iron-rich lining during **menses**. In the textbook model of the menstrual cycle, the first 14 days of the cycle are termed the follicular phase. It is a time of ovum development, estrogen domination, stable emotions, and the gradual increased synthesis of neuroactive chemicals that produce a sense of peace and well-being, especially **serotonin** and **endorphin**. At day 14, ovulation occurs from the peaking of FSH and LH, and the mood is best supported by the high concentrations of the "good mood" neurochemicals. The last 14 days of the menstrual cycle, termed the **luteal phase**, is characterized by increasing progesterone and diminishing concentrations of estrogen, serotonin, and endorphin. For some women, the combination of these hormonal changes causes unstable emotions, inflammation, and craving for sweet-fat treats. Eating carbohydrate and fats can increase the synthesis of the feel-good hormones serotonin and endorphin. Additionally, the effect of insulin is augmented by the hormonal shifts in the menstrual cycle, thus blood sugar levels are lower during the luteal phase as well. (See BioBeat 7.1 on page 349 for dietary tips that complement the variable biology of a female who is capable of reproduction.[H,139])

polycystic ovarian syndrome (PCOS) The most common female endocrine disorder that affects female fertility and may be caused by insulin resistance.

endocrine A type of gland that secretes hormones that are transmitted by the blood and act on cells of a different tissue.

ovaries The egg-producing reproductive organ, often found in pairs as part of the female vertebrate reproductive system.

prenatal A term used to refer to the period of pregnancy.

estrogen A group of steroid compounds, found in both males and females, but named for its importance in the estrous cycle; functions as the primary female sex hormone.

progesterone A steroid hormone that fluctuates along with estrogen, is largely manufactured in the ovaries and adrenal glands, and promotes the growth of the uterine lining during the luteal phase of the menstrual cycle.

luteinizing hormone (LH) One of the four important hormones of the menstrual cycle that surges at ovulation and helps release eggs from the ovaries.

follicle-stimulating hormone (FSH) One of the four important hormones of the menstrual cycle that stimulates egg development, surges at ovulation, and helps release eggs from the ovaries.

egg A mature female reproductive cell, also termed *ovum* or *ova*, or female gametes that normally contain 23 chromosomes.

menses The cyclic discharge of blood-rich uterine linings from the uterus of nonpregnant women, monthly from puberty to menopause.

serotonin A neurotransmitter, made from the essential amino acid tryptophan, that is involved in mood, sleep, appetite, pain, and memory.

endorphins Opioid protein compounds that are endogenously produced in the brain and produce an intense sense of well-being.

luteal phase The last 14 days of the menstrual cycle.

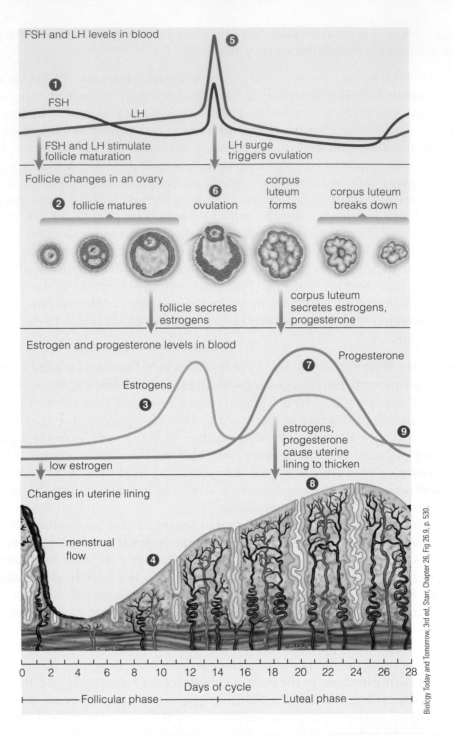

FSH and LH levels in blood

① FSH

LH

⑤

FSH and LH stimulate follicle maturation

LH surge triggers ovulation

Follicle changes in an ovary

② follicle matures

⑥ ovulation

corpus luteum forms

corpus luteum breaks down

follicle secretes estrogens

corpus luteum secretes estrogens, progesterone

Estrogen and progesterone levels in blood

Progesterone ⑦

Estrogens ③

estrogens, progesterone cause uterine lining to thicken

⑨

low estrogen

Changes in uterine lining

⑧

menstrual flow

④

0 2 4 6 8 10 12 14 16 18 20 22 24 26 28
Days of cycle

Follicular phase — Luteal phase

Biology Today and Tomorrow, 3rd ed, Starr, Chapter 26, Fig 26.9, p. 530.

FIGURE 7.1 Changes in the human female ovary and uterus, correlated with changing hormone levels. (1) Follicle stimulating hormone (FSH) causes (2) follicles (eggs) to mature. (3) With the rise of estrogens, (4) the uterine lining thickens, then (5) the surge of FSH and luteinizing hormone (LH) (6) triggers ovulation. (7) Without fertilization of an egg, progesterone and estrogen levels increase, causing (8) the lining of the uterus to thicken. (9) As the hormone levels of progesterone and estrogen diminish, the lining of the uterus does, too.

Energy and Nutrient Needs Any sexually active and reproductively capable female can become pregnant (see BioBeat 7.2). The specific energy and nutrient needs for reproductively capable women and adolescents are the same as for females of other various age groups. The Dietary Reference Intakes (DRIs) for energy and nutrients are established and shown in Appendix A. The goal of preconception nutrition is to acquire the nutrient stores to maintain a healthy pregnancy and to prevent maternal nutrient store depletion at birth.[69]

Calories Energy needs are determined using the Estimated Energy Requirement (EER) equations. Age- and gender-specific formulas are found in Appendix A. Calorie needs can be calculated based on body mass and physical activity. Adjustments of the calculated Calorie intake can be modified to promote a healthy body weight.

BioBeat 7.1

Tips for Premenopausal Cycling Females

To balance some of the biological changes that occur during a woman's menstrual cycle, especially during the luteal phase, women can consume small, frequent meals of low-glycemic-index, carbohydrate-rich foods, while eating a diet low in fat (less than 25 percent of Calories), with an omega-3 to omega-6 fatty acid ratio of 1:4. Protein intake needs to be adequate (0.8 to up to 1.6 grams of protein per kilogram of body weight). Total energy intake and energy expenditure balance needs to be employed to achieve a "physically fit" level of body fat (18 to 22 percent). A balanced diet that provides all of the essential nutrients and is antioxidant rich decreases the discomforts of the luteal phase. A balanced diet includes following the MyPlate food guidance system; eating the recommended amount of grains (at least half as whole grains); eating the recommended amount of vegetable variety and choosing from the cabbage family (brassica or cruciferous), dark green, and orange vegetables as frequently as possible; eating the recommended amount of fruit and choosing whole fruits rich in antioxidants (citrus fruits, melons, berries, kiwis, apricots, tropical fruits, and red grapes); and meeting the teaspoons of oils recommended by using flax seed meal, canola oil, fish oil, and walnuts. The dairy recommendation can be met by calcium- and protein-rich foods such as fortified soy products and yogurt with the live cultures added, which may have added biological benefits, especially during the luteal phase. Selecting from a wide variety of legumes, meat alternatives, and fish is needed to meet the protein-rich foods recommendation from the protein foods group.[8,31,38,39,66]

What is the best dietary approach for a woman to add stability to her variable biology?

The addition or reduction of about 250 Calories per day can promote gradual weight gain or weight loss, respectively, depending on the need for weight change.[69]

Carbohydrate, Protein, and Fat The DRI for preconception carbohydrate is the same as during adulthood. There is a minimal need of 130 grams per day of dietary carbohydrate for brain, central nervous system, and red blood cell functioning. The Acceptable Macronutrient Distribution Range (AMDR) for carbohydrate is the same as for adults, 45 to 65 percent of Calories with less than or equal to 25 percent of Calories from sugars. Fiber-rich sources are needed to promote gastrointestinal health and prevent obesity, type 2 diabetes, heart disease, and some cancers. The DRI is personalized at 1.4 grams per 100 Calories consumed. Based

sperm Male reproductive cells or male gametes, which contain 23 chromosomes in each.

mitosis A type of cell division in which the cell nucleus divides into identical nuclei containing the same number of chromosomes and produces the same type of cell.

chromosome An organized, threadlike strand of DNA in the cell nucleus where an individual's genes are encoded; the hereditary information can be transmitted from one generation to the next.

BioBeat 7.2

Basics of Conception and Genetics

There are roughly 48 hours during the menstrual cycle when conception can occur as a result of male climax during sexual intercourse. This means that viable **sperm** must be available at the right place at the correct time during the menstrual cycle (just before or at ovulation). Fertilization is the combination of two different cells: an egg and a sperm. In the body, the cell division process normally produces new cells for growth, repair, and the general replacement of older cells. This type of cell division is called **mitosis**. Typically, a body cell divides into two complete new cells that are identical to the original one. However, when an egg and sperm combine, each cell contributes 23 single-stranded **chromosomes**. **Meiosis** is a process of cell division where the number of chromosomes per cell is cut in half. The two sequential division processes of meiosis culminate in the production of **gametes** that have only half the number of chromosomes of body (somatic) cells. When the gametes fuse, the cell produced is called a **zygote**, and it has the combined genetic material from each gamete and the correct number of chromosomes: 46 total chromosomes, 23 pairs, with one gender- (sex-) determining pair. Then cellular division (mitosis) and cellular **differentiation** (the creation of different types of cells; less specialized cells become more specialized cell types) continue to occur during embryonic and fetal development.

What is the difference between meiosis and mitosis?

on the reference person, the DRI for reproductively capable women ranges from 25 to 26 grams per day. Fiber intake is increased by eating more whole fruits, vegetables, whole grains, and legumes.[69]

Adolescent females (ages 14 to 18) have a DRI for protein of 0.85 grams per kilogram of body weight, while women ages 19 to 50 require 0.80 grams of dietary protein per kilogram of body weight. Consuming adequate protein is very important for normal metabolism, tissue repair, and lean body mass maintenance. Actual protein needs of a reproductively capable adolescent will fluctuate depending on whether the adolescent is in a **growth spurt**.[69]

There is no DRI for total fat. However, there is a DRI for the essential fatty acids linoleic acid and alpha-linolenic acid. The DRI for linoleic acid ranges from 11 to 12 grams per day, while the DRI for alpha-linolenic acid is 1.1 grams per day (see Appendix A). The AMDR for total fat for adults is 20 to 35 percent of Calories (25 to 35 percent for adolescents), with the AMDR for the essential fatty acids being 5 to 10 percent of Calories for linoleic acid and 0.6 to 1.2 percent of Calories for alpha-linolenic acid. The tendency to overconsume saturated fat and cholesterol and underconsume the essential fatty acids is common. Consuming less fast food and processed food and more whole plant foods (including nuts and seeds) will support a healthy shift in fatty acid intake.[68,69]

Vitamins, Minerals, and Water   The DRIs for the vitamins and minerals are the adult female values. Many nutrient values are the same for males and females. The only nutrient DRI that is higher for females than males is iron. Making sure that iron stores are adequate (by a blood test that measures serum ferritin levels) before pregnancy is critical for preventing iron-deficiency anemia.[92] Adequate water intake that maintains positive water balance will support metabolic waste removal, cardiovascular function, thermal regulation, and metabolism. The DRI for water intake is 2.7 liters of water per day for women.

Folic acid is a nutrient that deserves special consideration for preconception and during pregnancy. It is commonly recommended in supplement form because some people poorly convert food folate to active folic acid. Preconception folic acid supplementation improves fertility and prevents spina bifida, a neural tube defect (NTD), and supports the rapid growth of the **placenta**, embryo, and **fetus**. It is needed to produce new DNA (genetic material) as cells multiply. Without adequate amounts of folic acid, cell division could be impaired, possibly leading to poor growth of the fetus or placenta.[166]

Women who consume adequate folic acid from supplements or fortified foods for at least one year before they become pregnant can reduce the risk for having a premature baby, as well as reduce the risk of NTDs. The March of Dimes and the Institute of Medicine recommend that all women who can become pregnant take a multivitamin that contains 400 micrograms of folic acid every day starting before pregnancy, as part of a healthy diet. Most prenatal vitamins contain 800 to 1,000 micrograms of folic acid. However, women should not take more than 1,000 micrograms (1 milligram) without their healthcare provider's advice. Because of the typical low dietary folate intake and thus inadequate dietary folate status, folic acid supplementation is recommended for all women of childbearing age, because about half of all pregnancies in this country are unplanned. Because spina bifida and NTDs originate in the first month of pregnancy, before many women know they are pregnant, it is important for a woman to have enough folic acid in her system before conception (see Figure 7.2).[138]

If a woman has already had a pregnancy affected by an NTD, it would be beneficial to take larger doses of folic acid daily. In this situation, supplementing 4,000 micrograms per day (4 milligrams) beginning at least one month before pregnancy and through the first trimester of pregnancy reduces the incidence of NTDs by about 70 percent.[30,40,69,92,107,153,166]

meiosis A process of cell division where the number of chromosomes per cell is cut in half after the egg and sperm fuse.

gametes Mature sexual reproductive cells (sperm and egg) that have a single set of unpaired chromosomes (half of the genetic material) necessary to form a complete human organism.

zygote The cell produced by the fusion of mature gametes (a sperm and an egg).

differentiation A biological process that causes a less specialized cell to become a different and more specialized cell, like when a fertilized egg engages in embryonic development, which precedes fetal development.

growth spurt A brief period of accelerated growth, noted by rapid increases in weight-for-age and height-for-age.

placenta A vascular structure that forms in the uterus of most mammals upon pregnancy that provides oxygen and nutrients to the developing fetus and transfers waste from the fetus.

fetus A term for the life form in the womb after the embryonic period.

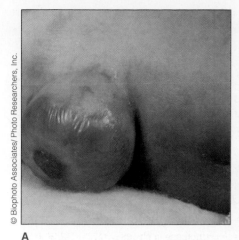

A

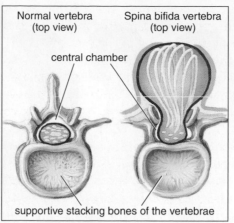

Normal vertebra (top view)   Spina bifida vertebra (top view)

central chamber

supportive stacking bones of the vertebrae

B Normally, the bony central chamber closes fully to encase the spinal cord and its surrounding membranes and fluid. In spina bifida, the two halves of the slender bones that should complete the casement of the cord fail to join.

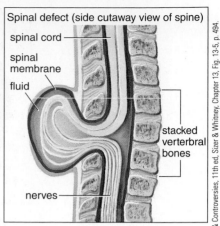

Spinal defect (side cutaway view of spine)

spinal cord

spinal membrane

fluid

stacked verterbral bones

nerves

C In the serious form shown here, membranes and fluid have bulged through the gap and nerves are exposed, invariably leading to some degree of paralysis and often to mental retardation.

Nutrition: Concepts & Controversies, 11th ed, Sizer & Whitney, Chapter 13, Fig. 13-5, p. 494.

Prenatal Nutrition

Prenatal nutrition includes Calories, building blocks from the macronutrients (proteins, carbohydrates, and fats), the essential micronutrients (vitamins and minerals), and the fiber and water that is necessary to maintain function, meet the maternal body changes that occur during pregnancy, and support fetal development. The major maternal and fetal changes that occur over the course of the 38 weeks of **gestation** are illustrated in Figure 7.3. The nutritional recommendations are based on the daily nutrient needs of both mother and fetus to stay alive, and to support the significant tissue changes made by both.[38,39,48]

Monitoring Good Nutrition During Pregnancy Many physiological changes take place in a pregnant woman's body. There are increases in the heart and respiratory rate, cardiac output, red blood cell and plasma volume, and oxygen consumption. The actual heart enlarges a little bit. With the need to support another life, there is an increase in appetite, thirst, metabolism, and body temperature. The kidneys filter waste more rapidly and retain more sodium, resulting in an increase in body water. The combination of increased body water, body fat, and lean body mass translates to maternal weight gain. Good nutrition is needed to support appropriate maternal weight gain as well as fetal growth and development. There are clear weight gain recommendations based on the woman's body mass before pregnancy, and there are clear markers of normal embryonic development.[16]

Maternal Weight Gain Maternal weight gain is a common method to monitor adequate nutrition of both mother and baby. Depending on the preconception weight and height of the mother, a recommended amount of weight to gain during the pregnancy is made. As seen in Table 7.2, an underweight woman (BMI less than 18.5) may gain 28 to 40 pounds (about 13 to 18 kilograms) during the pregnancy. A normal-weight woman (BMI 18.5 to 24.9) may gain 25 to 35 pounds (about 11 to 16 kilograms) during the pregnancy, as seen in Figure 7.3. Overweight women (BMI 25 to 29.9) should gain less weight during the pregnancy, 15 to 25 pounds (about 7 to 11 kilograms). Obese women (BMI 30 or greater) should control weight gain to an 11- to 20-pound increase (about 5 to 9 kilograms).[38,48,49,71,160]

FIGURE 7.2 Spina bifida, a neural tube defect. A: A photograph of a spina bifida lesion on a baby's back. B: The transverse section of the spinal vertebrae. C: The sagittal section of the spinal column.

gestation The period from conception to birth, usually noted in weeks when a woman carries a developing fetus in her womb.

A developing fetus between 32 and 36 weeks' gestation.

ancroft9/Shutterstock.com

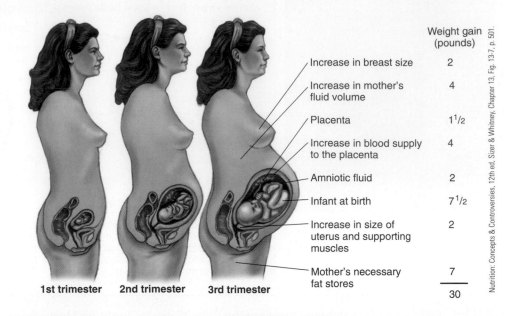

	Weight gain (pounds)
Increase in breast size	2
Increase in mother's fluid volume	4
Placenta	1½
Increase in blood supply to the placenta	4
Amniotic fluid	2
Infant at birth	7½
Increase in size of uterus and supporting muscles	2
Mother's necessary fat stores	7
	30

1st trimester 2nd trimester 3rd trimester

Nutrition: Concepts & Controversies, 12th ed, Sizer & Whitney, Chapter 13, Fig. 13-7, p. 501.

FIGURE 7.3 Weight gain during the first, second, and third trimesters of pregnancy.

TABLE 7.2

Recommended Weight Gain for Pregnancy

Prepregnancy BMI[1]	Pregnancy BMI[2]	Recommended Weight Gain
Underweight (BMI < 18.5)	Underweight (BMI < 19.8)	28–40 pounds
Normal weight (BMI < 25)	Normal weight (BMI < 26)	25–35 pounds
Overweight (BMI 25–29.9)	Overweight (BMI 26–29)	15–25 pounds
Obese (BMI > 30)	Obese (BMI > 29)	11–20 pounds

[1]Food and Nutrition Board, subcommittee on nutrition status during pregnancy and lactation.

[2]National Heart Lung and Blood Institute expert panel on identification, evaluation, and treatment of overweight and obesity in adults.

© Cengage Learning 2013

Fetal Growth and Development   The first eight weeks of gestation is considered the period of embryonic development. From week nine of gestation to birth is considered the period of fetal development. The main weekly changes taking place during fetal growth and development are illustrated in Figure 7.4.[H,OO]

The placenta forms from the mother's endometrium, the lining of the **uterus**. It begins to develop at two weeks of pregnancy and forms the site of embryo implantation. Although no mixing of blood occurs from mother to fetus, the placental membrane does allow for the transfer of substances, including medicines, alcohol, and food contaminants. Both nutrients and oxygen travel from the mother to the fetus through the placenta; in turn, metabolic waste from the fetus is excreted through the mother's kidneys and lungs.[H,OO]

The first six weeks of gestation is a period of rapid cellular differentiation and consequently a period where many defects in development can have lasting consequences. A period when tissues change and organs grow and develop rapidly is called a **critical period**. So naturally, the embryonic development time period is a critical period when any negative influence can have severe consequences. These negative influences, called **teratogens**, include alcoholic beverages of any

uterus A female muscular organ in the reproductive system in which the fetus develops until birth; also known as the womb.

critical period A crucial stage in human development when tissues change and organs grow and develop rapidly.

teratogens Any substances or agents that can interfere with normal embryonic development by causing malformations or defects.

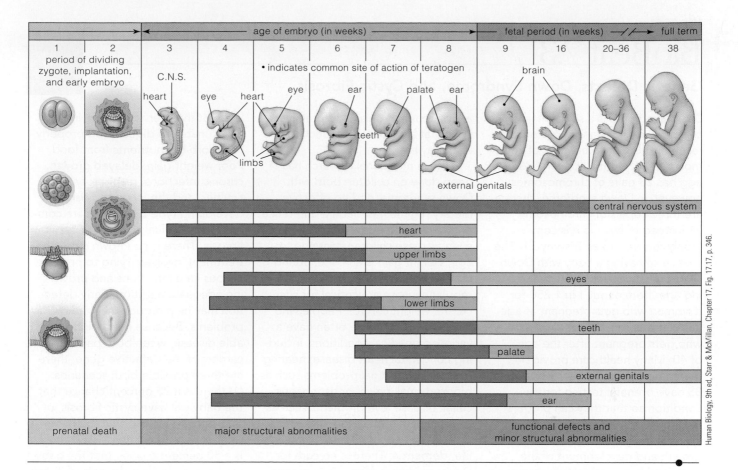

Human Biology, 9th ed, Starr & McMillan, Chapter 17, Fig. 17.17, p. 346.

FIGURE 7.4 Fetal growth, development, and suscepti- bility to birth defects.

sort, nutrient deficiencies such as folic acid, or the use of a vitamin A derivate, **isotretinoin** (Accutane™). Illicit drugs, medications, tobacco, and Calorie depri- vation can also negatively influence the developing embryo.[DD,31]

The children of pregnant women who are exposed to environmental agents such as **polycyclic aromatic hydrocarbons (PAHs)** may be predisposed to develop- ing cancers of the lung, liver, and skin. It is estimated that 3 percent of **birth defects** result from maternal exposure to teratogens. In addition, teratogens may play a role in increasing the rate of **miscarriages**, which may be as high as 50 percent of recog- nized and unrecognized pregnancies. Fetal exposure to the class of chemicals known as endocrine disruptors is believed to have potentially adverse effects on normal development and reproductive capability of the next generation. Incidental common endocrine disruptors include pesticide residues, steroid compounds in animal meat, and oral contraceptive agents. Exposure to some common teratogens, such as alcohol, cocaine, and prescription drugs, during pregnancy can be reduced or eliminated. In the case of **epileptics,** who depend on the medication phenytoin to control seizures, alternative drugs need to be considered during pregnancy in order to protect the fetus. Exposure to environmental teratogens is more difficult to control. However, certain foods in the diet may provide protection. The phytochemical indole-3-carbinol is abundant in cabbage family vegetables and green tea. However, other polyphenols may provide protection against PAH carcinogenesis by affecting maternal and/or fetal enzymes that metabolize the carcinogen or act as antioxidants.[31,38,39,48]

Fetal growth and development are monitored during pregnancy by measuring the uterus, listening to fetal heart tone, and taking ultrasounds. If the baby is at risk for genetic defects, genetic testing may be performed during the pregnancy, usually by **amniocentesis,** to discover genetic defects such as **Down syndrome** or **cystic fibrosis** (see BioBeat 7.3).[FF,LL,33,87]

isotretinoin An oral medication, chemically related to vitamin A, used for the treatment of moderate to severe acne; causes severe birth defects if taken dur- ing early pregnancy.

polycyclic aromatic hydrocar- bons (PAHs) Potentially toxic, organic compounds that contain only carbon and hydrogen in their chemical formula, but have at least two fused, six-sided rings in the chemical structures.

birth defects Permanent abnor- malities in an infant's structure or function detected at birth or shortly thereafter.

miscarriage The spontaneous ending of a pregnancy that occurs when the embryo or fetus is incapable of surviving in utero.

epileptics Individuals that have chronic neurological disorders characterized by a variety of recurrent, unprovoked seizures.

amniocentesis A prenatal diag- nostic procedure usually done between 16 and 20 weeks of a pregnancy to test the genetic traits of fetal cells in the amniotic fluid.

BioBeat 7.3

Genetic Defects, Down Syndrome, and Cystic Fibrosis

Down syndrome is a set of mental and physical signs and symptoms resulting from an extra copy of chromosome 21. Normally, a fertilized egg has 23 pairs of chromosomes; however, in Down syndrome, there are three copies of chromosome 21 instead of two, so it is commonly referred to as Trisomy 21. The chance of having a baby with Down syndrome increases with advancing age, from about 1 in 1,250 for a woman who gets pregnant at age 25 to about 1 in 100 for a woman who gets pregnant after the age of 40. Many healthcare providers recommend that women over age 35 have prenatal testing for the condition so that they can prepare for the child's special needs. Down syndrome significantly alters the growth and development of the body and brain; symptoms can range from mild to severe. Down syndrome people can be recognized by several physical signs, including a flat face with an upward slant to the eyes, a short neck, abnormally shaped ears, a deep crease in the palm of the hand, white spots on the iris of the eye, poor muscle tone, loose ligaments, and small hands and feet. Most people with Down syndrome have intellectual quotients (IQs) that fall in the mild to moderate range of intellectual and developmental disabilities, thus limiting intellectual abilities and adaptive behaviors such as conceptual, social, and practical skills that are needed to function in everyday life.

Much nutritional research has been done on children born with Down syndrome, and no effective dietary regime has made a difference on the children's development. However, as these children continue to age, weight gain is an issue. Sound dietary practices, Calorie control, and regular exercise are critical for successful weight control. People who have Down syndrome often have a variety of medical conditions, including congenital heart disease, hearing problems, intestinal problems such as blocked small bowel or esophagus, celiac disease, eye problems such as cataracts, thyroid dysfunctions, poor skeletal development, and in later life, **dementia**. There is no cure for Down syndrome, and management of the disorder addresses the medical issues that continue to develop throughout life.[LL,87,139]

Cystic fibrosis is an inherited chronic disease that affects the lungs and digestive system. A defective gene (CF gene) produces a protein product that causes the body to produce unusually thick, sticky mucus that clogs the lungs and obstructs the pancreas. These conditions lead to life-threatening lung infections and obstructions of the pancreas that prevent the release of digestive enzymes into the small intestine. Without these enzymes, the individual is unable to digest food properly or absorb the nutrients from food. Poor weight gain, delayed growth, chronic infections, respiratory difficulties, and the increased risk for developing type 2 diabetes are common to individuals born with cystic fibrosis. There is no known cure, and treatment involves trying to prevent serious lung infections and providing adequate nutrition. Early detection may help reduce the medical problems. Because it is an inheritable disease, when both parents are carriers of the defective gene, there are three possible birth scenarios: (1) there is a 25 percent chance that the baby will have cystic fibrosis, or a 75 percent chance that the baby will *not* have cystic fibrosis; (2) there is a 50 percent chance that the baby will be a carrier, but the baby will not have cystic fibrosis; and (3) there is a 25 percent chance that the baby will not have an abnormal copy of the CF gene, thus the inheritable trait is not present in the baby. Amniocentesis can be performed between 15 to 20 weeks of pregnancy to determine the presence of the defective CF gene. In any case, if both parents are carriers of the defective CF gene, newborn screening will be done.[S,FF,LL,33,97]

What is the role of nutrition in Down syndrome and cystic fibrosis?

Down syndrome A birth defect that is characterized by a variety of mental and physical signs and symptoms that exist throughout life, resulting from an extra copy of chromosome 21.

cystic fibrosis An inherited chronic disease that causes the body to produce thick mucus that obstructs the lungs and pancreas, leading to respiratory and digestive difficulties.

dementia The deterioration of cognitive function.

Birth Size, Nutrition, and Health Outcomes Critical periods take place at various points of pregnancy. The developing fetus has a specific need at each critical period, and the environmental conditions should be optimal to support the fetus. If the nutritional needs of the fetus are not met, the result can be low birthweight or premature birth.[2]

Birthweight is the first weight of the newborn measured immediately after birth. Birthweight of less than 5.5 pounds, or 2,500 grams, is considered low birthweight. A low-birthweight infant can be born too small as a result of intrauterine growth retardation, being born prematurely (before 37 weeks' gestation), or both. A low-birthweight baby that was born full term is a small for gestational age (SGA) baby. A low-birthweight baby that was born three weeks or more before the full-term 40-week gestation due date is a premature ("preemie") baby. Compared to infants of normal weight, low-birthweight infants may have an increased risk for illness through the first six days of life, infections, and impaired motor and social development and learning disabilities.[H]

Premature low birthweight has been—but is not necessarily—associated with exposure of the mother to lead, solvents, pesticides, PAHs, and air pollution during pregnancy. Exposure to these and other harmful substances may not always be avoidable. However, other avoidable risk factors that cause low-birthweight babies include maternal use of alcohol, low prenatal weight gain, and maternal smoking. Smoking during pregnancy can also result in placenta detachment, miscarriage, long-term disabilities for the baby, and premature birth.[BB,JJ]

Preemies have more health problems compared to SGA babies; the more premature the baby, the more severe his or her health problems are likely to be. Preemies have different growth charts to track their growth progress back to the standard growth charts.[71,139]

Energy and Nutrient Needs Nutrient requirements during pregnancy are increased to support all of the maternal and fetal changes. The nutrient increases are illustrated in Figure 7.5 on page 357. The macronutrient requirements to support tissue changes are significant, as are the increased micronutrient needs required to support metabolism and structure. The macronutrient recommendations are generally made for the overall pregnancy; however, many of the micronutrient recommendations are made according to the trimester of the pregnancy.[69]

Calories Energy needs during pregnancy are determined using the EER equations. The age- and gender-specific formulas are found in Appendix A. Calorie needs can be calculated based on body mass and physical activity. There are no additional Calorie needs during the first trimester. However, during the second trimester, the average daily increase in caloric intake is 340 Calories above the first trimester calculated EER, and 452 more Calories during the third trimester.

When a pregnant woman consumes too few Calories, a baby of low birthweight and low nutritional status is often the outcome of the pregnancy. This in turn can lead to lifelong risks in the developing fetus, including chronic diseases, obesity, short stature, lower intelligence, and other brain-functioning problems.[71]

When a pregnant woman consumes too many Calories, her baby is often born larger than normal and with more medical and surgical assistance needed at delivery. This baby is at greater risk for having heart and neural tube defects. The woman is at greater risk for diabetes (gestational, type 2), hypertension, and infections with too great of a weight gain during the pregnancy.[H,J,R,2]

When a woman is healthy, has achieved a healthy body weight, and is nourished (has adequate nutrient stores in her body) before becoming pregnant, then the best environment is created and available to support her baby. This starts with good general nutrition and a healthy uterus and progresses to forming a healthy placenta and fetus. Thus, good nutrition is critical before and during pregnancy for a good birth outcome.[2]

Carbohydrate, Protein, and Fat The DRI for carbohydrate during pregnancy is increased from 130 to 175 grams per day minimally. Carbohydrate is needed especially for brain, central nervous system, and red blood cell functioning. The AMDR for carbohydrate is the same as for adults, 45 to 65 percent of Calories, with less than or equal to 25 percent of Calories from sugars. Fiber-rich sources are needed to promote gastrointestinal health and prevent obesity, type 2 diabetes, heart disease, and some cancers. The DRI is personalized at 1.4 grams per 100 Calories consumed. Based on the reference person, the DRI for a pregnant woman, no matter what the trimester, is 28 grams per day. Fiber intake is increased by eating more whole fruits, vegetables, whole grains, and legumes.[69]

The DRI for protein is 1.1 grams per kilogram of body weight during pregnancy. The total gram amount of protein required will increase with the weight gain during the pregnancy. Consuming adequate protein is very important for fetal development, maternal tissue changes, normal metabolism, tissue repair, and lean body mass maintenance.[38,39,48]

There is no DRI for total fat. However, there is a DRI for the essential fatty acids linoleic acid and alpha-linolenic acid. The DRI for linoleic acid is increased to 13 grams per day, while the DRI for alpha-linolenic acid is increased to 1.4 grams per day, compared to adult females (see Appendix A). Linoleic acid can be converted to **arachidonic acid** (AA), while alpha-linolenic acid can be converted to eicosapentaenoic acid (EPA) and docosahexaenoic acid (DHA). The AMDR for total fat is the same as for adults at 20 to 35 percent of Calories, with the AMDR for the essential fatty acids being slightly higher compared to female adults. Consuming less fast food and processed food and more whole plant foods (including nuts and seeds) and fish will support adequate fatty acid intake. Standard prenatal vitamins do not include omega-3 fatty acids, which help promote so many aspects of the baby's central nervous system development. If the pregnant mother does not choose to consume fish or other foods high in omega-3 fatty acids, including EPA and DHA, a fatty acid supplement in addition to prenatal vitamins is beneficial (see BioBeat 7.4).[26,68,114]

Vitamins, Minerals, and Water   The necessary nutrients to support the pregnant mother and developing fetus can be consumed in a well-balanced diet by following the MyPlate recommendations, if foods that are rich in these nutrients are selected and consumed. Most healthcare providers recommend taking a prenatal vitamin and mineral supplement during the entire pregnancy and continuing to consume a healthy diet. Most prenatal vitamin and mineral supplements contain higher amounts of many of the essential nutrients compared to the DRIs. The essential micronutrients are required for new tissue formation for both maternal and fetal purposes. Thus, most of the DRIs for the vitamins and minerals are higher during pregnancy and **lactation** compared to the adult female values (see Figure 7.5). Some of these nutrients—folate, iron, vitamin A, calcium, iodine, zinc, and water—are explored in more detail below.[32,92,95,107]

arachidonic acid An omega-6 long-chained fatty acid that is 20 carbons long with four double bonds and can be made in the body from linoleic acid.

lactation The production and secretion of milk by the mammary glands; also known as breastfeeding or lactogenesis.

BioBeat 7.4

DHA, the Central Nervous System, and Cognition

Two polyunsaturated fatty acids, arachidonic acid (AA) and docosahexaenoic acid (DHA), are critical for embryonic, fetal, and infant central nervous system (CNS) growth and development. AA is involved in cellular metabolism, while DHA is incorporated in retinal and brain membrane phospholipids, where it is critical for visual and neural function and neurotransmitter metabolism. AA is easily made in the body from omega-6 fatty acids, which can be abundant in the diet. However, omega-3 fatty acids, and especially DHA, are difficult to consume in the diet if fish is not regularly eaten. During the last trimester of a pregnancy, the fetus accumulation of omega-3 fatty acids is particularly great. Many studies done in human models show that prenatal consumption of omega-3 fatty acids improves birthweight as well as visual and cognitive function of the infant. About 3 grams consumed in the maternal diet per day (one-third eicosapentaenoic acid [EPA] and two-thirds DHA) has been associated with infant development benefits. Because only about 4 to 11 percent of DHA is retroconverted to EPA in the adult body, pregnant women who just take DHA supplements, without any dietary EPA, may be unable to produce the right balance of eicosanoids and may limit the transport and uptake of DHA into fetal cells. Postnatal DHA has been shown to benefit infants' CNS development and intelligence. Mothers who breastfeed need to continue to include omega-3 fatty acids in their diets. As for formula-fed infants, most infant formulas include DHA in the formulation.[1,23,40,65,68,72,114,122]

What are the benefits and sources of DHA during and after pregnancy for the baby?

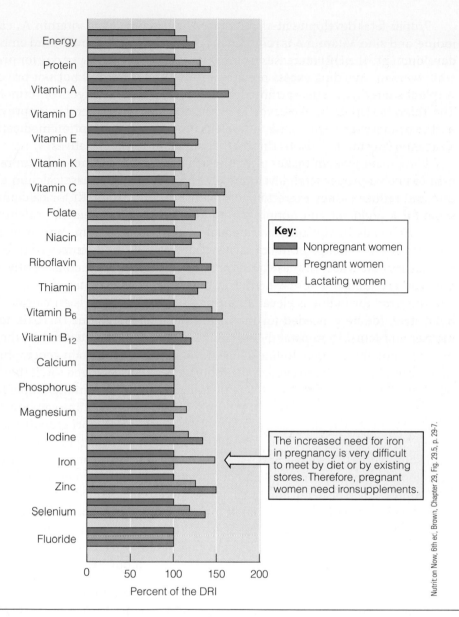

The increased need for iron in pregnancy is very difficult to meet by diet or by existing stores. Therefore, pregnant women need iron supplements.

Nutrition Now, 6th ed., Brown, Chapter 29, Fig. 29.5, p. 29-7.

FIGURE 7.5 Comparison of Dietary Reference Intakes (DRIs) for nonpregnant, pregnant, and lactating women.

Pregnancy requires dramatic new cell synthesis, so there is an increased need for folate and iron. Folate plays a critical role in new cell synthesis as well as development of the nervous system, which is well underway by the third week of pregnancy. To support this role, a pregnant woman needs to increase her consumption of folate from 400 to 600 micrograms per day. Excellent food sources of folate are dark green leafy and raw vegetables, orange juice, and whole and enriched grains. Highly fortified breakfast cereals providing 100 percent of the Reference Daily Intake for folic acid can provide this nutrient, which is associated with preventing spina bifida, a neural tube defect (see Figure 7.2 on page 351). The DRI for iron increases rather dramatically, from 18 to 27 milligrams per day, during pregnancy because of the need to generate a large number of new red blood cells to support the growth and development of the placenta and fetus. The red blood cells contain hemoglobin, a protein complex that contains iron. Oxygen binds to the iron component of hemoglobin and serves as the delivery protein carrier of oxygen to all cells in the body. Iron can cross the placenta to meet the fetal iron and oxygen delivery needs to support life. Excellent food sources of heme iron are clams, oysters, organ meats, and red muscle meats. Less bioavailable non-heme iron can be found in fortified cereals, beans, pumpkin seeds, and spinach. The absorption of non-heme iron is increased by vitamin C when consumed together.[CC,38,39,48,166]

Proper fetal development also requires higher intakes of vitamin A, calcium, iodine, and zinc. Vitamin A is required for gene expression, growth, and embryonic development. The DRI increases from 700 to 770 micrograms per day for pregnant adult women. Avoiding excess previtamin A (animal sources) but not provitamin A (plant sources) may be warranted to prevent teratogenic effects of retinoic acid. The Tolerable Upper Intake Level (UL) is 3,000 micrograms per day for previtamin A. The primary teratogenic risk, however, comes from prescription medications to treat acne (Accutane™, isotretinoin) that are vitamin A derivatives.

A pregnant woman makes physiological adjustments to her calcium metabolism to enable proper fetal development. The body will increase calcium absorption and reduce kidney excretion automatically. Thus, the DRI for calcium is the same for a pregnant and nonpregnant adult woman. To meet the calcium DRI, a woman needs to consume an adequate amount of calcium- and protein-rich foods each day—4 cups of milk (cow, soy, or yogurt) or 12 ounces of sardines or canned salmon. Many of these foods are difficult to consume consistently, so taking a dietary supplement may be the best way to meet the dietary need.

The need for iodine is elevated during pregnancy, which is an elevated metabolic state. Iodine is needed for the major metabolic hormone thyroxin for both mother and fetus. If prenatal dietary iodine levels are inadequate, then the infant may be mentally retarded. Iodine deficiency is common in certain geographic locations around the world, so expecting mothers need to use iodized salt if they do not regularly consume seafood. By selecting iodized salt or consuming ample seafood (at least two 4-ounce servings per week), adequate intake of iodine can be achieved.

Zinc has catalytic, structural, and regulatory roles to support growth and development. It functions as a cofactor for more than 100 different enzymes and is required for insulin synthesis, storage, and release. Thus, the DRI for zinc increases from 8 to 11 milligrams per day for pregnant adult women. Zinc can be consumed from the food supply in shellfish, red and organ meats, and some fortified cereals.[10,11,30,31,66]

Adequate water intake that maintains positive water balance will support metabolic waste removal, blood volume expansion, cardiovascular function, thermal regulation, and metabolism. The DRI for water intake is higher for pregnant women than for nonpregnant adult females. It is 3 liters (or about 3 quarts) per day.

Diet and Health Issues in Pregnancy Prenatal monitoring by a healthcare provider improves birth outcome. Pregnant women should learn as much as possible about the health risks associated with pregnancy and take responsibility for controlling any risk for jeopardizing the growth and development of the unborn child. Diet-related health issues addressed in this section include nausea, vomiting, constipation, heartburn, pica, gestational diabetes, preeclampsia, fetal alcohol syndrome, and the dilemma of consuming fish because of the fear of **methylmercury** toxicity.[TT,2,38,39,48]

Nausea, Vomiting, Constipation, and Heartburn Nausea and vomiting can occur early in the pregnancy, usually between gestational weeks 5 to 12. Approximately 70 percent of pregnant women will experience nausea, while 40 percent may experience nausea and vomiting. The exact cause is unknown, though hormonal changes are most likely the contributing factors. When the nausea and vomiting are not severe and do not persist past the first trimester, they are considered to be a normal part of pregnancy. When nausea and vomiting occur in the morning, they are commonly referred to as morning sickness. However, it is normal to experience nausea and vomiting throughout the day. The best way to control nausea and vomiting is by avoiding foods and beverages that are known triggers and separating the intake of solids and liquids. It is also helpful to keep something in the stomach by eating small, frequent meals. Consuming saltine crackers between meals may help.

methylmercury An organic form of mercury, created from metallic or elemental mercury by bacteria in sediments, that is easily absorbed into the living tissue of aquatic organisms and can bioaccumulate up the marine food chain.

Later in the pregnancy, constipation and heartburn become common complaints. Consuming enough fluids and increasing dietary fiber content promotes normal bowel movements. Iron supplements commonly cause constipation, so taking smaller doses of iron supplements multiple times per day can alleviate this problem. To control heartburn, eating small, frequent meals throughout the day that are low in fat is beneficial. Furthermore, avoiding foods that relax the lower esophageal sphincter (such as chocolate, caffeinated beverages, tomato products, citrus fruits, mustard, mint, and of course alcoholic beverages), along with sitting upright for an hour after eating, will reduce heartburn.[H]

Pica **Pica** is a term used to describe the consumption of non-nutritional foods such as starch and clay. When clay and dirt are consumed, it is also called geophagy. Pica is not well understood, but is most common in individuals with developmental disabilities, such as autism and mental retardation. However, eating starch frequently occurs in pregnancy, possibly as a treatment of morning sickness or as a manifestation of iron deficiency. Pica may also occur in adults who crave a certain texture in their mouth. The causes of pica frequently point to either underlying deficiencies of iron or zinc, or lead toxicity.[132,145] Inadequate iron or zinc nutritional status may trigger specific cravings for non-nutritional substances. Additionally, there may be cultural and familial factors associated with pica. Clay or laundry starch ingestion is regarded as acceptable by various social groups. Clay and starch eating are seen in the United States in some rural Southern African American communities, primarily among women and children.

Treatment for pica should first correct any nutrient deficiencies or address any other medical problems, such as lead exposure. For individuals with developmental disabilities, family education and behavior modification approaches may be required, such as associating the pica behavior with mild aversion therapy, followed by positive reinforcement for eating the right foods. Medications may help reduce the abnormal eating behavior, if pica occurs as part of a developmental disorder such as mental retardation.[16]

Gestational Diabetes **Gestational diabetes** is a transient type of diabetes that occurs in some pregnant women who did not have preexisting diabetes. Although gestational diabetes usually goes away after pregnancy, many women who experience gestational diabetes develop type 2 diabetes later in life, so for those who develop gestational diabetes, it is important to monitor blood sugar throughout life.

Routinely at the beginning of the third trimester, a modified fasting glucose tolerance test should be administered to all pregnant women to detect elevated blood sugar levels or rule out gestational diabetes (see Module 3, page 134). Maternal high blood sugar levels negatively impact the health of the developing fetus, promote excess growth of the fetus, and cause large-birthweight babies. Large babies are difficult to carry to term and deliver. Women who develop gestational diabetes must monitor their blood sugar daily. Good blood sugar control is the key to preventing problems during pregnancy or birth.

Treatment for gestational diabetes begins with making better dietary choices and incorporating regular exercise. Consuming several small meals of whole, fresh foods that have a low glycemic index and practicing Calorie control are great steps toward complementing an activity program (see Module 3, page 135). At least 30 minutes per day of moderate, low-impact exercise (like walking or water exercise during pregnancy) improves insulin sensitivity, which improves blood sugar control. Often, these lifestyle changes will keep the blood sugar level within a normal range. If lifestyle changes do not achieve acceptable blood sugar levels, then insulin administration may be required.[H,J,84,139]

pica A term used to describe people who persistently consume nonnutritional foods such as starch and clay.

gestational diabetes A transient type of diabetes that occurs in some pregnant women who did not have preexisting diabetes.

Preeclampsia　Several terms are used to describe hypertension during pregnancy: **preeclampsia**, toxemia, and pregnancy-induced hypertension (PIH). Toxemia is an older term based on a belief that the condition was the result of toxins (poisons) in the blood. PIH is a newer term that clearly describes the medical issue. However, preeclampsia is the umbrella term to cover all variants of hypertension during pregnancy.

Prevention strategies for preeclampsia are unclear because the etiology is unknown, even though several mechanisms for the **pathology** have been identified, stemming from both parents. Living a healthy lifestyle with adequate rest and exercise and consuming a well-balanced, antioxidant-rich diet have been shown to prevent hypertension. Additionally, calcium, magnesium, potassium, and fish oil supplementation, controlling excess dietary sodium intake, and daily aspirin administration during pregnancy appear to be beneficial in reducing blood pressure and its complications.

A successful birth outcome is always a healthcare goal during pregnancy. Proper prenatal care and the timing of the delivery of the baby are of utmost importance. Blood pressure monitoring as a part of regular prenatal care can achieve early detection, which is key to successful management of preeclampsia. Pharmacotherapy or prescription drug treatment is the medical intervention for the treatment of preeclampsia. Untreated preeclampsia increases the risk for heart disease, stroke, and type 2 diabetes later in life for the mother and restricted growth and respiratory distress for the baby. Besides having high blood pressure (hypertension, which can be asymptomatic), the pregnant woman with preeclampsia may also experience nausea, headache, blurred vision, and certain abnormal urine and blood tests.[H,2,15]

Alcohol Avoidance: Fetal Alcohol Syndrome　**Fetal alcohol syndrome (FAS)** is a group of problems in children born to mothers who drank alcohol during their pregnancy. FAS is completely preventable if alcohol abstinence is practiced. Alcohol consumption is most capable of producing fetal malformation between 15 and 25 days post-conception, a time when most women do not know that they may be pregnant. Three servings of alcoholic beverages during this period have caused FAS. Moderate to heavy alcohol drinking during this time of embryonic development results in a wide variety of lifelong physical and developmental issues for the baby.

FAS babies are usually small for gestational age and underweight. They often have a marked, visible characteristic of the head, forehead, nose, jaw, eyes, ears, and lips, as illustrated in Figure 7.6. They frequently are born with a variety of other birth defects, such as heart defects and vision or hearing problems and delayed development. Behavior problems often develop as they grow older. The most serious problem is mental retardation. First, the diagnosis of FAS is difficult and subjective. It is made by the healthcare provider based on his or her judgment, given some objective findings and a subjective impression. The best time for recognizing the syndrome is the period from 3 to 8 years, because the distinct pattern of FAS facial abnormalities is fully expressed. Additionally, cognitive and behavioral delays are clearly manifested. The practical diagnosis depends on knowledge of the mother's alcohol consumption during pregnancy. Occasional light drinking patterns, which are not on a daily or continuous basis during pregnancy, are not likely to increase the risk of FAS in offspring. However, caution is advised for daily drinkers. A safe level of alcohol consumption during pregnancy has not yet been established, so it is best to avoid drinking alcohol altogether. Heavy drinking during pregnancy (five drinks or more daily) places the fetus at high risk for FAS.[H,QQ,2,21,139]

Fish and Methylmercury　Methylmercury is a toxic, organic form of mercury, which is a heavy metal found mostly in water, soil, plants, and animals. Methylmercury is made by some bacteria from inorganic mercury through metabolic

preeclampsia An umbrella term used to refer to all variants of hypertension or high blood pressure during pregnancy.

pathology The scientific and medical study of disease.

fetal alcohol syndrome (FAS) A group of physical and mental problems in children born to mothers who drank alcohol heavily during their pregnancy; best diagnosed from ages 3 to 8.

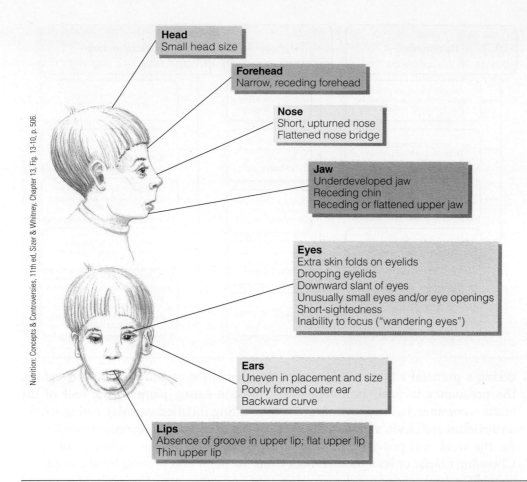

Head
Small head size

Forehead
Narrow, receding forehead

Nose
Short, upturned nose
Flattened nose bridge

Jaw
Underdeveloped jaw
Receding chin
Receding or flattened upper jaw

Eyes
Extra skin folds on eyelids
Drooping eyelids
Downward slant of eyes
Unusually small eyes and/or eye openings
Short-sightedness
Inability to focus ("wandering eyes")

Ears
Uneven in placement and size
Poorly formed outer ear
Backward curve

Lips
Absence of groove in upper lip; flat upper lip
Thin upper lip

FIGURE 7.6 Facial characteristics of fetal alcohol syndrome (FAS).

processes. Methylmercury contaminates water such as lakes, rivers, and streams and bioaccumulates, particularly in the fish food chain. Pregnant and breastfeeding women are advised of the health risks to their offspring associated with consuming high dietary levels of methylmercury. Methylmercury crosses the placenta and can be found in the baby's blood at levels higher than those in the mother. The baby's brain is the organ most sensitive to the effects of methylmercury exposure. The birth defects associated with methylmercury toxicity include small head size, cerebral palsy, developmental delay and/or mental retardation, blindness, muscle weakness, and seizures. Maternal blood and hair can be tested to determine exposure to methylmercury. Blood tests are better for detecting methylmercury immediately following exposures. Tests of hair may be able to detect long-term exposure.[W,DD,16]

Eating fish and seafood regularly is strongly recommended (at least two 4-ounce servings of low methylmercury-containing seafood servings per week). Good choices are shrimp, salmon, pollock, catfish, scallops, and sardines. Fish that are large and have long life spans are more likely to contain higher amounts of methylmercury than are small fish with short life spans. The following large fish have the highest levels of methylmercury and should be avoided during pregnancy: shark, swordfish, king mackerel, and tilefish.[38,39,48]

MyPlate for Pregnancy The MyPlate food guidance system is based on the Dietary Guidelines 2010 (see Module 2, page 61). Pregnant women can go to www.choosemyplate.gov and follow the links under Specific Audiences, Pregnant & Breastfeeding. The major MyPlate themes for pregnancy include visit your healthcare provider regularly, get your own daily food plan for moms, and include nutrient-rich foods from each basic food group (see Figure 7.7). The prenatal modifications for the consumption of the additional MyPlate serving equivalents are designed to meet the increased nutritional needs of mother and baby; however,

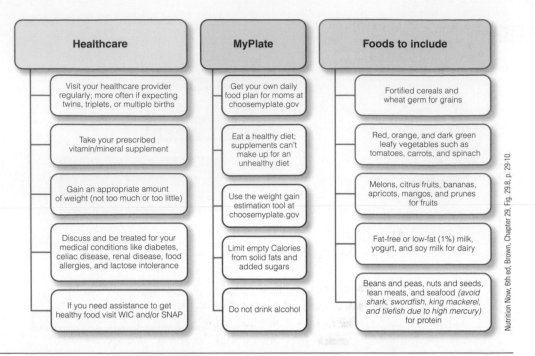

Healthcare	MyPlate	Foods to include
Visit your healthcare provider regularly; more often if expecting twins, triplets, or multiple births	Get your own daily food plan for moms at choosemyplate.gov	Fortified cereals and wheat germ for grains
Take your prescribed vitamin/mineral supplement	Eat a healthy diet; supplements can't make up for an unhealthy diet	Red, orange, and dark green leafy vegetables such as tomatoes, carrots, and spinach
Gain an appropriate amount of weight (not too much or too little)	Use the weight gain estimation tool at choosemyplate.gov	Melons, citrus fruits, bananas, apricots, mangos, and prunes for fruits
Discuss and be treated for your medical conditions like diabetes, celiac disease, renal disease, food allergies, and lactose intolerance	Limit empty Calories from solid fats and added sugars	Fat-free or low-fat (1%) milk, yogurt, and soy milk for dairy
If you need assistance to get healthy food visit WIC and/or SNAP	Do not drink alcohol	Beans and peas, nuts and seeds, lean meats, and seafood *(avoid shark, swordfish, king mackerel, and tilefish due to high mercury)* for protein

Nutrition Now, 6th ed, Brown, Chapter 29, Fig. 29.8, p. 29-10.

FIGURE 7.7 The MyPlate food guidance system core themes for pregnancy.

taking a prenatal vitamin and mineral supplement is recommended throughout the pregnancy, as well as meeting the MyPlate eating plan. Eating half of the grain recommendation as whole grains (including fortified cereals), eating whole, antioxidant-rich fruit, and meeting all of the vegetable subgroup recommendations for the week will provide the needed fiber and many of the required nutrients. Choosing nonfat or lean food choices from the dairy and protein food groups will provide several needed nutrients and protein, and will control the intake of saturated fatty acids and cholesterol. Choosing nuts, seeds, and plant oils will provide the needed essential fatty acids. Controlling the intake of empty Calories will increase the nutrient density of the diet and prevent excess weight gain. MyPlate serving equivalents can be explored in Appendix C. Remember, the MyPlate eating plan encourages the intake of whole and fresh plant foods, low-fat animal foods, and liquid oils; it discourages the intake of empty Calories from solid fats and sugars. Furthermore, daily exercise and fluid balance are needed.[XX,YY,84,139]

Prenatal Fitness and Lifestyle Management Pregnancy is a time of many body changes caused by hormonal shifts and the physiological and anatomical changes to accommodate fetal development. The physiological and psychological demands, the energy demands, and the fatigue of pregnancy can all be somewhat offset with better physical fitness and time set aside to exercise. Eating well, getting regular exercise, and getting adequate rest provide a triad of wellness that is valuable for a healthy pregnancy. Exercising during pregnancy will reduce the loss of muscular and cardiovascular fitness, prevent excessive maternal weight gain, and reduce the risk of gestational diabetes, preeclampsia, varicose veins, deep vein thrombosis, and complaints about low back pain, as well as the poor psychological adjustment to the physical changes of pregnancy. In the absence of medical or obstetric complications, 30 minutes or more of moderate-intensity physical activity every day is safe to do. Pregnant women should avoid activities that have a high risk of falling, causing abdominal trauma, or putting stress on the joints. In summary, exercise during pregnancy can help reduce the common discomforts of pregnancy and physically prepare the mother for childbirth (labor and delivery). Exercising after the birth will hasten the recovery process.[H,41,115]

Before an exercise program is initiated, it is a good idea to consult with a healthcare provider to ensure that exercise can be conducted safely. If you already exercise regularly, your doctor will probably tell you to continue your current regular exercise. A total fitness program is encouraged (see Module 4, page 190). After the first trimester of pregnancy, avoid doing any exercises on your back. Avoid brisk exercise in hot, humid weather or if you have a fever. When the core body temperature of the mother increases, stress on the baby is seen by the increase in fetal heart tone. Wearing light-colored, comfortable clothing will reduce thermal stress. A sports bra will support and protect the breasts. The consumption of plenty of fluids will ensure optimal thermal regulation and reduce the risk of overheating and dehydration. Exercising to the point of exhaustion is not recommended. During the third trimester, water exercise may be the key to being comfortable while exercising. The ability to keep cooler and not have the weight-bearing impact on the body is a wonderful combination to maintain the enjoyment of exercising. Warning signs that you should stop exercising and immediately contact your healthcare provider include vaginal bleeding, dizziness or feeling faint, increased shortness of breath, chest pain, headache, muscle weakness, calf pain or swelling, uterine contractions, decreased fetal movement, or fluid leaking from the vagina.[H,I,U,41,115]

Summary Points

- Celiac disease, PCOS, and preconception nutrition affect fertility, conception, implantation, and the development of the embryo; taking a folic acid supplement is recommended.
- The fertility diet is low in *trans* fatty acids and saturated fatty acids, higher in monounsaturated fatty acids, vegetable proteins, fiber, and low-glycemic-index carbohydrate sources, and adequate in all of the essential vitamins, minerals, amino acids, fatty acids, and water.
- The prenatal DRIs address the increased maternal and fetal nutritional needs to ensure proper growth, development, and function of the fetus.
- Prenatal weight gain and healthcare monitoring determine the risk for developing medical conditions of pregnancy, such as preeclampsia and gestational diabetes, and promote good nutrition. There are many prenatal dietary precautions to take to avoid a poor birth outcome.
- Prenatal fitness includes following the eating plan provided by MyPlate, taking a prenatal multivitamin and mineral supplement, engaging in regular exercise, getting plenty of rest, and being wise about lifestyle choices.

Take Ten on Your Knowledge Know-How

1. How does prenatal body weight influence the amount of weight a woman should gain during pregnancy?
2. When does the period of the embryo begin? When is the most likely time for a teratogen to affect an embryo? When does the fetal period begin?
3. What risks are there to a baby if the mother gains too little weight during the pregnancy? What risks are there to a baby if the mother gains too much weight during the pregnancy?
4. What beneficial effects are associated with sufficient intakes of EPA and DHA? Why should pregnant women eat fish with low methylmercury levels?
5. What are some of the common gastrointestinal distresses pregnant women experience? What remedies are available to manage them?
6. What is fetal alcohol syndrome? How can it be prevented? When can it be most accurately diagnosed?
7. How does the MyPlate food guidance system support a healthy pregnancy?
8. Why is a prenatal dietary supplement recommended throughout the pregnancy?
9. What benefits are there to exercising during pregnancy? How should exercise be modified during each trimester?
10. What is pica? Why is it believed to occur during pregnancy?

7.2 NUTRITION DURING INFANCY AND FOR LACTATION

T-Talk 7.2

To hear Dr. Turley talk about nutrition during infancy and for lacation, go to www.cengagebrain.com

Introduction

Infancy is the time of life from birth to one year. Specific nutrition needs should be met during this sensitive time of rapid growth and development. In the first four to six months, those needs are best met by breast milk but can also be met by proper infant formula feeding. The lactating mother also has unique nutrition needs to meet the demands of milk production for her baby. Between four and six months of age, the infant can begin eating specially prepared solid foods, along with breast milk or formula. Infants who have continued to receive breast milk are usually weaned from it by age 1, when they can begin drinking cow's milk or other calcium-fortified beverages, along with soft, nonchoking foods. As babies grow and develop, they need not only appropriate nourishment but also appropriate fitness opportunities. The family that grows together, eats nutritious foods together, and plays actively together will be happy and healthy.[B,Z,XX,ZZ,16]

The Physical, Cognitive, and Social Triad During Infancy

Diet, emotional and social factors, the environment, genetics, and ethnicity all influence growth and development. Nutrition not only supports physical growth and cognitive function, but it also plays a role in the acquisition of social skills. During the first year of life, breast or formula feedings nurture bonding. The introduction of solid foods provides an array of stimuli that contribute to development.[139]

Successful Growth and Development Infants are born totally dependent on their mother's nurturing. Most infants are born healthy and normal; however, occasionally an infant is born with health problems. In an effort to diagnose common problems early, newborn screening is commonly performed (see BioBeat 7.5). During the first six months of life, the growth and development of healthy newborns progresses quickly. They go from helplessness to gaining head and neck control, sitting up, crawling, walking, and self-feeding. To quantify how rapidly normal growth and development occur during this period, by four to six months of age the baby's body weight will double and length (height) will increase 25 percent. From 6 to 12 months, growth is a bit slowed. By one year of age, an infant's body weight has practically tripled, and length has increased 50 percent. If a 5-foot tall, 100-pound adult were to grow at this rate, that adult would be 6 feet 4 inches tall and weigh 200 pounds at the end of a six-month period.

Infants are not born with fully functional body systems; they could be considered a physiological work in progress, and they are in need of a healthy, nurturing environment. Table 7.3 provides milestones in normal infant development by body system.[OO,16,118,139] For example, their gastrointestinal tract and kidneys are immature, and their heads are disproportionately large compared to that of adults. The gut lining is thin and leaky, meaning that it can absorb intact macronutrients. The kidneys have a poor ability to concentrate waste and process electrolytes and nitrogenous waste. Part of this is evident with the seemingly constant need for diaper changing. They also progress from communication by crying to cooing and smiling within a few months, then to forming words and/or using sign language by six months.[B,Z,OO,118]

Monitoring Successful Growth and Development Growth charts are commonly used in medical settings to track and monitor physical development including

TABLE 7.3

Body Systems and Development

Organ/Site	Damage
Circulatory system	Three blood flow changes occur after birth and are mostly complete within hours after birth and totally complete within three weeks to several months. Newborns have more blood cells than are needed due to changes in respiration; therefore, the liver breaks down the red blood cells to bilirubin, which is normally excreted. Excess bilirubin results in jaundice or a yellowish appearance and can cause brain damage. Treatment involves exposure to ultraviolet light to break down bilirubin in the skin.
Digestive system	The smell of their mother's milk stimulates the flow of digestive juices in breastfed infants. Taste is influenced by the mother's diet. Gastric pH is alkaline at birth, then becomes acidic within 24 hours. Acidity then decreases for a few months. Some enzymes are equal to those of an adult (trypsin). Others are only 10 to 60% of adult levels (chymotrypsin, lipase, amylase). Intestines are proportionally larger than an adult's and may be leaky.
Endocrine system	Hormonal changes occur during infant growth to support the fastest rate of growth in the life span and blood sugar regulation.
Integumentary system	The sense of touch is fairly well developed at birth. The skin is soft and elastic.
Muscular system	Muscular development is largely controlled by the hormones governing growth and natural physical activity.
Nervous system	The brain is only 25% developed at birth (continues through childhood), and nerve fibers are not completely myelinated (myelin is a protective covering) until age 2.
Reproductive system	Reproductive capability is defined at birth.
Respiratory system	Respiration initiates at birth. During the final month of pregnancy, the neonate begins breathing-like motions, although the lungs are filled with amniotic fluid. The birthing process causes most of the amniotic fluid to be expelled from the lungs.
Skeletal system	Bone matrix increases and mineralizes with the nutrition delivered in breast milk or infant formula.
Urinary system	This system is immature until five months. By one month, the kidney nephrons are mature and water balance can be maintained.

© Cengage Learning 2013

BioBeat 7.5

Genetic Screening for Inborn Errors of Metabolism (Galactosemia, Phenylketonuria, Congenital Hypothyroidism)

There are at least 28 inheritable disorders (inborn errors) that can be tested for during newborn screening. The screening identifies disorders of amino acid, fatty acid, and organic acid metabolism, as well as endocrine, blood, respiratory, and metabolic disorders. Screening for **phenylketonuria (PKU)** is mandated in all 50 U.S. states. Two other common screenings include **galactosemia** and congenital hypothyroidism.

PKU is an inborn error in amino acid metabolism where the person cannot break down excess phenylalanine, causing levels to increase in the blood. Phenylalanine is an essential amino acid; however, when blood phenylalanine levels are high, development is delayed, especially in the

central nervous system. Infants born with PKU must follow a specific diet and be provided with a special formula in order to control their intake of phenylalanine. The child must be carefully monitored to reduce the risk of mental retardation. When blood levels of phenylalanine increase in a growing infant or child, mental retardation occurs.

Galactosemia is an inherited condition resulting in the inability of the body to break down galactose, one of the simple sugars, which comes from the lactose in breast and cow's milk. The resulting accumulation of galactose in the body can cause many problems, including liver damage, cataracts, mental retardation, and possible death. A lactose-free diet must be followed.

Congenital hypothyroidism occurs when a baby is born without enough thyroid hormone. Thyroid hormones require the trace mineral iodine and play a vital role in body growth and brain development. Mental retardation and stunted growth will occur without thyroid hormone supplementation when hypothyroidism is diagnosed from newborn medical screening. The screening of other inheritable disorders may be called for if the parents are carriers of altered genes.[75,86]

What is the role of nutrition in managing galactosemia, phenylketonuria, and congenital hypothyroidism?

weight, length, and head circumference. Monitoring at "well baby checkup" visits commonly occurs at three, six, nine, and twelve months of age. Growth charts can also be used personally for the same purpose. The Centers for Disease Control and Prevention (CDC) published the most recent growth charts in the year 2000. These are found in Appendix F and can be downloaded free at www.cdc.gov/growthcharts. The World Health Organization (WHO) released Child Growth Standards in the year 2006. These standards are the first of their kind and are based on a large, healthy, culturally diverse population. The WHO Child Growth Standards can be downloaded free at www.who.int/childgrowth/en. There are many growth and development assessment tools available, including length-for-age, height-for-age, weight-for-age, stature-for-age, weight-for-length, weight-for-height, body mass index-for-age, head circumference-for-age, arm circumference-for-age, subscapular skin fold-for-age, tricep skin fold-for-age, motor development milestones, and the rate of increase in weight, length, and head circumference. Before age 2, a child's height is measured as **recumbent length** (the child is measured while he or she is lying down). Head circumference measurements are taken to monitor brain growth. The head circumference measure is the only anthropometric measure not related to nutritional status, but it should be noted that a poor diet will not support normal brain development.[O,42,44,46,61]

A comparison study of the CDC and WHO standards found that the CDC charts reflect population samples of shorter and heavier individuals compared to the WHO standards. So use of the WHO standards would reflect less undernutrition (with the exception of the zero to six-month-old age group) and more overweight and obesity compared to the CDC charts. Another important finding is that the WHO standards after two months of age list healthy, breastfed babies as smaller in size, while the CDC charts would indicate that the smaller, breastfed baby may be underweight and at risk for **failure to thrive**. Percentile rankings compare the infant being assessed to other infants at the same age in the reference population group studied. A nine-month-old in the 75th percentile for weight is at the same or higher weight than 75 percent of the nine-month-olds in the reference population group and at a lower weight than 25 percent of the nine-month-olds in the reference population group.[O,42,44,124]

Babies tend to follow a consistent growth pattern. A dramatic shift in an infant's normal percentile ranking indicates a problem. Poor nutrition leads to poor growth and development, as well as behavioral problems. The degree to which malnutrition affects an infant is multifactorial; the timing, severity, and illness duration are all factors at play. For example, if a severe illness occurs during a growth spurt, then the outcome may be more predictably negative than if it occurs during a season of slower growth. The same thing would be true if the illness is long-lived rather than short-lived. When malnutrition is extreme enough to cause an infant to deviate from his or her normal growth pattern as measured by growth charts, then failure to thrive is medically implicated. This condition can be caused by disease, sickness, and even poor infant-caregiver interactions. Failure to thrive is official when an infant drops two or more grid lines within a six-month period in his or her height or weight for age. The role of nutrition in infancy is to support growth and development and promote physical readiness for increasing food experiences, **cognition**, and social interactions.[B,F,Z,JJ]

Cognitive Development
Infants have very large heads in proportion to the rest of their body and compared to adults. Infancy is a time of rapid intellectual development, and an adequate, healthy diet is important for normal cognitive development and skill acquisition. Cognitive development can also be supported by activities such as storytime, listening to and playing music, learning sign language, and playing with toys and games.[7,66,70,126,139]

phenylketonuria (PKU) An inborn error of metabolism that is screened in the United States and is managed by a special lifelong, low-phenylalanine diet.

galactosemia An inherited condition screened at birth causing the inability to break down galactose, resulting in many problems, including liver damage, cataracts, mental retardation, and possible death.

recumbent length Body length that is measured with the infant or child lying down flat on his or her back without the knees flexed.

failure to thrive A medical term describing poor weight gain and physical growth failure in which an infant or child experiences a weight-for-age percentile ranking less than two standard deviations within a six-month period; for example, a drop from the 50th to the 10th percentile weight-for-age within six months.

cognition The mental processes of thought, perception, reasoning, and learning.

The Psychosocial Sport of Eating Infants need a lot of special attention. They want the focus of the parent or caregiver to be on them. Mealtime is a great time to engage in healthy and enjoyable psychological and social behaviors. At first the meals are numerous and in the form of breast or bottle feeding. Feeding is provided on demand, and a trust relationship is established between infant and provider. When the baby is held close and snuggled as he or she is being fed, security and trust are built, and a deep bonding relationship is formed.

As infants mature and become more independent, mealtimes become more social, and provide an opportunity to expand sensory experiences as a variety of foods are introduced. Infants learn to express their preferences, desires, and needs. They may need snuggling and parental time or desire a favorite food. As an infant's senses develop, food becomes more pleasurable and interesting. The brain receives taste sensations (**gustation**) in the form of sweet, sour, bitter, and salty from the tastebuds on the tongue and smell sensations (**olfaction**) from neurochemical secretions. Pleasant smells promote feeding, whereas unpleasant smells impede it. Additionally, mouth feel, texture, and food temperature also play a role in the psychosocial sport of eating.[OO,53,96,125]

Nutrition Needs for the Lactating Mother

A woman who is breastfeeding has specific nutrition and lifestyle needs. Breastfeeding itself, along with specific exercises, can help women recover from pregnancy and delivery (see BioBeat 7.6). Proper nutrition supports both the mother's health and milk production for the newborn. DRI recommendations support the natural composition changes in mother's milk. The composition of breast milk and the changes that occur over time depend on whether the infant was born prematurely or full term and events that occur in the month **postpartum**. The DRIs for lactation are given in a variety of parameters. Some are based on age, some on the month postpartum, and others as a set amount that is independent of these factors. Macronutrient recommendations are generally made for lactation; micronutrient recommendations depend on the mother's age. In any case, the volume of milk production and the quality of the breast milk is affected by maternal nutrition.[A,H,AA,90]

Lactation Lactation is a maternal, **mammalian** biological process that has evolved over time and has provided optimal nutrition for the survival of offspring. In the last 50 years, medical science has learned a great deal about human milk, particularly in the area of immunology. After breastfeeding is initiated, the first fluids contain **colostrum**, a yellowish fluid that is loaded with antibodies that protect newborns from disease. Mature milk from a properly nourished lactating woman is high in fats, contains cholesterol, and is adequate in protein, carbohydrates, vitamins, and minerals.[A,AA,159]

During the first trimester of pregnancy, breasts enlarge because of the significant changes in hormone concentrations of estrogen, progesterone, **prolactin**, and others. The ducts and alveoli in the breast also multiply rapidly. The breast changes are occurring in preparation for breastfeeding.

Lactogenesis is the term for initiating lactation from the breasts (**mammary glands**), and occurs in three stages as described in Table 7.4. The process is initially controlled by hormonal changes. Progesterone levels fall, prolactin levels remain high, and **oxytocin** from the pituitary gland, as well as various hormones from the ovaries, thyroid, and adrenal glands, are at play (see Figure 7.8 on page 369). Milk production switches from endocrine (hormonal) control to **autocrine** control, and the supply and demand principle takes over. The more a mother nurses, the more milk she will produce. If she nurses less, milk production will slow down.[A,H,AA,55,86,155,159]

gustation The sense of taste or the ability to taste.

olfaction The sense of smell or the ability to smell (detect airborne molecules).

postpartum A period that begins immediately after delivering a baby, usually extending to the point of full recovery from childbirth.

mammalian A biological term used to refer to any warmblooded vertebrate having the skin more or less covered with hair; the young are born alive and nourished initially by mother's milk.

colostrum The first breast secretions from a lactating mother, which not only provides nourishment to the infant but also contains antibodies that protect the newborn from disease.

prolactin A hormone secreted by the pituitary gland; in females it stimulates growth of mammary glands during pregnancy and promotes lactation after birth.

lactogenesis The production and secretion of milk by the mammary glands; also known as breastfeeding or lactation.

mammary glands Exocrine glands that produce milk for the offspring of mammals.

oxytocin A hormone secreted by the pituitary gland that stimulates the uterine muscles to contract and the mammary glands to eject milk during lactation.

autocrine A form of cell signaling in which a cell secretes a hormone or chemical messenger that binds to receptors on the same cell and alters cell function.

BioBeat 7.6

Postpartum: Getting Back to Me!

During pregnancy, a woman's body undergoes several changes to accommodate the fetus, and childbirth causes abdominal and/or vaginal trauma. Exercise can speed a woman's recovery from both. Depending on the mode of delivery, most types of exercise can be continued or resumed in the postpartum period in about two weeks. However, with the physiological fatigue of delivery, the type of delivery, and the sleep-deprived lifestyle that newborn care requires, often new mothers may need to reduce the intensity, duration, and frequency of their exercise sessions and gradually increase the program over a three-month period. Women who have had **caesarean** delivery may slowly increase their aerobic and strength training, depending on their level of discomfort and other complicating factors such as anemia or wound infection. The six-week postpartum evaluation is an opportunity for women and their healthcare provider(s) to discuss these issues (see Module 4, page 196).

Some exercises can be initiated right after delivery. The initiation of pelvic floor exercises (**Kegel exercises**) in the immediate postpartum period may reduce the risk of future urinary **incontinence**, because pelvic floor muscles are weakened by the birth process. Kegeling is done by gently tightening, then relaxing the muscles of the perineum (pelvic floor muscles). Kegeling may be done lying down, sitting, or standing, and at first the exerciser should hold the contraction 2 to 3 seconds, often several times throughout the day. The goal for duration of a contraction is 20 seconds. This can be done by gradually increasing the duration of the contraction by 5 seconds each time. By two weeks postpartum, this goal should be met.

Another valuable exercise to do is pelvic tilts. They will help tone and strengthen the abdominal muscles and relieve backache. To do them, lie on the back, with the knees bent, and then tighten the stomach and buttock muscles to tilt the pelvis up. Before doing more advanced abdominal exercises, it is important to check for separation of the abdominal muscles. The abdominal muscles are divided by connective tissue that stretches apart during pregnancy. A hormone called **relaxin** allows all connective tissue to soften and stretch. If the gap between the abdominal muscles is greater than 2 inches, abdominal support must be applied while doing abdominal exercises. Engaging in power sport activities is not recommended until three months postpartum when relaxin levels have diminished, and tendons, ligaments, and connective tissues have had the time to tighten back up.[H,U,41,115]

How would you find the postpartum exercise classes in your community?

TABLE 7.4

The Three Stages of Lactogenesis

Stage	Description
1	Begins about 12 weeks before delivery when mammary glands begin to secrete colostrum, causing breast size to increase. The high maternal blood levels of progesterone inhibit the full production of milk until after birth.
2	Begins after the baby and the placenta have been delivered. Within two to three days postpartum, the "milk comes in." The amount of milk produced increases rapidly, and its composition gradually changes from colostrum to more mature milk. Breastfeeding often (and/or pumping if the baby cannot feed well) increases the number of prolactin receptors in the breast and improves milk production.
3	The mature milk supply is established. Continued milk production depends more on the ongoing removal of milk from the breasts than on the hormones circulating in the blood. The amount of milk the infant consumes is the amount of milk the mother produces.

© Cengage Learning 2013

caesarean An unnatural, surgical procedure used to deliver a baby by making an incision through the abdomen and wall of the uterus; also known as a C-section.

Kegel exercises A necessary postpartum pelvic floor exercise done by tightening then relaxing the muscles of the perineum, or the pelvic floor muscles.

Lactation and Nutrition Needs Nutrient requirements during lactation are increased to support milk production to sustain the expected growth of the infant.

Calories Calorie needs during lactation are determined using the age-appropriate EER equations. The age- and gender-specific formulas are found in Appendix A. Calorie needs can be calculated based on body mass and physical activity; generally speaking, for the first six months of breastfeeding, 330 additional Calories per day should be consumed, and during the second six months of nursing, 400 Calories should be added to the EER.[69,139]

Carbohydrate, Protein, and Fat The DRI for carbohydrate during lactation is increased to 210 grams per day, which is well above the adult minimum need of 130 grams per day of dietary carbohydrate. The additional carbohydrate is used for milk production and brain, central nervous system, and red blood cell functioning. The AMDR for carbohydrate is the same as for adults, 45 to 65 percent of Calories, with less than or equal to 25 percent of Calories from sugars. Fiber-rich sources (whole fruits, vegetables, whole grains, and legumes) are needed to promote gastrointestinal health and prevent obesity, type 2 diabetes, heart disease, and some cancers. The DRI is personalized at 1.4 grams per 100 Calories consumed. Based on the reference person, the DRI for a lactating woman, no matter the time postpartum, is 29 grams per day. The DRI for protein is 1.1 grams per kilogram of body weight during lactation. Consuming adequate protein is very important for milk production.

There is no DRI for total fat. However, there is a DRI for the essential fatty acids linoleic acid and alpha-linolenic acid. The DRI for linoleic acid is increased to 13 grams per day, while the DRI for alpha-linolenic acid is increased to 1.3 grams per day compared to adult females (see Appendix A). The AMDR for total fat is the same as for adults at 20 to 35 percent of Calories, with the AMDR for the essential fatty acids being slightly higher compared to female adults. Consuming less fast food and processed food and more whole plant foods (including nuts and seeds) and fish will support adequate and healthy fatty acid intake. Standard prenatal or vitamin and mineral supplements do not include omega-3 fatty acids, which help promote the development of so many aspects of the baby's central nervous system. If the lactating woman does not choose to consume fish or other foods high in omega-3 fatty acids, she can take a supplement.[23,66,68,69,122,133]

Vitamins, Minerals, and Water Most of the DRIs for vitamins and minerals are higher during lactation compared to the adult female values and are based on the age of the nursing mother (see Figure 7.5 on page 357). Continuing to take prenatal vitamin and mineral supplements or some other nutritionally complete supplement during lactation is recommended by most healthcare providers. Lactating women are also encouraged to consume adequate amounts of iodine and fluoride. The use of iodized salt and fluoridated water ensures adequate levels will occur in the breast milk. Furthermore, consuming a nutrient-rich diet by following the MyPlate food guidance system is necessary. Adequate water intake that maintains positive water balance will support milk production, metabolic waste removal, cardiovascular function, thermal regulation, and metabolism. The DRI for water intake is higher during lactation than for adult females or pregnant women. The need is 3.8 liters of water per day.[69]

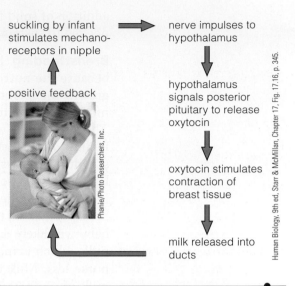

suckling by infant stimulates mechano-receptors in nipple → nerve impulses to hypothalamus

positive feedback

hypothalamus signals posterior pituitary to release oxytocin

oxytocin stimulates contraction of breast tissue

milk released into ducts

Human Biology, 9th ed, Starr & McMillan, Chapter 17, Fig. 17.16, p. 345.

Phanie/Photo Researchers, Inc.

FIGURE 7.8 Positive feedback keeps milk flowing to a suckling infant.

Nutrition Needs During Infancy

Based on the body weight of infants, nutritional needs are greater than at any other time during the life span. From zero to six months, infants will double their birthweight. By one year of age, infants triple their birthweight. The DRIs reflect these needs by establishing a birth to six-month category and a six-month to one-year category. As infant growth slows just a bit during the second six months of life, but the body weight has doubled, the amount of nutrients that infants need warrants the different recommendations. The normal, expected growth of infants can be sustained with breast milk or formula feedings only for the first few months of life; however, at around four to six months, the

incontinence Loss of bladder and/or bowel control.

relaxin A protein hormone that is secreted during pregnancy that relaxes ligaments (especially the pelvic ligaments) to prepare for labor and delivery.

addition of solid foods to the diet is required to supply the energy and nutrients needed to sustain expected growth and support development.[A,F]

Breastfeeding Breast milk is considered the gold standard for infant feeding because the nutrition provided in this single food supports optimal health and growth for infants. The hormone oxytocin causes the muscle cells around the mother's alveoli to contract, pushing milk down through the ducts to the nipple. This movement of milk down the ducts is called the milk-ejection reflex. Young infants will breastfeed frequently and should be fed on demand rather than on schedule. For best results, proper placement of the baby is needed to allow for latching on. A mother can be certain that her baby is being fed enough if the baby is growing and developing as expected and if the baby is soiling diapers as would be expected. As the baby grows, the feeding will become less frequent, but the baby will likely drink more milk at a time. Extended lactation produces toddler milk, which is more concentrated with immune factors as the toddler begins to nurse less. Milk that is produced after a premature birth is different from the milk of mothers whose babies are born full term, and those unique properties are beneficial for fragile, prematurely born infants. The natural changes in the nutritional composition of breast milk over time suit the changing needs of the infant. There are many more benefits of breastfeeding for mother and baby (see Table 7.5).[A,F,AA,77] Besides the benefits to the baby, breastfeeding benefits the mother. It has been shown to reduce the risk of breast and ovarian cancer and possibly osteoporosis later in life. The hormone oxytocin responsible for the milk-ejection reflex also causes the muscles around the uterus to contract, thus facilitating the recovery from pregnancy.

After birth, the mother's body produces antibody-laden colostrum. Within three days, more mature milk is produced. In this transition, the sodium, chloride, and protein levels in the milk decrease, and levels of lactose and other nutrients increase. The color gradually changes from the golden yellow typical of colostrum to a bluish white. Mature milk is high in fats, and the composition of fatty acids is related to the fatty acids consumed in the diet. For example, the more omega-3 fatty acids in the maternal diet, the more there will be in the breast milk. The high cholesterol concentration ensures the building blocks needed for nervous system structural development will be available. The proteins in breast milk are largely **casein** and **whey** proteins, but the concentration of **immunoglobulins** is notable. The predominant carbohydrate in breast milk is lactose, and essential vitamins and minerals are naturally present in adequate quantities as long as the maternal diet is adequate.[A,F,AA,25,77,86,155,159]

casein The dominant type of protein in mammalian milk.

whey A major mammalian milk protein, in addition to casein.

immunoglobulins Antibodies that are found in breast milk, blood, or other bodily fluids, and are used by the immune system to identify and neutralize foreign entities such as bacteria and viruses.

TABLE 7.5		
Benefits of Breastfeeding		
Infant benefits	• Highly bioavailable and balanced nutrient source	
	• Hormonal support for psychological development	
	• Aids cognitive development	
	• Strengthened immunity; possible protection against hypertension, type 1 diabetes, and allergies later in life	
Mother benefits	• Contraction of uterus speeds recovery from pregnancy	
	• Conserves iron by preventing menses (not a reliable method of contraception)	
	• Possible breast and ovarian cancer protection	
Other benefits	• Cost savings from improved health and not purchasing formula	
	• Environmentally responsible choice of nonmanufactured, unprocessed, unpackaged, and nonshipped food source	

© Cengage Learning 2013

Formula Feeding Breast milk or infant formula is recommended to be the diet or part of the diet for an infant up to the age of one year. A variety of infant formulas are available on the market. Most are fortified with DHA (because of the benefit to visual and cognitive development) and **taurine** (because it may be essential for **neonates**). Although the benefit of taurine has not been established, it is present in high concentrations in breast milk and is essential for prematurely born infants.[F,72]

The composition of breast milk is the standard used in the development of infant formulas. Infant formula feeding has supported comparable growth and development to breastfeeding. There are two types of protein-based infant formulas—cow or soy. The more expensive cow's milk–based formulas have had the protein **hydrolyzed** to partially digest it. Some infants who do not tolerate regular cow's milk–based formula do tolerate the hydrolyzed ones. If the infant develops mucus secretions or **colic**, a switch to soy formula may improve his or her discomfort because of the change in protein (casein to soy) and carbohydrate (lactose to glucose).[72,77,118]

Energy and Nutrient Needs for Baby Energy and nutrient needs are unique for infancy as with other life span groups. The DRIs for energy and nutrients are established for infancy. There are two age groups—zero to six months and seven to twelve months. DRIs are determined largely by the nutrient composition of breast milk, because breast milk has been used since antiquity to support infant life. Infants need a high amount of Calories compared to their body mass, a high percentage of Calories from fat compared to carbohydrate and protein, a sufficient amount of protein to support new tissue growth, and essential vitamins, minerals, and water.[F,69]

Calories Infant growth is remarkable. In order to support growth and development, sufficient Calories are needed, much more per kilogram (two to four times) of body weight than an adult maintaining existing body tissue. Infants need approximately 108 Calories per kilogram during the first six months of life. The exact Calorie need can be calculated as the EER from the life-stage-dependent formula in Appendix A.[69]

Carbohydrate, Protein, and Fat From age zero to six months, infants require 60 grams of carbohydrate each day, increasing to 95 grams between seven and twelve months of age. In the first four to six months, the carbohydrate source is primarily lactose in breast milk or infant formula. Infants do not have amylase enzyme activity until about six months of age. Thus, early introduction of starch from solid plant foods can cause gastrointestinal discomfort. There is no DRI for fiber during infancy.[69,118]

Because infants have underdeveloped kidneys, their protein needs are small. Although protein is needed for new tissue formation, too much can put a strain on the kidneys because of the need to process and eliminate nitrogenous waste with excess amino acid intake. Current research indicates that neonates require the same nine essential amino acids as adults and children, but also need arginine. Also, about half of the amino acids consumed should be essential (indispensable) amino acids. The DRI for protein is the highest per kilogram of body weight of any life span group. It is 1.52 grams per kilogram up to age six months and 1.5 grams per kilogram from seven to twelve months. For a reference infant, this translates to 9.1 and 11 grams of protein for the respective age groups.[B,F,Z,118]

Data GENEie Calculate the DRI for protein at ages six months (14 pounds) and one year (20 pounds). Note and discuss the change in DRI for protein and the factors influencing the grams of protein needed at each age. At what time in life was the protein need the highest, and why?

Dietary fat needs in infancy are high. Furthermore, infancy is the only life span time when there is a DRI for total fat. The DRI is 31 grams up to age six

taurine An amino acid intermediate that may be vital for the proper development and maintenance of the central nervous system during infancy, or for individuals with certain diseases or nutritional concerns.

neonate An infant age zero to four weeks.

hydrolyzed A chemical reaction in which water reacts with a compound (such as casein protein) to produce a more simple compound or protein structure.

colic A broad term used to describe abdominal pain or discomfort in infancy.

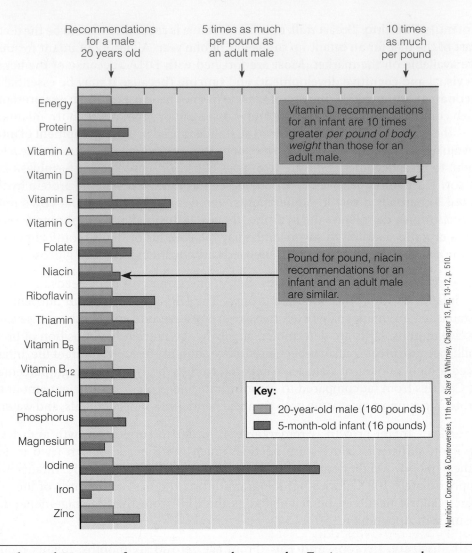

Recommendations for a male 20 years old
5 times as much per pound as an adult male
10 times as much per pound

Energy
Protein
Vitamin A
Vitamin D
Vitamin E
Vitamin C
Folate
Niacin
Riboflavin
Thiamin
Vitamin B₆
Vitamin B₁₂
Calcium
Phosphorus
Magnesium
Iodine
Iron
Zinc

Vitamin D recommendations for an infant are 10 times greater *per pound of body weight* than those for an adult male.

Pound for pound, niacin recommendations for an infant and an adult male are similar.

Key:
20-year-old male (160 pounds)
5-month-old infant (16 pounds)

Nutrition: Concepts & Controversies, 11th ed, Sizer & Whitney, Chapter 13, Fig. 13-12, p. 510.

FIGURE 7.9 Comparison of nutrient needs of a 5-month-old infant and a 20-year-old male.

months and 30 grams from seven to twelve months. Fat is very energy dense, and the Calories provided per gram are important for supporting growth and development. Essential fatty acids and other health-promoting, long-chain omega-3 and omega-6 polyunsaturated fatty acids are critical for cognitive development, brain and central nervous system growth and function, and maturation of the retina (recall BioBeat 7.4 on page 356). The DRIs for linoleic acid and alpha-linolenic acid are 4.4 to 4.6 grams and 0.5 grams, respectively (see Appendix A).[7,68,133]

Vitamins, Minerals, and Water Infants need the same essential vitamins and minerals as adults. The DRIs for vitamins and minerals for infants can be found in Appendix A. The nutrient needs for a five-month-old infant are compared to those for a 20-year-old adult male in Figure 7.9. Several nutrients that need to be monitored, along with fluid needs, will be addressed below.[18,54,63,92,95,107,117,130,137,167]

A vitamin K injection is given to all infants at birth. The vitamin K is needed for blood clotting. Adults receive much of their vitamin K from gastrointestinal tract bacteria synthesis. However, at birth, infants in essence have a sterile gastrointestinal tract, so they are devoid of this vitamin K resource.[137] If the baby is born prematurely, then vitamin E is administered to prevent hemolytic anemia. If the baby is full term, then blood vitamin E levels are usually adequate.

Vitamin D supplementation of 600 International Units per day is recommended if the infant is breastfed. Liquid multivitamin drops with vitamin D are available. Infant formula is adequate in vitamin D. Too little vitamin D can impede skeletal growth and weaken bones, which can eventually lead to rickets.[18,83,101,117,152]

Vitamin B$_{12}$ can be a nutrient of concern for infants being breastfed by strict vegan mothers. The vegan mother could have an inadequate body supply if she does not take vitamin B$_{12}$ supplements. Thus, her breast milk will also be lacking in this essential vitamin.[107]

Full-term infants have enough iron reserves in the body for the first six months of life. At six months of age, iron-fortified infant formula provides a significant iron source, usually 1 milligram per 100 Calories. Iron-fortified infant cereals, such as rice cereal, are introduced first, followed by oat cereals and then other grains like barley and then wheat. Then other iron-rich foods are incorporated into the baby's growing food choices, including baby food meats. Too much breast milk and not enough solid food after four to six months of age can promote a condition known as milk anemia. The infant develops iron-deficiency anemia because of a lack of iron intake when too much milk consumption displaces other age-appropriate, iron-containing food sources. Iron supplementation, in liquid form, is needed at 1 milligram per kilogram per day from age six to twelve months if the infant is breastfed. Preterm infants or low-birthweight infants need additional iron, 2 to 4 milligrams per kilogram per day, with supplementation starting at one month of age through twelve months.[10,92]

The need for iodine in infancy is relatively high in light of the DRI. For infants ages zero to six months, 110 micrograms per day are required. This increases to 130 micrograms per day from seven to twelve months of age. Iodine is critical for proper growth and development. A deficiency of iodine causes irreversible brain and central nervous system damage.[95]

Fluoride is necessary for tooth formation, which occurs during infancy. If the water supply is not fluoridated or does not contain adequate natural levels, some decisions need to be made to ensure nutrient adequacy for the infant. An amount of 0.25 milligrams per day starting at six months is required if there is no fluoride present in the water, if the water supply contains less than 0.3 parts per million, or if the water is purified or filtered, removing the fluoride present (see BioBeat 7.7). Women who breastfeed are not advised to supplement babies younger than six months of age with fluoride.[F]

Water has many important functions in the body. It is needed for nutrient and waste transport, to regulate body temperature, and as the medium for metabolism. During normal growth, a positive water balance is needed. This makes sense because the baby is increasing in size. The main source of fluid is breast milk or formula during the first six months, followed by the introduction of fruit juice and water. Care needs to be taken to limit fruit juice to 4 ounces per day or too many Calories will be consumed from sugar, and the larger volume will decrease the intake of other foods in the diet.

The body composition of an infant is about 75 percent water. By age 1, this has changed to about 60 percent water, the same as an adult. Water turns over more quickly in infants than at any other life stage. About 15 percent of total body water is lost and replaced (turned over) each day. The water losses occur through the urine, feces, skin, and breath. Water replacement occurs through breast milk or infant formula.

A normal, healthy baby does not need extra water. The DRI of 0.7 liters per day for infants ages zero to six months and 0.8 liters per day for infants ages seven to twelve months can be met under normal circumstances by proper breast and/or formula feeding. Fluid and electrolyte imbalance can occur with fever, vomiting, and diarrhea, and in a hot climate. During sickness or more extreme environmental conditions, it is critical to make sure the child doesn't get dehydrated. Fluid, electrolyte, and glucose replacement beverages designed for **pediatric** illness can be beneficial in preventing dehydrating during illness. If a baby stops urinating or the urine becomes reduced and concentrated, and the child is not taking in or retaining fluids, then medical attention is needed.[F,69,77,159]

pediatric The branch of medicine for infants and children.

BioBeat 7.7

Dietary Factors Important in Dental Health

Primary teeth develop two to three months in utero. By the last trimester, permanent teeth are forming. Humans usually form 52 teeth total (baby/primary and permanent). Prenatal nutrition profoundly affects all aspects of dentition. Furthermore, prenatal use of fluoride is effective (1 milligram per day), but the complete picture of its safety is not perfectly delineated. That is, it is not known whether the fetus can protect itself if there is too much fluoride taken by the mother.

Teeth are made up of calcium, phosphorus, and magnesium in a collagen or protein matrix called hydroxyapatite. Vitamins A and C play major roles in the formation of the hydroxyapatite, which is made into fluorohydroxyapatite in the presence of fluoride. This protein matrix is very resistant to the acids that cause tooth decay. The American Academy of Pediatrics recommends fluoride supplementation for the optimization of dental health after age six months. Fluoride is available in systemic and topical application. The amount of fluoride supplementation must be based on the fluoride present in the drinking water. Fluoride levels in water occur in parts per million (ppm). A level of 2 ppm yields white teeth, while 4 ppm can cause fluorosis, or brown teeth. Table 7.6 describes the recommendations for fluoride supplementation.

Salivary flow and microbial activity are also very important to tooth health. Protein and iron malnutrition affect the salivary flow and promote dental caries (cavities or tooth decay). Protein malnutrition will also disrupt tooth formation. *Streptococcus mutans* is one of the prevalent microbes that causes dental caries. Its growth in the mouth is supported by carbohydrates, especially sucrose. Thus, the intake of sugary foods promotes tooth decay. It is always advisable to avoid foods high in sugar and refined carbohydrates, especially before bedtime. Having children brush their teeth in the morning and at bedtime with a fluoride-containing toothpaste will also support healthy teeth.[B,F,Z]

What is the fluoride level in your drinking water, and how much fluoride should a six-month-old be supplemented with daily based on that amount?

TABLE 7.6

Age	Dosages Per Day (mg)		
	0–0.3 ppm	0.3–0.6 ppm	> 0.6 ppm
Birth to six months	0.0	0.0	0.0
Six months to three years	0.25	0.0	0.0
Three to six years	0.5	0.25	0.0
Six to sixteen years	1.0	0.5	0.0

Supplemental Fluoride Recommendations Based on Parts per Million (ppm) of Fluoride Content in the Drinking Water

Note 1: 1 ppm = 1 milligram per liter of water.

Note 2: The pink fluoride tablets provide 1 milligram of fluoride per tablet.
2.2 milligrams of sodium fluoride contain 1 milligram of fluoride.
© Cengage Learning 2013

Baby's First Foods A baby's first foods should be planned by the caregiver to match the baby's nutrition needs and developmental skills. The opportunity to establish healthy behaviors and instill the proper food-related building blocks for the rest of the baby's life should be embraced and practiced. Foods should be introduced gradually and in a specific sequence. Furthermore, they should be offered in an appropriate way and be safe to consume with a low risk of choking. Babies born into poverty situations can be supported through food assistance programs (see page 378). All of the aspects of baby's first foods will be discussed below, starting with the topic of the division of responsibility for the feeder and the baby being fed.[F,80,168]

in utero In the mother's womb.

Division of Responsibility   There is a system of feeding a child that starts in infancy. It is tied to understanding and respecting the division of responsibility between the caregiver and the consumer and the concepts of feeding and eating, respectively. The parent or caregiver has the responsibility of determining what to provide the infant to eat, while the infant decides how much, how often, and at what pace to eat. The caregiver needs to respect the infant's eating and sleeping cycles and continue to provide structured eating opportunities with age-appropriate foods to support proper growth and development (see Table 7.7). During infancy and up to about six months of age, the eating schedule is flexible. The caregiver should feed the infant when the infant demands to be fed. It is important to let the baby decide how to eat, such as the amount of food and the speed of intake. Once the baby begins to eat age-appropriate foods, he or she can join in family meals. The caregiver should take responsibility for providing regularly scheduled meals and snacks that are healthy and safe for the child.[80,118,130,131,168]

Sequence of Introducing Foods   The age of four to six months is a time when new foods are introduced gradually to the baby by spoon feeding. Solid foods are meant to supplement, not substitute, for breast milk or infant formula. These are still needed to meet the infant's high demand for energy and nutrients. Before four to six months of age, the baby is not ready to eat foods, either physically or physiologically. The indications that the baby is ready to start spoon feeding are when the developmental skills of sitting up with support and having good head and neck control are present.[F]

Iron-rich foods are important to feed four- to six-month-old infants. This is the time when the baby's iron status begins to decline as a result of rapid growth. So the first food that is usually offered is iron-fortified baby rice cereal, mixed with either iron-fortified formula or breast milk. The mixture should start out thin and may only be a few teaspoons to ¼ cup. Once the baby has mastered consuming thinner cereal, it can be mixed thicker. Other single-grain cereals can be introduced gradually. A good system is to add only one new food every three to four days. This gives the baby time to adjust to the food and the caregiver a chance to determine if the food is tolerated well by the baby.[92]

A complementary, widening variety of foods should be fed. It is a good idea to introduce vegetables before fruits; otherwise, the infant may immediately develop a preference for the sweet fruit and not want to eat the more bland vegetables. The baby who is being spoon fed should also continue to be breastfed or bottle-fed infant formula. This should be the case until age 1. Cow's milk can be introduced at one year of age and replace breast milk or infant formula. This progression is known as weaning. Weaning is a gradual process in which breast milk and/or formula and bottle feeding is decreased, a sippy cup is introduced, and the infant is offered increasingly more solid table foods.

When infants start self-feeding, they can be quite messy. Despite this, infants need the opportunity to experience and master self-feeding. So it is critical to allow the mess for developmental success to take place. (See Table 7.7 for more details on the progression of feeding, foods, and skills during infancy.[TT,34,53,96,125])

Inappropriate Foods, Beverages, and Feeding Methods There are several reasons why a food, beverage, or feeding method is considered inappropriate. Babies need to have the developmental skills to consume food and be protected from adverse eating experiences and developing nutrient deficiencies, food intolerances, and tooth decay.[80,168]

Feeding can be a messy experience, but that is part of the developmental process.

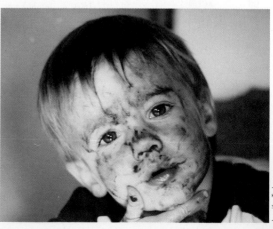

Jennifer Turley

TABLE 7.7

Growth, Development, and Feeding During Infancy

Age	Growth and Development	Feeding Skill	Foods to Introduce
0–1 month	Large head with soft spot on top; may hiccup, spit up, startle easily, sneeze, and tremble at the jaw; sleeps most of the time	**Rooting reflex; sucking reflex**; initially swallows liquids using back of tongue	Breast milk or iron-fortified infant formula every few hours on demand, with cuddling
2–4 months 	Lifts head briefly when lying on stomach; smiles, coos, and gurgles; whole body moves when lifted or touched; deliberate communication and movement begins; sleeps most of the time	Gradually begins to use front of tongue along with back of tongue; strong **extrusion reflex** (tongue-thrust) to push food out	Breast milk or iron-fortified infant formula every three to four hours or on demand, with cuddling, while alert and calm
4–6 months 	Weight nearly doubles and growth increases 3 to 4 inches since birth; follows objects with eyes and reaches for objects with both hands; grasps objects with palm of hand; puts fingers and objects in mouth; turns over, vocalizes, sits erect with support; sleeps six to seven hours at night and takes three to four naps per day	Extrusion reflex diminishes, ability to swallow nonliquid foods develops; indicates desire for food by opening mouth and leaning forward; opens mouth for spoon and closes lips over spoon; swallows semisolid food; begins chewing action and brings hand to mouth; indicates satiety by turning away and leaning back	Continue cuddling and feeding from breast or bottle; begin iron-fortified rice cereal mixed with breast milk or infant formula, and puréed vegetables and fruits according to baby's skills, using baby spoon and offering small bites; feed six to seven times per day
6–8 months 	Gains in weight and height are less rapid; appetite decreases; teething occurs; sits alone; stands up with help; takes three naps per day	Able to feed self with fingers; bites off food; chews with rotary motion; develops pincher (finger to thumb) grasp; begins to drink from cup; joins family meals	Continue cuddling and feeding from breast or bottle; begin wheat-free dry cereal, mashed vegetables, fruits, plain baby food meats, and unsweetened 100% fruit juices from cup; spoon feed thick/lumpy foods; introduce sippy cup
8–10 months 	Sits unsupported and crawls; explores objects with hands, eyes, and mouth; takes two naps per day	Begins to hold own bottle; curves lips around cup; reaches for and grasps food and spoon; chewing improves	Continue cuddling and feeding from breast or bottle, but at snack time only beginning around nine months; begin breads, cereals, crackers, yogurt, soft, chopped, cooked vegetables and fruits, finely cut meats and fish, casseroles, cheeses, eggs, and mashed, cooked legumes; serve from the table with hands, fingers, and/or spoon
10–12 months 	By twelve months has tripled birth weight, increased length by 50%, and begins to walk unassisted; grasps and releases objects with fingers; takes one to two naps per day	Chewing and cup drinking improves; finger feeds and masters spoon but still spills some food; by twelve months engages in family time, eating by self-feeding; is offered scheduled meals and snacks	Continue cuddling and feeding from breast or bottle, but at snack time only; add variety and increase portion sizes; at twelve months, offer all soft, non-choking-risk foods at the family table; may switch from breast milk or formula to whole, pasteurized milk; full transition from bottle to sippy cup with **weaning**

Jennifer Turley

Often, a caregiver may be in a rush to heat up a bottle for a crying baby. Heating liquid in a microwave can generate hot spots in the bottle that can scald the baby. Also, heating liquids in plastic containers can cause chemicals to leach into the plastic, creating a hazard for the baby. One such example is bisphenol A (BPA), a common chemical in plastics. BPA-free plastics and baby bottles are available.[31]

Improperly prepared infant formulas that are too dilute, too concentrated, or mixed with other substances are also inappropriate to offer babies. It is inappropriate to dilute infant formula because this leads to serious health problems, including failure to thrive, protein-energy malnutrition, and fluid and electrolyte imbalance, including hyponatremia. Caregivers may be tempted to fill a baby up more, or speed up the pace of eating. Thickening the baby's formula with cereal and slitting the nipple of the bottle to feed the infant with the hopes of the baby sleeping longer through the night, or jiggling the nipple to speed up a slow eater, are improper feeding methods. Likewise, putting a baby to bed with a bottle of anything besides water (i.e., formula, milk, or juice) can cause baby bottle tooth decay if the baby consistently falls asleep with the bottle in his or her mouth. Another concern with putting babies to bed by themselves with a feeding bottle is that babies need to be held while they are bottle-fed. This creates an intimate bond between the caregiver and the baby and helps the baby feel loved, nurtured, and satisfied physically and emotionally.[7]

Infants and young children should always be supervised while eating to avoid mishaps. Choking (from the baby putting too much food in his or her mouth) is always a danger. Foods that pose a risk for choking, such as popcorn, peanuts, hot dogs, hard candy, and whole grapes, should be avoided until the child has developed good chewing skills (closer to age 2).[34]

Foods and beverages that displace nutrient-rich food sources should not be offered to infants. In addition, a balanced diet needs to be thoughtfully managed. For example, at six months of age the baby may be offered up to 4 ounces of 100 percent real fruit juice per day. Too much fruit juice intake will displace the intake of calcium- and/or iron-rich food sources and could contribute to nutrient-related health problems like anemia. Excess fruit juice consumption can also cause tooth decay, cramping, gas, and diarrhea. Furthermore, allowing a child to drink too much juice, milk, or other sweet beverages between meals can lead to poor eating at meals. If a child is thirsty between meals, water can be a choice beverage to offer.[16,139]

Introducing foods too soon is believed to increase the risk of allergies. Protein can be absorbed intact in the underdeveloped leaky infant gut, promoting greater allergy risk. Allergy is a genetically based condition and increases with exposure to the top eight food allergens: cow's milk, eggs, soy, peanuts, tree nuts, wheat, fish, and shellfish. Signs and symptoms that a food is not tolerated well may include spitting up, diarrhea or constipation, increased fussiness, and allergic reactions such as skin rashes and hives, runny nose, diarrhea, and difficulty breathing. If a child has an allergic-like reaction to a food, it is a good idea to discuss it with a medical doctor and follow his or her advice. Usually, allergy testing is not done on infants or young children, because their immune systems are undergoing rapid transformation and development. It is possible for children to outgrow allergies. However, the general rule of thumb is that if the allergy persists after age 3, then the child is likely to have the allergy for life.[E,V]

The immune systems of babies are not developed enough to fight off foodborne illness risks. For this reason, honey and even foods with corn syrup are not advised for infant consumption. They may contain even a low amount of *Clostridium botulinum* spores, which cause the potentially fatal condition botulism. Furthermore, sweetening foods to make them more palatable for the baby is not recommended.[113]

Forcing a child to clean his or her plate is another inappropriate feeding tactic. The portion size of food given to a child should be appropriate. Too much food can

rooting reflex An innate instinct in newborns to turn their head toward an object that brushes their cheek, to promote and support feeding.

sucking reflex An innate reflex, linked to the rooting reflex, in which an infant sucks on a nipple touching the roof of its mouth to enable eating.

extrusion reflex An innate reflect in infants causing them to push food out of their mouths; diminishes between four to six months of age; also known as the tongue-thrust reflex.

weaning Gradually withdrawing breast milk and/or infant formula from an infant, along with the introduction of table foods, in the transition to a diet consisting of adult-type foods.

be intimidating, and making a child eat more than he or she wants can result in the child learning to override the basic hunger and satiety mechanisms and put the child at risk for obesity later in life. Other improper feeding methods include bribing a child to eat a new or disliked food with a favorite food or toy; allowing the child to eat while watching television or wandering around the house; allowing a child to graze all day long; making a child wait too long to eat or get too hungry before meals; not eating meals with the child or modeling appropriate eating patterns and manners; and making different foods for the child if the dinner meal wasn't accepted. MyPlate begins at two years of age. The format of consuming a well-balanced diet using whole, fresh foods daily from all of the food groups should be in place as early as possible. Each nutritious food that a baby consumes supports nutritional needs.[34,53,130,131,168]

Food Assistance Programs   The federally supported program Women, Infants, and Children (WIC) provides monies to states to help needy women, infants, and children to purchase specific nutritious foods. The support comes in the form of providing supplemental foods, healthcare referrals, and nutrition education. Low-income pregnant, breastfeeding, and nonbreastfeeding postpartum women, as well as their infants and children up to age 5 who are at nutritional risk, may qualify for the program.[UU,YY,99]

The Supplemental Nutrition Assistance Program (SNAP) is essentially the modern-day equivalent to the food stamp program. The goal of SNAP is to assist low-income families to buy needed, healthy foods. The program also helps people learn to make healthy eating and active lifestyle choices.[X,UU,VV]

Infant Fitness and Lifestyle Management   Newborn infants spend a majority of their time eating and sleeping. Up to 20 hours a day may be spent sleeping. However, during awake times and as developmental skills progress, infants do enjoy physical activity (see Table 7.7 on page 376 for developmental milestones by infant age). Caregivers can gently bicycle babies' legs while they lie on the floor, carry them around and explore the surroundings together, and carefully assist them in a more seated position. Other enjoyable activities are sitting in an infant swing or vibrating infant chair with dangling toys for entertainment. As infants develop head and neck strength, they will start to lift their heads and push themselves off the floor when lying on their stomachs. They will learn to roll over, sit up, and crawl in time. With more skill development, infants will want to walk in walkers and jump in jumpers. An active baby is a healthy and happy baby. Some physical exertion and stimulation can reduce boredom and fussiness. Caregivers need to provide some safe, age-appropriate, stimulating infant fitness opportunities on a daily basis and within a regular and normal daily schedule.[F,I,U]

Summary Points

- Successful growth and development requires adequate nutrition and cognitive and social stimulation.
- Expected infant growth is monitored using a variety of anthropometric growth charts.
- Feeding the infant through the first year of life includes providing breast milk or formula.
- Solid foods may be introduced sometime between four and six months, one food at a time, with the least allergy-risk foods introduced first.
- When feeding, the principles incorporated in the division of responsibility should be adopted, and appropriate feeding practices should be established.
- Infants should be introduced to an active lifestyle by providing age-appropriate opportunities for physical activity.

Take Ten on Your Knowledge Know-How

1. What are the anthropometric parameters that can be monitored during infant and childhood growth?
2. What is failure to thrive?
3. What inborn errors of metabolism are commonly screened after birth?
4. What benefits to the infant are provided by breastfeeding?
5. What benefits to the mother are provided by breastfeeding?
6. What nutrients are vital for supporting dental health in infants and children?
7. What is the division of responsibility for feeding (for infants and for caregivers)?
8. How do feeding skills progress during the first six months of life, and from six months to one year?
9. How should the first solid foods be introduced, like cereals, vegetables, and fruits?
10. How can inappropriate feeding methods be replaced with appropriate feeding methods? Give five examples.

7.3 CHILDHOOD NUTRITION

T-Talk 7.3
To hear Dr. Turley talk about childhood nutrition, go to www.cengagebrain.com

Introduction Childhood, the period from age 1 to age 12, is an important time for growth and development. In the early years, ages 1 to 3, the children are toddlers. From ages 4 to 8 they are young children, and from ages 9 to 12 they are older children. Between the ages of 2 and 5, children are also known as preschoolers. School-age children are between 6 to 12 years of age. During childhood, there are definitive physical, cognitive, and social changes that are supported by proper nutrition, education, and lifestyle. Nutrition needs are supported by meeting the DRIs and following MyPlate (MyPyramid) for children. School meal programs are also a source of nutrition for many American children. When children are not at school, parents and caregivers should take responsibility for guiding eating and activity and providing health care to support the child's well-being. Parents have the role of making meals fun and nutritious, offering foods at appropriate times, and providing appropriate eating environments. Children decide how much and which of the foods to eat. Parents should also promote health and well-being by modeling healthy behaviors and providing physical activity opportunities for their children. Taken together, such diet and lifestyle experiences will set the stage for a long and healthy life at a normal body weight and free of chronic disease.[XX]

The Physical, Cognitive, and Social Triad During Childhood Nutrition and lifestyle choices can greatly impact a child's physical, mental, and social development. The patterns established during childhood carry throughout the life span. Providing children with education, healthy foods, and healthy lifestyle experiences will ensure greater success throughout life in many ways.[OO,16,140,168]

Monitoring Good Nutrition During Childhood Growth and Development
Growth and development—along with the acquisition of feeding, behavioral, and social skills—promote and reflect good nutrition. When a child is fed properly and provided with a healthy lifestyle schedule, the child thrives.[118]

Successful Growth and Development Childhood growth and development is measurable by standardized growth charts and developmental milestones (see

Appendix F and Table 7.8).[OO,140,168] These growth charts portray an expectation of normal growth in terms of height and weight. The growth rate during childhood is slower than during infancy and adolescence. Because the growth rate is slower, the child's appetite decreases. Hunger and appetite will increase during times of more rapid childhood growth. Because of these fluctuating phases, parents need to trust the child's internal controls for feeding and food quantity. Force feeding a child is never advised.

A child's growth and size is influenced by genetics, diet, exercise, age, and gender. Differences from one child to the next are to be expected. Successful growth is measured on the gender-specific CDC growth charts. Children are assessed by their weight in pounds or kilograms and height in inches or centimeters. Their anthropometric values are plotted on the growth chart, and their percentile rankings are determined. If a child is less than the fifth percentile, then that child is underweight (but the child has a 5 percent chance of being a healthy weight at the fifth percentile ranking); less than the third percentile is interpreted as underweight and at risk for malnutrition. There is only a 3 percent chance the child is healthy weight at the third percentile. When a child is between the 85th and 95th percentile, that child is labeled as overweight; greater than the 95th percentile is interpreted as obese. However, there is a 5 percent chance the child could be healthy weight. During a transient growth period, a child may be temporarily overweight, but if the weight-for-height measures consistently indicate overweight, parents should become concerned. The number of overweight children is increasing nationwide and even worldwide in developed nations. This is a concern because overweight and obesity during childhood dramatically increases the likelihood of overweight and obesity persisting into adulthood. And the diseases and conditions associated with overweight and obesity during adulthood, such as type 2 diabetes, hypertension, atherosclerosis, and metabolic syndrome, are beginning to occur in childhood.[118,135,157,168]

TABLE 7.8

Growth, Development, and Feeding During Childhood

Age	Growth and Development	Feeding Skill and Food
1–2	Slower height and weight gains; appetite declines; soft spot on the head disappears; development of large muscles; pulls self up to stand; walks alone; baby teeth continue to come in; takes one long nap a day	Uses finger and thumb to pick things up; uses short-shanked spoon to help self-feed; lifts and drinks from a cup; helps scrub, tear, break, and dip foods
2–3	Slow and irregular height and weight gains; development of medium hand muscles; runs, climbs, pulls, pushes, walks upstairs, rides tricycle; has all 20 teeth; when hungry, expresses a need for food	Self-feeds with fingers, spoon, and cup (spills frequently); can spear food with a fork, wrap, pour, mix, shake, spread foods, and crack nuts (with supervision); eats independently; may have one favorite food at times
3–4	Continued growth and height gains; carries things without spilling; may substitute quiet time for nap	Self-feeder and drinks well from a cup
4–5	Continued growth and height gains; development of small finger muscles; hops, skips, and throws balls; good coordination; can manipulate buttons and shoe laces	Can use knife, fork, and a napkin; is a good self-feeder; can roll, juice, mash, or peel foods and crack egg shells
5–6	Continued growth; legs lengthen; develops fine coordination of fingers and hands; begins to lose front baby teeth; permanent molars appear	Can measure, grind, grate, and cut soft foods with a dull knife; may use hand mixer with supervision; may prefer plain, bland, and unmixed foods
6–9	Slower growth and height gains; additional permanent teeth form; 11 to 13 hours of sleep needed per day	Appetite may decline with slower growth; feeding skills are mastered; taste preferences broaden
9–12	Legs lengthen; growth comes in spurts; permanent teeth continue to form; 10 hours of sleep needed per day	Good appetite; hunger increases with growth spurts

© Cengage Learning 2013

The primary factors involved in childhood obesity are the lack of healthy eating and activity patterns. Parents and caregivers need to establish healthy eating and activity patterns.[B,Z,34,80]

Cognitive Development   Along with central nervous system development, children acquire intelligence and problem-solving abilities. Their ability to develop cognitively—that is, to think, learn, and perceive—is greatly influenced by their nutrition state, in conjunction with genetics and environment. The earliest nutrition environmental influences have lasting cognitive effects; these include the nutrition environment in the womb (in utero), during infancy, and during childhood. Poorly and undernourished preschoolers and school-age children are often sick more frequently, are less active (fatigued), and are less social and curious. Malnourished children experience overall delayed cognitive development that can have lasting effects into adulthood and affect their ability to lead a full and productive life. The longer the child's nutrition status is compromised, the greater the chance that the child will suffer some degree of cognitive impairment. Simply put, hunger can impact a child's ability to learn. A nutritious diet and the impact of eating breakfast have been shown to improve cognitive function in school-age children.[7,10,17,40]

As toddlers grow and develop, they become increasingly independent and mobile, yet continue to need the support and security of their parents and caregivers. They are especially vulnerable to accidents and injuries. Before toddlerhood, babies are totally focused on their own needs. At around age 2, children begin to become less self-centered and more interactive. They explore the environment around them, establish new relationships, and imitate others. They also begin to understand the family culture. Of course, they are still somewhat self-centered and may express their own will and throw tantrums. They may also develop rituals to deal with fears (separation from parents, darkness, loud sounds, and the like). Young children become more and more proficient with their language skills. Their verbal skills and their understanding of words increase dramatically. A typical child may have a 100-word vocabulary by age 2 and be combining words to make sentences by age 3.[53,70]

School-age children continue to experience maturation of motor skills and become even more independent. Their personalities blossom, their language skills expand to a few thousand words, and they speak in complete sentences. They develop the ability to limit behavior without external cues from parents and caregivers, but the strong desire to control situations may still persist. Parents need to set boundaries and provide guidance but allow room for the older child to separate from them more. School-age children are also more engaged in organized play, tend to cooperate better, and may exhibit enchanted-type self-centered thinking (where they may act like a character they have seen in a movie and think they deserve the treatment or benefits the character has).[7,17,126,140]

Division of Responsibility   Parents or caregivers decide what, when, and where to feed the child. The child decides whether to eat and how much to eat. Children will eat a variety of foods when they are hungry, and they know how much to eat. Because children have smaller stomachs than adults, they need to eat smaller portions more frequently throughout the day. Thus, providing nutritious snacks becomes an important part of the daily eating pattern. A child accepts as normal whatever foods are introduced or allowed, so making every food choice wise and healthy is important.[20,130,131]

Parents encourage healthy eating habits in their children when they provide nutritious, regular meals and snacks (see Figures 7.10 on page 382 and 7.11 on page 383) in an atmosphere that is conducive to eating, make the effort to prepare age-appropriate foods, are aware of feeding skills, and have realistic expectations about what their children will eat (see Table 7.8 on page 380). Children should not be allowed to drink too much juice or other sweetened beverages or graze on food throughout the day. The eating process will be much more successful if the child is hungry at the table during a regularly scheduled meal or snack. Having a calm and relaxed atmosphere at the table supports healthy eating, too. It is important for parents and caregivers to not get upset or give in to the demands of children and to maintain the correct division of responsibility for food.

Children should be taught that food is needed to nourish the body. Parents should avoid using food as a tool to control behavior or for emotional support. They should also resist forcing their children to clean their plates or to stay at the table until a defined amount of food is eaten. The ability of children to self-regulate their food intake is innate. Children should be allowed to pay attention to their own hunger and satiety cues. Parents should provide an age-appropriate serving size. Too much food on the plate can be intimidating. A general recommendation for serving size is 1 or 2 tablespoons per year of age. If children finish their meal, going back for seconds is okay. If children do not finish their meal, it's okay for the parents to decline to give the child dessert.[34,80,125,143]

Food, Friends, and Family   Eating should be a social and enjoyable time. Parents or caregivers should eat with the children, model healthy eating behaviors, teach acceptable behavior, avoid jumping up to get things, and provide an atmosphere free of distractions such as televisions, computers, and phones during meals. Mealtime should be quality family time. What children see their parents and siblings eat, they will usually eat too.[34,80,143]

Childhood Eating Behaviors   Children evaluate food by its color, texture, and taste. As toddlers, motor skills are developed enough to allow for self-feeding, although the result can often be messy. Independence increases along with interest in the environment. Thus, interest in food may decline.

Children are not concerned or even aware of foods' nutritional values. They tend to like foods that are sweet and a little bit salty and dense in Calories. Many children have not developed a liking for bitter and sour foods or strong-flavored foods (such as those containing garlic, onions, and spices). Because

Bribing a child to eat is inappropriate.

FIGURE 7.10 A regular feeding schedule of meals and snacks, with three sample menu ideas.

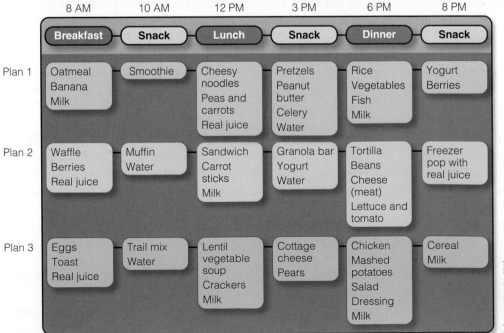

	8 AM Breakfast	10 AM Snack	12 PM Lunch	3 PM Snack	6 PM Dinner	8 PM Snack
Plan 1	Oatmeal Banana Milk	Smoothie	Cheesy noodles Peas and carrots Real juice	Pretzels Peanut butter Celery Water	Rice Vegetables Fish Milk	Yogurt Berries
Plan 2	Waffle Berries Real juice	Muffin Water	Sandwich Carrot sticks Milk	Granola bar Yogurt Water	Tortilla Beans Cheese (meat) Lettuce and tomato	Freezer pop with real juice
Plan 3	Eggs Toast Real juice	Trail mix Water	Lentil vegetable soup Crackers Milk	Cottage cheese Pears	Chicken Mashed potatoes Salad Dressing Milk	Cereal Milk

© Cengage Learning 2013

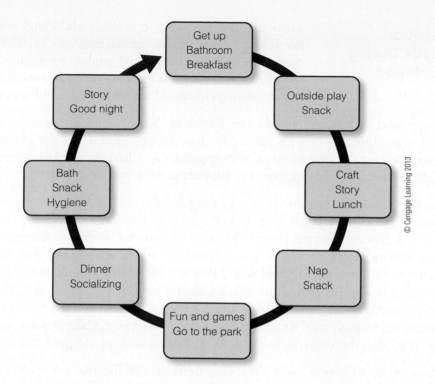

FIGURE 7.11 A lifestyle schedule for preschoolers.

© Cengage Learning 2013

of the growth and development occurring during childhood, parents and caregivers should realize that there is little room for unhealthy choices. They should delay introducing sugary foods as long as possible and satisfy the desire for sweetness with the natural sugars in whole fruits. Keeping foods simply prepared is a rule of thumb for providing meals that children will enjoy. Also, getting the child involved in the food preparation, selection, and even growing process has a big impact on the food consumption. Children who are involved in meal or snack preparation are eager to eat their food creations. When children request foods that are less nutritious, parents often want to eliminate or restrict the food. Too much restriction, however, can make the forbidden food more interesting, so it's a good idea to find a happy medium.[140]

Most children like to eat familiar foods, so introducing new foods can be a challenge. A child's interest for a new food can be encouraged by getting him or her involved. Perhaps reading a story about the food or about trying a new food (such as Dr. Seuss's *Green Eggs and Ham*), talking about attributes of the food (such as its color, shape, or smell), and manipulating the food (such as cutting it into fun shapes or combining it with a favorite food) can generate interest and help with the introduction of the new food. Introducing a new food alongside a familiar food and feeding children when they are hungry but not too hungry are tips to make mealtime successful. If children reject a new food, it often helps to introduce it again at a later date. It may take a few attempts over several months of exposing children to a new food before they accept it. As children age, it is also helpful to empower them by allowing them to make a few mealtime decisions, such as choosing which vegetable to cook for dinner. Although some food habits, likes, and dislikes established during childhood are transient, many form the basis for lifelong eating behaviors.[148]

Children can be opinionated and willful. They often go through eating phases that may involve different or altered food rituals. The ritual may be specific for the type of food, its shape, how it's cut, or how it goes with other foods. A food jag is usually a temporary eating pattern when a child will eat only one food for a period of time. Food rituals and jags are usually harmless and change over time. It is best if parents can understand, recognize, and go with the flow, realizing that as suddenly as the ritual came about, it can disappear just as suddenly.[34,57,80,143]

Nutrition Needs During Childhood

Many tools are available to evaluate childhood nutrition needs and to plan healthy meals and snacks. The DRIs, the MyPlate (MyPyramid) food guidance system, and the school food environment are areas that will be explored for understanding childhood nutrition needs and support.[168]

Energy and Nutrient Needs for Children Energy and nutrient needs are specific for children, as with other life span groups. The DRIs for energy and nutrients are established and shown in Appendix A. Children need a high number of Calories and adequate nutrients to support growth and development.

Calories Children are constantly gaining height and weight, and they need enough Calories to support their growth, development, and energy expenditure. There is a delicate balance, though, because too much Calorie consumption promotes obesity. A child's EER is a combination of total energy expenditure plus growth needs. Formulas to calculate a child's specific EER are found in Appendix A. A typical infant needs an additional 175 Calories per day to support growth, while a toddler needs an additional 20 Calories per day for growth. Calorie needs are also proportional to body size. As a child's body mass increases, Calorie needs increase because it takes more energy to maintain and move around a larger body.[W,69]

Carbohydrate, Protein, and Fat Children's needs for the energy-producing nutrients are similar to those of other life stage groups. Specific values are found in Appendix A. A few differences exist for protein and fat, but the carbohydrate recommendations for children are the same as for adults.

Carbohydrate is needed to support brain, central nervous system, and red blood cell functioning. Like adults, children need 130 grams per day, the minimum DRI for carbohydrate. The AMDR for carbohydrate for children is 45 to 65 percent of Calories, also the same as that for adults. After age 1, there is a DRI for fiber, which is 1.4 grams per 100 Calories consumed. For the reference child this translates to 19 grams per day for ages 1 to 3 and 25 grams per day for ages 4 to 8. Gender specificity begins at age 9. The fiber recommendation is then 26 grams per day for girls ages 9 to 13 and 31 grams per day for boys ages 9 to 13.

Just as for adults, there are many benefits of fiber, and there are negative effects of too much fiber (see Module 3, page 122). Children should consistently be offered fiber-rich food sources such as whole grains, legumes, whole fruits (rather than fruit juice), and vegetables. A diet that includes these foods will reduce the incidence of constipation and make bowel movements easier. If a child consumes too much fiber, he or she may get diarrhea and could further experience negative health effects if non-energy-providing fiber content in the diet displaces energy-dense foods. Additionally, the bioavailability of some minerals, such as iron and calcium, is compromised with a high-fiber diet.[69,111]

Children have higher protein needs than adults, but lower protein needs than infants. The difference is attributed to the demand for tissue repair, maintenance, and growth. As shown in Appendix A, the DRI for protein in childhood is 0.95 grams per kilogram of body weight, and the AMDR is 5 to 20 percent of Calories. Protein needs should be met by providing high-quality protein or the equivalent through ensuring complementation of plant protein sources (see Module 1, pages 18–19).[OO]

Fat provides an important source of energy to support childhood growth and development. Children can consume more fat than adults. The AMDR is 30 to 40 percent of Calories for children ages 1 to 3, and 25 to 35 percent of Calories for children and adolescents ages 4 to 18. Compare this to 20 to 35 percent of Calories during adulthood. Children have the same AMDR for the essential fatty acids as adults; for linoleic acid, 5 to 10 percent of Calories, and for alpha-linolenic acid, 0.6 to 1.2 percent of Calories. However, because children eat fewer Calories than adults, the actual DRI gram value for the essential fatty acids is lower. Depending on age, it is 7 to 12 grams per day for linoleic acid and 0.7 to 1.2 grams per day for alpha-linolenic acid (see Appendix A). Healthy fat sources such as nuts, seeds, and liquid plant oils will provide the essential fatty acids. Furthermore, from ages 1 to 2 it is appropriate to offer children whole milk because it provides needed Calories, protein, calcium, and fat.[69,124,133]

Vitamins, Minerals, and Water Children need the same essential micronutrients as adults. However, some nutrients are needed in greater amounts during growth. The DRI for vitamins, minerals, and water are provided in Appendix A. Vitamin and mineral needs can be met for a majority of children by consuming a variety of healthy foods and beverages daily. Getting essential nutrients from foods is preferred over supplements because whole plant foods provide other benefits, such as being high in fiber and providing phytochemicals. Vitamin and mineral supplements are recommended by the American Academy of Pediatrics when the child is undernourished, has poor eating habits, has a poor appetite or anorexia, has a fad or limited diet, or is a vegan (a vegetarian that does not consume any milk, eggs, or meat products). Special consideration for the use of iodized salt and fluoride supplements (up to the age of 16 years) may be needed, depending on soil quality and seafood intake for iodine and water fluoridation for fluoride. Highlights of a few nutrients of concern during childhood follow. General information on vitamins, minerals, and water can be revisited in Module 5, starting on page 225.[F,11,69,107,109]

Calcium needs are high during childhood because of the need to support bone growth. The DRI is 500 to 1,300 milligrams per day depending on the child's age (see Appendix A). To meet the need for calcium, children should consume 2 cups of milk or milk product (like yogurt or fortified milk alternatives such as soy or almond milk) each day. Calcium-fortified rice milk or orange juice will also provide needed calcium for the body, but are not rich sources of protein. A high intake of milk and juice leads to a low intake of iron-rich foods. Low iron intake eventually causes iron-deficiency anemia. Children are at increased risk for iron-deficiency anemia even without food intake pattern imbalances because of the increase in blood volume and supply that is required to support new tissue growth. Food sources of heme iron include red meat, poultry, and pork. Non-heme iron is available from iron-fortified cereals, whole grains, and legumes. Fluoride is another nutrient of concern during childhood because of its role in primary and secondary tooth formation (see BioBeat 7.7 on page 374).[18,54,92,95,107,124]

Water (fluid) needs are usually met by providing children with 2 cups of milk, ½ cup of real fruit juice, and water throughout the day, along with fluid-rich foods like whole fruits and vegetables. This daily pattern is usually sufficient for meeting a child's daily fluid needs. The DRI for water is 1.3 liters per day for children ages 1 to 3, 1.7 liters per day for children ages 4 to 8, and 2.4 liters per day for children ages 9 to 13. During times of sickness (vomiting and diarrhea) and with increased temperatures, altitude, and physical activity, fluid needs will increase. Dehydration (see Module 5, page 254) can be a serious and life-threatening condition. Therefore, sick children should be monitored carefully, and medical attention should be given when needed.[B,F,69]

MyPlate for Children The MyPlate food guidance system is designed for Americans starting at age 2 and continuing all the way through the life span to the ripe old age of 110. The model is based on the Dietary Guidelines 2010 (see Module 2, page 70). It can be accessed online at www.choosemyplate.gov and is available in Appendix C. The MyPlate materials for children continue to include the MyPyramid food guidance model. There is a MyPyramid model for preschoolers and school-age children. Figure 7.12 shows the model for school-age children. The food groups are the same regardless of age; however, the activities vary and are age appropriate. In general, the model provides a message that children should consume a variety

FIGURE 7.12 The MyPyramid food guidance system for children.

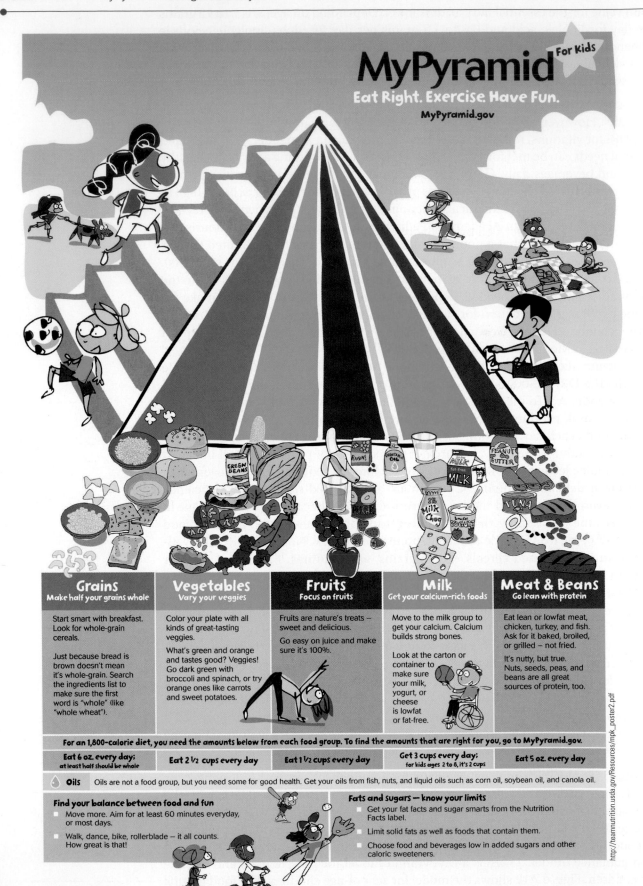

of whole, fresh foods from the different food groups, limit the intake of solid fats and added sugars, and be physically active. When the MyPlate (MyPyramid) food guidance system is followed, it becomes evident that there is very little room in a child's diet for unhealthy foods (including processed foods or confections) that provide an excess of empty Calories. The MyPlate system is designed to provide a balance of whole, fresh foods with Calorie control and nutrient density.[T,GG,119]

The School Food Environment Schools provide various food environments. In general, there are federal guidelines to meet basic nutrition standards. However, some schools minimally follow the standards, others exceed them, and others offset school expenses and budgets by permitting food chains and vending machines on their campuses. Just like the home eating environment, it takes an effort at school to promote childhood health and well-being and prevent overweight and obesity. The U.S. Department of Agriculture's (USDA) Food and Nutrition Service school meal programs are described briefly below.[UU,WW,81]

School Meals and Snacks The National School Lunch Program (NSLP) operates in public and not-for-profit private schools, as well as in residential childcare institutions. It is a federally assisted meal program that provides either low-cost or free meals (depending on the child's family's income) to children every school day. The lunch meal must meet federal nutrition requirements, which are to provide one-third of the DRI for Calories, vitamin A, vitamin C, calcium, and iron. Furthermore, the meal will provide no more than 30 percent of the Calories from fat and 10 percent of the Calories from saturated fat. The decision as to which foods are served and how the foods are prepared is determined by each local school food authority. Because many children continue to need food assistance during the summer months when school is out, schools can participate in the seamless summer program. Basically, this continues to provide the NSLP during the summer to low-income children.[TT,WW]

The School Breakfast Program operates similarly to the NSLP. One difference is that one-fourth of the DRIs for Calories, vitamin A, vitamin C, calcium, and iron is provided per meal. Schools that offer after-school activities may also participate in the after-school snack program that is a subcomponent of the NSLP.

The Fresh Fruit and Vegetable Program is controlled at the national level through the USDA's Food and Nutrition Service. The program strives to combat childhood obesity by giving children access to nutritious fruits and vegetables that they may not otherwise have available and helps children learn healthier eating habits.[VV,WW,111]

Childhood Fitness and Lifestyle Management

Dietary approaches to support childhood health are needed and vital. However, childhood fitness and lifestyle management are also crucial for complete health promotion. Active kids are healthy kids and tend to be free of diet- and health-related issues, including obesity and chronic disease.[I,U,47,91,127,154]

Active Kids are Healthy Kids Children need regular exercise and physical activity to promote physical, social, and psychological health. Children over the age of 2 should engage in at least 60 minutes of activity on most days of the week. Activities should be fun, and parents and caregivers should provide the opportunities, participate, and enjoy the benefits of physical activity together.

Activities for preschoolers include walking, playing chase and catch, ball-tossing, tumbling, running, jumping, dancing, riding tricycles or bicycles with training wheels, swimming, swinging, sledding, climbing on the playground or jungle gym, and sliding down slides. Activities for school-age children include those for preschoolers plus playing badminton, soccer, baseball, basketball, biking, martial arts, roller skating, skateboarding, skiing, playing tennis, and more.[B,I,U,Z,91,144]

Jennifer Turley

An active child is a healthy child.

Diet and Health Issues in Childhood Many diet-related health issues are applicable to children. A few of the more prevalent and pressing issues of today include obesity, type 2 diabetes, pediatric hypertension, heart disease, allergies, and toxin exposure.[24,73,79,140,146,168]

Obesity Sedentary lifestyles and unhealthy meal patterns and composition are contributing factors to the staggering rise in childhood obesity. Over the past 30 years, the incidence of childhood obesity has more than tripled in the United States. Now, approximately 20 percent of American children ages 6 to 11 are obese, compared to an expected rate of 5 percent. Obesity is commonly determined by having excess weight-for-height as measured by the body mass index (BMI). Figure 7.13 shows the BMI grid by gender for children and adolescents ages 2 to 18 by percentile. Normal weight-for-height occurs when a child is in the 10th to 85th percentile. A child is considered underweight when he or she is less than the 10th percentile, at risk for obesity when greater than the 85th percentile, and obese when greater than the 95th percentile. Because so many factors are involved in obesity and its treatment is so difficult, it is best to prevent it by adopting healthy lifestyle practices during childhood.[O,R,42,73]

Sedentary activities such as television watching and playing video games should be limited. A general recommendation is to allow no screen time for children under the age of 2 and to limit screen time to no more than two hours per day for children over age 2.[F,I,U]

FIGURE 7.13 Body mass index (BMI) charts for boys (left) and girls (right) ages 2 to 18.

The amount that children are allowed to eat and drink while watching television should be carefully controlled, because the distraction provided by television promotes mindless Calorie intake. Foods and beverages should be thoughtfully and mindfully consumed. The intake should be enjoyed and attention should be

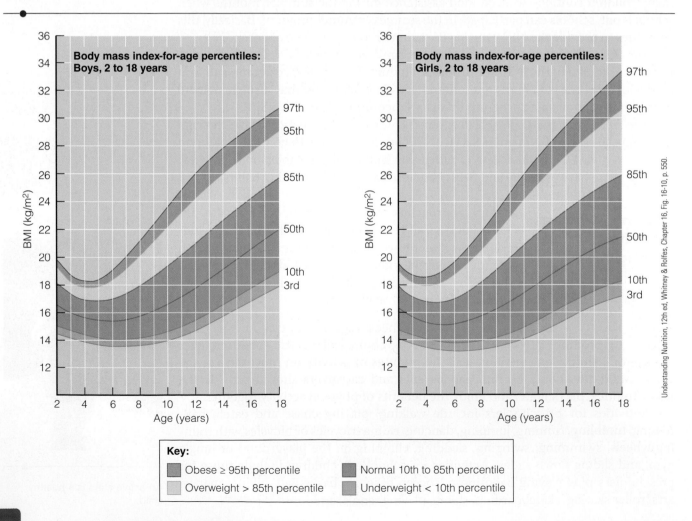

Key:
- Obese ≥ 95th percentile
- Overweight > 85th percentile
- Normal 10th to 85th percentile
- Underweight < 10th percentile

given to hunger, appetite, and satiety signals. Additionally, children should avoid drinking too many Calories. Drinking does not send the same satiety messages to the brain, so it's easy to consume an excess of Calories and promote weight gain by drinking liquid empty Calories from sodas and fruit drinks.[34,80,118,143,150,168]

Heart Disease, Hypertension, and Hyperglycemia   It may seem odd to mention heart disease, hypertension, and hyperglycemia during childhood. But because of the current childhood obesity problem, chronic disease risk is elevated. Heart disease is a chronic disease that most often comes to mind in later adulthood, and hypertension and hyperglycemia are contributors to atherosclerosis that in turn contribute to heart disease. The reality is that unhealthy diets and sedentary lifestyle practices taught or permitted during childhood set the child up for early death and loss of health and quality of life. These formerly adult diseases and conditions are now occurring in children with unhealthy lifestyle practices. The American Heart Association is a promoter of primary prevention that begins in childhood. Prevention is as simple as eating a well-balanced and nutrient-adequate diet, avoiding tobacco use, and being physically active.[73,91,93,118,146,158]

Food Allergies   Allergy is a genetic condition with dietary implications (see BioBeat 7.8). In order to assess a child's reaction to a new food, only one new food should be introduced at a time, and the child should be monitored for reaction for up to five days. Because allergy requires immune sensitization (see Module 3, page 131), a child should not have an allergic reaction to a food with the first exposure. However, allergy may develop with subsequent exposures. It is a good idea to minimize allergy risk by rotating foods in the diet, such as every three to four days, especially in children who are genetically predisposed to allergy. This builds in dietary variety as well. If a child does have an allergy to a food, then the nature and severity of the allergy should be assessed by a medical doctor. For severe and life-threatening reactions involving systemic anaphylaxis, complete dietary omission of the food and food ingredients is needed. As children grow and develop, their immune systems mature, and it is possible for a child to outgrow food sensitivities and allergies. However, if the allergy persists after age 3, it can be expected to persist for life.[E,V,KK,24,58]

BioBeat 7.8

The Evolution of Allergy

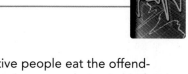

In humans, allergies are mediated primarily by one of five immunoglobulin (antibody) types, specifically IgE. Mammalian IgE and IgG evolved from an IgY-like ancestor millions of years ago. IgY is the primary antibody of amphibians, reptiles, and birds. The evolution of a specific antibody domain in IgY and thus IgE makes the antibody have a unique high affinity and slow dissociation rate for the antigen it is designed to bind. This evolutionary feature offers the natural selection advantage of improved immune defense against parasites. The bird (avian) IgY antibody co-functions to fight infection and to mediate anaphylaxis, which can occur in chickens.

Anaphylaxis is a systemic and life-threatening allergic reaction, which is IgE-mediated in humans. Allergies often begin in childhood. A child can be allergic to any food, although eight foods account for the vast majority of all food-allergic reactions: milk, eggs, wheat, soy, peanuts, tree nuts, fish, and shellfish. Even though 90 percent of allergic reactions are caused by the aforementioned foods, there are more than 160 other known allergic foods, plus the unknown potential for generating even more hypersensitivity by genetic engineering of plants in conventional food production. When allergic, intolerant, or chemically sensitive people eat the offending food—or a perceived safe food that is cross-contaminated with an allergic counterpart—their immune system responds immediately and dramatically, causing mild, moderate, or severe life-threatening reactions in the gastrointestinal tract, respiratory tract, and skin. Avoiding offending foods and seeking safe food alternatives is the only way for persons with allergies and/or intolerances to remain healthy.[V,151]

If a child you know has food allergies, what are the offending foods, the reactions, and some ways to cope with the allergy?

Toxin Exposure   A couple of incidental additives that are of concern are heavy metal exposure and residues that are in conventionally raised foods. Lead is a heavy metal and a toxin to the body that negatively affects brain, blood, and kidney functioning. It is also implicated in causing stunted growth, lower intelligence, and behavioral problems. Lead enters the body when children put lead-based objects into their mouths (such as paint chips from paint manufactured before the year 1950), from lead leaching into the water supply from lead-soldered pipes, from ceramic dishware with colorful lead paint that is not sealed and that leaches into the food on the dish, and from some canned goods produced in other countries. Iron deficiency can exacerbate lead toxicity, because free iron-binding receptor sites in the small intestine can bind lead and allow it to gain entrance into the body.[BB,OO,140]

Conventionally produced foods are exposed to toxins that can make their way into the body. One such example is the use of pesticides in plant crops and the bioaccumulation of such chemicals in the fat portions of conventionally produced animal milk and meat products. Children are especially vulnerable to the damage that pesticides can have on cells and body systems, especially the endocrine system. Switching a child's diet from conventional to organic foods will virtually eliminate pesticide exposure, along with eliminating exposure to hormones, antibiotics, and genetically modified organisms (see Module 6, page 321).[168]

Summary Points

- Childhood growth is measurable by standardized growth charts and developmental milestones.
- Cognitive development is greatly influenced by nutrition, education, genetics, and the environment.
- Parents or caregivers decide what, when, and where to feed the child, and the child decides whether and how much to eat.
- A daily routine with regular meals and snacks supports healthy eating.
- Getting children involved in mealtime preparation increases their interest in eating the meal.
- The DRIs, the MyPlate (MyPyramid) food guidance system, and school meals are beneficial for promoting proper childhood nutrition.
- Children need enough Calories and nutrients to support their growth, development, and energy expenditure.
- Dietary recommendations for protein and fat are higher for children than for adults.
- Calcium and iron are common nutrients of concern for children.
- Active kids are healthy kids and tend to be free of diet- and health-related issues, including obesity and chronic disease.
- Children with allergies may need to eat a modified and/or rotation diet for life if the allergies persist after age 3.
- Feeding children organically produced foods can almost eliminate their exposure to harmful components used in conventional food production practices.

Take Ten on Your Knowledge Know-How

1. How does nutrition support childhood growth?
2. How does nutrition support cognitive development in children?
3. What is the division of responsibility for feeding?
4. How would you outline a healthy daily schedule and meal/snack pattern to follow for a preschooler and a school-aged child?
5. What are the general nutrient recommendations for energy, carbohydrate, protein, and fat for children?

6. What are the general nutrient recommendations for vitamins, minerals, and water for children?
7. How does the MyPlate (MyPyramid) food guidance system support childhood nutrition?
8. Describe the school meal programs for children.
9. How can childhood obesity be prevented?
10. How can food allergies be prevented and managed in childhood?

7.4 ADOLESCENT NUTRITION

Introduction

Adolescence is the time of life between childhood and adulthood. There are dramatic changes in the body—weight, height, and composition—as well as reproductive ability. Additionally, adolescents develop their autonomy and identity and learn to function more self-sufficiently, ultimately becoming capable of independently caring for themselves in adulthood. Adolescent diet and lifestyle experiences determine their future health status and life quality.[OO,XX,118,158]

T-Talk 7.4
To hear Dr. Turley talk about adolescent nutrition, go to www.cengagebrain.com

The Physical, Cognitive, and Social Triad During Adolescence

Physical, cognitive, and social development continue throughout adolescence. Encouraging adolescents to practice healthy eating and lifestyle habits will support their growth and development, along with their acquisition of reproductive capability. The establishment of a healthy lifestyle during adolescence will carry them well through adulthood.[O,16,140]

Successful Growth and Development Both girls and boys experience a dramatic increase in growth during adolescence, and an increased appetite usually coincides with a growth spurt (see Figure 7.14). Growth takes place at the **epiphyseal plate**—the tissue near each end of each long bone that closes when growth terminates. At the point when growth stops, the tissue itself turns to bone. In girls, a growth spurt occurs commonly between the ages of 10 and 13 and is complete by two years after menarche (first period). In boys, the growth spurt usually occurs between 12 and 15 years, but growth may continue into the 20s. Girls usually gain 2 to 8 inches in stature, while boys gain 4 to 12 inches. Body weight increases during adolescence, too. Girls typically gain approximately 35 pounds, while boys gain 45 pounds. The type of body weight gained is different by gender. Girls gain more fat mass whereas boys gain more lean body mass. This is normal and necessary for successful reproductive health purposes.[54,118,167]

Boys are usually taller and heavier than girls throughout most of the life span. However, at the beginning of the growth spurt and between the ages of 9 and 13, girls are often taller and heavier than boys and begin menstruation and breast development. Boys experience rapid muscle growth during their growth spurt. Between the ages of 14 and 16, boys experience more rapid growth spurts and may have very large appetites. Their voices deepen, and they will develop more body and facial hair. Girls experience regular or irregular menstrual cycles. For both genders, acne is common, all of the permanent teeth are present except the wisdom teeth, and sleep takes on adult patterns.[140]

Puberty is the time during adolescence when the reproductive system matures. This process is hormonally driven. Although the age of onset of puberty can vary, obesity tends to trigger early reproductive development. Onset of puberty can also

epiphyseal plate The tissue near each end of each long bone, where growth takes place; it closes by turning to bone itself when growth terminates.

puberty A period in childhood growth that marks the transition to adulthood; the processes of physical change of a child's body that results in reproductive capability.

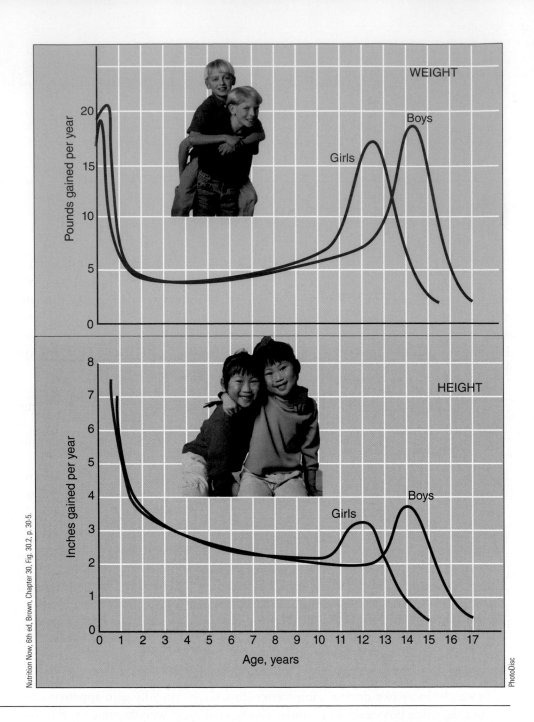

FIGURE 7.14 Growth rate from infancy through adolescence.

Nutrition Now, 6th ed, Brown, Chapter 30, Fig. 30.2, p. 30-5.

PhotoDisc

be negatively influenced by endocrine disruptors such as pesticide residues in food and bisphenol A in plastics.[OO,16,140]

Girls typically go through puberty earlier than boys. The usual age at which puberty starts is 9 years in girls and 11 years in boys. Early bloomers may enter puberty as early as ages 7 to 8 in girls and 9 to 10 in boys. Table 7.9 describes the stages of sexual maturation that can be expected once puberty starts.[OO,16,140] There are predictable changes that progress through five distinct stages of maturation. Figure 7.15 shows the typical timing of adolescent growth and development, including sexual maturity staging. In girls, the stages describe maturation and changes in the nipples, breasts, and pubic hair, along with the onset of menarche and acquisition of regular periods (menstruation). In boys, the stages describe maturation and changes in the testes, scrotum, penis, and pubic hair.[16,140]

TABLE 7.9

Sexual Maturity Rating for Girls and Boys

Stage	Girls (Breast, Pubic, Menarchy)	Boys (Genital, Pubic)
1	Nipple elevation; no pubic hair	No change in size or proportion of testes, scrotum, and penis; no pubic hair
2	Small raised breast bud; sparse growth of pubic hair along labia	Scrotum and testes enlarge; skin texture and color of scrotum change; little to no change in penis size; sparse growth of pubic hair at base of penis
3	Breast and areola enlarge; pubic hair pigments (darkens), coarsens, curls, and increases in amount; menarche occurs	Scrotum, testes, and penis (length then width) grow; pubic hair pigments (darkens), coarsens, curls, and increases in amount
4	Breast enlargement, with areola and nipple second mounding and projection; adult-type pubic hair but not spreading to thighs; irregular periods	Penis enlargement, development of glands, further scrotum and testes growth and darkening of scrotal skin; adult-type pubic hair but not spreading to thighs
5	Mature, adult-contoured breast and areola with only nipple projecting; adult-type and amount of pubic hair spreading to thighs; regular periods	Adult genitalia size and shape; adult type and amount of pubic hair spreading to thighs

© Cengage Learning 2013

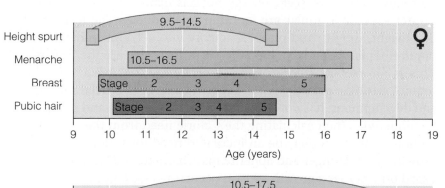

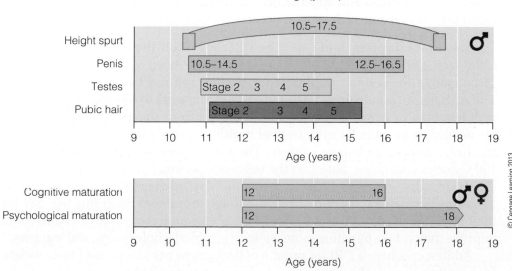

FIGURE 7.15 Typical timing of adolescent growth and development, including sexual maturity staging.

© Cengage Learning 2013

Cognitive and Social Changes During Adolescence Adolescents experience a maturity spurt cognitively and socially as well as physiologically. Slower rates of maturation may extend throughout young adulthood. Adolescents gain a greater ability to apply knowledge and think in multidimensional and abstract ways. Peer relationships can become the most important focus of an adolescent's life, even more so than academics and family. Gaining peer acceptance, which is important for an adolescent's self-esteem, is often based on his or her appearance. There are psychological and emotional changes, such as adolescents wanting more independence and possibly displaying more autonomy and even rebellious behaviors and attitudes. Parents still have a large influence over moral, religious, and educational issues and thought processes. As would be expected with the physical changes occurring during adolescence, adolescents become more interested in sexuality. This is accompanied by greater interest in intimacy, friendships, and relationships.[O,16,26,40,140]

Nutrition Needs During Adolecence

There are two DRI age groups for adolescences: ages 9 to 13 and 14 to 18. The DRIs in these age groups reflect the typical age of puberty by gender. Actual nutrient needs may be more adequately determined by the individual's developmental stage and physical growth rate. Respecting appetite, regulatory mechanisms, and continuing to eat a well-balanced diet are important to ensure adequate nutrition while discouraging overeating.[82,106,118,125,140]

Energy and Nutrient Needs for Adolescents Energy and nutrient needs are specific for adolescents, as with other life span groups. The DRIs for energy and nutrients are established and shown in Appendix A. Adolescents need a high number of Calories and adequate nutrients to support the final growth spurts and stages of life's growth and development, as compared to childhood stages.[118]

Calories   The adolescent energy need is expressed as the EER. The age- and gender-specific formulas are found in Appendix A. The formulas take age, gender, height, weight, and physical activity into account to determine daily caloric requirements. Essentially, the adolescent formulas are the EER plus 25 Calories. This is an average caloric increase over time, not what the adolescent would require during a growth spurt. Adequate energy is crucial for adolescent growth and development. If Calorie needs are not met, growth and sexual maturation may be hindered. When calculating the Calorie need from the EER formula, it should be noted that there may be an increased need for Calories during an adolescent growth spurt. Hunger and appetite may fluctuate during the years of adolescence based on whether a growth spurt is occurring (see Figure 7.16).[69]

Carbohydrate, Protein, and Fat   The DRI for carbohydrate during adolescence is the same as during adulthood. There is a minimal need of 130 grams per day of carbohydrate for brain, central nervous system, and red blood cell functioning. The AMDR for carbohydrate is the same as for adults, 45 to 65 percent of Calories, with less than or equal to 25 percent of Calories from sugars. Fiber-rich sources are needed to promote gastrointestinal health and prevent obesity, type 2 diabetes, heart disease, and some cancers. The DRI is personalized at 1.4 grams per 100 Calories consumed. Based on the reference person, the DRI for fiber for adolescent females ages 9 to 18 is 26 grams per day, whereas the DRI for males is 31 grams per day for the 9-to-13-year-old age group and 38 grams per day for the 14-to-18-year-old age group. Most adolescents underconsume fiber and could improve their intake by eating more fruits, vegetables, whole grains, and legumes.[69]

Adolescents have a DRI for protein of 0.95 grams per kilogram of body weight for ages 9 to 13 and 0.85 grams per kilogram of body weight for ages 14 to 18. The protein need is lower than in infancy and early childhood. Consuming adequate

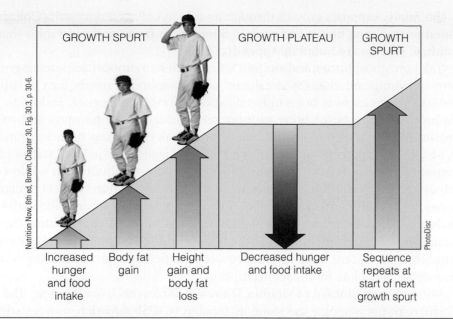

Nutrition Now, 6th ed, Brown, Chapter 30, Fig. 30.3, p. 30-6.

GROWTH SPURT			GROWTH PLATEAU	GROWTH SPURT
Increased hunger and food intake	Body fat gain	Height gain and body fat loss	Decreased hunger and food intake	Sequence repeats at start of next growth spurt

PhotoDisc

FIGURE 7.16 Growth spurts, hunger, and food intake during adolescence.

protein is very important for tissue repair and maintenance as well as supporting growth. Actual protein needs will fluctuate based on physiological demand, depending on whether the adolescent is in a growth spurt.

There is no DRI for total fat for adolescents. However, there is a DRI for the essential fatty acids linoleic acid and alpha-linolenic acid. The DRI for linoleic acid is 10 to 16 grams per day, while the DRI for alpha-linolenic acid is 1.0 to 1.6 grams per day. Both are based on age and gender (see Appendix A). The AMDR for total fat is slightly higher for adolescents than adults, at 25 to 35 percent of Calories, with the AMDR for the essential fatty acids being the same as for adults. Most adolescents consume fat near the upper range of the AMDR. Unfortunately, like adults, they tend to overconsume saturated fat and cholesterol and underconsume the essential fatty acids. Consuming less fast food and processed food and more whole plant foods (including nuts and seeds) will support a healthy shift in fatty acid intake.[125,133]

Vitamins, Minerals, and Water   The DRIs for the vitamins and minerals are approaching or at adult values. Many nutrient values are the same for males and females. Male adolescents do have a higher DRI than females for vitamin A, vitamin C, thiamin, riboflavin, niacin, vitamin B_6, choline, chromium, magnesium, manganese, zinc, and water. Females have a higher DRI than males for iron. Several nutrients are particularly important for adolescent growth and development; vitamin A, folate, zinc, and vitamin C will be discussed below.[28,29,69,129]

Vitamin A has many functions. It is primarily known to be required for vision. However, vitamin A is also needed for growth and reproduction. Some of the best food sources of provitamin A are from dark green leafy and deep orange pigmented fruits and vegetables. The typical male and female teen diet is low in vitamin A because it is low in fruits, vegetables, and fortified dairy products.

Like vitamin A, the intake of folate is typically deficient, especially in teen girls. This gives reason for concern with the onset of reproductive capability because folate is so important when it comes to preconception nutrition. Folate is also needed for new cell production because of its integral role in nucleic acid and protein synthesis. Food folate may not be converted properly to folic acid; therefore, a dietary supplement or fortified foods are important to consume.

Zinc is also required for adolescent growth and development in that it is needed to make DNA and proteins and is required for spermatogenesis. Some protein-rich foods, such as red meats and seafood, are good sources of zinc.

Milk and milk alternatives provide a good source of calcium to support growth and health.

Vitamin C supports growth through its role in collagen synthesis. Collagen is needed for building bones and tissues. Some fresh fruits and vegetables that are vitamin C sources are noted in Appendix C.

Calcium, phosphorus, and vitamin D are needed to support adolescent growth. There are significant changes in calcium and phosphorus metabolism to support the dramatic increase in bone mineralization during adolescence. Failure to provide proper nutrients for bone support can lead to low bone mass and related diseases, including osteoporosis in adulthood. Peak bone mass occurs around the age of 30, so adolescent nutrition affects bone health in the later stages of the life span. The need for calcium and phosphorus are the highest in adolescence compared to any other life stage group. The recommendation is 1,300 milligrams per day for calcium and 1,250 milligrams per day for phosphorus for males and females. Phosphorus is more abundant in the food supply than calcium. The best calcium sources are from milk and other dairy products. Calcium intake is usually adequate in adolescent boys, but inadequate to deficient in girls. This deficiency is attributed to a decline in milk consumption.[18,27,117,167]

An inadequate intake of vitamin D during adolescence is common. The DRI is 5 micrograms per day, the same as for adults. Calcium-rich sources such as milk and other dairy products are fortified with vitamin D, so when an adolescent reduces milk intake to consume sodas and sports drinks instead, both calcium and vitamin D status decline. Furthermore, lack of vitamin D leads to insufficient calcium absorption and retention in the body. Lack of sunlight exposure is also implicated in poor vitamin D status, so adolescents with poor self-synthesis from sunlight exposure should consider taking a vitamin D supplement, up to 1,000 International Units (25 micrograms) per day. Additionally, adolescents should consume 3 cups of milk each day as a source of calcium, phosphorus, and vitamin D. Vitamin K, largely produced by the bacteria in the gut, can be supported by consuming live cultured yogurt products.[18,36,101,117,167]

The need for iron increases during adolescence. In girls, the increase is needed to cover the losses of iron incurred during menstruation and the need for growth and new tissue formation. In boys, the needed increase in iron is less than that for girls in the 14-to-18-year-old life span group and is required to support intense growth. It is also needed to maintain normal body stores of iron. The DRI for iron is 8 milligrams per day for adolescents ages 9 to 13 and 11 to 15 milligrams per day for adolescent boys and girls ages 14 to18. If iron-deficiency anemia occurs, it can lead to increased fatigue and an inability to concentrate; these symptoms in turn can interfere with physical activities and intellectual pursuits in school. Food sources of iron include heme iron from animal meat and non-heme iron from plant sources, fortified foods, and animal products that do not have a blood supply, such as eggs and milk. As covered in Module 5 on page 268, heme iron is more bioavailable than non-heme iron, and the absorption of non-heme iron can be increased with vitamin C.[92]

Adolescents often have an unhealthy balance of sodium, potassium, and water intake. The DRI for sodium is the same for adolescents as for adults, 1,500 milligrams per day. The UL is 2,200 milligrams per day for adolescents ages 9 to 13 and 2,300 milligrams per day for ages 14 to 18. Many teens consume a high amount of sodium, which is above the UL from their processed and fast food intake.

Along with the excess intake of sodium usually comes an inadequate intake of potassium. Potassium is abundant in whole, fresh foods such as fruits, vegetables, meats, milk, grains, and legumes. Adolescents ages 9 to 13 have a DRI for potassium of 4,500 milligrams per day, while those ages 14 to 18 have a DRI of 4,700 milligrams per day. There is no UL for potassium. Fluid intake may also be low because of the lack of water intake and higher consumption of processed, caffeine-containing beverages. The DRI for water is 2.1 and 2.4 liters per day for girls and boys ages 9 to 13, respectively. For girls and boys ages 14 to 18, the DRI is 2.3 and 3.3 liters per day, respectively. Boys have a higher DRI for water because their muscle mass increases in the later adolescent age group. This is because muscle is 75 percent water. Fluid needs will be elevated above the DRI with exercise and hot climate conditions.[16,118,140]

Adolescent Fitness and Lifestyle Management

Fitness and lifestyle management are components for the healthy maturation of adolescents. The physical maturation processes improve adolescents' abilities to endure longer bouts of aerobic activity and respond better to resistive exercise programs (see Module 4, page 190). Unfortunately, some adolescents go astray and engage in behaviors and activities that are not health-promoting. Parents and caregivers can be good role models in their diet, fitness, and other lifestyle choices for the teens in their lives.[I,U,110,128]

Adolescent Fitness Activities Physical fitness and activity is needed throughout the life span, so fitness should be a part of the daily schedule of adolescents. As mentioned already, adolescents should engage in 60 minutes of physical activity daily. The types of fitness activities will become more like adult fitness activities. Adolescents can be physically challenged by playing organized sports (individual and team, recreational and competitive). Adolescents should explore different activities to find some that they enjoy doing both individually and with other teens. Some examples of excellent activities for adolescents include basketball, baseball, volleyball, football, track and field, soccer, swimming, running, hiking, biking, skiing, hockey, snowboarding, golf, tennis, racquetball, karate, badminton, and lacrosse. Physical activity, along with healthy eating, can help promote lean body mass weight gain rather than fat gain.[I,U]

Kayaking is an exciting form of recreational physical activity for adolescents.

Jennifer Turley

Diet and Health Issues in Adolescents There are many dietary and health-related issues facing today's adolescents. Weight gain, food intake shifts and imbalances, skin disorders, and disturbed and distorted eating practices will be discussed. The reliance on handy foods, such as packaged foods and fast foods, promotes the overconsumption of saturated fat, sodium, sugar, and Calories, with inadequate intakes of vitamins, minerals, and essential fatty acids. Focusing on eating whole, fresh foods improves nutrient density and the ability to meet nutrient needs.[158]

Weight Gain in Adolescents Although adolescents experience rapid growth that requires positive energy balance, many adolescents are overconsuming Calories compared to their actual need. So adolescence can often be a time in the life span when overweight begins. This is reflected in the rise in adolescent obesity across the nation.

Overweight or obese adolescents who begin weight loss programs should follow the guidelines set forth in Module 4 and not exceed losing 1 pound per week (see pages 186–188). Too-rapid weight loss can interfere with adolescent growth and development. If obesity persists, it can lead to premature health problems, including elevated blood cholesterol, blood glucose, and blood pressure, as well as risk for heart disease, cancer, and type 2 diabetes. MyPlate is a useful guide and resource for meeting nutrition needs and preventing excess weight gain during adolescence (see Module 2, page 61, and Appendix C).[R,14,73,91,93,146,150,162]

Food Intake Shifts and Imbalances Some teens have a healthy shift in behavior, such as eating a healthier diet than the rest of their family, engaging in more sports and organized physical activity events, or becoming more interested in intellectual activities. Some adolescents decide to become vegetarians in some form (see Module 1, page 10). Vegetarian diets can be health promoting, but like all diets they need to be appropriately planned. This is especially true for meeting the nutrient needs during adolescence.[OO,16,109,118,140]

Unfortunately, many teens begin to engage in unhealthy eating and lifestyle practices. They often eat more away-from-home foods, including more fast foods and processed foods. The teen diet often becomes devoid of fruits, vegetables, dairy products, and whole grains. Only one-fourth to one-third of American adolescents eat enough fruits and vegetables. Instead, they are eating more solid fat and added sugar and missing out on needed nutrients and fiber. Their diet is on the upper edge of the AMDR for total fat and frequently exceeds American Heart Association guidelines for saturated fat intake. Disproportionately higher consumption of sugary beverages and fatty foods is seen during adolescence compared to younger children.[111]

Teens often express dissatisfaction with their body image and are subject to peer influences. Teens are influenced greatly by their peers, role models, and the media. As a result, they may engage in unhealthy behaviors related to the use of alcohol, drugs, and tobacco. Sometimes it is just an experiment, but for many teens it becomes an unhealthy habit. Tobacco is addictive and interferes with nutrition and health. It is an appetite suppressant, so Calorie intake may decrease with its use. Tobacco also interferes with the metabolism of nutrients. Other negative lifestyle ramifications of tobacco use include reduced physical fitness, increased lung and respiratory damage, and increased chronic disease (such as heart disease, cancer, and osteoporosis) risk into adulthood.[16,19,89,110,118,140,146]

Adolescents often prefer to eat fast and processed food with other adolescents.

Disturbed, Distorted, and Disordered Adolescent Eating Sometimes unhealthy behaviors become extreme, as in the case of disordered eating. Distorted eating starts with body image dissatisfaction and progresses to dieting behavior and ultimately to clinical eating disorders in extreme cases (see Figure 7.17). Eating disorders are genetically, socially, emotionally, and psychologically based. They are also more common in girls than boys and typically develop during adolescence. At the heart of eating disorders are destructive behaviors, including fad dieting, skipping meals, using diet pills, laxatives, and/or diuretics, and overexercising. Restrictive dieting yields inadequate energy and nutrient intake, whereas overexercising leads to extreme negative energy balance. Both result in weight loss but also take an unhealthy toll on the body and impair bone health and the function of muscles, the brain, and many other body systems. Rapid and unhealthy weight loss often results in the loss of lean body mass and stored body fat. Low body fatness leads to lower hormone levels and the eventual cessation of menstruation in girls (amenorrhea). The reproductive development during adolescence can be halted and possibly permanently altered.[5,16,37,59,76,140]

Clinical eating disorders include anorexia nervosa, bulimia nervosa, and binge eating disorder (BED). These conditions are described in detail in the *Diagnostic and Statistical Manual of Mental Disorders* (DSM-IV).[3] The DSM-IV identifies many clinical signs and symptoms for meeting the criteria of an eating disorder. *Anorexia nervosa* is a condition in which the person engages in self-starvation and has an intense fear of becoming fat. *Bulimia nervosa* is characterized by periods of fasting followed by binging and then purging. The person consumes a large

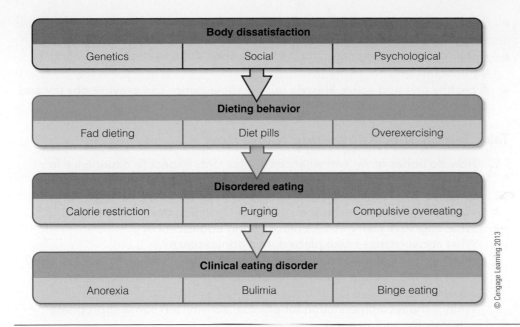

FIGURE 7.17 The progression of disordered eating.

© Cengage Learning 2013

amount of food and then voids it from his or her body by self-induced vomiting, laxative and diuretic use, enemas, and/or excessive overexercising. Individuals with anorexia are usually very thin, whereas those with bulimia are normal weight or slightly overweight. *BED* is uncontrollable binge eating. A person with BED consumes large quantities of foods in recurring episodes, at least twice a week. The person with BED has three or more of the following characteristics during an episode: rapid eating, eating without hunger, eating alone, eating until very full, and experiencing self-disgust about the binge.[HH,PP,59,120]

Most unhealthy approaches to weight loss are typically unsuccessful and leave adolescents with potential health issues, such as increased risk for osteoporosis resulting from bone demineralization, continued obsession with body weight and image, and feelings of failure from their lack of weight loss success.[103]

Skin Disorders Many teens are faced with battling acne. Acne may have dietary ties, but commonly during adolescence it is caused by hormonal changes, emotional stress, personal hygiene, and genetics. Other genetically influenced skin conditions include the skin form of celiac disease (dermatitis herpetiformis) and the skin reactions that occur with allergies. A powerful form of vitamin A, 13-*cis*-retinoic acid, can be prescribed as a topical medicine to cure acne. This medicine is known to cause birth defects shortly after conception and should be prescribed with great caution to females of reproductive age (see page 353). In addition, the person's lipid profile is dramatically changed to promoting heart disease, and depression is a common side effect of the medication.[B,Y,Z]

Summary Points

- Adolescence is a time of rapid growth and sexual maturation.
- Adolescent boys and girls progress through puberty in predictable stages as they transition from children to adults.
- The acquisition of abstract thinking and problem-solving skills and the increased emphasis on peer relationships are elements of the cognitive and social changes during adolescence.
- Proper adolescent nutrition will meet the increased demand for Calories and protein compared to adults and the need for unique vitamins and minerals to support growth and development.

- Adolescents need to be physically active by engaging in 60 minutes of exercise every day.
- Adolescence is a common time for food intake shifts to occur and eating disorders to begin.
- A healthy diet and active lifestyle during adolescence can prevent obesity and promote overall health and well-being throughout the life span.

Take Ten on Your Knowledge Know-How

1. How do bones grow, and what can adolescents expect to experience in regards to physical growth?
2. What do boys experience when going through puberty?
3. What do girls experience when going through puberty?
4. What cognitive changes take place during adolescence?
5. What social changes take place during adolescence?
6. What are the Calorie and energy-producing nutrient needs for adolescents?
7. What are the nutrient needs for vitamins, minerals, and water during adolescence?
8. What is the recommendation for physical activity and some appropriate activities for adolescents?
9. What are some food intake shifts that occur during adolescence?
10. What are some common health issues facing adolescents?

7.5 NUTRITION FOR THE OLDER ADULT

T-Talk 7.5
To hear Dr. Turley talk about nutrition for the older adult, go to www.cengagebrain.com

Introduction

Most of this textbook is devoted to nutrition for the young adult, ages 19 to 30 years, and the middle-aged adult, ages 31 to 50 years. This section of the module focuses on the nutrition needs of older adults. This group includes the young-old (ages 51 to 70) and the old-old (ages greater than 70 years). The term **geriatrics** is often used when describing the medical study of older adults. Here we will focus more on successful aging, but will explore common nutrition-related medical conditions of older adults and learn the physiology of aging.[K,M,98,147,165]

Achieving successful aging means adopting healthy eating and lifestyle habits that promote optimal physical, mental, and social well-being and support lean body mass, bone mass, nutrient status, and cognition. It also means preventing chronic diseases that are more prevalent in this period of the life span.

Each person will inevitably age, chronologically and biologically. Aging may be physiologically accelerated (premature aging) or delayed (maintaining functional capacity of all body systems), depending on lifestyle choices. Because of the large number of people born during the baby boom era (1940s to 1960s), the number of older American adults has increased. The current **life expectancy** from birth for Americans is 78 years. (See BioBeat 7.9 for information about the factors that are considered in estimating life expectancy.) The life span group over the age of 85 is the fastest growing segment of the American population. Thus, it is important to understand diet and lifestyle needs and resources to promote healthy aging and prevent disease (loss of body function) in older adults.[Q,22,52,88,100]

geriatrics The medical study of the physical, social, behavioral, and psychological diseases and issues of older adults.

life expectancy A statistical estimation of the expected number of years of life remaining at a given age.

The Physical, Cognitive, and Social Triad for Older Adults

Many changes occur later in life that affect the physical body, the cognitive process, and the social network. Living in a stimulating environment that is safe, being physically active, eating well, and limiting the exposure to toxic substances will preserve function of all body

BioBeat 7.9

The Life Expectancy Estimation

Life expectancy is the number of years a typical person lives. This number can be personalized, depending on a person's current age and other factors, such as genetics, family history, gender, personal lifestyle and behaviors, and access to available medical technologies, as well as future progress in life-extension sciences.

Life expectancy is increased (expected years are added to life) with more education, living in an advanced technological society, and having higher income. In addition, engaging in regular physical activity, being easy-going and mentally active, managing stress, controlling caloric intake, and having regular medical checkups all increase life expectancy. Life expectancy may be dramatically increased through scientific, medical, and pharmacological advances, hormone optimization, and proper nutrition.

Life expectancy is decreased (expected years are subtracted from life) with a family history of early

The number of candles yet to be put on a birthday cake is greatly influenced by diet and lifestyle choices.

death from chronic disease such as heart disease, stroke, diabetes, thyroid disorders, or cancer, especially if the descendant is taking no action to reduce these risks. In addition, being overweight or obese, having elevated blood pressure or blood cholesterol, eating fast foods and processed foods, skipping meals, eating in a hurry, and drinking excess alcohol

also decrease life expectancy. Smoking cigarettes dramatically reduces life expectancy. Living with a smoker or in a polluted environment and chronically using drugs therapeutically or recreationally also negatively impacts life expectancy.[Q,22,52,67,98,165]

What factors affect your estimated life expectancy?

systems, promote the ability to live well in old age, and prevent **senility**. As you learn about some theories of aging and the accompanying physiological changes, you can implement many strategies for successful aging.[40]

The Theories and Physiology of Physically Aging The physical body undergoes many predictable changes throughout life. Aging is growing old as a result of the combination of all the genetic, molecular, and physiological occurrences over time, from conception to death. In older adults, aging is characterized by declining physical abilities and increased pathology (incidence of disease). Pathology can be significantly reduced with proper diet and lifestyle management.[164]

Theories of Aging There are many theories of aging, including several that are evolutionary, molecular, cellular, and systems-based. Regarding evolution, one theory is that aging is caused by a decline in natural selection. Because evolution maximizes individual reproductive fitness, **longevity** may not be necessary for some species. Thus, theories with an evolutionary basis favor genetics for reproductive ability and not for longevity. However, at the molecular level, longevity has clearly been found to be associated with a person's family genetics. Having a gene pool that favors structural and functional endurance of body systems will increase longevity. Additionally, **senescence** (the aging process) may also result from changes in gene expression and molecular events. At the cellular level, senescence may occur from replication errors or stress. There are many forms of stress; oxidative stress is one form that can damage cells and thus trigger a decline in function (see BioBeat 5.4 in Module 5, page 234, for more information on oxidative

senility When the mind and body deteriorate in old age.

longevity Having a long duration of life or life span.

senescence The process of aging and physically showing the signs of increasing age, resulting from the cells ceasing to divide by mitosis and a decline in organ function.

stress and antioxidants). Oxidative stress causes DNA, proteins, and lipids to be chemically modified in a nonfunctional way. The cumulative damage to chemicals needed for structure or function promotes senescence. A systems-based explanation of aging refers to decreased function of neurological, endocrine, and immune systems as a person ages. The neurological system communicates internal and external responses of the entire body, while the endocrine system supports reproductive and other organ functions through hormonal control. As a person ages, the functions of both the neurological and endocrine systems decrease. The immune system wards off foreign threats and detects and destroys abnormal cells, all while sparing normal cells and tissues. The aging immune system is evidenced by increased infections, cancers, and autoimmune diseases in the elderly.[50,161]

Physiology of Physically Aging   The physiology of aging is described by body system in Table 7.10.[16,50,104,123,141,163,164] Each system has its own average rate of functional decline or aging. For example, many age-related changes affect food intake and the gastrointestinal system. With aging, saliva production decreases, making it more difficult to swallow foods (**dysphagia**) and increasing the risk of choking. Furthermore, diminished saliva increases tooth decay. This can lead to oral health changes, including missing teeth and the need for dentures.[16,98,141,165]

Teeth aid in the initial physical breakdown of food. About one-half of all Americans over the age of 65 do not have any teeth. Several dietary changes follow when a person is missing teeth. Pain and discomfort associated with altered dentition and improperly fitting dentures can lead people to avoid foods that need a lot of chewing, such as meats, raw vegetables, and fresh fruits. Additionally, dentures tend to be less effective than natural teeth in grinding and chewing food. Poor dentition can promote malnutrition.[C,M,NN,SS,16,141]

Sensory changes, including declining hearing, vision, taste, and smell, occur. Hearing impairment is very common, causing the need for hearing aids. Visual changes impair the ability to read food package labels, menus, expiration dates, and recipes. Loss of vision can also lead to loss of driving abilities, which in turn prevents older adults from going grocery shopping. Altered taste and smell can affect eating practices. The tastebuds at the back of the mouth are affected first; loss of these tastebuds reduces the sensations of sweet and salty and causes foods to taste more sour and bitter. And the pleasure associated with eating is reduced, which leads to malnutrition.

The intestinal villi are also negatively impacted by age. They are more often atrophied in older adults, resulting in inadequate nutrient absorption and a much greater incidence of lactose intolerance. Anorexia of aging causes reduced appetite and increased protein-energy malnutrition risk (see Module 3, page 117). Loss of appetite can result from delayed gastric emptying, so there is a longer feeling of fullness after eating.[6,112,123,164]

There is also a decline in kidney (renal), liver (hepatic), lung, and heart functioning with aging. Aging causes a reduction in renal function, leading to a reduced ability to concentrate urinary waste, as well as reduced bladder control. It is estimated that renal capacity diminishes about 1 percent per year after the age of 40. This promotes greater urine output and increased risk for dehydration. To complicate the matter, thirst sensations are reduced, leading to fluid and electrolyte imbalance and the signs and symptoms of dehydration, including headache, dizziness, fatigue, visual changes, disorientation, and mental confusion (see Module 5, page 253). Dehydration can also increase the risk for urinary tract infections. In a majority of men, urinary tract obstruction may result from benign prostatic hyperplasia by age 80. The result is an inability to urinate with urge and sphincter relaxation, causing incontinence.

In the hepatic system, the size and functional ability of the liver cells to regenerate and detoxify are reduced, resulting in a greater number of toxins circulating in

dysphagia Difficulty swallowing.

the body. From a respiratory point of view, lung function and airway size decrease, so ventilation and gas exchange decrease as well. Regarding the heart, the cardio-vascular system declines gradually, including a decline in maximum heart rate and an increase in heart diseases. These lung and heart changes result in a significant disability to exercise.[P,16,50,112,141]

Additional age-related physiological changes affect the bones, joints, muscles, nerves, and brain. Bone mineral density is lost, along with joint flexibility. The body undergoes changes in composition, including loss of muscle mass and increasing fat mass. Furthermore, core body temperature defense becomes compromised, and the ability to tolerate cold or hot environments is reduced. The older adult has slower nerve conductivity, as well as fewer neurons in the brain, a reduction in the blood supply to the brain, and fewer brain cells, all of which compromise cognitive function. Greater imbalance from loss of motor nerve function and poor posture are common physical signs associated with this aging phenomenon. All of these changes negatively impact muscle coordination, reflexes, and memory.[62,78,141]

TABLE 7.10

Body Systems and Aging

Organ/Site	Damage
Circulatory system	Blood vessel elasticity and cardiac output decline, while blood pressure steadily increases. Arteries harden, and blood clots are more common. The capillaries are more fragile.
Digestive system	Tooth loss, gum disease, loss of appetite, impaired taste and smell, difficulty swallowing, and reduced salivary and mucus production may cause a reduction in the amount or variety of foods eaten and create a fear of choking. The loss of intestinal strength slows motility and increases constipation. The reduction in stomach (gastric) acidity and villi atrophy causes impaired digestion and reduced nutrient absorption (especially for iron, zinc, calcium, folic acid, and vitamin B_{12}). There is a reduced sensation of thirst, which increases the risk for dehydration and associated confusion.
Endocrine system	Abnormal glucose metabolism occurs as the pancreas produces less insulin and cells become less responsive to insulin. Menopause and andropause occur, as evidenced by reduced estrogen and testosterone, respectively. Thyroid hormone production, growth hormone secretion, and vitamin D synthesis decrease.
Integumentary system	Drier skin and hair, reduced sweat production, loss of skin elasticity, thinning skin, more wrinkles, reduced collagen, bruising, hyperpigmented age spots on the face, hands, and wrists, graying hair, and hair loss occur. Vitamin D production from sunlight exposure to the skin is reduced, creating an increased need for dietary or supplemental vitamin D. Low vitamin D levels increase the susceptibility for poor calcium status.
Lymphatic system	Reduced immune functioning increases the risk for cancers, infectious diseases, and complications from foodborne illness.
Muscular system	Muscle capacity is reduced approximately 20 percent by age 45, 35 percent by age 65, and 50 percent by age 85. With the reduction in lean muscle mass, there is a reduction in metabolism, Calorie need, strength, and stability. Stamina and flexibility diminish. Physical inactivity leads to muscle loss and increased stiffness.
Nervous system	Diminished mental acuity and nerve conduction, including reduced peristalsis, are common. Reduced appetite, thirst, sight, smell, and taste negatively affect food selection and intake. Altered sleep patterns occur.
Reproductive system	Sex hormone production decreases. Reproductive capability ceases in women after menopause. Sperm count decreases in men.
Respiratory system	Respiration and lung capacity are reduced approximately 5 percent by age 45, 15 percent by age 65, and 20 percent by age 85. There is an increased risk of respiratory infections and pneumonia.
Skeletal system	Bone mass declines in men and women (women experience twice as much bone loss as men), but especially after menopause in women, increasing the risk of bone fractures and breaks. Fractures heal more slowly. Joints stiffen and are more painful.
Urinary system	Kidney function is reduced approximately 10 percent by age 45, 25 percent by age 65, and 45 percent by age 85. Blood filtration rate decreases and urine output increases. Bladder control is compromised.

© Cengage Learning 2013

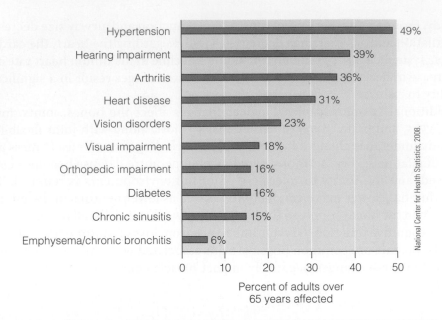

FIGURE 7.18 Leading chronic diseases and conditions in adults over age 65.

Figure 7.18 shows the leading chronic diseases and conditions in adults over the age of 65. Medications are commonly used by older adults to combat lifestyle-associated diseases and conditions such as heart disease, hypertension, cancer, type 2 diabetes, and arthritis. Medications often have side effects that interfere with the person's nutrition status, including altering taste and smell and impaired nutrient absorption.[104,112,163,165]

Cognition and Brain Preservation *Cognition* is a general term for the mental processes of thought, perception, reasoning, and learning, whereas *dementia* refers to the deterioration of cognition that can be progressive with aging. Dementia can have devastating effects on the individual, family, and friends. Some signs of dementia include agitated behavior, confusion, delusions, unclear thinking, loss of memory, loss of problem-solving skills, loss of familiarity with usual surroundings, and loss of interest in daily activities.[40,136,164]

Although many medical conditions promote dementia or decrease cognition, mental function is dependent on several nutrients, as shown in Table 7.11.[107,136] The deficiency of several nutrients can alter many aspects of cognitive function. A deficiency of vitamin B_{12} and insufficient omega-3 fatty acids are particularly noteworthy. Any of the nutritional anemias can reduce cognitive function, but of particular commonality in the elderly is vitamin B_{12} deficiency, which causes pernicious anemia. In this treatable condition, there are symptoms of anemia and dementia. The treatment involves adequate vitamin B_{12} administration, usually by intermuscular injection. Omega-3 fatty acids, especially highly unsaturated long-chain omega-3 fatty acids, support cognition and appear to slow age-related cognitive decline in older adults without dementia. Omega-3 fatty acids are usually orally supplemented from fish oil. Mental changes can also result from thyroid disease and depression, both of which are usually treatable with prescription medication. Increased depression can occur as a result of social isolation, loneliness, and the loss of loved ones, as well as from physical illness and the loss of independence.[16,51,104,112,141]

Dementia can impact the diet, because the person suffering from dementia may forget to eat, make limited food selections, and store food improperly. Beyond eating a healthy diet, cognition can be supported by being physically active, maintaining social relationships, being intellectually engaged, and maintaining a positive outlook on life.[107,136]

TABLE 7.11

The Role of Various Nutrients in Brain Function

Nutrient	Short-Term Memory	Problem Solving	Cognition	Neurotransmitters	Mental Health	Vision
Vitamin A						√
Vitamin E	√		√			
Thiamin					√	
Riboflavin		√				
Niacin					√	
Vitamin B_6			√			
Vitamin B_{12}	√	√	√			
Folate		√	√		√	
Choline				√		
Vitamin C	√	√				
Iron			√			
Zinc					√	
Tyrosine				√		
Tryptophan				√		
Linoleic acid						√
Alpha-linolenic acid						√

© Cengage Learning 2013

The Social Network The social network plays an important role in the health and vitality of older adults. The network may consist of family, friends, neighbors, health professionals, pets, and religious groups. The social network can provide emotional support, tangible aid, and valuable information that can improve the quality of life. Emotional support conveys love, reassurance, and belongingness and commonly comes from family and friends. Pets also provide emotional support. Any relationship that fosters some form of intimacy, whether physical, mental, or spiritual, will provide emotional support. Tangible aid may be needed for meals, shopping, and getting out for social events or medical appointments. Healthcare professionals are needed for providing informational health advice, while peers and family members may offer other advice, feedback, and suggestions, along with companionship at mealtime, that all improve the quality of life.[C,G,M,NN,SS]

Older adults may find social support at their local senior center, where there are often regular nutritious meals, crafts, games, experts available for various kinds of advice, travel opportunities, and exercise facilities. Many of the services, such as lunchtime meals, are either free or offered at reduced prices. Some seniors find it rewarding to serve their community and enlarge their social network by volunteering for some form of community service. The bottom line is that older adults often need the care, help, and support of their social network, and they need to be active in building and maintaining their social ties.[16,141]

Jeff Greenberg/Photo Edit

Communal mealtime is important for the health and vitality of older adults.

Nutrition Needs for Older Adults Like other life span groups, older adults have unique nutrition needs that can be supported by using the MyPlate food guidance system and Dietary Guidelines

2010 (see Module 2, page 70). Also, specific recommendations exist for Calories, carbohydrates, proteins, fats, vitamins, minerals, and water.[9,94,98,100,147]

Energy and Nutrient Needs for Older Adults Energy and nutrient needs are specific for older adults, as with other life span groups. The DRIs for energy and nutrients are established and shown in Appendix A. The needs for energy are reduced in older adulthood. We will focus on several vitamins, minerals, and water because of age-related, complicating medical conditions and changes in how these nutrients are absorbed and metabolized.

Calories   With increasing age in adulthood, caloric needs decline. The EER and Resting Energy Expenditure or Basal Metabolic Rate formulas for older adults take this decline into account. The EER formula has 10 fewer Calories per day for men and 7 fewer per day for women, for each year above 19 years. By the age of 69, this amounts to approximately 500 fewer Calories per day needed. There can be several reasons for the reduction in Calories required to maintain normal body weight and prevent weight gain with increasing age. Some reasons include reduced physical activity and more at-home leisure activities. A large contributing factor, however, is the reduction in lean body mass, especially muscle mass, and the increase in fat mass. Lean tissue is more metabolically active and requires more energy to maintain. The reduction in energy need can be prevented if the older person continues to exercise and be physically active, engaging in activities like walking and strength training.[69]

Carbohydrate, Protein, and Fat   Carbohydrate needs are the same for older adults as for younger adults. The DRI for carbohydrate continues to be 130 grams per day minimum, with an ADMR of 45 to 65 percent of Calories from total carbohydrate, with sugars limited to less than or equal to 25 percent of Calories. In looking at the DRIs, it appears on the surface that fiber should be reduced, but the lower gram recommendation (from 38 to 30 grams in men and from 25 to 21 grams in women) is a result of the reduced Calorie need, because the DRI for fiber is tied to the amount of Calories consumed. Carbohydrate-related concerns surface more commonly in the older adult years. Constipation is more common, resulting from a lack of adequate fiber and fluid intake, as well as a lack of physical activity. Reduced fluid intake is linked to the diminished thirst sensation and bladder control that occurs with aging. Consuming adequate fiber and fluids and continuing to be active can alleviate constipation in older adults. Excellent fiber sources include legumes, whole grains and whole-grain breads and cereals, vegetables, and fruits.[98,147]

The protein needs of older adults are also the same as for young and middle-aged adults. The DRI continues to be 0.8 grams per kilogram of body weight per day. Current research findings suggest, however, that the DRI for protein is probably set too low for most age groups, including older adults. Older adults often have reduced protein intake, which can contribute to reduced immunity. Legumes, whole grains, and vegetables, along with fish, lean meats, and low-fat dairy products, can support the protein needs of older adults.[108]

The recommendation for total fat intake continues to be 20 to 35 percent of Calories. The DRI for linoleic acid is reduced from 17 to 14 grams per day in men and 12 to 11 grams per day in women over the age of 50. The recommendation for alpha-linolenic acid is unchanged. Older adults should be mindful of reducing saturated fatty acid intake levels to below 7 percent of Calories and *trans* fatty acids to less than 1 percent of Calories to support heart health. Thus, high-fat animal meats and dairy products should be limited; lean and low-fat versions of these foods should be consumed in moderation instead. Omega-3 fatty acids from fatty fish or supplements can support brain function and reduce the symptoms of inflammatory diseases.[51,122]

Vitamins, Minerals, and Water   The vitamin and mineral needs of older adults are similar to those of young and middle-aged adults; some metabolic and health-related nutrient concerns are important to be aware of, however.[69,107]

The DRIs for vitamin A, vitamin C, vitamin E, and zinc are unchanged in older adults. The absorption of vitamin A is often increased, and the liver uptake of vitamin A is often reduced. Older adults are often advised to avoid excess vitamin A intake from supplements because of negative bone effects. Adequate vitamin A is needed to support epithelial cell and bone health. From food consumption survey results, vitamin C, vitamin E, and zinc intakes may be reduced. Low vitamin C and E status promotes more oxidative stress, and low vitamin C and zinc status leads to weakened immunity. Reduced antioxidant status may be implicated in accelerated senescence. Vitamin E is found in nuts, seeds, and plant oils, while vitamin C is abundant in citrus fruits, and zinc is present in lean meats.

Vitamin D deficiency is common in adults and becomes even more prevalent with aging. The DRI for vitamin D increases from 5 micrograms per day for adults to 10 micrograms per day for the young-old (ages 51 to 70) and 15 micrograms per day for the old-old (older than 70 years). Food sources of vitamin D are relatively scarce and include fortified milk, cereals, and fatty fish. Getting adequate sunlight exposure is one approach to help alleviate the problem; however, this is not always possible for older adults who may have health conditions and limited mobility, particularly if they live in an area with long winters (e.g., the Northeast) or reduced sunlight (e.g., Seattle). Furthermore, synthesis of active vitamin D from cholesterol upon ultraviolet light exposure to the skin is reduced with aging. Thus, individualized supplementation of vitamin D_3 is often a good idea to support normal vitamin D blood levels in older adults.[98,101,117]

The DRI for vitamin B_6 increases after age 50. For males, the DRI increases from 1.3 to 1.7 milligrams, while for females it increases from 1.3 to 1.5 milligrams per day. There are several reasons for the increase. Low vitamin B_6 status has been associated with reduced immune function in the elderly. Vitamin B_6 may also improve cognition and memory, because it is needed for the synthesis of neurotransmitters and functions along with folic acid and vitamin B_{12} to prevent hyperhomocystemia.[134,138]

Vitamin B_{12} status may be low in older adults, because there is reduced gastric acid production with aging, causing reduced vitamin B_{12} absorption. A higher dietary intake of vitamin B_{12} may not prevent pernicious anemia (an autoimmune disease) if the complicated B_{12} absorption processes are inadequate at any critical point; thus, an intramuscular injection of vitamin B_{12} may be needed (see Module 5, page 244). Low levels of vitamin B_{12} and/or folate cause macrocytic anemia and contribute to hyperhomocystemia, a risk factor for heart disease. Vitamin B_6 also functions in reducing hyperhomocystemia. Thus, there may be a need to consume more lean and low-fat animal products containing vitamins B_{12} and B_6 and more dark green leafy vegetables, whole grains, and citrus fruits for folate. Consuming fortified foods and supplements may also be warranted to achieve adequate intake of these vitamins.[134,138]

The DRIs for calcium and magnesium increase with age. After age 50, the DRI for calcium increases from 1,000 to 1,200 milligrams per day for men and women. After age 30, the DRI for magnesium increases from 400 to 420 and 310 to 320 milligrams per day in men and women, respectively. The increase in the amount of these nutrients supports bone mass during a time of potentially accelerated bone loss as a result of hormonal (see BioBeat 7.10 on page 412), dietary, and lifestyle changes. Calcium intake tends to decline in older adults, along with the body's ability to absorb it. Reduced absorption is attributed to **achlorhydria** (reduced gastric acid), and reduced intake is often linked to increased lactose

achlorhydria Low or absent production of gastric acid (hydrochloric acid) in the stomach.

intolerance with age. Medications, in addition to calcium and vitamin supplementation, may be required to support bone mineralization.[RR,98,147]

The DRIs for sodium, chloride, and chromium decrease for men and women with age, while the DRI for iron decreases in postmenopausal women. For sodium, the DRI decreases for men and women from 1,500 to 1,300 (at age 51) to 1,200 (at age 71) milligrams per day. For chloride, the DRI decreases for men and women from 2,300 to 2,000 (at age 51) to 1,800 (at age 71) milligrams per day. The need for chromium reduces from 35 to 30 micrograms and from 25 to 20 micrograms per day in men and women, respectively, at age 51. For iron in women, the DRI decreases from 18 milligrams per day to 8 milligrams per day at age 51; the risk of anemia is less common because menstruation ceases at menopause. However, reduced gastric acid levels with increasing age and/or overuse of antacids can lead to impaired iron absorption. So, it is common for iron status to be low in old age. Consuming vitamin C, along with iron-rich foods, remains an important way to increase iron absorption from non-heme (plant or fortified) sources.[92]

Water or fluid needs for older adults are the same as for younger adults: 3.7 liters per day for men and 2.7 liters per day for women. However, diminished thirst sensations may lead to underconsumption and dehydration. Fluid needs can be met by consuming beverages and fluid-rich foods such as fruits and vegetables.[69]

DA+ GENEie Go to Diet Analysis Plus and create your profile. Build a profile report to see your DRI recommendations. Now change your age to 15, 25, 35, 55, and 85. Note and discuss the changes in your Calorie and nutrient recommendations across the life span.

Food Assistance Programs Older adults are considered to be at risk for nutrition-related health problems. There are many reasons why older adults may be at risk, including physical and mental illness or disease, social isolation, multiple medication use, and food insecurity. Food insecurity means that there is a reduced ability to obtain, prepare, and/or purchase foods. Poverty is often the cause.[G,EE,UU]

The Department of Health and Human Services operates a food assistance program specifically designed for older people called Nutrition Services Incentive Program (NSIP). This program provides nutritious meals to any American age 60 or older; the program also includes the person's spouse, even if the spouse is younger. Another benefit of this program is that it does not have any income restrictions. NSIP makes meals available through neighborhood centers, provides transportation to the meal sites, and operates Meals on Wheels for homebound elderly people. The meals provided are hot or cold and are available at least once a day for five days out of the week. In addition to NSIP, poverty-stricken elderly people are eligible for food stamps by participating in the SNAP program (see page 378).[EE,UU]

The Senior Farmers' Market Nutrition Program is operated by the USDA Food and Nutrition Service. The program awards grants to states, U.S. territories, and federally recognized Indian tribal governments. The grant recipients then provide coupons to low-income seniors. The coupon holder uses the coupon like money to obtain food at farmers' markets, roadside stands, and community-supported agriculture programs. This is an exciting program because it provides fresh, locally grown fruits, vegetables, and herbs to older adults who otherwise may not have access to these nutritious and more sustainable foods.[SS]

Other food assistance programs exist, including the Child and Adult Care Food Program, Commodity Supplemental Food Program, and Emergency Food Assistance Program.[UU]

Older Adult Fitness and Lifestyle Management

Older adults can and should be physically active and manage their lifestyle to promote health and well-being. Staying physically fit is associated with less sickness and increased longevity. Although age-related physiological decline reduces the individual's ability for high-performance exercise, the ability to exercise age-appropriately continues throughout this phase of the life span.[I,U,102,149]

Fitness The American College of Sports Medicine, in conjunction with the American Heart Association, has basic recommendations for physical activity for older adults. There are recommendations for aerobic and resistive physical activity and fitness for adults over the age of 65, because both aerobic and muscle-strengthening activities are critical for healthy aging. The recommendations call for moderately intense aerobic exercise 30 minutes a day, five days a week, or vigorously intense aerobic exercise 20 minutes a day, three days a week. It also calls for 8 to 10 strength-training exercises (with 10 to 15 repetitions) at 60 to 70 percent of the one-repetition-maximum, and at least two sets of each exercise two to three times per week. Moderate-intensity aerobic exercise means working somewhat hard (perceived exertion rating of 70 percent of the maximum heart rate; Module 4, pages 193–194). The person exercising should still be able to carry on a conversation during exercise. The same recommendations also apply to adults ages 50 to 64 with chronic disease conditions, such as arthritis.[102]

Older adults who are not accustomed to exercising or who have chronic conditions should develop an activity plan with a health professional to manage their exercise risks and consider therapeutic needs. It is also wise to gradually increase physical activity or to start off easy and increase exercise time and intensity as skills and abilities increase. If the individual has a risk of falling, balance exercises should be done to improve core strength. Exercises such as yoga and t'ai chi have been shown to improve balance and mental acuity, while exercise in general preserves cardiovascular and respiratory function and reduces hypertension. Walking is by far the most common form of aerobic exercise for the aging population.[104,149,163]

Resistive exercise is very effective at preserving strength, because it targets the muscles, stimulates protein synthesis, and retains lean body mass. It also improves metabolism in every cell of the body and slows age-related decline. Maintaining strength can improve a person's independence by enabling that person to continue to do basic essential daily functions like climbing stairs and opening containers of food. Maintaining lean body mass will also reduce the risk for falls and injuries, thus preventing head trauma and hip fractures. A strength-training program for 12 weeks, in which the exercises are done at 80 percent of the one-rep-max (see Module 4, page 194), can increase strength 5 percent per session, with strength gains ranging between 100 and 200 percent.[62,102,149]

The take-home message for senior fitness is to have a physical activity plan and to make physical activity a regular part of daily life. Older adults will have a more youthful and functional status as a result of exercise, and an exercise program can be initiated anytime during the life span and result in improved function, health, and quality of life.[I,U]

Diet and Health Issues in Older Adults There are numerous diet and health issues for older adults. A few common conditions will be discussed here, including **Alzheimer's disease, arthritis, cataracts** and **macular degeneration**, osteoporosis, **sarcopenia**, and sleep disturbances.[12,98,147,163,165]

Alzheimer's Disease Alzheimer's disease is not a normal part of aging. It is a form of dementia referred to as early or presenile dementia. Memory loss may begin in the 40s or 50s. Memory loss progresses to loss of reasoning, loss of

Walking is a great form of exercise for older adults.

Alzheimer's disease Pre-senile dementia that starts with memory loss and leads to complete helplessness.

arthritis Breakdown of the cartilage surface in joints that leads to inflammation and pain.

cataracts The scar tissue that forms over the cornea as a result of eye surface damage.

macular degeneration Weakening of the major eye muscle due to oxidative stress.

sarcopenia The reduction of muscle mass and strength as a consequence of aging.

communication, loss of problem-solving skills, and loss of all physical abilities, resulting in complete helplessness. Individuals suffering from Alzheimer's disease may also exhibit behavior that is agitated, confused, or disinterested. This debilitating disease is believed to have a genetic component and to have diet and lifestyle implications. Maintaining good heart health and regular intake of fish may reduce the risk. Damage caused to DNA, proteins, and cell membranes in brain cells may result from oxidative stress and free radicals. Excess iron, copper, and aluminum may trigger oxidative stress, while low antioxidant nutrient status may compromise the ability to neutralize radical species. Damage may also occur from elevated blood pressure and blood glucose and reduced blood flow to the brain. Hyperhomocystemia is also implicated in Alzheimer's disease, especially in early dementia. Thus, adequate antioxidant intakes, along with folic acid, vitamin B_6, and vitamin B_{12}, may play supporting roles in healthy brain aging and slowing the progression of dementia. Additionally, maintaining brain activity by being mentally challenged and engaged may also slow progressive brain disorders.[D,9,12,51,136,149]

Arthritis   Arthritis is joint inflammation associated with damage of the cartilage surface, while **osteoarthritis** is deterioration of joint cartilage inducing inflammation, and **rheumatoid arthritis** is an autoimmune disease that causes joint damage, inflammation, and bone structure deformities. These conditions are common in older adults. They can be improved with exercise and worsened with obesity. A good antiinflammatory dietary approach is to consume highly unsaturated omega-3 fatty acids from fish oil in therapeutic amounts (2 to 4 grams per day) and consume antioxidant-rich fruits and vegetables.[N]

Cataracts and Macular Degeneration   Cataracts and macular degeneration are very common eye conditions in the elderly. Exposure to sunlight and low antioxidant intake promotes cataracts, which are thickenings over the lenses of the eyes. Wearing UV protective eyewear and consuming vitamin C–rich foods can reduce cataract formation. Another common eye disease is macular degeneration, which results from the deterioration of the macular muscle in the region of the retina. It is the leading cause of blindness in the United States. Diets low in vitamins C and E, zinc, and the phytochemicals lutein and zeaxanthin increase the risk of macular degeneration and cataracts. Consuming whole plant foods in a variety of colors will support vision by supplying the needed antioxidants and phytochemicals. Other controllable lifestyle factors include avoiding tobacco use, excessive oxidative stress, and excess sunlight exposure to the eyes.[L,MM,121]

Osteoporosis   Osteoporosis is a silent disease that is commonly diagnosed in postmenopausal women (see Module 5, BioBeat 5.7, page 261). Smaller, thin, Asian or white women with a family history of the disease are at the greatest risk. It is estimated that one-half of older women and one-fourth of older men have osteoporosis and will suffer from a broken bone from the condition. Fractures are not only painful, but they can affect quality of life, lead to loss of independence, and can be life threatening. Osteoporosis also causes loss of height, changes in posture, and back pain (Module 5, page 261).[RR,62]

It's impossible to feel bone being demineralized, so bone density scans, usually by dual energy x-ray absorptiometry (DEXA), are done on postmenopausal women to determine their potential disease status and to see if the person has osteopenia, or low bone mass. Osteoporosis is characterized by porous bone that has been depleted of major bone minerals, especially calcium. The osteoporotic bone is frail and breaks easily. The most common sites of bone fracture include the wrist, the spine, and the hip.[K]

osteoarthritis Deterioration of joint cartilage.

rheumatoid arthritis Autoimmune destruction of organs and tissues, especially joints, causing joint pain, inflammation, swelling, stiffness, and deformity.

When a person has osteoporosis, fall prevention is a major strategy. Falls can be caused by other medical conditions, certain medications, and unsafe surroundings. The environment can be made safer by keeping the house clean of clutter and area rugs, having adequate lighting, wearing nonslip shoes, installing grab rails, using a cane or walker for stability, avoiding slippery walking surfaces, and paying attention to each step taken.[64,163,165]

Both diet and lifestyle approaches are needed before the age of 25 to keep bones healthy and strong later in life. Consuming adequate calcium and vitamin D is essential. Older adults should continue to consume low-fat dairy products and choose lactose-reduced products when needed, as well as calcium-fortified dairy alternatives and other calcium-fortified foods and/or supplements. Other nutrients are needed for bone metabolism. For example, vitamin K is important for bone mineralization; thus, adequate levels are needed to prevent fractures. Good food sources of vitamin K include dark green leafy vegetables, such as spinach, kale, broccoli, and romaine lettuce, along with live cultured yogurt, which supports bone mineral density (see Module 5, page 235). Boron is another trace element that plays a role in calcium metabolism.[98,141]

Physical activity, especially weight-bearing exercise, is needed to promote bone mineral density. Furthermore, not using tobacco and not drinking alcohol in excess will support bone health. It's best to start taking care of bones early in life. The stronger they are before midlife hormonal changes occur, the lower the likelihood of demineralization that weakens the bone to the point of fracture.[62,64]

Demo GENEie Showcase the quantity difference between the amount of calcium found in the body of an active and healthy 25-year-old woman compared to a 75- and 95-year-old woman with osteoporosis. Place 1 pound of flour in each of three clear plastic bags to represent the amount of calcium in the young woman's bones. Take one of the 1-pound bags away to represent the amount of calcium in the bones of a 75-year-old woman with osteoporosis. Then take another 1-pound bag away to represent the amount of calcium in the bones of the 95-year-old woman with advanced osteoporosis.

Sarcopenia   Sarcopenia is the loss of muscle mass and strength that commonly occurs with aging. Adult body fatness peaks between 50 and 60 years old.[13,74,78,108] Lean body mass, especially skeletal muscle mass, decreases in older adulthood. This loss can be a result of hormonal changes (see BioBeat 7.10), diminished nerve function resulting in the loss of fast-twitch muscle fibers, living a sedentary lifestyle, eating a poor diet, and suffering from chronic disease. Along with decreased muscle mass and strength comes a reduction in metabolism and an increase in fat mass, especially visceral fat. Including a regular resistive exercise program will counteract some of the sarcopenia.

Sleep Disturbances   Sleep disturbances occur with aging, and sleep quality is an important marker of successful aging. Sleep disruption is commonly caused by medical and psychosocial diseases and conditions in older adults. Contributing factors to sleep disruption include decreased **melatonin** production, consuming foods and beverages containing stimulants, living a sedentary life, taking some medications, and some medical conditions, like thyroid disease, arthritis, depression, Parkinson's disease, and dementia. Sleep issues may include having a difficult time falling asleep or staying asleep, being awakened easily, and sleeping for a shorter period. When sleep isn't adequate or deep, then the person may not be alert during the day and may become sleepy early in the evening. Going to sleep too early can then cause the person to wake up very early in the morning, setting a pattern that doesn't follow a normal circadian rhythm.[43]

melatonin A hormone that controls the sleep and awake cycles; it is produced and secreted by the pineal gland when a person is in a dark environment, thus inducing sleep.

BioBeat 7.10

The Hormones of Aging

Many hormonal changes accompany normal aging. Here we will focus on the sex hormones. There is a decline in estrogen, progesterone, and testosterone production with age. The decline usually begins sometime between the ages of 45 and 55. Men go through **andropause**, while women go through menopause. Both are normal events in the aging process. Menopause is more commonly known than andropause because it has a clear physiological marker, the loss of menstruation.

The decline in testosterone is a major contributing factor to andropause in men. This results in physiological and psychological changes, including reduced sexuality and overall energy. In women, **perimenopause** precedes menopause. Perimenopause can last up to four years. It is a time of ovarian hormonal fluctuation, with many peaks and valleys leading to menstrual irregularity and menopause symptoms. The symptoms of menopause include hot flashes, night sweats, vaginal dryness, insomnia, and mood swings. Some women experience few or no menopause symptoms, whereas others suffer greatly from an arsenal of physical insults. Menopause is official when the woman has not had a menstrual cycle (period) for 12 months.

Hormone therapy is an option for men and women undergoing hormonal changes. Testosterone replacement is available for men, whereas estrogen, progesterone, and even testosterone replacement are available for women. Traditional hormone therapy includes estrogen compounds. This treatment alleviates menopause symptoms and severity. Medical risks of the loss of estrogen with menopause presently focus on cardiovascular disease caused by blood lipid changes and increased bone loss, so hormone replacement may reduce the risk of accelerated bone loss and thus osteoporosis, as well as sarcopenia or reduced muscle mass. The risks of using traditional hormone therapy, however, have been reported to include increased risk for cancer of the breast and uterus, heart attack, stroke, and blood clots.

The use of **bioidentical hormones**, herbal remedies, and lifestyle therapies have become popular to alleviate menopausal symptoms. Bioidentical hormones have the identical chemical structure as those made by the body. The risks of bioidentical hormone therapy are poorly defined, and it is believed that they may be similar to traditional hormone therapy. Herbal remedies commonly include the use of soy isoflavones, black cohosh, wild yam extracts, St. John's wort, valerian root, and melatonin. Soy, black cohosh, and wild yam all contain a variety of phytoestrogens, which mimic estrogen and help some women, but their efficacy is not routinely reported. St. John's wort is used to improve mood and can benefit those who experience mild to moderate depression, whereas valerian root and melatonin may alleviate sleep disturbances. Some individuals benefit from lifestyle therapies. Physical activity can decrease depression and improve the quality of sleep, and relaxation techniques such as deep breathing can improve sleep.[11,16,56,141,164]

What hormonal changes can you expect to experience when you become an older adult?

---

andropause A decline in male sex hormone levels, resulting in physiological and psychological changes and reduced sexuality and overall energy; also known as male menopause.

perimenopause A natural transition period of approximately four years before menopause when hormone levels decline, leading to menstrual irregularity and menopause symptoms.

bioidentical hormones Hormones that have the same chemical structure as those made by the body.

Aging Well The secret to immortality has been sought since antiquity. There have been no discoveries to prevent aging and death at the end of the life span, but controlling aging and maintaining optimal vitality are reasonable life pursuits. Individual genetics, plus social, behavioral, and environmental factors, all affect a person's life expectancy, but the individual can play a role in increasing his or her life expectancy by modifying personal behaviors and being responsible with his or her lifestyle.[163,165]

Although aging is inevitable, it can be done well (see Figure 7.19). Establishing lifelong habits that include eating a balanced diet, routinely being mentally and physically active, and abstaining from known behavioral risk factors are important elements of successful aging. Physical health is also supported by getting adequate rest, so repairing of the daily stresses in the body can occur. Emotional and social health, along with stress management, are all vital components of aging well.[98,147,156]

AGING WELL

Stress Busters
- Relax
- Go for a walk
- Breathe deeply
- Think positively

Emotional Health
- Reduce stress
- Learn relaxation techniques
- Cultivate a garden
- Seek out laughter
- Take time for spiritual growth
- Adopt and love a pet
- Take time off

Social Health
- Be socially active
- Volunteer for a special cause
- Make new friends
- Enroll in lifelong learning
- Be active in your community

Nutritional Health
- Choose nutrient-dense foods
- Eat at least 5 to 9 fruits and vegetables every day
- Drink plenty of water
- Keep fat intake at a healthy level
- Get adequate fiber

Physical Health
- Be physically active
- Get adequate sleep
- Challenge your mental skills
- Do aerobic and strength-training exercises
- Stretch for flexibility

Lifelong Habits for Successful Aging
- Cherish your personal values and goals
- Develop good communication skills
- Balance diet and exercise to maintain a healthy weight
- Practice preventive health care
- Develop skills and hobbies to enjoy for a lifetime
- Manage time
- Learn from mistakes
- Nurture relationships with family and friends
- Enjoy, respect, and protect nature
- Accept change as inevitable
- Plan ahead for financial security

Personal Nutrition, 7th ed, 3oyle & Long, Chapter 11, Fig. 11-10, p. 399.

FIGURE 7.19 The aging well pyramid.

Summary Points

- There are many theories of aging, including those tied to evolution, molecules, cells, and systems.
- The physiology of aging involves many changes to the body's senses and organ systems, including changes in hearing, vision, taste, smell, digestion, bone, immunity, skin, nerves, and renal, lung, liver, brain, and cardiac capacity.
- Mental changes associated with aging can often be prevented with diet, exercise, and lifestyle habits, such as being mentally challenged on a daily basis.
- A social network is a vital component to the health and well-being of older adults.
- Successful aging improves health and quality of life.
- Eating a healthy diet adequate and balanced in Calories, energy-producing nutrients, fiber, vitamins, minerals, and water will support successful aging.
- There are many food assistance programs to support healthy eating for older adults, including those who have food insecurity.
- Lifestyle practices, such as being physically active and not using tobacco or using alcohol in excess, will add significantly to the older adult's quality of life.
- Aerobic and strength-training exercises will preserve lean body mass and prevent age-related physical decline.
- Many common degenerative diseases seen in the elderly are preventable by eating a healthy diet, being active, engaging in social events, getting adequate rest, and avoiding unhealthy habits and situations.
- Aging well is possible and requires a good diet, exercise, stress management, mental stimulation, social and emotional support, and adequate sleep.

Take Ten on Your Knowledge Know-How

1. What happens physiologically during aging?
2. What is the role of diet and lifestyle in brain preservation?
3. What should the social network of an older adult ideally look like?
4. Why are Calorie needs reduced for the elderly?
5. What are some of the changes in nutrient needs for older adults?
6. How can a poverty-stricken older adult be assured of eating healthy meals on a regular basis?
7. What are the basic fitness recommendations for the elderly?
8. What are some common health issues of older adults? How many of these begin in young adulthood because of unhealthy lifestyle practices?
9. What is the role of nutrition in the common health issues of older adults?
10. How can a person age well?

 **SUMMARY**

CONTENT KNOWLEDGE

IN THIS MODULE, YOU HAVE LEARNED THAT:

- The quality and quantity of foods a person eats affect his or her health and physiological function from the womb to death.
- The effort an expectant mother makes to achieve a high level of reproductive fitness not only benefits her during the pregnancy, but also has a lifelong impact on the health of her baby.
- Introducing healthy foods and eating habits supports optimal growth and development of an infant and extends throughout the child's life.
- Nutrient requirements change significantly during infancy and childhood depending on age and gender and if the child is going through a growth spurt.
- Regulatory mechanisms for appetite are tuned to govern energy intake.
- To ensure nutrient adequacy, consuming a variety of whole, fresh foods and only a limited amount of solid fat and sugar is recommended in MyPlate food plans.
- In the later stages of the life span, age-related physiological decline is evident, but living a healthy lifestyle throughout the life span will promote successful aging, which is maintaining the highest level of function possible each day.

PERSONAL IMPROVEMENT GOALS

THE MANAGEMENT OF LIFESTYLE HAS IMPLICATIONS THROUGHOUT LIFE, SUCH AS THE FOLLOWING:

- Preconceptual fitness and nutritional status affect fertility, conception, implantation, and development of the embryo and fetus.
- Realizing that there are no second chances for embryonic development during the first eight weeks of gestation, nor for fetal development during the last 30 weeks of pregnancy, should motivate any woman to engage in exercise, follow her MyPlate eating plan, take a multivitamin and mineral supplement, and avoid environmental risks.
- Parents can be positive influences on their children's eating habits by supplying healthy food choices to the infant or child, allowing him or her to decide how much food to consume, making feeding environments pleasant, and using food appropriately (as something to be enjoyed, not as a reward).
- A sound foundation of healthy eating practices can last a lifetime.

- Regular, age-appropriate exercise and stimulation are also essential components of living well throughout the life span.
- Having a healthy lifestyle throughout your life can delay or avoid the common chronic diseases that manifest during the later stages of the life span.

Here is a tip for you: Make sound diet, exercise, and lifestyle choices for you and your children every day to promote optimal health and well-being from conception through old age.

You can assess if you met the learning objectives for this module by successfully completing the Homework Assessment and the Total Recall activities (sample questions, case study with questions, and crossword puzzle).

Homework Assessment

50 questions

1. A diet high in saturated fat supports normal fertility.
   A. True   B. False

2. The DRIs for a pregnant woman are increased compared to those for a nonpregnant female of the same age.
   A. True   B. False

3. The goal for preconceptual nutrition is to supply the energy, nutrients, and fiber needed for maternal structural changes and function.
   A. True   B. False

4. Preeclampsia is the term used to describe high blood pressure during pregnancy.
   A. True   B. False

5. Taking a prenatal supplement ensures dietary adequacy of the macronutrients.
   A. True   B. False

6. Normal cell division is called:
   A. Ovulation
   B. Conception
   C. Mitosis
   D. Meiosis
   E. Gametes

7. Prenatal weight gain is based on:
   A. Maternal girth circumference
   B. Maternal blood pressure
   C. Maternal height
   D. Maternal BMI
   E. Maternal fasting blood sugar

8. Which of the following is a health issue that may commonly develop during pregnancy?
   A. Osteoporosis
   B. Dementia
   C. Menopause
   D. Hearing loss
   E. Gestational diabetes

9. Regular prenatal exercise can:
   A. Promote bone loss
   B. Elevate LDL cholesterol
   C. Control weight gain
   D. Decrease muscle tone
   E. Cause constipation

10. Lactation is:
    A. Feeding the infant during the first year of life
    B. A process to feed mammalian offspring
    C. A nutritionally inadequate process to feed infants during the first few months of life
    D. A postpartum, avoidable event
    E. A process that extends the recovery process of pregnancy

11. Malnutrition causes failure to thrive.
    A. True   B. False

12. Babies should be held during breast or bottle feedings.
    A. True   B. False

13. Positive feedback keeps breast milk flowing to a suckling infant.
    A. True   B. False

14. Uterine contractions are a benefit to a mother who breastfeeds her baby.
    A. True   B. False

15. The dietary need for fat is low during infancy.
    A. True   B. False

16. At what age would an infant or child likely begin to transition from a bottle to a sippy cup?
    A. 6 months
    B. 9 months
    C. 12 months
    D. 24 months
    E. 36 months

17. In childhood, _____ is involved in the psychological and social aspects of eating.
    A. Olfaction
    B. Gustation
    C. Food texture
    D. A and B
    E. A, B, and C

18. During infancy, essential fatty acids and omega-3 fatty acids function in:
    A. Cognitive development
    B. Maturation of the retina
    C. Brain growth
    D. Central nervous system function
    E. All of the above

19. Which of the following statements is false regarding maternal and infant nutrition?
    A. Vitamin K injections are given at birth.
    B. The need for iodine is high.
    C. Breastfed infants may need vitamin D supplements.
    D. Vegan mothers produce the most nutritious breast milk.
    E. Fluoride is needed for tooth formation.

20. Which of the following is the best sequence for introducing foods to an infant?
    A. Breast milk, real fruit juice, wheat cereal, and then strained meat and vegetables
    B. Breast milk, iron-fortified rice cereal, strained, cooked vegetables, and then mashed fruit
    C. Breast milk, chopped meats, puréed fruits, and then diced, raw fruits and vegetables
    D. Cow's milk, beef, corn, and then honey and herbal teas
    E. Infant formula, fruit and vegetable juice, nuts and seeds, and then chopped meats

21. A child in the 25th percentile for weight-for-age is considered underweight.
    A. True   B. False

22. A lack of healthy eating and activity patterns promotes childhood obesity.
    A. True   B. False

23. Having a regular meal and snack schedule supports the eating process during childhood.
    A. True   B. False

24. Most food rituals and jags experienced during childhood persist into adolescence.
    A. True   B. False

25. Children should always be in positive energy balance to support growth and development.
    A. True   B. False

26. An undernourished child may experience:
    A. More frequent sickness
    B. Fatigue
    C. Less social behavior
    D. Less curiosity
    E. All of the above

27. In the division of responsibility, it is the child's job to determine:
    A. What to eat
    B. When to eat
    C. Where to eat
    D. How much to eat
    E. None of the above

28. All of the following affect childhood eating except the:
    A. Color of the food
    B. Texture of the food
    C. Taste of the food
    D. Nutritional value of the food
    E. Temperature of the food

29. Which of the following statements is false regarding nutrient needs during childhood?
    A. Adequate fiber will reduce constipation.
    B. Fat provides an important source of energy.
    C. Calcium needs are high to support bone growth.
    D. Fluid and electrolyte replacement drinks are needed to prevent dehydration.
    E. Low iron intake increases the risk of anemia.

30. Which of the following statements is false regarding childhood diet and health?
    A. Daily physical activity is needed.
    B. One-third of the DRI for several nutrients are provided by the National School Lunch Program.
    C. Adult health conditions like hypertension rarely occur during childhood.
    D. Food allergy is a genetic condition.
    E. Lead toxicity negatively affects brain function.

31. The protein need during adolescence may fluctuate based on physiological demand (growth spurts) rather than chronological age or the DRI for protein.
    A. True   B. False

32. It is normal for boys and girls to gain more than 50 pounds in body weight during adolescence.
    A. True   B. False

33. Adolescents should engage in 60 minutes of daily physical activity.
    A. True   B. False

34. Hormonal changes are most commonly the cause of acne in adolescents.
    A. True   B. False

35. Proper nutrition is needed to support adolescent growth and development.
    A. True   B. False

36. There is a decline in _____ consumption by adolescent girls.
    A. Milk
    B. Processed food
    C. Sweets
    D. B and C
    E. A, B, and C

37. Which of the following is an eating disorder that is characterized by self-induced starvation?
    A. Anorexia nervosa
    B. Bulimia nervosa
    C. Menarche
    D. Binge eating disorder
    E. Celiac disease

38. Which of the following statements is false regarding puberty?
    A. There are sexual maturity ratings by developmental stage.
    B. It is a time of reproductive maturation.
    C. It is hormonally driven.
    D. Girls usually mature earlier than boys.
    E. Most boys and girls have adult-type pubic hair by age 13.

39. Which of the following statements is false regarding energy and nutrients during adolescence?
    A. Insufficient Calorie intake can hinder growth and development.
    B. The Acceptable Macronutrient Distribution Range (AMDR) for total fat is 25 to 35 percent of Calories.
    C. Iron-deficiency anemia is more common in boys than girls.
    D. Vitamin D is needed to support bone mineralization.
    E. Zinc is important for growth and development.

40. Food and lifestyle choices during adolescence are largely influenced by:
    A. Parents
    B. Siblings
    C. Peers
    D. A and B
    E. A, B, and C

41. With aging, neurological function is reduced.
    A. True   B. False

42. Hypertension is a leading medical condition in older adults.
    A. True   B. False

43. Dentures are more effective than natural teeth in chewing food.
    A. True   B. False

44. The social network plays an important role in the health of older adults.
    A. True   B. False

45. Calorie needs increase after age 65.
    A. True   B. False

46. Which of the following is associated with increased life expectancy?
    A. Skipping meals
    B. Increasing Calorie intake
    C. Thyroid disorders
    D. Managing stress
    E. All of the above

47. Bone mass declines most significantly as a result of:
    A. Incontinence
    B. Menopause
    C. Sarcopenia
    D. Hypertension
    E. Macular degeneration

48. All of the following nutrients are needed for brain function except:
    A. Thiamin
    B. Iron
    C. Vitamin B_{12}
    D. Saturated fat
    E. Vitamin E

49. Which of the following is a food assistance program specifically for older people?
    A. Women, Infants, and Children
    B. The Fresh Fruit and Vegetable Program
    C. The Nutrition Services Incentive Program
    D. The Supplemental Nutrition Assistance Program
    E. All of the above

50. Which of the following statements is false regarding diet and health issues in older adults?
    A. Stimulating foods and beverages disturb sleep.
    B. Herbal remedies help alleviate menopause symptoms in some women.
    C. Oxidative stress and low antioxidant nutrient status may be implicated in Alzheimer's disease.
    D. There are no dietary recommendations to prevent macular degeneration.
    E. Omega-3 fatty acids from fish oil may improve arthritis.

Total Recall

SAMPLE QUESTIONS

True/False Questions

1. A woman with a normal body mass index (BMI) before pregnancy should gain 11 to 20 pounds during pregnancy.

2. Caregivers should provide regularly scheduled meals and snacks that are healthy and safe for children.

3. Diet and exercise influence a child's growth.

4. Substantial increases in body weight and height are experienced during adolescence.

5. There is a nutritional approach to reducing age-related visual problems.

Multiple Choice Questions: Choose the best answer.

6. Which of the following conditions negatively affects conception?
   A. Celiac disease
   B. Obesity
   C. Genetic mutation of MTRR or MTHFR genes
   D. Polycystic ovarian syndrome
   E. All of the above

7. All of the following nutrient needs are high during infancy except:
   A. Vitamin D
   B. Fiber
   C. Iodine
   D. Protein
   E. Vitamin B_{12}

8. At what age would a child be expected to use a fork and dull knife?
   A. 1–2 years
   B. 3–5 years
   C. 4–6 years
   D. 6–9 years
   E. 9–12 years

9. Which of the following nutrients is not directly involved in supporting adolescent growth?
   A. Vitamin A
   B. Sodium
   C. Zinc
   D. Folate
   E. Vitamin C

10. In older adults, physical activity is implicated for improving:
    A. Arthritis
    B. Sleep disturbances
    C. Sarcopenia
    D. Osteoporosis
    E. All of the above

CASE STUDY

Gabe is a healthy and active 6-year-old boy. He is a member of a middle-income, soccer-loving family, so as long as he's been aware, he has been around the sport. He was breastfed through infancy and switched to whole cow's milk at age 1. He is in first grade now at a public school and is progressing well with his cognitive and social skills. He is active at recess and after school. Throughout his life, he has usually tracked in the 75th percentile for weight-for-age and in the 50th percentile for height-for-age. His CDC growth chart is provided on the next page.

2 to 20 years: Boys
Stature-for-age and Weight-for-age percentiles

NAME Gabe

RECORD # _____

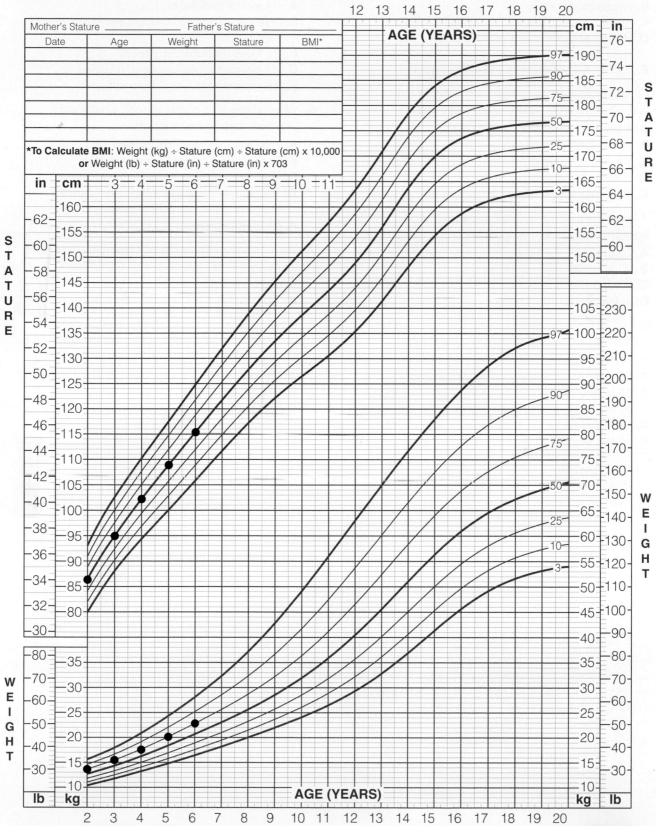

SAFER · HEALTHIER · PEOPLE™

7.6 SUMMARY

419

1. What is the earliest nutrition exposure that Gabe experienced?
   A. Breast milk
   B. Whole cow's milk
   C. His mother's nutritional status at conception
   D. His father's zinc status enabling spermatogenesis
   E. His first bite of organic baby rice cereal

2. Which of the following is a health benefit that Gabe has from being breastfed as an infant?
   A. Reduced cancer risk
   B. Strengthened immunity
   C. Improved blood pressure
   D. Greater bone mass
   E. None of the above

3. If Gabe maintains the same growth patterns, approximately how much will he weigh at age 12?
   A. 70 pounds
   B. 80 pounds
   C. 90 pounds
   D. 100 pounds
   E. 110 pounds

4. If Gabe maintains the same growth patterns, approximately how tall will he be at age 12?
   A. 56 inches
   B. 58 inches
   C. 60 inches
   D. 62 inches
   E. 64 inches

5. What can Gabe expect to be experiencing at his current age regarding his growth and development?
   A. He learns to drink from a cup
   B. His sleep drops to 10 hours per night with no naps
   C. He loses his front baby teeth
   D. He develops medium hand muscles
   E. He experiences a rapid growth spurt

6. When Gabe is at public school, what food program may he be involved in?
   A. Women, Infants, and Children
   B. The National School Lunch Program
   C. The Nutrition Services Incentive Program
   D. The Supplemental Nutrition Assistance Program
   E. All of the above

7. Gabe is reducing his risk of chronic disease later in life by being physically active and eating a well-balanced diet as a child.
   A. True   B. False

8. During Gabe's life span, his highest nutritional need for protein based on his body weight will be during:
   A. Infancy
   B. Childhood
   C. Adolescence
   D. Adulthood
   E. Older adulthood

9. If Gabe experiences puberty at a typical age, he can be expected to have mature adult sexual characteristics at age 12.
   A. True   B. False

10. When Gabe is old, if he remains as social as he is now, he will be healthier than if he were antisocial.
    A. True   B. False

FUN-DUH-MENTAL PUZZLE

ACROSS

4. A nutrient needed during pregnancy, for growth, and to cover losses from menstruation.
7. Body system affected by declining bone mass.
9. In the _____ week of pregnancy, the fetal central nervous system is developing.
11. Taste preference that children have.
12. Type of infant cereal usually fed as baby's first food.
13. A girl's first period.
15. Plate where bone growth takes place.
17. Normal aging process in which testosterone levels decline in men.
18. Food that poses a choking and allergy risk to infants and young children.
19. The period of time between birth and death.
22. Type of destructive, food-related behavior that many adolescent girls engage in.
23. Type of food production that may improve male fertility.
24. Sedentary lifestyle and unhealthy meal patterns and composition contribute to this rising childhood and adolescent health problem.

DOWN

1. Nutrient important for the brain and central nervous system from conception through old age.
2. State of energy balance that infants and children are typically in.
3. Type of milk that is best for infants.
4. The only life span time when there is a DRI for total fat.
5. Type of division that needs to be established between parents and children for successful eating to take place at meals and snacks.
6. Nutrient important for dental health that may be present in drinking water.
8. The number of minutes that children, adolescents, and older adults should exercise each day.
10. Deterioration of cognition that may occur with aging.
14. Good _____ is critical before and during pregnancy.
16. Acronym for a common condition impairing female reproduction.
20. The time when the reproductive system matures.
21. The loss of muscle mass and strength that occurs with aging.

References

Web Resources

A. Academy of Breastfeeding Medicine: www.bfmed.org
B. Action for Healthy Kids: www.actionforhealthykids.org
C. Administration on Aging: www.aoa.gov
D. Alzheimer's Association: www.alz.org
E. American Academy of Allergy, Asthma and Immunology: www.aaaai.org
F. American Academy of Pediatrics: www.aap.org
G. American Association of Retired Persons: www.aarp.org
H. American Congress of Obstetricians and Gynecologists: www.acog.org
I. American Council on Exercise: www.acefitness.org
J. American Diabetes Association: www.diabetes.org
K. American Geriatrics Society: www.americangeriatrics.org
L. American Macular Degeneration Foundation: www.macular.org
M. American Society on Aging: www.asaging.org
N. Arthritis Foundation: www.arthritis.org
O. Centers for Disease Control and Prevention, Growth Charts: www.cdc.gov/growthcharts
P. Centers for Disease Control and Prevention, Health Information for Older Adults: www.cdc.gov/aging/aginginfo/index.htm
Q. Centers for Disease Control and Prevention, Life Expectancy: www.cdc.gov/nchs/fastats/lifexpec.htm
R. Centers for Disease Control and Prevention, Overweight and Obesity: www.cdc.gov/obesity
S. Cystic Fibrosis Foundation: www.cff.org
T. Dietary Guidelines for Americans: www.health.gov/dietaryguidelines
U. Fitness, Sports, and Nutrition (President's Council): www.fitness.gov
V. Food Allergy and Anaphylaxis Network: www.foodallergy.org
W. Food and Nutrition Information Center: fnic.nal.usda.gov
X. Food Stamp Program: frac.org/federal-foodnutrition-programs/snapfood-stamps
Y. Healthfinder.gov: www.healthfinder.gov
Z. KidsHealth: kidshealth.org
AA. La Leche League: www.llli.org
BB. Lead Hazard Control: portal.hud.gov:80/hudportal/HUD?src=/program_offices/healthy_homes
CC. Linus Pauling Institute: lpi.oregonstate.edu
DD. March of Dimes: www.marchofdimes.com
EE. Meals on Wheels Association of America: www.mowaa.org
FF. MedlinePlus, Newborn Screening: www.nlm.nih.gov/medlineplus/newbornscreening.html
GG. MyPlate: www.choosemyplate.gov
HH. National Alliance on Mental Illness: www.nami.org
II. National Center for Complementary and Alternative Medicine: nccam.nih.gov
JJ. National Center for Health Statistics: www.cdc.gov/nchs
KK. National Digestive Diseases Information Clearinghouse, Celiac Disease: digestive.niddk.nih.gov/ddiseases/pubs/celiac
LL. National Down Syndrome Society: www.ndss.org
MM. National Eye Institute: www.nei.nih.gov
NN. National Institute on Aging: www.nia.nih.gov
OO. National Institute of Child Health and Human Development: www.nichd.nih.gov
PP. National Institute of Mental Health: www.nimh.nih.gov/index.shtml
QQ. National Organization on Fetal Alcohol Syndrome: www.nofas.org
RR. National Osteoporosis Foundation: www.nof.org
SS. NIH Senior Health: nihseniorhealth.gov
TT. Nutrient Rich Foods Coalition: www.nutrientrichfoods.org

UU. Nutrition Assistance Programs: www.fns.usda.gov/fns
VV. Supplemental Nutrition Assistance Program (SNAP): www.fns.usda.gov/snap
WW. USDA Food and Nutrition Service, School Meals: www.fns.usda.gov/cnd
XX. Wellness Council of America: www.welcoa.org
YY. Women, Infants, and Children: www.fns.usda.gov/wic
ZZ. World Health Organization, Child Growth Standards: www.who.int/childgrowth/en

Works Cited

1. Abad-Jorge, A. (2008). The role of DHA and ARA in infant nutrition and neurodevelopmental outcomes. *Today's Dietitian, 10*(10), 66–70.
2. American Dietetic Association. (2008). Position of the American Dietetic Association: Nutrition and lifestyle for a healthy pregnancy outcome. *Journal of the American Dietetic Association, 108*(3), 553–561.
3. American Psychiatric Association. (2000). *Diagnostic and statistical manual of mental disorders* (4th ed.). Washington, DC: Author.
4. Anderson, K., Norman, R. J, & Middleton, P. (2010). Preconception lifestyle advice for people with subfertility. *Cochrane Database of Systematic Reviews*, 4.
5. Anton, L. G. (2010). Disordered eating in adolescent athletes: Prevalence and risk factors. *SCAN'S (A Publication for Sports, Cardiovascular, and Wellness Nutritionists) PULSE, 29*(1), 9–12.
6. Benelam, B. (2009). Satiety and anorexia of aging. *British Journal of Community Nursing, 14*(8), 332–335.
7. Benton, D. (2010). The influence of dietary status on the cognitive performance of children. *Molecular Nutrition and Food Research, 54*, 457–470.
8. Berghella, V., Buchanan, E., Pereira, L., & Baxter, J. K. (2010). Preconception care. *Obstetrical and Gynecological Survey, 65*(2), 119–131.
9. Bernstein, M., & Luggen, A. S. (2010). *Nutrition for the older adult.* Sudbury, MI: Jones and Bartlett Publishing.
10. Black, J. L., Pinero, D. J., & Parekh, N. (2009). Zinc and cognitive development in children: Perspectives from international studies. *Topics in Clinical Nutrition, 24*(2), 130–138.
11. Black, M. M. (2009). Micromineral deficiencies and child development. *Nutrition Today, 44*(2), 71–76.
12. Blennow, K., de Leon, M. J., & Zetterberg, H. (2006). Alzheimer's disease. *Lancet, 368*, 387–403.
13. Boirie, Y. (2009). Physiopathological mechanism of sarcopenia. *Journal of Nutrition Health and Aging, 13*, 717–723.
14. Boutelle, K. N., Libbey, H., Neumark-Sztainer, D., & Story, M. (2009). Weight control strategies of overweight adolescents who successfully lost weight. *Journal of the American Dietetic Association, 109*(12), 2029–2935.
15. Briceño-Pérez, C., Briceño-Sanabria, L., & Vigil-De Gracia, P. (2009). Prediction and prevention of preeclampsia. *Hypertension and Pregnancy, 28*(2), 138–155.
16. Brown, J. (2008). *Nutrition through the life cycle* (3rd ed.). Belmont, CA: Wadsworth.
17. Bryan, J., Osendarp, S., Hughes, D., Calvaresi, E., Baghurst, K., & van Klinken, J. (2004). Nutrients for cognitive development in school-aged children. *Nutrition Reviews, 62*, 295–306.
18. Bueno, A. L., & Czepielewski, M. A. (2008). The importance for growth of dietary intake of calcium and vitamin D. *Journal of Pediatrics (Rio J), 84*(5), 386–394.
19. Burrowes, J. D. (2010). Preventing heart disease today and tomorrow in youth. *Nutrition Today, 45*(1), 33–42.
20. Byrd-Bredbenner, C., & Abbot, J. M. (2008). Food choice influencers of mothers of young children. *Topics in Clinical Nutrition, 23*(3), 198–215.
21. Calhoun, F., & Warren, K. (2007). Fetal alcohol syndrome: Historical perspectives. *Neuroscience and Biobehavioral Reviews, 31*(2), 168–171.

22. Cannella, C., Savina, C., & Donini, L. M. (2009). Nutrition, longevity and behavior. *Archives of Gerontology and Geriatrics, 49* (Suppl. 1), 19–27.

23. Carlson, S. E. (2009). Docosahexaenoic acid supplementation in pregnancy and lactation. *American Journal of Clinical Nutrition, 89*(2), 678S–684S.

24. Carney, L. N. (2009). Pediatric food allergies. *Today's Dietitian, 11*(7), 48–54.

25. Carney, L. N. (2010). Fortified breast milk: Vulnerable infants need safe administration of nature's ideal food. *Today's Dietitian, 12*(6), 46–53.

26. Casey, B. J., Jones, R. M., & Hare, T. A. (2008). The adolescent brain. *Annals of the New York Academy of Sciences, 1124*, 111–126.

27. Cashman, K. D. (2007). Vitamin D in childhood and adolescence. *Postgraduate Medicine Journal, 83*, 230–235.

28. Caudill, M. A. (2010). Pre- and postnatal health: Evidence of increased choline needs. *Journal of the American Dietetic Association, 110*(8), 1198–1206.

29. Cena, E. R., Joy, A. B., Heneman, K., Espinosa-Hall, G., Garcia, L., Schneider, C., . . . & Zidenberg-Cherr, S. (2008). Folate intake and food-related behaviors in nonpregnant, low-income women of childbearing age. *Journal of the American Dietetic Association, 108*(8), 1364–1368.

30. Cetin, I., Berti, C., & Calabrese, S. (2010). Role of micronutrients in the periconceptional period. *Human Reproduction Update, 16*(1), 80–95.

31. Chapin, R. E., Robbins, W. A., Schieve, L. A., Sweeney, A. M., Tabacova, S. A., & Tomashek, K. M. (2004). Off to a good start: The influence of pre- and periconceptional exposures, parental fertility, and nutrition on children's health. *Environmental Health Perspectives, 112*, 69–78.

32. Checkley, W., West, K. P., Wise, R. A., Baldwin, M. R., Wu, L., LeClerq, S. C., . . . & Sommer, A. (2010). Maternal vitamin A supplementation and lung function in offspring. *New England Journal of Medicine, 362*(19), 1784–1794.

33. Com, G. (2010). Cystic fibrosis newborn screening. *Journal of the Arkansas Medical Society, 106*(9), 210–212.

34. Cooke, L. (2007). The importance of exposure for healthy eating in childhood: A review. *Journal of Human Nutrition and Dietetics, 20*, 294–301.

35. Cooper, C. C. (2009). Intergenerational programs: Uniting young and old for good nutrition, physical activity, and wellness. *Today's Dietitian, 11*(6), 44–49.

36. Cooper, C. C. (2010). Probiotics in pediatrics: Using friendly bacteria to treat health conditions. *Today's Dietitian, 12*(1), 24–27.

37. Cortese, S., Falissard, B., Pigaiani, Y., Bazato, C., Bogoni, G., Pellegrino, M., . . . & Maffeis, C. (2010). The relationship between body mass index and body size dissatisfaction in young adolescents: Spline function analysis. *Journal of the American Dietetic Association, 110*(7), 1098–1102.

38. Cox, J. T., & Phelan, S. T. (2008). Nutrition during pregnancy. *Obstetrics and Gynecology Clinics of North America, 35*(3), 369–383.

39. Cox, J. T., & Phelan, S. T. (2009). Prenatal nutrition: Special considerations. *Minerva Ginecologica, 61*(5), 373–400.

40. Dauncey, M. J. (2009). New insights into nutrition and cognitive neuroscience. *Proceedings of the Nutrition Society, 68*, 408–415.

41. Davies, G. A., Wolfe, L. A., Mottola, M. F., MacKinnon, C., & Society of Obstetricians and Gynaecologists of Canada. (2003). Exercise in pregnancy and the postpartum period. *Journal of Obstetrics and Gynaecology of Canada, 25*(6), 516–522.

42. de Onis, M., Garza, C., Onyango, A. W., & Borghi, E. (2007). Comparison of the WHO Child Growth Standards and the CDC 2000 Growth Charts. *Journal of Nutrition, 137*, 144–148.

43. Driscoll, H. C., Serody, L., Patrick, S., Maurer, J., Bensasi, S., Houck, P. R., . . . & Reynolds, C. F. III. (2008). Sleeping well, aging well: A descriptive and cross-sectional study of sleep in "successful agers" 75 and older. *American Journal of Geriatrics and Psychiatry, 16*(1), 74–82.

44. Duggan, M. B. (2010). Anthropometry as a tool for measuring malnutrition: Impact of the new WHO Growth Standards and reference. *Annals of Tropical Paediatrics, 30*(1), 1–17.

45. Eisner, O., & Maltin, V. (2009). Celiac disease, infertility and pregnancy complications. *Women's Health Report, 1*, 3–4.

46. El Mouzan, M. I., Foster, P. J., Al Herbish, A. S., Al Salloum, A. A., Al Omar, A. A., Qurachi, M. M., & Kecojevic, T. (2009). The implications of using the World Health Organization Child Growth Standards in Saudi Arabia. *Nutrition Today, 44*(2), 62–70.

47. Faigenbaum, A. D. (2007). Resistance training for children and adolescents: Are there health outcomes? *American Journal of Lifestyle Medicine, 1*(3), 190–200.

48. Fall, C. (2009). Maternal nutrition: Effects on health in the next generation. *Indian Journal of Medical Research, 130*(5), 593–599.

49. Fernandez, I. D., Olson, C. M., & Dye, T. D. V. (2008). Discordance in the assessment of prepregnancy weight status of adolescents: A comparison between the Centers for Disease Control and Prevention sex- and age-specific body mass index classification and the Institute of Medicine–based classification used for maternal weight gain guidelines. *Journal of the American Dietetic Association, 108*(6), 998–1002.

50. Ferrari, A. U., Radaelli, A., & Centola, M. (2003). Physiology of aging: Invited review: Aging and the cardiovascular system. *Journal of Applied Physiology, 95*, 2591–2597.

51. Fotuhi, M., Mohassel, P., & Yaffe, K. (2009). Fish consumption, long-chain omega-3 fatty acids and risk of cognitive decline or Alzheimer disease: A complex association. *Nature Clinical Practice: Neurology, 5*(3), 140–152.

52. Foundation for Infinite Survival, Inc., & Life-Extension and Control of Aging Program. (n.d.). *A life-expectancy calculation.* Berkeley, CA: Author. Accessed at: http://fis.org/LE-Calc.

53. Fu, M., Cheng, L., Tu, S., & Pan, W. (2007). Association between unhealthful eating patterns and unfavorable overall school performance in children. *Journal of the American Dietetic Association, 107*(11), 1935–1943.

54. Gat-Yablonski, G., Yackobovitch-Gavan, M., & Phillip, M. (2009). Nutrition and bone growth in pediatrics. *Endocrinology Metabolism Clinics of North America, 38*(3), 565–586.

55. Gedney, L. (2010). Working mothers' challenge: Finding a way to pump throughout the day. *Today's Dietitian, 12*(5), 33–35.

56. Geller, S. E., & Studee, L. (2006). Soy and red clover for mid-life and aging. *Climacteric, 9*(4), 245–263.

57. Getz, L. (2009). No peas for me! Helping parents combat kids' picky eating behaviors. *Today's Dietitian, 11*(9), 40–44.

58. Getz, L. (2009). Proper planning: Schools need food allergy strategies to keep students safe. *Today's Dietitian, 11*(8), 44–53.

59. Gonzalez, A., Kohn, M. R., & Clarke, S. D. (2007). Eating disorders in adolescents. *Australian Family Physician, 36*(8), 614–619.

60. Grassi, A. (2008). Recognition and treatment approaches for polycystic ovary syndrome. *Women's Health Report, 1*, 3–4.

61. Grummer-Strawn, L. M., Reinold, C., & Krebs, N. F. (2010). Use of World Health Organization and CDC Growth Charts for children aged 0–59 months in the United States. *Morbidity and Mortality Weekly Report, 59*(RR-9), 1–14.

62. Guadalupe-Grau, G., Fuentes, T., Guerra, B., & Calbet, J. A. L. (2009). Exercise and bone mass in adults. *Sports Medicine, 39*(6), 439–468.

63. Guilarte, T. R. (2009). Vitamin B_6 and cognitive development: Recent research findings from human and animal studies. *Nutrition Reviews, 51*, 193–198.

64. Haber, D. (2007). *Health promotion and aging: Practical applications for health professionals* (4th ed.). New York: Springer.

65. Hawthorne, K. M., Abrams, S. A., & Heird, W. C. (2009). Docosahexaenoic acid (DHA) supplementation of orange juice increases plasma phospholipid DHA content of children. *Journal of the American Dietetic Association, 109*(4), 708–712.

66. House, S. H. (2009). Schoolchildren, maternal nutrition and generating healthy brains: The importance of lifecycle education for fertility, health and peace. *Nutrition and Health, 20*(1), 51–76.

67. Houston, D. K., Nicklas, B. J., & Zizza, C. A. (2009). Weighty concerns: The growing prevalence of obesity among older adults. *Journal of the American Dietetic Association, 109*(11), 1886–1895.

68. Innis, S. M. (2007). Dietary (n-3) fatty acids and brain development. *Journal of Nutrition, 137*, 855–859.

69. Institute of Medicine. (2006). *Dietary Reference Intakes: The essential guide to nutrient requirements.* Washington, DC: The National Academies Press.

70. Isaacs, E., Oates, J., & ILSI Europe A.I.S.B.L. (2008). Nutrition and cognition: Assessing cognitive abilities in children and young people. *European Journal of Nutrition, 47* (Suppl. 3), 4–24.

71. Jevitt, C. (2009). Pregnancy complicated by obesity: Midwifery management. *Journal of Midwifery and Women's Health, 54*(6), 445–451.

72. Joeckel, R. J., & Phillips, S. K. (2009). Overview of infant and pediatric formulas. *Nutrition in Clinical Practice, 24*(3), 356–362.

73. Jolliffe, C. J., & Janssen, I. (2006). Vascular risks and management of obesity in children and adolescents. *Vascular Health and Risk Management, 2*(2), 171–187.

74. Jones, T. E., Stephenson, K. W., King, J. G., Knight, K. R., Marshall, T. L., & Scott, W. B. (2009). Sarcopenia: Mechanisms and treatments. *Journal of Geriatric Physical Therapy, 32*(2), 39–45.

75. Kemper, A. R., Brewer, C. A., & Singh, R. H. (2010). Perspectives on dietary adherence among women with inborn errors of metabolism. *Journal of the American Dietetic Association, 110*(2), 247–252.

76. Kline, D. A. (2010). 'Under-the-radar' eating disorders: Know the signs to improve care. *Today's Dietitian, 12*(11), 68–74.

77. Koletzko, B., Baker, S., Cleghorn, G., Neto, U. F., Gopalan, S., Hernell, O., . . . & Zong-Yi, D. (2005). Global standard for the composition of infant formula: Recommendations of an ESPGHAN coordinated international expert group. *Journal of Pediatric Gastroenterology and Nutrition, 41,* 584–599.

78. Koopman, R., & van Loon, L. J. (2009). Aging, exercise, and muscle protein metabolism. *Journal of Applied Physiology, 106*(6), 2040–2048.

79. Kraak, V. I., & Story, M. (2010). A public health perspective on healthy lifestyles and public-private partnerships for global childhood obesity prevention. *Journal of the American Dietetic Association, 110*(2), 192–200.

80. Lanigan, J., & Singhal, A. (2009). Early nutrition and long-term health: A practical approach. *Proceedings of the Nutrition Society, 68*(4), 422–429.

81. LaRowe, T. L., Moeller, S. M., & Adams, A. K. (2007). Beverage patterns, diet quality, and body mass index of US preschool and school-aged children. *Journal of the American Dietetic Association, 107*(5), 1124–1133.

82. Larson, N. I., Neumark-Sztainer, D., Hannan, P. J., & Story, M. (2007). Family meals during adolescence are associated with higher diet quality and healthful meal patterns during young adulthood. *Journal of the American Dietetic Association, 107*(9), 1502–1510.

83. Leaf, A. A., & RCPCH Standing Committee on Nutrition. (2007). Vitamins for babies and young children. *Archives of Disease in Childhood, 92*(2), 160–164.

84. Lewis, B., Avery, M., Jennings, E., Sherwood, N., Martinson, B., & Crain, A. L. (2008). The effect of exercise during pregnancy on maternal outcomes: Practical implications for practice. *American Journal of Lifestyle Medicine, 2*(5), 441–455.

85. Libonati, J., & Libonati, C. (2009). Understanding celiac disease. *Today's Dietitian, 11*(6), 50–55.

86. Lipstein, E. A., Vorono, S., Browning, M. F., Green, N. S., Kemper, A. R., Knapp, A. A., Prosser, L. A., & Perrin, J. M. (2010). Systematic evidence review of newborn screening and treatment of severe combined immunodeficiency. *Pediatrics, 125*(5), 1226–1235.

87. Lott, I. T., & Dierssen, M. (2010). Cognitive deficits and associated neurological complications in individuals with Down's syndrome. *Lancet Neurology, 9*(6), 623–633.

88. Martin, G. M., Bergman, A., & Barzilai, N. (2007). Genetic determinants of human health span and life span: Progress and new opportunities. *PLoS Genetics, 3,* 1121–1130.

89. Martin, J. B. (2010). The development of ideal body image perceptions in the United States. *Nutrition Today, 45*(3), 98–110.

90. Matter, C. N., Fok, D., & Chong, Y. S. (2008). Common concerns regarding breastfeeding in a family practice setting. *Singapore Medical Journal, 49*(4), 272–279.

91. McCall, A., & Raj, R. (2009). Exercise for prevention of obesity and diabetes in children and adolescents. *Clinical Journal of Sports Medicine, 28*(3), 393–421.

92. McCann, J., & Ames, B. N. (2007). An overview of evidence for a causal relation between iron deficiency during development and deficits in cognitive or behavioral function. *American Journal of Clinical Nutrition, 85,* 931–945.

93. McMillen, I. C., Rattanatray, L., Duffield, J. A., Morrison, J. L., MacLaughlin, S. M., Gentili, S., & Muhlhausler, B. S. (2009). The early origins of later obesity: Pathways and mechanisms. *Advances in Experimental Medicine and Biology, 646*, 71–81.

94. Melanson, K. J. (2008). Exercise nutrition for adults older than 40 years. *American Journal of Lifestyle Medicine, 2*(4), 285–289.

95. Melse-Boonstra, A., & Jaiswal, N. (2010). Iodine deficiency in pregnancy, infancy and childhood and its consequences for brain development. *Best Practice and Research in Clinical Endocrinology and Metabolism, 24*(1), 29–38.

96. Meyer, R. (2009). Infant feed first year: Part 1: Feeding practices in the first six months of life. *Journal of Family Health Care, 19*(1), 13–16.

97. Mogayzel, P. J. Jr., & Flume, P. A. (2010). Update in cystic fibrosis 2009. *American Journal of Respiratory and Critical Care Medicine, 181*(6), 539–544.

98. Moritsugu, K. P. (2007). Healthy aging starts with healthful eating. *Journal of the American Dietetic Association, 107*(5), 723.

99. Murimi, M., Dodge, C. M., Pope, J., & Erickson, D. (2010). Factors that influence breastfeeding decisions among special supplemental nutrition program for women, infants, and children participants from central Louisiana. *Journal of the American Dietetic Association, 110*(4), 624–627.

100. National Institute on Aging, National Institutes of Health, & U.S. Department of Health and Human Services. (2010). *Healthy aging: Lessons from the Baltimore Longitudinal Study of Aging.* Publication No. 08-6440. Accessed at: http://www.nia.nih.gov/NR/rdonlyres/F1B25F15-BB89-4A73-842A-5E7A2A2F69C2/0/Healthy_Aging_2010.pdf.

101. National Institutes of Health Office of Dietary Supplements. (2011 update). *Dietary supplement fact sheet: Vitamin D.* Accessed at: http://ods.od.nih.gov/factsheets/vitamind.

102. Nelson, M. E., Rejeski, W. J., Blair, S. N., Duncan, P. W., Judge, J. O., King, A. C., Macera, C. A., & Castaneda-Sceppa, C. (2007). Physical activity and public health in older adults: Recommendation from the American College of Sports Medicine and the American Heart Association. *Circulation, 116*, 1094–1105.

103. Neumark-Sztainer, D., Wall, M., Haines, J. Story, M., & Eisenberg, M. E. (2007). Why does dieting predict weight gain in adolescents? Findings from Project EAT-II: A 5-year longitudinal study. *Journal of the American Dietetic Association, 107*(3), 448–455.

104. Ney, D. M., Weiss, J. M., Kind, A. J., & Robbins, J. (2009). Senescent swallowing: Impact, strategies, and interventions. *Nutrition in Clinical Practice, 24*(3), 395–413.

105. Niewinski, M. M. (2008). Advances in celiac disease and gluten-free diet. *Journal of the American Dietetic Association, 108*(4), 661–672.

106. Nisevich, P. M. (2008). Sports nutrition for young athletes: Vital to victory. *Today's Dietitian, 10*(3), 44–48.

107. Packer, L., Seis, M., Eggersdorfer, M., & Cadenas, E. (Eds.). (2010). *Micronutrients and brain health.* Parkway, NY: CRC Press.

108. Paddon-Jones, D., & Rasmussen, B. B. (2009). Dietary protein recommendations and the prevention of sarcopenia. *Current Opinion in Clinical Nutrition and Metabolic Care, 12*(1), 86–90.

109. Palmer, S. (2010). No meat, no problem: Vegetarian diets can support optimal health for infants and children. *Today's Dietitian, 12*(3), 28–34.

110. Pan, Y., & Pratt, C. A. (2008). Metabolic syndrome and its association with diet and physical activity in US adolescents. *Journal of the American Dietetic Association, 108*(2), 276–286.

111. Pearson, N., Biddle, S. J., & Gorely, T. (2009). Family correlates of fruit and vegetable consumption in children and adolescents: A systematic review. *Public Health Nutrition, 12*(2), 267–283.

112. Peppersack, T. (2009). Nutritional problems in the elderly. *Acta Clinica Belgica, 64*(2), 85–91.

113. Petri, W. A. Jr., Miller, M., Binder, H. J., Levine, M. M., Dillingham, R., & Guerrant, R. L. (2008). Enteric infections, diarrhea, and their impact on function and development. *The Journal of Clinical Investigation, 118*, 1277–1290.

114. Picone, O., Marszalek, A., Servely, J. L., & Chavatte-Palmer, P. (2009). Effects of omega-3 supplementation in pregnant women. *Journal of Obstetrics, Gynecology, and Reproductive Biology, 38*(2), 117–124.

115. Pivarnik, J., & Mudd, L. (2009). Oh baby! Exercise during pregnancy and the postpartum period. *ACSM's Health & Fitness Journal, 13*(3), 8–13.

116. Practice Committee of the American Society for Reproductive Medicine, in collaboration with the Society for Reproductive Endocrinology and Infertility. (2008). Optimizing natural fertility. *Fertility and Sterility, 90* (Suppl. 5), S1–S6.

117. Prentice, A., Goldberg, G. R., & Schoenmakers, I. (2008). Vitamin D across the lifecycle: Physiology and biomarkers. *American Journal of Clinical Nutrition, 88*(2), 500S–506S.

118. Prentice, A., Schoenmakers, I., Laskey, A. M., de Bono, S., Ginty, F., & Goldberg, G. R. (2006). Symposium on 'Nutrition and health in children and adolescents': Session 1: Nutrition in growth and development: Nutrition and bone growth and development. *Proceedings of the Nutrition Society, 65*(4), 348–360.

119. Reedy, J., & Krebs-Smith, S. M. (2010). Dietary sources of energy, solid fats, and added sugars among children and adolescents in the United States. *Journal of the American Dietetic Association, 110*(10), 1477–1484.

120. Ressler, A. (2008). Insatiable hungers: Eating disorders and substance abuse. *Today's Dietitian, 10*(10), 72–76.

121. Rhone, M., & Basu, A. (2008). Phytochemicals and age-related eye diseases. *Nutrition Reviews, 66*(8), 465–472.

122. Riediger, N. D., Othman, R. A., Suh, M., & Moghadasian, M. H. (2009). A systemic review of the roles of n-3 fatty acids in health and disease. *Journal of the American Dietetic Association, 109*(4), 668–679.

123. Ritz, P. (2000). Physiology of aging with respect to gastrointestinal, circulatory and immune system changes and their significance for energy and protein metabolism. *European Journal of Clinical Nutrition, 54* (Suppl. 3), S21–S25.

124. Roberts, S. B., & Dallal, G. E. (2001). The new childhood growth charts. *Nutrition Reviews, 59*(2), 31–32.

125. Rockett, H. R. H. (2007). Family dinner: More than just a meal. *Journal of the American Dietetic Association, 107*(9), 1498–1501.

126. Rosales, F. J., Reznick, J. S., & Zeisel, S. H. (2009). Understanding the role of nutrition in the brain and behavioral development of toddlers and preschool children: Identifying and overcoming methodological barriers. *Nutritional Neuroscience, 12*(5), 190–202.

127. Rowland, T., Carlin, S., & Nordstrom, L. (2007). Exercise prescriptions in the pediatrician's office: Feasibility or folly? *American Journal of Lifestyle Medicine, 1*(1), 48–53.

128. Sanchez, A., Normal, G. J., Sallis, J. F., Calfas, K. J., Cella, J., & Patrick, K. (2007). Patterns and correlates of physical activity and nutrition behaviors in adolescents. *American Journal of Preventative Medicine, 32*(2), 124–130.

129. Sanders, L. M., & Zeisel, S. H. (2007, July/August). Choline: Dietary requirements and role in brain development. *Nutrition Today, 42*(4), 181–186.

130. Satter, E. (1999). *Secrets of feeding a healthy family*. Madison, WI: Kelcy Press.

131. Satter, E. (2000). *Child of mine: Feeding with love and good sense*. Boulder, CO: Bull.

132. Schnuelle, P., Oberheiden, T., Hohenadel, D., Gottmann, U., Benck, U., Nebe, T., . . . & Birck, R. (2006). An unusual case of severe iron deficiency anaemia. *Gut, 55*(7), 1060.

133. Schuchardt, J. P., Huss, M., Stauss-Grabo, M., & Hahn, A. (2010). Significance of long-chain polyunsaturated fatty acids (PUFAs) for the development and behaviour of children. *European Journal of Pediatrics, 169*(2), 149–164.

134. Seshadri, S. (2006). Elevated plasma homocysteine levels: Risk factor or risk marker for the development of dementia and Alzheimer's disease? *Journal of Alzheimer's Disease, 9*(4), 393–398.

135. Sharma, A. J., Grummer-Strawn, L. M., Kalenius, K., & Smith, R. (2009). Obesity prevalence among low-income, preschool-aged children: United States, 1998–2008. *Morbidity and Mortality Weekly Report, 58*(28), 769–773.

136. Shatenstein, B., Kergoat, M., & Reid, I. (2007). Poor nutrient intakes during 1-year follow-up with community-dwelling older adults with early-stage Alzheimer dementia compared to cognitively intact matched controls. *Journal of the American Dietetic Association, 107*(12), 2091–2099.

137. Shea, M. K., & Booth, S. L. (2008). Update on the role of vitamin K in skeletal health. *Nutrition Reviews, 66*(10), 549–557.

138. Shelhub, J. (2008). Public health significance of elevated homocysteine. *Food and Nutrition Bulletin, 29* (Suppl. 2), S116–S125.

139. Shepherd, A. A. (2008). Nutrition through the life-span: Part 1: Preconception, pregnancy and infancy. *British Journal of Nursing, 17*(20), 1261–1268.

140. Shepherd, A. A. (2008). Nutrition through the life-span: Part 2: Children, adolescents and adults. *British Journal of Nursing, 17*(21), 1332–1338.

141. Shepherd, A. A. (2009). Nutrition through the life-span: Part 3: Adults aged 65 years and over. *British Journal of Nursing, 18*(5), 301–302, 304–307.

142. Shuaibi, A. M., House, J. D., & Sevenhuysen, G. P. (2008). Folate status of young Canadian women after folic acid fortification of grain products. *Journal of the American Dietetic Association, 108*(12), 2090–2094.

143. Sigman-Grant, M., Christiansen, E., Branen, L., Fletcher, J., & Johnson, S. L. (2008). About feeding children: Mealtimes in child-care centers in four Western states. *Journal of the American Dietetic Association, 108*(2), 340–346.

144. Silence, J. (2009). Physical activity for toddlers. *Today's Dietitian, 11*(11), 50–54.

145. Sing, D., & Sing, C. F. (2010). Impact of direct soil exposures from airborne dust and geophagy on human health. *International Journal of Environmental Research and Public Health, 7,* 1205–1223.

146. Singhal, A. (2009). The early origins of atherosclerosis. *Advances in Medicine and Biology, 646,* 51–58.

147. Stanner, S. (2009). Diet and lifestyle measures to protect the aging heart. *British Journal of Community Nursing, 14*(5), 210–212.

148. Stein, K. (2007). Playing with food: Promoting food play to teach healthful eating habits. *Journal of the American Dietetic Association, 107*(8), 1284–1285.

149. Sumic, A., Michael, Y. L., Carlson, N. E., Howieson, D. B., & Kaye, J. A. (2007). Physical activity and the risk of dementia in oldest old. *Journal of Aging and Health, 19,* 242–259.

150. Swinburn, B. (2009). Obesity prevention in children and adolescents. *Child and Adolescent Psychiatric Clinics of North America, 18*(1), 209–223.

151. Taylor, A. I., Gould, H. J., Sutton, B. J., & Calvert, R. A. (2008). Avian IgY binds to a monocyte receptor with IgG-like kinetics despite an IgE-like structure. *The Journal of Biological Chemistry, 283,* 16384–16390.

152. Taylor, J. A. (2008). Defining vitamin D deficiency in infants and toddlers. *Archives of Pediatrics and Adolescent Medicine, 162*(6), 583–584.

153. Tessema, J., Jefferds, M. E., & Cogswell, M. (2009). Motivators and barriers to prenatal supplement use among minority women in the United States. *Journal of the American Dietetic Association, 109*(1), 102–108.

154. Timmons, B. W. (2007). Exercise and immune function in children. *American Journal of Lifestyle Medicine, 1*(1), 59–66.

155. Tufts-New England Medical Center Evidence-Based Practice Center. (2007). *Breastfeeding and maternal and infant health outcomes in developed countries.* AHRQ Publication No. 07-E007. Accessed at: http://www.ahrq.gov.

156. Tufts University. (2009). To live to a biblical old age, stay physically active. *Tufts University Health & Nutrition Letter, 27*(10), 1–2.

157. Ventura, E. E., Davis, J. N., Alexander, K. E., Shaibi, G. Q., Lee, W., Byrd-Williams, C. E., . . . & Goran, M. I. (2008). Dietary intake and the metabolic syndrome in overweight Latino children. *Journal of the American Dietetic Association, 108*(8), 1355–1359.

158. Wadha, P. D., Buss, C., Entringer, S., & Swanson, J. M. (2009). Developmental origins of health and disease: Brief history. *Seminars in Reproductive Medicine, 5,* 358–368.

159. Walker, A. (2010). Breast milk as the gold standard for protective nutrients. *Journal of Pediatrics, 156* (Suppl. 2), S3–S7.

160. Ward, E. M. (2009). Pregnancy by the numbers: The IOM updates weight gain guidelines. *Today's Dietitian, 11*(10), 47–50.

161. Wardwell, L., Chapman-Novakofski, K., Herrel, S., & Woods, J. (2008). Nutrient intake and immune function of elderly subjects. *Journal of the American Dietetic Association, 108*(12), 2005–2012.

162. Weigensberg, M. J., & Goran, M. I. (2009). Type 2 diabetes in children and adolescents. *Lancet, 373*(9677), 1743–1744.

163. Weil, A. (2005). *Healthy aging.* New York: Anchor Books.

164. Weinert, B. T., & Timiras, P. S. (2003). Physiology of aging: Invited review: Theories of aging. *Journal of Applied Physiology, 95,* 1706–1716.

165. Wellman, N. S. (2007). Prevention, prevention, prevention: Nutrition for successful aging. *Journal of the American Dietetic Association, 107*(5), 741–743.

166. Wilson, R. D., Johnson, J. A., Wyatt, P., Allen, V., Gagnon, A., Langlois, S., . . . & Genetics Committee of the Society of Obstetricians and Gynaecologists of Canada and The Motherisk Program. (2007). Pre-conceptional vitamin/folic acid supplementation 2007: The use of folic acid in combination with a multivitamin supplement for the prevention of neural tube defects and other congenital anomalies. *Journal of Obstetrics and Gynaecology Canada, 29*(12), 1003–1026.

167. Wosje, K. S., & Specker, B. L. (2009). Role of calcium in bone health during childhood. *Nutrition Reviews, 58*(9), 253–268.

168. Wu, T. C., & Chen, P. H. (2009). Health consequences of nutrition in childhood and early infancy. *Pediatrics and Neonatology, 50*(4), 135–142.

169. Zeisel, S. H. (2006). The fetal origins of memory: The role of dietary choline in optimal brain development. *Journal of Pediatrics, 149* (Suppl. 5), S131–S136.

Reference Tables (DRI, UL, AMDR, EER, DRV, and RDI) and Chemical Structures of Nutrients

FIGURE A1 The chemical structure of essential vitamins.

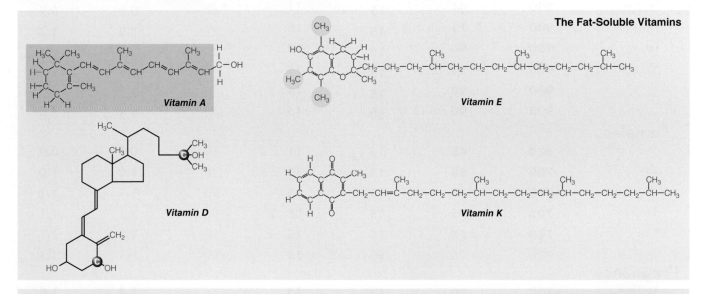

The Fat-Soluble Vitamins

Vitamin A

Vitamin E

Vitamin D

Vitamin K

The Water-Soluble Vitamins

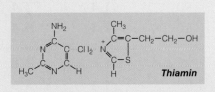

Thiamin

Riboflavin

Niacin   *Choline*

Biotin

B₁₂

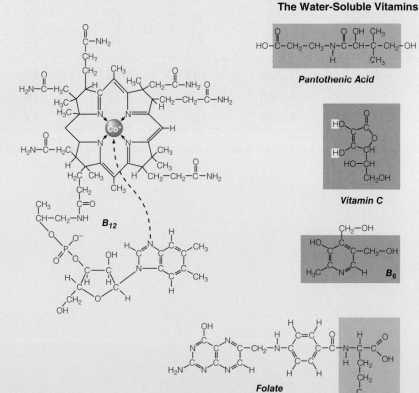

Pantothenic Acid

Vitamin C

B₆

Folate

Understanding Nutrition, 12th ed, Whitney & Rolfes, Appendix, pp. C5-C9.

Dietary Reference Intakes (DRIs): Vitamins

Life Stage Group	[1]Vit. A (μg/day)	Vit. C (mg/day)	Vit. D (μg/day)	Vit. E (mg/day)	Vit. K (μg/day)	Thiamin (mg/day)	Riboflavin (mg/day)
Infants							
0–6 mo.	400	40	10	4	2.0	0.2	0.3
7–12 mo.	500	50	10	5	2.5	0.3	0.4
Children							
1–3 y	**300**	**15**	**15**	**6**	30	**0.5**	**0.5**
4–8 y	**400**	**25**	**15**	**7**	55	**0.6**	**0.6**
Males							
9–13 y	**600**	**45**	**15**	**11**	60	**0.9**	**0.9**
14–18 y	**900**	**75**	**15**	**15**	75	**1.2**	**1.3**
19–30 y	**900**	**90**	**15**	**15**	120	**1.2**	**1.3**
31–50 y	**900**	**90**	**15**	**15**	120	**1.2**	**1.3**
51–70 y	**900**	**90**	**15**	**15**	120	**1.2**	**1.3**
> 70 y	**900**	**90**	**20**	**15**	120	**1.2**	**1.3**
Females							
9–13 y	**600**	**45**	**15**	**11**	60	**0.9**	**0.9**
14–18 y	**700**	**65**	**15**	**15**	75	**1.0**	**1.0**
19–30 y	**700**	**75**	**15**	**15**	90	**1.1**	**1.1**
31–50 y	**700**	**75**	**15**	**15**	90	**1.1**	**1.1**
51–70 y	**700**	**75**	**15**	**15**	90	**1.1**	**1.1**
> 70 y	**700**	**75**	**20**	**15**	90	**1.1**	**1.1**
Pregnancy							
≥ 18 y	**750**	**80**	**15**	**15**	75	**1.4**	**1.4**
19–30 y	**770**	**85**	**15**	**15**	90	**1.4**	**1.4**
31–50 y	**770**	**85**	**15**	**15**	90	**1.4**	**1.4**
Lactation							
≤ 18 y	**1,200**	**115**	**15**	**19**	75	**1.4**	**1.6**
19–30 y	**1,300**	**120**	**15**	**19**	90	**1.4**	**1.6**
31–50 y	**1,300**	**120**	**15**	**19**	90	**1.4**	**1.6**

[1]Vit. is an abbreviation for vitamin.
Boldfaced values represent Recommended Dietary Allowance (RDA) values. Non-boldfaced values represent Adequate Intake (AI) values.
© Cengage Learning 2013

Niacin (mg/day)	Vit. B$_6$ (mg/day)	Folate (μg/day)	Vit. B$_{12}$ (μg/day)	Pantothenic Acid (mg/day)	Biotin (μg/day)	Choline (mg/day)
2	0.1	65	0.4	1.7	5	125
4	0.3	80	0.5	1.8	6	150
6	**0.5**	**150**	**0.9**	2	8	200
8	**0.6**	**200**	**1.2**	3	12	250
12	**1.0**	**300**	**1.8**	4	20	375
16	**1.3**	**400**	**2.4**	5	25	550
16	**1.3**	**400**	**2.4**	5	30	550
16	**1.3**	**400**	**2.4**	5	30	550
16	**1.7**	**400**	**2.4**	5	30	550
16	**1.7**	**400**	**2.4**	5	30	550
12	**1.0**	**300**	**1.8**	4	20	375
14	**1.2**	**400**	**2.4**	5	25	400
14	**1.3**	**400**	**2.4**	5	30	450
14	**1.3**	**400**	**2.4**	5	30	450
14	**1.5**	**400**	**2.4**	5	30	450
14	**1.5**	**400**	**2.4**	5	30	450
18	**1.9**	**600**	**2.6**	6	30	450
18	**1.9**	**600**	**2.6**	6	30	450
18	**1.9**	**600**	**2.6**	6	30	450
17	**2.0**	**500**	**2.8**	7	35	550
17	**2.0**	**500**	**2.8**	7	35	550
17	**2.0**	**500**	**2.8**	7	35	550

TABLE A2

Dietary Reference Intakes (DRIs): Minerals

Life Stage Group	Calcium (mg/day)	Chloride (mg/day)	Chromium (µg/day)	Copper (µg/day)	Fluoride (mg/day)	Iodine (µg/day)	Iron (mg/day)	Magnesium (mg/day)
Infants								
0–6 mo.	200	180	0.2	200	0.01	110	0.27	30
7–12 mo.	260	570	5.5	220	0.5	130	**11**	75
Children								
1–3 y	**700**	1,500	11	**340**	0.7	**90**	**7**	**80**
4–8 y	**1,000**	1,900	15	**440**	1	**90**	**10**	**130**
Males								
9–13 y	**1,300**	2,300	25	**700**	2	**120**	**8**	**240**
14–18 y	**1,300**	2,300	35	**890**	3	**150**	**11**	**410**
19–30 y	**1,000**	2,300	35	**900**	4	**150**	**8**	**400**
31–50 y	**1,000**	2,300	35	**900**	4	**150**	**8**	**420**
51–70 y	**1,000**	2,000	30	**900**	4	**150**	**8**	**420**
> 70 y	**1,200**	1,800	30	**900**	4	**150**	**8**	**420**
Females								
9–13 y	**1,300**	2,300	21	**700**	2	**120**	**8**	**240**
14–18 y	**1,300**	2,300	24	**890**	3	**150**	**15**	**360**
19–30 y	**1,000**	2,300	25	**900**	3	**150**	**18**	**310**
31–50 y	**1,000**	2,300	25	**900**	3	**150**	**18**	**320**
51–70 y	**1,200**	2,000	20	**900**	3	**150**	**8**	**320**
> 70 y	**1,200**	1,800	20	**900**	3	**150**	**8**	**320**
Pregnancy								
≤ 18 y	**1,300**	2,300	29	**1,000**	3	**220**	**27**	**400**
19–30 y	**1,000**	2,300	30	**1,000**	3	**220**	**27**	**350**
31–50 y	**1,000**	2,300	30	**1,000**	3	**220**	**27**	**360**
Lactation								
≤ 18 y	**1,300**	2,300	44	**1,300**	3	**290**	**10**	**360**
19–30 y	**1,000**	2,300	45	**1,300**	3	**290**	**9**	**310**
31–50 y	**1,000**	2,300	45	**1,300**	3	**290**	**9**	**320**

Sulfate (a form of sulfur) is an essential major mineral. There is no specific DRI for sulfur because it is provided by two sulfur-containing amino acids (cysteine and methionine) that make up proteins. The amount of sulfur in protein is approximately 0.0085 grams per gram of protein. See Table A5 to determine the DRI for protein. To determine approximate sulfur intake based on that protein amount: Protein grams × 0.0085 = Sulfur grams × 1000 = milligrams.

Cobalt is a component of vitamin B_{12}. There is no specific DRI for cobalt; its need is met through the DRI for vitamin B_{12} (see Table A1).

Boldfaced values represent Recommended Dietary Allowance (RDA) values. Non-boldfaced values represent Adequate Intake (AI) values.
© Cengage Learning 2013

Manganese (mg/day)	Molybdenum (μg/day)	Phosphorus (mg/day)	Potassium (mg/day)	Selenium (μg/day)	Sodium (mg/day)	Water (L/day)	Zinc (mg/day)
0.003	2	100	400	15	120	0.7	2
0.6	3	275	700	20	370	0.8	3
1.2	**17**	**460**	3,000	**20**	1,000	1.3	**3**
1.5	**22**	**500**	3,800	**30**	1,200	1.7	**5**
1.9	**34**	**1,250**	4,500	**40**	1,500	2.4	**8**
2.2	**43**	**1,250**	4,700	**55**	1,500	3.3	**11**
2.3	**45**	**700**	4,700	**55**	1,500	3.7	**11**
2.3	**45**	**700**	4,700	**55**	1,500	3.7	**11**
2.3	**45**	**700**	4,700	**55**	1,300	3.7	**11**
2.3	**45**	**700**	4,700	**55**	1,200	3.7	**11**
1.6	**34**	**1,250**	4,500	**40**	1,500	2.1	8
1.6	**43**	**1,250**	4,700	**55**	1,500	2.3	9
1.8	**45**	**700**	4,700	**55**	1,500	2.7	8
1.8	**45**	**700**	4,700	**55**	1,500	2.7	8
1.8	**45**	**700**	4,700	**55**	1,300	2.7	8
1.8	**45**	**700**	4,700	**55**	1,200	2.7	8
2.0	**50**	**1,250**	4,700	**60**	1,500	3.0	13
2.0	**50**	**700**	4,700	**60**	1,500	3.0	11
2.0	**50**	**700**	4,700	**60**	1,500	3.0	11
2.6	**50**	**1,250**	5,100	**70**	1,500	3.8	14
2.6	**50**	**700**	5,100	**70**	1,500	3.8	12
2.6	**50**	**700**	5,100	**70**	1,500	3.8	12

© Cengage Learning 2013

FIGURE A2 The periodic table of elements with nutritional application for the essential minerals in human nutrition. The essential minerals are boldfaced and highlighted, including the major minerals [Calcium (**Ca**), Magnesium (**Mg**), Phosphorus (**P**), Sodium (**Na**), Potassium (**K**), Chloride (**Cl**), Sulfur (**S**)] and the trace minerals [Iron (**Fe**), Zinc (**Zn**), Iodine (**I**), Selenium (**Se**), Chromium (**Cr**), Molybdenum (**Mo**), Copper (**Cu**), Manganese (**Mn**), Fluoride (**F**), Cobalt (**Co**)].

REFERENCE TABLES (DRI, UL, AMDR, EER, DRV, AND RDI) AND CHEMICAL STRUCTURES OF NUTRIENTS

Tolerable Upper Intake Levels (UL) for Vitamins

Life Stage Group	[1]Vit. A (μg/day)	Vit. C (mg/day)	Vit. D (μg/day)	Vit. E (mg/day)	Niacin (mg/day)	Vit. B_6 (mg/day)	Folate (μg/day)	Choline (g/day)
Infants								
0–6 mo.	600	ND	25	ND	ND	ND	ND	ND
7–12 mo.	600	ND	25	ND	ND	ND	ND	ND
Children								
1–3 y	600	400	50	200	10	30	300	1.0
4–8 y	900	650	50	300	15	40	400	1.0
Males and Females								
9–13 y	1,700	1,200	50	600	20	60	600	2.0
14–18 y	2,800	1,800	50	800	30	80	800	3.0
19–70 y	3,000	2,000	50	1,000	35	100	1,000	3.5
> 70 y	3,000	2,000	50	1,000	35	100	1,000	3.5
Pregnancy								
≤ 18 y	2,800	1,800	50	800	30	80	800	3.0
19–50 y	3,000	2,000	50	1,000	35	100	1,000	3.5
Lactation								
≤ 18 y	2,800	1,800	50	800	30	80	800	3.0
19–50 y	3,000	2,000	50	1,000	35	100	1,000	3.5

[1]The UL for vitamin A is for retinyl esters (retinol, retinal, retinoic acid) from animal sources, not carotenoid plant precursors of vitamin A.
ND means the UL is not determined due to lack of data.
Essential vitamins with UL not determined for any life stage: vitamin K, thiamin, riboflavin, vitamin B_{12}, pantothenic acid, and biotin.
Nonessential substances with UL not determined for any life stage: carotenoids.
© Cengage Learning 2013

Tolerable Upper Intake Levels (UL) for Minerals

Life Stage Group	Boron (mg/day)	Calcium (g/day)	Chloride (mg/day)	Copper (μg/day)	Fluoride (mg/day)	Iodine (μg/day)	Iron (mg/day)	[1]Magnesium (mg/day)
Infants								
0–6 mo.	ND	ND	ND	ND	0.7	ND	40	ND
7–12 mo.	ND	ND	ND	ND	0.9	ND	40	ND
Children								
1–3 y	3	2.5	2,300	1,000	1.3	200	40	65
4–8 y	6	2.5	2,900	3,000	2.2	300	40	110
Males and Females								
9–13 y	11	2.5	3,400	5,000	10	600	40	350
14–18 y	17	2.5	3,600	8,000	10	900	45	350
19–70 y	20	2.5	3,600	10,000	10	1,100	45	350
> 70 y	20	2.5	3,600	10,000	10	1,100	45	350
Pregnancy								
≤ 18 y	17	2.5	3,600	8,000	10	900	45	350
19–50 y	20	2.5	3,600	10,000	10	1,100	45	350
Lactation								
≤ 18 y	17	2.5	3,600	8,000	10	900	45	350
19–50 y	20	2.5	3,600	10,000	10	1,100	45	350

© Cengage Learning 2013

Tolerable Upper Intake Levels (UL) for Minerals (continued)

Life Stage Group	Manganese (mg/day)	Molybdenum (µg/day)	Nickel (mg/day)	Phosphorus (g/day)	Selenium (µg/day)	Sodium (mg/day)	Vanadium (mg/day)	Zinc (mg/day)
Infants								
0–6 mo.	ND	ND	ND	ND	45	ND	ND	4
7–12 mo.	ND	ND	ND	ND	60	ND	ND	5
Children								
1–3 y	2	300	0.2	3	90	1,500	ND	7
4–8 y	3	600	0.3	3	150	1,900	ND	12
Males and Females								
9–13 y	6	1,100	0.6	4	280	2,200	ND	23
14–18 y	9	1,700	1.0	4	400	2,300	ND	34
19–70 y	11	2,000	1.0	4	400	2,300	1.8	40
> 70 y	11	2,000	1.0	3	400	2,300	1.8	40
Pregnancy								
≤ 18 y	9	1,700	1.0	3.5	400	2,300	ND	34
19–50 y	11	2,000	1.0	3.5	400	2,300	ND	40
Lactation								
≤ 18 y	9	1,700	1.0	4	400	2,300	ND	34
19–50 y	11	2,000	1.0	4	400	2,300	ND	40

© Cengage Learning 2013

[1]The UL for magnesium is for supplements only.
ND means the UL is not determined due to lack of data.
Essential minerals with UL not determined for any life stage: chromium, potassium, and water.
Nonessential minerals with UL not determined for any life stage: arsenic and silicon.

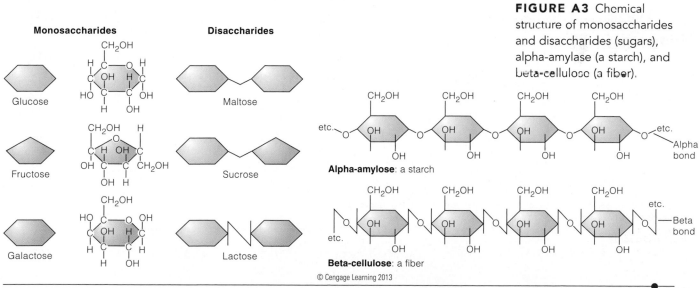

Monosaccharides

Glucose

Fructose

Galactose

Disaccharides

Maltose

Sucrose

Lactose

Alpha-amylose: a starch

Beta-cellulose: a fiber

© Cengage Learning 2013

FIGURE A3 Chemical structure of monosaccharides and disaccharides (sugars), alpha-amylase (a starch), and beta-cellulose (a fiber).

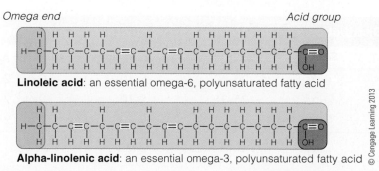

Omega end

Acid group

Linoleic acid: an essential omega-6, polyunsaturated fatty acid

Alpha-linolenic acid: an essential omega-3, polyunsaturated fatty acid

FIGURE A4 Chemical structure of the essential fatty acids linoleic and alpha-linolenic acid.

© Cengage Learning 2013

Dietary Reference Intakes (DRIs): Carbohydrate, Fiber, Fat, Fatty Acids, and Protein

Life Stage Group	Energy[1] (kCal/day)		Carbohydrate (g/day)	Fiber[2] (g/day)	Fat (g/day)	Linoleic Acid (g/day)	α-Linolenic Acid (g/day)	Protein (g/day)	Protein (g/kg/day)
Infants	♂	♀							
0–6 mo.	570	520	60	ND	31	4.4	0.5	9.1	1.52
7–12 mo.	743	676	95	ND	30	4.6	0.5	**11**	1.20
Children	♂	♀							
1–3 y	1,046	992	**130**	19	ND	7	0.7	**13**	**1.05**
4–8 y	1742	1,642	**130**	25	ND	10	0.9	**19**	**0.95**
Males									
9–13 y	2,279		**130**	31	ND	12	1.2	**34**	**0.95**
14–18 y	3,152		**130**	38	ND	16	1.6	**52**	**0.85**
19–30 y	3,067[3]		**130**	38	ND	17	1.6	**56**	**0.80**
31–50 y	3,067[3]		**130**	38	ND	17	1.6	**56**	**0.80**
51–70 y	3,067[3]		**130**	30	ND	14	1.6	**56**	**0.80**
> 70 y	3,067[3]		**130**	30	ND	14	1.6	**56**	**0.80**
Females									
9–13 y	2,071		**130**	26	ND	10	1.0	**34**	**0.95**
14–18 y	2,368		**130**	26	ND	11	1.1	**46**	**0.85**
19–30 y	2,403[3]		**130**	25	ND	12	1.1	**46**	**0.80**
31–50 y	2,403[3]		**130**	25	ND	12	1.1	**46**	**0.80**
51–70 y	2,403[3]		**130**	21	ND	11	1.1	**46**	**0.80**
> 70 y	2,403[3]		**130**	21	ND	11	1.1	**46**	**0.80**
Pregnancy									
1st trimester	+0		**175**	28	ND	13	1.4	**71**	**1.10**
2nd trimester	+340		**175**	28	ND	13	1.4	**71**	**1.10**
3rd trimester	+452		**175**	28	ND	13	1.4	**71**	**1.10**
Lactation									
1st 6 mos.	+330		**210**	29	ND	13	1.3	**71**	**1.30**
2nd 6 mos.	+400		**210**	29	ND	13	1.3	**71**	**1.30**

[1]The Estimated Energy Requirement values listed are for active reference persons. A personalized EER should be calculated from Table A8.
[2]The DRI for fiber can be personalized at 1.4 grams per 100 Calories consumed.
[3]Males subtract 10 and females subtract 7 Calories per day for each year above age 19.
ND means the value is not determined due to lack of data.
Boldfaced values represent Recommended Dietary Allowance (RDA) values. Non-boldfaced values represent Adequate Intake (AI) values.
© Cengage Learning 2013

Acceptable Macronutrient Distribution Ranges (AMDRs) for Healthy Diets as a Percent of Energy

Age (years)	Carbohydrate	Sugar	Total Fat	Linoleic Acid	α-Linolenic Acid	Protein
1–3	45–65	≤ 25	30–40	5–10	0.6–1.2	5–20
4–18	45–65	≤ 25	25–35	5–10	0.6–1.2	10–30
≥ 19	45–65	≤ 25	20–35	5–10	0.6–1.2	10–35

© Cengage Learning 2013

FIGURE A5 Chemical structure for all 20 amino acids grouped by shared chemical properties.

1. Amino acids with aliphatic side chains, which consist of hydrogen and carbon atoms (hydrocarbons):

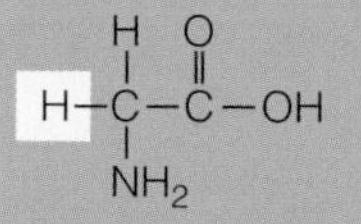

Glycine (Gly)

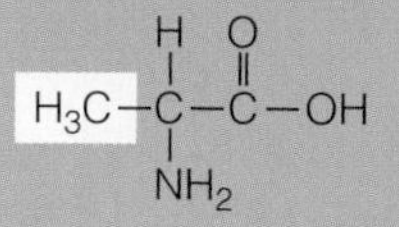

Alanine (Ala)

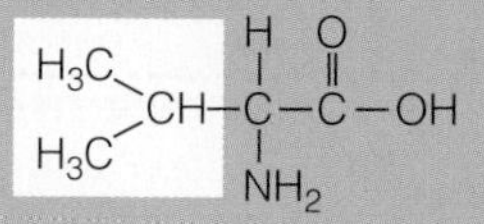

Valine* (Val)

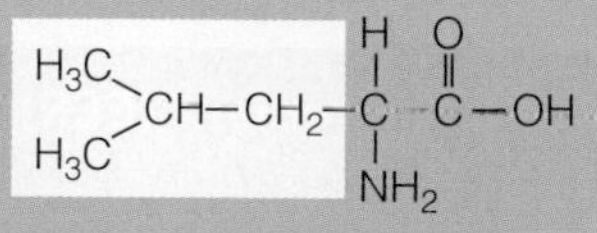

Leucine* (Leu)

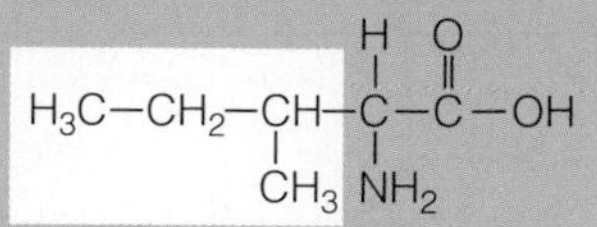

Isoleucine* (Ile)

2. Amino acids with hydroxyl (OH) side chains:

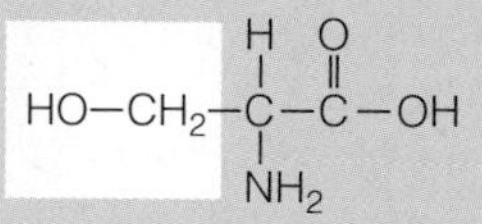

Serine (Ser)

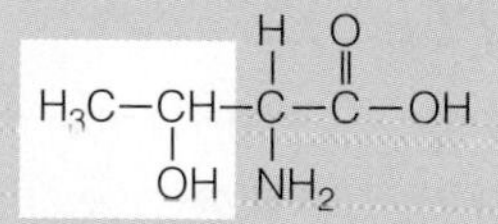

Threonine* (Thr)

3. Amino acids with side chains containing acidic groups or their amides, which contain the group NH_2:

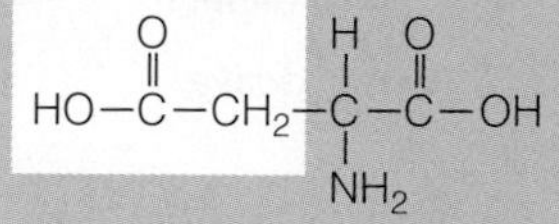

Aspartic acid (Asp)

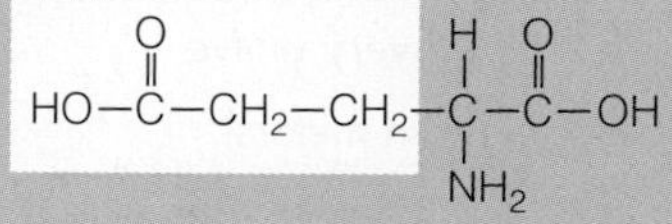

Glutamic acid (Glu)

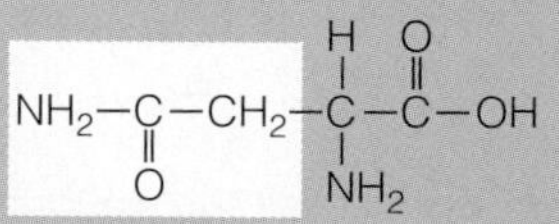

Asparagine (Asn)

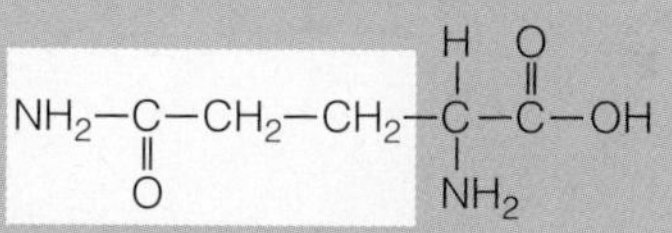

Glutamine (Gln)

4. Amino acids with basic side chains:

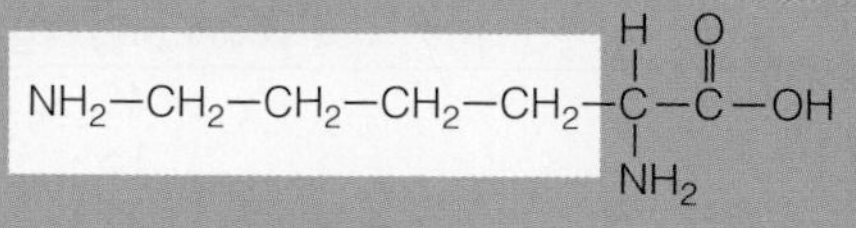

Lysine* (Lys)

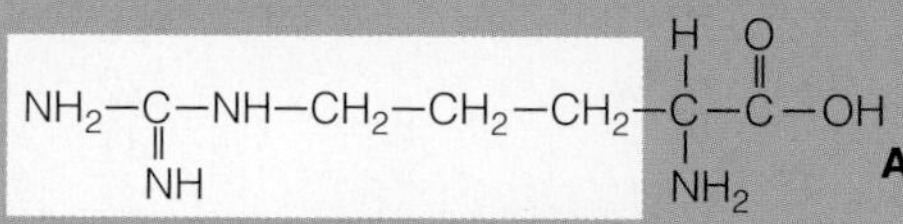

Arginine (Arg)

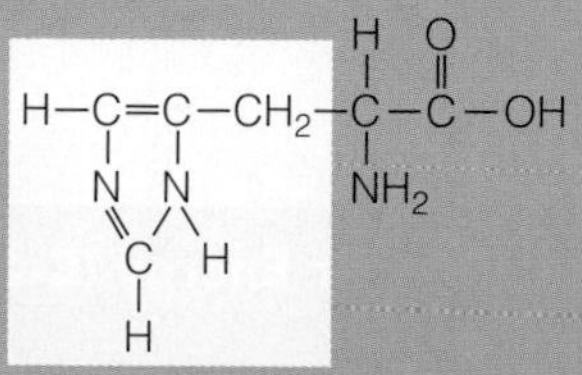

Histidine* (His)

5. Amino acids with aromatic side chains, which are characterized by the presence of at least one ring structure:

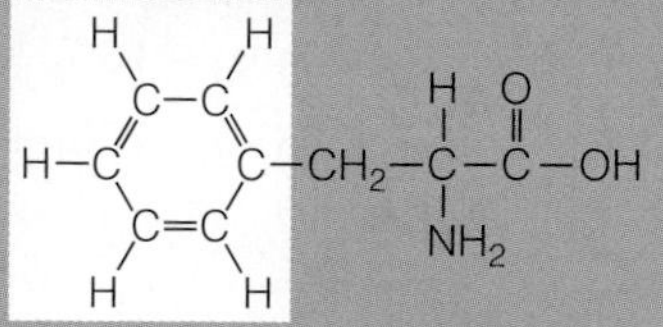

Phenylalanine* (Phe)

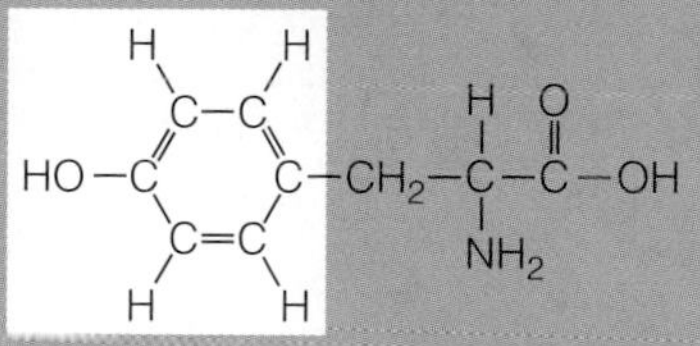

Tyrosine (Tyr)

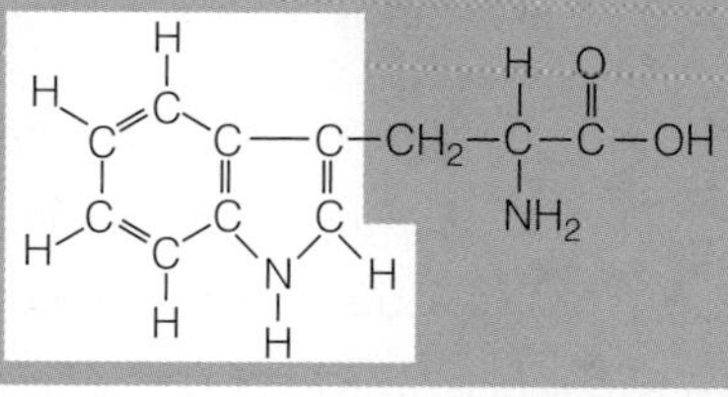

Tryptophan* (Trp)

6. Amino acids with side chains containing sulfur atoms:

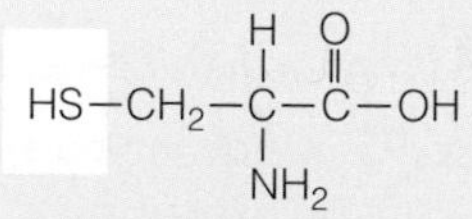

Cysteine (Cys)

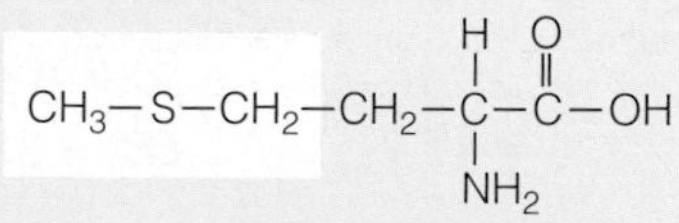

Methionine* (Met)

7. Imino acid:

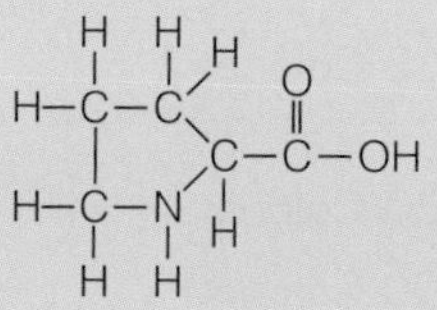

Proline (Pro)

Proline has the same chemical structure as the other amino acids, but its amino group has given up a hydrogen to form a ring.

*Essential amino acids

Understanding Nutrition, 12th ed. Whitney & Rolfes, Appendix, p. C-4.

Dietary Reference Intakes (DRIs) for Essential Amino Acids

Amino Acid	Estimated Average Requirement for Children 1–3 years (mg/kg/day)	Estimated Average Requirement for Adults (mg/kg/day)
Lysine	45	31
Histidine	16	11
Isoleucine	22	15
Leucine	48	34
Valine	28	19
Methionine + cysteine[1]	22	15
Phenylalanine + tyrosine[2]	41	27
Threonine	24	16
Tryptophan	6	4

[1]Tyrosine is produced from phenylalanine.
[2]Cysteine is produced from methionine.

© Cengage Learning 2013

Dietary Reference Intakes (DRIs) for Energy (Calories) Using the Estimated Energy Requirement (EER) Equations by Life Stage Group

Life Stage Group	EER Prediction Equation
Infants	
0–3 mo.	EER = (89 × Weight in kg − 100) + 175
4–6 mo.	EER = (89 × Weight in kg − 100) + 56
7–12 mo.	EER = (89 × Weight in kg − 100) + 22
Children	
1–2 y	
3–8 y	EER = (89 × Weight in kg − 100) + 20
Male	EER = 88.5 − (61.9 × Age in years) + PA (26.7 × Weight in kg + 903 × Height in meters) + 20
	PA (Physical Activity) = 1.0 if sedentary; 1.13 if low active; 1.26 if active; 1.42 if very active
Female	EER = 135.3 − (30.8 × Age in years) + PA (10.0 × Weight in kg + 934 × Height in meters) + 20
	PA (Physical Activity) = 1.0 if sedentary; 1.16 if low active; 1.31 if active; 1.56 if very active
Males	
9–18 y	EER = 88.5 − (61.9 × Age in years) + PA (26.7 × Weight in kg + 903 × Height in meters) + 25
	PA (Physical Activity) = 1.0 if sedentary; 1.13 if low active; 1.26 if active; 1.42 if very active
≥ 19 y	EER = 662 − (9.53 × Age in years) + PA (15.91 × Weight in kg + 539.6 × Height in meters)
	PA (Physical Activity) = 1.0 if sedentary; 1.11 if low active; 1.25 if active; 1.48 if very active
Females	
9–18 y	EER = 135.3 − (30.8 × Age in years) + PA (10.0 × Weight in kg + 934 × Height in meters) + 25
	PA (Physical Activity) = 1.0 if sedentary; 1.16 if low active; 1.31 if active; 1.56 if very active
≥ 19 y	EER = 354 − (6.91 × Age in years) + PA (9.36 × Weight in kg + 726 × Height in meters)
	PA (Physical Activity) = 1.0 if sedentary; 1.12 if low active; 1.27 if active; 1.45 if very active
Pregnancy	
1st trimester	Adult EER + 0 Calories
2nd trimester	Adult EER + 340 Calories
3rd trimester	Adult EER + 452 Calories
Lactation	
1st 6 mos.	Adult EER + 330 Calories
2nd 6 mos.	Adult EER + 400 Calories

Weight in pounds (lbs) ÷ 2.2 lbs/kg = Weight in kg
Height in inches ÷ 39.37 = Height in meters
See Table A9 for DRI for physical activity and PA factor description for EER equations.

© Cengage Learning 2013

TABLE A9

Physical Activity (DRI and PA Factors)

Physical Activity (PA)	Recommendation/Description (values based on an average-weight person)
DRI	A minimum of 60 minutes of moderate physical activity cumulative per day is recommended to prevent weight gain and reduce chronic disease. Moderate intensity is equivalent to walking at a 4-mile-per-hour (mph) pace.
Factor	
Sedentary	Only physical activities required for independent living.
Low active	Walking 1.5–3 miles per day at 2–4 mph pace.
Active	Walking 3–10 miles per day at 2–4 mph pace.
Very active	Walking 10 or more miles per day at 2–4 mph pace.

© Cengage Learning 2013

TABLE A10

Reference Person Values for Body Mass Index (BMI), Weight, and Height

Life Stage Group	BMI (kg/m²)	Weight kg (lbs)	Height cm (in)
Infants			
0–6 mo.	ND	6 (13)	62 (24)
7–12 mo.	ND	9 (20)	71 (28)
Children			
1–3 y	ND	12 (27)	86 (34)
4–8 y	15.3	20 (44)	115 (45)
Males			
9–13 y	17.2	36 (79)	144 (57)
14–18 y	20.5	61 (134)	174 (68)
19–30 y	22.5	70 (154)	177 (70)
Females			
9–13 y	17.4	37 (81)	144 (57)
14–18 y	20.4	54 (119)	163 (64)
19–30 y	21.5	57 (126)	163 (64)

© Cengage Learning 2013

TABLE A11

Daily Reference Values (DRVs) Based on a 2,000-Calorie Diet

Food Component	DRV
Fat	65 g
Saturated fat	20 g
Cholesterol	300 mg
Carbohydrate	300 g
Fiber	25 g
Protein	50 g high-quality protein or 65 g low-quality protein
Sodium (Na)	2,400 mg
Potassium (K)	3,500 mg

© Cengage Learning 2013

TABLE A12

Reference Daily Intakes (RDIs) Based on the Highest 1968 Recommended Dietary Allowance Value

Nutrient	Amount	Units	Nutrient	Amount	Units
Thiamin	1.5	mg	Calcium	1,000	mg
Riboflavin	1.7	mg	Iron	18	mg
Niacin	20	mg	Zinc	15	mg
Biotin	300	µg	Iodine	150	µg
Pantothenic acid	10	mg	Copper	2	mg
Vitamin B_6	2	mg	Chromium	120	µg
Folate	400	µg	Selenium	70	µg
Vitamin B_{12}	6	µg	Molybdenum	75	µg
Vitamin C	60	mg	Manganese	2	mg
Vitamin A	5,000	IU[1]	Chloride	3,400	mg
Vitamin D	400	IU[1]	Magnesium	400	mg
Vitamin E	30	IU[1]	Phosphorus	1,000	mg
Vitamin K	80	µg			

[1]IU: International Units
© Cengage Learning 2013

Facts, Formulas, Conversions, and Sample Calculations

B.1 GENERAL WEIGHTS AND MEASURES

Facts and Conversions These are standard universal conversions.

Length

1 inch (in) = 2.54 centimeters (cm)
1 foot (ft) = 30.48 cm or 12 in
1 meter (m) = 39.37 in

Volume

1 milliliter (mL) = 1/1,000 L
1 teaspoon (tsp or t) = 5 mL or 5 g
1 tablespoon (tbs or T) = 3 tsp or 15 mL
1 ounce, fluid (fl oz) = 2 tbs or 30 mL
1 cup (c) = 8 fl oz or 16 tbs or 240 mL
1 quart (qt) = 32 fl oz or 4 c or 0.95 L
1 deciliter (dL) = 100 mL
1 liter (L) = 1.06 qt or 1,000 mL
1 gallon (gal) = 16 c or 4 qt or 128 fl oz or 3.79 L

Weight

1 microgram (μg or mcg) = 1/1,000 mg
1 milligram (mg) = 1,000 μg or 1/1,000 g
1 gram (g) = 1,000 mg or 1/1,000 kg
1 ounce, weight (oz) = about 28 g or 1/16 lb
1 pound (lb) = 16 oz or about 454 g
1 kilogram (kg) = 1,000 g or 2.2 lbs

Formulas
Converting Pounds to Kilograms

$$1 \text{ lb body weight} \div 2.2 \text{ lb/kg} = \text{Weight in kg}$$

Sample calculation: Jerry weighs 150 pounds. How many kilograms is this?

150 lbs ÷ 2.2 lb/kg = 68.18 kg, 68.2 kg (rounded to the nearest tenth of a kg)

B.2 MYPLATE AND EXCHANGE LISTS

Facts, Formulas, and Conversions Tools to plan, manage, and evaluate diets.

Formulas

Determine the Number of Serving Equivalents by MyPlate Use Appendix C to look up what counts as a one serving equivalent by food group.

First, identify the food group the item is within. Second, solve for the number of equivalents = amount eaten ÷ amount of an equivalent.

Example: Pattern 12 fluid ounces of 2% milk.

Fact One cup of milk equals a 1 cup dairy group equivalent.

$$12 \text{ fl oz} \div 8 \text{ fl oz/c} = 1.5 \text{ c}$$

$$1.5 \text{ c milk} \div 1 \text{ c milk} = 1.5 \text{ dairy c equivalents}$$

Fact To figure the empty Calories from 2% milk:

1 cup of 2% milk provides 125 total Calories, of which 40 are empty.

$$1.5 \text{ c } 2\% \text{ milk} \times 40 \text{ empty Calories} = 60 \text{ empty Calories}$$

Determine the Number of Exchanges from the Exchange Lists Use Appendix D to look up what counts as an exchange by Exchange List.

The number of exchanges = amount eaten ÷ amount of an exchange

Example: Determine the number of starch exchanges in a 4-ounce bagel.

Fact One ounce of bagel equals one starch exchange.

$$4 \text{ oz} \div 1 \text{ oz} = 4 \text{ starch exchanges}$$

B.3 CARBOHYDRATE (INCLUDES ALCOHOL AND FIBER), PROTEIN, AND FAT

Facts and Conversions

Calories Yielded per Gram Weight:
- Carbohydrate: 4 Calories per gram
- Protein: 4 Calories per gram
- Fat: 9 Calories per gram
- Alcohol: 7 Calories per gram
- Fiber: 0 Calories per gram

Individual Protein Requirements The average American adult requires the DRI for protein (0.8 grams per kilogram of body weight). Individuals who strength train (lift weights) should consume 1.2 to 1.6 grams of protein per kilogram of body weight to promote an increase in lean body mass. Individuals who are endurance athletes should consume 1.8 to 2 grams of protein per kilogram of body weight to maintain lean body mass.

Sample calculation: Eric weighs 198 pounds. What is his DRI for protein? Assuming he is a long-distance runner, what is his maximum protein need?

Convert weight in pounds to kilograms: 198 lb ÷ 2.2 lb/kg = 90 kg

Determine the DRI: 90 kg × 0.8 g/kg = 72 g DRI for protein

Determine the high end of an endurance athlete's need: 90 kg × 2.0 g/kg = 180 g protein need based on activity

Formulas
Calculating Percentage of Calories from Carbohydrate, Protein, Fat, or Alcohol When Given Grams

First, calculate the Calories from the gram amount of the nutrient.

Grams of carbohydrate, protein, fat, or alcohol × Calories per gram factor = Calories from carbohydrate, protein, fat, or alcohol

Second, calculate the percent of Calories from the nutrient.

Calories from carbohydrate, protein, fat, or alcohol ÷ Total Calories × 100 = Percent of Calories from carbohydrate, protein, fat, or alcohol

This same principle can be applied to grams of a type of fatty acid, sugar, starch, amino acid, and so on.

Sample calculation: Margie ate 450 grams of digestible carbohydrate, 112 grams of protein, and 83 grams of fat in one day. She did not consume any alcohol. How many Calories from carbohydrate, protein, and fat did she eat? How many total Calories did she eat? What percentage of her Calories came from carbohydrate, protein, and fat?

First, calculate the number of Calories from carbohydrate, protein, and fat:

450 g carbohydrate	× 4 Calories/g	= 1,800 Calories
112 g protein	× 4 Calories/g	= 448 Calories
83 g fat	× 9 Calories/g	= 747 Calories

Second, sum the Calories from each of the energy-producing nutrients to get the total Calories:

1,800 carbohydrate Calories + 448 protein Calories + 747 fat Calories = 2,995 total Calories

Third, calculate the percentage of Calories from carbohydrate, protein, and fat:

1,800 Calories carbohydrate ÷ 2,995 total Calories × 100 = 60.1% carbohydrate

448 Calories protein ÷ 2,995 total Calories × 100 = 14.9% protein

747 Calories fat ÷ 2,995 total Calories × 100 = 25% fat

Calculating Grams When Given Percent Total Calories

Total Calories × Percent of Calories provided by the nutrient (divided by 100) = Fraction of Calories

Fraction of Calories ÷ Calories/g = grams

Sample calculation: A diet for an adult should provide 5 to 10 percent of total Calories from the essential omega-6 polyunsaturated fatty acid linoleic acid to meet the Acceptable Macronutrient Distribution Range. Sam is an adult male who consumed 3,100 Calories in one day. How many grams of linoleic acid (minimum) should he consume to minimally meet his AMDR?

3,100 Calories × 5% ÷ 100 = 155 Calories ÷ 9 Calories/g = 17.2 g linoleic acid

Polyunsaturated:Saturated Fat Ratio (P:S)   When the P:S ratio is greater than or equal to 3:1 and fat content is high in the diet, cancer risk is increased. When the P:S ratio is less than or equal to 0.33:1 and fat content is high in the diet, heart disease risk is increased. A P:S ratio of 1:1 is often recommended. Calculate the P:S ratio using dietary intakes of polyunsaturated fatty acids (PUFAs) and saturated fatty acids (SFAs) from the dietary analysis below. Calculate as follows:

PUFA (g) ÷ SFA (g)

The number calculated is assigned to the PUFA. The SFA is assigned the number 1.

Sample calculation: Blaine ate 8 grams of PUFAs and 48 grams of SFAs and had 37 percent of Calories from fat. What is his P:S ratio? Is the ratio increasing disease risk?

8 g ÷ 48 g = 0.17. This value is placed in the P position of the ratio.
The S position is always assigned the number 1.
The P:S ratio is 0.17:1.
This P:S ratio is increasing risk for heart disease.

Calculating Fiber DRI Based on Calorie Intake and Calculating Grams of Fiber per 100 Calories Consumed Based on Fiber Intake and Calorie Intake

To calculate the fiber DRI: Total Calories consumed ÷ 100 × 1.4 g fiber/ 100 Calories = Fiber DRI (g)

To calculate the fiber g/100 Calories: Fiber (g) ÷ Calorie intake ÷ 100 = Fiber g/100 Calories

To calculate the percent DRI for fiber: Grams of fiber consumed ÷ Gram DRI for fiber based on Calories consumed × 100

Sample calculation: An adult female ate 2,850 Calories and 23 grams of fiber in one day. What is her DRI for fiber? How many grams of fiber per 100 Calories eaten did she consume? Was her fiber intake deficient?

To calculate the DRI for fiber based on Calories consumed: 2,850 Calories ÷ 100 × 1.4 = 39.9 g DRI for fiber

To calculate the grams of fiber per 100 Calories consumed: 23 g ÷
2,850 Calories ÷ 100 = 0.8 g fiber per 100 Calories consumed

To calculate the percent DRI: 23 g ÷ 39.9 g × 100 = 57.6%. This is deficient.

B.4 VITAMINS AND MINERALS

Facts and Conversions Recommended nutrient intake is guided by the Dietary
Reference Intakes (DRIs).

Deficient Intake Nutrient deficiency is promoted when the nutrient is consumed at less than 66 percent of the DRI.

Inadequate Intake Nutrient inadequacy is promoted when the nutrient is consumed at 66 percent of the DRI to less than 100 percent of the DRI.

Adequate Intake Nutrient adequacy is promoted when the nutrient is consumed at levels greater than or equal to the DRI (100 percent) for the nutrient and
less than or equal to the Tolerable Upper Intake Level (UL).

Excessive or Possible Toxic Intake Nutrient toxicity is determined for each
individual nutrient. The UL tables in Appendix A should be used as a guide for
safe intake. The ULs are values that are likely to pose no risk of toxicity or adverse
reactions when consumed on a daily basis. The greater the intake is above the UL,
the greater the risk of toxicity.

International Units (IU) IU values are used in the RDIs on food and supplement
package labels for vitamins A, D, and E. However, the DRIs for these nutrients are
not expressed as IU any longer. Thus, conversion is required to compare nutrient
values on food and supplement labels to current dietary needs (DRI values).

> **Vitamin A:** Convert IU to microgram Retinol Equivalents (RE, vitamin A) for
> animal sources, divide by 3.33; for vegetable and fruit sources, divide by 10.
> **Vitamin D:** Convert IU to micrograms of vitamin D, divide by 40.
> **Vitamin E:** Convert IU to milligrams of alpha-tocopherol (α-TE, vitamin E),
> divide by 1.5.

Sodium The goal is based on limiting dietary intake to less than (or equal to) the UL
of 2,300 milligrams of sodium per day. The DRI for sodium is set at 1,500 milligrams.
Sodium intake can also be compared to the average American's sodium intake of
4,000 to 6,000 milligrams per day. Salt is 40 percent sodium and 60 percent chloride.

To convert grams of salt to milligrams of sodium: Salt (g) = mg sodium ÷ 400

Sample calculation: Jack consumed 4,500 milligrams of sodium in one day.
How many grams of salt did he consume?

To convert milligrams of sodium to grams of salt: 4,500 mg sodium ÷ 400
= 11.25 g salt

Formulas
Calculating Percent of DRI See Appendix A for these values.

Percent DRI = Intake amount of a nutrient ÷ DRI for the nutrient × 100

Sample calculation: Eric is a 20-year-old male. Dietary analysis showed that he consumed 122 milligrams of vitamin C. His DRI for vitamin C is 90 milligrams (see Appendix A). What percent of the DRI is this? Was his intake adequate, inadequate, or deficient?

$$122 \text{ mg} \div 90 \text{ mg} \times 100 = 136\%. \text{ His intake was adequate.}$$

Bone Health and Calcium:Phosphorus (Ca:P) Ratio   To reduce the risk of osteoporosis and achieve an optimal Ca:P ratio, two factors need to be considered: (1) the DRI for calcium and phosphorus (see Appendix A) needs to be met, and (2) the Ca:P ratio must be greater than or equal to 1:1. For adults 19 to 30 years old, the DRI for calcium and phosphorus is 1,000 milligrams per day and 700 milligrams per day, respectively. Calculate the Ca:P ratio using dietary intake of calcium and phosphorus from a dietary analysis report. Calculate as follows:

$$\text{Calcium (mg)} \div \text{Phosphorus (mg)}$$

The number calculated is assigned to the calcium. The phosphorus is assigned the number 1.

Sample calculation: Janet consumed 870 milligrams of calcium and 1,230 milligrams of phosphorus in one day. What is her Ca:P ratio for this day? Is the ratio optimal for bone health?

$$870 \text{ g} \div 1,230 \text{ g} = 0.71. \text{ The Ca:P ratio is 0.71:1. This ratio is not optimal}$$
$$\text{and she did not meet her DRI for calcium.}$$

B.5 FOOD LABELS

Facts and Conversions

Nutrient Density   Twenty percent or more of the RDI provided per food serving is nutrient dense.

Fat Content Assessment

When the total percent of Calories from fat in a food is less than 25 percent, the food is low fat.
When the total percent of Calories from fat in a food is between 25 and 35 percent, the food is moderate fat.
When the total percent of Calories from fat in a food is greater than 35 percent, the food is high fat.

The fat content by weight is the number used on the front of the food package label for dairy and meat products. It is the ratio of the gram weight of the fat in the food serving in the context of the gram weight of the food serving.

Sample calculation: The Nutrition Facts panel of whole milk shows that 1 cup of milk has 244 total grams weight, 8 grams fat, 9 grams protein, and 12 grams carbohydrate. What is the percentage of fat by Calories and the percentage of fat by weight?

$$\text{Amount of Calories from fat: 8 g} \times 9 \text{ Calories/g} = 72 \text{ Calories}$$

Amount of Calories from protein: 9 g × 4 Calories/g = 36 Calories

Amount of Calories from carbohydrate: 12 g × 4 Calories/g = 48 Calories

Amount of total Calories: 72 + 36 + 48 = 156 Calories

Fat content by weight: 8 g fat/serving ÷ 244 g total weight of the serving
× 100 = 3.3%

Amount of Calories from fat based on energy:
72 fat Calories ÷ 156 total Calories × 100 = 46%

Classification: Whole milk is a high-fat food.

Formulas

Calculating Percent of RDI or DRV   See Appendix A for these values.

Percent RDI = Intake amount of nutrient ÷ RDI for the nutrient × 100

Percent DRV = Intake amount of nutrient ÷ DRV for the nutrient × 100

Converting Percent RDI to the Nutrient Weight

RDI for the nutrient × Percent of the nutrient provided in one serving
(divided by 100) = Amount of the nutrient provided per serving

Sample calculation: The Nutrition Facts panel of a food shows that it provides
25 percent of the RDI for vitamin C per serving. How much vitamin C is this?
Appendix A shows the RDI for vitamin C is 60 milligrams.

60 mg × 25% ÷ 100 = 15 mg vitamin C/serving

B.6 BLOOD CHOLESTEROL, GLUCOSE, PRESSURE, AND METABOLIC SYNDROME

Facts and Conversions

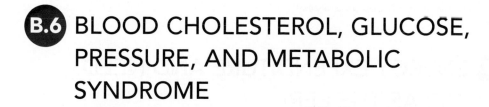

TABLE B1

Interpreting Blood Cholesterol Levels

Total Cholesterol	LDL Cholesterol	HDL Cholesterol
< 200 Desirable 200–239 Borderline high ≥ 240 High	< 100 Optimal 100–129 Near optimal 130–159 Borderline high 160–189 High ≥ 190 Very high	< 40 Low (indicates risk) > 60 High

All values mg/dL.
© Cengage Learning 2013

TABLE B2

Interpreting Blood Glucose Levels

Metabolic State	Indicates Diabetes	Prediabetes	Normal
Fasting	≥ 126 mg/dL	100–125 mg/dL	70–99 mg/dL
Fed (2 hours post-prandial)	≥ 200 mg/dL	140–199 mg/dL	< 140 mg/dL

© Cengage Learning 2013

TABLE B3

Interpreting Metabolic Syndrome: If Three or More of These Risk Factors Are Met

Factor	Value
Waist circumference	> 35 in ♀ > 40 in ♂
Fasting blood triglycerides	≥ 150 mg/dL
HDL cholesterol	< 50 mg/dL ♀ < 40 mg/dL ♂
Blood pressure	≥ 130/85 mmHg
Fasting blood glucose	≥ 110 mg/dL

© Cengage Learning 2013

TABLE B4

Interpreting Blood Pressure Measurements

	Systolic[1]	Conjunction	Diastolic[2]
Optimal	< 120	and	< 80
Prehypertension	120–139	or	80–89
Stage 1 hypertension	140–159	or	90–99
Stage 2 hypertension	> 160	or	> 100

[1]Systolic blood pressure in millimeters of mercury (Hg)

[2]Diastolic blood pressure in millimeters of mercury (Hg)

© Cengage Learning 2013

B.7 ENERGY EXPENDITURE AND NEED (DRI AS THE EER)

Formulas

Mifflin-St. Jeor Equation   Determining Resting Energy Expenditure (REE) Calories per day, which is the number of Calories required in a day to sustain life:

For men: REE (Calories per day) = $(10 \times W) + (6.25 \times H) - (5 \times A) + 5$

For women: REE (Calories per day) = $(10 \times W) + (6.25 \times H) - (5 \times A) - 161$

(W = Weight in kilograms; H = Height in centimeters; A = Age in years)

Sample calculation: Jane is a female, age 36, who weighs 133 pounds (60.5 kg) and is 5 feet 5 inches (165 cm) tall. What is her REE using the Mifflin-St. Jeor equation?

Female REE (Calories per day) = $(10 \times 60.5) + (6.25 \times 165) - (5 \times 36) - 161$

$$= 605 + 1{,}031 - 180 - 161 = 1{,}295 \text{ Calorie REE}$$

TABLE B5

Calculating Calorie Burn Based on Energy Expenditure Levels

Level	Description of Activities at Each Level	Activity Factor
A Resting (*Reclined activity*)	Resting: Sleeping, reclining at rest *Level A is at complete rest. The energy required at this state is enough to sustain basic function in the body.*	1.0
B Sedentary (*Very light activity*)	Very light activity: Sitting, studying, watching TV, doing computer work, playing cards, waiting in line, cooking, ironing, printing, counter clerking, doing lab work, photocopying, playing a musical instrument *Level B activities include those that are done standing in a small space or sitting.*	1.5
C Easy (*Low activity*)	Low activity: Wallpapering, house painting, washing windows, raking leaves, strolling, doing housework, playing volleyball, caregiving, providing daycare, playing golf, canoeing, walking at a 2.5–3-mph pace *Level C activities include those that require moving body parts and not sweating while engaged in the activities.*	2.5
D Rigorous (*Moderate activity*)	Moderate activity: Playing competitive volleyball, doing aerobics, backpacking, playing singles tennis, chopping wood, shoveling, power-walking, rowing, playing racquetball, boxing, skin diving, power-walking at a 4-mph pace *Level D activities include those that require moving body parts rigorously and induce sweating while engaged in the activities.*	5.0
E Strenuous (*Extreme activity*)	Extreme activity: Running for a touchdown in football, running the bases in baseball, doing circuit training, sprint swimming, digging trenches, doing judo, running intervals, cross-country skiing uphill *Level E activities include all-out efforts or intense activity that last less than a minute, which may be experienced at times during competition or in training.*	7.0

© Cengage Learning 2013

Formulas
Calculating the Calories Burned Within Each Activity Level

$$\text{Calories burned} = \text{Hours spent in activity} \times \text{REE} \times \text{Activity factor} \div 24 \text{ hours/day}$$

$$\text{Total energy expenditure in a day} = \text{Calories burned; Resting} + \text{Physical activity} + \text{Specific Dynamic Action of food}$$

Sample calculation: Jane's REE based on the Mifflin-St. Jeor equation is 1,295 Calories. One day she spent 9 hours of activity at level B, 4 hours at level C, 2 hours at level D, and 5 minutes at level E. How many total Calories did she burn in physical activity on this day?

First, calculate the Calories burned in physical activity.

Level B = 9 × 1,295 × 1.5 ÷ 24 = 728.44 or **728 Calories**

Level C = 4 × 1,295 × 2.5 ÷ 24 = 539.58 or **540 Calories**

Level D = 2 × 1,295 × 5 ÷ 24 = 539.58 or **540 Calories**

Level E = 5 minutes ÷ 60 minutes/hour = 0.08 hours × 1,295 × 7 ÷ 24 = **31 Calories**

Second, sum the Calories burned in each level of activity to determine total Calories burned in physical activity (PA).

Total PA = B + C + D + E = 728 + 540 + 540 + 31 = **1,839 Calories burned in physical activity**

Sample calculation: Jane's REE based on the Mifflin-St. Jeor equation is 1,295 Calories. She slept 8.91 hours on the same day she expended the activity in the sample calculation above, and diet analysis showed she consumed 3,135 Calories this day. How much energy did she expend while sleeping and in processing her food? What is her total energy expenditure for the day?

Resting (level A) = 8.91 × 1,295 × 1 ÷ 24 = 480.76 or **481 Calories**

Specific Dynamic Action of food = 0.1 × 3,135 Calories/day = **314 Calories**

Total energy expenditure = 481 + 1,839 + 314 = **2,634 Calories**

Average Energy Expenditure Formula

Sum of Calories expended each day ÷ Total number of days analyzed = Average energy expenditure

Example from a 2-day energy expenditure analysis: Day 1 + Day 2 ÷ 2 = 2-day average energy expenditure

Energy Balance Formulas
Maintaining body weight (energy balance or isocaloric conditions)
Calories consumed = Calories expended
Weight loss (negative energy balance)
Calories consumed < Calories expended
Weight gain (positive energy balance)
Calories consumed > Calories expended

Determining Calorie Difference (Intake versus Expenditure)

Calories consumed − Calories expended = + or − Calorie difference

The Calorie difference can be a positive, negative, or zero number.

Converting Calories in 1 Pound of Fat

1 lb = 3,500 Calories

Change in Body Weight Formula

Calorie difference ÷ 3,500 Calories per pound fat = Fat weight change

+ Calorie difference is weight gain (positive energy balance)
− Calorie difference is weight loss (negative energy balance)

Sample calculation: Stephanie consumed an average of 2,916 Calories over 2 days and expended an average of 2,477 Calories over the same 2 days. What state of energy balance was she in? How many pounds did she gain or lose in that time?

To calculate Calorie difference: 2,916 Calorie intake − 2,477 Calorie expenditure = + 439 Calories

She was in positive energy balance.

To calculate weight change: + 439 Calories ÷ 3,500 Calories/lb fat = 0.125 lb

She gained 0.125 pounds of fat.

Calculating Estimated Energy Requirement   The EER is also known as the DRI for Calories.

Look up the appropriate DRI formula based on age, gender, and physical activity using Appendix A. Look up the PA factor from Appendix A also.

Sample calculation: Eric is 20 years old, weighs 198 pounds, and is 6 feet (72 inches) tall. He is a distance runner (very active). What is his Estimated Energy Requirement (EER, which is the DRI for energy)?

$$EER = 662 − (9.53 \times A) + PA (15.91 \times W + 539.6 \times H)$$

(W = Weight in kilograms; H = Height in meters; A = Age in years)

Weight in kg = 198 lb ÷ 2.2 lb/kg = 90 kg

Height in meters = 72 in ÷ 39.37 = 1.83 m

PA (physical activity) = 1.48 (very active)

$$EER = 662 − (9.53 \times 20) + 1.48 (15.91 \times 90 + 539.5 \times 1.83)$$

$$EER = 662 − 190.6 + 1.48 (1,431.9 + 987.5)$$

$$EER = 662 − 190.6 + 3,580.7 = 4,052.1 \text{ Calories}$$

B.8 BODY MASS INDEX AND BODY COMPOSITION

Facts and Conversions
Central Adiposity   Fat distribution is a risk factor for cardiovascular disease, hypertension, and diabetes. A person has central adiposity/obesity when:
His waist (girth) measurement is greater than 40 inches.
Her waist (girth) measurement is greater than 35 inches.

Fact
Body Mass Index (BMI)   The BMI indicates the healthiness of the body weight in relation to height, and the BMI value is a screening tool to determine if a person is overweight. If the person has a BMI greater than 25 and central adiposity, then

that person is at risk for type 2 diabetes, hypertension, dyslipidemia, and cardio-vascular (heart) disease.

Formulas

$$BMI = \text{Weight in kilograms (kg)} \div \text{Height in meters squared (m}^2)$$

$$\text{Weight in kg} = \text{Weight in lbs} \div 2.2$$

$$\text{Height in m} = \text{Height in inches} \div 39.37$$

Sample calculation: Melissa weighs 132 pounds and is 5 feet 5 inches tall. What is her BMI? How is her BMI interpreted?

$$132 \text{ lbs} \div 2.2 \text{ lbs/kg} = 60 \text{ kg}$$

$$5 \text{ ft} \times 12 \text{ in/ft} = 60 \text{ in} + 5 \text{ in} = 65 \text{ in tall}$$

$$65 \text{ in} \div 39.37 \text{ m/in} = 1.65 \text{ m}$$

$$1.65^2 = 2.72 \text{ m}^2 \text{ (m}^2 \text{ could also be calculated as } 1.65 \times 1.65 = 2.72 \text{ m}^2)$$

$$BMI = 60 \text{ kg} \div 2.72 \text{ m}^2 = 22.1$$

This is interpreted as a normal BMI.

TABLE B6

Interpreting Body Mass Index (BMI)

BMI	Risk
< 18.5	Underweight
18.5–24.9	Normal
25–29.9	Overweight
30–34.9	Obesity (class I)
35–39.9	Obesity (class II)
≥ 40	Obesity (class III)

If BMI is 25 or greater, then percent body fat and central adiposity should be measured.

© Cengage Learning 2013

Fact

Body Composition   Body composition is measured in the exercise physiology lab to provide a value label to make the interpretation of body fatness. In general, females have more body fat than males.

TABLE B7

Interpreting Body Composition by Percent Body Fat

Body Fat Categories	% Body Fat (Male)	% Body Fat (Female)
Essential fat	3	12
Very lean	10	13
Lean	11–15	14–19
Not fat (average)	16–19	20–25
Fat*	20–27	26–32
Obese	> 28	> 33
Physically fit**	12–15	18–22

*Strong recommendations for fat cell reduction are made when males exceed 20 percent body fat, and females exceed 26 percent body fat.

**If you are male and have less than 12 percent body fat or a female with less than 18 percent body fat, you may interpret the result as being very lean, lean, and physically fit.

© Cengage Learning 2013

The MyPlate Food Guidance System

C.1 THE MYPLATE FOOD GUIDANCE SYSTEM

The USDA MyPlate system provides many options to help Americans consume a healthy diet and to be active every day. MyPlate has an individualized tool that considers age, gender, and physical activity level to determine a person's Calorie need. Calorie control, adequacy, moderation, variety, and balance are all integrated into this food guidance system.

The MyPlate food guidance system is built on the understanding that a single diet doesn't fit all. On the Web site www.choosemyplate.gov, consumers can access in-depth information about each major food group (grains, vegetables, fruits, dairy, and protein foods), oils, empty Calories, and physical activity. They can generate their personal MyPlate plan, find tips to improve intake or make wise choices, access resources to help them implement their food plan, learn what counts as a serving from each food group, see what foods are in each group, read tips for increasing whole, fresh food consumption and limiting solid fats and sugars, learn functions of nutrients in foods, and visualize foods in the photo gallery to help identify portion sizes. Understanding what constitutes one MyPlate equivalent and learning to pattern foods puts in perspective the extreme portion distortion to which Americans have become accustomed to over time. What is perceived as a serving on a typical dinner plate may be in reality two to three MyPlate serving equivalents.

Consumers can pattern foods using this simple formula:

$$\text{Amount eaten} \div \text{Amount per serving equivalent} = \text{Number of serving equivalents eaten}$$

MyPlate food groups will not list all foods available to consume. Thus, users should apply the concepts and examples to pattern their individual diet.

Additionally, the Web site supports a feature for users to track their progress and make gradual improvements. In building a personalized MyPlate plan online, it is possible, though optional, to enter individual weight and height. If a person's weight for height puts him or her in an overweight or obese body mass index category, then a modified plan can be generated at the choosemyplate.gov Web site to help that person move toward a healthier body weight. Typically, the modified plan is subtracting 200 Calories per day from the plan that would help the person maintain their current weight. A 200-Calorie deficit is typically achieved by reducing empty Calories from solid fat and sugar by approximately 155 Calories and reducing the number of teaspoons of oil by one (approximately 45 Calories).

The following tables C1 and C2 show the detailed information that the MyPlate system is based on without considering body weight and height. See Tables C3 through C7 for the food groups. See Table C8 for oils and Table C9 for empty Calories. See Table C10 for a detailed patterning example. See Table C11 for physical activity.

TABLE C1

MyPlate Food Intake Pattern Calorie Levels

Age	Males			Age	Females		
	Sedentary < 30 minutes PA + daily activity	Moderately Active 30 to < 60 minutes PA + daily activity	Active ≥ 60 minutes PA + daily activity		Sedentary < 30 minutes PA + daily activity	Moderately Active 30 to < 60 minutes PA + daily activity	Active ≥ 60 minutes PA + daily activity
2	1,000	1,000	1,000	2	1,000	1,000	1,000
3	1,000	1,400	1,400	3	1,000	1,200	1,400
4	1,200	1,400	1,600	4	1,200	1,400	1,400
5	1,200	1,400	1,600	5	1,200	1,400	1,600
6	1,400	1,600	1,800	6	1,200	1,400	1,600
7	1,400	1,600	1,800	7	1,200	1,600	1,800
8	1,400	1,600	2,000	8	1,400	1,600	1,800
9	1,600	1,800	2,000	9	1,400	1,600	1,800
10	1,600	1,800	2,200	10	1,400	1,800	2,000
11	1,800	2,000	2,200	11	1,600	1,800	2,000
12	1,800	2,200	2,400	12	1,600	2,000	2,200
13	2,000	2,200	2,600	13	1,600	2,000	2,200
14	2,000	2,400	2,800	14	1,800	2,000	2,400
15	2,200	2,600	3,000	15	1,800	2,000	2,400
16	2,400	2,800	3,200	16	1,800	2,000	2,400
17	2,400	2,800	3,200	17	1,800	2,000	2,400
18	2,400	2,800	3,200	18	1,800	2,000	2,400
19–20	2,600	2,800	3,000	19–20	2,000	2,200	2,400
21–25	2,400	2,800	3,000	21–25	2,000	2,200	2,400
26–30	2,400	2,600	3,000	26–30	1,800	2,000	2,400
31–35	2,400	2,600	3,000	31–35	1,800	2,000	2,200
36–40	2,400	2,600	2,800	36–40	1,800	2,000	2,200
41–45	2,200	2,600	2,800	41–45	1,800	2,000	2,200
46–50	2,200	2,400	2,800	46–50	1,800	2,000	2,200
51–55	2,200	2,400	2,800	51–55	1,600	1,800	2,200
56–60	2,200	2,400	2,600	56–60	1,600	1,800	2,200
61–65	2,000	2,400	2,600	61–65	1,600	1,800	2,000
66–70	2,000	2,200	2,600	66–70	1,600	1,800	2,000
71–75	2,000	2,200	2,600	71–75	1,600	1,800	2,000
> 75	2,000	2,200	2,400	> 75	1,600	1,800	2,000

PA = physical activity
© Cengage Learning 2013

MyPlate Food Intake Patterns

Daily Amount of Food from Each Group

Calorie Level	1,000	1,200	1,400	1,600	1,800	2,000	2,200	2,400	2,600	2,800	3,000	3,200
Grains	3 oz.	4 oz.	5 oz.	5 oz.	6 oz.	6 oz.	7 oz.	8 oz.	9 oz.	10 oz.	10 oz.	10 oz.
Vegetables	1 c	1.5 c	1.5 c	2 c	2.5 c	2.5 c	3 c	3 c	3.5 c	3.5 c	4 c	4 c
Fruits	1 c	1 c	1.5 c	1.5 c	1.5 c	2 c	2 c	2 c	2 c	2.5 c	2.5 c	2.5 c
Dairy	2 c	2 c	2 c	3 c	3 c	3 c	3 c	3 c	3 c	3 c	3 c	3 c
Protein foods	2 oz.	3 oz.	4 oz.	5 oz.	5 oz.	5.5 oz.	6 oz.	6.5 oz.	6.5 oz.	7 oz.	7 oz.	7 oz.
Oils	3 tsp	4 tsp	4 tsp	5 tsp	5 tsp	6 tsp	6 tsp	7 tsp	8 tsp	8 tsp	10 tsp	11 tsp
Empty Calories	165	171	171	132	195	267	290	362	410	426	512	648

c = cup, oz. = ounce equivalent, tsp = teaspoon

© Cengage Learning 2013

Grains Inside the MyPlate Food Guidance System

© Polara Studios, Inc.

Grains (G)
Make at least half of your grains whole grain (WG)

Includes: Whole grains such as amaranth, barley, brown rice, buckwheat, bulgur (cracked wheat), cornmeal, millet, oatmeal, popcorn, quinoa, rye, sorghum, triticale, whole wheat, wild rice; and whole-grain bread, cereal, tortilla, and pasta products. **Refined grains** may include products such as breads, crackers, cereals, flour tortillas, noodles, processed grains, and bakery goods. There may be whole grain versions available. Consumers need to read the food label ingredients and look for the word "whole" by the grain type used to make the product.

Servings in General: A 1-ounce MyPlate serving equivalent of grain could be 1 slice of bread, 1 cup of ready-to-eat cereal, or ½ cup of cooked rice, pasta, or cereal (*approximately 80 Calories*).

Health Benefits: Grains reduce heart disease, high blood pressure, some cancers, type 2 diabetes, neural tube defects during fetal development, and both constipation and obesity (useful in weight management) when eaten as whole grains.

Nutrients: Grains provide many nutrients, including several B vitamins (thiamin, riboflavin, niacin, and folate), minerals (iron, magnesium, and selenium), carbohydrate, fiber (as whole grains), and protein.

	Amount that counts as a 1-ounce equivalent
Bagels	1 mini bagel (*for WG choose whole wheat*) 1 large (3-inch diameter) is 4-ounce equivalents
Biscuits	1 small (2-inch diameter)
Breads	1 regular slice; 1 small slice French; 4 snack-size slices rye (*for WG choose 100% whole wheat*)
Bulgur	½ cup cooked (*cracked wheat is WG*)
Cornbread	1 small piece (2½ inch × 1¼ inch × 1¼ inch)
Crackers	5 whole-wheat crackers; 2 rye crisp breads; 7 square or round crackers (*for WG choose 100% whole wheat or rye*)
English muffins, buns	½ muffin; ½ hot dog or hamburger bun
Muffins	1 small (2½-inch diameter) (*for WG choose whole wheat*)
Oatmeal	½ cup cooked; 1 packet instant; 1 ounce dry, regular or quick (*oatmeal is WG*)
Pancakes	1 pancake (4½-inch diameter); 2 small pancakes (3-inch diameter) (*for WG choose 100% whole wheat or buckwheat*)
Popcorn	3 cups popped (*popcorn is WG*)
Ready-to-eat breakfast cereal	1 cup flakes or rounds; 1¼ cup puffed (*for WG choose toasted oat or whole-wheat flakes*)
Rice	½ cup cooked; 1 ounce dry (*for WG choose brown or wild*)
Pasta (spaghetti, macaroni)	½ cup cooked; 1 ounce dry (*for WG choose whole wheat*)
Tortillas	1 flour or corn tortilla (6-inch diameter) (*for WG choose whole wheat, whole grain, or whole corn*)

© Cengage Learning 2013

Vegetables Inside the MyPlate Food Guidance System

Vegetables (V)
Make half your plate
fruits and vegetables
Choose a variety of vegetables
from the five subgroups

Includes: All fresh, frozen, canned, and dried vegetables and vegetable juices.

Servings in General: A 1 cup MyPlate serving equivalent of vegetables could be 1 cup raw or cooked vegetables or vegetable juice, or 2 cups of raw leafy greens (*approximately 50 Calories or 120 Calories for starchy vegetables*).

Health Benefits: Vegetables reduce heart disease, heart attack, high blood pressure, stroke, type 2 diabetes, some cancers, kidney stones, obesity, and bone loss. Eating vegetables that are low in Calories instead of higher-Calorie foods may be useful in helping to lower Calorie intake.

Nutrients: Vegetables provide many nutrients, including potassium, vitamin A, vitamin C, folate (folic acid), carbohydrate, fiber, and protein. Most are low in fat and Calories. None have cholesterol.

Subgroups: Dark green, red-orange, beans and peas, starchy, and other vegetables.

Amount that counts as 1 cup of vegetables	
Dark-Green Vegetables	
Broccoli[Vit. C]	1 cup chopped or florets; 3 spears 5 inches long, raw or cooked
Greens[Vit. A] (collards, mustard greens, turnip greens, kale)	1 cup cooked
Spinach[Vit. A]	1 cup cooked; 2 cups raw
Raw leafy greens (spinach[Vit. A], romaine, watercress, dark green leafy lettuce, endive, escarole)	2 cups raw
Red-Orange Vegetables	
Carrots[Vit. A]	1 cup strips, slices, or chopped, raw or cooked; 2 medium; 1 cup baby carrots (about 12)
Pumpkin[Vit. A]	1 cup mashed, cooked
Sweet potatoes[Vit. A]	1 large baked (2¼-inch or more diameter); 1 cup sliced or mashed, cooked
Tomatoes	1 large raw whole (3 inches); 1 cup chopped or sliced, raw, canned, or cooked
Tomato or mixed vegetable juice	1 cup
Winter squash (acorn, butternut, hubbard)[Vit. A]	1 cup cubed, cooked
Beans and Peas	
Dried beans and peas (black, white, garbanzo, kidney, navy, pinto, soybeans, split peas, lentils)	1 cup whole or mashed, cooked
Tofu	1 cup ½-inch cubes (about 8 ounces)
Starchy Vegetables	
Corn	1 cup; 1 large ear (8 inches to 9 inches long)
Green peas, field peas, black-eyed peas, lima beans, water chestnuts	1 cup
White potatoes, cassava, plantains, green bananas, taro	1 cup diced or mashed; 1 medium boiled or baked potato (2½-inch to 3-inch diameter); French fried, 20 medium to long strips (2½ inches to 4 inches long; has empty Calories)
Other Vegetables	
Artichokes	1 whole
Asparagus, bean sprouts, beets, Brussels sprouts, eggplants, green or wax beans, okra, parsnips, turnips	1 cup cooked
Cabbage[Vit. C]	1 cup chopped or shredded, raw or cooked
Cauliflower[Vit. C]	1 cup pieces or florets, raw or cooked
Celery	1 cup diced or sliced, raw or cooked; 2 large stalks (11 inches to 12 inches long)
Cucumbers	1 cup raw, sliced or chopped
Green or red peppers[Vit. C]	1 cup chopped, raw or cooked; 1 large pepper (3-inch diameter, 3¾ inches long)

Vegetables Inside the MyPlate Food Guidance System (*continued*)

Other Vegetables

Lettuce, iceberg or head	2 cups raw, shredded or chopped
Mushrooms	1 cup raw or cooked
Onions	1 cup chopped, raw or cooked
Summer squash or zucchini	1 cup raw or cooked, sliced or diced

Vegetable Subgroup Amounts in Cups per Week

Calories	1,000	1,200	1,400	1,600	1,800	2,000	2,200	2,400	2,600	2,800	3,000	3,200
Dark green	1	1.5	1.5	2	3	3	3	3	3	3	3	3
Red-Orange	0.5	1	1	1.5	2	2	2	2	2.5	2.5	2.5	2.5
Beans and peas	0.5	1	1	2.5	3	3	3	3	3.5	3.5	3.5	3.5
Starchy	1.5	2.5	2.5	2.5	3	3	6	6	7	7	9	9
Other	3.5	4.5	4.5	5.5	6.5	6.5	7	7	8.5	8.5	10	10

Vit. C A good food source of vitamin C.

Vit. A A good food source of provitamin A, beta-carotene.

© Cengage Learning 2013

TABLE C5

Fruits Inside the MyPlate Food Guidance System

© Polara Studios, Inc.

Fruits (F)
Make half your plate
fruits and vegetables
Choose a variety of whole, fresh fruit

Includes: All fresh, frozen, canned, and dried fruits and fruit juices.

Servings in General: A 1 cup MyPlate serving equivalent of fruit could be 1 cup of fruit or 100% fruit juice, or ½ cup of dried fruit (*approximately 100 Calories*).

Health Benefits: Fruits reduce heart disease, heart attacks, high blood pressure, stroke, type 2 diabetes, some cancers, kidney stones, obesity, and bone loss.

Nutrients: Fruits provide many nutrients, including potassium, vitamin C, folate (folic acid), carbohydrate, and fiber. Most are low in fat, sodium, and Calories. None have cholesterol.

	Amount that counts as 1 cup of fruit
Apple	½ large (3.25-inch diameter); 1 small (2.5-inch diameter); 1 cup sliced or chopped, raw or cooked
Applesauce, berries Vit. C (blueberries, blackberries, rasp-berries, marionberries), cherries, kiwis Vit. C, mangos, fruit cocktail	1 cup
Bananas	1 cup sliced; 1 large (8 inches to 9 inches long)
Cantaloupe Vit. C and Vit. A, honeydew Vit. C, crenshaw Vit. C	1 cup diced or melon balls
Grapes	1 cup whole or cut up; 32 seedless grapes
Grapefruit Vit. C	1 medium (4-inch diameter); 1 cup sections
Mixed fruit (fruit cocktail)	1 cup diced or sliced, raw or canned, drained
Oranges Vit. C, mandarin oranges Vit. C, tangerines Vit. C	1 large (3 1/16-inch diameter); 1 cup sections; 1 cup canned, drained
Peaches	1 large (2¾-inch diameter); 1 cup sliced or diced, raw, cooked, or canned, drained; 2 halves, canned
Pears	1 medium; 1 cup sliced or diced, raw, cooked, or canned, drained
Pineapple	1 cup chunks, sliced or crushed, raw, cooked or canned, drained
Apricots Vit. A, plums, figs, lemons, limes	3 medium or 2 large; 1 cup sliced, raw or cooked
Strawberries Vit. C	About 8 large berries; 1 cup whole, halved, or sliced, fresh or frozen
Watermelon	1 small wedge (1 inch thick); 1 cup diced or balls
Dried fruit (raisins, prunes, apricots Vit. A, etc.)	½ cup raisins; ½ cup prunes; ½ cup dried apricots
100% fruit juice (orange Vit. C, apple, grape, grapefruit Vit. C, etc.)	1 cup

Vit. C A good food source of vitamin C.

Vit. A A good food source of provitamin A, beta-carotene.

© Cengage Learning 2013

Milk Inside the MyPlate Food Guidance System

Dairy (D)
Choose fat free
or low fat (1%)

Includes: All fluid milk products and many foods made from milk that retain their calcium. Foods made from milk that have little to no calcium, such as cream cheese, cream, and butter, are not part of this group.

Servings in General: A 1 cup MyPlate serving equivalent of dairy could be 1 cup of milk or yogurt, 1½ ounces of natural cheese, or 2 ounces of processed cheese (*approximately 90 Calories when fat free or low fat*).

Health Benefits: Dairy products reduce the risk of low bone mass throughout the life cycle and may prevent osteoporosis.

Nutrients: Dairy products provide calcium, potassium, vitamin D, and protein. Low-fat or fat-free forms provide little or no solid fat.

Amount that counts as 1 cup of dairy	
Milk	1 cup; 1 half-pint container; ½ cup evaporated milk (*choose fat-free or low-fat milk most often*)
Yogurt	1 regular container (8 fluid ounces); 1 cup (*choose fat-free or low-fat yogurt most often*)
Cheese	1½ ounces hard cheese (cheddar, mozzarella, Swiss, parmesan); ⅓ cup shredded cheese; 2 ounces processed cheese (American); ½ cup ricotta cheese; 2 cups cottage cheese (*choose low-fat cheeses most often*)
Milk-based desserts	1 cup pudding made with milk; 1 cup frozen yogurt; 1½ cups ice cream (*choose fat-free or low-fat types most often*)

Calcium choices for those who do not consume dairy products include: calcium-fortified juices, soybeans and other soy products (soy-based beverages, soy yogurt, tempeh), calcium-fortified rice and almond beverages, canned fish (sardines, salmon with bones), some calcium-fortified cereals and breads, other dried beans, and leafy greens (collard and turnip greens, kale, bok choy). The amount of calcium that can be absorbed from these foods varies.
© Cengage Learning 2013

Protein Foods Inside the MyPlate Food Guidance System

Protein Foods (PF)
Choose lean or low-fat meats;
choose fish (8 ounces per week), nuts, and
seeds frequently instead of meat or poultry

Includes: All foods made from meat, poultry, fish, beans or peas, eggs, nuts, and seeds are considered part of this group. Beans and peas can be counted either as vegetables (beans and peas subgroup), or in the protein foods group. Generally, individuals who regularly eat meat, poultry, and fish would count beans and peas in the vegetable group. Individuals who seldom eat meat, poultry, or fish (vegetarians) would count some of the beans and peas they eat in the protein foods group.

Servings in General: A 1-ounce MyPlate serving equivalent of protein foods could be 1 ounce of lean meat, poultry, or fish, 1 egg, 1 tablespoon peanut butter, ¼ cup cooked beans, or ½ ounce of nuts or seeds (*approximately 55 Calories when lean*).

Health Implications: Foods in the protein foods group provide nutrients that are vital for health and body maintenance. However, choosing foods from this group that are high in saturated fat and cholesterol may increase the risk for heart disease.

Nutrients: Protein foods provide many nutrients including protein, B vitamins (niacin, thiamin, riboflavin, and B_6), vitamin E, iron, zinc, and magnesium.

Amount that counts as a 1-ounce equivalent	
Meats	1 ounce cooked lean beef, pork, or ham
Poultry	1 ounce cooked chicken or turkey (without skin); 1 sandwich slice of turkey (4⅓ inch × 2½ inch × ⅛ inch)

© Cengage Learning 2013

	Amount that counts as a 1-ounce equivalent
Fish	1 ounce cooked fish or shellfish
Eggs	1 egg, 1½ egg whites
Nuts and seeds	½ ounce nuts (12 almonds, 24 pistachios, 7 walnut halves); ½ ounce seeds (pumpkin, sunflower, or squash seeds, hulled, roasted); 1 tablespoon peanut butter or almond butter
Beans and peas	¼ cup cooked beans (black, kidney, pinto, or white); ¼ cup cooked peas (chickpeas, cowpeas, lentils, or split peas); ¼ cup baked beans or refried beans; ¼ cup (about 2 ounces) tofu; 1 ounce cooked tempeh; ¼ cup roasted soybeans; 1 falafel patty (2¼ inches, 4 ounces); 2 tablespoons hummus

TABLE C8

Oils Inside the MyPlate Food Guidance System

© Matthew Farruggio

Oils (O) Category
Consume the recommended amount of healthy liquid fats

Includes: Oils that are liquid at room temperature, like vegetable oils. Liquid oils come from plants (except coconut and palm) and from some fish. Foods that are mainly oil include mayonnaise, certain salad dressings, and soft margarine with no *trans* fats.

Servings in General: A 1 teaspoon MyPlate serving equivalent of oil could be 1 teaspoon of liquid plant or fish oil at room temperature (*approximately 40 Calories*). Most Americans consume enough oil in the foods they eat, such as nuts, fish, cooking oil, and salad dressing. Some oil is needed for health. Because it is a fat source, the amount should be limited to the recommendation to balance total Calorie intake.

Nutrients: Oils provide vitamin E, monounsaturated fatty acids, and polyunsaturated fatty acids, which contain essential fatty acids.

Health Benefits and Implications: Plant and fish oils promote heart health; however, overconsuming linoleic acid, which is dominate in most plant oils, can increase cancer risk.

	Amount of oils in common foods	
Oils	**Amount of Food**	**Amount of Oil**
Vegetable oils (canola, corn, cottonseed, olive, peanut, safflower, soybean, sunflower), fish oils	1 tablespoon	3 teaspoons/15 grams
Foods Rich in Oils	**Amount of Food**	**Amount of Oil**
Margarine, soft (*trans* fat free)	1 tablespoon	2½ teaspoons/11 grams
Mayonnaise	1 tablespoon	2½ teaspoons/11 grams
Mayonnaise-type salad dressing	1 tablespoon	1 teaspoon/5 grams
Italian dressing	2 tablespoons	2 teaspoons/8 grams
Thousand Island dressing	2 tablespoons	2½ teaspoons/11 grams
Olives, ripe, canned	4 large	½ teaspoon/2 grams
Avocados	½ medium	3 teaspoons/15 grams
Peanut butter	2 tablespoons	4 teaspoons/16 grams
Peanuts, dry roasted	1 ounce	3 teaspoons/14 grams
Mixed nuts, dry roasted	1 ounce	3 teaspoons/15 grams
Cashews, dry roasted	1 ounce	3 teaspoons/13 grams
Almonds, dry roasted	1 ounce	3 teaspoons/15 grams
Hazelnuts	1 ounce	4 teaspoons/18 grams
Sunflower seeds	1 ounce	3 teaspoons/14 grams
Flax seeds	1 ounce or 1 tablespoon	3 teaspoons/14 grams

Empty Calories Inside the MyPlate Food Guidance System

Empty Calorie Foods Category (EC)
Limit foods and beverages with solid fat and added sugars
Empty Calorie foods are discretionary food choices that promote malnutrition

Includes: Solid fats and added sugars that, when consumed in excess, promote obesity, which is associated with heart disease, type 2 diabetes, and cancer.

Solid Fats: Solid fats are solid at room temperature, like butter and shortening. Solid fats come from many animal foods, can be made from vegetable oils through hydrogenation, and are found naturally in coconut and palm plant foods.

- Common solid fats are butter, shortening, stick margarine, animal fat, and coconut oil.
- Foods high in solid fats include many cheeses, creams, ice creams, well-marbled cuts of meats, regular ground beef, bacon, sausages, poultry skin, and many baked goods (such as cookies, crackers, doughnuts, pastries, and croissants).

Added Sugars: Added sugars are sugars and syrups that are added to foods or beverages during processing or preparation. This does not include naturally occurring sugars, such as those that occur in milk and fruits.

- Foods that contain added sugars are regular soft drinks, candy, cakes, cookies, pies, fruit drinks (such as fruitades and fruit punch), milk-based desserts and products (such as ice cream, sweetened yogurt, and sweetened milk), and grain products (such as sweet rolls and cinnamon toast).
- Ingredients shown on the food labels of processed foods that indicate added sugar are brown sugar, corn sweetener, corn syrup, dextrose, fructose, fruit juice concentrates, glucose, high-fructose corn syrup, honey, invert sugar, lactose, maltose, malt syrup, molasses, raw sugar, sucrose, sugar, and syrup.

Allowance: The empty Calorie allowance is the remaining amount of Calories needed to meet the food intake pattern (after accounting for the Calories needed for all food groups and oils, using forms of foods that are fat free or low fat and with no added sugars). The empty Calorie allowance can also be used to eat more whole, fresh foods from the major food groups.

Food	Amount	Total Calories	Empty Calories
Dairy			
Fat-free milk	1 cup	85	0
1% milk	1 cup	100	20
2% milk (reduced fat)	1 cup	125	40
Whole milk	1 cup	145	65
Low-fat chocolate milk	1 cup	160	75
Cheddar cheese	1½ ounces	170	90
Nonfat mozzarella cheese	1½ ounces	65	0
Whole-milk mozzarella cheese	1½ ounces	130	45
Fruit-flavored low-fat yogurt	1 cup (8 fluid ounces)	240 to 250	100 to 115
Frozen yogurt	1 cup	220	140
Ice cream, vanilla	1½ cups	435	308
Cheese sauce	¼ cup	120	75
Protein Foods			
95% fat-free, extra-lean meat	1 ounce, cooked	55	0
90% fat-free meat	1 ounce, cooked	65	10
80% fat-free meat (such as regular ground beef)	1 ounce, cooked	77	22
Turkey roll, light	1 slice (1 ounce each)	42	0
Roasted chicken breast (skinless)	1 ounce	47	0

© Cengage Learning 2013

Empty Calories Inside the MyPlate Food Guidance System (*continued*)

Protein Foods			
Roasted chicken thigh with skin	1 ounce	70	23
Fried chicken with skin and batter	1 wing (1 ounce each)	158	112
Beef sausage	1 ounce, cooked	115	60
Sausages and hot dogs, precooked	1 ounce, cooked	97	42
Beef bologna	1 slice (1 ounce each)	88	33
Grains			
Whole-wheat bread	1 slice (1 ounce)	70	0
White bread	1 slice (1 ounce)	70	0
English muffin	½ muffin (1 ounce)	68	0
Blueberry muffin	½ small (1 ounce)	93	23
Croissant	½ medium (1 ounce)	115	48
Biscuit, plain	1 to 2-inch diameter	100	45
Cornbread	1 piece (2½ inch × 2½ inch × 1¼ inch)	190	50
Graham crackers	2 large pieces	120	50
Whole-wheat crackers	5 crackers	90	20
Round snack crackers	7 crackers	105	35
Chocolate chip cookies	2 large	135	70
Cake-type doughnuts, plain	2 mini doughnuts (1½-inch diameter)	120	50
Glazed doughnut, yeast type	1 medium (3¾-inch diameter)	240	165
Cinnamon roll	½ roll	155	50
Vegetables			
French fries	1 medium order (1 cup)	460	325
Potato or corn chips	1 ounce (1 cup)	150	100
Onion rings	1 order (8–9 rings, 1 cup)	275	160
Extras			
Regular soda	1 can (12 fluid ounces)	155	155
Diet soda	1 can (12 fluid ounces)	5	5
Ketchup and mustard	2 tablespoons	25	25
Fruit punch or sports drinks	1 cup	115	115
Jelly/jam and sugar	1 tablespoon	50	50
Table wine	5 fluid ounces	115	115
Beer (regular)	12 fluid ounces	145	145
Beer (light)	12 fluid ounces	110	110
Distilled spirits (80 proof)	1½ fluid ounces	95	95
Butter, stick margarine	1 teaspoon	35	35
Candy bar	1 bar	260	260
Cream cheese	1 tablespoon	50	50
Cream and creamer	1 tablespoon	50	50
Dessert topping, frozen, semisolid	1 tablespoon	15	15
Gravy, canned	¼ cup	30	30

TABLE C10

Sample 1-Day Diet Patterned According to MyPlate

April, a 20-year-old, moderately active female of normal body weight, kept the following food record and patterned her foods according to the MyPlate food guidance system.

Put an asterisk by foods that provide high biological value protein. Circle foods that are high in beta-carotene.

Highlight foods that are high in vitamin C.

Food and Drink Consumed		Number of Servings in the Food Groups						
Food and Drink	Amount	G (ounce equivalent)	V (cup equivalent)	F (cup equivalent)	D (cup equivalent)	PF (ounce equivalent)	O (teaspoon equivalent)	EC (Calories)
Breakfast:								
2% milk	8 fluid ounces				1*			40
Peanut butter	2 tablespoons					1	4	
Strawberry jelly	2 tablespoons							100
White bread	2 slices	2						
Banana	1 each			1				
Herbal tea	8 fluid ounces							
Lunch:								
Beef hot dog	2 ounces					2*		83
Ketchup	1 tablespoon							13
Mustard	1 tablespoon							12
White bun	1 each	2						
Fruit punch	12 fluid ounces							173
Snacks:								
Raw baby carrots	0.5 cups		(0.5)					
Cereal bar	1 medium	1						~ 50
Tap water	16 fluid ounces							
Dinner:								
Spaghetti								
Pasta	1.5 cups	3						
Regular ground beef	4 ounces					4*		87
Tomato sauce	0.75 cups		0.75					
Parmesan cheese	1 ounce				0.67*			60
French bread	2 slices	2						
Butter	2 teaspoons							70
Salad								
Iceberg lettuce	1.5 cups		0.75					
Cherry tomatoes	0.25 cups		0.25					
Ranch dressing	2 tablespoons						2	
Chocolate ice cream	1.25 cups				0.83*			171
Gender—Female	Total	10	2.25	1	2.5	7	6	859
Age—20	Recommend	7 ounces	3 cups	2 cups	3 cups	6 ounces	6 teaspoons	290
Pregnant—No	Evaluation *(circle one)*	**A** or D **not** **½ WG**	A or **D** **low** **variety**	A or **D**	A, **D**, or E	A, D, or **E** **not low** **fat**	**A**, D, or E	NE or **E**
Lactating—No								

Food Groups: G (Grains), V (Vegetables), F (Fruit), D (Dairy), PF (Protein Foods), O (Oils), EC (Empty Calories)

Indicate whole grains (WG). Evaluation by MyPlate: "A" = Adequate; "D" = Deficient; "E" = Excessive; "NE" = Not Excessive.

Otto Greule Jr./Time Life
Pictures/Getty Images

Physical Activity
Expend energy through body movement and exercise

Physical Activity: Physical activity and nutrition work together for better health. Being active increases the amount of Calories burned. As people age their metabolism slows, so maintaining energy balance requires moving more and eating less. Physical activity simply means movement of the body that uses energy. Walking, gardening, briskly pushing a baby stroller, climbing the stairs, playing soccer, or dancing the night away are all good examples of being active. For health benefits, physical activity should be moderate or vigorous and amount to *30 minutes* minimally each day. Increasing the intensity or the amount of time of activity can have additional health benefits and may be needed to control body weight. Some physical activities, like walking at a casual pace while grocery shopping and doing light household chores, are not intense enough to help meet the recommendations. Although the body is moving, these activities do not increase the heart rate, so they are not counted toward the 30 or more minutes a day that should minimally be achieved. About *60 minutes* a day of moderate physical activity may be needed to prevent weight gain. For those who have lost weight, at least *60 to 90 minutes* a day may be needed to maintain the weight loss. At the same time, Calorie needs should not be exceeded. Children and teenagers should be physically active for *at least 60 minutes* every day, or most days.

- Moderate physical activities include walking briskly (about 3½ miles per hour), hiking, gardening/yard work, dancing, golf (walking and carrying clubs), bicycling (less than 10 miles per hour), and weight training (general light workout).
- Vigorous physical activities include running/jogging (5 miles per hour), bicycling (more than 10 miles per hour), swimming (freestyle laps), aerobics, walking very fast (4½ miles per hour), heavy yard work such as chopping wood, weight lifting (vigorous effort), and basketball (competitive).

Health Benefits: Regular physical activity can produce long-term health benefits. People of all ages, shapes, sizes, and abilities can benefit from being physically active. The more physical activity you do (within the 2008 Physical Activity Guidelines for Americans), the greater the health benefits. Being physically active can help you increase your chances of living longer, feel better about yourself, decrease your chances of becoming depressed, sleep well at night, move around more easily, have stronger muscles and bones, stay at or get to a healthy weight, be with friends or meet new people, and enjoy yourself and have fun.

Health Implications: When you are not physically active, you are more likely to get heart disease, get type 2 diabetes, have high blood pressure, have high blood cholesterol, or have a stroke.

© Cengage Learning 2013

The Exchange Lists

D.1 THE EXCHANGE LISTS

The Exchange System and its Lists were created by the American Diabetes Association in conjunction with the Academy of Nutrition and Dietetics. They were revised in 2003. The Exchange System sorts foods based on carbohydrate, protein, fat, and Calorie content. The energy (Calorie)–producing nutrients (carbohydrate, protein, and fat) are considered in determining the portion size of each food within each Exchange List. Then, the Calories are naturally controlled as well. The Exchange Lists, along with the macronutrient contents, are shown in Table D1. The idea behind the system is that any food on a specific Exchange List can be exchanged for any other food on the same list. Portion control is paramount. The foods and the amount listed within each Exchange List table are viewed as equivalent in carbohydrate, protein, fat, and Calories. A sample of a healthy diet plan using the Exchange List System is shown in Table D2.

Exchange Lists exist for:

1. Starches (grains, cereals, pasta, breads, crackers, some snacks, starchy vegetables, dried beans, peas, and lentils; see Table D3)
2. Fruits (see Table D4)
3. Nonstarchy vegetables (see Table D5)
4. Milk (see Table D6)
5. Meat and meat substitutes (see Table D7)
6. Fats (see Table D8)
7. Sweets, desserts, and other carbohydrates (see Table D9)
8. Free foods (see Table D10)
9. Combination foods (see Table D11)

To pattern foods, use this simple formula:

Amount eaten ÷ Amount per exchange = Number of exchanges eaten

TABLE D1

Overview of the Exchange Lists

Lists	Carbohydrate (grams)	Protein (grams)	Fat (grams)	Calories
Carbohydrate-Rich Foods				
Starch	15	3	0–1	80
Fruit	15	—	—	60
Milk				
• Fat free, low fat	12	8	0–3	90
• Reduced fat	12	8	5	120
• Whole	12	8	8	150
Other Carbohydrates	15	varies	varies	varies
Nonstarchy Vegetables	5	2	—	25
Protein-Rich Foods				
Meat and Meat Substitutes				
• Very lean	—	7	0–1	35
• Lean	—	7	3	55
• Medium fat	—	7	5	75
• High fat	—	7	8	100
Fat-Rich Foods				
Fat (polyunsaturated, mono-unsaturated, and saturated)	—	—	5	45
Combination and Free Foods				
Free	< 5	—	—	< 20
Combination	15	varies	varies	varies

© Cengage Learning 2013

TABLE D2

A Sample Diet Plan

Provides 1,735 Calories with 55 to 60 percent of its Calories from carbohydrate, 15 to 20 percent from protein, and 20 to 30 percent from fat.

Accompanying Meal Plan: Breakfast of cereal with fruit and milk; lunch of a ham sandwich and fruit; snack of popcorn and juice; dinner of pasta with meat sauce, salad sprinkled with seeds and dressing, green beans, and corn on the cob; and snack of milk and crackers.

Exchange	Calories	Breakfast	Lunch	Snack	Dinner	Snack
9 starches	720	2	2	1	3	1
4 vegetables	100				4	
3 fruits	180	1	1	1		
6 lean meats	330		2		4	
2 fat-free milks	180	1				1
5 fats	225		1		4	

© Cengage Learning 2013

TABLE D3

The Starch List

One starch exchange is equal to:

15 grams of carbohydrate, 3 grams of protein, 0–1 gram of fat, and 80 Calories

Bread	Exchange Amount
Bagel, 4 ounce	¼ (1 ounce)
Bread, reduced-Calorie	2 slices (1½ ounces)
Bread, white, whole-wheat, pumpernickel, rye	1 slice (1 ounce)
Bread sticks, crisp, 4 inch × ½ inch	4 (⅔ ounce)
English muffin	½ each
Hot dog bun or hamburger bun	½ (1 ounce)
Naan, 8 × 2 inch	¼ each
Pancake, 4 inches across, ¼-inch thick	1 each
Pita, 6 inches across	½ each
Roll, plain, small	1 each (1 ounce)
Raisin bread, unfrosted, 1 slice	(1 ounce)
Tortilla, corn, 6 inches across	1 each
Tortilla, flour, 6 inches across	1 each
Tortilla, flour, 10 inches across	⅓ each
Waffle, 4 inches square or across, reduced-fat	1 each
Cereals and Grains	**Exchange Amount**
Bran cereals	½ cup
Bulgur	½ cup
Cereals, cooked	½ cup
Cereals, unsweetened, ready-to-eat	¾ cup
Cornmeal, dry	3 tablespoons
Couscous	⅓ cup
Flour, dry	3 tablespoons
Granola, low-fat	¼ cup
Grape-Nuts®	1 cup
Grits	½ cup
Kasha	½ cup
Millet	⅓ cup
Muesli	¼ cup
Oats	½ cup
Pasta	⅓ cup
Puffed cereal	1½ cups
Rice, white or brown	⅓ cup
Shredded Wheat®	½ cup
Sugar-frosted cereal	½ cup
Wheat germ	3 tablespoons

continued

The Starch List (*continued*)

Starchy Vegetables	Exchange Amount
Baked beans	⅓ cup
Corn	½ cup
Corn on cob, large	½ cob (5 ounces)
Mixed vegetables with corn, peas, or pasta	1 cup
Peas, green	½ cup
Plantain	½ cup
Potato, boiled	½ cup or ½ medium (3 ounces)
Potato, baked	¼ large (3 ounces)
Potato, mashed	½ cup
Squash, winter (acorn, butternut, pumpkin)	1 cup
Yam, sweet potato, plain	½ cup

Crackers	Exchange Amount
Animal crackers	8 each
Graham cracker, 2½-inch square	3 each
Matzoh	¾ ounce
Melba toast	4 slices
Oyster crackers	24 each
Popcorn, popped, no fat added, or low-fat microwave	3 cups
Pretzels	¾ ounce
Rice cakes, 4 inches across	2 each
Saltine-type crackers	6 each
Snack chips, fat-free or baked (tortilla, potato)	15–20 (¾ ounce) each
Whole-wheat crackers, no fat added	2–5 (¾ ounce) each

Beans, Peas, and Lentils (*1 starch exchange, plus 1 very lean meat exchange*)	Exchange Amount
Beans and peas (garbanzo, pinto, kidney, white, split, etc.)	½ cup
Lima beans	⅓ cup
Lentils	½ cup
Miso (*400 mg or more sodium per exchange*)	3 tablespoons

Starchy Foods Prepared with Fat (*1 starch exchange plus 1 fat exchange*)	Exchange Amount
Biscuit, 2½ inches across	1 each
Chow mein noodles	½ cup
Corn bread, 2-inch cube	1 piece (2 ounces)
Crackers, round butter type	6 each
Croutons	1 cup
French-fried potatoes, oven-baked (see fast foods list)	1 cup (2 ounces)
Granola	¼ cup
Hummus	⅓ cup
Muffin, 5 ounces	⅕ (1 ounce)
Popcorn, microwaved	3 cups
Sandwich crackers, cheese or peanut butter filling	3 each

The Starch List (continued)

Starchy Foods Prepared with Fat (*1 starch exchange plus 1 fat exchange*)	Exchange Amount
Snack chips (potato, tortilla)	9–13 (¾ ounce)
Stuffing, bread, prepared	⅓ cup
Taco shell, 6 inches across	2 each
Waffle, 4 inches square or across	1 each
Whole-wheat crackers, fat added	4–6 each (1 ounce)

© Cengage Learning 2013

TABLE D4

The Fruit List

One fruit exchange is equal to:

15 grams of carbohydrate, 0 grams of protein, 0 grams of fat, and 60 Calories

Fruit (whole)	Exchange Amount
Apple, unpeeled, small	1 each (4 ounces)
Applesauce, unsweetened	½ cup
Apples, dried	4 rings
Apricots, fresh	4 whole (5½ ounces)
Apricots, dried	8 halves
Apricots, canned	½ cup
Banana, small	1 (4 ounces)
Blackberries	¾ cup
Blueberries	¾ cup
Cantaloupe, small	⅓ melon (11 ounces) or 1 cup cubes
Cherries, sweet, fresh	12 each (3 ounces)
Cherries, sweet, canned	½ cup
Dates	3 each
Figs, fresh	1½ large or 2 medium (3½ ounces)
Figs, dried	1½ each
Fruit cocktail	½ cup
Grapefruit, large	½ (11 ounces)
Grapefruit sections, canned	¾ cup
Grapes, small	17 each (3 ounces)
Honeydew melon	1 slice (10 ounces) or 1 cup cubes
Kiwi	1 each (3½ ounces)
Mandarin oranges, canned	¾ cup
Mango, small	½ fruit (5½ ounces) or ½ cup
Nectarine, small	1 each (5 ounces)
Orange, small	1 each (6½ ounces)
Papaya	½ fruit (8 ounces) or 1 cup cubes
Peach, medium, fresh	1 each (4 ounces)

© Cengage Learning 2013

continued

The Fruit List (continued)

Fruit (whole)	Exchange Amount
Peaches, canned	½ cup
Pear, large, fresh	½ each (4 ounces)
Pears, canned	½ cup
Pineapple, fresh	¾ cup
Pineapple, canned	½ cup
Plums, small	2 each (5 ounces)
Plums, canned	½ cup
Plums, dried (prunes)	3 each
Raisins	2 tablespoons
Raspberries	1 cup
Strawberries	1¼ cups whole berries
Tangerines, small	2 each (8 ounces)
Watermelon	1 slice (13½ ounces) or 1¼ cups cubes

Fruit Juice, Unsweetened	Exchange Amount
Apple juice/cider	½ cup
Cranberry juice cocktail	⅓ cup
Cranberry juice cocktail, reduced-Calorie	1 cup
Fruit juice blends, 100% juice	⅓ cup
Grape juice	⅓ cup
Grapefruit juice	½ cup
Orange juice	½ cup
Pineapple juice	½ cup
Prune juice	⅓ cup

© Cengage Learning 2013

TABLE D5

The Nonstarchy Vegetable List

One nonstarchy vegetable exchange is equal to:

5 grams of carbohydrate, 2 grams of protein, 0 grams of fat, and 25 Calories

½ cup cooked vegetables or vegetable juice or 1 cup raw vegetables is one exchange

Artichokes	Cauliflower
Artichoke hearts	Celery
Asparagus	Cucumbers
Beans (green, wax, Italian)	Eggplant
Bean sprouts	Green onions or scallions
Beets	Greens (collard, kale, mustard, turnip)
Broccoli	Kohlrabi
Brussels sprouts	Leeks
Cabbage	Mixed vegetables (without corn, peas, potatos, or pasta)
Carrots	Mushrooms

© Cengage Learning 2013

The Nonstarchy Vegetable List (continued)

½ cup cooked vegetables or vegetable juice or 1 cup raw vegetables is one exchange

Okra	Tomatoes
Onions	Tomatoes, canned
Pea pods	Tomato sauce*
Peppers (all varieties)	Tomato/vegetable juice*
Radishes	Turnips
Salad greens (endive, escarole, any variety of lettuce, spinach)	Water chestnuts
Sauerkraut*	Watercress
Spinach	Zucchini
Summer squash (crookneck)	

*Provides 400 mg or more sodium per exchange.
© Cengage Learning 2013

TABLE D6

The Milk List

One milk exchange is equal to:

12 grams of carbohydrate, 8 grams of protein, varying amounts of fat and Calories

Fat-Free/Low-Fat Milk (*12 grams of carbohydrate, 8 grams of protein, 0–3 grams of fat, and 90 Calories*)	Exchange Amount
Fat-free milk	1 cup
½% milk	1 cup
1% milk	1 cup
Fat-free or low-fat buttermilk	1 cup
Evaporated fat-free milk	½ cup
Fat-free dry milk	⅓ cup dry
Soy milk, low-fat or fat-free	1 cup
Yogurt, fat-free or low-fat, flavored, sweetened with nonnutritive sweetener and fructose	⅓ cup (6 ounces)
Yogurt, plain, fat-free	⅔ cup (6 ounces)
Reduced-Fat Milk (*12 grams of carbohydrate, 8 grams of protein, 5 grams of fat, and 120 Calories*)	**Exchange Amount**
2% milk	1 cup
Soy milk	1 cup
Sweet acidophilus C	1 cup
Yogurt, plain, low-fat	¾ cup
Whole Milk (*12 grams of carbohydrate, 8 grams of protein, 8 grams of fat, and 150 Calories*)	**Exchange Amount**
Whole milk	1 cup
Evaporated whole milk	½ cup
Goat's milk	1 cup
Kefir	1 cup
Yogurt, plain, made from whole milk	¾ cup

© Cengage Learning 2013

TABLE D7

The Meat and Meat Substitutes List

One meat exchange is equal to:

0 grams of carbohydrate, 7 grams of protein, varying amounts of fat and Calories

Very Lean Meat and Substitutes (*0 grams of carbohydrate, 7 grams of protein, 0–1 grams of fat, and 35 Calories*)	Exchange Amount
Poultry: Chicken, turkey, or Cornish hen (white meat, no skin)	1 ounce
Fish: Fresh or frozen cod, flounder, haddock, halibut, trout, lox (smoked salmon), or tuna, fresh or canned in water	1 ounce
Shellfish: Clams, crab, lobster, scallops, shrimp, imitation shellfish	1 ounce
Game: Duck or pheasant (no skin), venison, buffalo, ostrich	1 ounce
Cheese with 1 gram or less of fat per ounce: Fat-free or low-fat cottage cheese	¼ cup (1 ounce)
Other: Processed sandwich meats with 1 gram of fat or less per ounce (such as deli-thin shaved meats, chipped beef*, turkey, ham)	1 ounce
Egg whites	2 each
Egg substitutes, plain	¼ cup
Hot dogs with 1 gram of fat or less per ounce* and sausages with 1 gram of fat or less per ounce	1 ounce
Kidneys (high in cholesterol)	1 ounce
Beans, peas, lentils (cooked) (1 very lean meat + 1 starch exchange)	½ cup

Lean Meat and Substitutes (*0 grams of carbohydrate, 7 grams of protein, 3 grams of fat, and 55 Calories*)	Exchange Amount
Beef: USDA Select or Choice grades of lean beef trimmed of fat (round, sirloin, and flank steak), tenderloin, roast (rib, chuck, rump), steak (T-bone, porterhouse, cubed), ground round	1 ounce
Pork: Lean pork (fresh ham), canned, cured, or boiled ham, Canadian bacon*, tenderloin, center loin chop	1 ounce
Lamb: Roast, chop, leg	1 ounce
Veal: Lean chop, roast	1 ounce
Poultry: Chicken, turkey (dark meat, no skin), chicken white meat (with skin), domestic duck or goose (well drained of fat, no skin)	1 ounce
Fish:	
Herring, uncreamed or smoked	1 ounce
Oysters	6 medium
Salmon (fresh or canned), catfish, tuna (canned in oil, drained)	1 ounce
Sardines, canned	2 medium
Game: Goose (no skin), rabbit	1 ounce
Cheese:	
4.5%-fat cottage cheese	¼ cup
Grated parmesan	2 tablespoons
Cheeses with 3 grams of fat or less per ounce	1 ounce

© Cengage Learning 2013

The Meat and Meat Substitutes List (continued)

Lean Meat and Substitutes (continued) (0 grams of carbohydrate, 7 grams of protein, 3 grams of fat, and 55 Calories)	Exchange Amount
Other: Hot dogs with 3 grams of fat or less per ounce*	1½ ounces
Processed sandwich meat with 3 grams of fat or less per ounce (turkey pastrami)	1 ounce
Liver, heart (high in cholesterol)	1 ounce

Medium-Fat Meat and Substitutes (0 grams of carbohydrate, 7 grams of protein, 5 grams of fat, and 75 Calories)	Exchange Amount
Beef: Most beef products (ground beef, meatloaf, corned beef, short ribs), prime grades of meat trimmed of fat (prime rib)	1 ounce
Pork: Top loin, chop, Boston butt, cutlet	1 ounce
Lamb: Rib roast, ground	1 ounce
Veal: Cutlet, ground or cubed, unbreaded	1 ounce
Poultry: Chicken (dark meat, with skin), ground turkey or ground chicken, fried chicken (with skin)	1 ounce
Fish: Any fried fish product	1 ounce
Cheese with 5% or less fat	
Feta	1 ounce
Mozzarella	1 ounce
Ricotta	¼ cup (2 ounces)
Other:	
Egg (high in cholesterol, limit to 3 per week)	1 each
Sausage with 5 grams of fat or less per ounce	1 ounce
Soy milk	1 cup
Tempeh	¼ cup
Tofu	½ cup (4 ounces)

High-Fat Meat and Substitutes (0 grams of carbohydrate, 7 grams of protein, 8 grams of fat, and 100 Calories)	Exchange Amount
Pork: Spareribs, ground pork, pork sausage	1 ounce
Cheese: All regular cheeses (American, cheddar, Monterey jack, Swiss)	1 ounce
Other:	
Processed sandwich meats with less than 8 grams of fat per ounce (bologna, pimento loaf, salami)	1 ounce
Sausages (bratwurst, Italian sausage, knockwurst, Polish sausage, smoked sausage)	1 ounce
Pork bacon	2 slices or 1 ounce each before cooking
Turkey bacon	3 slices or ½ ounce each before cooking
Peanut butter (contains unsaturated fat)	1 tablespoon
Hot dogs: Beef, pork, turkey, chicken, or combination (1 high-fat meat + 1 fat exchange)	1 (10 per 1 pound-sized package. The Exchange Lists specify the 10 per 1 pound-sized package.)

*Provides 400 mg or more of sodium per exchange.

TABLE D8

The Fat List

One fat exchange is equal to:

0 grams of carbohydrate, 0 grams of protein, 5 grams of fat, and 45 Calories

Monounsaturated Fats	Exchange Amount
Avocados	2 tablespoons (1 ounce)
Oil (canola, olive, peanut)	1 teaspoon
Olives, ripe, black	8 large
Olives, green, stuffed*	10 large
Almonds, cashews, mixed nuts (50% peanuts)	6 nuts
Peanuts or pistachios	10 nuts
Pecans	4 halves
Peanut butter, smooth or crunchy	½ tablespoon
Sesame seeds	1 tablespoon
Tahini or sesame paste	2 teaspoons

Polyunsaturated Fats	Exchange Amount
Walnuts (English or black)	4 halves
Margarine, stick, tub, or squeeze	1 teaspoon
Margarine, lower-fat spread (30%–50% vegetable oil)	1 tablespoon
Mayonnaise, regular	1 teaspoon
Mayonnaise, reduced-fat	1 tablespoon
Oil (corn, safflower, soybean)	1 teaspoon
Salad dressing, regular*	1 tablespoon
Salad dressing, reduced-fat	2 tablespoons
Mayonnaise-type salad dressing, regular	2 teaspoons
Mayonnaise-type salad dressing, reduced-fat	1 tablespoon
Seeds (pumpkin, sunflower)	1 tablespoon

Saturated Fats‡	Exchange Amount
Bacon, cooked	1 slice (20 slices per pound)
Bacon grease, shortening, or lard	1 teaspoon
Butter, stick (whipped) [reduced-fat]	1 teaspoon (2 teaspoons) [1 tablespoon]
Chitterlings, boiled	2 tablespoons (½ ounce)
Cream, half and half	2 tablespoons
Coconut milk	1 tablespoon
Coconut, sweetened, shredded	2 tablespoons
Cream cheese, regular	1 tablespoon (½ ounce)
Cream cheese, reduced-fat	1½ tablespoons (¾ ounce)
Sour cream, regular	2 tablespoons
Sour cream, reduced-fat	3 tablespoons

*Provides 400 mg or more sodium per exchange.
‡Saturated fats can raise blood cholesterol levels.
© Cengage Learning 2013

The Sweets, Desserts, and Other Carbohydrates List

One sweets, desserts, and other carbohydrates exchange contains carbohydrates and often fat.
1 carbohydrate = 15 grams of carbohydrates and 1 fat = 5 grams of fat

Other Carbohydrates	Exchange Amount	Number of Exchanges
Angel food cake, unfrosted	1/12 cake	2 carbohydrates
Brownies, small, unfrosted	2-inch square (1 ounce)	1 carbohydrate, 1 fat
Cake, unfrosted	2-inch square (1 ounce)	1 carbohydrate, 1 fat
Cake, frosted	2-inch square (2 ounces)	2 carbohydrates, 1 fat
Cookies or sandwich cookies with creme filling	2 small (1/3 ounce)	1 carbohydrate, 1 fat
Cookies, sugar-free	3 small or 1 large (1 ounce)	1 carbohydrate, 1–2 fats
Cranberry sauce, jellied	1/4 cup	1 1/2 carbohydrates
Cupcake, frosted	1 small (2 ounces)	2 carbohydrates, 1 fat
Doughnut, plain cake	1 medium (1 1/2 ounces)	1 1/2 carbohydrates, 2 fats
Doughnut, glazed	3 3/4 inches across (2 ounces)	2 carbohydrates, 2 fats
Energy, sport, or breakfast bar	1 bar (1 1/3 ounces)	1 1/2 carbohydrates, 0–1 fat
Energy, sport, or breakfast bar, large	1 bar (2 ounces)	2 carbohydrates, 1 fat
Fruit cobbler	1/2 cup (3 1/2 ounces)	3 carbohydrates, 1 fat
Fruit juice bar, frozen, 100% juice	1 bar (3 ounces)	1 carbohydrate
Fruit snacks, chewy (pureed fruit concentrate)	1 roll (3/4 ounce)	1 carbohydrate
Fruit spreads, 100% fruit	1 1/2 tablespoons	1 carbohydrate
Gelatin, regular	1/2 cup	1 carbohydrate
Gingersnaps	3 each	1 carbohydrate
Granola or snack bar, regular or low-fat	1 bar (1 ounce)	1 1/2 carbohydrates
Honey	1 tablespoon	1 carbohydrate
Ice cream	1/2 cup	1 carbohydrate, 2 fats
Ice cream, light	1/2 cup	1 carbohydrate, 1 fat
Ice cream, low-fat	1/2 cup	1 1/2 carbohydrates
Ice cream, fat-free, no sugar added	1/2 cup	1 carbohydrate
Jam or jelly, regular	1 tablespoon	1 carbohydrate
Milk, chocolate, whole	1 cup	2 carbohydrates, 1 fat
Pie, fruit, with 2 crusts	1/8 of 8-inch commercially prepared pie	3 carbohydrates, 2 fats
Pie, pumpkin or custard	1/8 of 8-inch commercially prepared pie	2 carbohydrates, 2 fats
Pudding, regular (made with reduced-fat milk)	1/2 cup	2 carbohydrates
Pudding, sugar-free or fat-free (made with fat-free milk)	1/2 cup	1 carbohydrate
Reduced-Calorie meal replacement (shake)	1 can (10–11 ounces)	1 1/2 carbohydrates, 0–1 fat
Rice milk, low-fat or fat-free, plain	1 cup	1 carbohydrate
Rice milk, low-fat, flavored	1 cup	1 1/2 carbohydrates
Salad dressing, fat-free*	1/4 cup	1 carbohydrate
Sherbet, sorbet	1/2 cup	2 carbohydrates
Spaghetti or pasta sauce, canned*	1/2 cup	1 carbohydrate, 1 fat

continued

The Sweets, Desserts, and Other Carbohydrates List (*continued*)

Other Carbohydrates	Exchange Amount	Number of Exchanges
Sports drinks	8 ounces (1 cup)	1 carbohydrate
Sugar	1 tablespoon	1 carbohydrate
Sweet roll or Danish	1 each (2½ ounces)	2½ carbohydrates, 2 fats
Syrup, light	2 tablespoons	1 carbohydrate
Syrup, regular	1 tablespoon	1 carbohydrate
Syrup, regular	¼ cup	4 carbohydrates
Tortilla chips	6–12 each (1 ounce)	1 carbohydrate, 2 fats
Vanilla wafers	5 each	1 carbohydrate, 1 fat
Yogurt, frozen	½ cup	1 carbohydrate, 0–1 fat
Yogurt, frozen, fat-free	⅓ cup	1 carbohydrate
Yogurt, low-fat, with fruit	1 cup	3 carbohydrates, 0–1 fat

*Provides 400 mg or more of sodium per exchange.
© Cengage Learning 2013

TABLE D10

The Free Foods List

One free foods exchange provides:
≤ 5 grams of carbohydrate and 20 Calories

Fat-Free or Reduced-Fat Foods	Exchange Amount
Cream cheese, fat-free	1 tablespoon (½ ounce)
Creamers, nondairy, liquid	1 tablespoon
Creamers, nondairy, powdered	2 teaspoons
Margarine spread, fat-free	4 tablespoons
Margarine spread, reduced-fat	1 teaspoon
Mayonnaise, fat-free	1 tablespoon
Mayonnaise, reduced-fat or salad dressing type	1 teaspoon
Mayonnaise-type salad dressing, fat-free	1 tablespoon
Nonstick cooking spray	NA
Salad dressing, fat-free or low-fat	1 tablespoon
Salad dressing, fat-free, Italian	2 tablespoons
Sour cream, fat-free, reduced-fat	1 tablespoon
Whipped topping, regular	1 tablespoon
Whipped topping, light or fat-free	2 tablespoons
Sugar-Free Foods	**Exchange Amount**
Candy, hard, sugar-free	1 piece
Gelatin dessert, sugar-free	NA
Gelatin, unflavored	NA
Gum, sugar-free	NA

© Cengage Learning 2013

The Free Foods List (*continued*)

Sugar-Free Foods	Exchange Amount
Jam or jelly, light	2 teaspoons
Sugar substitutes	NA
Syrup, sugar-free	2 tablespoons

Drinks	Exchange Amount
Bouillon, broth, consommé*	NA
Bouillon or broth, low-sodium	NA
Carbonated or mineral water	NA
Club soda	NA
Cocoa powder, unsweetened	1 tablespoon
Diet soft drinks, sugar-free	NA
Drink mixes, sugar-free	NA
Tea	NA
Tonic water, sugar-free	NA

Condiments	Exchange Amount
Catsup	1 tablespoon
Horseradish	NA
Lemon juice, lime juice	NA
Mustard	NA
Pickle relish	1 tablespoon
Pickles, dill*	1 medium
Pickles, sweet (bread and butter)	2 slices
Pickles, sweet (gherkin)	¾ ounce
Salsa	¼ cup
Soy sauce, regular or light*	1 tablespoon
Taco sauce	1 tablespoon
Vinegar	NA
Yogurt	2 tablespoons

Seasonings	Exchange Amount
Flavoring extracts	NA
Garlic	NA
Herbs, fresh or dried	NA
Hot pepper sauces	NA
Pimento	NA
Spices	NA
Wine, used in cooking	NA
Worcestershire sauce	NA

Limit those with serving sizes to less than 3 per day. Eat freely those without serving sizes (NA).

*Provides 400 mg or more of sodium per exchange.

TABLE D11

The Combination Foods List

Combination foods can contain varying amounts of carbohydrates, proteins, and fats. 1 carbohydrate = 15 grams of carbohydrates and 1 fat = 5 grams of fat. See Table D7 for protein and fat amounts in meat and meat substitutes.

Fast Foods	Amount	Number of Exchanges
Burrito, with beef*	1 each (5–7 ounces)	3 carbohydrates, 1 medium-fat meat, 1 fat
Chicken nuggets*	6 each	1 carbohydrate, 2 medium-fat meats, 1 fat
Chicken breast and wing, breaded and fried*	1 each	1 carbohydrate, 4 medium-fat meats, 2 fats
Chicken sandwich, grilled*	1 each	2 carbohydrates, 3 very lean meats
Chicken wings, hot	6 each (5 ounces)	1 carbohydrate, 3 medium-fat meats, 4 fats
Fish sandwich, with tartar sauce*	1 each	3 carbohydrates, 1 medium-fat meat, 3 fats
French fries*	1 medium serving (5 ounces)	4 carbohydrates, 4 fats
Hamburger, regular	1 each	2 carbohydrates, 2 medium-fat meats
Hamburger, large*	1 each	2 carbohydrates, 3 medium-fat meats, 1 fat
Hot dog, with bun*	1 each	1 carbohydrate, 1 high-fat meat, 1 fat
Individual pan pizza*	1 each	5 carbohydrates, 3 medium-fat meats, 3 fats
Pizza, cheese, thin crust*	¼ of 12-inch (about 6 ounces)	2½ carbohydrates, 2 medium-fat meats
Pizza, meat, thin crust	¼ of 12-inch (6 ounces)	2½ carbohydrates, 2 medium-fat meats, 1 fat
Soft serve cone	1 small (5 ounces)	2½ carbohydrates, 1 fat
Submarine sandwich*	1 sub (6-inch)	3 carbohydrates, 1 vegetable, 2 medium-fat meats, 1 fat
Submarine sandwich, less than 6 grams fat*	1 sub (6-inch)	2½ carbohydrates, 2 lean meats
Taco, hard or soft shell	1 (3–3½ ounces)	1 carbohydrate, 1 medium-fat meat, 1 fat

Soups	Amount	Number of Exchanges
Bean*	1 cup	1 carbohydrate, 1 very lean meat
Cream, made with water*	1 cup (8 ounces)	1 carbohydrate, 1 fat
Instant*	6 ounces prepared	1 carbohydrate
Instant, with beans/lentils*	8 ounces prepared	2½ carbohydrates, 1 very lean meat
Split pea, made with water*	½ cup (4 ounces)	1 carbohydrate
Tomato, made with water*	1 cup (8 ounces)	1 carbohydrate
Vegetable beef, chicken noodle, or other broth-type*	1 cup (8 ounces)	1 carbohydrate

Entrées	Amount	Number of Exchanges
Tuna noodle casserole, lasagna, spaghetti with meatballs, chili with beans, macaroni and cheese*	1 cup (8 ounces)	2 carbohydrates, 2 medium-fat meats
Chow mein, without noodles or rice	2 cups (16 ounces)	1 carbohydrate, 2 lean meats
Tuna or chicken salad	½ cup (3½ ounces)	½ carbohydrate, 2 lean meats, 1 fat

Frozen Entrées and Meats	Amount	Number of Exchanges
Dinner-type meal*	Generally 14–17 ounces	3 carbohydrates, 3 medium-fat meats, 3 fats
Entrée or meal with less than 340 Calories	About 8–11 ounces	2–3 carbohydrates, 1–2 lean meats
Meatless burger, soy-based	3 ounces	½ carbohydrate, 2 lean meats
Meatless burger, vegetable- and starch-based	3 ounces	1 carbohydrate, 1 lean meat
Pizza, cheese, thin crust*	¼ of 12-inch (6 ounces)	2 carbohydrates, 2 medium-fat meats, 1 fat
Pizza, meat topping, thin crust*	¼ of 12-inch (6 ounces)	2 carbohydrates, 2 medium-fat meats, 2 fats
Pot pie*	1 each (7 ounces)	2½ carbohydrates, 1 medium-fat meat, 3 fats

*Provides 400 mg or more of sodium per exchange.
© Cengage Learning 2013

Nutrition Resources

Nutrition Organizations

Academy of Nutrition and Dietetics
www.eatright.org
216 West Jackson Boulevard,
Suite 800
Chicago, IL 60606-6995
(800) 877-1600
(312) 899-0040

American Society for Nutrition
www.nutrition.org
9650 Rockville Pike
Bethesda, MD 20814
(301) 634-7050

National Academy of Sciences/Institute
of Medicine/National Research Council
(NAS/IOM/NRC)
www.nas.edu
500 Fifth Street
Washington, DC 20001
(202) 334 2000

Society for Nutrition Education
www.sne.org
9100 Purdue Road,
Suite 200
Indianapolis, IN 46268
(800) 235-6690

Health and Disease

Alzheimer's Association
www.alz.org
225 North Michigan Avenue,
17th Floor
Chicago, IL 60601
(800) 272-3900

Alzheimer's Disease Education
& Referral Center
www.nia.nih.gov/Alzheimers
P.O. Box 8250
Silver Spring, MD 20907-8250
(800) 438-4380

American Academy of Allergy, Asthma,
and Immunology
www.aaaai.org
611 East Wells Street,
Suite 1100
Milwaukee, WI 53202
(414) 272-6071

American Cancer Society
www.cancer.org
National Home Office
1599 Clifton Road NE
Atlanta, GA 30329-4251
(800) 227-2345

American Council on Science
and Health
www.acsh.org
1995 Broadway, 2nd Floor
New York, NY 10023-5860
(212) 362-7044

American Dental Association
www.ada.org
211 East Chicago Avenue
Chicago, IL 60611
(312) 440-2500

American Diabetes Association
www.diabetes.org
1701 North Beauregard Street
Alexandria, VA 22311
(800) 342-2383

American Heart Association
www.heart.org
7272 Greenville Avenue
Dallas, TX 75231
(800) 242-8721

American Institute for Cancer Research
www.aicr.org
1759 R Street NW
Washington, DC 20009
(800) 843-8114

American Medical Association
www.ama-assn.org
515 North State Street
Chicago, IL 60654
(800) 621-8335

American Public Health Association (APHA)
www.apha.org
800 I Street NW
Washington, DC
20001-3710
(282) 777-2742

American Red Cross
www.redcross.org
National Headquarters
2025 E Street NW
Washington, DC 20006
(202) 303-4498

Arthritis Foundation
www.arthritis.org
P.O. Box 7669
Atlanta, GA 30357
(800) 283 7800

Celiac Disease Foundation
www.celiac.org
13251 Ventura Boulevard,
Suite 1
Studio City, CA 91604
(818) 990-2354

The Food Allergy & Anaphylaxis Network
www.foodallergy.org
11781 Lee Jackson Highway,
Suite 160
Fairfax, VA 22033-3309
(800) 929-4040

Gluten Intolerance Group
www.gluten.net
31214 124th Avenue SE
Auburn, WA 98092
(253) 833-6655

Mayo Clinic Health Oasis
www.mayoclinic.com
13400 East Shea Boulevard
Scottsdale, AZ 85259
(480) 301-8000

National Council Against Health Fraud,
Inc. (NCAHF)
www.ncahf.org
P.O. Box 141
Fort Lee, NJ 07024
(212) 723-2955

National Osteoporosis Foundation
www.nof.org
1150 17th Street NW, Suite 850
Washington, DC 20036
(800) 231-4222

Aging

American Association of Retired Persons
(AARP)
www.aarp.org
601 E Street NW
Washington, DC 20049
(800) 424-3410

The Gerontological Society of America
(GSA)
www.geron.org
1220 L Street NW, Suite 901
Washington, DC 20005
(202) 842-1275

National Council on Aging, Inc. (NCOA)
www.ncoa.org
1901 L Street NW, 4th Floor
Washington, DC 20036
(202) 479-1200

Infancy and Childhood

American Academy of Pediatrics
www.aap.org
141 Northwest Point Boulevard
Elk Grove Village, IL 60007-1098
(847) 434-4000

Birth Defect Research for Children, Inc.
www.birthdefects.org
976 Lake Baldwin Lane,
Suite 104
Orlando, FL 32814
(407) 895-0802

National Center for Education in
Maternal and Child Health
www.ncemch.org
2000 15th Street North, Suite 701
Arlington, VA 22201-2617
(703) 524-7802

Pregnancy and Lactation

The American Congress of Obstetricians
and Gynecologists
www.acog.org
P.O. Box 96920
Washington, DC 20090
(202) 638-5577

La Leche International, Inc.
www.llli.org
957 N. Plum Grove Road
Schaumburg, IL 60173
(800) 525-3243

March of Dimes
www.marchofdimes.com
1275 Mamaroneck Avenue
White Plains, NY 10605
(914) 997-4488

U.S. Government

Administration on Aging
www.aoa.gov
1 Massachusetts Avenue NW
Washington, DC 20001
(202) 619-0724

Centers for Disease Control and
Prevention (CDC)
www.cdc.gov
1600 Clifton Road NE
Atlanta, GA 30333
(800) 232-4636

Federal Trade Commission (FTC)
www.ftc.gov
600 Pennsylvania Avenue NW
Washington, DC 20580
(202) 326-2222

Food and Drug Administration (FDA)
www.fda.gov
10903 New Hampshire
Avenue
Silver Spring, MD 20993
(888) 463-6332

Food and Nutrition Information Center,
National Agricultural Library
fnic.nal.usda.gov
10301 Baltimore Avenue,
Room 105
Beltsville, MD 20705-2351

National Cancer Institute
www.cancer.gov
31 Center Drive,
MSC 2580
Bethesda, MD 20892
(800) 422-6237

National Digestive Diseases Information
Clearinghouse (NDDIC)
www.digestive.niddk.nih.gov
2 Information Way
Bethesda, MD 20892-3570
(800) 891-5389

National Heart Lung and Blood Institute
www.nhlbi.nih.gov
NHLBI Health Information Center
P.O. Box 30105
Bethesda, MD 20824-0105
(301) 592-8573

National Institute on Aging
Public Information Office
www.nia.nih.gov
31 Center Drive, MSC 2292
Bethesda, MD 20892
(800) 222-4225

National Institute of Allergy and
Infectious Diseases
www.niaid.nih.gov
6610 Rockledge Drive,
MSC 6612
Bethesda, MD 20892-6612
(301) 402-1663

National Institute of Dental and
Craniofacial Research (NIDCR)
www.nidcr.nih.gov
9000 Rockville Pike
Bethesda, MD 20892
(866) 232-4528

National Institute of Diabetes and
Digestive and Kidney Diseases
www2.niddk.nih.gov
31 Center Drive, MSC 2560
Bethesda, MD 20892-2560
(301) 496-3583

National Institutes of Health (NIH)
www.nih.gov
9000 Rockville Pike
Bethesda, MD 20892
(301) 496-4000

National Marine Fisheries
www.nmfs.noaa.gov
1315 East West Highway
Silver Spring, MD 20910

NIH Osteoporosis and Related Bone
Diseases National Resource Center
www.niams.nih.gov/Health_Info/
Bone
2 AMS Circle
Bethesda, MD 20892-3676
(800) 624-2663

Nutrition Assistance Programs
www.fns.usda.gov
3101 Park Center Drive
Alexandria, VA 22302

Office of Dietary Supplements
ods.od.nih.gov
6100 Executive Boulevard,
Room 3B01, MSC 7517
Bethesda, MD 20892
(301) 435-2920

Physical Activity Guidelines
for Americans
www.health.gov/PAGuidelines
P.O. Box 1133
Washington, DC 20013-1133
(301) 565-4167

Substance Abuse and Mental Health
Services Administration
www.samhsa.gov
P.O. Box 2345
Rockville, MD 20847-2345
(800) 487-4889

U.S. Department of Agriculture (USDA)
Food and Nutrition Service
www.usda.gov
14th Street and Independence
Avenue SW
Washington, DC 20250
(202) 720-2791

U.S. Department of Agriculture (USDA)
Nutrient Data Laboratory
www.nal.usda.gov/fnic/foodcomp/
search
Building 005, Room 105,
BARC-West
Beltsville, MD 20705

U.S. Department of Education (DOE)
www.ed.gov
400 Maryland Avenue SW
Washington, DC 20202
(800) 872-5327

U.S. Department of Health and Human
Services
www.hhs.gov
200 Independence Avenue SW
Washington, DC 20201
(877) 696-6775

U.S. Environmental Protection Agency
(EPA)
www.epa.gov
1200 Pennsylvania Avenue NW
Washington, DC 20460
(202) 272-0167

U.S. National Library of Medicine
www.nlm.nih.gov
8600 Rockville Pike
Bethesda, MD 20894

International Agencies

Canadian Council of Food and Nutrition
www.nin.ca
2810 Matheson Boulevard East,
1st Floor
Mississauga, Ontario, Canada
L4W 4X7
(905) 625-5746

Dietitians of Canada
www.dietitians.ca
480 University Avenue,
Suite 604
Toronto, Ontario, Canada M5G 1V2
(416) 596-0857

Food and Agriculture Organization of
the United Nations (FAO)
www.fao.org
Liaison Office for North America
2175 K Street, Suite 300
Washington, DC 20437
(202) 653-2400

Food Insight: International Food
Information Council Foundation
www.foodinsight.org
1100 Connecticut Avenue NW,
Suite 430
Washington, DC 20036
(202) 296-6540

Food Standards Agency
www.food.gov.uk
125 Kingsway
London WC2B 6NH
(020) 7276-8000

Health Canada
www.hc-sc.gc.ca
Address Locator 0900C2
Ottawa, Ontario, Canada
K1A OK9
(613) 957-2991

International Life Sciences Institute
www.ilsi.org
1156 Fifteenth Street NW, Suite 200
Washington, DC 20005
(202) 659-0074

UNICEF
www.unicef.org
3 United Nations Plaza
New York, NY 10017
(212) 326-7000

World Health Organization (WHO)
www.who.int/en
Regional Office
525 23rd Street NW
Washington, DC 20037
(202) 974-3000

Alcohol and Drug Abuse

Al-Anon Family Groups, Inc.
www.al-anon.alateen.org
1600 Corporate Landing Parkway
Virginia Beach, VA 23454-5617
(888) 425-2666

Alcoholics Anonymous (AA)
www.aa.org
P.O. Box 459
New York, NY 10163
(212) 870-3400

Narcotics Anonymous (NA)
www.na.org
P.O. Box 9999
Van Nuys, CA 91409
(818) 773-9999

National Council on Alcoholism and
Drug Dependence (NCADD)
www.ncadd.org
244 East 58th Street, 4th Floor
New York, NY 10022
(212) 269-7797

Weight Control and Eating Disorders

National Association of Anorexia Nervosa
and Associated Disorders (ANAD)
www.anad.org
P.O. Box 640
Naperville, IL 60566
(630) 577-1330

Obesity Society
www.obesity.org
8757 Georgia Avenue, Suite 1320
Silver Spring, MD 20910
(301) 563-6526

Overeaters Anonymous (OA)
www.oa.org
6075 Zenith Court NE
Rio Rancho, NM 87144-6424
(505) 891-2664

TOPS Club Inc. (Take Off Pounds Sensibly)
www.tops.org
4575 South Fifth Street
P.O. Box 070360
Milwaukee, WI 53207-0360
(414) 482-4620

Weight Watchers, Inc.
www.weightwatchers.com
11 Madison Avenue,
17th Floor
New York, NY 10010
(800) 651-6000

Fitness

American Alliance for Health, Physical
Education, Recreation & Dance
www.aahperd.org
1900 Association Dr.
Reston, VA 20191-1598
(800) 213-7193

American College of Sports Medicine
(ACSM)
www.acsm.org
401 West Michigan Street
Indianapolis, IN
46202-3233
(317) 637-9200

American Council on Exercise (ACE)
www.acefitness.org
4851 Paramount Drive
San Diego, CA 92123
(858) 576-6500

National Association for Health & Fitness
www.physicalfitness.org
65 Niagara Square,
Room 607
Buffalo, NY 14202
(716) 583-0521

National Institute of Sport and Fitness
www.nifs.org
250 University Boulevard
Indianapolis, IN 46202
(317) 274-3432

National Strength and Conditioning
Association (NSCA)
www.nsca-lift.org
1885 Bob Johnson Drive
Colorado Springs, CO 80906
(719) 632-6722

Food Production

Blue Ocean Institute
www.blueocean.org
250 Lawrence Hill Road
Cold Spring Harbor,
NY 11724
(631) 659-3746

Center for Science in the Public Interest
(CSPI)
www.cspinet.org
1220 L Street NW, Suite 300
Washington, DC 20005
(202) 332-9110

Center for Sustainable Systems
css.snre.umich.edu
University of Michigan
3012 Dana Building
440 Church Street
Ann Arbor, MI 48109-1041
(734) 764-1412

Community Food Security Coalition
www.foodsecurity.org
3830 SE Division Street
Portland, OR 97202
(503) 954-2970

Food Routes
www.foodroutes.org
439 Phinney Drive
Troy, PA 16947
(570) 673-3398

Institute for Agriculture and Trade Policy
www.iatp.org
1100 15th Street NW,
11th Floor
Washington, DC 20005
(202) 222-0749

Monterey Bay Aquarium
www.montereybayaquarium.org
886 Cannery Row
Monterey, CA 93940
(831) 648-4800

National Pesticide Information Center
npic.orst.edu
Oregon State University
333 Weniger Hall
Corvallis, OR 97331-6502
(800) 858-7378

NCAT Sustainable Agriculture Project
https://attra.ncat.org
P.O. Box 3657
Fayetteville, AR 72702
(800) 346-9140

Slow Food USA
www.slowfoodusa.org
20 Jay Street, Suite M04
Brooklyn, NY 11201
(718) 260-8000

Sustainable Table
www.sustainabletable.org
215 Lexington Avenue,
Suite 1001
New York, NY 10016
(212) 991-1930

United Fresh Produce Association
www.unitedfresh.org
1901 Pennsylvania Avenue NW,
Suite 1100
Washington, DC 20006
(202) 303-3400

USA Rice Federation
www.usarice.com
4301 North Fairfax Drive,
Suite 305
Arlington, VA 22203
(703) 351-8161

World Hunger

Bread for the World
www.bread.org
50 F Street NW, Suite 500
Washington, DC 20001
(800) 822-7323
(202) 639-9400

Freedom from Hunger
www.freedomfromhunger.org
P.O. Box 2000
1644 DaVinci Court
Davis, CA 95616
(530) 758-6241

World Hunger Year
www.worldhungeryear.org
505 Eighth Avenue,
Suite 2100
New York, NY 10018-6582
(212) 629-8850

Worldwatch Institute
www.worldwatch.org
1776 Massachusetts Avenue NW
Washington, DC 20036
(202) 452-1999

Nutrition-Related Journals

The American Journal of Clinical Nutrition
www.ajcn.org

CA: A Cancer Journal for Clinicians
caonline.amcancersoc.org

Cell
cell.com

Circulation
circ.ahajournals.org

Diabetes
diabetes.diabetesjournals.org

European Journal of Clinical Nutrition
www.nature.com/ejcn/index.html

Hypertension
hyper.ahajournals.org

International Journal of Behavioral
Nutrition and Physical Activity
www.ijbnpa.org/home

International Journal of Obesity
www.nature.com/ijo/index.html

Journal of the American College of
Nutrition
www.jacn.org

Journal of the American Dietetic
Association
www.adajournal.org

The Journal of the American Medical
Association
www.jama.org

The Journal of Nutrition
jn.nutrition.org

Journal of Nutrition Education and
Behavior
www.jneb.org

Morbidity and Mortality Weekly Report
www.cdc.gov/mmwr

Nature
www.nature.com/nature/index.html

The New England Journal of Medicine
www.nejm.org

Nutrition Journal
www.nutritionj.com/home

Nutrition Reviews
www.ilsi.org/Pages/
NutritionReviews.aspx

Nutrition Today
www.nutritiontodayonline.com

Obesity Journal
www.obesity.org/publications/
obesity-journal.htm

PubMed (National Library of Medicine
published abstract search engine)
www.ncbi.nlm.nih.gov/sites/
entrez?db=pubmed

Science
www.sciencemag.org

Scientific American
www.sciam.com

CDC Growth Charts

FIGURE F1 Infant boys, birth to 36 months, length-for-age and weight-for-age.

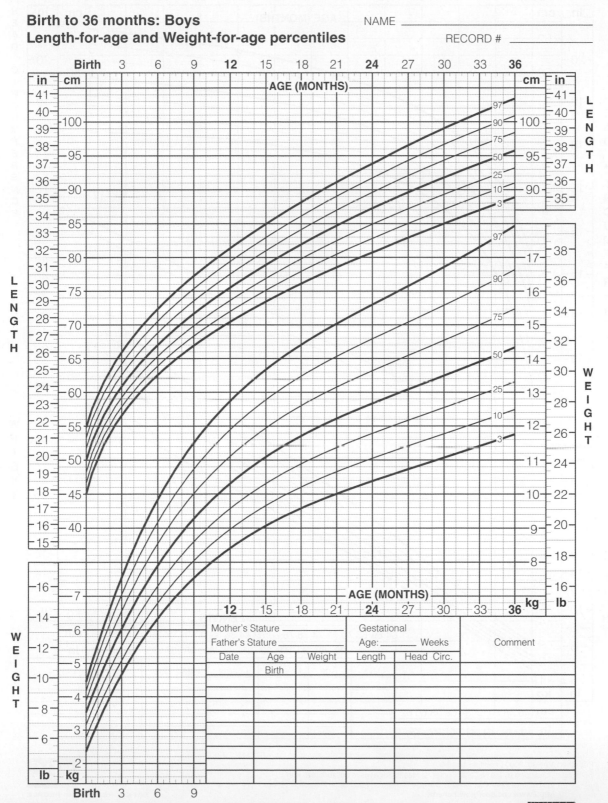

Birth to 36 months: Boys
Length-for-age and Weight-for-age percentiles

NAME _____

RECORD # _____

Published May 30, 2000 (modified 4/20/01).
SOURCE: Developed by the National Center for Health Statistics in collaboration with
the National Center for Chronic Disease Prevention and Health Promotion (2000).
http://www.cdc.gov/growthcharts

SAFER · HEALTHIER · PEOPLE™

FIGURE F2 Infant boys, birth to 36 months, head circumference-for-age and weight-for-length.

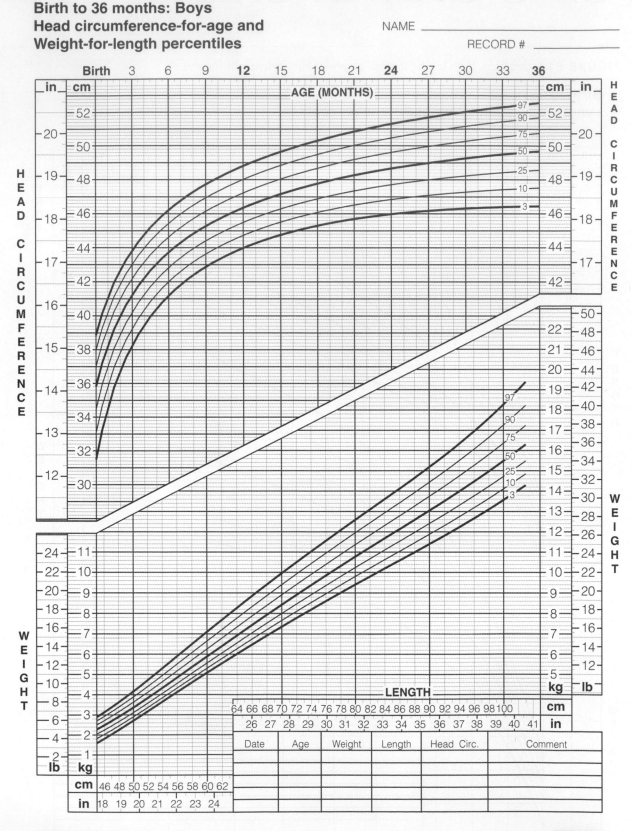

Birth to 36 months: Boys
Head circumference-for-age and
Weight-for-length percentiles

NAME _____

RECORD # _____

Published May 30, 2000 (modified 10/16/00).
SOURCE: Developed by the National Center for Health Statistics in collaboration with
the National Center for Chronic Disease Prevention and Health Promotion (2000).
http://www.cdc.gov/growthcharts

SAFER · HEALTHIER · PEOPLE™

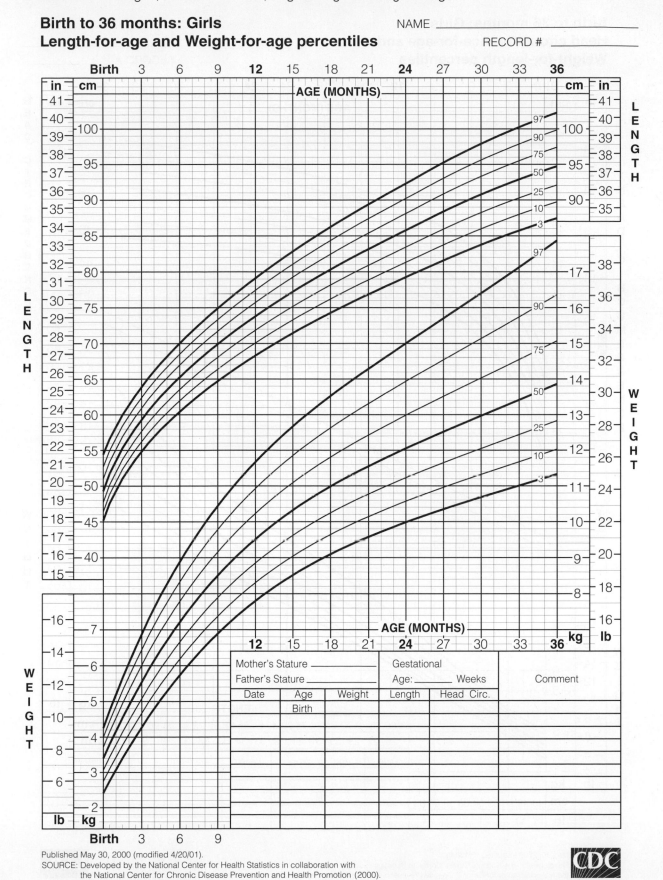

Birth to 36 months: Girls
Length-for-age and Weight-for-age percentiles

NAME _____

RECORD # _____

Published May 30, 2000 (modified 4/20/01).
SOURCE: Developed by the National Center for Health Statistics in collaboration with
the National Center for Chronic Disease Prevention and Health Promotion (2000).
http://www.cdc.gov/growthcharts

FIGURE F4 Infant girls, birth to 36 months, head circumference-for-age and weight-for-length.

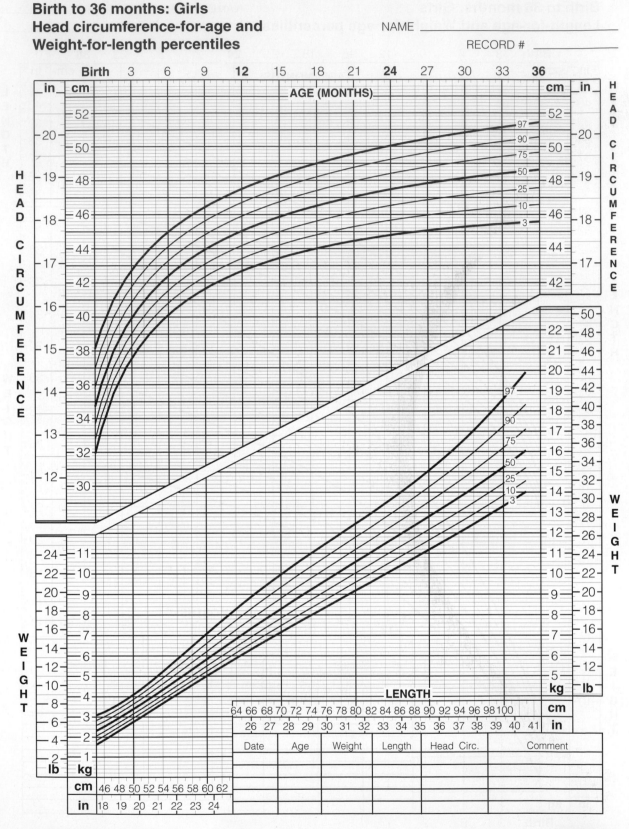

Birth to 36 months: Girls
Head circumference-for-age and
Weight-for-length percentiles

NAME _____

RECORD # _____

Published May 30, 2000 (modified 10/16/00).
SOURCE: Developed by the National Center for Health Statistics in collaboration with
the National Center for Chronic Disease Prevention and Health Promotion (2000).
http://www.cdc.gov/growthcharts

SAFER · HEALTHIER · PEOPLE™

FIGURE F5 Children and adolescent boys, 2 to 20 years, stature-for-age and weight-for-age.

2 to 20 years: Boys
Stature-for-age and Weight-for-age percentiles

NAME _____
RECORD # _____

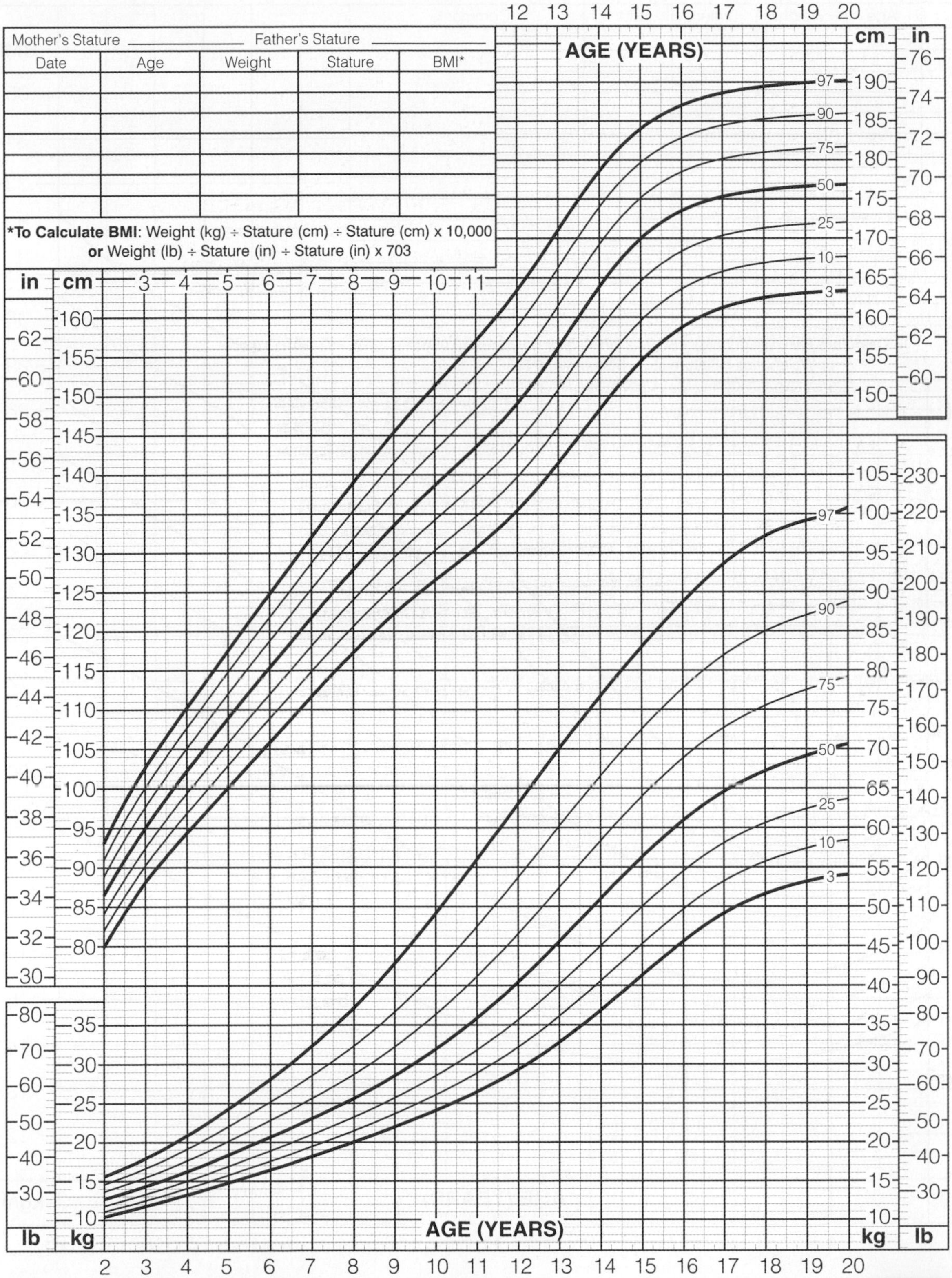

*To Calculate BMI: Weight (kg) ÷ Stature (cm) ÷ Stature (cm) x 10,000
or Weight (lb) ÷ Stature (in) ÷ Stature (in) x 703

Published May 30, 2000 (modified 11/21/00).
SOURCE: Developed by the National Center for Health Statistics in collaboration with
the National Center for Chronic Disease Prevention and Health Promotion (2000).
http://www.cdc.gov/growthcharts

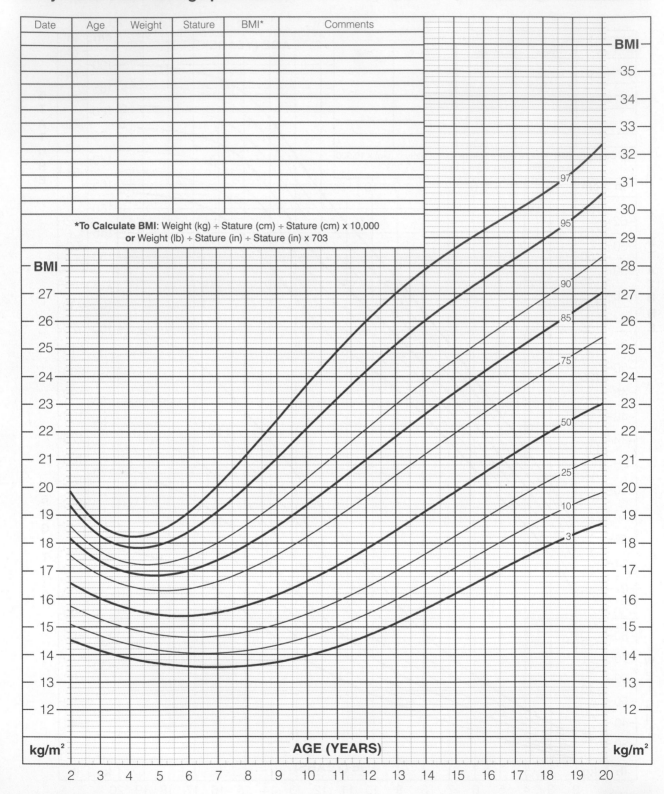

2 to 20 years: Boys
Body mass index-for-age percentiles

NAME _____

RECORD # _____

*To Calculate BMI: Weight (kg) ÷ Stature (cm) ÷ Stature (cm) x 10,000
or Weight (lb) ÷ Stature (in) ÷ Stature (in) x 703

Date	Age	Weight	Stature	BMI*	Comments

AGE (YEARS)

Published May 30, 2000 (modified 10/16/00).
SOURCE: Developed by the National Center for Health Statistics in collaboration with
the National Center for Chronic Disease Prevention and Health Promotion (2000).
http://www.cdc.gov/growthcharts

SAFER • HEALTHIER • PEOPLE™

FIGURE F7 Children and adolescent girls, 2 to 20 years, stature-for-age and weight-for-age.

2 to 20 years: Girls
Stature-for-age and Weight-for-age percentiles

NAME _____

RECORD # _____

Published May 30, 2000 (modified 11/21/00).
SOURCE: Developed by the National Center for Health Statistics in collaboration with
the National Center for Chronic Disease Prevention and Health Promotion (2000).
http://www.cdc.gov/growthcharts

CDC GROWTH CHARTS

493

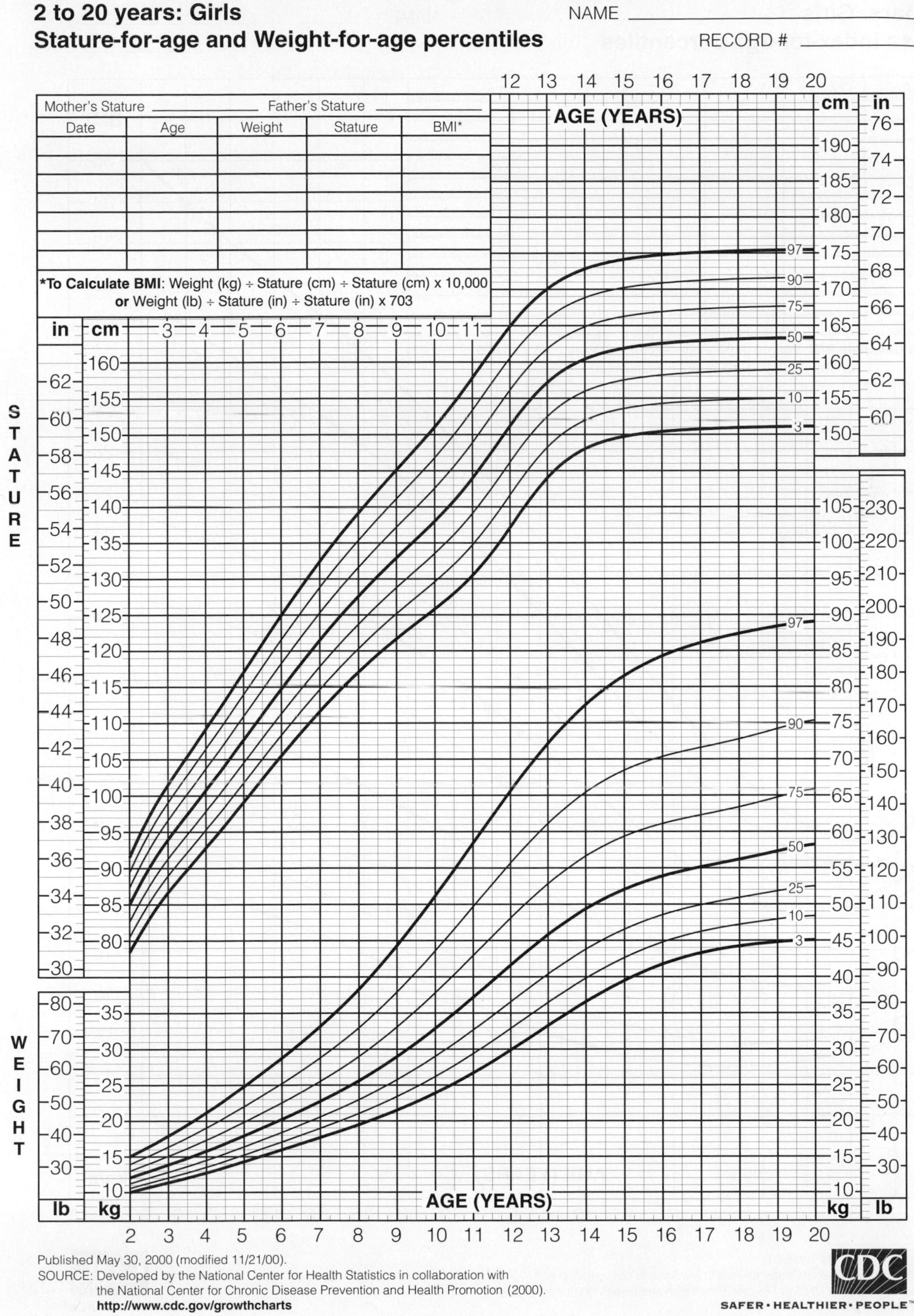

2 to 20 years: Girls
Body mass index-for-age percentiles

NAME _____

RECORD # _____

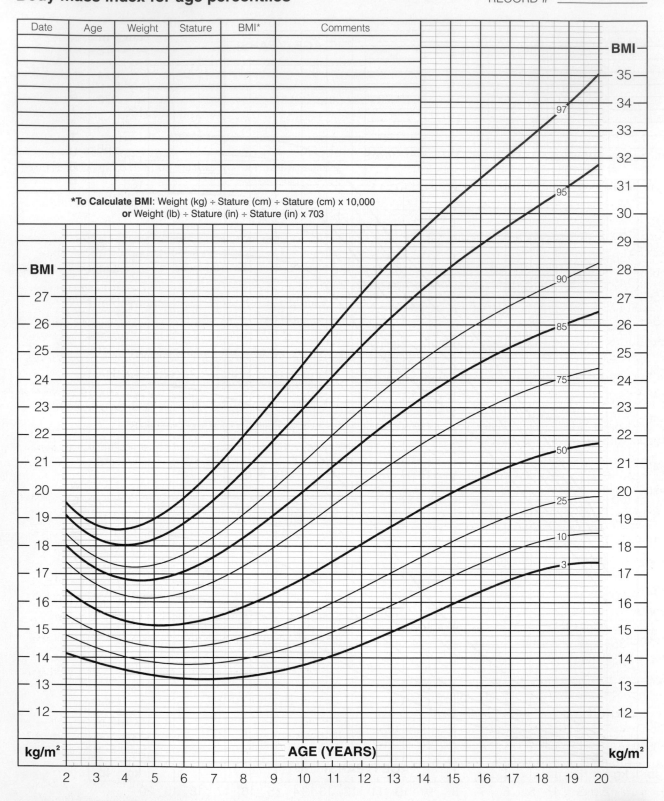

*To Calculate BMI: Weight (kg) ÷ Stature (cm) ÷ Stature (cm) x 10,000
or Weight (lb) ÷ Stature (in) ÷ Stature (in) x 703

Published May 30, 2000 (modified 10/16/00).
SOURCE: Developed by the National Center for Health Statistics in collaboration with
the National Center for Chronic Disease Prevention and Health Promotion (2000).
http://www.cdc.gov/growthcharts

SAFER • HEALTHIER • PEOPLE™

Glossary

1,25-dihydroxycholecalciferol The active form of vitamin D, also called 1,25-dihydroxyvitamin D₃ and calcitriol.

absorption The transport of nutrients from the intestinal lumen into small intestine cells, and then into either the blood or the lymph.

Acceptable Macronutrient Distribution Range (AMDR) A recommended range of Calories expressed as percents for carbohydrate, sugar, protein, fat, and essential fatty acid dietary intake.

accessory organs Relating to digestion, the liver, gallbladder, and pancreas.

acetylcholine A neurotransmitter made from the water-soluble vitamin choline.

achlorhydria Low or absent production of gastric acid (hydrochloric acid) in the stomach.

acrodermatitis enteropathica An inborn error in metabolism that results in zinc deficiency.

active transport A selective absorption mechanism in which nutrients cross cell membranes by binding to a specific carrier protein or receptor that requires ATP energy to cross the lipid bilayer.

acutane A prescription drug made from a form of vitamin A (retinoic acid) that is used to treat acne.

added sugars Simple sugars and syrups used as an ingredient in the preparation of processed foods or added to foods by an individual.

adenosine triphosphate (ATP) The ultimate form of energy and a short-term energy store generated by converting adenosine diphosphate (ADP) to ATP.

adequacy A diet that provides essential nutrients, fiber, and energy (Calories) in amounts adequate or sufficient to sustain life and maintain health.

Adequate Intake (AI) The average dietary amount of a nutrient that appears sufficient for health; used when an RDA cannot be determined.

aerobe A living organism that requires oxygen.

aerobic capacity The ability of the respiratory and cardiovascular systems to deliver oxygen to working muscles.

aerobic exercise Physical activity where the heart rate can be elevated and maintained steadily for at least 10 minutes and there is increased oxygen uptake and delivery to the body and muscles.

affiliation A partnership between two parties, such as an employee and employer; can also be the professional body of which you are a member.

alcohol A central nervous system depressant, consumable drug, and carbohydrate-related compound providing 7 Calories per gram.

algae A large and varied group of chlorophyll-containing photosynthetic organisms that are primarily aquatic and include seaweed and kelp.

allergen A substance that can cause allergy or an allergic response; also known as an *antigen*.

allergic foods Milk, eggs, fish, crustacean shellfish, tree nuts, peanuts, wheat, and soybeans are the most common foods that people have allergies to.

allergy An immune-mediated reaction, usually to a protein component in food.

alpha-linolenic acid An essential fatty acid with 18 carbons and three double bonds, with the first double bond occurring at the third carbon from the methyl end, making it an omega-3 fatty acid.

alpha-tocopherol A form of vitamin E most commonly used in dietary supplements and food fortification.

Alzheimer's disease Pre-senile dementia that starts with memory loss and leads to complete helplessness.

amenorrhea Absence of the normal menstrual cycle for three months, or no period within the last six months.

amino acids Organic and nitrogen containing compounds that can be essential and nonessential; there are 20 used to make proteins.

ammonification Conversion of ammonia to ammonium salts.

amniocentesis A prenatal diagnostic procedure usually done between 16 and 20 weeks of a pregnancy to test the genetic traits of fetal cells in the amniotic fluid.

amylase An enzyme that hydrolyzes the starch amylose (cleaves the alpha bonds between the glucose molecules).

anabolic When small molecules are put together to build larger ones through metabolic reactions; requires a condensation chemical reaction.

anabolism When small molecules are put together to build larger ones through metabolic condensation reactions (water is released).

anaerobe A living organism that does not require oxygen.

anaphylaxis A sudden, life-threatening, systemic or multisystem hypersensitivity allergic reaction.

andropause A decline in male sex hormone levels, resulting in physiological and psychological changes and reduced sexuality and overall energy; also known as male menopause.

anemia An inability for the red blood cells to deliver oxygen to body cells, causing the individual to feel tired, weak, breathless, apathetic, and often experience headache; presents in many types, such as macrocytic, microcytic, and hemolytic.

antagonistic muscles A pair of opposing muscles positioned adjacent to an articulating joint where, when one set of muscles contracts, the angle of the joint decreases, and when the other set contracts, the angle of the joint increases.

antagonists A chemical substance that counteracts the effects of another chemical substance.

antibiotics A chemical substance that is administered as a drug to kill bacteria.

antibodies Proteins that are produced by the immune system in response to antigens, which are recognized as foreign substances that invade the body.

antigen A substance that elicits the formation of antibodies by the immune system.

antioxidant A chemical compound that can donate an electron without becoming chemically reactive itself, and thus can inhibit oxidation and reduce the damage that electron-deficient chemicals cause.

appetite The psychological desire for food; affected by sensory inputs like seeing, smelling, or thinking about food.

aqueous Of or pertaining to water and water-soluble body secretions.

arachidonic acid An omega-6 long-chained fatty acid that is 20 carbons long with four double bonds and can be made in the body from linoleic acid.

ariboflavinosis The condition caused by a deficiency of the water-soluble vitamin riboflavin.

arteries The largest blood vessels that carry blood from the heart to the tissues.

arthritis Breakdown of the cartilage surface in joints that leads to inflammation and pain.

artificial sweeteners Artificial chemicals such as aspartame, saccharin, sucralose, and Aceulfame-K that have been approved by the Food and Drug Administration to replace sugar in processed foods.

atherosclerosis A disease of the arteries that is characterized by the accumulation of lipid-containing material called *plaque* on the inner walls of the arteries, particularly the coronary and cranial arteries.

autocrine A form of cell signaling in which a cell secretes a hormone or chemical messenger that binds to receptors on the same cell and alters cell function.

autoimmune response When the immune system targets self and destroys the cells, tissues, organs, and systems within the body.

autonomic nervous system The part of the central nervous system that regulates involuntary vital functions.

autotrophs Organisms that produce their own food using light or heat energy.

bacteria A large group of single-celled organisms of various shapes, lacking a nucleus, that are everywhere on earth, important for nutrient cycling, and abundant on and in the human body.

balance Foods and nutrients provided in the diet in proportion to each other and the body's needs.

Basal Metabolic Rate (BMR) The number of Calories expended in a 24-hour period to sustain vital function under basal conditions.

behavior modification Techniques used to promote desired behavior changes, where thoughts and activities can be changed to achieve personal goals.

beriberi The condition caused by a deficiency of the water-soluble vitamin thiamin; can present in a dry or wet form.

beta-carotene A phytochemical (plant-derived, health-promoting food component) that is a precursor or a "provitamin" for vitamin A.

bile An emulsifier that is synthesized by the liver, stored in the gallbladder, and released into the small intestine with fat consumption. It then enables fat-soluble substances to integrate into water for digestion by lipase enzymes.

bioaccumulate The buildup (increased concentration) of substances in living tissues that accumulate up food chains and within food webs.

biodiversity The genetic variation among all life forms on earth.

bioidentical hormones Hormones that have the same chemical structure as those made by the body.

biological classification The scientific organization of living organisms on earth into the categories kingdom, phylum, class, order, family, genus, and species.

biomolecule Organic, biological molecules produced by a living organism that make up cells, tissues, organs, and systems and include proteins, carbohydrates, lipids, and nucleic acids.

biotin A water-soluble essential vitamin that functions in carboxylation reactions, and is present in food in a free form or bound to protein.

birth defects Permanent abnormalities in an infant's structure or function detected at birth or shortly thereafter.

blind study A placebo-controlled study design where the subjects do not know if they are taking a placebo or the treatment, but the investigator does know.

blood lipids A variety of carrier lipoproteins in blood, including high-density lipoproteins (HDLs), low-density lipoproteins (LDLs), very-low-density lipoproteins (VLDLs), and chylomicrons.

body composition The percentages of fat mass, mineral mass, and lean body mass contributing to the total body weight of a person.

body mass index (BMI) A calculated number based on weight and height, used to interpret the healthiness of body weight.

bolus A chewed, softened mass of ingested food that is swallowed and propelled through the esophagus to the stomach.

bone mineralization The process in which minerals such as calcium, phosphorus, magnesium, and fluoride crystallize on the collagen matrix of bone.

buffer A water-based solution that resists pH change.

cadmium A heavy metal causing slow and irreversible damage to the kidneys and liver.

caesarean An unnatural, surgical procedure used to deliver a baby by making an incision through the abdomen and wall of the uterus; also known as a C-section.

calcium Ninety-nine percent of this major mineral is found in bone and teeth, but it is also important for blood clotting, muscle contraction, and nerve conduction.

calorically dense An energy-producing substance that contains a lot of energy per gram weight, such as triglycerides (9 Calories per gram) and alcohol (7 Calories per gram), compared to starches, sugars, and proteins (4 Calories per gram).

Calorie The unit used to measure energy in food; technically a kilocalorie (Kcal).

Calorie control Managing the intake of food energy in the diet to achieve a healthy body weight.

cancer A disease in which cells become engaged in uncontrolled cellular division and growth through the processes of initiation, promotion, and progression.

carbohydrates Organic compounds composed of carbon, oxygen, and hydrogen that provide the preferred fuel of the body; categorized as simple and complex; many provide 4 Calories per gram (except fibers, which are indigestible and thus noncaloric).

cardiomyopathy A medical condition where the contractility of the heart muscle is weakened, and thus the pumping action of the heart muscle is inadequate to supply oxygen to the cells.

cardiorespiratory systems Systems of the body, including the heart and lungs and associated blood vessels.

cardiovascular disease A variety of diseases that affect the heart and blood vessels.

cardiovascular system Of or pertaining to the heart and blood vessels or the circulatory system of the body.

carnitine A vitamin-like compound that helps long-chain fatty acids get into the mitochondria for the generation of ATP energy.

casein The dominant type of protein in mammalian milk.

case studies Noncontrolled medical investigation that may be published in medical literature to provide suggestive, scientific information.

catabolic When large molecules are broken apart to yield smaller ones through metabolic reactions; requires a hydrolysis chemical reaction.

catabolism When large molecules are broken apart to yield smaller ones through hydrolysis reactions, which require water.

catalyze To facilitate a biochemical reaction to occur.

cataracts The scar tissue that forms over the cornea as a result of eye surface damage.

celiac disease An inheritable autoimmune disease that damages the villi in the small intestine when the person eats wheat, barley, or rye.

cell The smallest structural and functional unit of life in all known living organisms.

central adiposity A waist circumference greater than 40 inches in men and greater than 35 inches in women.

cerebral spinal fluid Extracellular fluid that circulates around the spinal cord.

certification insignia A marker of quality for displaying product assurance or guaranteed potency.

chemical poisons Toxic substances like lead, mercury, and pesticides that have the potential to kill.

chloride A major mineral that is part of salt, but also used to make HCl (hydrochloric acid).

chlorophyll Large, magnesium-containing molecules that can be seen as green pigments found in life forms capable of photosynthesis.

cholecalciferol Vitamin D_3, which is the form of vitamin D that is used to fortify food and is added to dietary supplements.

cholecystokinin (CCK) A hormone produced by cells of the duodenum that signals the pancreas to release pancreatic juice, the gallbladder to release bile, and satiety in the brain.

choline A water-soluble vitamin that is used to make the neurotransmitter acetylcholine.

chromium A trace mineral that potentiates the action of insulin and may improve the glucose tolerance test.

chromosome An organized, threadlike strand of DNA in the cell nucleus where an individual's genes are encoded; the hereditary information can be transmitted from one generation to the next.

chronic diseases Diseases, such as heart disease, cancer, diabetes, and osteoporosis, that progress slowly and have a long duration.

chylomicrons A post-prandial (after eating), temporary lipoprotein that transports lipids from intestinal cells to the liver for repackaging into the stable lipoprotein fractions: HDL, LDL, IDL, and VLDL.

chyme A liquefied, partly digested substance that is made and released from the stomach into the small intestine.

circadian rhythm A daily cycle of biochemical, physiological, and/or behavioral processes of living organisms, greatly influenced by the exposure to 24-hour cycles of light and dark.

cis fatty acids The natural configuration of unsaturated fatty acids, in which the hydrogen atoms are on the same side of the double bond in the carbon chain.

citric acid cycle A series of oxygen-dependent enzymatic reactions taking place inside mitochondria, in which covalent bonds are broken and the energy is captured to produce ATP; also known as the Krebs cycle or the tricarboxylic acid cycle.

clinical trials Controlled medical investigation where statistical analyses can be applied to determine the significance of the study results.

clonal expansion Asexual reproduction of one organism, producing a population of genetically equivalent organisms.

cobalamin The chemical form of vitamin B_{12}, which contains cobalt.

cobalt A trace mineral that is essential for vitamin B_{12} structure; cobalamin.

coenzyme A substance, such as a vitamin or mineral, that participates in metabolism as a structural part of an enzyme; also called a cofactor.

cognition The mental processes of thought, perception, reasoning, and learning.

colic A broad term used to describe abdominal pain or discomfort in infancy.

colostrum The first breast secretions from a lactating mother, which not only provides nourishment to the infant but also contains antibodies that protect the newborn from disease.

complementation Combining protein-containing plant foods so that all the essential amino acids are present with the food combination; examples include rice and black beans, whole-wheat bread and peanut butter, and soy milk and cereal.

complete protein A high-quality or high biological value protein that is from the animal kingdom and provides all the essential amino acids.

complex carbohydrates Polysaccharides composed of straight or branched chains of monosaccharides.

concentric When the muscle shortens and it is contracting, usually when it is opposing resistance.

conception The moment in time when an egg (ova) is fertilized by a sperm.

condensation reaction An anabolic chemical reaction producing water and a larger molecule.

control group A group in a study that is not treated or is given a placebo and is used as a comparison to the experimentally treated group.

conventional food production A food production system used in many developed nations that employs the use of pesticides, synthetic fertilizers, antibiotics, hormones, and/or genetically modified organisms—all used to increase yields.

copper A trace mineral; Menkes disease is associated with a deficiency and Wilson's disease is associated with a toxicity.

core body temperature The temperature of the body at the deep central part.

credentials The qualification or competence issued to an individual by a third party, such as a degree title earned by a graduate of a university or a national credentialing association.

cretinism Mental retardation at birth, which can be caused by an iodine deficiency.

critical period A crucial stage in human development when tissues change and organs grow and develop rapidly.

cross-hybridization Genetically mixing different species to produce hybrids; cross-breeding.

cross-over study A study design where the subjects are tested against themselves, because every participant will be studied on the placebo and on the treatment.

cystic fibrosis An inherited chronic disease that causes the body to produce thick mucus that obstructs the lungs and pancreas, leading to respiratory and digestive difficulties.

Daily Reference Values (DRVs) Daily Values for fat, saturated fat, cholesterol, carbohydrate, fiber, protein, sodium, and potassium that are based on a 2,000-Calorie diet and expressed as percents (except protein, sugar, and potassium) in the Nutrition Facts panel on food package labels.

Daily Values (DVs) Reference values including Daily Reference Values (DRVs) and Reference Daily Intakes (RDIs) used on food labels.

danger zone A range of food storage temperature from 40 to 140 degrees Fahrenheit where microorganisms can replicate quickly enough to cause foodborne illness.

deamination The removal of a nitrogen group from an amino acid by a vitamin B_6 coenzyme.

deficient Nutrient intake amount below the level recommended for a healthy person to prevent deficiency-related signs, symptoms, and diseases.

dehydration A reduction of total body water when thermal regulation, cardiovascular function, and electrolyte balance are all impaired.

dehydroascorbic acid A reduced form of vitamin C, or L-ascorbic acid, that is commonly caused by oxidation but can be restored by gluthione peroxidase.

dementia The deterioration of cognitive function.

denaturation The change in shape of proteins and loss of their function when exposed to heat, acids, bases, and/or heavy metals.

dental caries Tooth decay, cavities.

deoxyribonucleic acid (DNA) A linear, double-stranded polymer of the nucleic acids (nucleotides) adenine, guanine, cytosine, and thymine, found in the cell's nucleus, that makes up the cell's transmittable genetic information and encodes bioactive molecules.

diabetes A chronic disorder of carbohydrate metabolism, typically from insufficient or ineffective insulin.

diet The kind and amount of foods and beverages a person eats and drinks.

dietary fiber The nonstarch polysaccharides in plant foods that are not digested by human digestive enzymes, although some are digested by bacteria in the gastrointestinal tract.

dietary guidelines Modern society's instructional messages for reducing the risks for diet-related diseases, including U.S. Dietary Guidelines 2010; American Heart Association guidelines; American Cancer Society guidelines; and Healthy People 2020.

Dietary Guidelines for Americans Science-based advice for Americans to promote health and to reduce risk for major chronic diseases through diet and physical activity; updated every five years by the Department of Health and Human Services in conjunction with the U.S. Department of Agriculture.

Dietary Reference Intake (DRI) A recommended intake value for an essential nutrient, fiber, and Calories; also exists for physical activity.

dietary supplement Processed food component(s) from any of the following categories: essential or nonessential mineral, essential or nonessential vitamin, herbal, glandular, enzyme, nutritional substance, or fiber.

differentiation A biological process that causes a less specialized cell to become a different and more specialized cell, like when a fertilized egg engages in embryonic development, which precedes fetal development.

digestion The processes by which food and its components are broken down chemically and mechanically into units that can be absorbed.

digestive system Collectively, the organs and processes associated with the ingestion, digestion, absorption, and excretion of food.

disaccharide Double units of sugar including sucrose, maltose, and lactose; two monosaccharides chemically bound together.

diverticular disease Diverticulosis, a condition where diverticula (out-pocketing) are present in the colon as a result of weak colon wall structures and increased pressure against them; when the diverticula are inflamed, it is termed *diverticulitis*.

docosahexaenoic acid (DHA) An omega-3 polyunsaturated fatty acid with 22 carbons and six double bonds with health-promoting properties, present in fish and synthesized in limited amounts in the body from the essential omega-3 fatty acid alpha-linolenic acid (18 carbons with three double bonds).

double-blind, cross-over study A placebo-controlled study design where the subjects do not know if they are taking a placebo or the treatment, nor does the investigator know who is receiving the placebo, but every participant will be studied on the placebo and on the treatment.

double-blind study A placebo-controlled study design where the subjects do not know if they are taking a placebo or the treatment, nor does the investigator know who is receiving the placebo.

Down syndrome A birth defect that is characterized by a variety of mental and physical signs and symptoms that exist throughout life, resulting from an extra copy of chromosome 21.

duodenum The first segment of the small intestine, where most of the digestion and absorption of nutrients from chyme takes place.

dyslipidemia Any combination of abnormal circulating lipoprotein fractions of hypercholesterolemia, high LDL cholesterol, low HDL cholesterol, and hypertriglyceridemia.

dysphagia Difficulty swallowing.

eccentric When the muscle is lengthening while it is contracting, and when it is opposing resistance; it is known as a negative lift.

ecology The interrelationship of living organisms and the environment.

ecosystem The interaction of living organisms with each other and their environmental surroundings.

edema Fluid retention by body tissues as a result of protein deficiency (kwashiorkor) and other conditions that cause excessive amounts of fluid to be retained in interstitial (in between cells) spaces.

editorial board A group of individuals who are responsible for the quality and content of a frequent publication.

egg A mature female reproductive cell, also termed *ovum* or *ova*, or female gametes that normally contain 23 chromosomes.

eicosanoids Biologically active compounds, such as prostaglandins, derived from long-chain polyunsaturated fatty acids that help to regulate blood pressure, blood clotting, blood lipids, inflammation, and other body functions.

eicosapentaenoic acid (EPA) An omega-3 polyunsaturated fatty acid with 20 carbons and five double bonds with health-promoting properties, present in fish and synthesized in limited amounts in the body from the essential omega-3 fatty acid alpha-linolenic acid (18 carbons with three double bonds).

electrolyte balance The equilibrium or normal ratio of charged particles in a solution that function as electrolytes (such as sodium, potassium, and chloride) in the body.

electrolytes Ions such as sodium, potassium, and chloride that govern fluid balance across semipermeable membranes (between the inside and outside of cells).

electron A negatively-charged particle in a molecular structure.

electron transport system A membrane-bound system of proteins in the mitochondria of cells that couples an electron donor and an electron acceptor to create a membrane potential that generates ATP energy.

embryo The product or stage after the fertilization of the egg when rapid cellular differentiation occurs; precedes the fetal stage of development.

empty Calorie allowance The Calories allotted for solid fat and added sugars in a person's energy allowance after consuming enough nutrient-dense foods from the MyPlate food groups to meet all nutrient needs for a day.

empty Calories Calories provided by solid fat and added sugars, neither of which are health promoting; an allowance is given based on Calorie need.

emulsifier A substance with aqueous and lipid affinities (water-soluble and fat-soluble chemical attractions), such as bile and lecithin, that promotes the formation of a stable mixture, or emulsion, of oil and water.

emulsifying agent (or emulsifier) A substance that associates water-soluble and fat-soluble substances such as water and oil together.

endocrine A type of gland that secretes hormones that are transmitted by the blood and act on cells of a different tissue.

endocrine disruptors Certain exogenous, organic chemical compounds that, when taken into the body, alter or disrupt the hormonal balance or homeostasis.

endogenous From inside the body.

endorphins Opioid protein compounds that are endogenously produced in the brain and produce an intense sense of well-being.

energy The ability to do work (chemical, mechanical, or osmotic).

energy balance The consumption of the same number of Calories as expended; Calories eaten equals Calories burned; isocaloric; body weight is maintained.

energy expenditure The number of Calories used by the body in a 24-hour period in the categories of Basal Metabolic Rate, Specific Dynamic Action of food, and physical activity.

energy-producing Substances that provide Calories or energy, including digestible carbohydrates (4 Calories per gram), proteins (4 Calories per gram), fats (4 Calories per gram), and alcohol (7 Calories per gram).

enrichment The term associated with the process of adding iron, thiamin, riboflavin, niacin, and folic acid (but not vitamin B_6, magnesium, zinc, or fiber) back to processed grains at approximate levels of the original whole grain.

enterohemorrhagic colitis Intestinal bleeding causing bloody diarrhea.

enterotoxin A protein toxin that can be released by a microorganism and affects the intestine.

enzyme A protein synthesized by cells to catalyze, or facilitate, a specific chemical reaction involving other substances without itself being altered.

epidemiology Retrospective statistical investigation of populations to provide supportive data and ideas for further research.

epileptics Individuals that have chronic neurological disorders characterized by a variety of recurrent, unprovoked seizures.

epiphyseal plate The tissue near each end of each long bone, where growth takes place; it closes by turning to bone itself when growth terminates.

epithelial A type of cell that covers the cavities, surfaces, and structures of the body, notably the respiratory tract, skin, and the lining of the GI tract.

ergogenic Any agent that, when applied, improves the capacity to exercise when tested against a placebo.

esophagus A muscular tube that carries the bolus and fluids from the mouth to the stomach.

essential amino acids Amino acids that are needed by the body to build proteins but cannot be synthesized endogenously; thus they are needed from food sources.

essential fat mass The percentage of total body weight from fat that is minimally required for structure and function, also known as essential body fat.

essential fatty acids (EFAs) Linoleic acid and alpha-linolenic acid, which are needed by the body but are not made by the body in amounts sufficient to meet physiological needs.

essential nutrients Substances that are found in food and are needed by the body, but are not made by the body in amounts sufficient to meet physiological needs.

Estimated Average Requirement (EAR) The average dietary amount of a nutrient that will maintain adequate function in half of the healthy people of a given age and gender group.

Estimated Energy Requirement (EER) Calorie level that is calculated and intended to maintain energy balance and good health in a person of a given age, gender, weight, height, and level of physical activity; also called the DRI for Calories.

estrogen A group of steroid compounds, found in both males and females, but named for its importance in the estrous cycle; functions as the primary female sex hormone.

evolution The change in heritable genetic composition of a population, such as by gene mutation and as a result of natural selection, which leads to adaptation over successive generations.

Exchange Lists A diet-planning tool created by the American Diabetes Association and Academy of Nutrition and Dietetics that organizes foods by their proportions of carbohydrate, fat, and protein (and thus Calories); there are nine lists: Starch, Nonstarchy Vegetables, Fruit, Milk, Meat and Meat Alternatives, Fat, Free Foods, Combination Foods, and Other Carbohydrates, that are used to plan diets.

excretion The elimination of waste from cellular metabolism from the body by respiration (breathing), skin secretions (sweating), and urinary output controlled by the kidneys, in addition to the elimination of solid waste from the GI tract (feces).

exogenous From outside the body.

extracellular Outside cells.

extracellular fluid (ECF) The fluid that is outside of cells.

extra lean Five grams of fat or less, 2 grams of saturated and *trans* fat or less, and 95 milligrams or less of cholesterol per 3.5-ounce serving of meat.

extrusion reflex An innate reflect in infants causing them to push food out of their mouths; diminishes between four to six months of age; also known as the tongue-thrust reflex.

facilitated diffusion A selective absorption mechanism in which nutrients cross cell membranes by binding to a specific carrier protein or receptor to cross the lipid bilayer; facilitated by a concentration gradient that creates osmotic pressure from the forces of migration from an area of high concentration to a lower one.

failure to thrive A medical term describing poor weight gain and physical growth failure in which an infant or child experiences a weight-for-age percentile ranking less than two standard deviations within a six-month period; for example, a drop from the 50th to the 10th percentile weight-for-age within six months.

fake fat An artificial chemical compound (such as olestra) used in processed food that is indigestible and noncaloric, but provides the properties of fat.

fasting The voluntary lack of food and beverage intake, promoting the catabolism of lean and fat body masses for energy and the depletion of stored nutrients.

fat mass The pounds of fat estimated from the percentage of total body weight from fat; includes essential fat mass and stored body fat.

fats Triglycerides or dietary fat.

fat-soluble nutrients The nutrients that are soluble in oil.

fat substitutes A wide variety of chemical compounds (such as starches, gums, gels, and proteins) used in processed food to reduce the amount of fat and Calories in the food.

fauna All of the animal life in a region, time period, or environment.

feasting The continuation of food intake beyond satiety, promoting fat synthesis.

feces Solid waste that is compacted in the large intestine (colon) and then released from the anus; composed of bacteria, fiber, sloughed-off intestinal epithelial cells, undigested food, and small amounts of GI tract juices.

ferritin An iron-containing protein that is used to evaluate total body iron stores.

fertility The quality and ability of the natural capability to produce offspring.

fetal alcohol syndrome (FAS) A group of physical and mental problems in children born to mothers who drank alcohol heavily during their pregnancy; best diagnosed from ages 3 to 8.

fetus A term for the life form in the womb after the embryonic period.

fiber Plant polysaccharides, such as cellulose, that are composed predominately of repeating units of glucose hooked together by beta bonds, and are indigestible by humans.

fibrocystic breast disease The presence of lumpy cysts in the breast that may be treated with vitamin E.

fibrous protein Proteins that are uniform in their structure; either in a helical structure or a pleated sheet.

fission The splitting of one unicellular organism to become two.

flavin adenine dinucleotide (FAD) A coenzyme made from riboflavin that carries two electrons.

flavin mononucleotide (FMN) A coenzyme made from riboflavin that carries one electron.

flora The collective bacteria and other microorganisms in an ecosystem, in a host, or the gastrointestinal tract of the host.

fluorhydroxyapatite The mineralized protein matrix of bone where the phosphorous has been replaced because of the presence of fluoride.

fluoride A trace mineral that functions to strengthen tooth enamel, resist dental caries, and stabilize bone.

fluorosis Tooth discoloration caused by too much exposure to fluoride.

folate A generic term used to describe a collection of chemical compounds found in food that can be converted to folic acid in the body and function in the formation of nucleic acids; deficiency results in macrocytic normochromic anemia.

folic acid The form of a water-soluble vitamin (pteroylmonoglutamic acid) that synergizes with vitamin B_{12} and has been associated with reducing the risk of spina bifida.

follicle-stimulating hormone (FSH) One of the four important hormones of the menstrual cycle that stimulates egg development, surges at ovulation, and helps release eggs from the ovaries.

food Any ingestible substance that nourishes the body.

food additives Chemicals that the Food and Drug Administration has approved to be added to processed foods from the following categories: antimicrobial agents, antioxidants, nutrients, stabilizers and thickening agents, artificial colors and flavors, bleaching agents, emulsifiers, or chelating agents.

foodborne illness Sickness caused by the consumption of food contaminated by pathogenic microorganisms.

food chains The linking of plants and animals by their food relationships and transfer of food energy.

food composition The chemical composition of nutrients in foods, including carbohydrate, starch, fiber, sugars, fat, fatty acids, protein, amino acids, vitamins, minerals, water, and other bioactive substances such as phytochemicals, alcohol, and caffeine.

food intolerance An adverse reaction to a food or food component that is not mediated by the immune system.

food label The Nutrition Facts panel, which provides information according to law about the manufacturer, nutrients, ingredients, terms, health claims, and allergic foods in that item.

food webs A network of interacting food chains within an ecosystem.

for-profit An organization in business to make money.

free Negligible amounts of fat, cholesterol, sodium, sugar, or Calories per serving in a food product.

fructose A monosaccharide, sometimes called fruit sugar.

functional fiber Indigestible dietary components that have been isolated from natural sources or synthetically made, and have beneficial physiological effects in humans.

functional foods Foods that may have additional health-promoting properties beyond providing basic nutrition needs (Calories and essential nutrients); may also be called medicinal foods.

fungi A taxonomic kingdom that includes yeast, molds, mushrooms, and other species similar to plants but lacking chlorophyll.

galactose A monosaccharide that is part of the disaccharide lactose.

galactosemia An inherited condition screened at birth causing the inability to break down galactose, resulting in many problems, including liver damage, cataracts, mental retardation, and possible death.

gallbladder An accessory organ of digestion that stores bile produced by the liver.

gametes Mature sexual reproductive cells (sperm and egg) that have a single set of unpaired chromosomes (half of the genetic material) necessary to form a complete human organism.

gaseous Of or pertaining to gas; a substance in vapor form.

gastric acid An acidic secretion containing hydrochloric acid produced by the cells of the stomach.

gastric emptying time The amount of time it takes chyme to pass through the stomach.

gastrin A hormone secreted by cells of the stomach that induces the production and secretion of gastric juice.

gastrointestinal tract The long, muscular tube extending from the mouth to the anus.

gene expression The transcription of the nucleotide bases making up the genetic code of a gene into RNA and then the translation into a functional protein.

genetically modified organisms (GMOs) Portions of DNA from one living organism are introduced into another living organism (transgenic), or the expression of a gene is blocked (antisense); also called genetic engineering or biotechnology.

genetics The study of an organism's genetically transmitted heredity (familial or inherited) defined by deoxyribonucleic acid (DNA), the gene sequences, or genetic material.

geriatrics The medical study of the physical, social, behavioral, and psychological diseases and issues of older adults.

Germany's Commission E The governmental regulatory agency in Europe that controls the production, distribution, and use of herbal medicines.

gestation The period from conception to birth, usually noted in weeks when a woman carries a developing fetus in her womb.

gestational diabetes A transient type of diabetes that occurs in some pregnant women who did not have preexisting diabetes.

globular protein Proteins with inconsistent and loosely interacting protein strands in their structure; the protein conformation can be part helical structure, part pleated sheet, part random, or completely random.

glucagon A protein hormone that is secreted by pancreatic alpha cells in response to low blood glucose levels, stimulates glycogen breakdown in the liver, and causes increases in blood glucose levels from the release of glucose into the blood from the liver glycogen stores.

glucose A monosaccharide, sometimes called blood sugar.

glutathione peroxidase A selenium-dependent liver enzyme.

glycemic index A number assigned to a food based on the blood-glucose-raising potential of the given food compared to glucose; has a standard of 100.

glycemic response The rise and fall of blood glucose levels after the consumption of a food.

glycerol backbone The three-carbon sugar alcohol that forms the three sites for the three fatty acids in a triglyceride to attach to (in an esterification process).

glycogen A glucose-based storage molecule, sometimes called animal starch, that is synthesized and stored in the liver and muscle tissues.

glycolysis The energy pathway where glucose is either broken down anaerobically (without oxygen) to pyruvate to release ATP for the body, or aerobically (with oxygen) to acetyl CoA and to enter the citric acid cycle.

goiter An enlargement of the thyroid gland caused by an iodine deficiency or toxicity, malfunction of the gland, or overconsumption of goitrogens.

governmental agency A not-for-profit, public sector agency that is supported by tax dollars—"by the people, for the people."

growth spurt A brief period of accelerated growth, noted by rapid increases in weight-for-age and height-for-age.

gustation The sense of taste or the ability to taste.

gut-associated lymphoid tissue A significant portion of the immune system (40 percent) that is largely integrated into the intestines of the digestive system.

health Complete physical, mental, and social well-being; not just the absence of infirmity.

health claims Statements approved by the Food and Drug Administration (FDA) linking the nutrition profile of a food to a reduced risk of a particular disease or health-related condition.

healthy flora The collective bacteria (such as *lactobacillus* and *bifidobacterium*) that colonize the gastrointestinal tract and prevent or reduce infection caused by pathogenic microorganisms.

heme A portion of hemoglobin and myoglobin proteins; a portoporphrine, which is a chemical ring structure that holds iron.

hemochromatosis A genetic disease that causes high absorption rates of iron, resulting in iron toxicity; classically diagnosed by the triad of bronze-colored skin, diabetes (hyperglycemia), and liver disease.

hemoglobin A large, protein-containing iron that is essential for oxygen transport and carbon dioxide removal by red blood cells.

hemolysis When red blood cells break.

hemolytic anemia A type of anemia resulting from red blood cell destruction, which can be caused by insufficient vitamin E levels in prematurely born infants or by vitamin E toxicity.

hemorrhage Uncontrolled, excessive bleeding that can also occur from a reduced ability for the blood to clot.

hemosiderosis An iron toxicity condition, characterized by excess iron storage, that is commonly genetically linked and is more prevalent in men and often tied to hemochromatosis.

high-density lipoprotein (HDL) A class of lipoproteins made of lipids including cholesterol, phospholipids, triglycerides, and protein that is known as the "good" kind of cholesterol because it scavenges cholesterol from tissues and returns it to the liver for processing.

high-fat food Provides more than 35 percent of the Calories from fat.

homeostasis A balanced state (within a tolerable range) of the collection of anabolic biochemical reactions (synthesis) and catabolic biochemical reactions (degradation) occurring inside the body.

homocysteine An intermediary compound in the metabolism of the amino acids methionine and cysteine that can build up in the blood stream when the diet is inadequate in folic acid, vitamin B_6, and vitamin B_{12}, which are all needed for the metabolic conversions.

homocysteine transmethylase A folic acid-dependent enzyme that is necessary for the detoxification of homocysteine.

hormone An active chemical substance formed in one part of the body and carried to another part of the body to alter cellular behavior or activity.

hunger The physiological need for food.

hydration Of or pertaining to water; the status of body water content assessed by fluid and electrolyte balance in the intracellular and extracellular fluids.

hydrochloric acid An acid secreted by the stomach that causes protein denaturation and aids digestion.

hydrogenation A food processing technique that chemically forces the addition of hydrogen atoms by saturating monounsaturated or polyunsaturated fatty acids (the double bonds are eliminated, making the fats saturated, solid, resistant to oxidation, shelf stable, and less healthy).

hydrolysis A chemical reaction that uses water to break the bonds in chemical compounds, in which water (H_2O) is split, generating a free hydrogen atom (H) and a hydroxyl group (OH), which are used to chemically balance the products of catabolism.

hydrolyzed A chemical reaction in which water reacts with a compound (such as casein protein) to produce a more simple compound or protein structure.

hydroxyapatite The mineralized protein matrix of bone and teeth that is rich in calcium, phosphorus, and magnesium.

hydroxylated The addition of a hydroxyl (OH) group to a compound, which often results in an electron-deficient chemical.

hypercalcemia A state of the blood that is characterized by high blood levels of calcium.

hypercarotenemia A state of the blood that is characterized by high blood levels of beta-carotene.

hypercholesterolemia A fasting total blood cholesterol level of 240 mg/dL or greater.

hyperglycemia A blood glucose concentration that is greater than or equal to 126 mg/dL, which is both above normal (70 to 99 mg/dL) and prediabetes (100 to 125 mg/dL).

hyperhomocystemia High levels of homocysteine in the blood.

hyperinsulinemia High levels of insulin in the blood.

hyperkalemia A state of the blood that is characterized by high blood levels of potassium.

hyperphosphatemia A state of the blood that is characterized by high blood levels of phosphorus that leads to metastatic calcification.

hyperplasia Increase in cell number.

hypertension Elevated blood pressure above 120/80 millimeters of mercury; the prehypertension stage is established by the range 120–139/80–89, stage 1 of hypertension is given by the range 140–159/90–99, and stage 2 is greater than or equal to 160/100.

hypertriglyceridemia A fasting serum triglyceride level of 150 mg/dL or greater.

hypertrophy Increase in cell size.

hypoglycemia A blood glucose concentration that is below normal (70 to 99 mg/dL) and becomes symptomatic at about 60 mg/dL.

hypogonadism A condition in which decreased production of gonadal hormones leads to below-normal function of the gonads and retardation of sexual growth and development.

hypokalemia A state of the blood that is characterized by low blood levels of potassium.

hyponatremia A state of the blood that is characterized by low blood levels of sodium.

hypophosphatemia A state of the blood that is characterized by low blood levels of phosphorus and causes general debility.

hypothesis A proposed scientific belief based on facts and previous observations that needs additional validation by the scientific method.

hypothyroidism Low blood levels of thyroid hormone fractions.

hypovitaminosis A A condition of vitamin A deficiency characterized by impaired growth, night blindness, xerophthalmia, diarrhea, depressed immunity, and hyperkeratosis.

ileum The third and last segment of the small intestine.

immunity The body's immune system defense mechanisms against invasion of foreign entities.

immunoglobulins Antibodies that are found in breast milk, blood, or other bodily fluids, and are used by the immune system to identify and neutralize foreign entities such as bacteria and viruses.

impotence A male sexual dysfunction characterized by the inability to develop or maintain an erection of the penis sufficient for satisfactory sexual performance.

inborn error of metabolism A disease caused by an inherited genetic defect causing a specific abnormality in metabolism.

incomplete protein A low-quality or low biological value protein from the plant kingdom that is missing or limited in one or more of the essential amino acids.

incontinence Loss of bladder and/or bowel control.

indirect calorimetry A measurement of Resting Energy Expenditure using oxygen consumption, carbon dioxide production, and mathematical conversions to determine the Calories used by a person.

infant A term used to refer to the very young offspring of humans; usually includes the time from birth to at least age 1.

infertility The state of being unable to produce offspring; usually in a woman, it is an inability to conceive; in a man, it is an inability to impregnate.

inflammation A cellular response to injury that is characterized by pain, swelling, redness, and heat.

ingredient A component of a processed food product.

ingredients list A list of the components of a processed food product given in descending order by gram weight or volume.

inorganic Does not contain carbon and is not a living thing.

insoluble fibers Indigestible, mostly plant-derived dietary components that do not soften in water, including cellulose, hemicelluloses, and lignin.

insulin A protein hormone secreted by pancreatic beta cells in response to elevated blood glucose, which changes the cell membrane permeability to allow glucose to enter liver and muscle cells, thus lowering blood glucose levels.

intermediate-density lipoprotein (IDL) Lipoproteins that are transiently formed by the degradation of very-low-density lipoproteins to low-density lipoproteins and do not promote or regress coronary artery disease.

intermediate storage A substance that is stored but only lasts for a few hours, such as liver glycogen.

International Units (IU) A standardized unit of measure used on food package labels for vitamins A, D, and E.

interstitial fluid One of the extracellular fluid compartments, which refers to fluid between cells.

intervention trial A prospective statistical comparison between populations that compares the effect of performing an intervention program in one population against the other (similar) population that had no intervention.

intracellular Inside cells.

intracellular fluid (ICF) The fluid inside of cells.

intrinsic factor A vitamin B_{12} binding protein that is needed for absorption.

in utero In the mother's womb.

in vivo Refers to studies conducted, or any processes that occur, in living organisms.

iodine A trace mineral that is essential for the structure of thyroid hormone.

iron A trace mineral that is a component of several functional proteins, including hemoglobin and myoglobin.

irradiation The exposure of food to high-energy particle bombardment, which destroys microbial and insect contamination.

isocaloric When energy intake is equal to energy expenditure; the Calorie amount consumed meets the energy need.

isokinetic The muscle force exerted or muscle resistance applied during limb movement is at a fixed speed or constant velocity.

isometric When the muscle remains the same length while contracting or the muscle is static when tension occurs.

isotretinoin An oral medication, chemically related to vitamin A, used for the treatment of moderate to severe acne; causes severe birth defects if taken during early pregnancy.

jejunum The second segment of the small intestine, after the duodenum and before the ileum.

Kegel exercises A necessary postpartum pelvic floor exercise done by tightening then relaxing the muscles of the perineum, or the pelvic floor muscles.

Keshan disease The deficiency disease for selenium that causes cardiomyopathy.

kinetic The motion of molecules, related to the ability to do work.

kwashiorkor Protein malnutrition resulting when a person consumes adequate Calories but an inadequate amount of protein, which is required to sustain growth and/or repair vital tissues.

L-ascorbic acid The chemical name for vitamin C.

laboratory experiments Well-controlled, experimental conditions taking place in a laboratory setting, often using animals, tissues, cells, insects, microbes, or other living systems, where the results of the study produce strong, supportive scientific evidence.

lactase An enzyme that breaks the bond between a glucose and a galactose molecule, or digests lactose.

lactation The production and secretion of milk by the mammary glands; also known as breastfeeding or lactogenesis.

lactoferrin An iron-containing protein that is produced in mother's milk and is well absorbed.

lactogenesis The production and secretion of milk by the mammary glands; also known as breastfeeding or lactation.

lactose A disaccharide that is made up of glucose and galactose, commonly called milk sugar.

lactose intolerance An inborn error of metabolism in which a person cannot make lactase and thus has no ability to digest the milk sugar lactose; signs include severe gas, bloating, and diarrhea after lactose ingestion.

lactose maldigestion A condition in which a person has a limited ability to make lactase and thus has a compromised ability to digest the milk sugar lactose, resulting in the signs of gas, bloating, and diarrhea after lactase ingestion.

large intestine The large bowel portion of the GI tract that completes the digestive process and includes the ascending, transverse, descending, and sigmoid colon.

lead A heavy metal that displaces calcium, iron, zinc, and other minerals from their action in the bones, kidneys, and liver, resulting in organ damage.

lean Ten grams of fat or less, 4.5 grams of saturated and *trans* fat or less, and 95 milligrams or less of cholesterol per 3.5-ounce serving of meat.

lean body mass Animal tissue, including mostly vital organs and muscle mass, but also connective tissue.

lecithin A phospholipid used by the food industry as an emulsifier in processed foods and by the body in cell membranes.

legumes Plant foods from the bean and pea family, with seeds that are rich in protein compared to other plant-derived foods.

leukopenia A condition of the blood that is characterized by lowered white blood cell count.

life expectancy A statistical estimation of the expected number of years of life remaining at a given age.

life span The period of time between birth and death; the maximum years alive attained by a member of a species.

light or lite One-third fewer Calories per serving as compared to the original product; one-half the fat or sodium as compared to the original product; or light in color or texture compared to the original product.

lignin An insoluble, indigestible organic molecule that holds cellulose fibers together in plants.

linoleic acid An essential fatty acid with 18 carbons and two double bonds, with the first double bond occurring at the sixth carbon from the methyl end, making it an omega-6 fatty acid.

lipids A family of fat-soluble organic compounds that includes triglycerides, phospholipids, and sterols.

lipoproteins Clusters of lipids associated with proteins that serve as transport vehicles for fat-soluble substances in the lymph and blood, including chylomicron, HDL, LDL, IDL, and VLDL.

liver An organ with multiple functions, including the synthesis of cholesterol and bile and detoxification of chemicals.

longevity Having a long duration of life or life span.

long-term storage Substances that are stored for days, months, or years, such as body fat and vitamin B_{12}.

low 140 milligrams or fewer of sodium per serving; 20 milligrams or fewer of cholesterol per serving; 40 Calories or fewer per serving.

low-density lipoprotein (LDL) A class of lipoproteins made of lipids including cholesterol, phospholipids, triglycerides, and protein that is known as the "bad" kind of cholesterol because it delivers cholesterol to tissues and can promote atherosclerosis and contribute to heart disease when elevated in the blood.

low-fat food Provides less than 25 percent of the Calories from fat.

luteal phase The last 14 days of the menstrual cycle.

luteinizing hormone (LH) One of the four important hormones of the menstrual cycle that surges at ovulation and helps release eggs from the ovaries.

lymph A fluid containing lymphocytes that travels through the lymphatic vessels and is the first conduit for the transport of fat-soluble substances, packaged as chylomicrons, after digestion and intestinal cell absorption.

lymphatic vessels The tubular system of vessels where lymph fluid circulates.

macrocytic normochromic anemia An anemia that can be used to describe the red blood cells as large and normal color.

macronutrient Any of the categories of energy-producing nutrients: carbohydrates, proteins, and fats.

macular degeneration Weakening of the major eye muscle due to oxidative stress.

magnesium A major mineral, abundant in plant foods, needed to mineralize bones and teeth, as well as to participate in muscle contractions, nerve conduction, and blood clotting.

major minerals Essential minerals found in the adult reference body (150 pounds) in quantities greater than 5 grams, including calcium, phosphorus, magnesium, sodium, potassium, chloride, and sulfur.

malnutrition Bad nutrition; can include over- and undernutrition related to the intake of too much or too little of a nutrient, energy, and/or Calories.

maltose A disaccharide that is composed of two units of glucose; can be referred to as malt sugar.

mammalian A biological term used to refer to any warm-blooded vertebrate having the skin more or less covered with hair; the young are born alive and nourished initially by mother's milk.

mammary glands Exocrine glands that produce milk for the offspring of mammals.

manganese A trace mineral, different from the major mineral magnesium, that plays a role in bone formation and the metabolism of amino acids, carbohydrates, and cholesterol.

marasmus Protein-energy malnutrition resulting from both protein and energy (Calorie) deficiencies.

mastication The act of mechanical digestion, accomplished by chewing food to soften it for swallowing.

maximum heart rate (MHR) The highest number of times the heart can beat competently or contract per minute, given in beats per minute.

medium-chain triglycerides Fatty acids that have 6 to 12 carbons in the chain and are esterified to glycerol.

meiosis A process of cell division where the number of chromosomes per cell is cut in half after the egg and sperm fuse.

melatonin A hormone that controls the sleep and awake cycles; it is produced and secreted by the pineal gland when a person is in a dark environment, thus inducing sleep.

menarche The central event of female puberty, signaled by the first menstrual cycle.

Menkes disease An inborn error in metabolism that causes a copper deficiency, which is characterized by osteoporosis in infants and children, normocytic hypochromic anemia, and a low white blood cell count.

menopause The time of a woman's life when her menstrual cycle ceases permanently; 12 months of amenorrhea (no menstrual cycle).

menses The cyclic discharge of blood-rich uterine linings from the uterus of nonpregnant women, monthly from puberty to menopause.

mercury A heavy metal that poisons the nervous system, especially in fetuses.

messenger ribonucleic acid (mRNA) A long polymer of the nucleic acids (nucleotides) adenine, cytosine, guanine, and uracil that are made inside the nucleus of the cell; upon the expression and transcription of deoxyribonucleic acid, mRNA then migrates to the ribosome in the cytosol of the cell, where it is translated for protein synthesis.

metabolic acidosis A blood pH less than 7.35 that results from too many hydrogen ions in the blood.

metabolic alkalosis A blood pH greater than 7.45 that results from not enough hydrogen ions in the blood; leads to paralytic ileus and can be caused from magnesium toxicity.

metabolic syndrome A lifestyle disease where the diagnosis is made by having three or more of the five following metabolic risk factors: central adiposity, hypertension, hyperglycemia, hypertriglyceridemia, or low HDL cholesterol.

metabolism The conversion (anabolic or catabolic) of a substance from one form to another by a living organism.

metabolites The products (molecules) of biochemical reactions or metabolism.

metalloenzymes Enzymes containing minerals that are metals.

microbes A general term for microorganisms such as bacteria, protozoa, parasites, fungi, and algae.

microbial growth The many ways microorganisms perpetuate themselves, which are affected by temperature, protein, pH, and moisture.

microcytic hypochromic anemia An anemia that can be used to describe the red blood cells as small and low in color.

micronutrient Any of the categories of non-energy-producing nutrients: vitamins and minerals.

microorganisms A wide variety of life forms, also called microbes, that are so small that a microscope is needed to see them.

microvilli Tiny hair-like projections of cell membranes on epithelial cells of the small intestine that increase the absorptive surface area.

Mifflin-St. Jeor equation Mathematical equation, sensitive to gender, body mass, and age, used to estimate Resting Energy Expenditure.

mineral mass The noncombustible, inorganic substances incorporated in animal bodies; the ash after cremation.

minerals Inorganic elements naturally found in the earth that are categorized as major and trace in human nutrition; some minerals are essential nutrients needed in small amounts (milligrams or micrograms) by the body to function in a structural capacity, as coenzymes, and in fluid and pH balance.

miscarriage The spontaneous ending of a pregnancy that occurs when the embryo or fetus is incapable of surviving in utero.

mitochondria The cellular organelles that generate most of the cell's supply of adenosine triphosphate (ATP), which is the ultimate source of chemical energy used by cells to do work.

mitosis A type of cell division in which the cell nucleus divides into identical nuclei containing the same number of chromosomes and produces the same type of cell.

moderate exercise An intensity of physical activity that causes the body to sweat.

moderate-fat food Provides 25 to 35 percent of the Calories from fat.

moderation A diet that provides no unwanted substance in excess and provides enough (but not too much of) essential nutrients and Calories.

modified atmosphere packaging A food preservation technique that replaces oxygen in packaging with carbon dioxide or nitrogen to extend shelf life.

molecule Two or more atoms (the smallest component of an element) held together or stabilized by a chemical bond.

molybdenum A trace mineral that is a coenzyme for molybdoenzymes.

monosaccharide Single units of sugar that typically form a single ring and include glucose, fructose, and galactose.

monounsaturated fatty acids (MUFAs) A fatty acid with one double bond in the carbon chain, found in abundance in olive and canola oils, almonds, and avocados.

mouth The opening of the oral cavity where food ingestion occurs.

multicellular organisms Complex living systems that have several types of cells in the life form.

mutagenic A physical or chemical agent that causes a mutation or a physical change to the genetic material.

myoglobin An iron-containing protein of the muscle cells that is needed for oxygen uptake.

MyPlate An icon identifying the portions of the major food groups to consume each day; accompanied by a Web-based, Calorie-controlled, personalized food guidance system based on the age, gender, height, weight, and physical activity level of a person; sponsored by the U.S. Department of Agriculture.

naphthoquinones A family of vitamin K chemical compounds known as vitamin K_2, which are made by the bacteria in the gut but are present in small amounts in dairy and meat products.

natural selection The evolution of a species through genetic alterations in heritable traits that enable the species to adapt and survive in its environment over time.

natural toxicants Chemical compounds naturally occurring in foods that can cause illness or death.

negative energy balance The consumption of fewer Calories than expended; Calories eaten are less than the Calories used; body weight decreases.

neonate An infant age zero to four weeks.

neural tube defect The malformations of the brain and/or spinal cord occurring during embryonic development that can be caused by insufficient folate status.

neurochemicals A variety of organic molecules that activate functions in the central nervous system.

neurotoxin A toxin that can be produced by bacteria and targets nerve cells.

neurotransmitter An active chemical substance that allows transmission of messages in the nervous system.

neutropenia A condition of the blood that is characterized by lowered neutrophil cell count.

neutrophil An abundant type of white blood cell (leukocyte) that is an important phagocyte (engulfing and destroying cell).

niacin A water-soluble vitamin that plays a role as a coenzyme in NAD, NADP, energy metabolism, and steroid synthesis; deficiency causes pellegra.

nicotinamide The chemical form of niacin that is fortified in foods and used for most dietary supplements; also known as niacinamide.

nicotinamide adenine dinucleotide (NAD) A coenzyme made from niacin.

nicotinamide adenine dinucleotide phosphate (NADP) A phosphorylated coenzyme made from niacin.

nicotinic acid The chemical form of niacin that is used to reduce total blood cholesterol but causes flushing if an aspirin has not been taken before the supplementing of this water-soluble vitamin.

nitrification Conversion of ammonium to nitrite and nitrate.

nitrogen balance Nitrogen loss, compared to nitrogen intake from protein sources.

nonessential amino acids Amino acids that are needed by the body to build proteins but are synthesized from nitrogen and carbohydrate intermediates and can also be provided by food sources.

normocytic hypochromic anemia The characterization of the red blood cells that are normal in size but low in color.

not-for-profit An agency or organization that exists to serve the public good and not for the purpose of making money; also called *nonprofit*.

nourish To provide with food, nutrients, or other substances necessary for life and growth.

nucleic acids The building blocks of genetic material (DNA and RNA) used by all cells to divide, differentiate, or synthesize protein constituents; also called *nucleotides*.

nutrient adequacy The dietary intake of the needed amount of energy, nutrients, and fiber to support optimal health and function; numerically defined as 100 percent of the DRI up to the Tolerable Upper Intake Level.

nutrient deficiency The dietary intake of much less than the needed amount of energy, nutrients, and fiber to support optimal health and function; numerically defined as less than 66 percent of the DRI.

nutrient density A high amount of nutrients relative to the number of Calories.

nutrient excess The dietary intake of much more than the needed amount of energy, nutrients, and fiber to support optimal health and function; numerically defined as above the Tolerable Upper Intake Level for vitamins and minerals.

nutrient inadequacy The dietary intake of less than the needed amount of energy, nutrients, and fiber to support optimal health and function; numerically defined as 66 to 99 percent of the DRI.

nutrients Organic and inorganic chemicals in food that may provide energy, structural materials, and/or be regulatory agents that support life, growth, maintenance, and repair of the body's tissues; can be essential or nonessential, caloric or noncaloric.

nutrigenomics The study of the relationships between dietary influences and components and gene expression.

nutrition The science of foods and the nutrients and other substances they contain, and of their actions within the body, as well as the social, economic, cultural, and psychological implications of food and eating.

nutritional anemias The types of anemia caused by nutrient deficiencies.

nutritional sciences The study of nutrition, including dietary components and metabolism.

Nutrition Facts panel An area on the food package that shows the serving size, servings per container, Calories per serving, Calories from fat per serving, percent of the Daily Value (DV) including the DRVs and RDIs, and the ingredients.

nutritionist Term for a person who advises people on dietary matters, with or without appropriate credentials.

nutritious A food that provides nourishment to a high degree.

obese Body fatness in excess of what is conducive for maintaining good health.

obesity epidemic A rapidly growing number of people within a population with excesses of stored body fat (BMI of 30 or greater) that leads to increased morbidity and mortality.

oils Dietary fats that are liquid at room temperature.

olfaction The sense of smell or the ability to smell (detect airborne molecules).

omega-3 fatty acid Long-chained, polyunsaturated fatty acids (PUFAs) in which the first double bond is three carbons away from the methyl (CH_3) end of the carbon chain.

omega-6 fatty acids Polyunsaturated fatty acids in which the first double bond is after the sixth carbon, counting from the methyl (CH_3) end of the carbon chain.

omnivore An eating style that includes consuming plant and animal foods.

oral hypoglycemic agents Drugs used to treat type 2 diabetes that restore the functioning of self-produced insulin.

organic A carbon-containing substance or molecule; in lay terms or on food labels, means organically produced.

organic food production A food production system that follows specific standards that do not permit the use of pesticides, synthetic fertilizers, antibiotics, hormones, and/or genetically modified organisms and also strives to enhance the environment and conserve earth's resources.

osmolarity A measure of the concentration of particles in a solution.

osmotic pressure The pressure exerted against a semipermeable membrane that is created by the concentration of particles in a solution.

osteoarthritis Deterioration of joint cartilage.

osteoblasts The type of bone cell that builds or mineralizes bone.

osteoclasts The type of bone cell that breaks down or demineralizes bone.

osteomalacia A softening of the bone caused by a vitamin D deficiency in adults.

osteopenia Low bone density that can be caused by not consuming enough calcium in the diet or a toxicity of vitamin A.

osteoporosis Fragile bones that break with low stress, usually resulting from bone loss over time.

ovaries The egg-producing reproductive organ, often found in pairs as part of the female vertebrate reproductive system.

overload principle A greater than normal workload demand placed on the cardiorespiratory or skeletal-muscular systems that leads to increased functional capacity.

overnutrition Excess intake of energy and/or nutrients.

overweight Having excess body weight in relation to height than what is considered normal.

ovulation The moment in time when egg(s) are released from the ovaries.

oxidation The process of a substance combining with oxygen, resulting in the loss of an electron and the creation of a chemically unstable (more reactive) molecule.

oxidative stress A measurable shift to a more electron-deficient state in a biological system that results in injury.

oxidizable substrate Any chemical compound capable of losing an electron.

oxytocin A hormone secreted by the pituitary gland that stimulates the uterine muscles to contract and the mammary glands to eject milk during lactation.

pancreas An organ that has exocrine (secreting digestive enzymes and juices into the duodenum) and endocrine (secreting hormones into the blood that help to maintain glucose homeostasis) functions.

pantothenic acid A water-soluble vitamin that is a component of coenzyme A and widespread in foods; deficiency is poorly documented in humans.

parasites Living organisms that live in or on another living organism (the host), and take their nourishment from the host without benefiting the host.

partial hydrogenation A food processing technique that chemically forces partial saturation (addition of hydrogen atoms and the removal of the double bonds) of monounsaturated or polyunsaturated fatty acids, which can generate *trans* fatty acid configurations.

passive diffusion Absorption mechanism in which nutrients cross the intestinal cell membrane freely by a concentration gradient that creates osmotic pressure from the forces of migration from an area of high concentration to a lower one.

pathogenic Disease-causing.

pathology The scientific and medical study of disease.

patterning Determining how an amount of food eaten quantifies as an equivalent amount from the MyPlate food guidance system or the Exchange List system.

peak bone mass The highest attainable bone density for an individual that is developed typically up to approximately ages 25 to 35.

pediatric The branch of medicine for infants and children.

peer-reviewed When colleagues professionally evaluate the work of other colleagues; also called *refereed*; required for publication of scholarly work in prestigious journals.

pellagra The deficiency disease of the water-soluble vitamin niacin, characterized by dermatitis, diarrhea, and dementia.

pepsin A gastric enzyme that specifically breaks peptide bonds between specific amino acids in a protein.

peptide bonds The type of chemical bond that joins two amino acids together.

perceived exertion Standard words that are used to describe how hard you feel you are working during exercise.

percent body fat The percent of total body weight from fat, which can be estimated using many body composition evaluation techniques.

perimenopause A natural transition period of approximately four years before menopause when hormone levels decline, leading to menstrual irregularity and menopause symptoms.

peripheral neuropathy Decreased sensation of the nerves that go to the limbs.

peristalsis Waves of circular muscular contractions in the GI tract, which are controlled by the autonomic nervous system and propel food along the entire GI tract.

pernicious anemia The deficiency of vitamin B_{12}, characterized by macrocytic normochromic anemia and central nervous system impairment.

personal preference The food likes and dislikes of an individual.

pesticide A heterogeneous category of chemical compounds (including fungicides, herbicides, insecticides, and rodenticides) that can be applied to conventionally produced agriculture to exterminate or ward off insect, microbial, plant, and animal pests; may also bioaccumulate in conventional animal foods; many are endocrine disruptors.

petechia Pinpoint hemorrhaging (bleeding) under the skin.

pH Acid-base balance measured on a scale of 1 to 14, with low numbers being acidic, middle numbers being neutral, and high numbers being basic (alkali).

phenylketonuria (PKU) An inborn error of metabolism that is screened in the United States and is managed by a special life-long, low-phenylalanine diet.

phosphatidylcholine The chemical term for the phospholipid lecithin, which is made out of choline.

phospholipids A compound (such as lecithin) composed of a glycerol backbone and a phosphate group with choline (a B vitamin), which is water-soluble, and two fatty acids, which are fat-soluble; these compounds are used as emulsifying agents, to build cell membranes, and as a precursor for acetylecholine; provides 0 Calories per gram.

phosphorus Eighty-five percent of this major mineral is in bone and teeth; if too much is consumed, bones demineralize to increase the calcium levels in the blood.

phosphorylation When a phosphate group is added to a molecule, such as in the chemical reaction converting adenosine diphosphate (ADP) to adenosine triphosphate (ATP).

photosynthesis The process by which green plants use the sun's energy to make carbohydrates from carbon dioxide and water.

phylloquinones A family of vitamin K chemical compounds known as vitamin K_1, which are in dark green and cabbage-family vegetables.

physical activity Any form of body movement.

physical fitness Having good cardiovascular, respiratory, and muscular capacity assessed by body composition, aerobic activity, and muscular performance and flexibility.

physiological Relating to physiology, the study of the function of living systems.

phytochemical A plant-derived, nonnutrient chemical that has biological activity and health-promoting properties in the body.

pica A term used to describe people who persistently consume nonnutritional foods such as starch and clay.

placebo A blank treatment given to subjects in a control group, sometimes referred to as the "sugar pill."

placenta A vascular structure that forms in the uterus of most mammals upon pregnancy that provides oxygen and nutrients to the developing fetus and transfers waste from the fetus.

plant sterols and stanols Heart health-promoting chemicals from plant foods, including grains, vegetables, nuts, seeds, cereals, and legumes.

plasma The largest of the extracellular fluid volumes; provides a medium for the red and white blood cells to be transported through the circulatory system.

polychlorinated biphenyls Organic compounds causing long-lasting skin eruptions, eye irritation, growth retardation, anorexia, and fatigue.

polycyclic aromatic hydrocarbons (PAHs) Potentially toxic, organic compounds that contain only carbon and hydrogen in their chemical formula, but have at least two fused, six-sided rings in the chemical structures.

polycystic ovarian syndrome (PCOS) The most common female endocrine disorder that affects female fertility and may be caused by insulin resistance.

polysaccharides Organic compounds composed of many monosaccharides chemically linked together; also referred to as complex carbohydrates (such as starch and fiber).

polyunsaturated fatty acids (PUFAs) A fatty acid with more than one double bond in the carbon chain; found in abundance in nuts, seeds, and most liquid plant oils.

positive energy balance The consumption of more Calories than expended; Calories eaten are greater than the Calories used; body weight increases.

postpartum A period that begins immediately after delivering a baby, usually extending to the point of full recovery from childbirth.

post-prandial In a fed state, or after eating.

potassium The major mineral in intracellular fluid that maintains cell volume.

potential Stored energy, in the context of the ability to do work.

prediabetes When fasting blood glucose levels are in the range of 100 to 125 mg/dL, which is higher than normal (70 to 99 mg/dL) but not yet high enough to be considered diabetic (greater than or equal to 126 mg/dL).

preeclampsia An umbrella term used to refer to all variants of hypertension or high blood pressure during pregnancy.

pregnancy The period from conception to birth when a woman houses a developing embryo and then fetus in her womb.

prenatal A term used to refer to the period of pregnancy.

prion An infectious protein believed to cause mad cow disease and be transmittable to humans.

private sector The part of the economy that is self controlled and can be for-profit or not-for-profit.

probiotics Live microorganisms that, when administered in adequate amounts, confer a health benefit on the host.

processed food A food that has been manipulated to change its physical, chemical, microbiological, or sensory properties.

product assurance standards Standards used to ensure batch-to-batch consistency, potency, dissolution, and disintegration of dietary supplements or drugs.

professional affiliation The public, private for-profit, or private not-for-profit agency a professional represents.

progesterone A steroid hormone that fluctuates along with estrogen, is largely manufactured in the ovaries and adrenal glands, and promotes the growth of the uterine lining during the luteal phase of the menstrual cycle.

prolactin A hormone secreted by the pituitary gland; in females it stimulates growth of mammary glands during pregnancy and promotes lactation after birth.

protein character The unique aspects of an individual protein, determined by the amino acid sequence and the folding and interacting of the protein strand that makes a three-dimensional functional molecular structure.

protein digestion When the peptide bonds between the amino acids in a protein strand are broken by a protease enzyme, causing the amino acids to be released for absorption, transportation, utilization, and excretion by body cells.

protein-energy malnutrition A condition known as *marasmus*, which results from dietary deficiencies of both protein and energy (Calories).

protein excess The intake of too much protein, causing dehydration, increased calcium and zinc excretion, liver and spleen enlargement, and long-term reduced liver and kidney function.

proteins Organic, energy-producing compounds made of amino acids for tissue repair and maintenance, as well as for growth; classified as complete or incomplete, high-quality or low-quality, or high or low biological value.

protein sparing When carbohydrates provide adequate energy so that the amino acids from protein can be used for tissue repair, maintenance, and growth purposes, rather than energy production.

protein synthesis A series of condensation reactions between amino acids that build a protein, directed by the genetic code, which occurs by transcription of DNA and translation of mRNA.

protozoa Microscopic animals made of one cell or a group of like cells that live predominately in water; some are human pathogens.

provitamin A Beta-carotene and other antioxidant carotenoids that can be converted to vitamin A inside the body and are found in yellow/orange or dark green fruits and vegetables.

psychological Relating to psychology, the study of the mind.

puberty A period in childhood growth that marks the transition to adulthood; the processes of physical change of a child's body that results in reproductive capability.

public sector The part of the economy that is owned and controlled by the public through the government and is not-for-profit and "by the people, for the people."

pulmonary surfactant A chemical substance produced by lung cells that is critical for the structure and function of the very small lung sacs.

pyridoxal (PL) The name for the coenzyme made from vitamin B_6 that carries the aldehyde group and is found in animal foods.

pyridoxal phosphate (PLP) The name for the most important form of the coenzyme made from vitamin B_6 that is phosphorylated and carries the aldehyde group, and is found in animal foods.

pyridoxamine (PM) The name for the coenzyme made from vitamin B_6 that carries an amine group and is found in animal foods.

pyridoxamine phosphate (PMP) The name for the coenzyme made from vitamin B_6 that is phosphorylated and carries an amine group, and is found in animal foods.

pyridoxine (PN) The name for the coenzyme made from vitamin B_6 that carries a pyrimidine and is found in plant foods.

pyridoxine phosphate (PNP) The name for the coenzyme made from vitamin B_6 that is phosphorylated and carries a pyrimidine, and is found in plant foods.

reactive oxygen species Highly unstable and reactive molecules produced as a byproduct of oxygen metabolism, which damage cell structures unless stabilized by an antioxidant.

recombination The breaking down of DNA (usually), which then rejoins or crosses over to a different DNA molecule and begins a generational transfer of the genetic change.

Recommended Dietary Allowances (RDAs) The dietary amount of a nutrient considered adequate to meet the known nutrient needs of practically all healthy people; also a nutrient intake goal for individuals.

recumbent length Body length that is measured with the infant or child lying down flat on his or her back without the knees flexed.

reduced Twenty-five percent less of a nutrient is present as compared to the original food product; a food label must specifically state which nutrient is reduced (e.g., Calories, fat, or sodium).

Reference Daily Intake (RDI) The highest level of the essential vitamins and minerals for men or women based on the 1968 Recommended Dietary Allowances (RDAs); values are expressed as percentages on the Nutrition Facts panel of food labels, and only percents for vitamin A, vitamin C, calcium, and iron are required by law to be shown.

references Another term for citations, which are published sources recognizing the source of information or quotes used to support the author's work.

refined Removing components of a food, such as the bran and germ from whole wheat, to increase shelf life and consumer appeal.

Registered Dietitian A person who has studied diet and nutrition at an Academy of Nutrition and Dietetics (formerly American Dietetic Association)–approved university program and has passed a standardized exam.

relaxin A protein hormone that is secreted during pregnancy that relaxes ligaments (especially the pelvic ligaments) to prepare for labor and delivery.

resistive exercise A form of physical activity done to increase the size and strength of the skeletal muscles.

Resting Energy Expenditure (REE) An estimated number of Calories needed to sustain vital functions under resting conditions in a 24-hour period.

Retinol Activity Equivalent (RAE) The conversion of vitamin A activity from provitamin A and previtamin A, also known as Retinol Equivalents (REs).

Retinol Equivalent (RE) The conversion of vitamin A activity from provitamin A and previtamin A, also known as Retinol Activity Equivalents (RAEs).

rheumatoid arthritis Autoimmune destruction of organs and tissues, especially joints, causing joint pain, inflammation, swelling, stiffness, and deformity.

riboflavin A water-soluble vitamin that plays its role as a coenzyme for FMN and FAD in fatty acid energy metabolism; deficiency causes ariboflavinosis.

rickets The deficiency condition of vitamin D as it manifests in children; seen by bowed legs and a pigeon chest.

rooting reflex An innate instinct in newborns to turn their head toward an object that brushes their cheek, to promote and support feeding.

saliva The secretions from the salivary glands in the mouth that contain the enzyme salivary amylase, which begins starch digestion.

salt A crystalline mineral compound of sodium and chloride.

sarcopenia The reduction of muscle mass and strength as a consequence of aging.

satiety The physiological feedback mechanism that terminates food intake; a feeling of fullness after eating a meal.

saturated fatty acids (SFAs) A fatty acid with no double bonds between the carbon molecules; found in abundance in animal meats, high-fat dairy products, tropical oils (coconut and palm), and hydrogrenated oils.

science A way of knowing and explaining the phenomena and processes of the natural world.

scientifically literate Capable of proposing and evaluating evidence based on scientific methods.

scientific knowledge The enduring understanding of the natural world when tested repeatedly using the scientific method.

scientific method The scientific investigation process used by scientists to produce evidence for understanding the natural world; involves identifying a problem to solve, formulating a hypothesis, testing the hypothesis, collecting and analyzing the data, and making conclusions.

scurvy The deficiency condition of vitamin C (or L-ascorbic acid) that has signs of swollen, red, bleeding gums, petechia, and follicular hyperkerotosis.

sebaceous Of or pertaining to fat and fat-soluble body secretions.

secretin A protein hormone produced by duodenal cells, causing the stomach to make pepsin, the liver to make bile, and the pancreas to make its digestive juices (enzymes and sodium bicarbonate).

selenium A trace mineral that functions through selenoproteins and is an antioxidant.

selenosis The toxicity condition of selenium commonly caused by selenomethionine that causes changes in connective and nervous tissue, garlic breath, and GI distress.

semipermeable membrane A membrane made of a lipid bilayer that will allow certain molecules or ions to pass through it by diffusion.

senescence The process of aging and physically showing the signs of increasing age, resulting from the cells ceasing to divide by mitosis and a decline in organ function.

senility When the mind and body deteriorate in old age.

serotonin A neurotransmitter, made from the essential amino acid tryptophan, that is involved in mood, sleep, appetite, pain, and memory.

serum cholesterol The level of total cholesterol in the blood stream, including mostly LDL and HDL; it is at a desirable level when it is less than 200 mg/dL.

serum triglycerides The level of total triglycerides in the blood stream, including chylomicrons and VLDL; it is at a desirable level when it is less than 150 mg/dL.

set point theory The concert of psychological and physiological regulatory mechanisms that maintain a comfortable body weight range.

short-term storage When a substance is stored for seconds or minutes, such as ATP (energy).

signs Measurable markers or marked visual changes that indicate a medical or nutritional ailment.

simple sugars Carbohydrates that are monosaccharides or disaccharides.

small intestine The part of the digestive tract located between the stomach and the large intestine (colon) that has three sections—the duodenum, jejunum, and ileum—and is the part of the GI tract where most digestion and absorption of nutrients occurs.

sociological Relating to sociology, the study of society.

sodium A major mineral that maintains electrolyte balance in the extracellular fluid and makes up 60 percent of salt.

sodium sensitivity A person who has a significant elevation in blood pressure associated with sodium intakes above the Tolerable Upper Intake Level (UL).

soluble fibers Indigestible, mostly plant-derived dietary components that soften in water, including pectins, gums, and mucilages.

Specific Dynamic Action of food The energy expended to ingest, digest, assimilate, and excrete food; also called the *thermic effect of food*.

sperm Male reproductive cells or male gametes, which contain 23 chromosomes in each.

spina bifida A neural tube defect present at birth that can be caused by poor preconceptual folic acid status.

splenic shunt An increase in blood flow to the working muscle caused by the increased oxygen demands of exercise, thus diverting the blood supply away from the vital organs to the skeletal muscles.

stanol esters A group of sterol compounds (chemically modified by the addition of hydrogen atoms) that are found in plants and reduce the level of low-density lipoproteins in blood and thus are heart-healthy.

starch Plant polysaccharides, such as amylose and amylopectin, that are made up of repeating units of glucose hooked together by alpha bonds, that are digestible by humans.

sterol esters A group of chemically esterified phytosterol compounds found in plants that reduce the level of low-density lipoproteins in blood and thus are heart-healthy.

sterols Fat-soluble compounds containing a four-ring carbon structure with any of a variety of side chains attached that provide 0 Calories per gram.

stomach A large, muscular, sac-like organ of the digestive system where chemical and mechanical digestion occurs; the mixing of the bolus with gastric acid and enzymes and the churning due to muscle contractions produces the liquid mixture called *chyme.*

stored body fat The difference between total body fat and essential fat mass.

stroke The rupture or blockage of a blood vessel supplying oxygen to the brain, resulting in brain damage.

subcutaneous fat Fat mass under the skin, which can be used to estimate total body fat.

substrate A chemical compound that is acted on by enzymes that support life.

sucking reflex An innate reflex, linked to the rooting reflex, in which an infant sucks on a nipple touching the roof of its mouth to enable eating.

sucrose A disaccharide that is made up of glucose and fructose; commonly called table sugar.

sugar alcohols Simple sugars that don't promote tooth decay, which may include any of the following chemicals: xylitol, mannitol, maltitol, lactitol, and erythritol.

sugar substitutes A wide variety of chemical compounds that are perceived as sweet to the human tastebuds and are added to processed foods to reduce the sucrose content.

sulfur A major mineral that helps stabilize protein structures; deficiency is poorly documented in humans; the need is met by consuming a protein-adequate diet.

supplements Products, such as vitamins, minerals, amino acids, glandular extracts, and herbs, that are taken by mouth in addition to a normal diet.

sustainable Able to meet current needs without compromising the ability to meet future needs.

symptoms A subjective complaint about the state of wellness.

taurine An amino acid intermediate that may be vital for the proper development and maintenance of the central nervous system during infancy, or for individuals with certain diseases or nutritional concerns.

temperature adjustment The process of chilling, freezing, and heating foods to minimize microbial growth or kill microbes.

teratogens Any substances or agents that can interfere with normal embryonic development by causing malformations or defects.

testimonials Nonscientifically-based, personal testaments that something is effective.

tetany Involuntary muscle contraction without relaxation.

theory Phenomena believed to be true based on scientific evidence, but lacking conclusive evidence, thus not taken as fact.

thermal injury Damage to the body as a result of elevated temperature above 104 degrees Fahrenheit, causing reduced functional capacity and performance.

thermic effect of food The energy expended to ingest, digest, assimilate, and excrete food; also called the *Specific Dynamic Action of food.*

thermodynamics The processes involved in the conversion of energy to work (mechanical, chemical, or osmotic) and the energy lost as heat.

thiamin A water-soluble vitamin needed for the enzyme thiamin pyrophosphate (TPP), which is needed for ATP production from carbohydrate, normal appetite, and nervous system functioning; the deficiency causes beriberi, and toxicity is poorly documented in humans.

thirst A craving for water when there is a need to rehydrate tissues.

thyroxine A hormone produced by the thyroid gland that regulates the metabolic rate and requires iodine in its chemical structure.

tinctures Alcohol extracts of biologically active plant compounds that are concentrated and used in a liquid form to remedy illness.

tocopherols A class of chemical compounds, along with tocotrienols, that make up the vitamin E family.

tocotrienols A class of chemical compounds, along with tocopherols, that make up the vitamin E family.

Tolerable Upper Intake Level (UL) The maximum dietary amount of a nutrient that can be consumed daily with little risk of illness; however, an intake at a higher level increases the risk of adverse health effects.

trace minerals Essential minerals found in the adult reference body (150 pounds) in quantities less than or equal to 5 grams, including iron, zinc, iodine, selenium, chromium, molybdenum, copper, manganese, fluoride, and cobalt.

transamination The transfer of a nitrogen group from an amino acid by a vitamin B_6 coenzyme.

***trans* fatty acids** Fatty acids with hydrogens on opposite sides of the double bond on the carbon backbone; generated by partial hydrogenation of unsaturated fatty acids and to a lesser extent by bacterial action on animal fats.

transferrin An iron-containing transport protein in the body.

transit time The amount of time it takes food to pass through the entire GI tract from mouth to anus.

transport The movement of molecules across a cell membrane boundary.

transportation The movement of nutrients from one site to another within the body, usually within the blood stream, but can occur within the lymphatic system for fat-soluble substances after their absorption.

triglyceride An organic, energy-producing compound (commonly called fat) that is made up of three fatty acids attached to a glycerol backbone and provides 9 Calories per gram.

trimester One of the three three-month periods into which human pregnancy is divided.

trophozoites Protozoa that are in a growing phase rather than a reproductive or resting phase.

tryptophan An essential amino acid that can be used to synthesize niacin in the body.

tumorigenic Cancer or a tumor caused by alterations in the DNA from many things, such as tobacco, drinking alcohol, being sedentary, eating high-fat animal foods, not eating a plant-based diet, and so forth.

type 1 diabetes Diabetes in which the pancreas fails to produce any insulin, usually because of autoimmune destruction of insulin-producing pancreatic beta cells.

type 2 diabetes The most prevalent type of diabetes, in which resistance to self-produced insulin occurs because of poor diet and lifestyle, obesity, and genetic factors.

undernutrition A lack of (or deficiency in) energy and/or essential nutrients.

underweight Having a lower body weight in relation to height than what is considered normal; a BMI less than 18.5.

urea A toxic nitrogen-containing organic compound that is a byproduct of amino acid catabolism and is excreted in the urine.

uric acid A nitrogen-containing waste product generated from purine catabolism that contributes to gout and other illnesses.

uterus A female muscular organ in the reproductive system in which the fetus develops until birth; also known as the womb.

variety A diet in which different foods are used for the same purpose.

vegan An eating style that includes consuming only plant foods.

very-low-density lipoprotein (VLDL) The type of lipoprotein cholesterol made primarily by liver cells to transport triglycerides to body cells.

villi Finger-like projections within the small intestine that increase the absorptive surface area, each with its own blood and lymph supply.

virion A single, complete virus particle.

viruses Encapsulated genetic material that require a host in order to perpetuate.

visceral fat Abdominal fat mass that indicates unhealthy accumulated fat around vital organs.

vitamin A A fat-soluble vitamin in the active forms retinol, retinal, and retinoic acid that functions in vision, epitheial cell maintenance, and gene expression; deficiency causes blindness.

vitamin B$_6$ Various forms of this water-soluble B vitamin (PN, PL, PM, PNP, PLP, and PMP) function as coenzymes in the metabolism of amino acids, glycogen, and some lipids.

vitamin B$_{12}$ A B vitamin called cyanocobalamin that functions in nucleic acid synthesis; deficiency results in pernicious anemia.

vitamin C A water-soluble vitamin that functions as an antioxidant, in collagen synthesis, and in the maintenance of connective tissue; deficiency results in scurvy; also known as L-ascorbic acid.

vitamin D A fat-soluble vitamin that is synthesized in the body from cholesterol and functions in the mineralization of bone by regulating calcium and phosphorus; deficiency causes rickets and osteomalacia (see also **1,25-dihydroxycholecalciferol** and **cholecalciferol**).

vitamin E The family of fat-soluble vitamins that functions in antioxidant activity and is also known as tocopherols and tocotrienols.

vitamin K A fat-soluble vitamin that functions in blood clotting; most of the dietary source is produced by the bacteria in the gut.

vitamins Organic, essential nutrients (categorized as fat- and water-soluble) that are needed in small amounts (milligrams or micrograms) by the body for health.

VO$_2$ max The maximum amount of oxygen that can be taken up by the body, determined under maximum exercise stress and given in milliliters of oxygen per kilogram of body weight per minute.

water An inorganic compound that is made of hydrogen and oxygen (H_2O) and is essential for life; the medium for metabolism and nutrient transport; dietary sources include all fluids and fluid-rich foods, such as fruits and vegetables.

water balance When the amount of water consumed equals the amount lost by the body, and equilibrium is achieved.

water intoxication The result of drinking too much water.

water-soluble nutrients Nutrients that dissolve in an aqueous solution or water.

weaning Gradually withdrawing breast milk and/or infant formula from an infant, along with the introduction of table foods, in the transition to a diet consisting of adult-type foods.

weight control Diet and lifestyle measures taken to achieve or maintain a healthy body weight.

weight gain An increase in total body weight caused by fluid retention, fat mass accumulation, or an increase in lean body mass.

weight loss A decrease in total body weight caused by dehydration or loss of fat or lean body mass.

Wernicke-Korsakoff syndrome The name of the syndrome associated with a thiamin deficiency resulting in wet beriberi, caused by excessive alcohol intake, manifesting with encephalopathy and psychosis.

whey A major mammalian milk protein, in addition to casein.

Wilson's disease The name of the disease caused by a genetic disorder resulting in copper overload, largely causing GI tract distress and liver damage.

xerophthalmia The name of the eye condition causing blindness resulting from a vitamin A deficiency.

zinc A trace mineral that is a structural component in more than 100 metalloenzymes; deficiency is caused by a genetic disorder resulting in acrodermatitis enteropathica.

zygote The cell produced by the fusion of mature gametes (a sperm and an egg).

Index

American Home Products, 299
American Journal of Clinical Nutrition, 291
American Medical Association, 292
Amines
    in amino acids, 18
    nitrates and nitrites with, 306
Amino acids, 6, 17, 18, 113–114
    aspartame and, 306
    ATP energy production and, 199
    categories of, 17–18
    essential amino acids, 17, 19, 114–115
    folate and, 245
    food composition information, 77–80
    functions of, 116
    glucose and, 128
    molybdenum and, 273
    nonessential amino acids, 114–115
    pantothenic acid and, 247
    phenylketonuria (PKU) and, 365
    in sweat, 205
    translation process, 115
    vitamin B_6 and, 243
    vitamin B_{12} and, 244
    weight control and, 187–188
Ammonification, 328, 329
Amniocentesis, 353
    cystic fibrosis, diagnosis of, 354
Amylase, 103
    in pancreatin, 304
    salivary amylase, 101
Amylopectin, 13
Amylose, 13, 127
Anabolic reactions, 30, 31
Anabolism, 109
Anaerobes, 311
Anaphylaxis, 130, 131
    food allergies and, 389
Andropause, 412
Anemia, 225. *See also* Hemolytic anemia; Iron-deficiency anemia; Pernicious anemia
    aging and, 404
    macrocytic normochromic anemia, 244, 245
    microcytic hypochromic anemia, 266
    normocytic hypochromic anemia, 274
    nutritional anemias, 266
    vitamin B_6 and, 243
Aneurin, 238
Animal Drug Amendment to the FD&C Act, 1968, 296
Animal foods
    and cholesterol, 26
    and phospholipids, 26
    proteins in, 19
    and triglycerides, 22
Anorexia
    of aging, 402
    hypophosphatemia and, 262

nervosa, 398–399
    thiamin deficiency and, 240
ANS (autonomic nervous system), 101
Antacids and magnesium toxicity, 263
Antagonistic muscles, 194
Antagonists of vitamin K, 236
Anthocyanins, 152
Antibiotics, 313
    in food production, 321, 323
    organic food production and, 327
Antibodies
    classes of, 131
    proteins and, 116
Antigens, 130
Antimicrobial agents, 305
Antioxidants, 144–145
    Alzheimer's disease and, 410
    cancer and, 150
    as food additives, 305
    food sources for, 234
    functions of, 234
    phytochemicals and, 68
    vitamin E, 233
Antiproliferation and vitamin D, 230
Anti-raw egg white factor, 248
Anus, 102, 110
Appendix, 110
Appetite
    chloride deficiency and, 256
    factors affecting, 9
    food choices and, 8
    set point theory and, 184
    thiamin and, 238
Aquaculture and Dietary Guidelines 2010, 71
Aqueous, 101
Aqueous secretions, 100
Aqueous waste, 110
Arachidic acid, 23
Arachidonic acid (AA), 23, 356
Arginine, 17
Ariboflavinosis, 241
Arrhythmia and potassium deficiency, 257
Arteries, 137
    coronary arteries, 136–137
Arthritis, 409–410. *See also* Rheumatoid arthritis
    aging and, 404
    hereditary hemochromatosis (HHC) and, 267
    HUFAs (highly unsaturated fatty acids) and, 147
    obesity and, 175
    sleep disturbances and, 411
Artificial colors and flavors, 305, 306
Artificial sweeteners, 307–309
Ascending colon, 110
Ascorbic acid, 246. *See also* Vitamin C
Asparagine, 17
Aspartame, 306, 308
Aspartic acid, 17

Asperigillus flavus, 311, 312
Asthma
    allergens and, 131
    HUFAs (highly unsaturated fatty acids) and, 147
    oxidative stress and, 234
Atherosclerosis, 137–138
    serum cholesterol and, 140
Athletes
    iron deficiencies in, 266–267
    performance of, 191
    vitamin E and, 234
ATP (adenosine triphosphate), 6, 7, 107–108, 306
    carbohydrates and, 198
    minerals and, 30
    phosphorus and, 261
    physical activity and, 199–200
    proteins and, 198
    water-soluble vitamins and, 239
Autism and pica, 359
Autocrine cells, 367
Autoimmune disease. *See also* Rheumatoid arthritis
    type 1 diabetes and, 133
    vitamin D deficiency and, 232
Autotrophs, 328, 329
Availability and food choices, 8–9
Azodicarbonamide, 306

B

Babies. *See* Infants; Newborns
Bacteria, 311, 312
    growth of, 317
    protein-rich foods and, 318
    toxins, bacterial, 312
Balanced diet, 7
Balance lifts, 195
Basal Metabolic Rate (BMR). *See* BMR (Basal Metabolic Rate)
Beef tallow, 24
Behavior modification, 176
    weight control and, 188–189
Behenic acid, 23
Belladonna, 320
Bench press, 192
Benzoyl peroxide, 306
Beriberi, 240
Beta-alanine, 303
Beta-carotene, 29
    as antioxidant, 144
    fiber and binding of, 123
    food label health claims on, 55
    food sources of, 230
    functions of, 228
    hypercarotenemia, 229
Betaine hydrochloric, 304
Bifidobacterium bifidus, 313, 315
Bilberry, 302